ENGINEERING CHALLENGES FOR SUSTAINABLE FUTURE

PROCEEDINGS OF THE 3RD INTERNATIONAL CONFERENCE ON CIVIL, OFFSHORE & ENVIRONMENTAL ENGINEERING (ICCOEE2016), KUALA LUMPUR, MALAYSIA, 15–17 AUGUST 2016

Engineering Challenges for Sustainable Future

Editor

Noor Amila Wan Abdullah Zawawi
Universiti Teknologi PETRONAS, Seri Iskandar, Perak, Malaysia

CRC Press is an imprint of the
Taylor & Francis Group, an **informa** business

A BALKEMA BOOK

Cover photos information (clockwise)
Oil and gas platform with gas burning, Power energy.
Copyright: curraheeshutter

Water recycling on sewage treatment station
Copyright: Kekyalyaynen

Tubes abstract background
Copyright: Eugene Sergeev

Aerial view of Hoover Dam
Copyright: Andrew Zarivny

All photos courtesy of Shutterstock (http://www.shutterstock.com)

CRC Press/Balkema is an imprint of the Taylor & Francis Group, an informa business

© 2016 Taylor & Francis Group, London, UK

Typeset by MPS Limited, Chennai, India
Printed and bound in Singapore by Markono Ltd, Singapore

All rights reserved. No part of this publication or the information contained herein may be reproduced, stored in a retrieval system, or transmitted in any form or by any means, electronic, mechanical, by photocopying, recording or otherwise, without written prior permission from the publishers.

Although all care is taken to ensure integrity and the quality of this publication and the information herein, no responsibility is assumed by the publishers nor the author for any damage to the property or persons as a result of operation or use of this publication and/or the information contained herein.

Published by: CRC Press/Balkema
P.O. Box 11320, 2301 EH Leiden, The Netherlands
e-mail: Pub.NL@taylorandfrancis.com
www.crcpress.com – www.taylorandfrancis.com

ISBN: 978-1-138-02978-1 (Hbk)
ISBN: 978-1-4987-8151-0 (eBook PDF)

Table of contents

Preface   XI
Organising committee   XIII

Coastal and offshore engineering

Numerical simulation of the flow passing a pipeline close to a flat seabed   3
F. Namazi-saleh, V.J. Kurian & Z. Mustaffa

Coastal vulnerability index at a RAMSAR site: A case study of Kukup mangrove island   9
A.K.A. Wahab, D.S.M. Ishak & M.H. Jamal

Effects of mooring lines and water depth parameters on the dynamic motion of a turret-moored FPSO   15
M.O. Ahmed & A. Musaad

VIV of four cylinders in side by side configurations – experimental and CFD simulations   21
A.M. Al-Yacouby, V.J. Kurian, M.S. Liew & V.G. Idichandy

Investigating the potential of soil-nail retaining structures as a slope strengthening remedy by capitalizing on reliability analysis technique   27
S. Qasim, I.S.H. Harahap, S. Baharom, M. Imran & D. Kazmi

Response of spar-mooring-riser system under random wave, current and wind loads   33
M. Jameel, A.E. Ibrahim & M.Z. Jumaat

Effects of current coexisting with random wave on offshore spar platforms   39
V.J. Kurian, S.N.A. Tuhaijan & C.Y. Ng

Nonlinear dynamic analysis of jacket-type offshore platform and optimization of leg batter   45
R. Senthil, S. Ishwarya & M. Arockiasamy

FPSO response and water depth: A study using model tests   51
V.J. Kurian, N. Rini, M.S. Liew & A. Whyte

Real time kinematic-GNSS methods for monitoring subsidence of offshore platform   55
K.C. Myint, A.N. Matori & A. Gohari

Incremental wave analysis method: An alternative to load factor method in SACS pushover analysis   59
M.M.A. Wahab, A.R.M.M. Kamal & V.J. Kurian

FRP shear contribution in externally bonded reinforced concrete beams with stirrups   63
H. Fazli & W. Teo

Coupled analysis for the effect of wave height on FPSO motions   69
N. Rini, V.J. Kurian, M.S. Liew & A. Whyte

The effect of short-crested waves on the dynamic responses of semisubmersible: A comparison of long and short-crested waves with current   75
C.Y. Ng & S.J. Ng

Decommissioning decision criteria for offshore installations and well abandonment   81
T.M.Q. Al-Ghuribi, M.S. Liew, N.A. Zawawi & M.A. Ayoub

Hydraulic efficiency of a pile-supported free surface breakwater subjected to random waves   87
S.S. Hassan, H.M. Teh & V.J. Kurian

Development of ultimate strength versus plate index diagram for initially deflected steel plate under axial compression *D.K. Kim, B.Y. Poh, J.R. Lee, K.S. Park & J.K. Paik*	93
Rheological behavior of tsunami run-up water containing alluvial deposit at coastline of Kedah, Malaysia *A.K. Philip, H.M. Teh, S.H. Shafiai, A. Jaafar & A.H.M. Rashidi*	101
An optimised model of pipe-in-pipe installation for subsea field *M.Z. Jiwa, D.K. Kim, Z. Mustaffa & H.S. Choi*	107
Method for installation of marine HDPE pipeline *Y.T. Kim, K.S. Park, H.S. Choi, S.Y. Yu & D.K. Kim*	113
Dynamic behaviour of tension leg platform in short-crested directional seas *D.O. Oyejobi, M. Jameel & N.H.R. Sulong*	119
A simplified prediction method for subsea pipeline expansion *J.H. Seo, H.S. Choi, K.S. Park, S.Y. Yu & D.K. Kim*	125
Tsunami risk assessment: How safe is Malaysia? *D.B. Shahruzzaman & A.M. Hashim*	131
Wave transmission characteristics of sand containers used as submerged breakwaters *S. Gupta, H.M. Teh, A.M. Hashim & R.B.R. Sulaiman*	137
The need of early response system to HNS accident based on case analysis *J.W. Ryu, J.M. Kim, S.M. Shin & H.K. Park*	143

Construction and project management

Coordination process in construction projects management *W.S. Alaloul, M.S. Liew & N.A. Zawawi*	149
Effectiveness of preventive safety management in construction *I. Othman, M. Napiah, M.F. Nuruddin & M.M.A. Klufallah*	155
Highway project performance evaluation: A study *W. Sutrisno*	159
Ranking of principal causes of construction waste for Malaysian residential project *U.A. Umar, N. Shafiq, A. Malakahmad, M.F. Nuruddin & I.U. Salihi*	163
Attributes of coordination process in construction projects *W.S. Alaloul, M.S. Liew & N.A. Zawawi*	169
Structural relationship of success factors for Small Medium Enterprises (SME) contractors in PLS-SEM model *I.A. Rahman, N.I. Rahmat & S. Nagapan*	173
Categorization of Saudi Arabia's construction delay factors using factor analysis technique *I.A. Rahman, N. Al-Emad & S. Nagapan*	177
The awareness of green building ratings among university students *G. Hayder, O.A. Ahmed & S.B.C. Selia*	183
Factors affecting the embodied carbon footprint potential—Assessment of conventional Malaysian housing habitat *S.S.S. Gardezi, N. Shafiq, M.F. Nuruddin, N.A. Zawawi & F.B. Khamidi*	187
Optimization of residential roof design using system dynamics and building information modelling *S.A. Farhan, N. Shafiq, K.A.M. Azizli, F.K. Soon & L.C. Jie*	193
The development of embodied carbon emission benchmark model for purpose built offices in Malaysia *M.M.A. Klufallah, M.F. Nuruddin, I. Othman & M.F. Khamidi*	199

Environmental and water resource engineering

Influence of environmental factor on the performance of *Pistia Stratiotes* in a surface sequencing baffled steep flow constructed wetland — 207
S.R.M. Kutty, E.H. Ezechi & A.T.A. Nagum

Influence of organic loading rate on a submerged two-stage integrated bioreactor without external carbon addition — 211
E.H. Ezechi, S.R.M. Kutty, M.H. Isa, A. Malakahmad & E. Olisa

Performance of a bench scale anaerobic baffled reactor operated at ambient temperatures — 215
U.A. Abubakar, M.M. Muhammad, F.B. Ibrahim, M.A. Ajibike & A. Ismail

Use of low frequency ultrasound for solids solubilization in Palm Oil Mill Effluent — 221
L.-P. Wong, M.H. Isa & M.J.K. Bashir

The performance investigation of three glass cover solar stills using different basin absorbents — 225
K.W. Yusof, A. Riahi, N. Sapari, M.H. Isa, N.A.M. Zahari, E. Olisa, K.M. Khouna & B.S.M. Singh

Artificial neural network approach for modeling of Cd (II) adsorption from aqueous solution by incinerated rice husk carbon — 229
T. Khan, M.H. Isa & M.R. Mustafa

Implementation of attached growth system in Malaysia: An overview — 235
G. Hayder, N.F.S.Bt. Mohamad Fu'ad & P.A./L. Perumulselum

Comparative study of polymeric adsorbent for copper removal from industrial effluents — 241
S.M. Laghari, M.H. Isa, Z.M. Ali & A.J. Laghari

Analysis of mould growth contamination in library building — 245
S.N.A. Wahab, N.I. Mohammed, M.F. Khamidi, Z.M. Noor & M. Ismail

Petroleum sludge thermal treatment and use in cement replacement – A solution towards sustainability — 251
E.N. Pakpahan, N. Shafiq, M.H. Isa, S.R.M. Kutty & M.R. Mustafa

Residential landscape minimize indoor temperature in tropical climate: Myth or reality — 257
N. Kamarulzaman, N.A. Zawawi & N.I. Mohammed

Enhancement of waste activated sludge disintegration and dewaterability by H_2O_2 oxidation — 263
G.C. Heng, K.W. Chen & M.H. Isa

Application of integrated bioreactor system (i-SGBR) for simultaneous treatment of wastewater and excess sludge degradation — 267
S.R.M. Kutty, N. Aminu, M.H. Isa & I.U. Salihi

Hydraulic assessment of grassed swale as bioengineered channel — 273
M.M. Muhammad, K.W. Yusof, M.R. Mustafa & A.Ab. Ghani

Sulfide precipitation as treatment for iron rich groundwater — 279
N. Sapari, H. Jusoh, E. Olisa & E.H. Ezechi

Climate change impact on water resources of Bakun hydroelectric plant in Sarawak, Malaysia — 283
M.R. Mustafa, M. Hussain & K.W. Yusof

Investigation of the influence of particle size on the migration of DNAPL in unsaturated sand — 289
M.Y.D. Alazaiza, S.K. Ngien, M.M. Bob & S.A. Kamaruddin

Comparison of support vector machines kernel functions for pore-water pressure modeling — 293
N.M. Babangida, K.W. Yusof, M.R. Mustafa & M.H. Isa

Toxicological studies of Perak River water using biological assay — 299
T.S. Abd Manan, A. Malakahmad & S. Sivapalan

Predicting CMIP5 monthly precipitation over Kuching using multilayer perceptron neural network — 305
M. Hussain, K.W. Yusof & M.R. Mustafa

Drought analysis of Cheongmicheon in Korea based on various drought indices — 311
K.J. Won, S.H. Kim, E.S. Chung, S.U. Kim & M.W. Son

Introduction of decision support system for design of LID based on SWMM5.1: A case study in Korea — 317
J.Y. Song, E.S. Chung, S.H. Kim & S.-H. Lee

Geotechnical engineering and geoinformatics

Effects of tunnel face distance on surface settlement — 323
A. Marto, H. Sohaei, M. Hajihassani & A.M. Makhtar

Characterization of Pb and Cd contaminated sandy soil by dielectric means — 327
H.M. Al-Mattarneh, R.M.A. Ismail, M.F. Nuruddin, N. Shafiq & M.A. Dahim

Microbially induced cementation to improve the strength of residual soil — 331
M. Umar, K.A. Kassim & Z. Ibrahim

The behaviour of electrical resistivity correlated with converted SPT-N results from seismic survey — 335
S.B. Syed Osman, H. Jusoh & K. Chai

Development of new distributed optical fibre sensor as borehole extensometer — 339
H. Mohamad

Pile set up investigation for improvement on pile capacity — 345
I.S.H. Harahap & L.C. Fai

Feasibility study of P-wave monitoring of selective bioplugging caused by biofilms — 351
D.-H. Noh & T.-H. Kwon

Prioritizing the criteria for urban green space using AHP-multiple criteria decision model — 355
R. Ahmad & A.N. Matori

Effects of DEMs from different sources in deriving stream networks threshold values — 361
N.A. Ishak, M.S.S. Ahamad, S.K.M. Abujayyab & A.Ab. Ghani

Spatial compliances study of present landfill siting to Malaysian standard — 365
S.Z. Ahmad, M.S.S. Ahamad, S.K.M. Abujayyab, M.S. Yusoff & N.Q. Omar

Monitoring deformation of liquefied natural gas terminal with persistent scatterer interferometry — 371
A.S. Ab Latip, A.N. Matori & D. Perissin

Highway and transportation engineering

An evaluation of 85th operating speed and posted speed limit based on horizontal, vertical alignments and traffic conditions: Case study of two lane urban roadway — 377
A.M.A. Bakar, M.S.N.A. Rani, R. Mohamed, N.S. Sani, N. Jalal, M.A. Adnan & T.B.H.B.T. Besar

Mechanical performance of temperature reduced mastic asphalt — 381
M. Dimitrov & B. Hofko

Influence of factors to shift private transport users to Park-and-Ride service in Putrajaya — 385
I.A. Memon, M. Napiah, M.A. Hussain & M.R. Hakro

Statistical interpretations about the use of footbridges by diverse groups of pedestrians in Malaysia — 391
R. Hasan & M. Napiah

Physical and storage stability properties of linear low density polyethylene at optimum content — 395
N. Bala & I. Kamaruddin

Feasibility of reclaimed asphalt pavement in rigid pavement construction — 401
G.D. Ransinchung.R.N, S. Singh & S.M. Abraham

Optimal maintenance planning for Trans-West Africa coastal highway infrastructure using HDM-4 — 405
O. Oloruntobi, N. Madzlan, K. Ibrahim & O. Johnson

The effect of water on the performance of polymer fibre-reinforced road bituminous mixtures — 409
I. Kamaruddin, M. Napiah & M.H. Nahi

Structures and materials

3D nonlinear finite element analysis of HPFRC beams containing PVA fibers — 417
N. Shafiq, T. Ayub & S.U. Khan

Displacement-based design for precast post-tensioned segmental columns with different aspect ratios — 423
E. Nikbakht & M.S. Mahzabin

Carbon emission analysis of double storey reinforced concrete house in Malaysia — 427
M.S.A. Hamid, N. Shafiq, N.A. Zawawi, M.F. Nuruddin, M.F. Khamidi & M.S.H.M. Shaharmi

Addition of natural lime in incinerated sewage sludge ash concrete — 433
T.Y. Yu, D.S. Ing & C.S. Choo

Properties and structural behavior of sawdust interlocking bricks — 437
B.S. Mohammed, M. Aswin & Vethamoorty

Development of nano silica modified solid rubbercrete bricks — 443
N. Mahamood, B.S. Mohammed, N. Shafiq & S.M.B. Eisa

Analysis layers method of strengthening reinforced concrete T-beams wire rope moment negative (attention reinforcement slab) — 447
D.L.C. Galuh & H.P. Riharjo

Concrete panel from polystyrene waste — 451
D. Sulistyorini & I. Yasin

Fault tree analysis for reinforced concrete highway bridge defect — 457
W.S. Wan Salim, M.S. Liew & A. Shafie

Effect of high volume fly ash in ultra high performance concrete on compressive strength — 463
N.M. Azmee & M.F. Nuruddin

Behaviour of structural properties of fiber reinforced high strength concrete beams subjected to static loads — 467
N. Shafiq, M.F. Nuruddin & A.E.A. Elshekh

Single strap pull test of carbon fibre reinforced polymer plated steel member under fatigue loading — 473
N. Osman, A.Y.M. Yassin & I. Akbar

Experimental study for stepped reinforcement concrete beams — 477
M.A. Hussain & N. Safiq

Nano silica modified roller compacted rubbercrete – An overview — 483
M. Adamu, B.S. Mohammed & N. Shafiq

Contribution of natural pozzolan to sustainability of well cements for oil and gas industry — 489
A. Talah, F. Kharchi & R. Chaid

Costs of urbanization and its impacts on the urban density — 493
B. El Kechebour & S. Haddad

Numerical study on shear and normal stress variation of RC wall with L shaped section — 497
A. Ahmed-Chaouch

Effects of alkaline solution on the microstructure of HCFA geopolymers — 501
M.F. Nuruddin, A.B. Malkawi, A. Fauzi, B.S. Mohammed & H.M. Al-Mattarneh

Influence of pozzolan on sulfate resistance of concrete — 507
A. Merida & F. Kharchi

Presentation of a helping tool for the sizing and the optimization of 'beam' in reinforced concrete — 511
A. Boukhaled & F. Mendaci

Sustainable waste management of bottom ash as cement replacement in green building — 517
N.L.B.M. Kamal, G. Hayder, O.A. Ahmed, S.B. Beddu, M.F. Nuruddin & N. Shafiq

Numerical analysis of the composite "ball-and-socket" carcass design for unbonded flexible pipe — 521
A.F. Billah, Z. Mustaffa & B.T.M.B. Albarody

Milling time influence of ultrafine treated rice husk ash to pozzolanic reactivity in portlandite 527
M.F. Nuruddin, S.A. Saad, N. Shafiq & M. Ali

Mix design proportion for strength prediction of rubbercrete using artificial neural network 531
A. Awang, B.S. Mohammed & M.R. Mustafa

Roof insulation material from low density polyethylene (LDPE), kapok fibre and silica aerogel 537
N.H.A. Puad, M.F. Nuruddin, J.J. Lian & I. Othman

Application of new codes for fatigue assessment of older bridges 543
A.Q. Ayilara, T. Wee & M.S. Liew

Author index 551

Preface

This book contains papers presented at the 3rd International Conference on Civil, Offshore & Environmental Engineering (ICCOEE2016) under the banner of World Engineering, Science & Technology Congress (ESTCON2016) held from 15th–17th August 2016 at Kuala Lumpur Convention Centre (KLCC), Malaysia. The ICCOEE series of conferences started in Kuala Lumpur, Malaysia 2012, and the second event of the series took places in Kuala Lumpur, Malaysia 2014.

The main objective of the ICCOEE Conferences is to provide a platform for academia and industry to showcase their latest advancements and findings in civil, offshore and environmental engineering areas. The conference also provides great opportunities for participants to exchange new ideas and experiences as well as to establish research and business relations with global partners for future collaborations.

The aim is to promote engineering knowledge for sustainable future to be more efficient, environmentally friendly, reliable and safe using the latest methods and procedures design and optimisation. The 103 papers are categorized in the following themes and areas of research:

- Coastal and offshore engineering
- Construction and project management
- Environmental and water resources engineering
- Geotechnical engineering and geoinformatics
- Highway and transportation engineering
- Structures and materials

The articles in this book were accepted after a review process, based on the full text of the papers. Thanks are due to the Technical Programme Committee and to the Advisory Committee who had most of the responsibility for reviewing the papers. We are also grateful to the additional reviewers who helped the authors deliver better papers by providing them with constructive comments. We hope that this process contributed to a consistently good level of the papers included in the book.

N.A. Zawawi

Organising committee

CONFERENCE CHAIR

Associate Professor Dr. Noor Amila Wan Abdullah Zawawi

CONFERENCE CO-CHAIR

Associate Professor Dr. Amirhossein Malakahmad

SECRETARY

Dr. Zahiraniza Mustaffa
Dr. Nurul Izma Mohammed

TREASURER

Associate Professor Dr. Abdul Nasir Matori
Dr. Ng Cheng Yee

TECHNICAL COMMITTEE

Associate Professor Dr. Mohamed Hasnain Isa
Dr. Muhammad Raza Ul Mustafa
Dr. Teo Wee
Ir. Mohamed Mubarak Abdul Wahab
Professor Dr. Nasir Shafiq

PUBLICATION COMMITTEE

Associate Professor Dr. Bashar S. Mohammed
Dr. Montasir Osman Ahmed Ali
Dr. Do Kyun Kim

LOGISTIC COMMITTEE

Associate Professor Dr. Ahmad Mustafa Hashim

IT & MEDIA COMMITTEE

Dr. Airil Yasreen Mohd Yassin

PUBLICITY & PROTOCOL COMMITTEE

Associate Professor Dr. Shamsul Rahman Kutty
Ms. Niraku Rosmawati Ahmad

F&B COMMITTEE

Mrs. Aslinda Jamaluddin

EVENT MANAGEMENT COMMITTEE

Dr. Teh Hee Min

SPONSORSHIP COMMITTEE

Ir. Dr. Idris Othman

Coastal and offshore engineering

Numerical simulation of the flow passing a pipeline close to a flat seabed

Fatemeh Namazi-saleh, Velluruzhathil John Kurian & Zahiraniza Mustaffa
Department of Civil and Environmental Engineering, Universiti Teknologi PETRONAS, Seri Iskandar, Perak, Malaysia

ABSTRACT: In this study, the vortex-shedding pattern and hydrodynamic force acting on the surface of a marine pipe near a flat boundary are simulated numerically. The Navier-Stokes equations with the use of Large Eddy Simulation (LES) model are numerically solved to predict flow pattern and to calculate the pressure distribution on the surface of the pipe. The governing equations are solved based on 2D Galerkin finite volume algorithm. Due to the high flexibility of unstructured triangular meshes for modeling any complex geometry, this type of meshing is selected in the present study. Evaluation of the numerical modeling shows that this method is able to predict the hydrodynamic force and flow pattern over the pipe properly. In addition, the results prove that the vicinity of the bed shear layer to cylinder shear layer would significantly influence the vortex-shedding pattern.

1 INTRODUCTION

Simulation of flow around a circular cylinder in the vicinity of a flat boundary is needed in offshore engineering, especially in design of marine pipelines placed on the seabed. The proximity of the cylinder to the seabed can affect the flow pattern around. Also, it causes a large alternation on the hydrodynamic forces direction and magnitude. To investigate the unsteady nature of the hydrodynamic force acting on the pipe and the effects of bed vicinity on the pipe loads, an accurate numerical simulation of flow pattern around the pipe is necessary.

Simulation of the flow pattern over a circular cylinder exposed to turbulent flow and consequently the pressure distribution on the surface of the pipe is complex and depends on the various factors. These factors are the effect of bed vicinity, which is a shown in term of the gap (h) to pipe diameter ratio (h/D), Reynolds number, and the ratio of boundary layer thickness to the pipe diameter (δ/D) (Gao et al., 2006, Ong et al., 2010, Namazi-Saleh et al., 2014). It is well known that offshore pipelines are usually exposed to flow with high Reynolds number around $1 \times 10^4 - 1 \times 10^6$, covering sub-critical flow, $300 < Re < 2 \times 10^5$, to trans-critical flow, $Re > 3.6 \times 10^6$. Here $Re = u_0 D/\vartheta$ where u_0 is the free-stream velocity, D is the cylinder diameter and ϑ is the kinematic viscosity.

Many researchers have investigated the effect of various turbulent models using numerical simulation of flow around an isolated cylinder and cylinder near a wall. In all these researches, the force coefficients are estimated based on computational fluid dynamics (CFD) equations. These numerical estimations significantly depend on the correct simulation of the flow pattern at the computational domain. In order to simulate the 2D and 3D of fluid flow over a pipe in the vicinity of a rigid seabed, Li et al. (Li et al., 1997) and Lee et al. (Lee et al., 1994) employed finite element method to solve the incompressible Navier-Stokes equations and Smagorinsky subgrid scale model to simulate turbulent flow passing a circular cylinder near a rigid wall. The accuracy of the numerical results was approved by qualitative comparisons of Bearman and Zdravkovich (Barad et al.) measurements. They showed that both 2D and 3D simulations results are compatible with the measured data. Dongfang and Liang (Liang and Cheng, 2005) stated that the 2D SGS model over predict the fluctuation velocities at high Reynolds number ($Re = 7000$) in comparison with Bearman and Zdravkovich (Barad et al.) measurements. They also indicated that the obtained results for the SGS turbulence model significantly depend on employing computational mesh while the standard $k - \varepsilon$ turbulence model could simulate the time average velocity reasonably well regardless of computational mesh. In Kazeminezhad et al. (Kazeminezhad et al., 2010) numerical simulation, the standard $k - \varepsilon$ turbulent model was applied at $Re = 1.5 \times 10^4$. Their results showed very good qualitative agreement with reported experimental data.

In the current study, flow modeling around a circular cylinder near a plane boundary is investigated numerically by using Galerkin Finite Volume method and the Large Eddy simulation. The ability of this numerical simulation of simulating of flow pattern over a flat plate and isolated cylinder is proved by author in previous studies (Namazi-Saleh et al., 2016, Sabbagh-Yazdi and Namazi-Saleh, 2009, Namazi-Saleh and Sabbagh-Yazdi, 2009). In the present study, flow with different

Reynolds number of 9500 and 15000 and gap ratio of $h/D = 0.2$ and 0.4 is selected. In addition, the effect of mesh resolution on quality of results is discussed. Then the model is evaluated for the velocity profile at different location of flow around the cylinder. Moreover, flow pattern and mechanism of vortex shedding for a circular cylinder near a plane boundary are discussed.

2 MATHEMATICAL FORMULATION

The incompressible Navier-Stokes equations are used to model the flow field around a horizontal pipe, which can be written in the following:

$$\frac{\partial u_i}{\partial x_i} = 0 \qquad (1)$$

$$\frac{\partial u_i}{\partial t} + u_j \frac{\partial u_i}{\partial x_j} = -\frac{1}{\rho}\frac{\partial p}{\partial x_i} + \vartheta \frac{\partial}{\partial x_j}\left(\frac{\partial u_i}{\partial x_j} + \frac{\partial u_j}{\partial x_i}\right) - \frac{\partial \overline{u_i' u_j'}}{\partial x_j} \qquad (2)$$

where $x_i (i = 1, 2)$ point to the Cartesian coordinate system for horizontal and vertical directions. In the cited equations, u_j and u_i denote the velocities in vertical and horizon direction, p is the pressure, t is time marching, $\overline{u_i' u_j'}$ denote Reynolds stress tensor, and v is the kinematic viscosity. The Reynolds stress tensor is modeled based on Boussinesq assumption by introducing ϑ_t as eddy viscosity, following formulation is arrived at:

$$-\overline{u_i' u_j'} = \vartheta_t \left(\frac{\partial u_i}{\partial x_j} + \frac{\partial u_j}{\partial x_i}\right) - \frac{2}{3} k \delta_{ij} \qquad (3)$$

in which k is dimensionless turbulent kinetic energy and δ_{ij} is the Kronecker delta. The Sub-Grid Scale (SGS) model is used for turbulence modeling. The eddy viscosity $\vartheta_t = \vartheta_{SGS}$ is calculated as follows:

$$\vartheta_{SGS} = (C_s \Delta)^2 \left[1/2\, \overline{s_{ij}}\, \overline{s_{ij}}\right]^{1/2} \qquad (4)$$

For definition of ϑ_{SGS}, the Sub-Grid Scale model is utilized, where Δ is the area of a triangular cell and C_s is assumed as 0.10 (Sohankar, 2008).

In this study, the discretization process is based on "Galerkin Finite Volume Method" to derive the discrete formulas of the governing equations on triangular meshes. The following formula can be obtained:

$$\frac{\Delta W}{\Delta t} = -\frac{P}{\Omega}\left[\sum_{k=1}^{N_{face}}[\overline{F_c}(\Delta y) - \overline{H_c}(\Delta x)]\right] - \frac{P}{A}\left[\frac{2}{3}\sum_{k=1}^{N_{cells}}[\overline{F_d}(\Delta y) - \overline{H_d}(\Delta x)]\right] \qquad (5)$$

where, $\overline{H_c}, \overline{F_c}$, represent the average values of convective fluxes at the faces of control volume boundary and $\overline{H_d}, \overline{F_d}$ are the mean values of viscous fluxes determined at each triangle cell.

To develop the computational efficiency, various numerical techniques such residual smoothing, edge-base algorithm and Runge-Kutta multi-stage time stepping are applied (Hino et al., 1993).

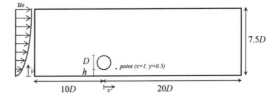

Figure 1. Definition sketch of computational domain.

In present study the laboratory boundary conditions of Oner et al. (Oner et al., 2008) and Jensen (Jensen, 1988) are modeled using rectangular domain, $7.5D$ in high and $30D$ in length. Hence, the ratio of gap (h) to the pipe diameter (D), in this study, is so small that the vortex shedding at downstream side of the pipe near rigid seabed is not as strong as isolated pipe. In order to reduce computational cost, the height of $7.5D$ is appropriate for this simulation. Figure 1 shows schematically the computational domain.

The boundary condition for the inflow current was specified to be equal to the unit free stream velocity, U_∞, and outflow boundary was specified as the unit free pressure. The non-slipping condition for solid wall nodes of the cylinder and the plane boundary layer was considered by setting tangential and normal components of velocity to zero. Moreover, the wall function is employed at the plane boundary layer.

3 RESULTS AND DISCUSSION

The appropriate time step length and the mesh size are the first and most important step in the simulation. Therefore, different test cases are done to examine the sensitivity analysis for the time step and mesh size in the present model. To do so, simulation of flow around a circular cylinder near a plane boundary has done in a case of $h/D = 0.4$ and $Re = 15000$. According to a similar study of Kazeminezhad et al. (Kazeminezhad et al., 2010), dimensionless time step of 0.02 is selected on three different mesh system such as coarse mesh, fine mesh and very fine mesh. Table 1 shows the number of nodes, number of cells, circumferential nodes number and number of nodes on the plain bed. The local meshes around the cylinder are also shown in Figure 2.

In order to explore the sensitivity of model on mesh size, the value of horizontal velocity (u/u_0) of specific point, at $x = 1$ and $y = 0.5$ see Figure 1, for three different mesh is illustrated in Figure 3. It can be seen that even though the coarse mesh cannot predict flow characteristic accurately, the predicted velocity obtained by fine mesh and very fine mesh are very close.

Further investigations in mesh resolution, using instantaneous lift force for three different mesh. Accordingly, the lift force is calculated as the following equation (Kazeminezhad et al., 2010):

$$F_L = \int_0^{2\pi} p \sin(\phi)\, r_0\, d\theta \qquad (6)$$

Table 1. Characteristics of the different mesh system.

Mesh name	number of nodes	number of cells	Circumferential nodes	Nodes on the plain bed
Coarse	2570	4810	160	80
Fine	5565	10690	200	140
Very fine	8788	13132	240	180

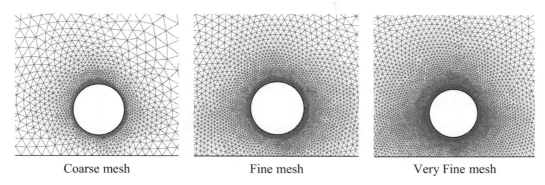

Coarse mesh   Fine mesh   Very Fine mesh

Figure 2. Computational mesh.

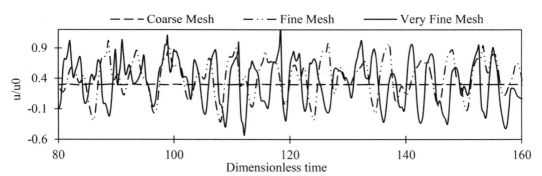

Figure 3. Instantaneous horizontal velocity of wake flow at specific point ($x = 1$ and $y = 0.5$) using different mesh resolution at $Re = 1500$.

Table 2. Lift force at different mesh system.

Mesh name	Coarse	Fine	Very fine
Lift force	0.023	0.121	0.143

Where p is pressure on the surface of the cylinder and θ is the angle in horizon and r_0 represents diameter of cylinder. The results show that the lift force on coarse mesh is much less than fine and very fine mesh, whereas there are no any remarkable difference between the results obtained by fine mesh and very fine mesh (see Table 2). As a result, it can be concluded that to reduce any computing costs, using fine mesh is sufficient to obtain the results, which are not dependent on mesh resolution.

For accurate prediction of hydrodynamic loads applied on the surface of the pipeline, the model should be able to simulate fluid flow around the pipe correctly. Hence, the present model is validated by comparison with the reported numerical data and experimental measurements of velocity and pressure. The simulation was performed based on Oner et al. (Oner et al., 2008) laboratory conditions to validate the present numerical simulation. Oner et al. (Oner et al., 2008) modeled a horizontal pipe near a plane boundary and measures the velocity field in a steady flow. The present model is evaluated for the two test cases with different gap ratio, $h/D = 0.2$ and 0.4, and two Reynolds number of 9500 and 1.5×10^4. In this study the important flow parameter such as velocity profile, pressure coefficient for this case study will be discussed.

Figure 4 illustrates the simulated velocity profile around a pipe with gap ratio of 0.2 at various cross sections and measured with numerical simulation of Kazeminezhad et al. (Kazeminezhad et al., 2010) and experimental data reported by Oner et al. (Oner et al., 2008). Kazeminezhad et al. (Kazeminezhad

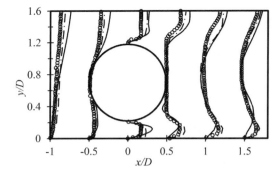

Figure 4. Comparison of mean horizontal velocity at different location of present simulation (——) with numerical-URANS simulation (— - —) (Kazeminezhad et al., 2010), and experimental reports (o) (Oner et al., 2008) ($h/D = 0.2$, $Re = 9500$).

et al., 2010) used two-dimensional numerical simulation to model the flow passing a pipe near the rigid seabed. Their simulation was based on finite volume model for RANS equations. As can be seen in Figure 4, the numerical modeling successfully captured the mean horizontal velocity profile around the cylinder. The model could simulate the wake at downstream of the pipe well, though it overpredicts the velocity captured near the pipe. This overestimation of the velocity profile is also reported by Kazeminezhad et al. (Kazeminezhad et al., 2010), Smith and Foster (Smith and Foster, 2005) and Liang et al. (Liang and Cheng, 2005), where all of them used 2D numerical simulation with $k - \varepsilon$ turbulence modeling.

Vortex shedding at the downstream side of the pipe is the most important fact, which strongly affects the hydrodynamic load applied to the pipe frequently. This oscillation causes a periodic variation of force on the pipe. Figure 5 shows the instantaneous the velocity and pressure distribution due to hydrodynamic force around the pipe near rigid seabed for the gap ratio of 0.4 at $Re = 15000$. The instantaneous streamline are visualized the simulated flow and shown in Figure 5(a). Since the selected gap ratio, selected in this simulation, is larger than the critical gap to diameter ratio and the vortex shedding is simulated at downstream side of the cylinder successfully.

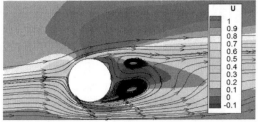

a) Instantaneous color-coded map of velocity with streamline around a circular cylinder.

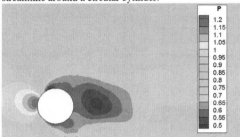

b) Instantaneous color-coded map of pressure around a circular cylinder.

Figure 5. Instantaneous (a) velocity (b) pressure distribution with vortex shedding pattern around.

of Oner et al. (Oner et al., 2008). This confirms that the model is capable of predicting the hydrodynamic pressure acting on the pipe properly. The following conclusions can be obtained.

- By using the Galerkin finite volume flow solver, there is no need to apply any reconstruction method and cumbersome matrix computations for explicit computation of the laminar and turbulent flow at the nodal points for unstructured triangular meshes.
- The LES turbulence model provides much deeper insight in modeling turbulent flow and interaction of wake and boundary layer compared with RANS model.
- The appropriate time step and mesh size are the most important step in numerical simulation to obtain accurate result with the lowest computing cost.

4 CONCLUSIONS

Flow around a circular cylinder near to a rigid boundary at subcritical Reynolds number has been simulated using the incompressible Navier-Stokes equations and SGS turbulence model. The discretization method was based the on artificial compressibility method and Galerkin finite volume approach. The hydrodynamic quantities such as pressure distribution and velocity profile around the cylinder have been predicted. The predictions are in fair agreement with numerical results reported by Kazeminezhad et al. (Kazeminezhad et al., 2010) and experimental measurement

ACKNOWLEDGMENT

The authors are grateful for the funding and facilities support by Universiti Teknologi PETRONAS.

REFERENCES

Barad, S. G., Ramaiah, P., Giridhar, R. & Krishnaiah, G. 2012. Neural network approach for a combined performance and mechanical health monitoring of a gas turbine engine. *Mechanical Systems and Signal Processing*, 27, 729–742.

Gao, F.-P., Yang, B., Wu, Y.-X. & Yan, S.-M. 2006. Steady current induced seabed scour around a vibrating pipeline. *Applied Ocean Research*, 28, 291–298.

Hino, T., Martinelli, L. & Jameson, A. A finite-volume method with unstructured grid for free surface flow simulations. Sixth International conference on numerical ship hydrodynamics, 1993.

Jensen, B. Large-scale vortices in the wake of a cylinder placed near a wall. Laser Anemometry-Advances and Applications, 1988. 153–163.

Kazeminezhad, M., Yeganeh-Bakhtiary, A. & Etemad-Shahidi, A. 2010. Numerical investigation of boundary layer effects on vortex shedding frequency and forces acting upon marine pipeline. *Applied Ocean Research*, 32, 460–470.

Lee, Y.-G., Hong, S.-W. & Kang, K.-J. A numerical simulation of vortex motion behind a circular cylinder above a horizontal plane boundary. The Fourth International Offshore and Polar Engineering Conference, 1994. International Society of Offshore and Polar Engineers.

Li, Y., Chen, B. & Lai, G. The numerical simulation of wave forces on seabed pipeline by three-step finite element method and large eddy simulation. The Seventh International Offshore and Polar Engineering Conference, 1997. International Society of Offshore and Polar Engineers.

Liang, D. & Cheng, L. 2005. Numerical modeling of flow and scour below a pipeline in currents: Part I. Flow simulation. *Coastal Engineering*, 52, 25–42.

Namazi-Saleh, F., John, K. V. & Mustaffa, Z. B. 2016. Numerical evaluation of galerkin finite volume solver for laminar/turbulent flow over flat plate *ARPN Journal of Engineering and Applied Sciences*, 11, 2393–2399.

Namazi-Saleh, F., Kurian, V. & Mustaffa, Z. 2014. Investigation of vortex induced vibration of offshore pipelines near seabed. *Applied Mechanics and Materials*. Trans Tech Publications Ltd.

Namazi-Saleh, F. & Sabbagh-Yazdi, S.-R. 2009. Iterative imposing near wall logarithmic velocity profile in turbulent flow pressure modeling over cylindrical for unstructured triangular meshes. *11th WSEAS International Conference on Mathematical and Computational Methods in Science and Engineering, MACMESE '09*. World Scientific and Engineering Academy and Society (WSEAS).

Oner, A. A., Kirkgoz, M. S. & Akoz, M. S. 2008. Interaction of a current with a circular cylinder near a rigid bed. *Ocean Engineering*, 35, 1492–1504.

Ong, M. C., Utnes, T., Holmedal, L. E., Myrhaug, D. & Pettersen, B. 2010. Numerical simulation of flow around a circular cylinder close to a flat seabed at high Reynolds numbers using a k–ε model. *Coastal Engineering*, 57, 931–947.

Sabbagh-Yazdi, S. & Namazi-Saleh, F. Compression of 2D Simulation Turbulent Flaw with/without Logarithmic Law Near Flat Plate. at 8th Internation Hydraulic conference in The Tehran Univesity, 2009.

Smith, H. D. & Foster, D. L. 2005. Modeling of flow around a cylinder over a scoured bed. *Journal of waterway, port, coastal, and ocean engineering*, 131, 14–24.

Sohankar, A. 2008. Large eddy simulation of flow past rectangular-section cylinders: Side ratio effects. *Journal of wind engineering and industrial aerodynamics*, 96, 640–655.

Coastal vulnerability index at a RAMSAR site: A case study of Kukup mangrove island

A.K.A. Wahab, D.S.M. Ishak & M.H. Jamal
Center for Coastal and Ocean Engineering, Research Institute for Sustainable Environment, Universiti Teknologi Malaysia, Kuala Lumpur, Malaysia

ABSTRACT: The coastal zone of Pulau Kukup is characterized by muddy substrates and is classified as a microtidal coast. The mangroves of Pulau Kukup are potentially threatened by accelerated Sea Level Rise (SLR). The study evaluates the Coastal Vulnerability Index (CVI) through the assessment of physical and environmental variables. The data sources were computed and ranked with a value of 1 to 5 based on the criteria stated for each variable. The ranking categories range from very low (1) to very high (5) vulnerability indices. From the data analysis, Pulau Kukup shorelines were dominated by the highly vulnerable sectors (42%), concentrated at the northern and the southern part of the island. Gentle coastal slopes were found at the northern shoreline due to the presence of a 800 ha mudflat. The least vulnerable sectors (16%) were found to be at the eastern part of Pulau Kukup, facing the mainland. The CVI studies revealed the role of dense mangroves around the study area. The existence of dense mangrove vegetation around the shoreline reduced the impacts from SLR to the shorelines. As a conclusion none of the sectors was analysed as having very high vulnerability to SLR.

1 INTRODUCTION

Natural threats originating from the ocean waters have become a major concern to coastal communities. Sea-Level-Rise (SLR) phenomenon was identified as one of the expected problem faced by intertidal communities resulting from global climate change (IPCC, 2001).The mean annual temperature for selected monitoring stations in Malaysia showed significant increase in warming trend in the past 40 years (Ng et al., 2005). The projection of global sea level rise by several models indicated an increase from 12–22 cm in seawater surface during the 20th century (Solomon et al., 2007). The long term effects from SLR will reduce the quality of intertidal ecosystems due to habitat deterioration (Gilman et al., 2007). However, many initiatives at the local and global levels have been developed to understand the shoreline behavior in view of impending SLR.

The Coastal Vulnerability Index (CVI) to SLR is defined as the degree to which a system is susceptible to, and unable to cope with the adverse effects of SLR (Rameiri et al., 2011).This approach was among the first vulnerability assessment studies developed and to date, it was the most widely used in the world. In Malaysia, the coastal vulnerability study was first embarked in 2007 (DID, 2007). As a pilot study, it was conducted at two sites with contrasting coastal characteristics namely Pulau Langkawi in Kedah and Tanjung Piai in Johor. The Malaysian National Coastal Vulnerability Index (NCVI) integrated the physical, socio-economic and environmental factors in the assessment tools. The data handling and ranking for the Physical Vulnerability Index (PVI) was adopted from the USGS method. The Socio-economic Vulnerability Index (SVI) was developed based on Cutter et al., (2000), while the Environmental Vulnerability Index (EVI) was computed based on the procedure applied by the South Pacific Applied Geo-Science Commission (SOPAC, 2005). Finally, the Total Composite Vulnerability Index (TCVI) is developed to integrate the data ranking from parameters identified in each physical, socio-economic and environment/biological vulnerability indices.

This paper was aimed to study and quantify the Coastal Vulnerability Index along the island perimeter to understand the shoreline behavior and the degree of exposure to the impacts of SLR. The coastal vulnerability index study at Pulau Kukup will be conducted based on the methodology adopted in the NCVI study (DID, 2007). However, for this study, only two vulnerability indices were considered. The status of Pulau Kukup as an uninhabited mangrove forest limited the socio-economic activities along the shoreline (i.e. no human population or demography survey can be evaluated at the island). From observations, only two SVI parameters were relevant to the study area. As a conclusion, only PVI and EVI were determined in this study to compute the vulnerability indices of the area.

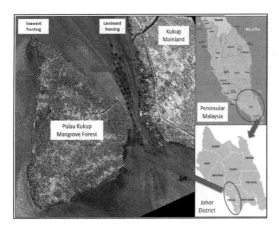

Figure 1. Location of the study area and the coastal sectors.

2 METHODOLOGY

2.1 Study area

Pulau Kukup mangrove forest was designated as RAMSAR site number 1287 in 2003 (Ramsar list, 2015). Pulau Kukup (Figure 1) is a unique example of an intact island which is a rarity in Malaysia. It was reported that Pulau Kukup is home to the critically endangered mangrove species, *Brugueira hainaseii* or commonly known as *Berus Mata Buaya* (IUCN, 2014). Moreover, despite its size, Pulau Kukup contained approximately 18 true mangrove species which represents the very rich species diversity (Ramsar, 2003). Pulau Kukup is significant as a flood control and physical protection to Kukup town. It helps to block the wave energy reaching Kukup town and provides a favorable condition for maricultural activities located close to the island. It is an uninhabited mangrove island located 1km offshore from the south–western tip of the state of Johor, Malaysia.

2.2 Data collection and analyses

The data collection for Coastal Vulnerability Index (CVI) is divided into 2 components which are physical and biological variables. Physical variables consists of seven parameters namely geomorphology, geologic material, regional coastal slope, shoreline changes rates, mean wave height, tidal range and relative sea level change (Table 1). Whereas, biological variables are defined by the percentage of vegetation cover and the number of endangered species present at the study area. The percentage of vegetation covers was identified from visual observation during the field survey and supported by the topography map published by Department of Survey and Mapping Malaysia. The number of endangered species at Pulau Kukup was gathered from literature survey and records from IUCN red list. The regional slope of the coastal zone is calculated from the grid of topographic map and bathymetric chart extending 5 km seaward of the shoreline. Digitial Shoreline Analysis System (DSAS) was adopted to run the End Point Rate (EPR) analysis for the identification of the historical shoreline movement. It was used to compute the statistical analyses from the distance of shoreline movement by the time elapsed between earliest and latest shoreline.

Table 1. Description of data collection for physical parameters at the study area.

No.	Parameter	Item	Data Source
1	Geomorphology	a. Field observation b. Topography map 　Sheet : 174 　Scale : 1: 50000 　Published year : 2003	Department of Survey and Mapping Malaysia
2	Geologic Materials	a. Field observation b. Topography map 　Sheet : 174 　Area : Johor 　Scale : 1: 50000 　Published year : 2003	Department of Survey and Mapping Malaysia
3	Regional coastal slope	a. Bathymetry chart 　Sheet : MAL 5129 　Scale : 1:75000 　Published Year : 2008	National Hydrographic Centre, Royal Malaysian Navy
4	Tidal range	Tidal level at standard port (Kukup station) published in Buku Jadual Pasang Surut Malaysia, Vol 1 (2014) Year : 1989-2003	National Hydrographic Centre, Royal Malaysian Navy
5	Mean wave height	The value was adopted from National Coastal Vulnerabilities Index Studies (DID, 2007)	Department of Irrigation and Drainage (DID)
6	Shoreline changes rate	a. SPOT Image 　Type: SPOT - 1xs. (Year : 2000) 　Resolution: 10m 　Type : SPOT-5 PNC (Year : 2005) 　Resolution : 2.5m b. World View 2 Image 　Type : 8-band Multispectral (Year 2011) 　Resolution : 2m	Malaysian Remote Sensing Agency and Digital Globe
7	Relative sea level change	The local rate was adopted from NAHRIM report	NAHRIM

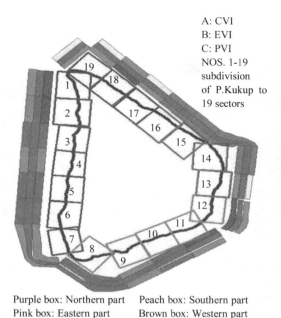

A: CVI
B: EVI
C: PVI
NOS. 1-19 subdivision of P.Kukup to 19 sectors

Purple box: Northern part   Peach box: Southern part
Pink box: Eastern part   Brown box: Western part

Figure 2. Distribution of Coastal Vulnerability Index (CVI) for Pulau Kukup.

The multidisciplinary data sources related to the physical and environmental parameters were evaluated and quantified to develop the Physical Vulnerability Index (PVI) and Environmental Vulnerability Index (EVI). The shorelines were divided into 19 coastal sectors of 0.5 km × 0.5 km each (Figure 2). The PVI and EVI score were integrated to form a single indicator in color-coded ranking for Coastal Vulnerability Index (CVI), (Table 2). The PVI and EVI parameters were calculated based on the standard equation developed by the USGS (Gornitz et al., 1997) and SOPAC (SOPAC, 2005) (Table 3).

3 RESULT AND DISCUSSION

From the analysis of the results, Pulau Kukup shorelines can be categorized into four vulnerability indices. The majority of the shorelines were classified into high-risk (42%), 26% was low-risk and 16% was classified as having moderate and very low-risk vulnerability (Figure 3). However, none of the sectors were classified as having a very high-risk index category. The majority of Pulau Kukup coastline was identified as being highly vulnerable, indicating that the Pulau Kukup shorelines has a high potential to be impacted by SLR.

Comparison was made with the output of CVI analysis from Tanjung Piai, Johor reported by DID, (2007). It was found that Tanjung Piai area was dominated by – very low vulnerability index (37.5%), followed by – high vulnerability index (20.8%). The area was also reported as highly vulnerable to the impact of SLR by 4.2%. The distribution of highly vulnerability index

Table 2. Color-coded Coastal Vulnerability Index Parameters.

Parameter	Very low (1)	Low (2)	Moderate (3)	High (4)	Very high (5)
Geomorphology	Rocky cliff coast	Medium cliff	Low cliff, alluvial plain	Cobbles	Mudflat, mangrove
Average Coastal Slope (%)	>1.2	1.2-0.9	0.9-0.6	0.6-0.3	<0.3
Mean Tidal Range (m)	>6.0	4.0-6.0	2.0-4.0	1.0-2.0	<1.0
Shoreline Changes Rate (m/yr)	>2.0	1.2 to 2.0	-1.0 to 1.0	-2.0 to 1.0	<-2.0
Mean Wave Height (m)	<0.55	0.55-0.85	0.85-1.05	1.05-1.25	>1.25
Relative Sea Level Change (mm/yr)	<1.8	1.8-2.5	2.5-3.0	3.0-3.4	>3.4
Vegetation Cover (%)	>80	60-79	40-59	20-39	<20
No. of Endangered Species	0	1	2	3	>3

Table 3. Computation of the Physical, Environmental and Coastal Vulnerability Indices.

Vulnerability Index	Parameter	Equation
Physical	- Shoreline changes rate, a - Mean tidal range, b - Geomorphology, c - Geologic material, d - Average coastal slope, e - Mean significant wave height, f - Relative sea level rise, g	$\sqrt{\dfrac{a \times b \times c \times d \times e \times f \times g}{7}}$ (1)
Environmental	- Vegetation cover, EVI-1 - Number of endangered species, EVI-2	$\dfrac{(EVI\text{-}1 + EVI\text{-}2)}{2}$ (2)
Coastal	- Physical Vulnerability Index, (PVI) - Environmental Vulnerability Index, (EVI)	$\dfrac{(PVI + EVI)}{2}$ (3)

was concentrated mainly at Tanjung Pelepas Port area. This is might be due to the developed area, where the vegetation has been replaced by man-made structures which classified to be moderate to highly vulnerable under EVI analysis. They accounted for 54.2% of the shoreline.

The northern and southern part of Pulau Kukup was dominated by highly vulnerable sectors. Sixty percent (60%) of the shorelines was categorized as highly vulnerable at both locations (sectors 2, 8, 9, 11, 18 and 19). Meanwhile, the eastern part of Pulau Kukup was the least vulnerable to SLR due to the fact that the shorelines face the mainland and are not directly exposed to the forces of the open sea. Seventy-five percent (75%) of the eastern shorelines are covered by the very low vulnerability index sectors which are sectors 14, 15 and 16.

The existence of mangrove vegetation around Pulau Kukup is significant in the CVI study. From the analysis, none of the sectors in Pulau Kukup was classified with very high risk index. It is believed that the dense mangrove forest had protected the coastal shorelines from the worst outcomes. Pulau Kukup shorelines are highly populated by mangrove vegetation in every sector. The presence of coastal wetland plant protects the shorelines through direct or indirect actions. The aboveground plant structure will directly dampen the wave energy while the roots systems will indirectly maintain the shoreline function by trapping sediment (Gedan et al., 2010).

Figure 4 shows the variability of the regional coastal slope around the study area based on sectors. The high vulnerability indices in the northern sectors recorded gentle slopes ranging from 0.04–0.37%. Whereas, the low vulnerability indices in the eastern part depicted the steepest slopes ranging from 0.45–1.98%. The gentle slope found in the northern part could be explained by a 800-ha mudflat deposition. This phenomenon contributed to the gentle sloping characteristics at sectors 1, 2, 17, 18 and 19. The average coastal slope indicated the relative vulnerability to inundation and the potential rapidity of the shoreline retreat (Brunel and Sabatier, 2009). A low-slope coastal regional is more exposed to the risk of flood inundation due to SLR. In gentle sloping coasts, any rise in seawater level would increase the inundation impacts to the mangrove island.

Based on the study, it showed that the region with gentler slopes will retreat faster and the effect of SLR was more prominent when compared to the steeper regions. The development of CVI for the study area is one of the significant approaches to provide the basic information on the degree of the shoreline susceptibility to sea level rises. The information on CVI will facilitate the future planning and management strategies to the prone area towards the impending SLR.

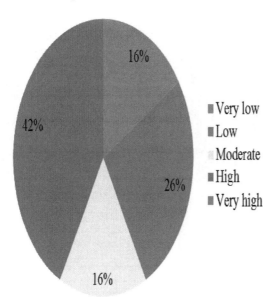

Figure 3. Percentage distribution of CVI at Pulau Kukup.

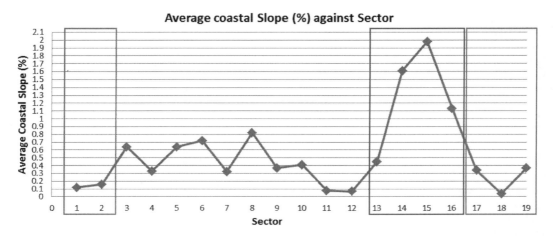

Figure 4. Average coastal slopes in every sector. Red box indicate the northern part, Green box indicate the eastern part.

4 CONCLUSION AND RECOMMENDATION

The Coastal Vulnerability Index at Pulau Kukup was classified into 4, -very low, -low, -moderate and -high vulnerability indices. None of the sectors was classified as having very high vulnerable index, which could be related to the role of extensive mangrove trees around the shoreline. Overall, CVI around Pulau Kukup shoreline can be classified into 42% (high vulnerability index), 26% (low vulnerability index) and 16% (moderate and very low vulnerability index). The variability of the regional coastal slopes, influence the distribution of CVI at the study area. The eastern part was found to have the steepest slopes which contribute to the least vulnerability index whilst the northern part recorded with the gentlest slope was identified as the most prone area to the projected SLR. Other than that, the incident waves, around the study area might also play a role in the data ranking classification.

ACKNOWLEDGEMENT

This paper has been made possible through the research work granted by Ministry of Education to Research Management Centre, Universiti Teknologi Malaysia through the Fundamental Research Grant Scheme (FRGS) using vote no. Q.J130000.7822.4F607.

REFERENCES

Brunel, C & Sabatier, F (2009), 'Potential influence of sea-level rise in controlling shoreline position on the French Mediterranean Coast', *Geomorphology*, vol.107, pp. 47–57.

Cutter, S.L., Mitchell J.T. and Scott M.S., (2000) Revealing the vulnerability of people and places: A case study of Georgetown County, South Carolina, *Annals of the Assoc. of American Geographers*, 90 (4), 713–737.

DID, (2007), *The National Coastal Vulnerability Index Study-Phase 1 Report*. Available from: Drainage and Irrigation Department (DID), Malaysia. Report submitted by Coastal and Offshore Engineering Institute [December 2007].

Gedan, KB, Matthew, L, Wolanski, E & Barbier EB (2010), 'The present and future role of coastal wetland vegetation in protecting shorelines: answering recent challenges to the paradigm', *Climatic Change*, vol. 106, pp. 7–29.

Gilman, EL, Ellison, J & Coleman, R (2007), Assessment of mangrove response to projected relative sea-level rise and recent historical reconstruction of shoreline position', *Environmental Monitoring Assessment*, vol.124, pp. 105–130.

Gornitz, VM, Beaty, TW & Daniels, RC (1997), *A coastal hazards data base for the U.S. West coast*. Available from: U.S Department of Energy, Environmental Sciences Division. [December, 1997].

IPCC (2001), Climate *change 2001: The scientific basis*. Cambridge University Press, Cambridge, United Kingdom, pp. 881.

IUCN, (2014), International *Union for Conservation Nature red list*. Available from: <http://www.iucnredlist.org/search> [27 December 2014].

Ng, PKL & Sivatoshi, N (1999), *A Guide to the mangrove of Singapore 1: The Ecosystem and Plant Diversity*, Singapore Science Centre, Singapore, pp. 101–103.

Rameiri, E. A. Hartley, B. Andrea, D. Filipe Santos, G. Ana and H.Mikael (2011). Method for Assessing Coastal Vulnerability to Climate Change. ETC CCA Technical Paper 1/2011.

Ramsar List. Ramsar.org [Retrieved on 13 Nov 2015].

Ramsar, Information Sheet on Ramsar Wetlands (RIS). (2003). Categories approved by Recommendation4.7, Ramsar Convention Bureau, Rue Mauverney 28, CH-1196 Gland, Switzerland. (pp. 7).

Solomon, S., Quin, D., Manning, M., Chen, Z., Marquis, M., Averyt, K.B., Tignor, M. & Miller, H.I. (2007), Climate change 2007: The physical science basis. Cambridge University Press, Cambridge, United Kingdom.

SOPAC., (2005). *Building Resilience in SIDS: The Environmental Vulnerability Index* Report by South Pacific Applied Geoscience Commission (SOPAC). United Nations Environment Programme (UNEP).

Effects of mooring lines and water depth parameters on the dynamic motion of a turret-moored FPSO

Montasir Osman Ahmed & Ashraf Musaad
Department of Civil and Environmental Engineering, Universiti Teknologi PETRONAS, Seri Iskandar, Perak, Malaysia

ABSTRACT: This research studies the effects on the dynamic responses of a turret-moored FPSO with respect to the changes in the mooring lines' configurations that include pretension forces and azimuth angles as well as the changes in the water depth. Numerical calculation is carried out based on a time domain analysis using the computational fluid dynamics' software STAR-CCM+. The dynamic responses in the heave, surge and pitch motions are studied when the FPSO is subjected to an environmental conditions based on a 100-yr hurricane condition for the Gulf of Mexico (GoM).

1 INTRODUCTION

The oil and gas industry nowadays is considered one of the fastest growing industries in the world, and that is due to the increasing demand of oil and gas all over the globe, which is proportional to the population growing economy. To meet this demand different exploitation and production alternatives are studied. FPSO (Floating, Production, Storage and Offloading vessel) is one of these alternatives that provides a low cost, reliable and more flexible offshore exploitation in various water depth. Consequently, the dynamic responses of the FPSO in both translational and rotational directions need to be studied in order to obtain a good station keeping of the vessel.

Many researchers have studied the dynamic responses of FPSOs under various environmental conditions and various water depth. Lingzhi et al. (2015), for example studied the dynamic responses of FPSO in shallow water and concluded that the accurate prediction of the dynamic responses is more difficult in shallow water because of the very low frequency motions caused by waves in shallow water. Moreover, it was found that the reduction of the water depth causes an increment in the sway, surge and yaw motion while it causes a decrease in the heave, roll and pitch motions.

A similar research by Li et al. (2003) on the responses of an FPSO in shallow water at a depth ranging from 21 to 26 m showed a reduction in the responses of the platform in the roll, pitch and heave directions as the water depth is reduced. Moreover, a reduction in the wave frequency motions of the platform was recorded as the water depth is reduced.

In addition, in shallow water a significant increase in the low frequency wave drift forces occurred causing an increase in the surge low frequency responses. These changes in the low wave drift forces can be a result of the variations in the waves natural frequency as well as the nonlinearity of the waves in shallow water (Xiao et al., 2007).

On the other hand, several researchers have been studying the dynamic responses of coupled and non-coupled FPSOs to understand the effects of the mooring systems on the global performance of the FPSO in order to provide a good station keeping of the vessel. When uncoupled analysis is used and the drag forces on the mooring systems and risers are neglected, mooring/riser dynamic tension could be hugely over-predicted and a great increase is witnessed in the sway and surge responses (Tahar and Kim, 2003).

This research shall contribute in filling the gap by studying the effects on the dynamic responses of turret-moored FPSOs when operating in shallow water, which will aid the industry to study the feasibility of using floating structures in shallow water, to reduce the costs associated with using fixed platform especially for production in marginal oil and gas fields.

Moreover, it will aid in the understanding of the optimum mooring lines' configuration for a turret-moored FPSO to provide more stability and safety during operation.

This study is determining the dynamic responses of a turret-moored Floating, Production, Storage, and Offloading (FPSO) platform by the means of numerical modeling in the time domain using the computational fluid dynamics' software STAR-CCM+, to investigate the heave, surge and pitch dynamic motions. The different environmental loadings and conditions that are used are wind, current and waves. The diffraction theory is used for computing the wave forces, and the wave profile of random one-direction waves is generated using JONSWAP wave spectrum. A similar data to the FPSO in the DeepStar study used by Wichers and Devlin (2001) is used for the modeling of the FPSO and the mooring system for validation purposes.

2 DESCRIPTION OF FPSO AND MOORING LINES

The FPSO and the mooring lines particulars in this research are similar to those used in the DeepStar study as mentioned earlier. The FPSO produces 120,000 bpd and has a storage capacity of 1,440,000 bbls, with an LBP of 310 m, a depth of 28.04 m, a beam of 47.17 m and a size of 200 KDWT.

The mooring system is a single point mooring (turret-mooring) consisting of a 12 chain-polyester-chain mooring lines arranged into fours groups of lines where each group is consisting of 3 lines that are 5o apart. The FPSO and the mooring lines' particulars are shown in Table 1 and Table 2 respectively.

3 ENVIRONMENTAL DATA

The environmental conditions will be based on a 100-yr hurricane condition for the Gulf of Mexico (GoM) with non-parallel current, wind and wave. The numerical data obtained from OCIMF (Oil Company International Marine Forum, 1994) will be used for the calculation of current and wind forces, and the wave profile will be generated using JONSWAP spectrum. The environmental loadings conditions are defined in Table 3.

Table 1. FPSO main particulars.

Description	Symbol	Unit	Quantity
Vessel size	–	kDWT	200.0
Length between perpendiculars	Lpp	m	310.0
Breadth	B	m	47.17
Depth	H	m	28.04
Draft	T	m	18.9
Length beam ratio	L/B		6.57
Beam draft ratio	B/T		2.5
Displacement		MT	240,869
Block coefficient	C_b		0.85
Water plane area	A	m^2	13,400
Water plane coefficient	C_w		0.9164
Center of gravity above base	KG	m	13.32
Metacentric height transverse	MGt	m	5.78
Metacentric height longitudinal	MGl	m	403.83
Transverse radius of gyration in air	K_{xx}	m	14.77
Longitudinal radius of gyration in air	K_{yy}	m	77.47
Yaw radius of gyration in air	$K_{\psi\psi}$	m	79.30
Wind area frontal	A_f	m^2	1,012
Wind area side	A_b	m^2	3,772
Turret in centerline behind Fpp (20.5% Lpp)	–	m	63.55
Turret elevation below tanker base		m	1.52
Turret diameter	–	m	15.85

Table 2. Mooring systems particulars.

Designation	Unit	Quantity
Water depth	m	914
Pre-tension	kN	1201
Number of lines	–	4 × 3
Degree between the three lines	deg	5
Length of mooring line	m	2088
Radius of location of chain stoppers on turn table	m	7.0
Segment 1 (ground section): chain		
Length at anchor point	m	914.4
Diameter	mm	88.9
Dry weight	N/m	1617.10
Weight in water	N/m	1406.90
Stiffness AE	kN	794484
Mean breaking load (MBL)	kN	6512
Segment 2: polyester rope		
Length	m	1127.80
Diameter	mm	88.9
Dry weight	N/m	412.23
Weight in water	N/m	349.75
Stiffness AE	kN	689858
Mean breaking load (MBL)	kN	6418
Segment 1 (ground section): chain		
Length at anchor point	m	45.7
Diameter	mm	88.9
Dry weight	N/m	1617.09
Weight in water	N/m	1406.89
Stiffness AE	kN	794484
Mean breaking load (MBL)	kN	6512

4 NUMERICAL MODELLING

4.1 *Wave theory*

Determining the wave kinematics such as the particle velocity, dynamic pressure and the acceleration can be carried out through various wave theories. The selection of the appropriate wave theory depends on certain environmental parameters, e.g., wave height, wave period and water depth. All the water waves are assumed to be periodic and uniform, having a height H and period T.

According to Chakrabarti (1987) Stokes nonlinear theory is more appropriate in deep water ($d/T^2 > 1.0$) compared to the Linear Airy wave theory, which is mostly used in shallow water.

Stokes used the perturbation technique to obtain further approximations of the theory, such that the

Table 3. Environmental loadings used for simulation.

Description	Unit	Quantity
Wave		
Significant wave height, H_s	m	12.19
Peak periods, T_p	s	14
Wave spectrum	JONSWAP (GAMMA = 2.5)	
Direction	deg	180[a]
Wind		
Velocity at 10 m above MWL	M/s	41.12
Spectrum	NPD	
Direction	deg	150[a]
Current		
Profile		
at free surface 0 m	m/s	0.9144
at 60.96 m	m/s	0.9144
at 99.44 m	m/s	0.0914
on the sea bottom	m/s	0.0914
Direction		210[b]

higher-order solutions are in order of magnitude smaller than the immediate lower one.

The dynamic pressure is given by

$$p = \rho g \frac{H}{2} \frac{\cosh ks}{\cosh kd} \cos\theta + \frac{3}{4} \rho g \frac{\pi H^2}{L} \frac{1}{\sinh 2kd}$$
$$\times \left[\frac{\cosh 2ks}{\sinh^2 kd} - \frac{1}{3}\right] \cos 2\theta - \frac{1}{4}\rho g \frac{\pi H^2}{L} \frac{1}{\sinh 2kd} \times [\cosh 2ks - 1] \quad (1)$$

The wave velocity or celerity given to third order is

$$c^2 = C_o^2 \left\{1 + \left(\frac{\pi H}{L}\right)^2 \left[\frac{5 + 2\cosh 2kd + 2\cosh^2 kd}{8\sinh^4 kd}\right]\right\} \quad (2)$$

where c_o is the celerity given by linear wave theory ($c_o^2 = (g/k)\tanh kd$) and the term within the bracket is the correction term.

The fifth-order velocity potential is written in a series form as

$$\Phi = \frac{c}{k}\sum_{n=1}^{5} \lambda_n \cosh nks \sin n\Theta \quad (3)$$

where the non-dimensional coefficient, λ_n are written as

$$\lambda_1 = \lambda A_{11} + \lambda^3 A_{13} + \lambda^5 A_{15}$$
$$\lambda_2 = \lambda^2 A_{22} + \lambda^4 A_{24}$$
$$\lambda_3 = \lambda^3 A_{33} + \lambda^5 A_{35}$$
$$\lambda_4 = \lambda^4 A_{44}$$
$$\lambda_5 = \lambda^5 A_{55}$$

The wave height is given by

$$H = \frac{2}{k}[\lambda + B_{33}\lambda^3 + (B_{35} + B_{55})\lambda^5] \quad (4)$$

According to stokes fifth-order wave theory, the celerity is given by

$$c^2 = c^2{}_o(1 + \lambda^2 c_1 + \lambda^4 c_2) \quad (5)$$

where λ is an unknown along with the wave number. The quantities λ and k are determined from equations (4) and (5) using an iterative technique. After knowing the potential Φ, the kinematics of the water particles and the dynamic pressure from Bernoulli equation can be found.

4.2 Equation of motion

In order to determine the motions of the structure, equations of motion in different degrees of freedom must be solved. These equations are generally non-linear due to the existence of non-linear damping and exciting force, in addition to a nonlinear restoring force. As a result of all the nonlinearities, these equations generally can only be solved by using numerical means.

The equation of motion for a system with mass, m and spring constant, K, which is linearly damped with a damping coefficient, C and an external force of amplitude, F_0 and frequency, ω is given by:

$$mx''(t) + Cx'(t) + Kx(t) = F_0 \sin \omega t \quad (6)$$

4.2.1 Added mass and damping coefficients

Floating structures' motions are influenced by the damping introduced by the motion of the structure in the water and by the added mass effect in the water. Prior to conducting a motion analysis these quantities must be known. Usually, these quantities are obtained from experiments when small structures are analyzed, but on the other hand, when the structure size is large these coefficients can be analytically obtained.

When using the experiment case, the damping coefficient and the added mass may be obtained from (7) and (8) respectively.

$$\zeta = \frac{C}{C_c} \quad (7)$$

where C_c is the critical damping coefficient given by $C_c = \sqrt[2]{Km}$ and ζ is called the damping factor.

$$\omega_n = \sqrt{\frac{K}{m}} \quad (8)$$

where ω_n is the natural frequency of the system.

4.2.2 Stiffness

The variation of the net weight and the buoyancy load with the changes in position from the datum is specified by the hydrostatic stiffness matrix. The components of stiffness can be specified for the roll, pitch and heave degrees of freedom, while it is considered as zero for the rest of the degrees of freedom.

The definition of the stiffness matrix for a system of M mooring lines is

$$k = \sum_{m-1}^{M} K^m = \sum_{m=1}^{M} -\frac{\partial F_m}{\partial x} \quad (9)$$

where K^m is the mooring line m stiffness matrix, while X is the translational and rotational vector which is given by $X^T = [X_o Y_o Z_o \theta_X \theta_Y \theta_Z]$ where $X_o Y_o Z_o$ are the displacement of the center of gravity and $\theta_X \theta_Y \theta_Z$ are the rotations of the floating body around the X, Y and Z axes. F_m is the vector of the reaction forces and the corresponding reaction moments that the mooring line m exerts on the floating structure.

4.2.3 Wave forces

The size of the FPSO is large, compared to the length of the wave, and due to that reason Morison's equation is no longer applicable. Moreover, Morison's equation in an empirical formula based on cylindrical members which is another limitation why it can't be used for computing the wave forces in this project. Instead the diffraction theory is considered when calculating the wave forces. According to Chakrabarti (1987), when a vessel is subjected to incident waves, there are two distinct effects of these waves as they meet the vessel. These waves are reflected or diffracted as they hit the barge and scattered in all directions. Moreover, these waves will set the body in motion, since the vessel is free to move in surge. The body's motion will produce waves that will radiate in all directions. Forces that are proportional to the amplitude of motion (or velocity) of the body are produced by the radiated waves. If this force is broken up into two components, then the added mass will be given by the component in phase with the body acceleration, while the damping force coefficient will be given by the component in phase with the body velocity. The force in a particular direction is computed by integrating the pressure in that direction over the submerged surface. The fluid pressure is calculated using Bernoulli's equation,

$$p = -p\frac{\partial \Phi}{\partial t} - \frac{1}{2}p(\nabla \Phi)^2 \quad (10)$$

where

$$p = \sum_{n=1}^{\infty} \varepsilon^n p_n \quad (11)$$

Substituting the value of Φ, the first and the second order dynamic fluid pressures are given by

$$p_1 = p\frac{\partial \Phi_1}{\partial t} \quad (12)$$

$$p_2 = p\frac{\partial \Phi_2}{\partial t} + \frac{1}{2}p(\nabla \Phi_1)^2 \quad (13)$$

thus,

$$F_{nj} = \varepsilon^n \iint p_n n_j \, dS \quad (14)$$

where F_{nj} is the nth-order force in the jth direction, S is the submerged surface, and n_j is the direction normal in the jth direction of interest.

5 RESULTS AND DISCUSSION

In each simulation only the parameter under the study is changed while the rest of the parameters are kept constant in order to examine the effects of the variation of that parameter on the dynamic motion of the FPSO.

5.1 Validation

Prior to the start of the parametric study, a full simulation is conducted in order to verify the validity of the modeling process by comparing the results obtained from the simulation to the results obtained from the DeepStar study (Wichers and Devlin, 2001). The results obtained from the simulation showed comparable values in the heave, surge and pitch motions, which are shown in figures 1, 2 and 3 respectively.

Table 4. The simulation cases based on the changing variable.

Variable	Case	Value
Water depth	Case 1	914 m
	Case 2	80 m
Pretension	Case 1	1200 kN
	Case 2	1500 kN
	Case 3	2200 kN
Azimuth angle	Case 1	30 degrees rotation
	Case 2	60 degrees rotation
	Case 3	90 degrees rotation

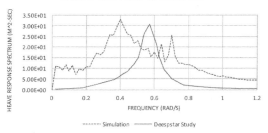

Figure 1. Comparison of simulation and the DeepStar study heave motion.

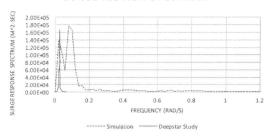

Figure 2. Comparison of simulation and the DeepStar study Surge motion.

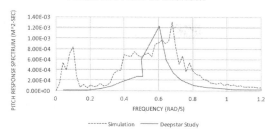

Figure 3. Comparison of simulation and the DeepStar study pitch motion.

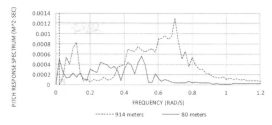

Figure 6. Comparison of pitch motion at shallow and deep water.

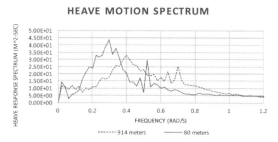

Figure 4. Comparison of heave motion at shallow and deep water.

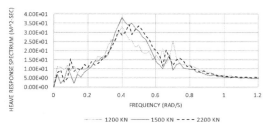

Figure 7. Comparison of heave motion at different mooring-lines' pretension forces.

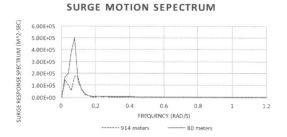

Figure 5. Comparison of surge motion at shallow and deep water.

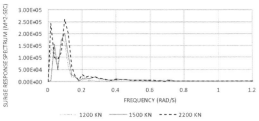

Figure 8. Comparison of surge motion at different mooring-lines' pretension forces.

5.2 Water depth

The Change in the water depth to 80 meters showed a significant increase in the surge motion energy density to about 5×10^5 m²-sec at a frequency of 0.1 rad/sec as shown in Fig. 5. Likewise, the heave motion energy density increased to 45 m²-sec at a frequency of 0.3 rad/sec as shown in Fig. 4. Contrarily, the pitch motion energy spectrum deceased with the reduction in the water depth yielding a maximum value of 6×10^{-4} m²-sec at a frequency of 0.5 rad/sec as shown in Fig. 6.

5.3 Pretension

Three simulations were conducted to study the effects of increasing the mooring lines' pretension forces on the dynamic motion of the FPSO, which were at 1200 kN, 1500 kN and 2200 kN. This increment yielded a significant rise in the pitch motion energy density to about 1.5×10^{-3} m²-sec at a frequency of 0.1 rad/sec as shown in Fig. 9, this low frequency motion is caused by the low frequency wave forces which are known as the second order drift forces. On the other hand, the heave and surge motion spectrum witnessed a minimum increment with the increase in the pretension forces as shown in figures 7 and 8 respectively.

5.4 Azimuth angle

The effects of the variation in the azimuth angle on the dynamic motion of the FPSO are studied at 30 degrees and 60 degrees with respect to the wave direction axis, which is represented by the 90 degrees' graph in the following figures. The increment in the azimuth showed an increase in all the motions spectrum, but most significantly in the heave motion with an energy density of 60 m²-sec at a frequency of 0.3 rad/sec as shown in Fig. 10.

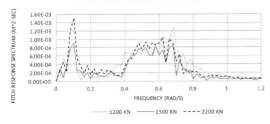

Figure 9. Comparison of pitch motion at different mooring-lines' pretension forces.

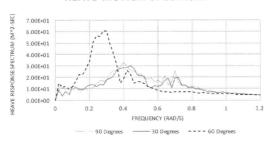

Figure 10. Comparison of heave motion at different azimuth angles' configurations.

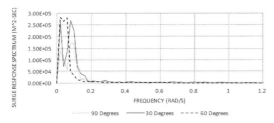

Figure 11. Comparison of surge motion at different azimuth angles' configurations.

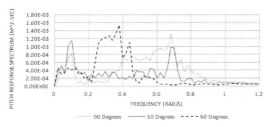

Figure 12. Comparison of pitch motion at different azimuth angles' Configuration.

6 CONCLUSIONS

This research studied the effects on the dynamic motion of a turret-moored FPSO with respect to the changes in the mooring lines configurations as well as the water depth, and the following conclusions can be drawn:

- Analysis in shallow water showed a significant increment in the surge motion, a minimum increment in the heave motion, while the pitch motion decreased notably.
- Increasing the mooring lines' pretension increases the dynamic responses in the heave, pitch and surge motions, with a significant increase in the pitch motion.
- Increasing the azimuth angle of the mooring lines with respect to the wave direction significantly increases the heave motion and a minimum increment is witnessed in the surge and pitch motions.

REFERENCES

Chakrabarti, S.K. (1987). Hydrodynamics of Offshore Structures. Computational Mechanical Publications. Southampton Boston.

Lingzhi Xiong, Haining Lu, Jianming Yang, Wei Zhang and Guang Yang. (2015). Experimental Investigation on Global Responses of a Large Floatover Barge in Extremely Shallow Water. In: Proc. 25th Int. Offshore Polar Engr. Conf., ISOPE. USA: pp. 1098–6189.

Tahar, A. and Kim, M. H. (2003). Hull/mooring/riser Coupled Dynamic Analysis and Sensitivity Study of a Tanker-based FPSO. Applied Ocean Research, 25(6), pp. 367–382.

Wicher, J.E.W. and Devlin, P.V. (2001). Effect of Coupling of Mooring Lines and Risers on the Design Values for a Turret Moored FPSO in Deep Water of Gulf of Mexico. In: Proc. of 11th Intl. Offshore & Polar Engr., ISOPE. Norway: pp. 480–487.

Xiao Lomg-fie, Yang Jian-min and Li Xin, (2007). Shallow Water Effects on Surge Motion and Load of Soft Yoke Moored FPSO. China Ocean Engineering, pp. 187–196.

Xin Li, Jainmin Yang and Longfei Xiao. (2003). Motion Analysis on a large FPSO in Shallow Water. In: Proc. 13th Int. Offshore Polar Engr. Conf., ISOPE. France: pp. 1098–6189.

VIV of four cylinders in side by side configurations – experimental and CFD simulations

A.M. Al-Yacouby, V.J. Kurian & M.S. Liew
Department of Civil and Environmental Engineering, Universiti Teknologi PETRONAS, Seri Iskandar, Perak, Malaysia

V.G. Idichandy
Department of Ocean Engineering, IIT Madras, India

ABSTRACT: Vortex Induced Vibrations (VIVs) of circular cylinders is a complex phenomenon encountered in many engineering fields such as marine and civil applications. This paper presents experimental and Computational Fluid Dynamic (CFD) simulation results for group of four cylinders, arranged in 2×2 rectangular arrays, subjected to VIVs. The physics involved in VIV is not fully understood, especially when a cylinder is falling in the wake of another cylinder. The objective of this study is to determine the effect of center to center spacing between the cylinders on VIV forces and response amplituds. The CFD simulation was conducted using Large Eddy Simulation (LES) method. The model test was carried out in the offshore engineering laboratory at Universiti Teknologi PETRONAS, Malaysia. The range of Reynolds (Re) Number achieved in the present study varied between 4.96E+03-2.48E+04. The comparison of the results shows that CFD simulations outcome is in good agreement with the experimental results.

1 INTRODUCTION

Group interference of slender structures is one of the complex area of research in ocean engineering and fluid dynamics due to its vast engineering applications. For instance, in marine and civil applications, production and drilling risers, tensioned leg platform (TLP) tendons, subsea group pipelines, transmission lines, and heat exchangers, group interference of VIV has been the subject of some successful studies in the past due to its great influence on several characteristics of fluid forces (Sanaati and Kato 2013). Generally, estimation of VIV forces and the corresponding responses on array of tubular cylinders with sufficient accuracy is a complex task, as the shielding effects on VIV are not fully understood. Due to flow separation and shedding of vortices, oscillating fluid forces develop and result in structural vibrations (Tofa, Maimun et al. 2014). The review of the literature indicates that during the last decade, successful research have been conducted to study the VIV of circular cylinders after the pioneering work of Feng (Feng 1968). For instance, overviews on the topic was presented by (Sarpkaya 1979), (Bearman 1984), (Blevins and Coughran 2009) and (Williamson and Govardhan 2004). For the latest experimental studies on VIVs, one can refer to several original research studies available in the literature such as (Vandiver 2012), (Roshko 2012, Zhao and Cheng 2014) and (Narendran, Murali et al. 2015). However, the majority of earlier studies focused on single degree of freedom cylinders where the model is allowed to oscillate along the cross-flow direction. On the other hand, CFD simulation for VIV of circular cylinders has become a promising area of research, especially in the last decade due to the advanced high performance computing facilities. However, many of these investigations are limited to 2D simulations at low Re Number as reported by (Saltara, D'Agostini Neto et al. 2011). Further, efficient examples of CFD simulations are presented by (Willden and Graham 2001), (Meneghini, Saltara et al. 2004) and (Bao, Huang et al. 2012). Many of the CFD computation studies available in the literature depend on $k-\varepsilon$ turbulence model, however, the present study is based LES at comparatively high Re Number.

2 CFD SIMULATIONS

CFD calculates numerical solutions using the equations governing fluid flow. In this study, CFD simulation was done using ANSYS Fluent V. 15.0.7. The flow is governed by the well-known Navier–Stokes equations. The turbulent flow was solved using LES. This method was initially proposed by Smagorinsky (Smagorinsky 1963) to simulate atmospheric air currents. In this model, the large-scale structure of the turbulent flow is computed directly, while the effects of the smallest isotropic eddies are modeled (Pletcher, Tannehill et al. 2012). The reason is that largest

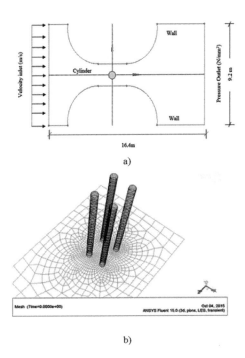
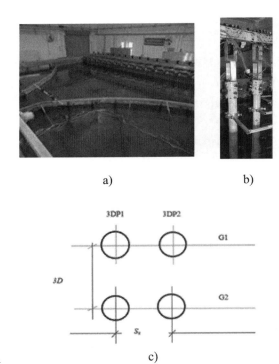

Figure 1. a) Flow domain, b) meshing details of the test section and the group of cylinders.

Figure 2. Details of the model setup: a) wave tank, b) cylinders fitted with VIV force sensors, c) model configuration for 2 × 2 rectangular arrays, with $S_x = $ 3D, 4D and 5D.

eddies are stimulated by the boundary conditions, and because it carry most of the Reynolds stress, it has to be computed. But the small scale turbulences are weaker and their contribution to the Reynolds stress are weaker and less critical (Wilcox 2006). The CFD solution was processed using High Performance Computers. The processing was performed using 10 supercomputers in parallel. The time was selected to be transient with a time interval of 1 millisecond i.e. ($\Delta t = 0.001$ seconds). The flow domain, and the meshing of the cylinders are presented in Figs. 1a and 1b respectively.

3 EXPERIENTIAL SET-UP

3.1 Wave tank details

The VIV tests were conducted in the offshore engineering laboratory in UTP. Details of the tank is depicted in Fig. 2a. The models used are fixed flexible cylinder. Instead of using the conventional spring setup normally adopted in VIV model tests, the VIV force sensor was developed custom-made for this purpose. The VIV force sensor depicted in Fig. 2b has 2 mm thick walls that act as spring, allowing the cylinder to oscillate in x and y directions.

The top portion of the sensor acts as a spring permitting motions in x direction, while the bottom block allow the system to move in the y direction. The motion of the cylinder was captured using Optic-Track system with three cameras. The cylinders are arranged in a 2 × 2 rectangular array. Thus, to study the effect of cylinders proximity on VIV responses, the center to center spacing along the flow direction was varied as $S_x = $ 3D, 4D and 5D, while keeping the spacing along the transverse direction constant i.e. $S_y = $ 3D as presented in Fig. 2c.

3.2 Model detail

In the present study, cylinders with outer diameter $D = 42$ mm with a total length of 1.32 m and wall thickness of 3 mm were used.

4 RESULTS AND DISCUSSIONS

Figure 3 shows the results of CFD simulation for four cylinders, with $S_x = $ 4D, subjected to uniform flow with $U = 0.43$ m/s. The graph indicate that the two cylinders which are falling in the wake of the leading cylinders are subjected to low current velocity (Fig. 3a) due to the blockage effects. In addition, one can observe that although the spacing between the cylinders presented in Fig. 3 corresponds to $S_x = $ 4D, the vortices developed in the wake of the leading cylinders influence the incident flow as presented in Fig. 3b.

The variation of flow velocity around the cylinder causes pressure differences as shown in Fig. 3c. The highly unstable boundary layer which separate from the surface of the cylinder forms the vortex shedding, which consequently forces the cylinder to oscillate in the transverse direction.

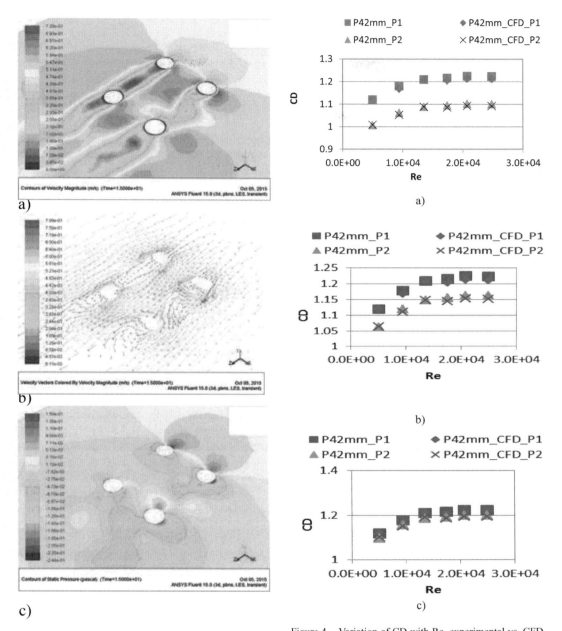

Figure 3. CFD Simulation results: a) velocity contour, b) velocity vectors, c) pressure distribution.

Figure 4. Variation of CD with Re, experimental vs. CFD, a) $S_x = 3D$, b) $S_x = 4D$, and c) $S_x = 5D$.

4.1 Drag coefficient

Drag coefficient is a non-dimensional form of the drag force. Generally, the total drag coefficient is estimated as $C_D = F_D/0.5\rho U^2 DL$, where F_D is the drag force, ρ is fluid density, U is the fluid velocity and D is the cylinder diameter. The total drag force was determined experimentally using the wave tank model test results. Figures 4a–4c show the variation of C_D with Re Number for a group of cylinders arranged in different center to center spacing, subjected to uniform flow. The graphs also provide comparison of CFD plotted against the experimental results of drag coefficients. In all the three cases, where the center to center spacing between the cylinders was varied as $S_x = 3D$, 4D and 5D, the values of C_D determined using CFD are comparatively smaller than the experimentally determined results, although the difference is not exceeding 5%. In addition, the results indicate that the values of C_D for the leading cylinders are generally higher than that of the trailing cylinders. For $S_x = 3D$, the maximum value for C_D for P1 determined experimentally and numerically are 1.23 and 1.21 respectively, recorded at Re = 2.07E+04, while the values of C_D corresponding to the second cylinder (P2) were 1.1025 (Exp.) and 1.09 (CFD) respectively, showing a difference of 9% between P1 and P2 (refer G1, Fig. 2c). Further, increasing the value of S_x to 4D, one can noticed that the values of C_D corresponding to P2 were determined as 1.16 (Exp.) and 1.15 (CFD), showing a slight decrease in the drag coefficients as compared to the previous case. Similarly, from Fig. 4c, where $S_x = 5D$, one can observe that the difference between the C_D values of the different tests configurations become negligible as the spacing between the cylinders decreases. This indicates that as the center to center distance between the cylinders increases, the influences of cylinder proximity decreases, and the variance becomes insignificant. Although, the array of cylinders presented in Fig. 2c consists of four cylinders, only the results corresponding to cylinders in group 1, denoted by G1 have been discussed and presented in this paper due to geometric similarity between G1 and G2. Generally, for an intermediate range of separations i.e. 3D to 20D for two flexibly mounted cylinders in tandem arrangement, several investigators (Bokaian and Geoola 1984), (Hover and Triantafyllou 2001) reported the existence of different regimes for the vibration of downstream cylinder, such as the VIV, upstream wake-excited galloping, or their combinations.

4.2 Lift coefficients

The lift coefficient (C_L) is a dimensionless parameter that relates the lift generated by a lifting body to the fluid density around the body, the fluid velocity and an associated area. The fluctuating lift coefficients provide an accessible record for estimation of vortex shedding frequency. The lift coefficient can be estimated as $C_L = F_L/0.5\rho U^2 D$, where F_L represents the lift force, ρ is the fluid density, D is the cylinder diameter, U is the flow velocity. In the present study, the total lift force was measured using a special force sensor developed for measuring the instantaneous time series of the forces. Figs. 5a–5c illustrate the variation of lift coefficients with $S_x = 3D$, 4D and 5D respectively. Figure 5a presents a comparison showing the variation of C_L with Re for P1 and P2, both experimentally and numerically using CFD simulation. In contradiction to the case of drag coefficients discussed in the previous section, here one can observe that the trailing cylinder has higher lift coefficients as compared to the leading cylinder. This observation is applicable for $S_x = 3D$, 4D and 5D, although at $S_x = 5D$, the influences of the cylinder proximity was negligible, and not exceeding 5% as depicted in Fig. 5c.

4.3 Strouhal number

This parameter is expressing the frequency of vortex shedding non-dimensionally and can be determined as $St = f_s D/U$, where f_s is the frequency of vortex shedding, D is the cylinder diameter and U is the flow velocity.

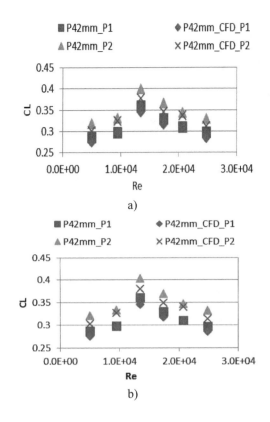

Figure 5. Variation of CL with Re, experimental vs. CFD, a) $S_x = 3D$, b) $S_x = 4D$ and c) $S_x = 5D$.

Fig. 6, shows the variation of St with Re Number. The graphs indicate that the values of St Number fluctuates between 0.18 to 0.22, which is very close to 0.2, the recommended value of St for the subcritical flow regime (Sumer and Fredsøe 2010). Although, the graphs presented in Fig. 6 are for $S_x = 3D$, one can observe that the influence of the wake on the trailing cylinder is obvious, as the values of St Number for the cylinder P2 increased by about 6%. This observation is also applicable for the CFD simulation results, where the values of St Number are found to be comparatively higher. This change on the values of St Number

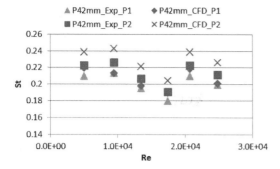

Figure 6. Variation of St with Re number.

can be associated with the group interference and the vortex shedding developed on the wake of the leading cylinder, P1.

4.4 Concluding remarks

In this study, the VIV of four flexible cylinders with outer diameter $D = 42$ mm, at different center to center spacing have been investigated. The range of Re Number achieved in this study varied between 4.96E+03-2.48E+04. This study consists of 3D LES, with Smagorinsky subgrid scale turbulence model. The CFD simulation was validated using, wave tank model tests. From this study, the following concluding remarks can be drawn:

- The analysis of the results indicates that, in a group of four cylinders in tandem arrangement, with different center to center spacing, the cylinders which are in the wake of the leading cylinders have comparatively smaller drag coefficients and comparatively higher lift coefficients and St Numbers.
- The analysis of the results also indicates that the center to center spacing between the cylinders has a major influence on drag and lift coefficients as well as on St Number.
- Generally, the CFD simulation results are in good agreement with the experimental results.

ACKNOWLEDGMENT

The authors would like to gratefully acknowledge their gratitude to Universiti Teknologi PETRONAS for support and encouragement.

REFERENCES

Bao, Y., C. Huang, D. Zhou, J. Tu and Z. Han (2012). "Two-degree-of-freedom flow-induced vibrations on isolated and tandem cylinders with varying natural frequency ratios." Journal of Fluids and Structures **35**: 50–75.

Bearman, P. W. (1984). "Vortex shedding from oscillating bluff bodies." Annual Review of Fluid Mechanics **16**(1): 195–222.

Blevins, R. D. and C. S. Coughran (2009). "Experimental investigation of vortex-induced vibration in one and two dimensions with variable mass, damping, and Reynolds number."Journal of Fluids Engineering **131**(10):101–202.

Bokaian, A. and F. Geoola (1984). "Wake-induced galloping of two interfering circular cylinders." Journal of Fluid Mechanics **146**: 383–415.

Feng, C. (1968). "The measurement of vortex induced effects in flow past stationary and oscillating circular and d-section cylinders."

Hover, F. and M. Triantafyllou (2001). "Galloping response of a cylinder with upstream wake interference." Journal of Fluids and Structures **15**(3): 503–512.

Meneghini, J. R., F. Saltara, R. de Andrade Fregonesi, C. T. Yamamoto, E. Casaprima and J. A. Ferrari (2004). "Numerical simulations of VIV on long flexible cylinders immersed in complex flow fields." European Journal of Mechanics-B/Fluids **23**(1): 51–63.

Narendran, K., K. Murali and V. Sundar (2015). "Vortex-induced vibrations of elastically mounted circular cylinder at Re of the O (10 5)." Journal of Fluids and Structures.

Pletcher, R. H., J. C. Tannehill and D. Anderson (2012). Computational fluid mechanics and heat transfer, CRC Press.

Roshko, A. (2012). "On the wake and drag of bluff bodies." Journal of the Aeronautical Sciences (Institute of the Aeronautical Sciences) **22**(2).

Saltara, F., A. D'Agostini Neto and J. Lopez (2011). "3D CFD Simulation of Vortex-induced Vibration of Cylinder." International Journal of Offshore and Polar Engineering **21**(3): 192–197.

Sanaati, B. and N. Kato (2013). "Hydroelasticity of four flexible cylinders in square arrangement subjected to uniform cross-flow." Journal of Offshore Mechanics and Arctic Engineering **135**(2): 021103.

Sarpkaya, T. (1979). "Vortex-induced oscillations: a selective review." Journal of Applied Mechanics **46**(2): 241–258.

Smagorinsky, J. (1963). "General circulation experiments with the primitive equations: I. the basic experiment*." Monthly weather review **91**(3): 99–164.

Sumer, B. M. and J. Fredsøe (2010). Hydrodynamics around cylindrical structures, World Scientific.

Tofa, M. M., A. Maimun, Y. M. Ahmed, S. Jamei and N. Khairuddin (2014). Numerical Studies of Vortex Induced Vibration of a Circular Cylinder at High Reynolds Number. Offshore Technology Conference-Asia, Offshore Technology Conference.

Vandiver, J. K. (2012). "Damping parameters for flow-induced vibration." Journal of Fluids and Structures **35**: 105–119.

Wilcox, D. C. (2006). Turbulence modeling for CFD.

Willden, R. and J. Graham (2001). "Numerical prediction of VIV on long flexible circular cylinders." Journal of Fluids and Structures **15**(3): 659–669.

Williamson, C. and R. Govardhan (2004). "Vortex-induced vibrations." Annu. Rev. Fluid Mech. **36**: 413–455.

Zhao, M. and L. Cheng (2014). "Vortex-induced vibration of a circular cylinder of finite length." Physics of Fluids (1994-present) **26**(1): 015111.

Investigating the potential of soil-nail retaining structures as a slope strengthening remedy by capitalizing on reliability analysis technique

Sadaf Qasim
NED University of Engineering & Technology, Karachi, Pakistan

I.S.H. Harahap, Syed Baharom & Muhammad Imran
Department of Civil Engineering and Environmental Engineering, Universiti Teknologi PETRONAS, Seri Iskandar, Perak, Malaysia

Danish Kazmi
NED University of Engineering & Technology, Karachi, Pakistan

ABSTRACT: Landslide is one of the major geo-hazard escorted by uncertainties that give escalation to hundreds of deaths every year. The strategy of slope remediation or stabilization is a common treatment of landslides. At present, safety factor method is agreed to soil nailing design and analysis. In one way it seems to be the positive assumption. Thus if reliability analysis method is introduced into soil nailed wall design and analysis, a lot of consequences can be mitigated. In this study, the author has targeted the uncertainties; poses threats to soil-nail retaining structure's reliability.

1 INTRODUCTION

Landslide is one of the major geo-hazard escorted by uncertainties that give escalation to hundreds of deaths every year. Problems of landslides often occur, due to instability of slopes, distressed slopes, and cut slopes (Cheung and Tang, 2005, Lo and Lam, 2013).

The appraisal of the stability of slopes, mostly natural slope, is one class of problems that is subjugated by uncertainties. Geological incongruities, material properties, environment conditions and analytical models are all factors participating to uncertainty. Conventional slope design practices do not report for these uncertainties, and fully believed with the adequacy of predictions. On the other hand, reliability analysis for slope stability proffers a proficient framework for rational methodical inclusion of uncertainty, thus given that a more logical basis for design.

Regarding Factor of safety, one of the researcher (El-Ramley, 2001) clearly exemplifies the limitations of conventional practices. For example the Lodalen slope (Sevaldson, 1956) designed for a factor of safety of 1.33 has a zero probability of unsatisfactory performance while Congress Street cut (Irland, 1954) on a factor of safety of 1.44 carries a probability of unsatisfactory performance of 2×10^{-2}. In other words, slope designed for a lower factor of safety is safer than that designed for higher factor of safety. Such discrepancy is a straight consequence of disbelieving uncertainty; the input parameters of Lodalen slope are highly reliable as compared to the input parameters of Congress street cut (Irland, 1954).

The strategy of slope remediation or stabilization is a common treatment of landslides (Gue and Fong, 2003). Soil nailed wall has been used generally for slope stability and foundation excavation. At present, safety factor method is agreed to soil nailing design and analysis. In one way it seems to be the positive assumption that without measuring the impact of uncertainties and its intensities, structure is assumed to be stable.

Thus, if reliability analysis method is introduced into soil nailed wall design and analysis, a lot of consequences can be mitigated. Therefore, some sages at home and abroad have done some work on introducing reliability method into soil nailed wall design and analysis (Jian-Xin et al., 2003, Babu and Singh, 2009, Lazarte and Baecher, 2003). In this study, the author has targeted the random variables; poses threats to soil-nail retaining structure's reliability. According to reliability theory, performance function was established for strength failure and stability failure of soil nailed wall. The reliability index of every failure mode was computed by design point method through first order second moment (FORM). Furthermore, theory of structural reliability analysis was also replicated by Monte-Carlo method and Importance sampling technique to put forward the inaccuracies of FORM to some extent.

Soil Nail Walls: Design and construction of slope restorative works masquerade high risk to both geotechnical designers and constructors, as the slope is vulnerable to further failure during the execution of the remedial works. Soil nails, if properly configured, has confirmed to be capable and cost efficient slope stabilization measure. Various slopes stabilized using soil nails have been carried out in Malaysia and all over the world on this pattern, but the method of designing still requires some improvement. This shows that proper geotechnical design is in need to prevent face failure, pull-out failure, nail tendon failure and overall slope failure.

Soil nailing is an in-situ earth reinforcement technique. Due to numerous gains such as prompt construction, no difficulty in application, less environmental impact etc. Slope engineers prefer soil nailing as a feasible substitute to the other earth retaining systems (Gue and Fong, 2003). Soil Nail is frequently used in Malaysia particularly in cut slope.

The use of soil nails for slope strengthening works has been gaining reputation in Malaysia since its first application in 1980s. In view of easy and swift installation, soil nail walls are commonly used as reinforcement but still the cases of slope failures persists there.

In order to ensure the design adequacy, a soil-nailed system must have the capacity to fulfil the stability criteria. Broadly speaking, design should carry the surety of safety against various failure modes. Failure modes of soil nail walls are mainly classified into three diverse groups: external failure modes, internal failure modes and facing failure modes (Babu and Singh, 2009).

Like the case of a failed soil nailed slope of Chainage 6100 of Federal Route 59 Cameron Highland, Pahang. The slope was stabilized using 100 mm diameter nailed with bar size 25 mm diameter. The penetration length is 12 m length with spacing 1 m c/c. The slope failed during construction stage of the project (JKR, 2010). Jabatan Kerja Raya, statistically showed that inadequate factor of safety is the prominent cause (JKR, 2009).

Similarly, Sri Plentong Industrial Park slope, located at a hilly rolling terrain in Johor Bahru having many industrial buildings at its toe. It is a steep slope of 20 m with an angle of around 70°. This is basically a cut slope. The face of the slope is close to the Public Utility Board pipeline. The stability of the slope is alarmed, and it needs proper strengthening/design as it is quite elevated (Hut, 2010). In this connection author has taken this case for further explorations.

A 20 m high soil nail wall is supporting a vertical cut but the stability is at the verge. The properties of the in-situ soil and other soil nail wall parameters are shown in Table 1. Influence of variability of in-situ soil is studied over a range of coefficients of variation (COV) of soil parameters (cohesion, angle of internal friction and unit weight of the soil), in accordance with the values reported by past researchers (Phoon and Kulhawy, 1999, Duncan, 2000).

2 RELIABILITY ANALYSIS

This section is basically the persuasion of reliability analysis, carried out with reference to a soil nail wall failure modes. The properties of the soil and other soil nail wall parameters are shown in Table 1. The influence of variability of soil is studied over a range of coefficients of variation (COV) of soil parameters in accordance with the values reported in literature (Phoon and Kulhawy, 1999, Duncan, 2000). The reliability analysis of five failure modes of the soil nail wall is carried out using FORM (through Microsoft Excel optimization program Solver), Monte Carlo simulation (MCS) and Importance Sampling (IS).

Table 1. Soil nail wall configurations (Hut, 2010, Duncan, 2000, Phoon and Kulhawy, 1999).

Description	Symbol	Value
Height of the wall	H	20 m
Soil cohesion	c	12 kPa
Friction angle	φ	38 degree
Unit weight of soil	γ	19 kN/m^3
Diameter of the nail	d	25 mm
Length of the nail	L	10 m
Drill hole diameter	DDH	100 mm
Nail spacing	Sh, Sv	1.5 m × 1.5 m
Yield strength of nail	fy	415 MPa

Table 2. Influence of cohesion on reliability indices (β) of soil nail failure modes.

	Soil nail failure modes				
(COVs)	Global	Sliding	Tensile	Pullout	Punching
10%	3.6	4.71	x	4.5	x
15%	3.57	4.70	x	4.5	x
20%	3.37	4.67	x	4.5	x
25%	3.08	4.66	x	3.9	x
30%	2.76	4.64	x	3.6	x

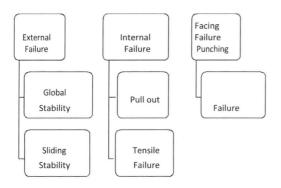

Figure 1. Soil nailing principal failure modes (Babu and Singh, 2009).

Table 3. Influence of angle of friction on reliability indices (β) of soil nail failure modes.

(COVs)	Soil nail failure modes				
	Global	Sliding	Tensile	Pullout	Punching
2%	4	17	6.52	7.51	3.07
4%	3.86	6.99	3.89	3.72	1.87
6%	3.37	4.67	2.82	2.56	1.34
8%	2.27	3.43	2.14	1.93	1.01
10%	2.26	2.73	1.73	1.51	0.82

Table 4. Influence of unit weight on reliability indices (β) of soil nail failure modes.

(COVs)	Soil nail failure modes				
	Global	Sliding	Tensile	Pullout	Punching
3%	3.49	4.68	2.98	2.61	1.41
4%	3.43	4.68	2.91	2.59	1.38
5%	3.37	4.67	2.82	2.56	1.34
6%	3.3	4.67	2.73	2.52	1.3
7%	3.22	4.67	2.64	2.49	1.26

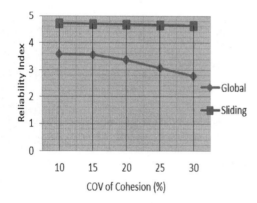

Figure 2. Influence on reliability index of external stability due to different COVs of cohesion.

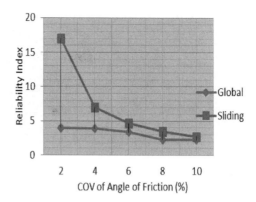

Figure 3. Influence on reliability index of external stability due to different COVs of angle of friction.

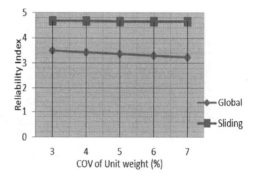

Figure 4. Influence on reliability index of external stability due to different COVs of unit weight.

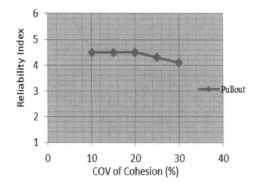

Figure 5. Influence on reliability index of internal stability due to different COVs of cohesion.

3 RESULTS AND DISCUSSION

When referring to external failure modes, reliability index of sliding capacity remains unaffected by cohesion and unit weight of the soil. In context with angle of friction, reliability index seems to be very high at 2% of COV but at 4% of COV it decreases drastically. Further at different COVs of 6%, 8% and 10% reliability index continuously decrease in one followed pattern. In case of, global failure all the three soil properties, cohesion, angle of friction and unit weight marginally affects the reliability indices (Figure 1, Figure 2, and Figure 3).

Referring the plots of internal failure modes (Figure 4, Figure 5, Figure 6 and Figure 7) tensile and pull out failures are line up to consider with. In relation with tensile and pull out failure modes, angle of friction has shown almost the same order of control on reliability indices. Clearly speaking reliability indices of both failure categories lower down from 6.51 to 1.73 and 7.51 to 1.51 at COVs of 2% to 10% respectively. For unit weight of soil, reliability indices of tensile and pullout failure modes decreases from 2.98 to 2.64 and 2.61 to 2.49 respectively. This explains that angle of friction and unit weight of the soil is equally important for both categories of internal failure modes but angle of friction creates a significant role in overall stability of the slope. In case of Punching failure angle of friction also poses its influence on reliability indices

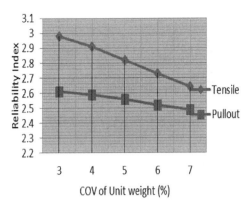

Figure 6. Influence on reliability index of internal stability due to different COVs of unit weight.

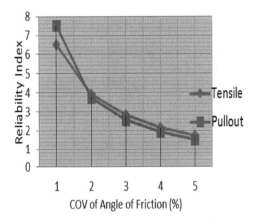

Figure 7. Influence on reliability index of internal stability due to different COVs of angle of friction.

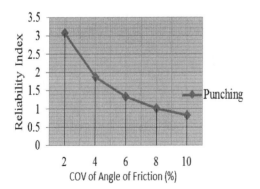

Figure 8. Influence on reliability index of facing stability due to different COVs of angle of friction.

(Figure 8 and Figure 9). If refers to variation in unit weight reliability indices also differs but as compared to angle of friction no prominent change has been observed.

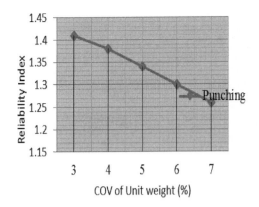

Figure 9. Influence on reliability index of facing stability due to different COVs of unit weight.

Table 5. Probability of failure of soil nail failure modes.

Failure modes	Monte Carlo simulation		Importance sampling technique	
	No of cycles	Probability of failure	No of cycles	Probability of failure
Global	100000	0.06247	600	0.051667
Sliding	100000	0.0179	900	0.0177
Tensile	100000	0.0035	1000	0.00318
Pull out	100000	0.0045	1000	0.004
Punching	100000	0.0218	600	0.02291

A credible justification for this observation might be extracted from the basic soil mechanics theory concerning influence of variability of in-situ soil parameters on the failure modes of a soil wall. In slope stability features, soil cohesion and angle of friction are the strength parameters contributing significantly to the resisting component of soil slope against failure. Conversely, unit weight of the soil participates to the loading element. From the results it is validated that in comparison to cohesion and unit weight, a small variation of angle of friction poses remarkable influence on the stability of all the focused failure modes of soil nail wall.

In the second phase of the previous section, probability of failure has been concluded for all five failure modes. The probability of failure is also an indirect measure of the reliability index but in some cases due to non-linearity of the performance functions, both the quantities differs with each other. Like in reliability index method (FORM), sliding category of external failure mode gives the value of 4.67 for the original case which is good according to the U.S. Army Corps of Engineers but when refer to Monte Carlo simulation (MCS) and Importance Sampling (IS) the computed probability of failure is 0.0148 and 0.0110. The computed probability of failure according to U.S. Army Corps of Engineers lies between below average and poor. This is principally, due to the approximations (by default) made in the method of FORM to make it simple and less tedious. When discussing about

Importance Sampling and Monte Carlo simulations both the methods perfect in their accuracies however they differ in their computational costs (Table 5). When pinpointing about pull out and punching failure, more or less reliability indices computed through FORM are in accordance with the probability of failure computed through MCS and IS technique. This proves that FORM is not always inaccurate; it depends upon the complexity of the performance function that results drawn from FORM are conservative but not misleading.

4 CONCLUSION

The following are the conclusion drawn based on reliability analysis:

- Considering External failure modes, reliability index of sliding capacity remains unaffected by cohesion and unit weight of the soil.
- The Global failure of all three soil properties including cohesion, angle of friction and unit weight marginally affects the reliability indices.
- The angle of friction and unit weight of the soil is important for both of the internal failure modes but angle of friction contributes to a significant role in overall stability of the slope.
- In case of Punching failure, angle of friction also poses its influence on reliability indices.

The probability of failure is an indirect measure of the reliability index but in some cases due to non-linearity of the performance functions, both the quantities differs with each other. The FORM is not always inaccurate; but depends upon the complexity of the performance function that results drawn from FORM are conservative.

REFERENCES

Babu, G. L. S. & Singh, V. P. 2009. Reliability Analysis of Soil Nail Walls. *Georisk: Assessment and Management of Risk for Engineered Systems and Geohazards*, 3, 44–54.

Cheung, W. M. & Tang, W. H. 2005. Realistic Assessment of Slope Reliability for Effective Land-Slide Hazard Management. *Geotechnique*, 55, 85–94.

Duncan, J. M. 2000. Factors of Safety and Reliability in Geotechnical Engineering. *Journal of geotechnical and geoenvironmental engineering*, 126, 307–316.

El-Ramley. 2001. *Probabilistic Analyses of Landslide Hazards and Risks: Bridging Theory and Practice*. Doctor of Philosphy, University of Alberta.

Gue, S. S. & Fong, C. C. 2003. Slope Safety: Factors and Common Misconceptions. *Buletin Ingenieur*, 19.

Hut, L. 2010. Slope Remedial Works at Sri Plentong. Universiti Teknologi Malaysia, Johor Bahru.

Irland, H. O. 1954. Stability Analysis of the Congress street Open Cut in Chicago. *Geotechnique*, 4, 163–168.

Jian-Xin, Y., Yuwen, Y., Leslie George, T., Peter Kai Kwong, L. & Yuet, T. 2003. New Approach to Limit Equilibrium and Reliability Analysis of Soil Nailed Walls. *International Journal of Geomechanics*, 3, 145–151.

JKR 2009. Jabatan Kerja Raya. National Slope Master Plan. Sectoral Report Research and Development Malaysia.

JKR 2010. Jabatan Kirja Raya, Soil Nail slope/wall Collapsed in Malaysia – Never ending story. Landslide Forensic Unit, Slope Engineering Branch, Public Works Department of Malaysia, ed.

Lazarte, C. & Baecher, G. B. LRFD for Soil Nailing Design and Specifications. LSD2003: International Workshop on Limit State design in Geotechnical Engineering practice 2003 Cambridge, Massachusetts.

Lo, D. O. K. & Lam, H. W. K. Value of Landslide Investigation to Geotechnical Engineer-ing Practice in Hong Kong. 18th International Conference on Soil Mechanics and Ge-otechnical Engineering, 2013 Paris.

Phoon, K.-K. & Kulhawy, F. H. 1999. Characterization of Geotechnical Variability. *Canadian Geotechnical Journal*, 36, 612–624.

Sevaldson, R. A. 1956. The Slide in Lodalen, October 6, 1954. *Geotechnique*, 6, 162–167.

Response of spar-mooring-riser system under random wave, current and wind loads

Mohammed Jameel, Abdulrahman Eyada Ibrahim & Mohd Zamin Jumaat
Department of Civil Engineering, University of Malaya, Kuala Lumpur, Malaysia

ABSTRACT: As a result of excessive consumption of oil and gas products in the twentieth century and the depletion of most of the hydrocarbon reservoir available on land, the attention has turned towards finding alternative sources of energy. One of the most important alternatives in recent decades is to exploit the hydrocarbon reservoir discovered under deep seas. Several types of floating structures have been developed and used in deep-water oil and gas production. The spar platform is the latest type which is designed and used in deep and ultra-deep waters. In this paper, stochastic analysis of spar-mooring-riser system has been conducted. Finite element analysis of the Spar-mooring-riser system simulated as a single fully coupled integrated model using ABAQUS/AQUA code was carried out to obtain the system responses under the long crested random wave with current and wind load inclusion. Surge, heave and pitch motion in addition to the maximum top tension of mooring lines and riser were obtained for steady state condition, as well as PSDFs for the same period were investigated.

1 INTRODUCTION

The offshore oil and gas exploration activities are rapidly expanding towards deeper waters, therefore, conventional fixed platforms cannot meet the requirement of the new fields. As a result, many countries and oil companies are investing in this vital area. Several approaches have been devised and developed for drilling and exploration of oil and gas in the marine environment. Spar platform is the latest type of floating structures that are designed, developed and used for oil and natural gas production in deep waters. Due to its high efficiency and stability in harsh environmental conditions, it can be considered as one of the best option for deep water exploration. During the past two decades spar platforms has been successfully installed around the world, especially in the Gulf of Mexico and North Sea. Numerous studies have been recently carried out so as to evaluate the hydrodynamic and aerodynamic loads influence on spar platform. Zaheer & Islam (2008) reported that wave and wind loads have significant influence on offshore. Yang & Kim (2010) carried out TLP coupled hull/tendon/riser dynamic analysis. Tendons were simulated as an elastic rods connected through linear and rotational springs to the hull. (SUN et al. 2011) presented coupled dynamic analysis of deep draft spar platform and its mooring lines under wave and current influence in time domain. (Jameel et al. 2013) investigated coupled spar responses taking into account essential nonlinearities. A time domain second order approach has been developed for the hydrodynamic load calculation. (de Pina et al. 2013) employed artificial neural networks (ANNs) tool to analyse mooring lines and risers. Deep water risers instability mechanism was investigated and analytically modeled by Khorasanchi & Huang (2014) to predict the instability conditions, considering 2D problem case. (Girón et al. 2014) carried out fully integrated mooring-riser system design approach for deep-water FPS. Yang & Xu (2015) investigated the spar platform parametric instability due to multi-frequency waves under realistic sea states. (Chen et al. 2015) explored the top end motion impact on the riser vortex-induced vibration (VIV). Although implementation of spar platforms is growing all over the world, there is limited understanding in precise modeling of nonlinear behavior of coupled spar-mooring-riser system. Moreover, the mooring lines and risers contribution in terms of damping, drag and inertia are not fully included, which is noticeable in deep water conditions. The present study aims to idealize the spar-mooring-riser system as an integrated structure under random wave, current and wind loading.

2 MATHEMATICAL FORMULATION

The deterministic model development includes the non-linear stiffness matrix formulation taking into consideration tension fluctuations of mooring line resulted by variable buoyancy along with other nonlinearities. The model takes into consideration selection and solution of wave theory which represents the kinematics of water particles to estimate the inertia and drag force of the 6 DOF. The static problem is solved

Table 1. Mechanical, geometrical and hydrodynamic properties of spar, mooring and riser.

Spar (Classic JIP Spar)		Mooring lines and Riser	
Length	213.044 [m]	No. of Moorings	4
Diameter	40.54 [m]	Mooring Length	2000.0 [m]
Draft	198.12 [m]	Riser Length	819.88 [m]
Mass	2.515276E8 [kg]	No. of Mooring Nodes	101
Mooring Point	106.62 [m]	No. of Riser Nodes	42
No. of Nodes	17	Mooring Pre-tension	1.625E+07 [N]
No. of Elements	16	Riser Pre-tension	1.20E+07 [N]
Type of Element	Rigid beam element	Mooring Mass	1100 [Kg/m]
Drag	0.6	Element Type	Hybrid beam element
Drag (vertical)	3.0	Drag	1.0
Inertia	2.0	Inertia	2.2
Added mass	1.0	Added mass	1.2

using Newton's method. Newmark time integration approach with iterative convergence is used in coupled dynamic model solution. The motion equation representing the Spar-mooring-riser system equilibrium of damping, restoring, inertia and exciting forces is assembled as follows:

$$[M]\{\ddot{X}\} + [C]\{\dot{X}\} + [K]\{X\} = \{F(t)\} \quad (1)$$

Where, [M], [C], [K] represent the total mass, damping and stiffness matrices of the Spar-mooring-riser system respectively. {X} represent the six degree of freedom structural displacements and the dot symbolizes differentiation with respect to time. The total Spar-mooring-riser system mass matrix consists of structural mass and added mass components. The Spar-mooring-riser systems structural mass comprises of moorings-riser elemental consistent mass matrices and the rigid Spar hull lumped mass properties. The total stiffness matrix element [K] consists of the geometrical stiffness matrix KG and the elastic stiffness matrix KE. The total damping to the system is provided by structural and hydrodynamic damping. The main damping is caused by the hydrodynamic influence. It is calculated from velocity term in the Morison equation force vector. The structural damping is simulated by Rayleigh damping.

3 FINITE ELEMENT MODEL

The finite element approach was used to idealise and discretize spar platform. The platform is modelled as a rigid body, whose mass and moment of inertia are located at the centre of gravity. The mooring lines and risers are modelled as hybrid beam elements and these are well coupled to the platform so as to ensure the platform behaves as a single system. The coupling was achieved with the aid of six springs. Hence, the platform motion is given by the rigid body reference point, (centre of gravity). In Abaqus/Aqua module, hydrodynamic and aerodynamic loadings are applied on the beam elements. The Morison equation was used for calculation of hydrodynamic loading while API spectrum model was used for aerodynamic forces. The mass and damping forces from the mooring lines were also included in the solved equation of motion. The structural damping was achieved with Rayleigh model while the hydrodynamic damping is recognised as the major source of damping.

4 NUMERICAL STUDY

Spar-mooring-riser system in 1018 m deep water has been studied numerically under random wave environment. The geometric characteristics and other related data for the present study are given in Table 1. Spar-mooring-riser system responses are obtained under long crested random sea state with current and wind forces. The sea state case study is (6 m) significant wave height "Hs", (14 sec) zero up crossing period "Tz" with current velocity 1.0 m/sec and wind speed 26.28 m/sec. Nonlinear dynamic analysis in time domain with step-by-step integration technique has been applied, sufficient time period enough to attain steady state has been taken. System response recorded for a period length of 5000 sec to 6000 sec to obtain the statistical analysis avoiding the initial transient phase.

5 RESULTS AND DISCUSSION

The realistic simulation of marine environment is take into account along with hydrodynamic and aerodynamic loading. Current velocity of 1.0 m/sec and 26.28 m/sec mean wind velocity under 6 m significant wave height "Hs" and 14 sec zero up crossing period "Tz" in deep water conditions are investigated to predict spar-mooring-riser system response for adequate period of time enough to attain steady state. The following sections explain the system response under random wave with current and wind.

Surge response. The predictability of surge response is of significant importance since it influences all the activities performed during the operation time. Current and wind loading, in random wave environment, are important parameter affecting the surge motion.

It is observed that entire platform displaced 44 m away from its mean position as shown in Fig. 1. The predictability of the spar platform surge response is acquired great importance because its influence upon all the activities carried out during the platform operation time. Figure 1 shows that entire platform displaced about 44 m away from its mean positions and started oscillation. The maximum and minimum values of oscillation are presented in Table 1. Figure 2 presents the power spectrum graph depicting the several distinct peaks compatible with surge and pitch frequencies. Maximum peak occurs at 0.1981 rad/sec.

Heave response. Heave response have direct effect on the mooring tension, riser tension and other associated operations. Figure 3 shows the time history response of heave motion. The platform oscillated randomly about the mean value of −0.032 m. Table 2 presents the statistical results of spar-mooring-riser system. More than one peak is observed in power spectrum graph as shown in Figure 4. Maximum peak value of 0.3387 rad/s is found while other peak have less energy content.

Pitch response. Figure 5 shows the time history response of spar-mooring-riser system under combination of random wave, current and wind loads. Time history shows cluster of reversals occurring at varying time intervals showing random behavior. The pitch response fluctuate about the mean position oscillating from smaller to greater heights and repeating the same tend onwards all through the time history. In Figure 6,

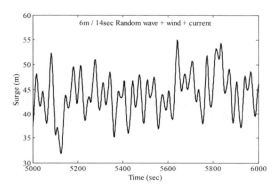

Figure 1. Surge time history.

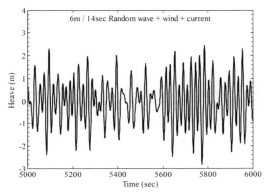

Figure 3. Heave time history.

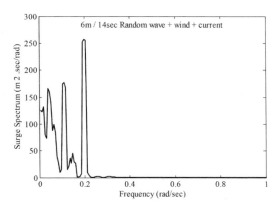

Figure 2. PSD of surge response.

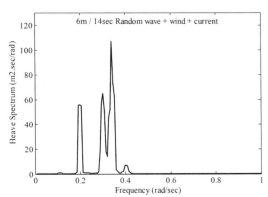

Figure 4. PSD of heave response.

Table 2. Statistical response of Spar-Mooring-Riser system.

Hs:6 m/ Tz:14 s Random Wave + Current + Wind	Max	Min	Mean	Standard deviation
Surge [m]	55.030	31.872	43.934	4.196
Heave [m]	2.452	−2.808	−0.032	1.001
Pitch [rad.]	0.0833	−0.1014	0.00003	0.0182
Top Tension in CML1 [N]	2.184E+07	1.603E+07	1.955E+07	1.102E+06
Top Tension in CML3 [N]	1.694E+07	1.327E+07	1.534E+07	7.250E+05
Top Tension in Riser [N]	3.066E+08	3.931E+07	1.681E+08	4.909E+07

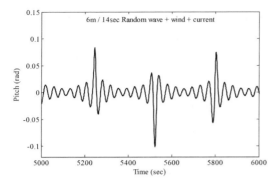

Figure 5. Pitch time history.

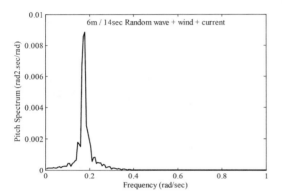

Figure 6. PSD of pitch response.

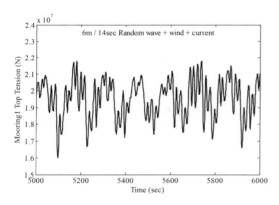

Figure 7. Mooring line 1 top tension time history.

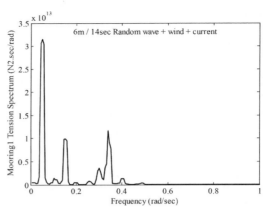

Figure 8. PSD of Maximum Tension of mooring line1.

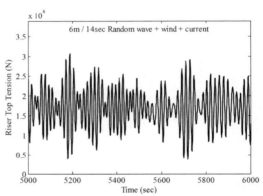

Figure 9. Riser top tension time history.

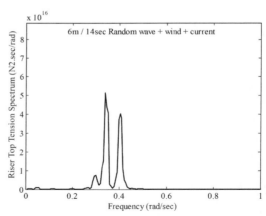

Figure 10. PSD of riser top tension.

maximum peak of 0.179 rad/s is observed which is later on diminished.

Mooring line response. Mooring lines 1 and 3 are located in the wave propagation direction while the mooring lines 2 and 4 are located in perpendicular direction. Therefore, they take the main portion of the tensile forces resulted by the hydrodynamic and aerodynamic loads. Initial mooring line pretension of 1.625E7 [N] is applied in current study. Mooring line 1 time history shows that top tension oscillate randomly around mean value od 1.95E+7[N] (Figure 7). The statistical analysis shows the max. and min. values of top tension as 2.184E+07 [N] and 1.603E+07 [N] respectively (Table 2). Power spectrum in Figure 8 depict that maximum peak occurs at 0.0511 rad/s. Mooring line 3, which is in the same direction of the external loads, experienced the slacking effects due to hydrodynamic and aerodynamic impact.

Riser response. Riser are physically connected with the spar keel and hinged at the seabed. It is expected that heave will significantly influence the riser stress and top tension response. Repeated oscillation mode is observed about the mean value of 1.681E+08N in comparison with the pre-tension value of 1.20E+07N, as shown in Figure 9. The power spectrum of riser top tension response time history shows three prominent peaks, in which maximum peak value occurs at 0.3387 rad/s, as illustrated in Fig. 10. Additional details of statistical analysis are presented in Table 2.

6 CONCLUSION

The present study concludes that; current and wind loads inclusion have a significant influence on the spar-mooring-riser system responses. The platform has been displaced away from its initial place to an approximate distance of 44 m. This displacement is reflected on the responses of mooring lines and riser as shown through the results above. The top tension increased obviously in mooring line 1, in conjunction with that mooring line 3 shows slacking and decreasing in top tension. At the same time the riser top tension increased significantly.

ACKNOWLEDGMENT

This work was supported by the University of Malaya Research Grant [RP004E-13AET].

REFERENCES

Chen, W., Li, M., Zheng, Z., Guo, S., Gan, K., 2015. Impacts of top-end vessel sway on vortex-induced vibration of the submarine riser for a floating platform in deep water. Ocean Engineering 99, 1–8.

de Pina, A.C., de Pina, A.A., Albrecht, C.H., de Lima, B.S.L.P., Jacob, B.P., 2013. ANN-based surrogate models for the Analysis of mooring lines and risers. Applied Ocean Research 41, 76–86.

Girón, A.R.C., Corrêa, F.N., Hernández, A.O.V., Jacob, B.P., 2014. An integrated methodology for the design of mooring systems and risers. Marine Structures 39, 395–423.

Jameel, M., Ahmad, S., Islam, A.S., Jumaat, M.Z., 2013. Nonlinear dynamic analysis of coupled spar platform. Journal of Civil Engineering and Management 19, 476-491. Khorasanchi, M., Huang, S., 2014. Instability analysis of deepwater riser with fairings. Ocean Engineering 79, 26–34.

Sun, Z.-c., Wang, X.-g., Liang, S.-x., Shu-Xue, L., Si, L., 2011. Dynamic response analysis of ddms platform subjected to actions of wave groups and current fources. Journal of Hydrodynamics, Ser. B 23, 697–708.

Yang, C.K., Kim, M., 2010. Transient effects of tendon disconnection of a TLP by hull–tendon–riser coupled dynamic analysis. Ocean Engineering 37, 667–677.

Zaheer, M.M., Islam, N., 2008. Aerodynamic response of articulated towers: state-of-the-art. Wind and Structures 11, 97–120.

# Effects of current coexisting with random wave on offshore spar platforms

V.J. Kurian, S.N.A. Tuhaijan & C.Y. Ng
Department of Civil and Environmental Engineering, Universiti Teknologi PETRONAS, Seri Iskandar, Perak, Malaysia

ABSTRACT: The wave and current coexist in the open sea. The existence of current in the water body changes the dynamic behavior of the sea wave and thus applies different hydrodynamic loading on the offshore structures. In this paper, the effect of the co-existence of wave and current in the water body is presented for both classic and truss spars. The purpose of this study is to suggest the better performing spar platform for the oil and gas operation to be installed in Malaysian offshore locations. In order to achieve the objective of this study, model tests have been done in the Universiti Teknologi PETRONAS Offshore Laboratory. The results are presented in terms of Response Amplitude Operator (RAO) for surge, heave, and pitch. The results show that the truss spar has lower responses compared to the classic spar and thus is preferable to be installed in the Malaysian offshore region where the current and wave act jointly.

1 INTRODUCTION

Spar is a floating structure stabilized by mooring lines attached to the seafloor and has been installed in ultra deepwater up to 3,000 m of water depth. It has natural stability since it has a large counterweight at the bottom. It also has the ability to move horizontally and to position itself at some distance from the main platform location by adjusting the mooring line tensions.

There are three types of spars which are classic spar, truss spar, and cell spar (Welton & Guez 2003). The difference among these spars is in the structure design. Generally, the spars consist of several elements such as topside, hull shell, buoyancy tank, centerwell, risers, and mooring lines (Halkyard 2001). The motions and loads of spars are controlled by two parts. The primary part is controlled by the hull configurations which consists of draft and heave plates. The secondary part is controlled by the mooring system which consists of taut and synthetics cables. Thus, the design of the spar is very important for the stability when it is installed in the water. The spar is having long periods in heave, pitch, and rolls which makes it insensitive to the wave frequencies and their height harmonics. In addition, it does not undergo any type of springing and ringing response in severe storms as tension leg platforms may undergo. It is also insensitive with the water depth since it is mainly a floating cylinder, thus, it can be relocated to another spot in the ocean regardless of the water depth or the deck load. The effect of wave drift damping of the spar structure is small. However, it improves slightly the response amplitude at the natural frequencies particularly at the early stages of the analysis.

The effect of the hydrodynamic forces may be critical on the connection of the offshore structures (Luo et al. 2003). For both classic and truss spars, the connections between the topsides and spar hull are critical locations for fatigue design due to the motion characteristics of the structures. In line with the global development, Malaysia had constructed and installed Kikeh spar which is the first deepwater floating platform in its water region and the first spar ever installed outside the Gulf of Mexico. It is operated in 1,330 m water depth of deepwater offshore Sabah, Malaysia. An innovative configuration of floating platform was required for the exploration of the hydrocarbon reservoir under the seabed in the very deepwater. There are some studies done on spar responses by using frequency and time domain analysis (Mekha et al. 1996, Zhang et al. 2008, Kurian et al. 2009, Kurian et al. 2008, Mekha et al. 1995, Wang et al. 2008, Agarwal & Jain 2003).

This research is required to understand the hydrodynamic interactions between the structure, the wave and the quantification of the nonlinear component of this interaction. In this paper, the dynamic responses of the truss spar and classic spar are presented. The structural responses of these two structures have been compared and the structure with a lower response is assumed to have higher stability for the oil and gas production. Thus, a more economical and productive structure can be designed in this industry.

2 MODEL TEST SETUP

2.1 Introduction

Two typical classic and truss spars were selected and analyzed in this model test. The models have been designed as rigid bodies attached to the sea floor by four-mooring lines. The random wave model spectrum was used for computing the incident wave kinematics and the wave forces. The metocean data in Malaysia was used to determine the model test environmental data. The purpose of this study is to gain general understanding on the effect of the current co-existed with wave in the water body to the response of the classic and truss spars. From the results, it is found that the structure with a lower response is more stable to be installed in the Malaysian water region.

2.2 Prototype description

The classic and truss spars model as shown in Figures 1 and 2 were used in this study. The models were designed by adopting a scale ratio of 1:100 and were fabricated by using galvanized steel. The hull diameter is 300 mm and the total length is 900 mm for the classic spar while for the truss spar the hull diameter and length are 300 mm and 430 mm. The total length of the truss spar is 900 mm.

2.3 Test facilities and instrumentation

The physical modeling study had been conducted in the Offshore Laboratory of Universiti Teknologi PETRONAS, Malaysia. The model testing facility consists of 23 m long, 12 m wide, and 1.5 m deep wave tank equipped with six glass windows and two movable remote control bridge platforms to support the testing personnel and equipment instruments. The wave and response devices such as wave probes, vectrino velocimeters, pressure transducers, accelerometers, Qualisis and the optical tracking system are also available in this laboratory. The wave generator is capable of generating regular and irregular waves and currents.

2.4 Metocean data

This study was focusing on the Malaysia offshore region especially at the South China Sea. Those values of the metocean data were taken from the PETRONAS Technical Standard (PTS) (Petronas 2010). The wave height, wave period, and current velocity ranges were taken from the lowest to the highest values occurred at this area and these values had been adopted in this study. The PTS stated that the wave height is ranging between 3.1 to 4.9 m, the wave period is ranging from 6 to 8.1 s, and the current velocity is ranging from 0.7 to 1.6 m/s. In this study, the values of the wave height, wave period, and current velocity were taken between these ranges.

2.5 Generation of wave and current

The spar models were subjected to current and random wave. The wave was generated by using the wave paddle, and the current was generated by using the water pump installed near to the wave paddle. The current reading was taken by using the vectrino velocimeters and the structure responses were recorded by using Opti-Tracking System (OTS). In this model test, the

Figure 1. Classic spar model.

Figure 2. Truss spar model.

model responses were recorded for 300 s. The JONSWAP spectrum was used to generate the random wave. The specified current and random wave conditions for the wave test are shown in Tables 1 and 2. In this paper, the results of combination of random wave (RD1) and current (C1) for both spars are presented.

2.6 Model test setup and procedure

The model was connected to the sea floor by using the horizontal mooring system comprising of four wires attached to linear springs. When the wave and current were run, the model was free to respond to the wave loading in all six degree of freedom within the constraints of the mooring system. Figures 3 and 4 show the model test arrangement from the section and top view.

2.7 Model test data processing

The Opti–Tracking System (OTS) was used to capture and record the responses of the models in surge, heave, and pitch when subjected to wave and current while the wave probe was used to obtain the wave height. A data post-processing program which is MATLAB was used to convert the recorded response time series to the response spectra by using Discrete Fast Fourier Transform (DFFT). The RAO values were obtained by dividing the response spectrum to the wave spectrum. Figures 5 to 7 show the time series of the structural responses in surge, heave, and pitch respectively. While Figure 8 shows the time series of the wave height profile, and Figure 9 shows the converted response spectral density.

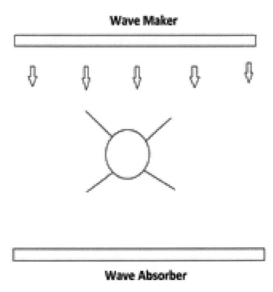

Figure 4. Model test arrangement (top view).

Table 1. Current conditions.

Test No	Current (m/s)
C1	0.1317
C2	0.1293
C3	0.1050
C4	0.0986
C5	0.0855
C6	0.0727
C7	0.0590
C8	0.0347

Table 2. Random wave conditions.

Test No	Wave Height m	Wave Period s
RD1	0.05	1.0
RD2	0.07	1.2

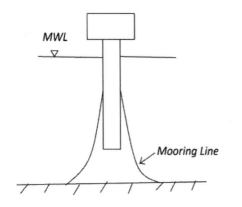

Figure 3. Model test arrangement (section view).

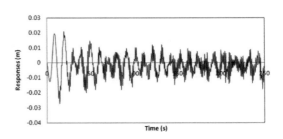

Figure 5. Time series for surge response.

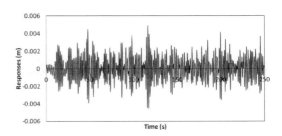

Figure 6. Time series for heave response.

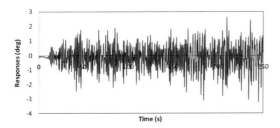

Figure 7. Time series for pitch response.

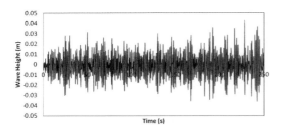

Figure 8. Time series for random wave (RD1) and current (C1).

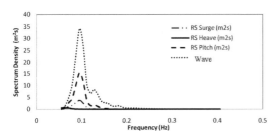

Figure 9. Spectra density.

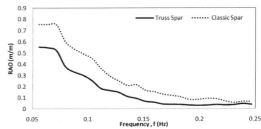

Figure 10. Surge RAO comparison for classic and truss spars.

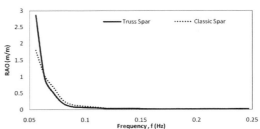

Figure 11. Heave RAO comparison for classic and truss spars.

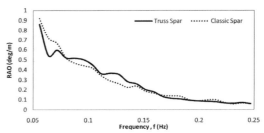

Figure 12. Pitch RAO comparison for classic and truss spars.

3 RESULTS AND DISCUSSION

In this paper, the comparison of structure responses in surge, heave, and pitch between classic and truss spars in terms of Response Amplitude Operator (RAO) are presented. As mentioned earlier, the results of the combination of random wave (RD1) and current (C1) for both spars are presented.

The comparison of surge RAO for the classic and truss spar is shown in Figure 10. It can be observed that the surge RAO for classic spar is higher compared to the surge RAO for truss spar through out all frequencies, and nearly the same as the frequency is increased. The truss spar is having higher stability due to its design. The truss part has increased the structural stability, thus reducing the structural responses. There is slightly different in magnitude at the lower frequency region by 25%, this might be due to the low frequency effect. However, this effect is neglected in this study.

Figure 11 shows the comparison of heave RAO for the classic and truss spars. It can be observed that the heave RAOs for both classic and truss spars are nearly the same for all frequencies. In this study, the heave responses for both spars is having not much difference, since the structures were subjected to the horizontal lateral forces only (random wave and current). Hence, the difference of heave responses for both classic and truss spars are not too significant.

The comparison of pitch RAO for the classic and truss spar is shown in Figure 12. It can be observed that the pitch RAO for truss spar and classic spar is having small difference in magnitude for all frequencies, and as the frequencies increased, the pitch RAO values for both spars are nearly the same. There is slightly different in magnitude at the lower frequency region, this might be due to the low frequency effect. As mentioned earlier, this effect is neglected in this study. Although for both type of spars the responses are having not much different, still the pitch response has to be considered in the preliminary design stage of the offshore structure.

4 SUMMARY

In this study, both classic and truss spars were subjected to random wave coexisting with current. Model tests had been done in order to determine the structural responses. Based on the results, the structural responses of truss spar for surge, heave, and pitch are lower compared to the classic spar. As a conclusion, the truss spar has proved to result in lower responses compared to the classic spar for the Malaysian offshore condition of random wave coexisting with current, thus is the most preferable structure to be installed in the Malaysian offshore regions.

REFERENCES

Agarwal, A. K & Jain, A. K. 2003. Dynamic Behavior of Offshore Spar Platforms under Regular Sea Waves. *Ocean Engineering*, 487–516.

Halkyard, J. 2001. Spar Global Response. The Coflexip Stena Offshore Group. Available: http://info.ogp.org.uk/metocean/FloatingSystems/presentations/Halkyard.pdf

Kurian, V. J., Montasir, O. A. A. & Narayanan, S. P. 2009. Numerical and Model Test Results for Truss Spar Platform. International Society of Offshore and Polar Engineers, 1–9.

Kurian, V. J., Wong, B. S. & Montasir, O. A. A. 2008. Frequency Domain Analysis of Truss Spar Platform. International Conference on Construction and Building Technology, (pp. 235–244).

Luo, M. Y., Wang, J. J. & Lu, R. R. 2003. Spar Topsides-to-Hull Connection Fatigue–Time Domain vs. Frequency Domain. International Offshore and Polar Engineering Conference (pp. 280–284). Hawaii, USA: The International Society of Offshore and Polar Engineers.

Mekha, B. B., Johnson, C. P. & Roesset, J. M. 1995. Nonlinear Response of a Spar in Deep Water: Different Hydrodynamic and Structural Models. International Offshore and Polar Engineering Conference (pp. 462–469). The Netherlands: The International Society of Offshore and Polar Engineers.

Mekha, B. B., Weggel, D. C., Johnson, C. P. & Roesset, J. M. 1996. Effects of second Order Diffraction Forces on the Global Response of Spars. International Offshore and Polar Engineering Conference (pp. 273–280). Los Angeles: The International Society of Offshore and Polar Engineers.

PTS, "PETRONAS Technical Standard," ed. Malaysia: PETRONAS, 2010.

Wang, Y., Yang, J., Hu, Z. & Xiao, L. 2008. Theoretical Research on Hydrodynamics of a Geometric Spar in Frequency- and Time-Domains. *Journal of Hydrodunamics*, 30–38.

Welton, G. C. & Guez, D. A. 2003. *Spar Overview*. Technip–Coflexip. Available: http://phx.corporate-ir.net

Zhang, F., Yang, J., Li, R. & Chen. G. 2008. Coupling Effects for Cell–Truss Spar Platform: Comparison of Frequency- and Time-Domain Analysis with Model Tests. *Journal of Hydrodynamics*, 424–432.

Nonlinear dynamic analysis of jacket-type offshore platform and optimization of leg batter

R. Senthil & S. Ishwarya
Division of Structural Engineering, Department of Civil Engineering, Anna University, Chennai, India

M. Arockiasamy
Department of Civil, Geomatics and Environmental Engineering, Florida Atlantic University, USA

ABSTRACT: Offshore steel jacket structures have been commonly used for oil and gas fields for decades. These structures contribute to about 90% of the world's offshore platforms and are appropriate for relatively shallow water depth. In this study, nonlinear dynamic analysis is performed on a 3-D model of a jacket-type fixed offshore platform for North Sea conditions. The structure is modeled, analyzed and designed using finite element software-Structural Analysis Computer System (SACS). The behaviour of jacket under wave load and seismic load are examined separately. Further, by varying the leg batter values of the platform, weight optimization is carried-out. Soil-structure interaction effect is considered in the analyses and these results are compared with the hypothetical fixed-support end condition. From the analyses, it is found that the optimum leg batter value for the chosen jacket structure is 16 and 2% of weight saving is achieved.

1 INTRODUCTION

Fixed jacket-type offshore structures is a type of oil platform with facilities to drill wells, to extract and process oil and natural gas, and to temporarily store product until it can be brought to shore for refining and marketing. These structures are subjected to various environmental loads during their life time. A jacket-type platform consists of three parts: Jacket, Pile and Topside. The primary function of a jacket structure is to transfer the lateral loads to the foundation and also support the weight of the topside structure.

Nonlinear dynamic analysis and design of jacket-type platform has been carried out by many researchers in last decades. Venkataramana et al. (1988) proposed a method for dynamic response analysis of offshore structures subjected to simultaneous wave and seismic loading and concluded that the sea waves act as a damping medium and reduce the amplitude of seismic response of offshore structures. Yasser Mostafa et al. (2003) investigated the response of fixed offshore platforms supported by clusters of piles and concluded that the top soil layers have an important role in the response of the structure. Mohsen Mohammad Nejad et al. (2010) investigated the effect of leg batter on overall cost of the Persian Gulf platform and found the optimum batter as 11 which lead to weight reduction of steel by 4%. Bargi et al. (2011) performed nonlinear dynamic analysis of a Persian Gulf platform and showed that the maximum displacement response of platform under the combination of earthquake and wave loads were more than the maximum displacement response of earthquake load alone. Narayanan Sambu Potty et al. (2013) assessed the ultimate strength of steel jacket offshore platform and found that X-bracing contributes highest rigidity to the whole platform. Taha Nasseri et al. (2014) applied genetic algorithm for optimizing the design of fixed offshore structure subjected to environmental loads and determined the contribution of member types in optimization. Duy-Duan Nguyen and Chana Sinsabvarodom (2015) investigated the nonlinear behaviour of a fixed-jacket offshore platform with different bracing systems and concluded that X-bracing system has high seismic performance. From the literature, it is evident that the difference between response of jacket platform under fixed support condition and the structure with Pile-Soil Interaction (PSI), under In-place and seismic analysis, need to be addressed.

In this study, at first, a four legged jacket-type platform is investigated for its structural response to wave and seismic loads separately, by performing non-linear dynamic analyses. In-place and seismic analyses have been performed on the chosen jacket for North Sea environment, using a finite element software, SACS. Secondly, weight optimization has been carried out

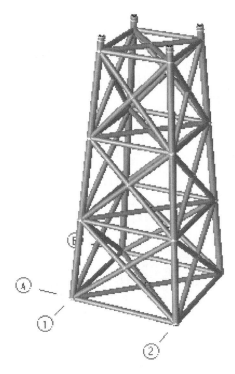

Figure 1. SACS 3D model of jacket.

by analyzing the platform with different values of leg batter, varying from 4 to 18.

2 METHODOLOGY

2.1 Model

The analysis and design are done using SACS software. The platform considered in the study is a typical four legged production platform made from tubular steel. Water depth at the North Sea location considered is 75 m. Total height of the platform (jacket and deck) is 90 meters and has a square cross-section in plan. The platform is designed based on the API RP 2A-WSD and DNV standards. The dimension at the foot is 32 m by 32 m in plan, major deck framing is 20 m by 20 m, and the jacket legs are battered at 1 to 12.5 (horizontal to vertical) in both broad side and end-on framing. The jacket is subdivided into 3 bays of 25 meters each with X-bracing. The 3D model of jacket is shown in Figure 1. It is assumed that the jacket structure is analyzed with the load from topsides acting as equivalent concentrated vertical load on the piles and the wind effect is neglected as the jacket is under water.

2.2 Material and geometric properties

The material and geometric properties of the jacket are shown in Tables 1 and 2 respectively. The length of the pile used is 100 m below mud-line.

Table 1. Material properties and mass of deck (Duy–Duan Nguyen, 2015).

Parameter	Value
Young's Modulus	2×10^8 kN/m^2
Poisson's ratio	0.3
Steel density	78.5 kN/m^3
Yield stress	3.2×10^5 kN/m^2
Ultimate stress	4×10^5 kN/m^2
Mass of the deck	4800 Tonnes

Table 2. Geometric properties (Duy–Duan Nguyen, 2015).

Member type	Diameter (cm)	Thickness (cm)
Horizontal bracings	1500	20
Diagonals in horizontal plane	1500	20
Diagonals in vertical plane	1600	30
Legs	2000	50
Piles	1900	40

Table 3. Wave load data (Yasser E. Mostafa, 2004).

Return Period	Wave height (m)	Height above MSL (m)	Wave period (s)
1	22.5	12.8	13.8
10	25.3	14.2	14.6
100	28.5	16.1	15.3
1000	36	20.4	17.1

Table 4. Soil layering data (Yasser E. Mostafa, 2004).

Return Period	Wave height (m)	Height above MSL (m)	Wave period (s)
1	22.5	12.8	13.8
10	25.3	14.2	14.6
100	28.5	16.1	15.3
1000	36	20.4	17.1

2.3 Loading data

- Dead Load
  The total deck load of 4800 tonnes is applied as concentrated vertical load on the 4 piles.
- Wave and current load The maximum directional wave heights for the 100-year return period are given in the Table 3. Surface current velocity of 1 m/s is considered for operating condition and 2 m/s for extreme condition
- Soil data The following soil layering data (Table 4) is used for the analysis.
- Load combinations

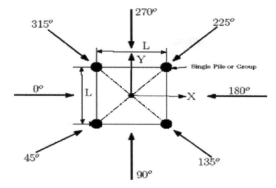

Figure 2. Wave load directions as per API guidelines.

The analysis is done for both operating and extreme conditions in eight directions as specified by the API code (Figure 2).

3 RESULTS AND DISCUSSIONS

3.1 In-Place Analysis

The global integrity of the structure against premature failure is checked using In-Place Analysis. The following steps are performed in the In-Place analysis.

1) Static analysis of the structure under wave load acting in eight directions, as per API guidelines
2) Free vibration analysis to obtain natural time period for corresponding modes.
3) Dynamic wave response analysis using the significant wave height and time period.

3.1.1 Static analysis

The combined utilisation ratio for the members showed that the stresses are within allowable limits and also the maximum Unity Check (UC) ratio of critical members was lesser than one for both fixed support condition and for PSI condition. The maximum horizontal joint displacements are within allowable limits (33% of length of the member) as per API.

3.1.2 Free vibration analysis

- Free vibration analysis is performed to obtain the natural time period of the structure.
- The natural time period of the structure under fixed support condition is 2.61 s in the 1st and 2nd modes.
- The natural time period of the structure with PSI is 2.95 s in the 1st and 2nd modes.
- The values are well within the range for fixed offshore structures.

3.1.3 Dynamic wave response analysis

Dynamic wave response analysis is carried out with Pierson–Moskowitz (P-M) wave spectrum. Maximum hydrodynamic force obtained is 7880 kN in X-direction. Table 5 shows the results of dynamic wave response analysis.

Table 5. Results of dynamic wave response analysis.

Response	Fixed support condition (m)	PSI condition (s)
Max. overturning moment	0.81×10^6	0.56×10^6
Base shear	13252	9795

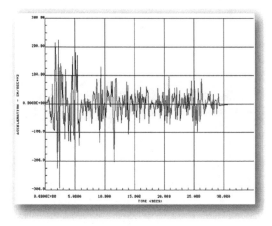

Figure 3. Time history of El-Centro (1940).

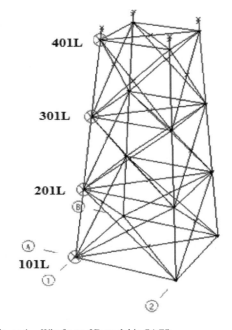

Figure 4. Wireframe 3D model in SACS.

3.2 Earthquake analysis

Seismic analysis is carried out using El-Centro Time History (Figure 3) which has duration of 31.18 sec and maximum acceleration of 312.76 cm/s^2.

The location of joints considered for joint displacement is shown in the Figure 4. The results of time

Responses	Fixed support condition	PSI condition
Base shear	53661 kN at 11.2 s	63857 kN at 10.3 s
Overturning moment	2.97×10^6 kN m at 5.7 s	3.26×10^6 kN m at 11.6 s
Joint displacement (401L)	32 cm at 7 s	42 cm at 13.2 s

Figure 5. Results of time history analysis.

	Fixed support condition	PSI condition
Base shear (kN)	0.46×10^5	0.52×10^5
Overturning moment (kN m)	0.24×10^7	0.26×10^7
Joint (401L) displacement (cm)	50.1	56.8

Figure 6. Results of response spectrum analysis.

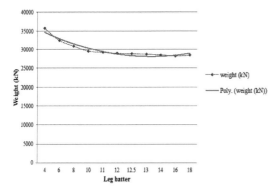

Figure 7. Optimization of leg batter.

history earthquake analysis under both fixed support condition and with PSI are as tabulated in Figure 5 and the results of response spectrum analysis are given in Figure 6.

3.3 *Optimization of leg batter*

The chosen jacket type platform is investigated for optimum weight. The jacket leg batter is an important parameter which influences the total weight of the jacket. Therefore, eleven jackets with a range of 3° to 14° (4 to 18 Vertical to Horizontal) slope in leg batter have been modelled and analysed by a finite element program, SACS. Totally, 11 jackets are modelled and analysed. Members are redesigned until the stress ratio reaches 0.8 in the critical members. The variation of jacket weight vs. leg batter is given in Figure 7 and the optimum leg batter obtained is 16.

4 CONCLUSION

The following conclusions are drawn based on the analytical investigations carried out using SACS.

- The natural frequency of the structure is around 2.9 seconds which is well within the range for fixed offshore structures.
- The maximum base shear obtained from dynamic wave response analysis for fixed support condition is about 35% more than that of analysis with soil-structure interaction.
- Base shear and overturning moment values due to earthquake forces for jacket with soil condition are about 25% more in time history analysis than response spectrum method.
- Maximum responses are increased by considering PSI for seismic analysis. Hence, consideration of soil condition is necessary for designing a safe structure.
- The weight of the chosen jacket has been reduced by 2% with optimized leg batter of 16.

ACKNOWLEDGEMENT

The authors wish to thank National Institute of Ocean Technology, Chennai, India for providing SACS software facility.

REFERENCES

Ahmed A. E., Mahmoud R. H., Marzouk. H (2009), 'Dynamic response of offshore jacket structures under random loads', Marine Structures, vol. 22, pp. 504–521.

API Recommended Practice 2A-WSD (1996), 'Recommended Practice for Planning, Designing and Constructing Fixed Offshore Platform-Working Stress Design,' 20th Edition, Official Publication, US.

API Recommended Practice 2EQ-WSD (2014), 'Seismic Design Procedures and Criteria for Offshore Structures', first edition.

Det Norske Veritas (DNV-1977), 'Result for the Design, Construction and Inspection of Offshore Structures,' Oslo.

Duy-Duan Nguyen, Chana Sinsabvarodom (2015), 'Nonlinear behavior of a typical oil and gas fixed-jacket offshore platform with different bracing systems subjected to seismic loading', The 20th National Convention on Civil Engineering, Thailand.

Khosro B., Reza Hosseini S., Mohammad H. and Hesam. S. (2011), 'Seismic Response of a Typical Fixed Jacket-Type Offshore Platform (SPD1) under Sea Waves', Open Journal of Marine Science 1, pp. 36–42.

Mehrdad Kimiaei, Mohsen Ali Shayanfar, Hesham El Naggar. M, Ali Akbar Aghakouchak (2004), 'Nonlinear response analysis of offshore piles under seismic loads', 13th World Conference on Earthquake Engineering Vancouver, B.C., Canada, Paper No. 3056.

Mohsen Mohammad Nejad et al. (2010), 'Optimization of Legs Batter in Fixed Offshore Platforms', Proceedings of the Twentieth International Offshore and Polar Engineering Conference, ISBN 978-1-880653-77-7; ISSN 1098-6189.

Narayanan Sambu Potty and Ahmad Fawwaz Ahmad Sohaimi (2013), 'Ultimate strength assessment for fixed steel offshore platform', Malaysian Journal of Civil Engineering, vol. 25, No. 2, pp. 128–153.

Taha Nasseri, Naser Shabakhty and Mohammad Hadi Afshar (2014), 'Study of Fixed Jacket Offshore Platform in the Optimization Design Process under Environmental Loads', International Journal of Maritime Technology, vol. 2, pp. 75–84.

Venkataramana K. (1988), 'Seismic response of offshore structures in Random seas' Proceedings of 9th world conference on earthquake engineering, Vol. 6.

Yasser Mostafaa and Hesham El Naggarb. M (2004), 'Response of fixed offshore platforms to wave and current loading including soil–structure interaction', Elsevier, Soil Dynamics and Earthquake Engineering, vol. 24, pp. 357–368.

FPSO response and water depth: A study using model tests

V.J. Kurian, N. Rini & M.S. Liew
Department of Civil and Environmental Engineering, Universiti Teknologi PETRONAS, Seri Iskandar, Perak, Malaysia

A. Whyte
Department of Civil Engineering, Curtin University, Australia

ABSTRACT: The modern era necessities and overuse of non-renewable energy by mankind have depleted the natural resources to a scaring extent. The effect of water depth to the FPSO response is an important aspect to be investigated, as today, new floating structures are being installed in deep waters of up to 3000 m depth. Hence in this paper, a study on the effect of FPSO response by varying the water depth has been presented by conducting model tests in the Offshore Laboratory, Universiti Teknologi PETRONAS. The results give an insight on the variation of surge, heave and pitch responses of the FPSO and mooring line tensions to the varying water depth. The results show an increase in the response as the water depth is increased and it is finally concluded that for minimizing the FPSO motions, the mooring line tension has to be increased as the water depth increases.

1 INTRODUCTION

Floating Production Storage and Offloading (FPSO) systems are producing oil since the late 1970s. The operating water depth for the FPSO ranges from 50 m to 3000 m in the Malaysian region. In the South East Asia, Malaysian waters is a prominent region in supplying the energy resource for the global needs. Model tests and software simulations aid the engineers in designing suitable platforms with optimum configuration for the various design considerations. Even though Malaysian water is calm when compared to the North Sea, proper consideration should be given to the various criteria like water depth and metocean data. If designed and conducted appropriately, model testing proves to be a reliable method for the prediction of platform responses and mooring line forces.

Pinkster et al. gives an outline of model tests with single point mooring systems emphasizing the purpose of model tests, scope and possible sources of errors, the required information to set up the model tests and the characteristics which can be simulated (Pinkster et al. 1975). Wichers conducted a comprehensive study for numerical simulations of a turret-moored FPSO in irregular waves with wind and current, derived the equations of motion of such model in the time domain using an uncoupled method and solved rigid body and mooring line dynamics separately (Wichers 1988). The model test of the FPSO was carried out by Lu et al. to evaluate the applicability of linearised spectral technique in combined low/high frequency motion reponses (Lu et al. 1990). Heurtier et al. compared the coupled and uncoupled analysis for a moored FPSO in harsh environments and suggested that the uncoupled analysis results are efficient to be used in the early design phase of the mooring system. There was relatively good agreement between the uncoupled and coupled analysis values even though the maximum values were different (Heurtier et al. 2001). Kim et al. developed a program in time domain for simulating the global motion of a turret moored FPSO. They also conducted physical model tests to study the vessel motion and mooring tension for non-parallel wind, wave, current and 100 year hurricane condition in Gulf of Mexico (Kim et al. 2005).

In this paper, the observations made by conducting model tests on the FPSO motion response and mooring line tension for different water depths are presented. The model tests were conducted in the Universiti Teknologi PETRONAS Offshore Laboratory. The spread-moored FPSO was free to move in surge, sway, heave, roll, pitch and yaw directions while it was subjected to unidirectional random waves. The model scale 1:100 was chosen for the ease of handling the FPSO and the mooring line connecting the FPSO to the seabed was considered as a linear spring with negligible mass. The results help in predicting the motion response of FPSO when the water depth and wave frequency are varied for a maximum water depth of 100 m since the maximum water depth allowed in the wave tank is 1 m.

2 MODEL TEST OF FPSO

The experimental investigation of the FPSO responses were carried out in a 20 m long, 10 m wide, and 1.5 m deep wave tank in the offshore laboratory, Universiti

Teknologi PETRONAS, Malaysia. In the present study, the spread moored FPSO model with the main dimensions as shown in Table 1 was kept at the center of the wave tank as shown in Figure 1. A horizontal mooring system which consists of four soft springs of stiffness 9 N/m connected with a cable was used to hold the FPSO on position. Soft springs were used to minimise the inertia effect of springs on the FPSO motions thereby minimising the coupling effects. Load cells were connected between the FPSO and mooring lines to measure the tension in the mooring lines. The spread-moored FPSO was thus oriented along the centre line of the wave tank with its bow facing the wave paddles to simulate the head sea condition. The weight of the FPSO model in air was 168 N. The model was ballasted with additional weight to achieve 50% of the dead weight tonnage (DWT) loading condition with a draft of 6.3 m.

All the waves were calibrated in the wave tank at the desired water depth before the FPSO model was kept in the tank. Qualysis Oqus 500+p 4 high speed motion capture system with SLR optics was used to measure the FPSO response in 6 degrees of freedom. Free decay test was conducted to measure the natural period of the system and the surge natural period was found to be 103.3 s.

Table 1. Main particulars of Berantai FPSO.

Designation	Quantity	Unit
Overall Length	207.43	m
Beam of Ship	32.25	m
Depth of Ship	16.75	m
Draft to baseline	12.6	m
Free board	4.15	m
DWT	55337	ton

Figure 1. Berantai FPSO model.

3 RESULTS AND DISCUSSION

The experiments were carried out in head sea condition under the action of regular waves. The wave height was 4 m and time period ranged from 8 s–16 s. The calibrated wave for the wave height 4 m and time period 8 s is given in Figure 2.

The water depth was varied from 62 m to 100 m. The incident wave elevation, mooring line force and motions of the FPSO were recorded. The typical time series for surge, heave and pitch motion responses are given in Figure 3.

The variation of the FPSO motion responses in the surge, heave and pitch degrees of freedom were studied by varying the water depth from 62 m to 100 m. At each water depth, keeping the wave height as 4 m, the wave periods were changed from 8 s to 16 s. The variations in surge, heave and pitch RAO are plotted for different water depths and shown in Figure 4. The plot shows that the mean values of surge and heave RAOs increase as the water depth increases. At a wave period of 9.09 s, the surge RAO is minimum and then it increases until a wave period of 12.5 s and then decreases at the same wave period a bit, again to increase until the wave period of 16 s. This pattern of deviation in surge RAO

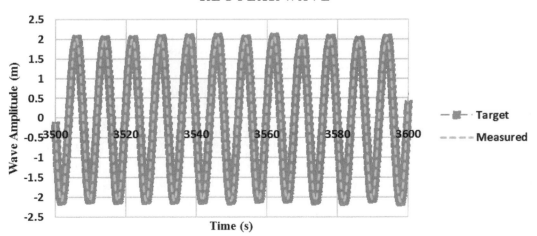

Figure 2. Calibrated regular wave.

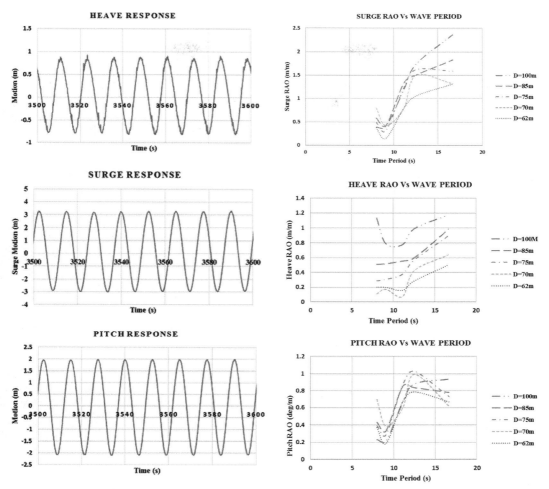

Figure 3. Typical time series of Surge, Heave and Pitch motion.

Figure 4. Variation in motion responses for different water depths.

is seen as the same for all the water depths. The heave RAO shows a general trend in increase of its value from the wave period of 8 s and a slight decline in the value for the RAO at the wave period of 11.11 s. The general trend in the variation of heave RAO for all the wave periods for different water depths is the same.

The pattern of change in pitch RAO for different wave periods in the range 9.09 s to 12.5 s is seen to be same as the surge RAO. The pitch RAO is minimum at the wave period of 9.09 s and then increases until 12.5 s and then takes a sudden dip. The variation in pitch RAO for different wave periods is also same for different water depths. The majority of the wave momentum is directed in the horizontal direction, which results in an increase of FPSO motion in the horizontal direction as the time period (or wave length) increases. The tanker, in effect becomes a particle floating on the surface to the wave as an effect of the wave length getting much larger than the length of the tanker. The tanker thus, does not offer much resistance and results in large amplitude of the motion (Gernon et al. 1987). The decrease in response during the time period of 8 s to 9 s is due to the half-length of the ship becoming an odd or even multiple of the wave lengths. If the length of a ship is half the waves generated, the resulting wave will be very small due to cancellation, and if the length is the same as the wavelength, the wave will be large due to enhancement. At 8.5 s, the wave length is half of the ship length which results in decrease in wave momentum and hence results in decrease in response. And hence the curves generated can be used for predicting the change in responses for different water depths and different wave periods for the similar configured FPSOs.

Load cells were connected between the FPSO hull and mooring lines to measure the tension in the mooring line. The variation in the mooring line tension for different water depths is plotted and shown in Figure 5. The mooring tension increases at a wave period of 8 s and declines at a wave period of 9.09 s until the wave period of 11 s and then again increases. The mooring line tension again decreases for the wave period range of 12.5 s to 16.6 s.

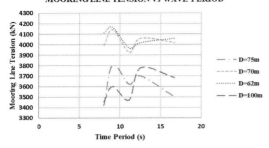

Figure 5. Variation in Mooring Line Tension for different water depths.

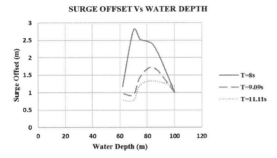

Figure 6. Variation in Surge Offset of FPSO for different frequencies.

The variation in the mean offset of the FPSO in surge direction for different wave frequencies is plotted and shown in Figure 6. The mean offset increases as the water depth decreases from 100 m to 75 m and then decreases when the water depth again decreases to 62 m.

4 CONCLUSIONS

Based on the present experimental study of a spread moored FPSO, the following conclusions are derived.

1. For a wave height of 4 m, the surge and pitch motions vary in the same pattern for different wave periods. Both the RAOs decrease when the wave period changes from 8s to 9.09 s and then rapidly increase when the wave period changes from 9.09 s to 12.5 s. The observed pattern remains true for different water depths.
2. The mooring line tension decreases for a wave period range of 9.09 s to 11 s and for 12.5 s to 16.6 s.
3. For both heave and surge degrees of freedom, the mean RAO increases as the water depth increases.

Hence it is finally concluded that for minimizing the FPSO motions, either the water depth should be minimum or the mooring line tension has to be increased or the wave periods and wave lengths should be minimum. Since controlling the metocean parameters and water depth is not a practical solution, proper tensioning of the mooring lines has to be adopted. The response of the FPSO should be kept as low as possible for a safe and comfortable working environment. Instead of pretensioning the mooring line with constant tension for all water depths, an increase in the pretension value for corresponding increase in water depth has to be ensured for minimizing the FPSO motions. This can be considered for the future research works. Since there was limitation in increasing the water depth, these results will not be applicable for deep water scenarios.

REFERENCES

Gernon, B. J. & Lou, J. Y. K. 1987. Dynamic Response of a Tanker Moored to an Articulated Loading Platform. *Ocean Engineering*, 14(6), 489–512.

Heurtier, J. M., Buhan, P., Fontane, E., Cunff, C., Biolley, F. & Berhault, C. 2001. Coupled Dynamic Response of Moored FPSO with Risers. *Proc. of the Eleventh ISOPE Conference*, 1, 319–326. Norway.

Kim, M. H., Koo, B. J., Mercier, R. M. & Ward, E. G. 2005. Vessel/ Mooring/ Riser Coupled Dynamic Analysis of a Turret – Moored FPSO Compared with OTRC Experiment. *Ocean Engineering*, 32, 1780–1802.

Lu, D. M. & Zhiliang, M. 1990. Analysis of Motion Response of Offshore Petroleum Floating Production Storage Offloading System. *Proc. of the First Pacific Asia Offshore Mechanics Symposium*, 295–303. Korea.

Pinkster, J. A. & Remery, G. F. M. 1975. The Role of Model Tests in the Design of Single Point Mooring Terminals. *OTC*, 1–24. Texas.

Wichers, J. E. W. 1988. A simulation Model for a single Point Moored Tanker, PhD Dissertation, Delft University.

Real time kinematic-GNSS methods for monitoring subsidence of offshore platform

Khin Cho Myint, A.N. Matori & Adel Gohari
Department of Civil and Environmental Engineering, Universiti Teknologi PETRONAS, Seri Iskandar, Perak, Malaysia

ABSTRACT: Reservoir compaction or shallow gas phenomena may cause offshore platforms to experience subsidence and deformation which could affect their structural integrity. Excessive subsidence may also cause loss of production as well as loss of live. However, the physical location of many offshore platforms which are typically hundreds of kilometers offshore poses a challenge to measure the subsidence rate accurately and in real-time. Over the years GNSS in the form of GPS has been used to determine such subsidence rate with great accuracy (cm level). The present GPS technology would not provide the monitoring in real-time mode. Present RTK GPS however has very serious limitation to be utilized in offshore due to its short range (approximately 50 km baselines) capabilities; typical offshore platforms are 150 km from shore. This paper therefore develops a new or improve RTK algorithm for longer range to be applied in real-time offshore platform monitoring.

1 INTRODUCTION

Offshore deformation monitoring has been a reality for the last few decades. The offshore deformations are usually situated at the ocean at a particular depth (Peuchen, 2013). Shallow gas or reservoir compaction phenomena may cause offshore platforms to experience subsidence and deformation which could affect their structural integrity. Moreover excessive subsidence may also cause loss of infrastructure as well as loss of life (Badellas & Savvaidis, 1990). Therefore, it is very crucial for shallow gas prone gas field platform. Subsidence monitoring is carried out regularly to anticipate their consequences. However the physical location of many offshore platforms which are typically hundreds of kilometers offshore poses a challenge to measure the subsidence rate accurately and in real-time.

Over the years GNSS in the form of GPS has been used to determine such subsidence rate with great accuracy (cm level). However the present GPS technology would not provide the monitoring in real-time mode. This paper carry out the new GPS monitoring that could be done in real-time mode using Real-Time-Kinematic (RTK) technique. Present RTK GPS however, has very serious limitation to be utilized in offshore due to its short range (approximately 50 km baseline) capabilities; typical offshore platforms are 150 km from shore.

The RTK performance is much depends upon baseline length defined as the distance between the user receiver and the base station. In the case of RTK with a short baseline under 10 km, errors of satellite ephemerides, effects of ionosphere and troposphere are almost eliminated by forming DD (double-difference) measurement equations. With a medium length baseline less than 100 km, in particular ionosphere effects are hard to be cancelled by DD (double-difference). This network-RTK technique has been well verified and demonstrated by a lot of field experiments in previous researches (H.J. Euler, et al. 2001) Global Navigation Satellite Systems (GNSS) as are collectively known have been in operation while more are still being developed. At this moment, several commercial precise positioning services with GNSS based upon such the network-RTK systems have been already started and widely utilized by many users.

2 REAL-TIME KINEMATIC GLOBAL POSITIONING SYSTEM (RTK-GPS)

RTK (real-time kinematic) is one of the most precise positioning technique which can obtain cm-level accuracy of user positions in real-time by using the measurements of GNSS. (Tomoji Takasu & Akio Yasuda, 2010). Focusing on optimal ways of RTK GPS processing reference receiver data and then in real-time, providing "correction" information to users receiver and at the base station. Nowadays, navigation technique surveying is enables the utilization of a static base station at a recognized spot and for real-time data collection utilizes mobile rover unit in the RTK positioning with GNSS. The measurements of RTK GPS is required to ensure the accuracy of the position to reach up to centimeter level is one reference station within 10 km to 15 km (Rizos & Han, 2003). The processes of data are transmitted from the reference station, wherein the roving receiver, the computer processor combines its measurement with the data. RTK positioning is a system that allows centimeter level accuracy positioning in real-time. However, it can cause efficiently differing a way similar errors and biases which is atmospheric effects, GNSS satellite orbit errors and clock bias in the carrier phase interpretation of the receivers at baseline, a reference station and a rover.

RTK-GPS positioning system typically comprises of a single reference station which transmits code and carrier phase observations information to one or mobile rover units in the field conventionally. Therefore, the data of reference station is combined with local measurements collected at the rover using proprietary differential processing techniques to yield precise relative estimates coordinate. The reference station and the rover station is equipped with dual frequency receivers. Moreover, reference receiver has a radio transmitter to the rover receiver sending phase observation, a correction which is equipped with radio modem to ascertain a link with the reference station (Loomis et al. 1991). The data volume sent by the reference receiver increases because of the density if the data update rate; as a result of this, RTK- GPS requires the data link to have an optimal capacity of 24,00 bps (bytes per second) or at least 9,600 bps or even in some cases 19,200 bps. According to Langley, this kind of data could be supported by the VHF of UHF bands which have a wide spectrum of bandwidths (GPS World 1997).

Traditionally, RTK is applied over short baselines involving one reference station and one roving receiver, using double-differencing of GNSS observations and employing some ambiguity fixing technique (Leick, 2004) As mentioned, conventional RTK range is limited because atmospheric and orbit errors grow with baseline length. Here in lies the primary motivation for using a network of base stations: to model and correct for distance-dependent errors that reduce the accuracy of conventional RTK.

3 GLOBAL NAVIGATION SATELLITE SYSTEMS (GNSS)

Global Navigation Satellite Systems (GNSS) applications commonly rely on the very precise GNSS carrier-phase observations with successfully fixed ambiguities [8]. GNSS has been used extensively by land surveyors since the late 1980s, primarily for geodetic control networks and for photo control. As systems have become more compact, most technologically advanced, easier to use and with a full complement of satellites enabling 24-hour usage, the diversity of surveying applications has increased significantly. GNSS systems are now available for many surveying and mapping tasks, including establishing control, setting out, real-time deformation monitoring, on-board camera positioning for aerial photography; the list is continually growing. Hence ambiguity resolution (AR) is the key for precision GNSS applications; it comprises ambiguity estimation and evaluation.

4 OFFSHORE DEFORMATION

The serious and potentially deceive consequences are one of the major problems in offshore deformation. According to fielding, extraction of hydrocarbon could lead to severe subsidence due to reservoir compaction and depletion (E.J. Fielding et al. 1998). Moreover, reported by A. Sroka and R. Hejmanowski (2006) it happened to Bedrige Oil Field that experienced subsidence of 50 cm/year, which lead to well failures and damages to exploration infrastructure that caused lost revenue of more than USD 20 million/year. Other damages of exploration and production infrastructures such as offshore platform, pipelines and building. Hence this underlines the importance of subsidence monitoring to an offshore platform to mitigate its adverse consequences. Due to their locations which are typically 120–250 km offshore, GPS technique was used to monitor their subsidence (A.N. Matori & H. Setan, 2000), (Peuchen, 2013). The implementing of a deformation monitoring system to maintain regular surveillance of the stability is a means to address both human safety and company profitability.

However, this study investigated into the effect of subsidence and deformation of an offshore platform due to factors such as reservoir compaction and depletion or shallow gas phenomena and its effect to their structural integrity. The approach developed in this study uses a precise relative Global Positioning System (GPS) which is advantageous for deformation monitoring in terms of long-baseline data as offshore platforms are located hundreds of kilometers from shore (widjajanti 2010).

5 RESEARCH METHODOLOGY

The method has been divided into two sections. The first part involves experimental work; conducting field measurement trials to study and collect Network RTK (NTRK) GNSS data- considering different navigation settings and measurement scenarios. The second part of the methodology is towards developing long-baseline. RTK GPS for real-time monitoring there will be three main steps involved which are:

- Correction Generation
- Correction Interpolation and extrapolation
- Correction transmission

In correction generation, the main work will be to solve ambiguity resolution in a long-baseline environment where the satellites observed are not common. This step will also involve residual error computation. During correction interpolation, this stage would explore the noble ways to interpolate the correction applied for long-baseline.

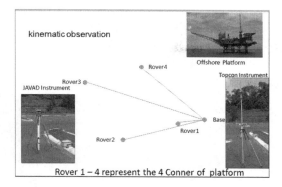

Figure 1. Kinematic observation using GNSS.

6 PP KINEMATIC DATA OBSERVATION AND PROCESSING

The PP kinematic survey of GNSS observation has been conducted using four receivers simultaneously. The survey of Static observation is conducted for 60 minutes time (start time: 02:30–03:45 in UTC time). Topcon GR-5 instrument were used as a base and 4 GNSS as a rover. During the time of observation, the data were received the position and type of satellite which is GPS and GLONASS satellites for all the four receivers. Therefore, the height of antenna was change three times during the observation. The processing is done by using the MAGNET tool.

7 ZERO BASELINE CALIBRATION USING GNSS RECEIVERS

Two (2) GNSS receivers were connected to a single antenna model JAVAD G5T using splitter and were installed on the rooftop connecting with 30 m long RG cable. Both GNSS receivers were simultaneously logged in the Netviews software with 15 seconds interval logging and the raw data was processed using Trimble Business Center 2.8. The zero baseline calibration indicates the ellipsoidal distanced and the ellipsoidal height difference between two GNSS positions is equal to zero. Both stations have been confirmed to be in good condition before the observations are made. The selections of control point are based on the criteria which are open sky view located on a hard structure, GNSS antenna located at safe area from any obstruction (Multipath) and optimum antenna location installed to the high point.

8 RESULTS AND DISCUSSION

Post-process Kinematic is to simulate the movement of deformation and to identify the movement of uplift and subsidence. However, the software (Magnet field) has limitation in term of virtual reference station and post processing. The results for the all receivers' position were accurate and reliable.

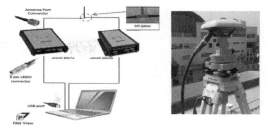

Figure 2. Process of zero baseline using GNSS receiver.

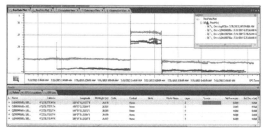

Figure 3. PP kinematic result.

Table 1. Result of zero-base-line test.

Calibration	GNSS Receiver S/N	Time Span	Latitude WGS 84 (EPSG Code 4326)	Longitude WGS 84 (EPSG Code 4326)	Ellipsoid Height (m)
#1	01804 (Delta)	3:00:00	3°07'30.45435" N	101°44'23.84448" E	53.818
	00660 (Delta)		3°07'30. 45435" N	101°44'23.84447" E	53.818
#2	00647 (Sigma)	1:37:30	3°07'30.44291" N	101°44'23.81832" E	51.809
	00548 (Sigma)		3°07'30.44295" N	101°44'23.81841" E	51.810

Zero baseline tests are used to examine the performance of GPS receivers. Since the two receivers are connected to the same antenna with a low-noise amplifier all common errors due to multipath, noise, satellite orbit perturbation, impacts from propagation media, etc. are eliminated in the GPS data processing. The results of obtained coordinate are shown in below. Hence, the maximum horizontal is identified was 0.00009" for Sigma model different in longitude. Difference height for and Sigma is 1 mm. The noise bias error is the major contribution for the position and height of the GNSS observation. The receiver is tested and working in good condition.

9 CONCLUSION

Real-Time-Kinematic (RTK) GPS has been developed since the middle of nineties to provide high accuracy positioning in real time. However RTK GPS has very severe limitation to use for offshore platform, real-time monitoring due to its short range operational capability which is typically 65 km as reported by (Grejner-Brzezinska, et al. 2009). Typical Malaysian Offshore platforms are located approximately 150–200 km away from the shore. Hence, obviously there is a need to develop new RTK technology for long baseline real-time offshore platform monitoring. The receiver tends to receive not only signals directly from the satellites, but also signal reflected from those reflecting surfaces (multipath effect). Moreover, the processing of long –based line GPS data algorithms, systematic errors such as GPS satellite orbits, multipath and atmospheric effects errors in GPS measurement can't be eliminated completely. Significantly, these systemic errors effect ambittiguity resolution processes and a potentially critical problem for GPS positioning high precision applications. Thus, it is necessary to customize a proper data processing algorithm which can effectively obtain a good result in offshore deformation.

REFERENCES

A.N. Matori & H. Setan. 2000. Role of global positioning system (GPS) in hydrocarbon exploration-subsidence monitoring of the offshore platform, J. Platform, vol. 1, no. 2, Universiti Teknologi PETRONAS, Malaysia.

A. Sroka & R. Hejmanowski. 2006. "Subsidence Prediction Caused by the Over oil and Gas Development," in 3rd *Int. Assoc. of IAG/12*th *Int.FIG Symp.*, Baden.

A. Badellas & P. Savvaidis. 1990. Monitoring of Deformation of Technical Works and Ground Landslides with Geodetic Methods. Papageorgiou Publ.Co., Thessaloniki.

D.A. Grejner-Brzezinska, Toth, C.K. Li, L. Park & X. Wang, et al. 2009. "Positioning in GPS challenged Environment dynamic sensor network with Distributed GPS Aperture and Inter-nodal Ranging of Signals, *Proceedings of the 22th Int Technical meeting of the satellite Division of the institute of navigation (ION GNSS 2009)*. 111–123.

E.J. Fielding et al. 1998. Rapid subsidence over oil fields measured by SAR interferometry Geophysical Research Letters, v. 25, no. 17, 3215–3218.

H.J. Euler, C.R. Keenan & B.E. Zebhause, Study of a simplified approach in utilizing information from permanent reference station arrays, ION GPS. 2001.

H. Setan & R. Othman. 2006. Monitoring of offshore platform subsidence: using permanent GPS stations," JGlobal Positioning System, vol. 5. no. 1–2, 17–21.

R.B. Langley. 1997. GPS receiver system noise. GPS World, 8(6): 40–45.

A. Leick. 2004. GPS Satellite Surveying, Third Ed., J. Wiley & Sons, N.J. http://igs.ifag.de/index_ntrip.htm

P. Loomis, L. Sheynblatt, & T. Mueller. 1991. Differential GPS Network Design. In Proceedings of the 4th International Technical Meeting of the Satellite Division of the Institute of Navigation (ION GPS-91), 511–520.

N. widjajanti. 2010. Deformation analysis of offshore platform using GPS technique and its application in structural integrity assessment. Phd thesis, universiti teknologi petronas.

J. Peuchen. 2013. Site characterization in near shore and offshore geotechnical projects, Volume 1, 83–111.

C. Rizos & S. Han. 2003. Reference Station Network Based RTK Systems - Concepts and Progress. School of Surveying and Spatial Information Systems. The University of New South Wales, Sydney NSW 2052 AUSTRALI.

Tomoji Takasu & Akio Yasuda. 2010. Kalman-Filter-Based Integer Ambiguity Resolution Strategy for Long-Baseline RTK with Ionosphere and Troposphere Estimation. Tokyo University of Marine Science and Technology, Japan.

Incremental wave analysis method: An alternative to load factor method in SACS pushover analysis

M.M.A. Wahab, A.R.M.M. Kamal & V.J. Kurian
Department of Civil and Environmental Engineering, Universiti Teknologi PETRONAS, Seri Iskandar, Perak, Malaysia

ABSTRACT: Many offshore jacket platforms in Malaysia are operating beyond their design life. In reassessment for life extension, Pushover Analysis is used to acquire the reliability of the platforms. Pushover Analysis imposes loadings onto platform model by incrementally increasing the Load Factor (LF) until the platform exceeds specified displacement or collapse limit. Recent study has established a new approach to perform Pushover Analysis. It is called the Incremental Wave Analysis (IWA) method. IWA increases wave height which coherently increases lateral load acting on the platform. Hence, to quantify the differences, this study compares both the methods; including analyzing Reserve Strength Ratio (RSR) and failure mechanism. In IWA apart from wave height other prospective significant parameters such wind speed and wave period are studied. The study concludes that IWA results in a lower RSR than LF, even when the wave period is included. But LF has more member failures compared to IWA.

1 INTRODUCTION

Although currently the market experiencing severe decline in oil price, life extension of the operating jacket platforms is inevitable (Bowler T 2015). Hence, it is essential to address the safety of the life extended platforms. Structural Reliability Analysis (SRA) is used to evaluate the safety of life extended platforms. In SRA, Pushover Analysis is commonly applied. Pushover Analysis evaluates platforms in commercial finite element programs such as SACS. The platform is modelled and subject to vertical and lateral loadings until the model reaches the displacement or collapse limit. The common practice in Pushover Analysis is termed Load Factor (LF) approach, where dead load, live load and environmental load acting on the platform are increased by a specified factor, at regular intervals to identify the failure load factor at the point of collapse (Kurian V et al. 2014). Recently a study has claimed that a new method in performing Pushover Analysis termed Incremental Wave Analysis (IWA) has been developed. IWA increases the forces acting on platform by increasing the wave height at every iteration anticipating the physical wave action location change on the platform rather than just the increase in magnitude. This method also allows incorporating the wave-in-deck phenomenon which is a serious threat to any platform (Golafshani A et al. 2011). Hence in this study a through comparison is planned on both methods.

To qualify the fitness for continued usage of a platform, a common response indicator is Reserve Strength Ratio (RSR). RSR is the ratio of collapse base shear to the design base shear (Eq. 1) (Krawinkler H 1996). In IWA method a unique response indicator, Collapse Wave Height (CWH) is proposed (Golafshani A et al. 2010).

$$RSR = \frac{BS_{collapse}}{BS_{design}} \qquad (1)$$

In lieu to the RSR, collapse mechanism of jacket platform is also a viable study. In conducting the reassessment of jacket platform, a study by George J et al. (2015) mentioned that LF method does not conform to at-site collapse condition. Commonly, pushover analysis produces results that show collapse behavior at the foundation. But, when compared to at-site collapse of platforms in the Gulf of Mexico during extreme storms, it is found that the platforms fail in the structure level (George J et al. 2015). Hence it is important to understand the collapse mechanism and how it is affected by the different methods being studied.

IWA introduces a method where the wave height is the parameter that acts as the force exerting on the platform. However wave height alone may not be sufficient to represent the wave action. Other wave parameters have not been studied specifically in the literature. It is essential to observe the effects of other parameters such as wind speed and wave period on the IWA method.

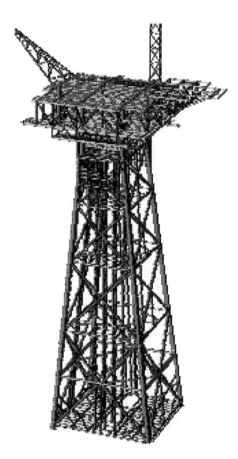

Figure 1. Platform A.

2 MODELLING CONFIGURATION

For the purpose of this research, an offshore jacket platform named as Platform A is utilized. It is a 4-legged platform in 94.8 meters water depth. The platform is located about 250 km offshore Bintulu near Sarawak. It is a drilling platform, with typical load combinations. The loading is applied from directions 0, 45, 90, 135, 180, 225, 270a, 270b and 315 degrees. The metocean load parameters utilized for this study are based on 100 year return period storm condition values. Figure 1 shows the configuration of platform A.

3 METHODOLOGY

The work procedure in achieving the objectives of this study is elaborated below:

Pushover Analysis was performed in SACS software. The utilized input files were the model input file (*sacinp*), pile-soil interaction input file (*psiinp*) and collapse input file (*clpinp*) of Platform A. In the collapse input file, the incremental factors for LF method were defined as in Table 1.

For LF the analysis is run for all nine directions separately. The results were taken from the collapse

Table 1. Adopted factors in collapse file.

Load Types	Factor		
	Initial	Final	Iteration
Dead	0	1	5
Live	0	1	5
Environmental			
LF	0	10	50
IWA *(end at 1* then wave height increased)	0	1	5

view of the collapse module. Design Base Shear was at factor 1 and Collapse Base Shear was at the last load step registered. And the failure mode was noted.

For IWA the platform model in the model file is subjected to individual incremental wave heights at each direction separately. The analysis was run for each wave height and the particular base shear value is recorded. Then the wave height is increased and the analysis is repeated and the new base shear is recorded. This is repeated until the platform reaches its collapse limit, i.e. collapses or reaching its maximum allowed displacement. At collapse, the base shear and wave height values are recorded. From the obtained values, the RSR and the collapse mechanism of the platform model was compared. Similar procedure was repeated to the other directions. For these analyses, the design wave heights from the design report were utilized. With a multi-directional approach, different sets of design wave heights are expected from different direction of a platform. In this study, for the IWA method, the wave heights were increased arbitrarily, 10% from the original design wave height at each step until collapse.

4 RESULTS AND DISCUSSION

4.1 *Collapse limit comparison between LF and IWA*

The significance of this comparison is attributed to the sensitivity of the methods used. From Figure 2, the general trend of RSR value from the LF is higher than IWA, except for direction 45 degree where both methods produce similar RSR values. Hypothetically LF does not represent the actual environmental loading imposed on a platform hence it is assumed that IWA gives a closer representation of the collapse limit for a platform. LF method should have been a conservative method as it uses approximation to produce the RSR. However, opposite to that it is observed that the values obtained from LF are less conservative than IWA. This may be attributed to the limits of LF, where its load pattern is undefined. The observation of the two load combinations in direction 270 found that the difference is in the vertical load acting on the platform. LF suggests that the axial load will result in tremendous change of collapse base shear. However, Irmawan et al. (2014) noted the dominant loads in offshore is the

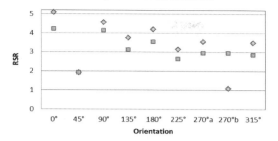

Figure 2. RSR comparison between LF and IWA.

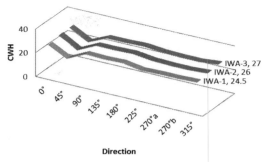

Figure 3. CWH comparison.

wave loading. The result obtained from the LF method in the 270b direction is rather objectionable.

4.2 Member failure comparison between load factor and incremental wave analysis

From the results, there is no general failure mode distinguishing the two methods. However, the results show that members fail at the bottom elevation in directions 45° and 90° in both methods; 135° for LF method; and member fails at the top elevation of the platform in directions 180°, 270° and 315° in both methods. Generally, it can be concluded that for specific platform the failure of members depends on direction rather than different nature of methods used.

From the number of member failure observed, it can deduce that members tend to fail more in the LF method than IWA method. Although most of the members fail at different sections of the same member, this is insensible as IWA did not show this failure behavior. One deduction may be attributed to the nature of the analysis. For each direction, the analysis is done only once. In the same analysis, the load factors are increased resulting in nodal displacement and change in stiffness constant. The members continue to undergo this until the platform reaches its collapse or displacement limit. IWA on the other hand, uses different analysis for different wave heights. This method does not take into account the member failure from the previous analysis. Thus, there is more member failure for LF as compared to IWA method.

4.3 Determination of wave parameters

The prediction of wave height and wave period will be based on a relationship between the said parameters with wind velocity. This method is taken from Chakrabarti S (2001). Sverdrup and Munk in Bretschneider's (SMB) formula (Bredmose H 2012; Feng X 2014; Michel W H 1968) for nearly developed sea is adopted as in Eq. 2–4:

$$\frac{gH_s}{U_w^2} = 0.254 \ (90\%) \qquad (2)$$

$$\frac{gH_s}{U_w^2} = 0.226 \ (80\%) \qquad (3)$$

$$\frac{gT_s}{U_w} = 4.764 \qquad (4)$$

Where Hs is the significant wave height in meter, Ts is the significant period in seconds, and Uw is the average wind speed in meter per second.

Base on the relationship from SMB formula, the wave height and wave period is input into the model for IWA method. In there, the change in wave height alone is termed IWA-1, the change in both wave height and wave period is termed IWA-2 and IWA-3 for 90% and 80% respectively. The result of the Collapse Wave Height (CWH) comparison among the considered cases is shown in Figure 3. The significance of CWH will determine the effects of other parameters to IWA method. Though the y-axis was provided in the range of 0–40, but the actual highest value recorded was 26.5 m.

From the values of the collapse wave height obtained for the IWA method, it is observable that wave period plays an important role in collapse limit of the platform. The increase in wave period will affect the frequency of the imposed loads on the platform, where the platform is hit at a slower frequency and the increase in wave height will affect the base moment due to increased moment arm. This behavior is reflected in the response of the platform where the collapse base shear and the Collapse Wave Height increases as a result of incorporating wave relationships.

4.4 RSR calculated for IWA and LF methods

From the Figure 4, generally the RSR value is highest for LF, followed by IWA with respect to wave period and wave height, IWA-3 followed by IWA-2, and lastly the conventional IWA. The exceptions to this generalization are storm condition 2, 8 and 9 which are at direction 45, 270 and 315 degrees. To put it in context, LF which should have been the conservative method gives a higher RSR value than the IWA methods which is more realistic. However, IWA-1 method gives the lowest RSR value which indicates that IWA-1 method is the most conservative of the four. IWA-2 and

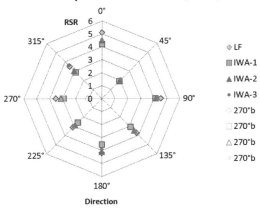

Figure 4. RSR comparison between LF, IWA-1, IWA-2 and IWA-3.

IWA-3 methods are in between LF and IWA-1 methods which indicate that these methods are used for optimal response. From this observation, it is apparent that wave period has a significant impact on Incremental Wave Analysis method. Furthermore, theoretically for the wave height to be increased the wave period has to be affected. This is to be further investigated.

5 CONCLUSION

This study has compared LF and IWA methods by utilizing RSR parameter and collapse mechanism. From the comparison, a general trend is seen in RSR. IWA results in a lower RSR value when compared to LF. When wave height increases, the internal stresses of members also increase and results in increased base shear. Base shear in LF method is attributed to the magnitude alone which explains why it is much higher at collapse for LF than IWA.

Collapse mechanism of the platform is compared at the location of the member failure. It is found that generally the member failure depends on the direction rather than the methods used. A notable difference between the two methods however falls in the number of member failure. LF has more member failures compared to IWA. This is attributed to the load pattern of LF method.

This research also studied the relationship of other parameters that affects IWA method. From the available scarce information, this research has studied wave period and wind speed parameters only. From the responses received, it is found that LF method still results in higher RSR value than IWA method, even when wave period is included in the analysis. Generally, when wave period is incorporated, the RSR value increases along with increase in wave period and wave height. Thus it is proven that wave period plays an important role in determining the RSR. It is concluded that IWA method increases wave height with respect to wave period.

REFERENCES

Bowler, T. (2015, january 19). BBC News. Retrieved June 5, 2015, from Falling oil prices: Who are the winners and losers?: http://www.bbc.com/news/business-29643612.

Bredmose, H., Larsen, S. E., Matha, D., Rettenmeier, A., Marino, E., & Saettran, L. (2012). D2.4: Collation of Offshore Wind-Wave Dynamics. Marinet.

Chakrabarti, S. (2001). Hydrodynamics of Offshore Structures. Southamptons Boston: WITpress; Computational Mechanics Publications.

Feng, X., Tsimplis, M., Yelland, M., & Quartly, G. (2014). Changes in significant and maximum wave heights in Norwegian Sea. Global and Planetary Change, 113, 68–76.

George, J., Kurian, V., & Wahab, M. (2015). Incorporation of Aging Effects of Piles into the Pushover Analysis of Existing Offshore Jacket Platforms in Malaysia.

Golafshani, A., Ebrahimian, H., & Baygi, V. (2010). A Technical Framework for Probabilistic Assessment of 5th National Congress on Civil Engineering.

Golafshani, A., Bagheri, V., Ebrahimian, H., & Holmas, T. (2011). Incremental Wave Analysis and Its Aplication to Performance-based Assessment of Jacket Platforms. Journal of Constructional Steel Research.

Irmawan, M., Piscesa, B., & Fada, I. A. (2014). Pushover Analysis of Jacket Structure in Offshore Platform Subjected to Earthquake With 800 Years Return Period. Conference: 2nd International Conference on Earthquake Engineering and Disaster Mitigation (ICEEDM). Surabaya, Indonesia: Research Gate.

Krawinkler, H. (1996). Pushover Analysis: Why, How, When and When Not To Use it. Stanford, California: Structural Engineers Association of California.

Kurian, V., Wahab, M., Voon, M., & Liew, M. (2014). Sensitivity Study and Safety Ratios.Pushover Analysis for Jacket Platform.

Michel, W. H. (1968). Sea Spectra Simplified. Marine Technology.

FRP shear contribution in externally bonded reinforced concrete beams with stirrups

H. Fazli & W. Teo
Department of Civil and Environmental Engineering, Universiti Teknologi PETRONAS, Seri Iskandar, Perak, Malaysia

ABSTRACT: Deterioration of marine concrete structures and its strengthening are challenges of the world of today. Recently, the increasing use of carbon fiber reinforced polymer (CFRP) composite appears to be a solution for strengthening of the concrete reinforced (RC) structures because of its physical and mechanical properties. This study aims to investigate the shear contribution of CFRP as a composite material in RC beams. In order to achieve this aim three RC beams investigated. One is un-strengthened and remainder two beams were externally shear strengthened in two scheme, U wrap and 2-side wrap with CFRP. Results indicate that CFRP increase the shear strength of the RC beams significantly. Moreover, the findings of this paper confirmed that the shear capacity predicting by design codes is higher and conservative than the experimental results.

1 INTRODUCTION

Reduction in strength and serviceability of reinforced concrete (RC) structures in marine environment occurred due to the moisture and chloride diffusion. Therefore, RC structures may need to be strengthened in order to prevent from deterioration and increase their ultimate shear or flexural capacity. For use the full potential ductility of the RC members, it is needed to make sure that RC member fail in flexure and not in shear. Externally bonded fiber reinforced polymer (EBFRP) and near surface mounted (NSM) systems are the most popular FRP strengthening techniques. Due to easy installation, lower cost and prevention of chloride ion penetration into concrete the EBFRP technique is more preferable than NSM technique for engineers. EBFRP with FRP sheets or plates is the most useful technique for improve and increase the shear strength of RC beams (Khalifa & Nanni 2000). Previous studies shown that strengthening concrete structures with FRPs lead to increase the ultimate load capacity of RC members (Soudki et al. 2007, ACI 440.2R-02 2008). Various factors contribute between FRP strengthening parameters such as shear span to depth ratio, the bond slip relationship and inclination of FRP (Khalifa & Nanni 2000, Chajes et al. 1995). Therefore, due to this complexity, the previous studies and also standards guidelines have not yet been sufficient to predict the shear strength.

This paper summarizes the concept of shear design provisions and experimental results of CFRP strengthened RC beams to investigate the CFRP shear contribution in RC beams. Furthermore, in order to evaluate reliability and accuracy of ACI 440-08, Fib, CNR and TR55 design codes the prediction of design guidelines compared with tested RC beams results.

2 DESIGN CODE PROVISIONS FOR PREDICTING SHEAR STRENGTH OF FRP STRENGTHENED RC BEAMS

2.1 *ACI 440-08 model*

The American concrete institute (ACI 440.2R-02 2008) code provision is based on a research study by Khalifa et al. (1998). The contribution of CFRP to the shear capacity of FRP strengthened concrete beam is given in Equation 1.

$$V_f = \left(\frac{A_{fv} \times f_{fe}(\sin \alpha + \cos \alpha) d_{fv}}{s_f} \right) \quad (1)$$

Where A_{fv} is the area of CFRP and determined by the total thickness of the applied CFRP sheets ($2nt_f$) times to the width of the CFRP strip (w_f). f_{fe} is the FRP effective stress and calculated by applying an effective strain (ε_{fe}) to the tensile modulus of elasticity of FRP (E_f). α is the angle between principal fiber orientation and longitudinal axis of member and d_{fv} is the effective depth of the FRP shear reinforcement. s_f is the span between each FRP sheet. Maximum strain of FRP systems that can be recorded at the failure load level is effective strain, ε_{fe}. Effective strain of FRP in U wraps or 2-side wraps calculated using a bond-reduction coefficient (k_v) as can be seen in Equations 2–7.

$$\varepsilon_{fe} = k_v \times \varepsilon_{fu} \leq 0.004 \quad (2)$$

$$k_v = \frac{(K_1 \times K_2 \times L_e)}{(11900 \times \varepsilon_{fu})} \leq 0.75 \quad \text{in SI units} \tag{3}$$

$$L_e = \left(\frac{23300}{(n_f \times t_f \times E_f)}\right)^{0.58} \quad \text{in SI units} \tag{4}$$

$$K_1 = \left(\frac{f'_c}{27}\right)^{\frac{2}{3}} \quad \text{in SI units} \tag{5}$$

$$K_2 = \left(\frac{d_{fv} - L_e}{d_{fv}}\right) \quad \text{For U wrap} \tag{6}$$

$$K_2 = \left(\frac{d_{fv} - 2L_e}{d_{fv}}\right) \quad \text{For 2side wrap} \tag{7}$$

2.2 Fib model

The European code fib (2006) shear strength predictions of FRP strengthened RC beams are based on the regression of experimental results carried out by Triantafillou & Antonopoulos (2000) and given in Equation 8:

$$V_{fd} = 0.9\varepsilon_{fd,e} \times E_{fu} \times \rho_f \times b_w \times d \,(\cot\theta + \cot\alpha)\sin\alpha \tag{8}$$

Where V_{fd} is the CFRP design shear contribution value, $\varepsilon_{fd,e}$ is the design value of effective FRP strain and E_{fu} is the elastic modulus of FRP in the principal fiber orientation. ρ_f is the FRP reinforcement ratio equal to $(2t_f \sin\alpha)/b_w$ for continuously distribution of FRP. However in FRP strip distribution b_w is equal to $(2t_f/b_w)(b_f/s_f)$ which s_f is strips spacing and b_f is width of FRP strip. b_w is the minimum width of cross section over the effective depth (d) and θ is the angle of diagonal cracks (assumed as 45°).

2.3 CNR model

Italian guidelines CNR (2004) determined the contribution of FRP in shear strength based on a research study by Monti et al. (2003). The proposed equations 9–10 are for U wrapped scheme and 2-side wrapped scheme, respectively.

$$V_{Rdf} = \left(\frac{1}{\gamma_{Rd}}\right) 0.9d \times f_{fed} \times 2 \times t_f (\cot\theta + \cot\beta)\left(\frac{w_f}{\rho_f}\right) \tag{9}$$

$$V_{Rdf} = \left(\frac{1}{\gamma_{Rd}}\right) \times \min\{0.9d, h_w\} \times f_{fed} \times 2 \times t_f \left(\frac{\sin\beta}{\sin\theta}\right)\left(\frac{w_f}{\rho_f}\right) \tag{10}$$

Where γ_{Rd} is the partial factor for resistance model and assumed as 1.20 and h_w is the beam stem depth. f_{fed} is the effective FRP design strength and given in Equation 11 for U wrapped scheme and Equation 12 for 2-side wrapped scheme. β, w_f and ρ_f are the angle, width and the spacing of the FRPs, respectively.

$$f_{fed} = f_{fdd}\left(1 - \frac{1}{3} \times \left(\frac{l_e \times \sin\beta}{\min\{0.9d, h_w\}}\right)\right) \tag{11}$$

$$f_{fed} = f_{fdd}\left(\frac{Z_{rid,eq}}{\min\{0.9d, h_w\}}\right) \times \{1 - 0.6\sqrt{Z_{rid,eq}}\}^2 \tag{12}$$

Where f_{fdd} is the ultimate design strength and l_e is the effective bond length of the FRP which determined by Equations 13–14. $Z_{rid,eq}$ is given in Equations 15–17 where, Z_{rid} is the vertical projected length of the strip.

$$f_{ffd} = \left(\frac{1}{\gamma_{fd}\sqrt{\gamma_c}}\right) \times \left(\frac{\sqrt{2E_f \Gamma_{FK}}}{t_f}\right) \tag{13}$$

Where $\gamma_{f,d}$ is the partial factor for FRP debonding and γ_c is the partial factor for concrete. The specific fracture energy, Γ_{FK}, is calculated as Equation 18.

$$l_e = \sqrt{\frac{E_f t_f}{2 f_{ctm}} E_f \Gamma_{FK}} \tag{14}$$

$$Z_{rid,eq} = Z_{rid} + l_{eq} \tag{15}$$

$$Z_{rid} = \min\{0.9d, h_w\} - (l_e \times \sin\beta) \tag{16}$$

$$l_{eq} = \left(\frac{s_f \times \sin\beta}{\frac{f_{dd}}{E_f}}\right) \tag{17}$$

Where s_f is the ultimate debonding slip and assumed as 0.2 mm.

$$\Gamma_{FK} = 0.3 k_b \times \sqrt{f_{ck} \times f_{ctm}} \tag{18}$$

Where f_{ck} is characteristic strength of concrete and f_{ctm} is average tensile strength of concrete. Geometric coefficient, k_b, is defined as below (Eq. 19):

$$k_b = \sqrt{\frac{2 - \frac{b_f}{b}}{1 + \frac{b_f}{400}}} \geq 1 \tag{19}$$

2.4 TR55 model

The British design code TR55 (2000) proposed the Equation 20 to determine the contribution of CFRP to the shear capacity of concrete beams:

$$V_{rd,sf} = \frac{A_{sw}}{s} \times z \times f_{ywd} \times \cot\theta + \frac{A_{fw}}{s_f} \times \left(df - \frac{n_s}{3} \times l_{t\,max} \times \cos\beta\right) \times E_{fd} \times \varepsilon_{fse} \times (\sin\beta + \cos\theta) \tag{20}$$

Where A_{sw} is cross sectional area of steel shear reinforcement and s is the spacing between the steel shear reinforcement. z is the lever arm between the longitudinal steel reinforcement and the centroid of the compression in the section. f_{ywd} is the design yield strength of the steel shear reinforcement. A_{fw} is area of FRP and s_f for continues FRP sheet is taken as 1.0. For fully wrapped, U wrapped and 2-side bonded n_s

value is 0, 1 and 2, respectively. L_{tmax} is the anchorage length and required to develop full anchorage capacity, β angel between the FRP and longitudinal axis of the member, E_{fd} is the design tensile modulus of the FRP laminate, ε_{fse} the effective strain in the FRP for shear strengthening.

3 EXPERIMENTAL PROGRAM

3.1 Preparation of test specimens and materials

An experimental study was conducted with a total of three reinforced concrete beams. The specimens were reinforced with longitudinal steel bars and steel shear reinforcement in rectangular-shaped with two-legged and hooks. All three beams had a rectangular cross section with dimension of 150 mm (W) × 2000 mm (L) × 250 mm height as shown in Figure 1.

One specimen was kept as a baseline specimen (control), whereas the remainder two specimens were externally strengthened with CFRP sheets. Table 1 presents the tested specimens details. The specimens were casted with concrete provided by a local ready-mix supplier. The 28 days compressive strength of the concrete was 22.13 MPa. The yield stress of the tension and compression steel reinforcing bars for T16 and T12 were 546 MPa and 554 MPa, respectively.

3.2 Specimens strengthening

Strengthening scheme consisted of U-wrap scheme (Beam-U) and 2-side wrap scheme (Beam-2S) as shown Figure 2. One layer of CFRP unidirectional Sika-wrap-231C carbon fiber fabric were applied following the manufacturer instructions and Sikadur-330 epoxy impregnating resin used as an adhesive. First concrete surface was prepared with hand grinder and then with sand paper to remove all the irregularities. To clean up the remained dust from surface acetone was used. First coat of saturate resin was applied follow by the CFRP sheets. A ribbed roller was used on the sheets to ensure impregnation of the fibers by the saturate. Table 2 shows the mechanical properties of the CFRP/Epoxy.

3.3 Ultimate load test setup

All the specimens instrumented with three linear variable displacement transducers (LVDTs) to determine mid-span deflections. For measure the strain at stirrups, waterproof strain gauges were used. Concrete beams tested under four-point loading with shear span to effective depth ration (a/d) equal to 3.

A gradually increasing load with rates of loading 0.10 was applied through a hydraulic actuator of universal testing machine (UTM) of 500 kN capacity. Applied loads were measured and recorded with a 500 kN capacity load cell. The simply supported RC beam specimens tested on the strong floor at the structural engineering research laboratory at Universiti Teknologi Petronas. Test setup and Instruments positions are shown in Figure 3.

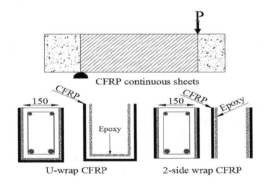

Figure 2. Beam strengthening schemes.

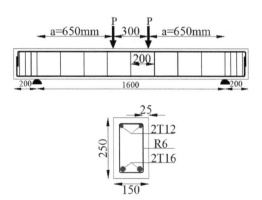

Figure 1. Test specimens details.

Table 1. Description of test program.

Beam	Wrapping schemes	Fiber distribution
Beam-0	Un-strengthened	–
Beam-U	U-wrapped	Continuous
Beam-2S	Sides-bonded	

Table 2. Mechanical and physical details of specimens and materials.

Type	Thickness (mm)	Tensile strength (N/mm^2)	Modulus of elasticity (N/mm^2)	Elongation at break (%)
Sika Wrap-231C	0.127	4,900	230,000 (Tensile)	2.1
Sikadur-330	–	30	3,800 (Flexural) 4,500 (Tensile)	0.9

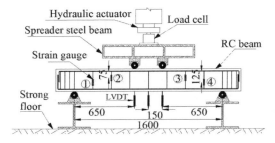

Figure 3. Test setup and instruments position.

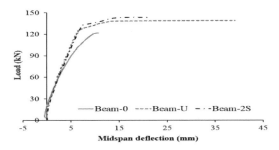

Figure 4. Beams load-deflection curve at mid-span.

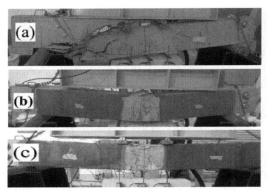

Figure 5. Tested beams failure mode.

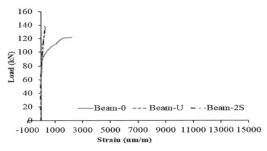

Figure 6. Load-strain curves of strain gauge 1.

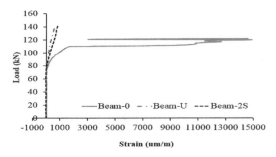

Figure 7. Load-strain curves of strain gauge 2.

4 RESULTS AND DISCUSSIONS

4.1 *Relationship between load and deflection*

In order to assess the shear capacity enhancement of a CFRP bonded RC beam, the load-deflection obtained results from experimental program was investigated. Mid-span load–deflection curves for the tested beams illustrated in Figure 4. Mid-span deflection for Beam-0 was 10.7 mm and for Beam-1 and Beam-2 were 21.24 and 39.17, respectively. Two modes of failure could be noticed for the beams, shear failure in the Beam-0 and flexural failure in strengthened beams.

Strengthening with CFRP composites could be effected on both failure mode and shear capacity. However, it is difficult to conclude from the results that a higher failure load obtain using specific strengthening scheme but results shown that U-wrap strengthening scheme has a significant effect on deflection behavior.

Figure 5a shows the mode of failure of Beam-0 which failed in shear. Initial cracks in Beam-0 started at shear spans at a load of 52 kN. The critical crack in the beam was the first shear crack which started to widen and propagated by increased the load and failed at a load of 121.85 kN.

Beam-U, failure was initiated by flexural cracks at mid-span of the beam as shown in Figure 5b. It was followed by flexural compression failure at a load of 138.54 kN with a 39.17 mm deflection. Externally U-wrap strengthening with CFRP resulted in a 14% increase in shear capacity compare to Beam-0. Beam-2S failure mode also flexural mode as shown in Figure 5c. The failure occurred at a load of 143 kN with 17 % increase in shear capacity compare to Beam-0. The failure load of beam-2S was slightly close and higher to that of beam-U.

4.2 *Comparison of strain response*

The strain gauges 1, 3 and 2, 4 were located at distance of 200 and 400 mm from supports, respectively. The distributions of strains in the tested beams stirrups shown in Figure 6–9. Results highlighted that control specimen was more strained compared to the strengthened beams. Therefore, it showed that CFRP relieved strain on the shear reinforcement due to shear strengthening. As shown in Figure 7 Maximum recorded vertical strain in stirrup observed in Beam-0 strain gauge 2 with 14591 um/m due to shear failure of the beam, while the maximum strain for beam-U

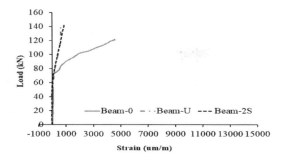

Figure 8. Load-strain curves of strain gauge 3.

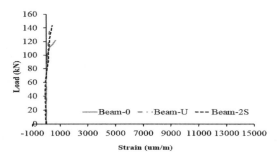

Figure 9. Load-strain curves of strain gauge 4.

and beam-2S at strain gauge 2 was 631 um/m and 824 um/m, respectively.

In Beam-0 the first shear cracks could be observed with the naked eye. The first shear cracks occurred at the early stages of strain in the stirrup steel. In fact, the observation of strain in shear reinforcements indicates that the significant contributions of the CFRP and the transverse steel in shear behavior of RC beams. It is observed that, the most strained stirrups were crossing the principal shear cracks. The strain distribution shows that the strains are much higher within 1.25d from the face of the support. This result is in good agreement with code specifications (ACI 318-02 2002) related to the location of the critical sections for design. First shear reinforcements which located in the critical sections yield. This verifies the results of formation and development of cracks in the Beam-0 during the test (Bousselham & Chaallal 2006, Carolin & Täljsten 2005) which shows that the first diagonal cracks observed in shear span between the support and the applied load.

By increasing the load the diagonal cracks become wider and developed simultaneously toward both the applied load point and the support. Yielding of stirrups occurred in high shear forces and lead to residual capacity of the beams decreased and become minimal. The lowest strain observed at strain gauge 4 of Beam-U with 229 um/m as shown in Figure 9. Beam-0 and Beam-2S strain value at strain gauge 4 was 695 um/m and 444 um/m, respectively.

Results indicated that strengthening by CFRP increased the shear capacity. The strains that recorded with four strain gauges shows that beam-U had almost lowest vertical strain which indicate that U wrap scheme reduce the strain in stirrups and lead to increase in shear capacity.

Table 3. Experimental FRP shear strength comparison with predicted FRP shear strength.

		Beam-0	Beam-U	Beam-2S
Experimental results	V_{exp}	60.92	69.27	71.52
	V_{frp},exp	–	8.35	10.60
$V_{frp,code}$	ACI 440-08	–	40.45	25
	fib	–	32	32
	TR55	–	44.64	38.57
	CNR	–	38	23.90
$V_{frp,code}/V_{frp,exp}$	ACI 440-08	–	4.84	2.26
	fib	–	3.83	3.02
	TR55	–	5.35	3.64
	CNR	–	4.55	2.25
COV (%)		–	13.62	23.88

4.3 Comparison between the test results and calculated values

The comparison between the test results and calculated shear strength using the code provisions are listed in Table 3. $V_{frp,exp}$ obtained by subtracting the shear strength of non-strengthened RC beams from CFRP strengthened one. The comparison indicated that the design codes approach gives conservative results than experimental results and the results acceptable. From the results it can be seen that fib design code prediction of $V_{frp,code}$ for U wrap scheme more closer to $V_{frp,exp}$ than other introduced codes. Also $V_{frp,code}$ of 2-side strengthening scheme in ACI and CNR has almost same predictions and more closer to the experimental results. This comparison evidenced a weakness that lies in the expression reported in the design codes. Furthermore, among the evaluated codes, TR55 design code consider the shear reinforcements effect on shear contribution and remainder of design codes not take into account.

5 CONCLUSION

In this study, CFRP shear contribution of CFRP strengthened RC beams with different scheme of strengthening were investigated. The study results showed that the externally strengthening CFRP can be used to increase the shear capacity of the concrete beams. In this study the experimental results achieved the 14%–17% increase in shear capacity of the beams without rupturing or debonding compare to the control specimen. Also using CFRP 2-sided scheme only gives less shear contribution compared to U-wrap. Comparison of the experimental results and design code predictions indicated that the current code provisions provides acceptable and conservative predictions.

ACKNOWLEDGEMENTS

Both authors would like to thank the UTP for providing financial support via Yayasan Universiti Teknologi PETRONAS (0153AA-A52).

REFERENCES

ACI 318-02 2002. *Building code requirements for reinforced concrete and commentary*. Farmington Hills, USA: American Concrete Institute.

ACI 440.2R-02 2008. *Guide for the design and construction of externally bonded FRP systems for strengthening concrete structures*. Farmington Hills, USA: American Concrete Institute.

Bousselham, A. & Chaallal, O. 2006. Effect of transverse steel and shear span on the performance of RC beams strengthened in shear with CFRP. *Composites Part B: Engineering* 37(1): 37–46.

Carolin, A. & Täljsten, B. 2005. Experimental study of strengthening for increased shear bearing capacity. *Journal of Composites for Construction* 9(6): 488–496.

Chajes, M. J., Januszka, T. F., Mertz, D. R., Thomson Jr, T. A. & Finch Jr, W. W. 1995. Shear strengthening of reinforced concrete beams using externally applied composite fabrics. *ACI Structural Journal* 92(3): 295–303.

CNR 2004. *Guide for the design and construction of externally bonded FRP systems for strengthening existing structures*. Rome, Italy: Advisory Committee on Technical Recommendations for Construction.

fib 2006. *Retrofitting of concrete structures by externally bonded FRPs: with emphasis on seismic applications*. Lausanne, Switzerland: International Federation for Structural Concrete.

Khalifa, A., Gold, W., Nanni, A. & Abdel Aziz, M. I. 1998. Contribution of externally bonded FRP to shear capacity of RC flexural members. *Journal of Composites for Construction* 2(4): 195–202.

Khalifa, A. & Nanni, A. 2000. Improving shear capacity of existing RC T-section beams using CFRP composites. *Cement and Concrete Composites* 22(3): 165–174.

Monti, G., Renzelli, M. & Luciani, P. FRP adhesion in uncracked and cracked concrete zones. *FRPRCS6*, 2003 Singapore. 183–192.

Soudki, K., El-Salakawy, E. & Craig, B. 2007. Behavior of CFRP strengthened reinforced concrete beams in corrosive environment. *Journal of Composites for Construction* 11(3): 291–298.

TR55 2000. *Design guidance for strengthening concrete structures using fibre composite materials*. Crowthorne, UK: The Concrete Society.

Triantafillou, T. & Antonopoulos, C. 2000. Design of concrete flexural members strengthened in shear with FRP. *Journal of Composites for Construction* 4(4): 198–205.

Coupled analysis for the effect of wave height on FPSO motions

N. Rini, V.J. Kurian & M.S. Liew
Department of Civil and Environmental Engineering, Universiti Teknologi PETRONAS, Seri Iskandar, Perak, Malaysia

A. Whyte
Department of Civil Engineering, Curtin University, Australia

ABSTRACT: Fully coupled time domain analysis is performed to study the effect of wave height on FPSO motions. The turret-moored FPSO used in the OTRC experimental studies were made use of to investigate the FPSO motion patterns. The study was conducted by varying the significant wave height for a particular peak period. The peak period ranges from 5–25 seconds, where the wave frequency response is significant. The results show significant variation in the horizontal and vertical motions of the FPSO, which are discussed in this paper. The results can be employed to ensure the safety and operability of the FPSO by predicting the FPSO motions where the sea state varies considerably.

1 INTRODUCTION

The number of operating Floating Production Storage and Offloading (FPSO) systems are significantly increasing in the present decade. The stability and station keeping of the FPSO are major contributors to the safety of the people onboard. The effect of the horizontal FPSO motion on the station keeping is significant and has to be studied thoroughly to avoid any future accidents. The high amplitude vertical motions significantly affect the operability of the FPSO followed by green water on the deck.

The environmental loads play an important role in the station keeping of the floating structures due to its effect on the FPSO horizontal motion. Wichers initiated the study on FPSO dynamic responses by conducting an uncoupled analysis to predict the FPSO motions and mooring line dynamics separately in the presence of wind, wave and current (Wichers 1988). Jiang et al developed a time domain program to predict the effect of these environmental loads on the LF and WF motions of the FPSO and detailed the possibility of low frequency slow drift oscillations in the horizontal degrees of freedom (Jiang et al. 1989). Heurtier et al did a comparative study on the FPSO motions using uncoupled analysis and coupled analysis and arrived at a conclusion that the uncoupled analysis is sufficient for the preliminary design of the mooring system (Heurtier et al. 2001), while Wichers et al showed that, for FPSO, uncoupled analysis will give large errors. Finally, Wichers et al concluded that fully coupled dynamic analysis is necessary to obtain accurate FPSO responses (Wichers et al. 2001a, b). Hence in this study, fully coupled time domain analysis is performed to investigate the FPSO motions. The non-linear time domain analysis of the wave frequency as well as low frequency force is carried out.

In this paper, the observations made by conducting time domain analysis of a turret moored FPSO in deep water using coupled analysis are presented. The turret moored FPSO used in the OTRC experiments is used for this study. The FPSO responses obtained from the commercial software is compared with the OTRC experimental results and the coupled model is calibrated and validated. The results obtained from simulation match well with the experimental results. The calibrated model is used for the parametric study using coupled analysis varying the wave height. The results can be used for predicting the variation of FPSO response with wave height in deep waters.

2 VALIDATION OF FPSO MODEL

The time domain simulation of the turret FPSO was carried out using fully coupled time domain analysis. The experimental studies for the same FPSO was conducted in the OTRC wave basin using truncated mooring system with a scale of 1:60 for 80% loading condition (Kim et al. 2005). The FPSO was free to move in all the six degrees of freedom. The prototype system had 12 mooring lines which were grouped into four sets with three mooring lines. In the experimental tests, four equivalent mooring lines at 45° apart were used. The main dimensions of the OTRC FPSO is given in Table 1. The mooring details and other parameters are given by Kim et al (Kim et al. 2001) and hence is not repeated here.

The free decay test was conducted to ensure that the mass distribution and hydrodynamic performance of the model and the prototype are the same. A specified force was applied on the respective degrees of freedom for 20–30 s and the analysis was carried out for small running time of 200–1000 s to obtain the free decay graphs. The natural periods obtained by performing free decay test using the commercial software is in the same range as the natural period obtained from the OTRC experiment. The values are compared and tabulated in Table 2.

Table 1. Main particulars of OTRC FPSO.

Designation	Quantity	Unit
Length Between Perpendicular	310	m
Beam of Ship	47.17	m
Depth of Ship	28.04	m
Draft (80% loaded)	15.121	m
Displacement (80% loaded)	186051	ton
Turret location (12.5% LPP)	38.73	m

Table 2. Natural periods of OTRC FPSO.

DOF	Present Work	OTRC Experiment
	Natural Periods (s)	Natural Periods (s)
Surge	204.2	206.8
Heave	11.2	10.7
Roll	12.4	12.7
Pitch	10.6	10.5

The experiments were carried out in head sea condition under the action of unidirectional random waves for the 100 year hurricane condition of Gulf of Mexico. The FPSO model along with the mooring system and the environment developed for conducting coupled analysis, is shown in Figure 1. The significant wave height was 12.19 m with a peak period of 14 s, represented by JONSWAP spectrum with gamma as 2.5. The water depth was 1829 m with a free surface current of 0.9144 m/s acting at 210°. The wind velocity at 10 m above MWL is 41.12 m/s represented using NPD spectrum acting at 150°. The same metocean data were used to simulate the environmental condition in coupled analysis and a 3-hour sea state simulation was carried out. The resulting motion spectrum in all the six degrees of freedom are compared with the published results obtained by conducting experiments in the OTRC wave basin and plotted as shown in Figure 2. The spectral densities match well except for the roll motion. This is due to the significant role of viscous effect in that mode. This can be effectively managed by providing additional roll damping in the analysis.

3 RESULTS AND DISCUSSION

Fully coupled time domain analysis was performed on the turret moored FPSO to obtain realistic motion amplitudes in the presence of unidirectional waves,

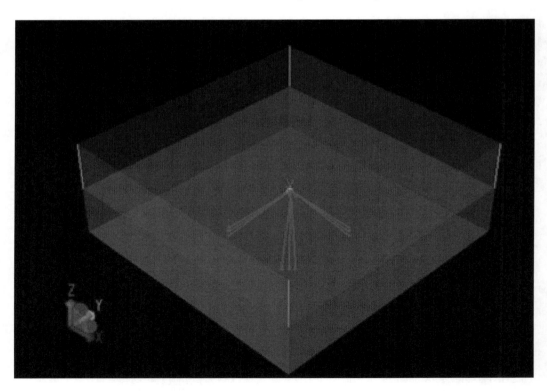

Figure 1. FPSO model and mooring system.

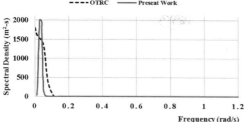
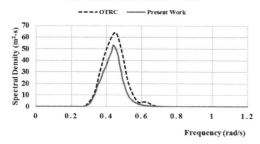
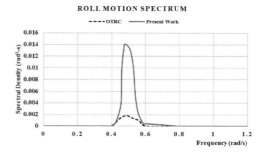
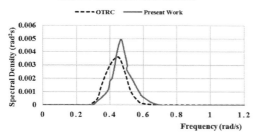
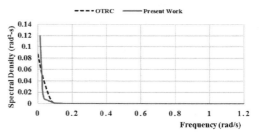

Figure 2. Motion spectrum comparison.

current and wind. The water depth was maintained at 1829 m and the significant wave height was varied from 4 m to 8 m for each peak period. Since Airy's linear wave theory is used in the analysis, wave height was not varied to higher values. The peak period ranges from 6 s to 25 s. The variation of RMS response with the wave height for all the six degrees of freedom were investigated. The variation of surge, sway and heave motions are presented in Figure 3.

Table 3 shows the variation of response height for all the degrees of freedom for significant wave height varying from 4 m to 8 m at peak period 20 s. The values clearly show the variation of the translation and rotation degrees of freedom.

The horizontal motions of the FPSO (surge, sway and yaw) decrease with an increase in wave height. This is because, the larger wave height causes an increase in velocity of the FPSO which in turn results in the increase of viscous damping in the waves (Ishihara et al. 2012, Faulkner et al. 1991). Comparing the motion of FPSO in horizontal plane, surge and heave are almost linear. For surge, heave and yaw degrees of freedom, the amplitudes of motion fluctuate significantly for the incoming waves with peak period 10 s–11 s. This is because, at this time period, the wave length of incoming waves becomes half the overall length of the FPSO.

This will result in cancellation of waves and fluctuation in its amplitudes. The amplitude of yaw motion is much higher compared to the amplitudes of roll and pitch motions. This is due to the fact that turret mooring allows rotation of the FPSO in yaw direction to avoid damages to the mooring system in adverse climates. For the FPSO motions in the vertical plane, all the three degrees of freedom, heave, roll and pitch show an increase in amplitude of motion with an increase in wave height. For heave, roll and pitch, the relationship between motion amplitude and significant wave height is linear at all the time periods. The results show that a rise in wave height will diminish the FPSO motion in horizontal plane, while the FPSO motion in the vertical plane escalates.

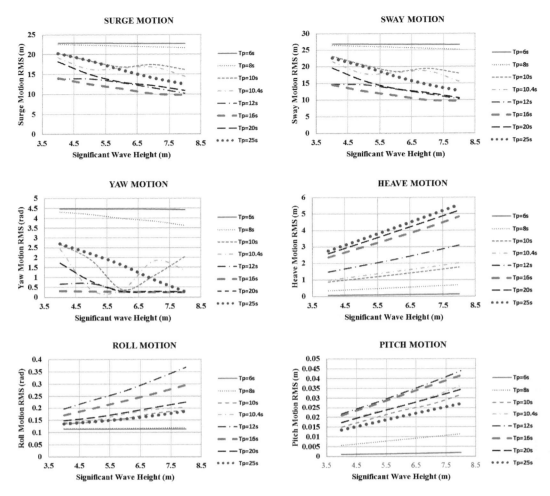

Figure 3. Response variation with wave height.

Table 3. Variation of FPSO motion with wave height.

Wave Height	Surge (m)	Sway (m)	Yaw (rad)	Heave (m)	Roll (rad)	Pitch (rad)
4	25.69926	27.6984	2.430427	3.657029	0.208768	0.024278
5	21.22945	22.19473	1.125258	4.584081	0.227398	0.030474
6	18.47142	18.81267	0.407849	5.509422	0.255424	0.036604
7	17.24777	17.10634	0.386488	6.431861	0.28769	0.04266
8	15.46272	14.85964	0.345204	7.361957	0.319547	0.048742

4 CONCLUSIONS

The conclusions drawn from the present study are

1. Surge, sway and yaw motions decrease as the wave height increases.
2. Heave, roll and pitch motions increase linearly with the wave height.
3. The FPSO motions in the horizontal plane fluctuate significantly when the incoming wave length is half of that of the FPSO length.

These results can be used to avoid green water on the decks as it is due to the increase in amplitude of the vertical motion of FPSO. Since high amplitude waves result in green water on the deck, proper precautions can be taken to minimize them and this, in turn will result in longer operation time for the FPSO. The operability of an FPSO is heavily dependent on the metocean data of the particular location and the results can used for predicting the FPSO motions to undertake proper precautions in adverse climates.

REFERENCES

Faulkner, D., Cowling, M.J. & Incecik, A. 1991. Integrity of Offshore Structures – 4, *Elsevier Science Publishers LTD*, London

Heurtier, J.M., Buhan, P., Fontane, E., Cunff, C., Biolley, F. & Berhault, C. 2001. Coupled Dynamic Response of Moored FPSO with Risers, *Proc. of the Eleventh ISOPE Conference*, 1: 319–326. Norway.

Ishihara, T., Phuc, P.V., Sukegawa, H., Shimada, K. & Ohyama, T. 2012. A study on the dynamic response of a semi-submersible floating offshore wind turbine system Part 1: A water tank test, *ICWE 12*, 2511–2518. CAIRNS.

Jiang, T. & Schellin, T. E. 1989. Motion Prediction of a Single-Point Moored Tanker Subjected to Current, Wind and Waves, *Journal of Offshore Mechanics and Arctic Engineering*, 112: 83–90.

Kim, M.H., Koo, B.J., Mercier, R.M. & Ward, E.G. 2005. Vessel/Mooring/Riser Coupled Dynamic Analysis of a Turret – Moored FPSO Compared with OTRC Experiment, *Ocean Engineering* 32: 1780–1802.

Wichers, J.E.W. 1988. A simulation Model for a single Point Moored Tanker, PhD Dissertation, Delft University.

Wichers, J.E.W. & Devlin, P.V. 2001. Effect of coupling of mooring lines and risers on the design values for a turret moored FPSO in deep water of the Gulf of Mexico, *Proc. of the Eleventh ISOPE Conference*, 480–487.

Wichers, J.E.W., Voogt, H.J., Roelofs, H.W. & Driessen, P.C.M. 2001. DeepStar-CTR 4401-Benchmark Model Test, Technical Report No. 16417-1-OB, MARIN, Netherlands.

The effect of short-crested waves on the dynamic responses of semisubmersible: A comparison of long and short-crested waves with current

C.Y. Ng & S.J. Ng
Department of Civil and Environmental Engineering, Universiti Teknologi PETRONAS, Seri Iskandar, Perak, Malaysia

ABSTRACT: As the development of oil and gas industry has progressed into deeper water depth, the research and development focused on this particular region structures have become essential. In the design of the offshore structures, wave force is always the most significant environmental force among. Literature has been reported that long-crested waves, which widely used in the design of offshore structures, are rarely found in the real sea. Whereby, this condition is well represented by the short-crested waves. The short-crested waves were defined as the waves formed by the linear summation of a series of long-crested waves that propagated to different directions, which is randomly varying in the magnitude and directions. Literature as well has been reported that design considering long-crested waves would be overestimated. Hence, in order to simulate the responses of the structures in real sea condition and provide an optimum design for the structures, short-crested waves are preferable. As per the knowledge of authors, research focused on the investigation of the dynamic responses of semi-submersible particularly focusing on short-crested waves and current concurrently has yet been found. In this study, a numerical comparative study on a typical eight-column semi-submersible model had been conducted. Frequency domain method, which adopted the Morison Equation and wave spectrum, was utilized to obtain the dynamic responses of a semi-submersible platform responses in surge, heave, and pitch in term of Response Amplitude Operator (RAO). The maximum amplitude of each motion was determined and compared. From the results, it has proven that the responses of the semi-submersible platform are smaller as compared to the one due to long-crested waves. Whereby, the results indicated that the dynamic responses for short-crested waves are approximately 50% lower.

1 INTRODUCTION

The offshore structures in Malaysia can be categorized into fixed and floating platforms. Semi-submersible is one of the typical type of floating production platform utilized in the ultra-deep water. Its primary characteristic is able to float at a stationary location and maintain good stability when encountering the natural environmental forces. One example of semi-submersible is the Gumusut-Kakap field in Malaysia with water depth up to 1200 m (Terry, 2008). Among the environmental force acting in real sea state considered in the design of offshore structures, wave forces are the most dominant factor governing the structural dynamic responses. Waves are categorized as long-crested and short-crested waves based on the wave propagation direction. Long-crested wave was defined as two-dimensional waves, which extend infinitely in lateral direction. On the other hand, short-crested wave was defined as a series of long-crested waves with random magnitude and directions by Kurian et al. (2013c). This type of wave is complex, three-dimensional and well representing the ocean waves.

In the past, studies focused on the wave forces have been conducted intensively by considering long-crested waves (LCW). Sun (1980) presented a study focused on regular wave forces including the viscous damping effect and viscous exciting effect subjected to semi-submersible. Maeda et al. (2000) investigated the estimation method of time series responses on floating structure by comparing analytical and measured results. Yilmaz and Incecik (1995) established a non-linear time domain analysis to investigate the responses of moored semi-submersible due to waves, wind, and current factors.

However, the properties and behaviour of real wind-generated waves were not accounted in those studies. Thus, short-crested waves (SCW) were simulated in later research to represent the real sea state. One of the most significant waves on a vertical circular cylinder conducted by Zhu (1993). He drawn a conclusion that the force applied on the cylinder is lower in the wave direction than plane incident wave.

Apart from that, Kurian et al. (2011) had conducted an experiment to investigate the motion responses of semi-submersible due to the effect of bi-directional waves. In the study, multi-element wave generator was used to simulate multidirectional wave condition in the real ocean subjected to the semi-submersible platform model. They concluded that the magnitude and trend of dynamic responses can be influenced significantly by wave crossing angle. Kurian et al. (2013a, 2013b, 2013c) also performed numbers of research regarding motion responses of floating offshore structures numerically and experimentally subjected to the short-crested waves. In their study, time domain analysis was adopted to solve the equation of motions for the structure dynamic equilibrium and linear airy wave principle in numerical calculation to attain the wave properties data. They have proved that short-crested waves yield smaller dynamic motions than that by long-crested waves in surge, heave, and pitch motions through both experiment and numerical studies. This finding can contribute to a more cost-effective and optimum design of deep water offshore platforms. In addition, they agreed that numerical simulation is applicable to offer similar results with experimental values. From the above reviews, it was found that most of the studies conducted by the previous researcher mainly focused on wave kinematics utilized Morison Equation to compute wave forces calculation on the flooded member of offshore structures, whereby it offers relatively more accurate outcomes. Apart from that, all of the analysis on the subject of short-crested waves yield back positive feedback that it helps to design for more economical offshore structures.

However, the dynamic responses of semi-submersible by comparing between short-crested and long-crested waves incorporated with current, have not been studied broadly. Whereby studies reported mainly focused only on directional wave spectrum, wave kinematics, and vertical circular cylinder. Furthermore, studies had been conducted on spar instead of semi-submersible and without considering the current environmental factor. Hence in this study, the effect of short-crested waves incorporated with current on the dynamic responses of semi-submersible is necessary to be quantified and qualified.

2 THE NUMERICAL ANALYSIS

An eight column semi-submersible model was used in this project with scale factor of 1:100 considering Froude Scaling Law. The principle dimensions of the model and environmental data input were tabulated in Table 1 and Table 2.

2.1 Wave energy spectrums

In this study, P-M Model was adopted to represent the wave density distribution at the chosen deep-water region. The model was expressed in terms of frequency (ranged from 0 Hz to 0.5 Hz) as below (Chakrabarti, 1994):

$$S(f) = \frac{\alpha g^2}{(2\pi)^4} f^{-5} \exp\left[-1.25\left(\frac{f}{f_0}\right)^{-4}\right] \quad \text{(Eq. 1)}$$

Where f_0 was peak frequency.

Table 1. Dimensions of model.

Parameters	Dimension
Centre of gravity	(0,0)
Diameter of column 1,4,5,8 (m)	0.10
Diameter of column 2,3,6,7 (m)	0.06
Height of pontoon (m)	0.10
Height of column (m)	0.26
Distance of column 1 & 5 from (0,0) (m)	−0.45
Distance of column 2 & 6 from (0,0) (m)	−0.15
Distance of column 3 & 7 from (0,0) (m)	0.15
Distance of column 4 & 8 from (0,0) (m)	0.45
Total mass (kg)	42.83

Table 2. Environmental data.

Parameters	Criteria
Significant Wave Height (m)	0.04
Peak Wave Period, Tp (s)	0.6
Water Depth (m)	12
Current (m/s)	1.00
Natural Period for Fx (s)	1
Natural Period for My (s)	0.6
Natural Period for Fz (s)	0.5

The short-crested wave spectrum spreading function can be described by a cosine power law, where cosine-squared distribution of the directional long-crested wave-energy density distribution in the range of ±90° from the average wave direction in the common form of short-crested wave considered. The energy density was about a mean heading angle of waves in a function of the frequency and direction is given as:

$$S(f, \theta) = \frac{2}{\pi} S(f) \cos^2 \theta, -\frac{\pi}{2} \le \theta \le \frac{\pi}{2} \quad \text{(Eq. 2)}$$

2.2 Wave force

The wave force calculation is the most important task in the design of offshore platforms. Dependent on the type and size of the members in an offshore platform, wave force calculation methods as follow are applicable:

- Morison equation
- Diffraction theory

In this paper, Morison equation was considered. Morison et al. (1950) derived the excited wave force on a vertical pile that composed of two components i.e. the inertia and drag. Inertia force, FI was derived by considering the case study of a water particle moving along the circular, where the force exerted on a small segment of the cylinder is proportional to the water particle acceleration at the centre of the cylinder. The inertia force is given as

$$df_I = C_M \rho \frac{\pi}{4} D^2 \frac{\partial u}{\partial t} ds \qquad (Eq.\ 3)$$

where df_I = inertia force on an incremental segment ds per unit length of the pile, ρ = seawater density (1.035 kg/m3), D = diameter of the cylinder, $\partial u/\partial t$ = local water particle acceleration and C_M = inertia coefficient.

Drag component in Eq. 4 is predominating the total force of Morison wave force, which was derived as the pressure difference at the wake region surrounding the cylinder and proportional to the square of water particle velocity. In this study, the current was considered present with waves, hence the Morison equation in terms of the total velocity including a steady current, U is given as

$$df_D = \frac{1}{2} C_D \rho D |u \pm U|(u \pm U) ds \qquad (Eq.\ 4)$$

where df_D = drag force on an incremental segment ds, u = instantaneous water particle velocity and C_D = drag coefficient.

In order to obtain the wave force, Eq. 5 to 7 were integrated along the wetted length of each column and over whole length of hull for semi-submersible model.

$$F_x = \sum_{i=3}^{7}(F_{Ii} + F_{Di})\cos\phi_w + F_{p_{x\,hull_{1,2}}} \qquad (Eq.\ 5)$$

$$F_z = F_{z\,hull\,1,2} \qquad (Eq.\ 6)$$

$$M_y = \sum_{i=3}^{7}(M_{Ii} + M_{Di})\cos\phi_w + M_{y\,p} \qquad (Eq.\ 7)$$

2.3 Dynamic motion response

The dynamic responses of semi-submersible in the three degree of freedom (DOF) i.e. surge, heave, and pitch motion in terms of response amplitude operator (RAO) were computed. The maximum wave forces and moment obtained from Morison Equation were considered to determine the potential behavior of an offshore structure would possess when operating at the real sea state. To obtain the motion response spectrum, for long-crested waves, $S(f)$ was represented by the wave spectrum from P-M model whereas for short-crested waves, $S(f)$ was represented by the directional wave spectrum covering the range of $\pm 90°$.

$$S_x(f) = RAO^2 \cdot S(f) \qquad (Eq.\ 8)$$

$$RAO = \frac{F/\left(\frac{Hs}{2}\right)}{\left[(K - mw^2 + Cw^2)\right]^{1/2}} \qquad (Eq.\ 9)$$

Where K is stiffness of structure; m is mass of the structure; C is structural damping ratio.

The maximum amplitude from motion response was determined for comparison of dynamic responses between long-crested and short-crested waves, with and without current effect by using the following formulae. $H(f)$ was obtained from either wave spectrum or directional wave spectrum depending on the type of waves computed.

$$n(x,t) = \sum_{n=1}^{N} \frac{RAO * H(f)}{2} \cos[k(n)x - 2\pi f(n)t + \varepsilon(n)] \qquad (Eq.\ 10)$$

$$H(f) = 2\sqrt{2 * S(f) * \Delta f} \qquad (Eq.\ 11)$$

3 RESULTS AND DISCUSSION

3.1 Wave energy spectrum

The long-crested wave and short-crested wave energy are illustrated in Fig. 1.

The long-crested wave spectrum was found to be approximately one fourth of the peak value of short-crested wave spectrum. The values of these spectrums affected considerably to the maximum amplitude of motion response when comparing between long-crested and short-crested waves.

3.2 Dynamic motion in responses

The dynamic motion responses due to the influence of LCW and SCW without current are illustrated in Figs. 2 to 4. Due to the insignificant effect of current on the heave motion of the model, hence the heave motion will not be highlighted here-after. The surge and pitch motion responses of the model due to influence of LCW and SCW with current are shown in Figs. 5 and 6. The maximum amplitude of motion response and the differences of the responses due to long-crested and short-crested waves are tabulated in Table 3.

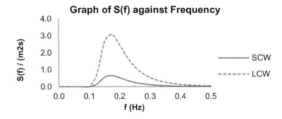

Figure 1. Comparison between wave spectrum and directional wave spectrum.

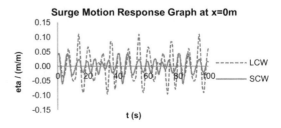

Figure 2. Comparison of surge motion response subjected to long-crested and short-crested waves.

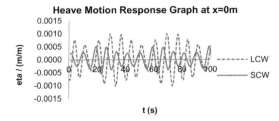

Figure 3. Comparison of heave motion response subjected to long-crested and short-crested waves.

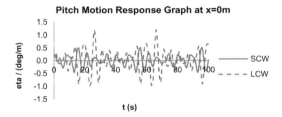

Figure 4. Comparison of pitch motion response subjected to long-crested and short-crested waves.

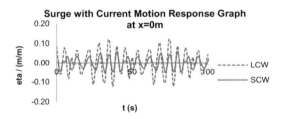

Figure 5. Comparison of surge motion response subjected to long-crested and short-crested waves with current.

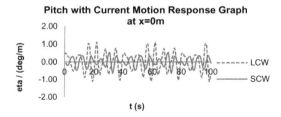

Figure 6. Comparison of pitch motion response subjected to long-crested and short-crested waves with current.

Generally, it can be observed that the responses yield by LCW are greater as compared to SCW in both the cases with and without current.

In Table 3, the maximum amplitude of the motion responses was summarized. In the comparison of LCW and SCW without current, about 34%, 50% and 45% smaller responses were found for surge, heave and pitch respectively due to SCW. This has proven that that the assumption that if the stretch of the platform subjected to the LCW, the effect will be on the whole stretch of the platform. However, in the real sea condition, SCW will be propagated to the stretch length in

Table 3. Summary of the maximum motion response of the models subjected to LCW and SCW without current.

Condition	Maximum Motion Responses		
	Surge	Heave	Pitch
LCW	0.114	0.00111	1.024
SCW	0.039	0.00055	0.470

Table 4. Comparison of the Maximum Motion Response of the model subjected to LCW and SCW with and without current.

Condition (LCW)	Maximum Motion Responses	
	Surge	Pitch
Wave only	0.114	1.024
Wave + Current	0.126	1.333
Condition (SCW)	Maximum Motion Responses	
	Surge	Pitch
Wave only	0.039	0.470
Wave + Current	0.084	0.547

Table 5. Comparison of the maximum motion responses subjected to LCW and SCW with current.

Condition (Wave + Current)	Maximum Motion Responses	
	Surge	Pitch
LCW	0.126	1.333
SCW	0.084	0.547

various angle and the net effect is expected to be less (Ng, 2014).

On the other hand, the comparison of the maximum motion responses with and without current for both LCW and SCW were shown in Table 4. It was found that differences about 10% and 23% greater responses for surge and pitch motions respectively were found for LCW with and without current. Meanwhile, 46% and 14% were found for surge and pitch motion respectively when the model subjected to the condition of SCW with and without current. Generally, the responses yielded 10% to 23% greater when the current was taken into account. However, a significant increment was found in the surge motion subjected to SCW a 46% increment was resulted.

In order to quantify the effect of SCW, Table 5 shows the comparison of the responses due to LCW and SCW with current. It was resulted that about 33% and 58% smaller value were found for surge and pitch motion when SCW was taken into consideration. Hence, the design considering LCW would be overestimated. In order to provide a cost and optimum design, it is recommended that SCW to be consider in the design of offshore structures.

4 CONCLUSION

In order to obtain an optimum and economical offshore structures design, directional wave statistics could be employed rather than conventional unidirectional wave data that would be overdesign. However, very few detailed research had been performed particularly focusing on short-crested waves and current concurrently to study the dynamic responses of semi-submersible. Therefore, a numerical comparative study on a typical eight-column semi-submersible model was conducted by using frequency domain analysis to study its dynamic responses subjected to long-crested and short-crested waves with current.

From this study, the following conclusion were drawn:

i. Without considering the current effect, about 34%, 50% and 45% smaller responses were found for surge, heave and pitch motion respectively subjected to SCW.
ii. Current effect was found not significant for heave motion.
iii. 10% and 23% increment on the responses were found for surge and pitch motion when model subjected to LCW with current.
iv. 46% and 14% increment in surge and pitch motion were found due to SCW with current.
v. It was proven that SCW yielded smaller responses even for condition with current. Hence, the application of short-crested waves is more effective and optimum in designing and costing of offshore structures.

ACKNOWLEDGEMENT

The authors would like to gratefully acknowledge their gratitude the Universiti Teknologi PETRONAS (UTP) and Offshore Engineering Center UTP (OECU) for their constant support and encouragement.

REFERENCES

Chakrabarti, S. K., 1994, *Hydrodynamic of Offshore Platforms*, Southampton: WIT Press.

Kurian, V. J., Ng, C. Y., Liew, M.S., 2011, Experiment Investigation for the Responses of Semi-Submersible Platform Subjected to Bi-directional Waves, *National Postgraduate Conference (NPC)*, Kuala Lumpur, 2011.

Kurian, V. J., Ng, C. Y., Liew, M.S., 2013a, A Numerical and experimental study on motion responses of semi-submersible platforms subjected to short-crested waves, *ICOVP-2013* Lisbon, 9–12 September 2013.

Kurian, V. J., Ng, C. Y., Liew M.S., 2013b, Dynamic Reponses Of Truss Spar Due To Wave Action, *Research Journal Of Applied Sciences, Engineering And Technology*, 5(3), 812–818.

Kurian, V. J., Ng, C. Y., Liew, M.S., 2013c, Effect of Short-Crested Waves On The Dynamic Responses Of Truss Spar Platform, *Proceedings of the Twenty-third (2013) International Offshore and Polar Engineering, Anchorage, Alaska, USA*, June 30–July 5, 2013.

Maeda, H., Ikoma, T., Masuda, K., Rheem, C.K., 2000, Time Domain Analyses Of Elastic Response And Second Order Mooring Force On A Very Large Floating Structure In Irregular Waves, *Marine structures* (13), 279–299.

Morison, J. R., O'Brien, M. P., Johnson, J. W., Schaaf, S. A., 1950, The Force Exerted by Surface Waves on Piles, *Petroleum Transactions, American Institute of Mining Engineers*, 189, 149–154.

Ng, C. Y., 2014, Comparative Study of the Dynamic Response of Floating Platforms Subjected to Long-crested and short-crested waves, *PhD Dissertation*, Universiti Teknologi PETRONAS.

Sun, F., 1980, Analysis of Motions of Semi-Submersible in Sea Waves, *Offshore Technology Conference*, 5–8 May, Houston, Texas.

Terry, F., 2008, Overview of Shell Deepwater Development in Malaysia, Retrieved from http://www.subseauk.com.

Yilmaz, O., Incecik, A, 1995, Dynamic Response Of Moored Semi-Submersible Platforms To Non-Colliner Wave, Wind And Current Loading, *The Fifth International Offshore and Polar Engineering Conference, The Hague, The Netherlands*, 11–16 June, 1995.

Zhu, S. P., 1993, Diffraction of Short Crested Waves, *Ocean Engineering*, 20(4), 389–407.

… Engineering Challenges for Sustainable Future – Zawawi (Ed.)
© 2016 Taylor & Francis Group, London, ISBN 978-1-138-02978-1

Decommissioning decision criteria for offshore installations and well abandonment

T.M.Q. Al-Ghuribi
Petroleum Engineering Universiti Teknologi PETRONAS, Tronoh, Perak, Malaysia

M.S. Liew
Faculty of Geoscience & Petroleum Engineering, Universiti Teknologi PETRONAS, Tronoh, Perak, Malaysia

N.A. Zawawi
Civil Engineering, Universiti Teknologi PETRONAS, Tronoh, Perak, Malaysia

M.A. Ayoub
Petroleum Engineering, Universiti Teknologi PETRONAS, Tronoh, Perak, Malaysia

ABSTRACT: Decommissioning of oil and gas offshore facilities is employed at the end of the life cycle of the field in compliance with national legislation, international laws and standards. The installation of oil and gas platforms keeps increasing worldwide necessitating the development of an innovative approach that determine the potential decommissioning option with least environmental impact and cost. This paper introduces an approach assessing financial and non-financial criteria of each decommissioning option by which financial analysis is assessed using net present value evaluation while non-financial factors are assessed through weighted evaluation technique. The economic analysis of removal option is considered inadequate in identifying the appropriate decommissioning option, so the assessment of the feasibility of environmental, health and safety, and public outcomes will be directive towards the best removal option with the lowest financial and environmental impacts combined.

Keywords: decommissioning; offshore platform; decision making model; environmental impact

1 INTRODUCTION

Decommissioning refers to the end of the operation of the facilities where the dismantling, disposal, and removal process of all upstream equipment and assets implemented. Undoubtedly, decommissioning in oil and gas industries has been recognized as a complex issue with a grown concern as the number of the fields approaching the end of their economic life surged considerably. Up to date, over 10000 offshore platforms have been installed all over the world, including fixed or floating platform, concrete or steel platforms, and offshore storage and loading facilities. The decommissioning approach must produce a consolidated method tackling all the resultant issues due to removal activities with the considerations of several factors namely, the type, shape and water depth of the installed platform (Anthony et al., 2000). Generally, most of the platforms are installed in water depth less than 75 meters. In Malaysia around 249 platforms have been installed across the Peninsular of Malaysia, Sabah, and Sarawak and all of these structures are fixed platforms with water depth not accessing 200 meters (Twomey, 2010). The upstream facilities disposal will cause a harmful impact upon the marine environment due to the high energy consumption during the removal process leading into unpleasant effects such as atmosphere emission, noise pollution, vibrations, and wasted substance deposited in the seabed (Fowler et al., 2014). Numerous studies have highlighted the necessity of the removal of old offshore structures and indicated that the abandonment of offshore installations is not necessarily to be the reverse of the installation process, but the main difference is the environmental impact induced throughout the abandonment process (Friederichs et al., 2015).

2 DECOMMISSIONING OVERVIEW

Decommissioning is purposely introduced to offshore facilities to free the field from the hazards developed from Hydrocarbon Asset as well as clearing the site of all the waste to ensure the restoration of the original conditions of the site. The process starts with engineering planning followed by obtaining the approval and then the implementation way after shutting down the production facilities when reservoir formation get depleted (Bernstein, 2015). The removal process will be risky without considering some factors such as ageing of the structures, the configuration and type of structure, way for shutting down the production and weight of the lift and soil strength (Ratnayake, 2012).

2.1 *Decommissioning process*

The decommissioning of offshore platforms involves a number of steps namely; project management, engineering planning; permitting and regulatory compliance; platform preparation; well plugging and abandonment; conductor removal; mobilization and demobilization of derrick barges; platform removal; pipeline and power cable decommissioning; materials disposal; and site clearance (Thornton and Wiseman, 2000). The removal of these structures commonly involves leaving in place, dismantling, complete or partial removal, and the toppling of disused facilities (Zawawi et al., 2012).With the complication associated with the removal of upstream asset there is always obstacle in predicating the best option of decommissioning and disposal of offshore facilities where a consistent approach needs to be set to produce the appropriate option, as this paper will be finalizing the best guidelines through a model approach that the author will develop in the methodology of this study. As illustrated in figure 1, decommissioning project should redeem regulatory, environmental, health and safety, and financial criteria to maintain a good reputation of the operators.

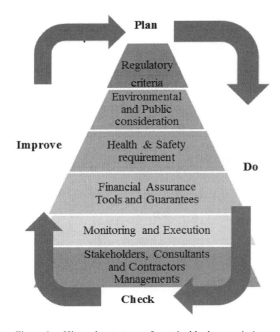

Figure 1. Hierarchy strategy of sustainable decommissioning program.

3 OVERVIEW ON PREVIOUS DECOMMISSIONING REGULATIONS

Over the past 50 years, global guidelines and regulations have been developed as the decommissioning activities rise significantly necessitating the establishment of a decommissioning approach to protect the marine environment and retain the right for other parties (local communities, industries) in accordance to consensus regulations (Tularak et al., 2007). There are several global conventions and guidelines that introduced to oil and gas industries purposely to minimize the induced risk and threats due to decommissioning operations. The Geneva Convention 1958 requires a complete removal of any oil and gas installation which has been abandoned or disused (Gutteridge, 1959). The 1972 London Convention prohibits the dumping any waste and materials in the sea unless it is confirmed, it has no harm to the marine environment, despite to some materials can be considered for dumping due to the limitations of the land disposal (de La Fayette, 1998). The United Nation Convention of the Law of the Sea (UNCLOS) in 1982 entails the removal of structures to guarantee the safety of other sea users (Nordquist and Nandan, 2011). In 1989 the International Maritime Organization (IMO) provided criteria and guidelines for the removal based on water depth and the weight of the structure which required a complete removal of all structure installed at water depth less than 100 meters with substructure weighting less than 4000 tonnes (Ebdon and Ellinas, 1989). However, for those structures installed at deeper water a partial removal will be implemented leaving 55 meter of clear water column for safety of navigation. The petroleum act 1998 requires the topsides of all structure to be moved to the shore for reuse or final disposal on land. Decommissioning of pipelines is enclosed at separate agenda from that of installations (Unit and House). The approval for pipeline decommissioning is basically obtained under marine and costal act (Appleby and Jones, 2012). In Malaysia, there is no national legislations to be imposed upon the operators, however the framework of decommissioning planning must be in compliance with at least eight laws: Merchant Shipping Ordinance, Continental Shelf Act, Exclusive Economic Zone Act, Environmental Quality Act, Fisheries Act, Occupational Safety and Health Act, Natural Resources and Environmental Ordinance and

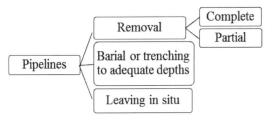

Figure 2. Pipeline removal options.

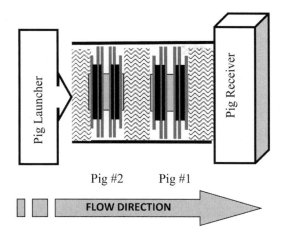

Figure 3. Pig Launcher & Pig Receiver General process flow diagram.

Conservation of Environment Enactment (Lun et al., 2012).

4 DECOMMISSIONING OPTIONS OF PIPELINES

There are several options of decommissioning of sub-sea pipelines and other sub-sea structures, and it is necessary to finalize the feasibility of each option with regard to weather conditions, water depth, and complexity of engineering plans.

Pipeline must undergo cleaning process to ensure it has reached the appropriate level of cleanliness regardless whether the pipeline left in place or lifted to the disposal yard. The operational pipeline must be freed from Hydrocarbon contents by purging the line system with chemicals or treated seawater followed by launching cleaning pigs which are being propelled through the line to reduce the Hydrocarbon deposit (Philip et al., 2014). The cleaning process of pipeline is performed as illustrated in figure 3:

- Pig launcher preparation
- Pig receiver preparation
- Pigging operations
- Pig arrival
- Pig receiver purging
- Pig execution

5 METHODOLOGY

The analysis of decision criteria of decommissioning of offshore platforms and oil and gas pipelines require data collection from the stakeholders and contractors through questionnaire. The questionnaire are planned to follow the literature planning and eligible standards, such as PETRONAS Technical Standards (PTS). The collected data will be utilized in the analysis through the model approach below.

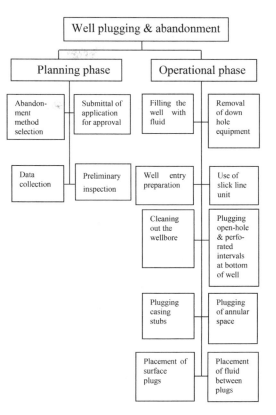

Figure 4. Well plugging process.

5.1 Well plugging and abandonment process

The method of performing well plugging and abandonment is presented in two phase namely planning phase and operational phase as summarized in figure 4.

6 IMPLEMENTATION OF THE MODEL APPROACH

The developed model is utilized to assess each decommissioning option in order to identify the preferable option that provide the ideal regulatory, environmental, health and safety, and pubic outcomes. Each criterion will be treated based on experts' opinions with regards to the standardized requirement. The feedback of experts on screening criteria will be examined through the model and then ranked in order of importance. The decision of each option will be further analysed in term of financial and non-financial considerations. As for financial criteria the assessment is determined through the calculation of Net Present Value (NPV) for each available decommissioning option by which the superior option will have the lowest NPV. In contrast, financial analysis is considered as inadequate in identifying the best removal option since it highlights the quantitative data only, yet the non-financial factors need to be screened in the decision making. The assessment of non-financial criterions is examined

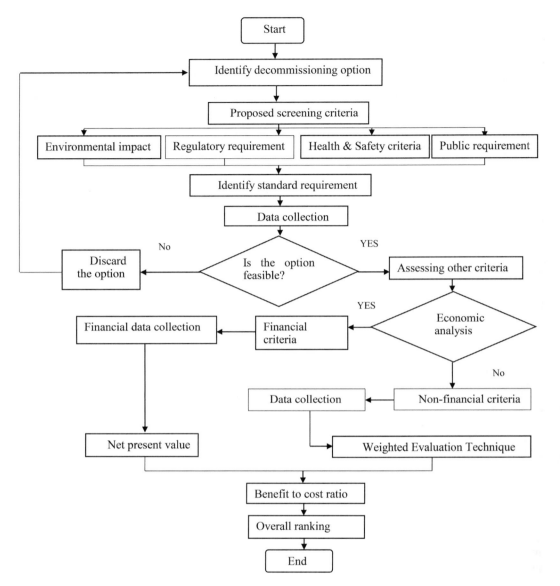

Figure 5. Model approach evaluating removal options.

by using Weighted Evaluation Technique (WET). The WET illustrates the chosen criteria and then establish the weights of importance of all criteria. The analysis is conducted through pair comparison of all criteria in order to finalize the significant one.

The Net present value for each option is determined through equation 1 as follows:

$$NPV = p*(1+d)^T \qquad (1)$$

Where, p = present cost of the option, d = the inflation rate, and T = the analysis period (years).

Benefit to Ratio is utilized to verify the derived advantages from all analysed options, so the ration can be obtained through equation 2:

$$A^* = A_i \,|BTC_i = V_{(i=1,n)} \; S_i/NPV \qquad (2)$$

Where A^* = the best alternative, S_i = the total score of alternatives, and BTC = benefit to ratio.

7 EVALUATIONS OF NON-FINANCIAL CRITERIONS

Non-financial criterions including environmental impact, regulatory, health and safety, and public analysis are attained using the weighting evaluation technique to determine the best alternative A^* as in equation 3:

$$A^* = A_i | S_i = V_{i=1..n} \sum_{i=1}^{m} W_i \, x \, S_{ii} \qquad (3)$$

Where W_i = the criterion final weighting, and S_{ii} = the scale of each option from 1 to 5 (poor to excellent).

The determination of decommissioning alternatives undergoes criteria identification and then weight score of their significance are established accordingly. Then the rate of removal options, namely complete removal, partial removal, and conversion to reef are assigned to a scale performance 1 to 5 (poor to excellent) in term of non-financial criterions. The total score of A^* will be determined for each removal option and the best alternative A^* would rather have the highest overall score. Finally, the BTC (benefit to cost ratio) technique is utilized involving both financial and Non-financial calculations to measure the effectiveness of each option where the best option should have the highest benefit to cost ratio.

8 CONCLUSION

The decommissioning and abandonment is still under development investigating all the possible alternatives for removal of topsides, substructures, and pipelines as well as the well abandonment, taking into account international regulatory constraints. This paper assessed several decommissioning options using the model approach involving all the necessary criteria in term of financial and non-financial criteria. The approach has defined all methodologies based on the collected data through survey assessment on decommissioning to guarantee the success and the congruity of the preferable option. The cost estimation of the removal was determined through cost parameters of vessel, equipment. The assessment of NPV and weighted evaluation technique analysis determined the ideal option with the highest benefit ratio. Meanwhile the placement of well plugging is controlled by the characteristic of the well itself.

REFERENCES

Anthony, N. R., Ronalds, B. F. & Fakas, E. 2000. Platform Decommissioning Trends. Society of Petroleum Engineers.

Appleby, T. & Jones, P. J. 2012. The marine and coastal access act—A hornets' nest? *Marine Policy*, 36, 73–77.

Bernstein, B. B. 2015. Evaluating alternatives for decommissioning California's offshore oil and gas platforms. *Integrated environmental assessment and management*, 11, 537–541.

De Le Fayette, L. 1998. The London Convention 1972: preparing for the future. *The International Journal of Marine and Coastal Law*, 13, 515–536.

Ebdon, R. & Ellinas, C. 1989. Decommissioning and removal of offshore structures: a state of the art.

Fowler, A., Macreadie, P., Jones, D. & Booth, D. 2014. A multi-criteria decision approach to decommissioning of offshore oil and gas infrastructure. *Ocean & Coastal Management*, 87, 20–29.

Friederichs, C., Dibello, F., Tignanelli, A. & Garzia, A. Reliable and Innovative Approach for Decommissioning Study of Oil & Gas Plants. Offshore Mediterranean Conference and Exhibition, 2015. Offshore Mediterranean Conference.

Gutteridge, J. 1959. 1958 Geneva Convention on the Continental Shelf, The. *Brit. Yb Int'l L.*, 35, 102.

Lun, N. K., Zawawi, N. A. W. A. & Liew, M. S. Conceptual framework: Semi-PSS for sustainable decommissioning of offshore platforms in Malaysia. Statistics in Science, Business, and Engineering (ICSSBE), 2012 International Conference on, 2012. IEEE, 1–5.

Nordquist, M. & Nandan, S. N. 2011. *United Nations Convention on the Law of the Sea 1982, Volume VII: A Commentary*, Martinus Nijhoff Publishers.

Philip, N. S., Wilde, S., Arshad, R., Washash, I. & Al-Sayed, T. A. R. 2014. Decommissioning Process for Subsea Pipelines. Society of Petroleum Engineers.

Ratnayake, R. C. Challenges in inspection planning for maintenance of static mechanical equipment on ageing oil and gas production plants: The state of the art. ASME 2012 31st International Conference on Ocean, Offshore and Arctic Engineering, 2012. American Society of Mechanical Engineers, 91–103.

Thornton, W. & Wiseman, J. Current trends and future technologies for the decommissioning of offshore platforms. Offshore Technology Conference, 2000. Offshore Technology Conference.

Tularak, A., Khan, W. A. & Thungsuntonkhun, W. Decommissioning Challenges in Thailand. SPE Asia Pacific Health, Safety, and Security Environment Conference and Exhibition, 2007. Society of Petroleum Engineers.

Twomey, B. G. 2010. Study assesses Asia-Pacific offshore decommissioning costs. *Oil & gas journal*, 108.

Unit, O. D. & House, A. Guidance Notes Decommissioning of Offshore Oil and Gas Installations and Pipelines under the Petroleum Act 1998. *Policy*, 1, 2.

Zawawi, N. A. W. A., Liew, M. & Na, K. Decommissioning of offshore platform: A sustainable framework. Humanities, Science and Engineering (CHUSER), 2012 IEEE Colloquium on, 2012. IEEE, 26–31.

Hydraulic efficiency of a pile-supported free surface breakwater subjected to random waves

S.S. Hassan, H.M. Teh & V.J. Kurian
Department of Civil and Environmental Engineering, Universiti Teknologi PETRONAS, Perak, Malaysia

ABSTRACT: Wave transmission, reflection and energy dissipation characteristics of a pile-supported Free Surface H-type Breakwater (FSHB) were experimentally studied under the action of random waves. Large scale laboratory experiments were conducted using a wave flume. The test models of different drafts were subjected to random waves of varying characteristics in different water depths. The experimental results revealed that the fixed FSHB attenuated the incident waves mainly by wave reflection. The wave suppression ability of the fixed FSHB was significantly enhanced when the breakwater was confronted by steeper waves. Most of the wave energy dissipation was associated with wave trapping between arms of the structure during wave overtopping and the eddies generated under keel of the structure during wave-structure interaction. Furthermore, the results indicated that increase in immersion depth of the structure increased its wave attenuation ability.

1 INTRODUCTION

Breakwaters are extensively used to protect coastlines and coastal structures from destructive impact of sea waves. Free surface breakwaters are also called open breakwaters. These structures are preferred over the gravity type breakwaters at environmentally sensitive sites where complete wave attenuation is not needed, especially in relatively deeper waters (Nishihata et al., 2012). The main body of such breakwater is located near the sea surface and is usually supported by piles fixed at the sea bottom. Free surface breakwaters pose minimal interruption to water circulation, fish migration and coastal sediment transport. They are mainly used to attenuate the shorter period waves in which the period is less than 5 s (Isaacson et al., 1995).

Hydraulic efficiency of the free surface breakwater is largely dependent upon geometric properties of the structure. The potential of the open breakwaters in small ports and harbours has motivated researchers and engineers in developing free surface breakwaters that are both hydraulically and economically efficient. These include solid-type, plate-type, caisson-type and multipart-type breakwaters (Teh, 2013). Teh (2013) outlined the pros and cons of each type of the breakwaters. The box type breakwater is the simplest form of design in the development of free surface breakwaters. Li et al. (2005) claimed that the rectangular barrier was a better wave attenuator compared to the cylindrical barrier at larger range of relative width (B/L). Some other breakwater configurations comprise of quadrant front face (Sundar & Subbarao, 2003), U-shape (Günaydın & Kabdaşlı, 2004), Π-shape (Günaydın & Kabdaşlı, 2007) and semi-circular (Teh et al., 2011).

Each of these breakwaters has a unique response to waves. He and Huang (2016) incorporated oscillating water column (OWC) in the pile-supported structure for both tapping of wave energy and wave height reduction at the lee side.

Teh et al. (2014) developed the H-type floating breakwater (H-Float) to provide wave protection to the coastal and offshore facilities. The breakwater was claimed to be hydraulically efficient when it was held by taut leg mooring in random waves. It optimized energy dissipation around the structure rather than reflecting a large amount of waves seaward. The H-type breakwater can be converted to a pile-supported structure for shallow water applications. Would hydraulic performance of the H-type breakwater improve if it is fixed by vertical piles or jacket structures? This research question motivates the present study to evaluate hydraulic efficiency of the pile-supported or fixed H-type breakwater in random waves via physical modelling.

In this study, hydraulic properties of the pile supported Free Surface H-type Breakwater (FSHB) were characterized by the coefficients of transmission ($C_t = H_t/H_i$), reflection ($C_r = H_r/H_i$) and energy dissipation ($C_l^2 = 1 - C_t^2 - C_r^2$), where H_i, H_r and H_t are incident, reflected and transmitted wave heights.

2 METHODOLOGY

2.1 Test model

The H-Float model of 1:15 scaling factor was used in the experiments. The dimensions of the test model

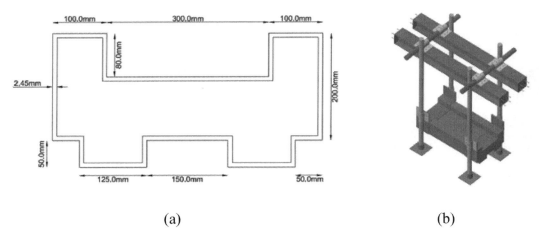

Figure 1. Test model: (a) H-Float model; (b) Model setup.

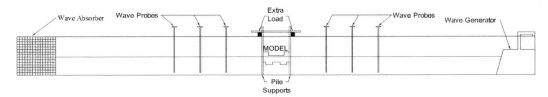

Figure 2. Side view of experimental setup.

were 500 mm in width, 1450 mm in length and 250 mm in height as shown in Figure 1(a). The model was constructed by mild steel and was coated with water proof paint. The breakwater has a pair of arms extended 80 mm from the main body set apart from each other at 300 mm, and a pair of legs projecting 50 mm downwards. The model was supported by four vertical cylindrical piles fixed to the flume bed by means of mounting plates. Clamps were used to attach the mounting plates and the model to the vertical piles at arbitrary levels as shown in Figure 1(b). To ensure the rigidity of the model, the upper end of the vertical piles were locked to rectangular beams placed across the width of the wave flume.

2.2 Experimental

Experimentation for investigation of performance of the pile supported H-type breakwater was conducted in a 25 m long, 1.5 m wide and 1.5 m deep wave flume at Offshore Engineering Laboratory of Universiti Teknologi PETRONAS, Malaysia. The facility was equipped with an active type wave paddle fabricated by Edinburgh Design Ltd. (UK) at one end of the flume. It can generate both regular and random waves of different characteristics. At the other end of the flume, a passive wave absorber was installed to eliminate the reflected waves from the test model during the experiments. Three-probe method proposed by Mansard and Funke (1980) was used to decompose the incident and reflected waves from the recorded wave series. As a result, two sets of three wave probes each were respectively placed at about 5 m distance from the test model at both sides as shown in Figure 2. The probes at the front were for measurement of incident and reflected waves, and those at the rear were for measurement of transmitted waves.

2.3 Test conditions

The model was tested in three water depths ($d = 0.6$, 0.7 and 0.8 m) representing three different tidal levels at sites. To optimize the breakwater efficiency, the immersion depths at each water depth were adjusted at $D = 0.025$, 0.065 and 0.115 cm from still water level. This yields a range of breakwater draft-to-water depth ratios (D/d), i.e. $0.031 < D/d < 0.164$. At each water depth, the model of a specific draft was tested in random waves of significant wave periods ranging from 0.8 to 2.0 s with interval of 0.2 s. The wave characteristics were subsequently represented by wave steepness parameter, H/L, values of 0.04, 0.05 and 0.06.

3 RESULTS AND DISCUSSIONS

Hydraulic characteristics of the pile-supported FSHB are presented in terms of energy coefficients, i.e. C_t, C_r and C_l^2. The coefficients are then graphically plotted relating to relative breakwater width, B/L and relative breakwater draft, D/d.

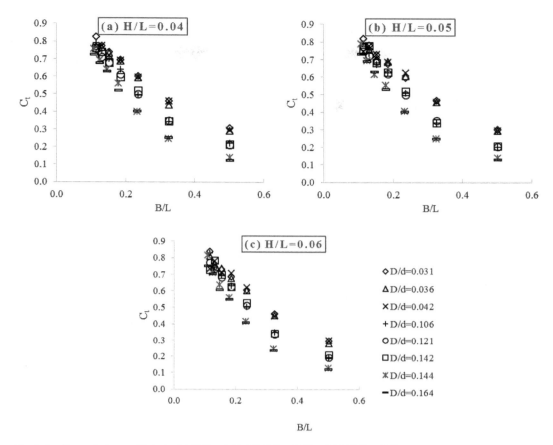

Figure 3. Transmission coefficients: (a) $H/L = 0.04$, (b) $H/L = 0.05$ and (c) $H/L = 0.06$.

3.1 Transmission coefficient

Figure 3 presents the transmission coefficients of the test model subjected to different wave climates, i.e. mild ($H/L = 0.04$), moderate ($H/L = 0.05$) and severe ($H/L = 0.06$) conditions. It can be seen that C_t of all D/d test cases rapidly decrease with an increase of B/L, regardless of their wave steepness. This indicates that the breakwater width is one of the important parameters in controlling the wave attenuation ability of the fixed FSHB. The larger the breakwater width with respect to the incident wavelength, the greater will be the wave suppression ability it demonstrates. The breakwater is capable of reducing the incident wave height by half when the breakwater width is beyond 25% of the incident wave length ($B/L > 0.25$) as shown in Figure 3. It is also observed from the figures that wave dampening ability of the breakwater can be further enhanced to 80% ($C_t < 0.20$) when the width of the largely immersed breakwater ($D/d > 0.15$) is extended to 50% of the wavelength, for all the tested wave steepness cases. Wave steepness does not seem to pose any significant influence on C_t of the breakwater in all D/d test cases.

3.2 Reflection coefficient

Figure 4 demonstrates reflectivity of the fixed FSHB subjected to different wave steepness. A drastic increase of C_r is found when B/L increases from 0.10 to 0.30, and the maximum C_r value recorded is approximately 0.80. At larger B/L range ($B/L \approx 0.5$), C_r values seem to hover around 0.80 for all test cases. The finding suggests that wave reflection characteristics of the fixed FSHB become dominant when the structure outsizes the incident waves (i.e. larger B/L). This explains why waves of smaller wave steepness are better intercepted and reflected by the breakwater, particularly by larger draft. In summary, the FSHB is a strong wave reflector ($C_r > 0.8$) when the barrier is designed at $B/L > 0.3$ and $D/d > 0.10$. It is important to ensure that the maritime operations at the front of the breakwater are not interfered by the confusing sea state as a result of wave reflection.

3.3 Energy loss coefficient

The scatter plots of energy dissipation with respect to B/L are presented in Figure 5. Despite weaker correlation between C_l^2 and B/L, it is still apparent that C_l^2,

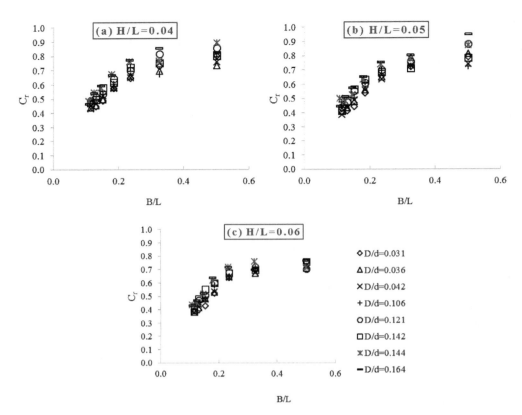

Figure 4. Reflection coefficients: (a) $H/L = 0.04$, (b) $H/L = 0.05$ and (c) $H/L = 0.06$.

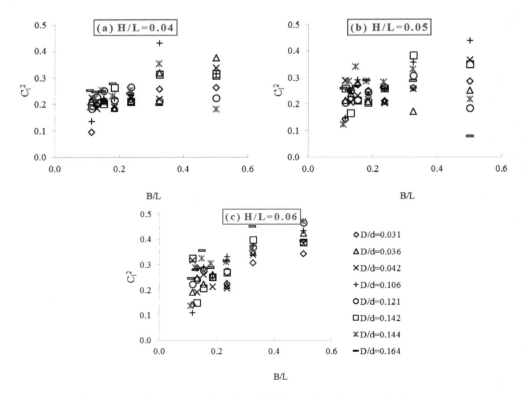

Figure 5. Energy dissipation coefficients: (a) $H/L = 0.04$, (b) $H/L = 0.05$ and (c) $H/L = 0.06$.

in general, increases with the increasing B/L. This relation becomes more prominent in larger wave steepness ($H/L = 0.06$). Nevertheless, the C_l^2 variations are rather small ($\Delta C_l^2 \approx 0.4$) compared with other energy coefficients. It is seen that breakwaters of higher D/d tend to pose more energy loss at the structures. As D/d increases, the increment of breakwater draft limits the freeboard. Waves overtop and interact with upper section of the breakwater resulting in energy dissipation. On the contrary, breakwaters of smaller D/d will trigger energy dissipation underneath the breakwater and the amount of energy loss is not comparable to the former. It is also seen from the figure that waves of higher steepness are prone to more energy dissipation when interacting with the breakwater. Overall, the FSHB is capable of dissipating the incident wave energy between 10 to 50%. The breakwater may serve as good energy dissipaters if both B/L and D/d are sufficiently large.

4 CONCLUSION

The pile-supported free surface H-type breakwater of different water depths and drafts was experimentally tested under random waves. The experimental results revealed that the H-type breakwater acts as a reflecting structure under fixed condition. Wave attenuation and energy dissipation abilities of the fixed FSHB were controlled by the relative breakwater width and the relative breakwater draft. Wave reflection by the breakwater was mainly affected by the relative breakwater width and was less depended upon the relative breakwater draft. Wave dissipation by the breakwater was largely dependent upon the relative breakwater draft. The experimental results revealed that the arms and legs of the breakwater played important part in wave energy dissipation.

ACKNOWLEDGMENT

This research was funded by *Universiti Teknologi PETRONAS* through YUTP Grant. The authors wish to thank the Center for Graduate Studies of UTP for financial support of this paper presentation.

REFERENCES

Günaydın, K. & Kabdaşlı, M. (2004). Performance of Solid and Perforated U-Type Breakwaters under Regular and Irregular Waves. *Ocean Engineering*, 31(11), 1377–1405.

Günaydın, K. & Kabdaşlı, M. (2007). Investigation of Π-Type Breakwaters Performance Under Regular and Irregular Waves. *Ocean Engineering*, 34(7), 1028–1043.

He, F. & Huang, Z. (2016). Using an Oscillating Water Column Structure to Reduce Wave Reflection from a Vertical Wall. *Journal of Waterway, Port, Coastal, and Ocean Engineering*, 04015021.

Isaacson, M., Whiteside, N., Gardiner, R. & Hay, D. (1995). Modelling of a Circular-Section Floating Breakwater. *Canadian Journal of Civil Engineering*, 22(4), 714–722.

Li, D., Panchang, V., Tang, Z., Demirbilek, Z. & Ramsden, J. (2005). Evaluation of an Approximate Method for Incorporating Floating Docks in Harbor Wave Prediction Models. *Canadian Journal of Civil Engineering*, 32(6), 1082–1092.

Mansard, E. P. & Funke, E. (1980). The Measurement of Incident and Reflected Spectra Using a Least Squares Method. *Coastal Engineering Proceedings*, 1(17).

Nishihata, T., Tajima, Y. & Sato, S. (2012). Numerical Analysis of Wave and Nearshore Current Fields around Low-Crested Permeable Detached Breakwaters. *Coastal Engineering Proceedings*, 1(33).

Sundar, V. & Subbarao, B. (2003). Hydrodynamic Performance Characteristics of Quadrant Front-Face Pile-Supported Breakwater. *Journal of Waterway, Port, Coastal, and Ocean Engineering*, 129(1), 22–33.

Teh, H. (2013). Hydraulic Performance of Free Surface Breakwaters: A Review. *Sains Malaysiana*, 42(9), 1301–1310.

Teh, H. M., John, K. V. & Hashim, A. M. (2014). *Hydraulic Investigation of the H-Type Floating Breakwater*. Paper presented at the OCEANS 2014-TAIPEI.

Teh, H. M., Venugopal, V. & Bruce, T. (2011). Hydrodynamic Characteristics of a Free-Surface Semicircular Breakwater Exposed to Irregular Waves. *Journal of Waterway, Port, Coastal, and Ocean Engineering*, 138(2), 149–163.

Development of ultimate strength versus plate index diagram for initially deflected steel plate under axial compression

D.K. Kim, B.Y. Poh & J.R. Lee
Department of Civil and Environmental Engineering, Universiti Teknologi PETRONAS, Seri Iskandar, Perak, Malaysia

K.S. Park
Steel Business Division, POSCO Center, Seoul, Republic of Korea

J.K. Paik
The Korea Ship and Offshore Research Institution, Pusan National University, Busan, Republic of Korea

ABSTRACT: An advanced method for the prediction of the ultimate strength of initially deflected steel plate subjected to axial compression is introduced. In order to predict the ultimate strength performance of a plate, a new concept so called the plate index (PI) which is a function of plate slenderness ratio (β) and coefficient of initial deflection is proposed. For the definition of PI, the ultimate strength analysis by the nonlinear finite element method using ANSYS was conducted. Totally selected 100 plate scenarios were adopted based on obtained probability density of β from three representative types of commercial ships at midship section. In the case of initial imperfection, only initial deflection was considered. The ultimate strength versus plate index diagram together with proposed formula which shows high accuracy was developed. The obtained method will be useful in predicting the ultimate performance of steel plate under axial compression.

1 INTRODUCTION

It is well known that plate and stiffened panel are the main components of ships and offshore structures. From a safety point of view, the ultimate limit state (ULS) design is being more highlighted since the Harmonised Common Structural Rule (CSR-H) has been recognised (IACS, 2015). During the design process of ships and ship-shaped offshore structures, the global (for hull girder collapse) and local (for buckling collapse) strength assessments are required.

Recently, Koh and Paik (2015) have investigated the structural failure of MSC Napoli, a post-Panamax class containership, due to the unexpected excessive hull girder load with the loss of structural stiffness and strength at 0.4L amidships. After this accident, the Marine Accident Investigation Branch figured out that the MSC Napoli collapsed due to several reasons such as 1) no sufficient buckling strength in the engine room part, 2) no requirement for buckling assessment outside 0.4L amidship (after this accident, the rule was revised by the class society), 3) dramatically increased hull girder loading by whipping and other. The importance of the ULS based structural design for a plate, stiffened panel, and hull girder is emphasised by above lesson.

A number of studies have been conducted to develop a simplified methodology (or empirical formula

method) to predict the ultimate strength of a plate based on effective width concept which was originally proposed by von Karman and showed good agreement with thin-plate. In order to improve the accuracy, Winter (1940) proposed a modified von Karman equation. Extensive reviews and researches to predict the ultimate strength performance of steel plate were conducted from the mid-1970s to early 1990s. The existing methods are briefly summarised in Table 1.

In the present study, axial compression which causes the buckling phenomenon due to the hull girder bending moment and different level of initial deflections is considered in proposing the advanced design formula. For the selection of reliable plate scenarios, three (3) representative midship sections from oil tanker, bulk carrier, and container ships in four (4) different sizes were investigated. Therefore, totally twelve (12) midship section data were analysed by probabilistic approaches. Through those probabilistic approaches, reliable plate slenderness ratio, plate width, thickness, and length were obtained. The selected hundred (100) cases of plate scenarios were employed for the ultimate strength analysis subjected to axial compression. The accuracy of obtained formulas by ANSYS (2014) nonlinear finite element analysis (FEA) was compared with the existing equations.

In addition, a new concept to predict the ultimate strength performance of initially deflected steel plate

Table 1. Existing design formulas for plate element.

Year	Author	Proposed design formulas ($=\sigma_u/\sigma_Y$)
1940	Winter	$=\begin{cases} 1 & \beta \leq 1.27 \\ 1.9/\beta - 0.8/\beta^2 & \beta > 1.27 \end{cases}$
1940	Frankland	$=\begin{cases} 1 & \beta \leq 1 \\ 2.25/\beta - 1.25/\beta^2 & 1 < \beta \leq 3.5 \\ 1.9/\beta & \beta > 3.5 \end{cases}$
1957	Gerard	$=\begin{cases} 1 & \beta \leq 1.51 \\ 1.42/\beta^{0.85} & \beta > 1.51 \end{cases}$
1975	Faulkner	$=\begin{cases} 1 & \beta < 1 \\ 2/\beta - 1/\beta^2 & \beta \geq 1 \end{cases}$
1975	Ueda et al.	$=\dfrac{1.338\omega_0^2 + 4.380\omega_0 + 2.647}{\beta + 6.130\omega_0 + 0.720} - 0.271\omega_0 - 0.088$
1977	Carlsen	$=\left(\dfrac{2.1}{\beta} - \dfrac{0.9}{\beta^2}\right)\left(1 - \dfrac{0.75\omega_0}{\beta}\right)\left(1 - \dfrac{\sigma_{rc}}{\sigma_Y}\right)$
1979	Ivanov & Rousev	$=\dfrac{1}{1 + (0.3\beta + 0.08)\omega_0}$
1983	Soreide & Czujko	$=1.42\left[1 - 2.19\left(\dfrac{\omega_0}{b/t}\right)^{0.129}\left(\dfrac{1.26}{\beta} - \dfrac{1.43}{\beta^2} + \dfrac{0.55}{\beta^3}\right)\left(\dfrac{2.74}{\beta} - \dfrac{2.56}{\beta^2} + \dfrac{0.92}{\beta^3}\right)\right]$
1983	Hughes	$=\dfrac{1}{4}\left(1.6 + \left(1 + \dfrac{2.75}{\beta^2}\right) - \sqrt{\left(1 + \dfrac{2.75}{\beta^2}\right)^2 - \dfrac{10.4}{\beta^2}}\right)$
1987	DNV	$=\begin{cases} 1 & \beta \leq 1 \\ 1.8/\beta - 0.8/\beta^2 & \beta > 1 \end{cases}$
1988	Smith	$=0.23 + \dfrac{1.16}{\beta} - \dfrac{0.48}{\beta^2} + \dfrac{0.09}{\beta^3}$
1992	Ueda	$=\begin{cases} (-2.431\omega_0^2 + 1.6826\omega_0 - 0.2961)(\beta^2 - 4.0) \\ +(7.2745\omega_0^2 - 4.7431\omega_0 + 0.6709)(\beta - 2.0) + z_1 \\ \qquad \text{for } 0.8 \leq \beta \leq 2.0 \\ \dfrac{-0.3597\omega_0^2 + 0.1748\omega_0 + 0.8598}{\beta + 2.2432\omega_0 - 0.6678} + 0.0373\omega_0 + 0.2481 \\ \qquad \text{for } 2.0 \leq \beta \leq 3.5 \end{cases}$
1998	Cui and Mansor	$=\begin{cases} 1 & \beta \leq 1.9 \\ 0.08 + 1.09/\beta + 1.26/\beta^2 & \beta > 1.9 \end{cases}$
2004	Paik et al.	$=\begin{cases} -0.032\beta^4 + 0.002\beta^2 + 1.0 & \beta \leq 1.5 \\ 1.274/\beta & 1.5 < \beta \leq 3.0 \\ 1.248/\beta^2 + 0.283 & \beta > 3.0 \end{cases}$

Note: σ_u = ultimate strength, σ_Y = yield strength, β = plate slenderness ratio, b = plate breadth, t = plate thickness, σ_{rc} = compressive residual strength, w_o = non-dimensional amplitude of initial deflection (=amplitude/t)

through the plate index method (so called, U-P diagram) was proposed. The complex relation between plate slenderness ratio (β) and coefficient of initial deflection considering buckling mode was also investigated. Throughout the proposed U-P diagram, ship designer may easily predict the ultimate strength performance of steel plate in the midship section.

2 PROCEDURE FOR THE DEVELOPMENT OF U-P DIAGRAM

The procedure for the development of U-P diagram is presented in Fig. 1. Once plate data was collected from

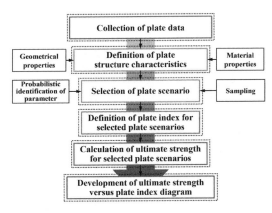

Figure 1. Proposed procedure for the development of U-P diagram.

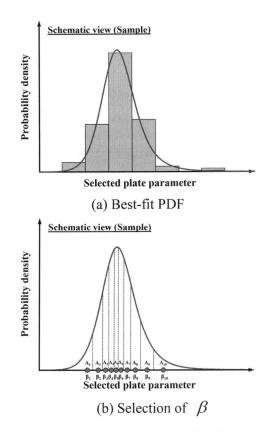

(a) Best-fit PDF

(b) Selection of β

Figure 2. Schematic view of probabilistic identification and sampling of selected plate.

the representative commercial ships, structural characteristics of the plate can be defined in terms of material and geometrical properties. Based on obtained information, plate slenderness ratio (β) can be calculated for all the collected plates. In order to select reliable plate scenarios, the probabilistic identification of selected β is required as shown in Fig. 2(a).

For the selection of the best-fit probability density function (PDF), the goodness-of-fit test (Anderson-Darling test) for plate slenderness ratio was performed. Once the best-fit PDF was selected, the area under PDF can be divided into the required number of plate scenario. Eq. (1) may give a better understanding for the selection of reliable plate scenario.

$$\frac{A_T}{N} = \begin{cases} 2\int_{\beta_i}^{\beta_{i+1}} f(\beta)d\beta & \text{for } i = 0 \\ \int_{\beta_i}^{\beta_{i+1}} f(\beta)d\beta & \text{for } 1 \leq i < N \end{cases} \quad \text{Eq. (1)}$$

Where, A_T = total area under selected PDF, N = total number of scenario, β = plate slenderness ratio, β_i = selected beta, if the total number of scenario (N) is 10, β_1 to β_{10} can be determined by Eq. (1) and Fig. 2(b).

The ultimate strength of plate for selected scenarios can be calculated by analytical, numerical, and experimental methods. In the present study, the ANSYS nonlinear finite element method (NLFEM) which is one of the famous numerical simulation codes is applied to compute the ultimate strength of plate subjected to axial compression. It is to ensure that other methods may also be applied for the ultimate strength analysis of plate structure.

In order to compare the accuracy of the existing methods, especially for the nonlinear FE method, a number of studies and verifications have been conducted by several researchers (ISSC 2012; ISSC 2015). In the case of model size and boundary condition, one-bay and one-span plate model was used with simply supported condition in the present study. For the consideration of initial imperfection, only initial deflection with buckling mode shape was considered.

$$PI = \left| \frac{c_1}{\beta} + \frac{c_2}{\beta^2} + \frac{c_3}{\beta^3} + c_4 \right| \quad (2)$$

$$C_{i, i=1 \text{ to } 4} = i(C_{ID})^i + k \quad (3)$$

$$w_o = A_{omn} \cdot \sin\left(\frac{m\pi x}{a}\right) \cdot \sin\left(\frac{n\pi y}{b}\right) \quad (4)$$

$$w_o = C_{ID} \cdot \beta^2 \cdot t \quad (5)$$

where, $C_{ID} = \begin{cases} 0.025 & \text{for slight level} \\ 0.1 & \text{for average level} \\ 0.3 & \text{for severe level} \end{cases}$ by Smith et al. (1988).

Once the ultimate strength analysis was done for the selected reliable plate scenarios, the plate index (PI) needs to be defined. In this study, PI proposed consists of four (4) different coefficients; C_1 to C_4 as shown in Eq. (2). In order to consider the effect of initial deflection, initial deflection coefficient (C_{ID}) with additional three (3) coefficients, i.e., ω, ξ, ψ, were applied as shown in Eq. (3). Based on obtained Eq. (3), the ultimate strength performance of plate with other initial deflections can also be predicted. The detail may be covered in the applied example section.

C_{ID} can be explained by Equations (4) and (5). Smith et al. (1988) proposed three types of initial deflection level in Eq. (5). In applying the initial deflection to the plate, a buckling mode shape is used in Eq. (4).

Previously, the damage index concept similar to the plate index concept was proposed by Paik et al. (2012). They targeted to develop the damage index for assessing the safety of damaged oil tankers by grounding, and container ships were also considered (Kim et al., 2013). In this study, the plate index concept is proposed to consider the initially deflected plate.

Based on the obtained ultimate strength performance of plate under axial compression and computed plate index for selected plate scenarios, the ultimate strength versus plate index (U-P) diagram, which is called the advanced design diagram, can be developed. The proposed concept will be verified by applied example in next section.

3 APPLIED EXAMPLE

The applicability of the proposed method is verified by applied example. Totally 3,077 plates with details such as geometrical (length, breadth, thickness) and material properties (yield strength, elastic modulus, poison ratio, etc.) were collected from the existing commercial midship section.

Before the selection of reliable plate scenarios, plate slenderness ratio (β) considered as an important parameter to distinguish plate properties was computed. Then, the probability density of obtained β was plotted based on the selected scenarios as shown in Fig. 3(a). The best interval was selected based on the maximum mean and minimum COV values as shown in Fig. 3(b). Totally hundred (100) cases of β were selected based on the sampling technique as illustrated in Eq. (1) and the selected plate scenarios are summarized in Table A1. 3-Parameter Loglogistic function was selected for the best-fit PDF based on the Anderson-Darling test as shown in Fig. 3(c).

Once the reliable plate scenarios were selected, PI can be defined. Through applied example, unknown parameters as presented in Eqs. (2) and (3) will be determined based on the results of the numerical simulation.

Once the reliable plate scenarios were selected, the ultimate strength analysis was conducted by the ANSYS nonlinear finite element method. For the modelling of the plate for the selected scenarios, boundary condition, range of model, initial imperfection, and applied loading should be discussed. Figures 4(a) and (b) show a sample of applied initial deflection and boundary condition. Based on the mesh convergence study in Fig. 4(b), 0.1 of mesh size and plate breadth ratio was selected for the numerical simulation.

For the safe design of plate, which was the main component of ships and offshore structure, simply supported boundary condition was applied in general. In

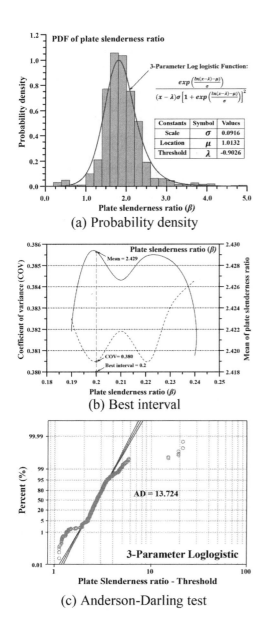

(a) Probability density

(b) Best interval

(c) Anderson-Darling test

Figure 3. Obtained probability density function of plate slenderness ratio with best-fit PDF.

the case of a model range, one-bay and one-span plate model was used based on ISSC (2012).

During the fabrication, initial imperfection naturally occurs in steel structure, especially for initial deflection and welding-induced residual stress which are taken into account in the ship and offshore construction. In this study, three different levels of initial deflection proposed by Smith et al. (1988) were considered as shown in Equations (4) and (5).

Once numerical simulations were done by the ANSYS for selected plate scenarios, plate index (PI) can be calculated as shown in Table A1. Simulation results were compared with other methods

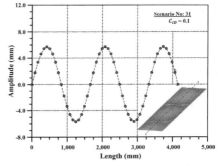

(a) Sample of initial deflection (Scenario No. 31)

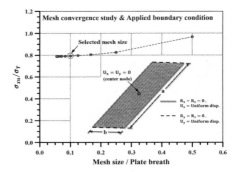

(b) Mesh convergence and boundary condition

Figure 4. Modelling for finite element analysis.

(ALPS/ULSAP, 2014) and existing design formulas illustrated in Table 1 are presented in Figs. 5(a) to (c). In addition, the advanced design formula which can cover a wide range of initial deflection level was also developed. PI is the main part of the proposed design formula, which requires computation of confidence (C_1 to C_4) in Eq. (3). In order to consider a wide range of initial deflection, additional empirical formulas are proposed in Fig. 6(a) from the results in Fig. 5(d).

Finally, the ultimate strength performance of plate versus plate index (U-P) diagram considering wide-ranges of initial deflections (C_{ID}) is proposed as illustrated in Fig. 6(b). Based on the obtained results from this study, the ultimate strength of initially deflected plate can be easily (using PI concept) and accurately ($R^2 = 0.99$) predicted.

4 CONCLUDING REMARKS

An advanced design formula by applying the plate index concept (U-P diagram) was proposed to predict the ultimate strength performance of plate under axial compression. The proposed concept was verified by the selected 300 plate scenarios and numerical simulations conducted by the ANSYS nonlinear FEM. The obtained design formulas were well fitted with the numerical simulation results ($R^2 = 0.99$) and compared with the existing design formulas. In the case of

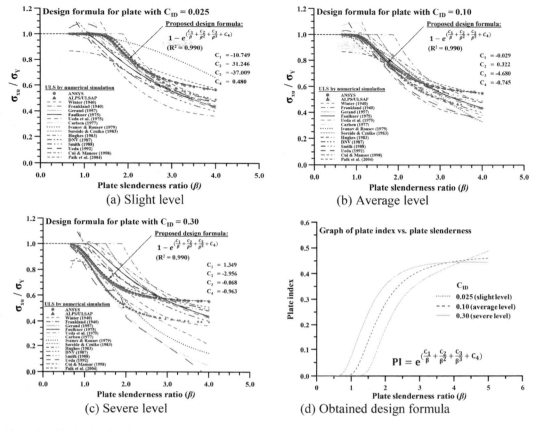

Figure 5. Obtained design formula based on simulation.

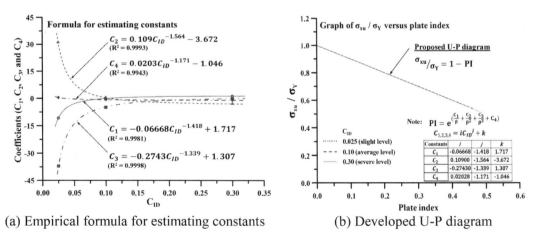

Figure 6. Development of U-P diagram.

initial deflection, three different levels such as slight, average, and severe were considered and general trends were formulated to predict other values of initial deflection. Other types of plate range and boundary condition will be considered for further study.

ACKNOWLEDGEMENTS

This research was supported by the Technology Innovation Program (Grant No.: 10053121) funded by the Ministry of Trade, Industry & Energy (MI) of Korea.

REFERENCES

ALPS/ULSAP. 2014. A computer program for ultimate limit state assessment of stiffened panels. Advanced Technology Center, DRS C3 Systems, Inc. Parsippany, NJ, USA (Available from: www.proteusengineering.com, www.maestromarine.com).
ANSYS. 2014. Version 13.0. ANSYS Inc., Canonsburg, PA, USA.
Carlsen CA. 1977. Simplified collapse analysis of stiffened plates. *Norwegian Maritime Research* 5(4): 20–36.
Chen, Y. 2003. *Ultimate strength analysis of stiffened panels using a beam-column method*. Ph.D. Dissertation, Virginia Polytechnic Institute and State University, Virginia, USA.
Cui, W. & Mansour, A. E. 1998. Effects of welding distortions and residual stresses on the ultimate strength of long rectangular plates under uniaxial compression. *Marine Structures* 11(6): 251–269.
DNV. 1987. Use of high tensile steel in ship structures. The Tanker Structure Co-operative Forum Meeting With Shipbuilders, Paramus, New Jersey, USA.
Faulkner, D. 1975. A review of effecting plating for use in analysis of stiffened plating in bending and compression. *Journal of Ship Research* 19(1): 1–17.
Frankland, J.M. 1940. The strength of ship plating under edge compression. *U.S. Experimental Model Basin Progress Report* 469, USA.
Gerard, G. 1962. Introduction to Structural Stability Theory. McGraw-Hill, USA.
Hughes, O.F. 1983. *Ship structural design: a rationally-based, computer-aided optimization approach*. New York: Wiley.
IACS. 2015. *Harmonised Common Structural Rules for Bulk Carriers and Oil Tankers*, International Association of Classification Societies (IACS), London, UK
Ivanov, L.D. & Rousev, S.H. 1979. Statistical estimation of reduction coefficient of ship's hull plates with initial deflections. *The Naval Architect* 4: 158–160.
ISSC. 2012. *Ultimate Strength (Committee III.1), 18th International Ship and Offshore Structures Congress (ISSC 2012)*, 9–13 September, Rostock, Germany.
ISSC. 2015. *Ultimate Strength (Committee III.1), 19th International Ship and Offshore Structures Congress (ISSC 2015)*, 7–10 September, Cascais, Portugal.
Kim, D.K., Pedersen, P.T., Paik, J.K., Kim, H.B., Zhang, X.M. & Kim, M.S. 2013. Safety guidelines of ultimate hull girder strength for grounded container ships. *Safety Science* 59: 46–54.
Koh, T.J. & Paik, J.K. 2015. Structural failure assessment of a post-Panamax class containership: lessons learned from the MSC Napoli accident. *Ships and Offshore Structures*, in-press (DOI:10.1080/17445302.2015.1074643).
Paik, J.K., Thayamballi, A.K. & Lee, J.M. 2004. Effect of initial deflection shape on the ultimate strength behavior of welded steel plates under biaxial compressive loads. *J Ship Res.* 48(1):45–60.
Paik, J.K., Kim, D.K., Park, D.H., Kim, H.B. & Kim, M.S. 2012. A new method for assessing the safety of ships damaged by grounding. *International Journal of Maritime Engineering* 154(A1): 1–20.
Smith, C.S., Davidson, P.C., Chapman, J.C. & Dowling, P.J. 1988. Strength and stiffness of ship's plating under in-plane compression and tension. *Trans RINA* 130: 277–296.
Soreide, T.H. & Czujko, J. 1983. Load-carrying capacity of plates under combined lateral load and axial/biaxial compression. *The proceedings of the 2nd International Symposium on Practical Design in Shipbuilding (PRADS 1983)*, October 17–22, Tokyo and Seoul, Japan and Republic of Korea.
Ueda, Y., Yasukawa, W., Yao, T., Ikegami, H. & Ominami, R. 1975. Ultimate strength of square plates subjected to compression (1st report) effects of initial deflection and welding residual stresses. *Journal of the Society of Naval Architects of Japan* 137: 210–221.
Ueda, Y., Yao, T., Nakacho, K. & Yuan, M.G. 1992. Prediction of welding residual stress, deformation and ultimate strength of plate panels. Engineering design in welded constructions. Oxford: Pergamon Press, 251–259.
Winter, G. 1940. Stress distribution in and equivalent width of flanges of wide thin-wall steel beams. *National Advisory Committee for Aeronautics (NACA) Technical Note No. 784*, NASA Technical Report Server (NTRS), Hampton, USA.

APPENDIX

Table A1. Plate information

No	a (mm)	b (mm)	t (mm)	σ_Y (MPa)	E (GPa)	β	Plate index 0.025	0.1	0.3
1	4150	830	44.50	315	205.8	0.73	0.00	0.00	0.01
2	4150	830	38.50	315	205.8	0.84	0.00	0.00	0.03
3	4150	830	36.00	315	205.8	0.90	0.00	0.00	0.04
4	4150	830	34.00	315	205.8	0.96	0.00	0.00	0.06
5	4150	830	32.50	315	205.8	1.00	0.00	0.01	0.07
6	4150	830	31.50	315	205.8	1.03	0.00	0.01	0.08
7	4150	830	31.00	315	205.8	1.05	0.00	0.01	0.09
8	4150	830	30.00	315	205.8	1.08	0.00	0.02	0.10
9	4150	830	29.50	315	205.8	1.10	0.00	0.02	0.11
10	4150	830	29.00	315	205.8	1.12	0.00	0.02	0.11
11	4150	830	28.50	315	205.8	1.14	0.00	0.03	0.12
12	4150	830	28.00	315	205.8	1.16	0.00	0.03	0.13
13	4150	830	27.50	315	205.8	1.18	0.00	0.03	0.14
14	4150	830	27.00	315	205.8	1.20	0.00	0.04	0.15
15	4150	830	26.50	315	205.8	1.23	0.00	0.05	0.15
16	4150	830	26.00	315	205.8	1.25	0.00	0.05	0.16
17	4150	830	25.50	315	205.8	1.27	0.00	0.06	0.17
18	4150	830	25.00	315	205.8	1.30	0.00	0.07	0.18
19	4150	830	24.50	315	205.8	1.33	0.00	0.07	0.19
20	4150	830	24.00	315	205.8	1.35	0.00	0.08	0.20
21	4150	830	23.50	315	205.8	1.38	0.01	0.09	0.21
22	4150	830	23.00	315	205.8	1.41	0.01	0.10	0.22
23	4150	830	22.50	315	205.8	1.44	0.01	0.11	0.23
24	4150	830	22.00	315	205.8	1.48	0.02	0.13	0.24
25	4150	830	21.50	315	205.8	1.51	0.03	0.14	0.25
26	4150	830	21.00	315	205.8	1.55	0.03	0.15	0.26
27	4150	830	20.50	315	205.8	1.58	0.04	0.16	0.27
28	4150	830	20.00	315	205.8	1.62	0.05	0.18	0.28
29	4150	830	19.50	315	205.8	1.67	0.07	0.19	0.29
30	4150	830	19.00	315	205.8	1.71	0.08	0.20	0.30
31	4150	830	18.50	315	205.8	1.76	0.10	0.22	0.31
32	4150	830	18.00	315	205.8	1.80	0.11	0.23	0.32
33	4150	830	17.50	315	205.8	1.86	0.13	0.25	0.33
34	4150	830	17.00	315	205.8	1.91	0.15	0.26	0.34
35	4150	830	16.50	315	205.8	1.97	0.17	0.28	0.35
36	4150	830	16.00	315	205.8	2.03	0.19	0.29	0.36
37	4150	830	15.50	315	205.8	2.09	0.21	0.30	0.37
38	4150	830	15.00	315	205.8	2.16	0.23	0.32	0.38
39	4150	830	14.50	315	205.8	2.24	0.25	0.33	0.38

Table A1. (*Continued*)

No	a (mm)	b (mm)	t (mm)	σ_Y (MPa)	E (GPa)	β	Plate index 0.025	0.1	0.3
40	4150	830	14.00	315	205.8	2.32	0.27	0.34	0.39
41	4150	830	13.50	315	205.8	2.41	0.29	0.35	0.40
42	4150	830	13.00	315	205.8	2.50	0.30	0.37	0.41
43	4150	830	12.50	315	205.8	2.60	0.32	0.38	0.41
44	4150	830	12.00	315	205.8	2.71	0.34	0.39	0.42
45	4150	830	11.50	315	205.8	2.82	0.35	0.40	0.42
46	4150	830	11.00	315	205.8	2.95	0.36	0.41	0.43
47	4150	830	10.50	315	205.8	3.09	0.38	0.42	0.43
48	4150	830	10.00	315	205.8	3.25	0.39	0.42	0.44
49	4150	830	9.50	315	205.8	3.42	0.40	0.43	0.44
50	4150	830	8.50	315	205.8	3.82	0.42	0.44	0.44
51	4150	830	42.00	235	205.8	0.67	0.00	0.00	0.00
52	4150	830	36.50	235	205.8	0.77	0.00	0.00	0.01
53	4150	830	34.00	235	205.8	0.82	0.00	0.00	0.02
54	4150	830	32.00	235	205.8	0.88	0.00	0.00	0.03
55	4150	830	30.50	235	205.8	0.92	0.00	0.00	0.05
56	4150	830	29.50	235	205.8	0.95	0.00	0.00	0.06
57	4150	830	29.00	235	205.8	0.97	0.00	0.00	0.06
58	4150	830	28.50	235	205.8	0.98	0.00	0.00	0.07
59	4150	830	27.50	235	205.8	1.02	0.00	0.01	0.08
60	4150	830	27.00	235	205.8	1.04	0.00	0.01	0.09
61	4150	830	26.50	235	205.8	1.06	0.00	0.01	0.09
62	4150	830	26.00	235	205.8	1.08	0.00	0.01	0.10
63	4150	830	25.50	235	205.8	1.10	0.00	0.02	0.11
64	4150	830	25.00	235	205.8	1.12	0.00	0.02	0.12
65	4150	830	24.50	235	205.8	1.14	0.00	0.03	0.12
66	4150	830	24.00	235	205.8	1.17	0.00	0.03	0.13
67	4150	830	23.50	235	205.8	1.19	0.00	0.04	0.14
68	4150	830	23.00	235	205.8	1.22	0.00	0.04	0.15
69	4150	830	22.50	235	205.8	1.25	0.00	0.05	0.16
70	4150	830	22.00	235	205.8	1.27	0.00	0.06	0.17
71	4150	830	21.50	235	205.8	1.30	0.00	0.07	0.18
72	4150	830	21.00	235	205.8	1.34	0.00	0.08	0.19
73	4150	830	20.50	235	205.8	1.37	0.01	0.09	0.21
74	4150	830	20.00	235	205.8	1.40	0.01	0.10	0.22
75	4150	830	19.50	235	205.8	1.44	0.01	0.11	0.23
76	4150	830	19.00	235	205.8	1.48	0.02	0.13	0.24
77	4150	830	18.50	235	205.8	1.52	0.03	0.14	0.25
78	4150	830	18.00	235	205.8	1.56	0.04	0.15	0.26
79	4150	830	17.50	235	205.8	1.60	0.05	0.17	0.28
80	4150	830	17.00	235	205.8	1.65	0.06	0.19	0.29
81	4150	830	16.50	235	205.8	1.70	0.08	0.20	0.30
82	4150	830	16.00	235	205.8	1.75	0.09	0.22	0.31
83	4150	830	15.50	235	205.8	1.81	0.11	0.23	0.32
84	4150	830	15.00	235	205.8	1.87	0.14	0.25	0.33
85	4150	830	14.50	235	205.8	1.93	0.16	0.27	0.34
86	4150	830	14.00	235	205.8	2.00	0.18	0.28	0.36
87	4150	830	13.50	235	205.8	2.08	0.21	0.30	0.37
88	4150	830	13.00	235	205.8	2.16	0.23	0.31	0.38
89	4150	830	12.50	235	205.8	2.24	0.25	0.33	0.38
90	4150	830	12.00	235	205.8	2.34	0.27	0.34	0.39
91	4150	830	11.50	235	205.8	2.44	0.29	0.36	0.40
92	4150	830	11.00	235	205.8	2.55	0.31	0.37	0.41
93	4150	830	10.50	235	205.8	2.67	0.33	0.38	0.42
94	4150	830	10.00	235	205.8	2.80	0.35	0.40	0.42
95	4150	830	9.50	235	205.8	2.95	0.36	0.41	0.43
96	4150	830	9.00	235	205.8	3.12	0.38	0.42	0.43
97	4150	830	8.50	235	205.8	3.30	0.39	0.43	0.44
98	4150	830	8.00	235	205.8	3.51	0.41	0.43	0.44
99	4150	830	7.50	235	205.8	3.74	0.42	0.44	0.44
100	4150	830	7.00	235	205.8	4.01	0.44	0.45	0.44

Rheological behavior of tsunami run-up water containing alluvial deposit at coastline of Kedah, Malaysia

A.K. Philip, H.M. Teh & S.H. Shafiai
Department of Civil and Environmental Engineering, Universiti Teknologi PETRONAS, Seri Iskandar, Perak, Malaysia

A. Jaafar
Department of Mechanical Engineering, Universiti Teknologi PETRONAS, Perak, Malaysia

A.H.M. Rashidi
National Hydraulic Research Institute of Malaysia, Seri Kembangan, Selangor, Malaysia

ABSTRACT: Tsunami run-up wave has been hypothesized as non-Newtonian fluid due to the presence of sediment and debris in its body. This paper assesses the rheology of run-up wave based on the established flow and viscosity curves. Soil sampling was conducted at shoreline of Kuala Teriang and Kota Kuala Muda, Kedah, Malaysia. Investigation on the geologic layers at four different trenches i.e. two from each sites showed that positive indication of tsunami deposits exist at Kuala Teriang. Observation at Kota Kuala Muda was rather challenging due to factors such non-discrepant grain size and sediment type between native and tsunami deposits, inadequate post sedimentation to preserve the tsunami soil, and weathering effect. Rheology experiment conducted used four different mixtures with varying volumetric concentration of tsunami fine sediment i.e. 10, 20, 30 and 40%. Results showed that all mixture exhibit non-Newtonian Bingham pseudoplastic behavior with threshold limit ranging from 0.0328 to 131.2 Pa.

1 INTRODUCTION

Tsunami, also known as 'harbor wave' in Japanese literature, is a natural disaster that causes massive damages to coastal regions. The Indian Ocean tsunami in 2004, a subsequent effect due to a fault rupture causing an underwater earthquake of magnitude 9.0 on the Richter scale, had brought wide-scale impacts to the coastal communities at the perimeter of the ocean (Colbourne, 2005). Although Malaysia is geographically situated within the shadow zone of tsunami, the diffracted tsunami waves had flooded the shoreline at the northern region of the peninsula i.e. Perak, Penang, Kedah and Perlis – see Figure 1. The casualties recorded a total loss of 68 lives and properties damage up to USD$25 million (Colbourne, 2005, Malaysia Department of Irrigation and Drainage [DID], 2005).

In light of the dire consequences of tsunami, coupled with the exponential population growth and rapid development at the coastal region, researchers around the globe had taken actions to gain deeper appreciation on the intricate behavior of tsunami phenomenon as part of the initiatives to provide mitigation solutions for similar future hazards. Extensive research have been conducted to study hydrodynamics of tsunami waves using both physical and numerical approaches, tsunami deposit characteristics, and the study of impacts toward socio-economics and policies of a country. Nonetheless, to the knowledge of the authors, very limited study have been undertaken to evaluate the rheology of the tsunami waves during the run-up, inundation and backwash phases. These phases are later referred as run-up water.

Rheology is the study of fluid deformation to classify the behavior into two general categories i.e. Newtonian and non-Newtonian. As defined by Lorenzini and Mazza (2004), Newtonian fluid refers to those fluid having a direct relationship between shear stress and shear rate e.g. water and oil. Non-Newtonian, on the other hand, is fluid that exhibit apparent change in viscosity (strain) with increasing shear rate, hence also affecting the applied shear stress. Two behaviors are observed in this class which are shear-thinning and shear-thickening. Shear-thinning or pseudoplastic fluid is characterize by the decrease in viscosity with increasing shear rate e.g. blood and paint. Contrary, shear-thickening or dilatant fluid shows an increase in viscosity with increasing shear rate e.g. starch mixture and quicksand. Additionally, there is also a non-Newtonian Bingham fluid, in which the fluid will flow as Newtonian, pseudoplastic or dilatant type after the yield stress was achieved (Lorenzini & Mazza, 2004). Other types of non-Newtonian fluid are thixotropy and rheopexy, a time-dependent change in viscosity.

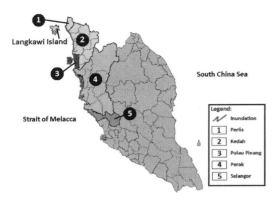

Figure 1. Shoreline of Malaysia affected by the 2004 tsunami waves. (DID, 2005).

For a debris or mudflow, the state of fluid at which it becomes non-Newtonian can be treated as single or multiphase (Santolo et al., 2010). When higher amount of fine fractions occurred for a particular debris flow i.e. 10% of the grain-size distribution, fluid deformation due to viscosity parameter occurred (Bin & Hulin, 2000). Alternatively, deformation due to collision will transpire when greater quantity of coarse particles exist.

It has been hypothesized that tsunami wave generated by submarine landslide behave as non-Newtonian due to the presence of soil debris within the body of water (Geist, 2009, Pudasaini et al., 2012). However, the theory made was only scoped at the source generation. The fluid characteristic during the tsunami run-up phase had yet been classified, regardless of the generation mechanism. Most comprehensive studies associated to mud and debris flows usually relates to slope failure or landslide regimes. Excluding the tsunami paradigm, the rheology of cohesive sediment in suspension in estuarine environment had been reported whereby the analysis used rotating cylinder viscometer (Krone, 1963).

The main objective of this paper is to classify the rheology of run-up tsunami water containing tsunami mud deposit. It is sensible to note that the rheological investigation in this study considers the fluid as single phase, where viscous dissipation and viscoplastic behavior are dominant.

2 SITE STUDY

2.1 Criteria for site selection

The determination of suitable study sites for soil sampling purpose are based on the following criteria which are: (1) availability of reference sources to determine the number of studied sites affected by the 2004 tsunami impact at Malaysia; (2) availability of reports on the characteristic of tsunami sediment; and (3) development status for minimal post tsunami site clean-up activities.

Substantial field data regarding site specific hydrodynamics and infrastructure damages had been reported (Koh et al., 2009, DID, 2005). On research that had investigated the tsunami deposit characteristic in the geographic layer, only two locations in Malaysia were studied by Hawkes et al. (2007) namely Kuala Teriang, Langkawi and Sungai Burong, Pulau Pinang.

Aerial observation via Google Earth software showed that out of the two studied areas by Hawkes et al. (2007), site at Sungai Burong (coordinate N 5°20'14.48" and E 100°11'46.90") had been densely vegetated and inaccessible. On the other hand, it had been cited that rapid reconstruction activities in Kuala Teriang to revert back to its former state had commenced in a matter of few months after the incident (Saleh, 2008). This signifies the possibilities that major on-land deposits had been cleaned up.

Additionally, five hectares of the muddy foreshore of Kuala Teriang had also been subjected to mangrove re-planting in year 2009. A total of 29500 mangrove seedlings were planted but only half of it survived a year later (National Audit Department Malaysia [NAD], 2010). Series of geotubes fronting the shoreline were installed as part of the replanting project package. Consequently, this implies that the upper soil strata, at area subjected to the mangrove replanting scheme, may have been altered. Nevertheless, an interview session with the locals had testified that the backshore area, particularly at the sandy berms, had minimal impacts from anthropogenic activities.

According to report by DID (2005), out of all the flooded shoreline in Malaysia, Kota Kuala Muda shoreline (stretching from Sungai Yu to Kampung Padang Salim) had encountered the most devastating impacts above all other, and thicker muddy soil deposited onshore were observed. However, no reports were available that comprehensively describe the characteristics of tsunami deposit at this particular site. Nonetheless, interview with the locals at Kota Kuala Muda had confirmed that the flat berm fronting the shoreline at Kampung Hujung Matang, and at an elevated open field at Kampung Padang Salim were subjected to minimal impact from human activities making it suitable for soil sampling purpose.

2.2 Site studies environmental settings

Two sites had been chosen for soil sampling namely Kuala Teriang (Langkawi Island) and Kota Kuala Muda, both located within the province of Kedah, Malaysia. For Kota Kuala Muda, the larger area were further narrowed down to Kampung Hujung Matang and Kampung Padang Salim.

Surrounded with village environment, the beach at Kuala Teriang and Kota Kuala Muda are categorized as muddy beach with scarce population of mangrove density growing at the mildly sloping foreshore. A limited strip of sandy beach berm was found between the muddy foreshore and the concrete pavement at Kuala Teriang study area.

Hawkes et al. (2007) reported that the native pre-tsunami soil at Kuala Teriang shoreline was silty clay sediment. He also had confirmed that the characteristic of tsunami sediment at Kuala Teriang was a medium to fine brown sand, and fining upward sequence was evident. At Kota Kuala Muda, the tsunami sediment reported was muddy soil at both sites (DID, 2005). However, no detailed reports on the tsunami nor native soil was found.

Survey reports stated that the tsunami wave run-up had achieved heights of 2.0 m LSD and inundation distance up to 400 m at Kota Kuala Muda area (DID, 2005). Meanwhile, Kuala Teriang was exposed to run-up height of 4.3 m MSL and subsequent inundation reaching 250 m inland (Hawkes, 2007). It was noted that discrepancies exist between each surveys available in terms of the measured wave run-up height and inundation distance (DID, 2005, Hawkes et al., 2007, Koh et al., 2009). Nonetheless, it was agreed that both shoreline were subjected to three series of tsunami waves (DID, 2005, Koh et al., 2009).

3 METHODOLOGY

Soil sampling activities were conducted on 5 May 2015 and four sampling trenches were made i.e. two from Kuala Teriang (T1-A and T1-B) and the other two from Kota Kuala Muda (T2-A and T2-B). Surface mud at the foreshore slope at both sites were also sampled to compare with the potential tsunami deposit. The geographical coordinates of each of the trenches are given in the Table 1.

3.1 Identification of tsunami deposit

In this study, qualitative analysis through physical observation was conducted to determine the potential tsunami deposit. For this purpose, guideline proposed by Peter & Jaffe (2010) was referred i.e. erosional contacts, deposit grading and thickness, landward thinning, boulder transport, number of layers, presence of mud cap and rip-up clast, sedimentary structure, and source of sediment. It is imperative to note that the observations conducted were point base i.e. one trench over the whole study area. Therefore, no attempt were made to derive the inland variation or flow condition. Thus, criteria such as landward thinning of deposit layer and boulder movement will not be used. The findings by Hawkes et al. (2007) was also set as the benchmarks in this analysis.

3.2 Particle size distribution method

Particle size distribution analysis for fine sediment was conducted at Geotechnical Laboratory, Universiti Teknologi PETRONAS. Two types of soil samples were analyzed which were (1) potential tsunami mud and (2) surface mud samples taken at the same site. The method follows the BS standard for hydrometer test (BS 1377: Part 2: 1990). A total dry weight of 50 g for each samples were prepared for the testing using soil that passes the 63 μm sieve. Dipersant solution (dilute solution of Sodium Hexametaphosphate mix with Sodium Carbonate) was used. The recorded results are presented in a semi-log graph.

3.3 Experimental procedure for rheology analysis

Rheometry test was conducted at Flow Assurance Laboratory, Universiti Teknologi PETRONAS. For this analysis, controlled stress rotational rheometer (AR-G2 TA instrument) equipped with parallel plate geometry was used to plot the flow and viscosity curves – see Figure 2 for the instrument setup.

The mixtures were imposed to varying shear rate ranging from 0.1 to 1000 s^{-1}, at constant temperature of 20° C. It is a requirement that the gap distance between each plates should be at ten times greater than the largest particle size. However, since only fine sediment was accounted, in which the solid sizes are less than 63 μm, the practical gap was set to 4 mm.

Four mixtures were prepared prior to the testing in which the tsunami mud samples acquired from the project sites were added with distilled water. The volumetric concentrations for each mixture were diversified to every 10% increment from mix 10% to mix 40%. Table 2 shows the mixing proportion for each volumetric concentration. The required mass for each mix were determined using density of water of 1000 Kg/m^3 and density of silty-clay marine sediment of 1690 Kg/m^3 (Nafe & Drake, 1961). The samples were prepared and stirred for 24 hours prior to

Table 1. Location of the trenches

Location	ID	Northing	Easting
Kg. Kuala Teriang, Langkawi, Kedah	T1-A	6° 21.669'	99° 42.767'
	T1-B	6° 21.674'	99° 42.776'
Kg. Hujung Matang, Kota Kuala Muda, Kedah	T2-A	5° 35.180'	100° 20.360'
Kg. Padang Salim, Kota Kuala Muda, Kedah	T2-B	5° 36.284'	100° 20.476'

Figure 2. (a) Controlled stress rheometer (b) Parallel plate geometry.

Table 2. Approximated mixing ratio

Ratio (Solid: Water)	10:90		20:80		30:70		40:60	
	S	W	S	W	S	W	S	W
Weight (g)	42	225	85	200	127	175	169	150

the rheometry testing to ensure homogeneity of the mixtures.

4 RESULT DISCUSSION

4.1 Evidence of tsunami deposit

Figure 4 depicts soil layers at four trenches i.e. T1-A, T1-B, T2-A and T2-B. The depth of the trenches at Kuala Teriang (i.e. T1-A and T1-B) were 30 and 45 cm, respectively. On the other hand, the depth of the trenches at Kota Kuala Muda (i.e. T2-A and T2-B) were 45 and 25 cm, respectively.

4.1.1 Identification of tsunami deposit at T1-A

For the case of soil at T1-A, four alternating layers between mud and sand were observed. The sandy sediments are mostly coarse materials mixed with shell fragments while the bottommost (0-10cm) dark colored mud layer is the pre-tsunami soil. Furthermore, mud cap and sharp contact between the sand and mud layers were also observed at position between 10–20 cm signifying features of tsunami residues as according to Peter & Jaffe (2010). Fining upward sequence of the coarse shell-mixed sand layer was also observed. No rip up clast were found in the strata. Although most of the indicator proposed by Peter & Jaffe (2010) were positive, the observed sediment did not contain fine sandy sediment as reported by Hawkes et al. (2007).

4.1.2 Identification of tsunami deposit at T1-B

Referring to trench T1-B, both the characteristic of tsunami sediment found by Hawkes et al. (2007) i.e. fine sand and fining upward sequence were observed in the 3 alternating layers of sand and mud soil. Mud cap was present, sandwiched between two sand layers. Erosional contact between the 10 cm mud deposit and sand layers was also recorded in the geologic layer. Nonetheless, rip-up clast was absent. Altogether, this confirms plausible existence of tsunami deposit at T1-B.

4.1.3 Identification of tsunami deposit at T2-A and T2-B

Contrary, no distinguishable strata were seen for trenches at Kota Kuala Muda i.e. T2-A and T2-B. Only clayey soil was found in T2-A and laterite soil in T2-B, in which both were the pre-tsunami sediment. Two justification can be derived for this case with regards to the post sedimentation at the elevated zone, and the

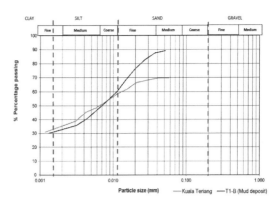

Figure 3. Grain size analysis of the tsunami and surface mud samples.

sediment characteristic between pre-tsunami and the new deposit brought by the wave. Firstly, lack of post sedimentation to these elevated flat zones causes the unconsolidated tsunami sediment to be easily eroded via nature forces. Peter & Jaffe (2010) mentioned the importance of posit sedimentation which act as natural preservative for the tsunami deposit. Secondly, the tsunami mud deposit at the site demonstrated a slight change in particle sizes with respect to the native soil hence complicates the identification process (Hawkes et al., 2007).

4.2 Hydrometer analysis

The analysis used the tsunami mud sample found in T1-B and the surface mud sample of the same site. The result shown in Figure 3 infers that the tsunami sediment is uniformly graded while the surface mud is well-graded. Higher fine sand amount was found at the foreshore.

The silt and clay distribution shows slight discrepancies between tsunami and surface mud of Kuala Teriang, where the clay content varies from 30–33%. The reduction of clay content seen in tsunami deposit as compared to the surface soil may be due to the loss in suspension transport during the final backwash.

The surface clay content shown by the hydrometer plot also contradict with the pre-tsunami sediment i.e. silty-clay where the reported silt and clay fractions by Hawkes et al. (2007) are at 70% and 20%, respectively. The differences observed may relate to the sampling points, depth of clay within the trench used for the analysis, and the change in sediment grain distribution within nine years duration since the past study. Moreover, it is important to note that the pre- and post-tsunami soil grain size distribution at Kuala Teriang that was described by Hawkes et al. (2007) were not sufficient to be used as the general distribution for the whole studies site. This is because only one trench was made by the previous researcher and uncertainty of the grain size distribution still exist for nearby area within the same study site.

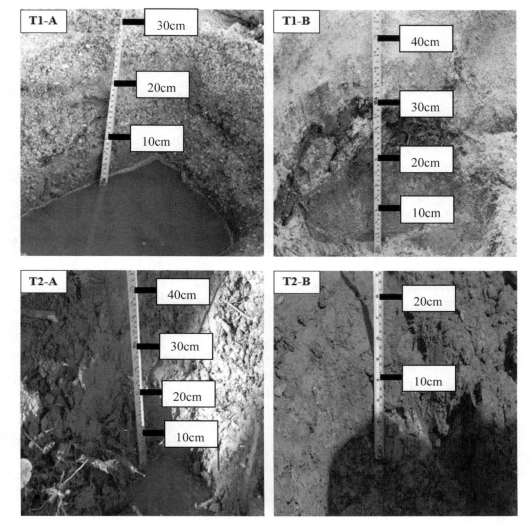

Figure 4. Observation of tsunami deposit layer at Kuala Teriang (T1-A, B) and Kota Kuala Muda (T2-A, B). At T1 trenches, clear layers of pre and post tsunami soils indicate possible tsunami deposit exist. At T2 trenches, no clear observation of soil lay-ers are observed.

4.3 Rheology of run-up water

Qualitative observation of flow curve shown in Figure 5 confirmed that all mixtures, in general, exhibit non-Newtonian Bingham pseudoplastic behavior, whereby an exponential rate of change in shear stress is observed within shear rate $0–200\,s^{-1}$. The viscosity curve shown in Figure 6 also showed that all mixture are in non-Newtonian behavior due to the decrease in viscosity with increasing shear rate. For a fluid to behave as Newtonian, a proportional relationship between shear stress and rate should be apparent in the flow curve, while a constant viscosity should be observed with increasing shear rate in viscosity curve.

For fluid to behave as Bingham would definitely suggest a threshold limit correspond to the concentration. Based on the flow curve, mix 10% had a mild Bingham behavior where the threshold limit is at 0.0328 Pa. On the other hand, mix 20% and 30% had threshold limits at 0.087 and 2.093 Pa, respectively. For mix 40%, higher stress was required to overcome the fluid strain induced by the viscosity of the fine sediment whereby the limit measured at 131.2 Pa. These data collectively suggest that tsunami run-up water will behave as non-Newtonian Bingham pseudoplastic provided that the volumetric concentration achieved 10% of the total volume and the flow stress is at 0.0328 Pa.

Through close observation of the viscosity curve, the initial viscosity for mixture 10% is at 10 Pa which is higher than mix 20 and 30%. The main reason for the error is due to the non-homogeneity of the 10% mixture. It was observed that during the rheological experimentation i.e. when sample was put onto the geometry plate, higher concentration of the fine solid

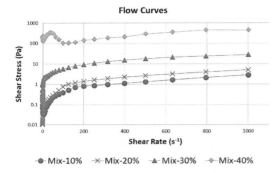

Figure 5. Flow curves for all the mixtures.

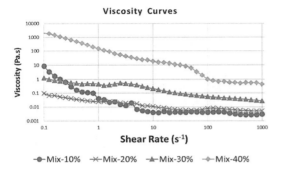

Figure 6. Viscosity curve for all the mixtures.

was taken thus increases the initial viscosity reading. Also, it was observed that the test sample loses fluidity at the end of the testing period and mud lump was observed on the plate.

5 CONCLUSION

Effort had been taken to identify the tsunami deposit and to determine the rheological behavior of tsunami run-up water based on qualitative assessment on the established flow and viscosity curves. Soil sampling was conducted at Kuala Teriang and Kota Kuala Muda, Kedah. Stratigraphy of soil at Kuala Teriang trenches showed positive indication of the 2004 Indian Ocean tsunami, especially for T1-B, whilst no indication of tsunami deposit was observed at Kota Kuala Muda. The main reason the identification of deposit at Kota Kuala Muda faced such result was due to the absence of contrasting sediment characteristic between tsunami and the native sediments, in addition to lack of post sedimentation at the elevated sampling zones and weathering effect. Experimental result from careful rheology analysis of four mixtures with volumetric concentration varied at 10, 20, 30 and 40% suggested that tsunami run-up wave will behave as non-Newtonian Bingham fluid provided that the solid concentration during its run-up phase achieve at least 10% of the volume. The threshold limits range from 0.0328 to 131.2 Pa for a range volumetric concentration from 10 to 40%.

ACKNOWLEDGEMENT

This research is funded by the Ministry of Science, Technology and Innovation (MOSTI) and Fundamental Research Grant Scheme.

REFERENCES

Bin, Y. & Huilin, Z. 2000. Direct Determination of Theological Characteristics of Debris Flow. *J. Hydraulic Engineering*, 126, 158–159.

Colbourne, F. W. 2005. Tsunami impact on the West Coast of Penang Island, Malaysia. Research Project Report, MS in Physical Sciences, Emporia State University, Emporia, Kansas, USA.

Coussot, P. 1997. Mudflow Rheology and Dynamics, *IAHR Monograph Series*, A. A. Balkema, Rotterdam, Brookfield.

Department of Irrigation and Drainage Malaysia. 2005. Investigation Report on Post Tsunami 26 December 2004.

Geist E. L. 2009. Chapter 3 – Phenomenology of Tsunamis: Statistical Properties from Generation to Runup. *Advances in Geophysics*, Elsevier, 51, 107–169.

Hawkes, A. D., Bird, M., Cowie, S., Grundy-Warr, C., Horton, B. P., Hwai, A. T. S. & Aik, L. W. 2007. Sediments Deposited by the 2004 Indian Ocean Tsunami Along the Malaysia–Thailand Peninsula. *Marine Geology*, 242(1), 169–190.

Koh, H. L., Teh, S. Y., Kew, L. M. & Zakaria, N. A. 2009. Simulation of Future Andaman Tsunami Into Straits of Malacca by TUNA. *Journal of Earthquake and Tsunami*, 3(02), 89–100.

Krone, R. B. A 1963. A Study of Rheologic Properties of Estuarial Sediments. Final Report No. 63-88, Hydraulic Engineering Lab and Sanitary Engineering Lab, University of California, Berkeley, CA.

Lorenzini, G. & Mazza, N. (2004). Debris Flow: Phenomenology and Rheological Modelling. *WITpress*, Southampton, Boston.

Nafe, J. E., & Drake, C. L. 1961. Physical Properties of Marine Sediments (No. TR-2). Lamont Geological Observatory Palisadesny.

National Audit Department Malaysia. 2010. Laporan ketua audit negara: aktiviti jabatan/agensi dan pengurusan syarikat kerajaan negeri Kedah tahun 2010. Wilayah Persekutuan Putrajaya.

Peters, R. & Jaffe, B. 2010. Identification of tsunami deposits in the geologic record; developing criteria using recent tsunami deposits (No.2010-1239). US Geological Survey.

Pudasaini, S. P., Miller, S. A., Simos, T. E., Psihoyios, G., Tsitouras, C. & Anastassi, Z. 2012. A Real Two-Phase Submarine Debris Flow and Tsunami. In *AIP Conference Proceedings-American Institute of Physics*, 1479(1), 197.

Saleh I. B. Final project evaluation: integrated coastal resources management in Pulau Langkawi (ICRM-PL), 2008.

Santolo S. D. A., Pellegrino, A. M. & Evangelista, A. 2010. Experimental Study on the Rheological Behavior of Debris Flow. *Natural Hazards and Earth System Science*, 10(12), 2507–2514.

An optimised model of pipe-in-pipe installation for subsea field

M.Z. Jiwa, D.K. Kim & Z. Mustaffa
Department of Civil and Environmental Engineering, Universiti Teknologi PETRONAS, Seri Iskandar, Perak, Malaysia

H.S. Choi
Graduate School of Engineering Mastership, Pohang University of Science and Technology, Pohang, Republic of Korea

ABSTRACT: One of the most important design challenges for engineers is the design for the installation of the pipeline including the pipe-in-pipe (PIP) system. A design underestimation could result in serious damage, while overestimation would result in high operation costs. An optimised modelling method is definitely required for the PIP system. The modelling of a finite element single pipe-lay simulation has been studied and discussed by many authors, and is very much understood by the people in the industry. However, modelling complex PIP systems for pipe laying simulations is quite challenging. To build the most economic finite element (FE) model for pipe-lay simulation of PIP systems, various modelling aspects of the installation were studied separately. In this work, three types of FE models which are two models with different simplifications and one actual PIP model were established and compared. Finally, a method for the optimization of performing PIP installation is presented.

1 INTRODUCTION

The temperature of production fluids inside pipelines is required to be kept above a critical wax and hydrate formation to prevent heat loss. Pipe-in-pipe (PIP) systems provide better insulation than conventional single pipelines (Kristofferson et al., 2012). PIP systems' high insulation properties minimize heat loss from the transported fluids to the environment better than traditional subsea coatings. This is achieved using thermal insulation of very low thermal conductivity that is encased in dry atmospheric conditions between the inner pipe that transports the fluid and the carrier pipe, which provides mechanical protection from the subsea environment.

The most common and efficient way used for laying offshore pipelines in Malaysia is the s-lay method. The method is suitable in shallow and intermediate water depths. The principle is that the pipeline is constructed in the vessel firing line through several processes which happen in the welding stations, NDT, repair station and field joint coating station. The pipeline is held by a series of tensioners and is eased off the stern of the vessel as the vessel moves forward. The pipe curves downward over a stinger mounted on the stern of the vessel to control the configuration of the pipeline into an "S" shape down to the seabed, as illustrated in Fig. 1.

The objective of this paper is to present an optimized installation design process for pipe-in-pipe systems. Many studies were conducted on single pipe stress analysis during its installation phase, but very few studies on pipe-in-pipe or pipe bundle systems have been published. Modelling the actual interaction of two or more pipes has proven to be a challenging process, regardless of the advancement of finite element modelling. Thus, the focus of this paper is on the comparison of modelling techniques for pipe-in-pipe systems' installation in s-lay configurations to come out with an optimized solution for PIP installation modelling.

2 PIPE-IN-PIPE SYSTEMS

A flowline system can be a single-pipe pipeline system, a pipe-in-pipe (PIP) system, or a bundled system (Bai & Bai, 2010). PIP systems can be categorised into three structural classifications which are sliding, fixed and restrained. Hausner and Dixon explained and described these systems using a compatibility matrix which show the possible combinations of insulation material type, field joint and installation method for the main structural categories (Hausner & Dixon 2002).

To date, only two pipe-in-pipe installation projects have been executed in Malaysia's offshore areas. The first installation is the Newfield Peninsular Malaysia Puteri field with a water depth of 60 m in 2011 (Dhillon 2014). The Puteri PIP replacement project comprised of a bundle of three (3) separate pipelines

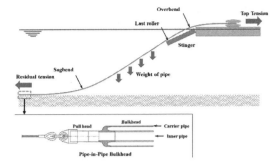

Figure 1. S-lay installation and PIP bulkhead.

Table 1. PIP case data.

No.	Parameter	Carrier pipe (mm)	Inner pipe (mm)
1	Outside diameter	304.8	213.4
	Wall thickness	12	12
2	Outside diameter	457.2	320
	Wall thickness	12.7	12.7

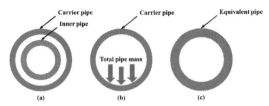

Figure 2. PIP modelling method.

welded together and installed in an *S-lay* method. The pipelines used are 8" inner pipe, 12" carrier pipe and 4" piggyback pipe. The second installation is the Murphy Oil Siakap North–Petai field with a water depth of more than 1190 m in 2014. This is considered the first *rigid-lay* installation of a pipe-in-pipe flowline in Asia-Pacific (Offshore magazine 2014).

This paper would discuss the PIP system using the sliding configuration since this is the most widely used system and is very flexible for installation. Sliding systems will usually have spacers for alignment and centralisation of its two pipes, but the system can work without spacers (Hausner & Dixon 2002).

3 MODELLING PROCESS

For model simplification, the distance and radius of vessel ramp and stinger are used for the initial estimation of lay stress, but it would not reflect the actual responses on the overbend region. These pipe-lay parameter configuration relates directly with water depth. Shallow water depths require short and large lay radii, whereas in deeper waters, longer and smaller lay radii are required. The stinger used in this study is a fixed type and has a total length of approximately 60 m.

The vessel ramp and stinger configuration are based on the actual pipeline installation design, with support rollers modelled in horizontal X and vertical Z coordinates to give a more realistic result (Orcina, 2013). In the case of laying in deeper waters, we may need to lower the stinger to reduce the stress on the last roller. Changing the stinger angle will also change the form of stinger radius. However in this study, the stinger angle and roller support parameters are fixed, which means that the ramp and stinger radius are constant. The cases are analysed in 50 m water depth.

The main PIP design data for modelling study are presented in Table 1. Only the bare pipe model is used for the study.

When modelling a PIP system for installation, it is a common industrial practice to model the PIP as a single pipe or as an equivalent pipe. The reason for this is normally due to its simplicity, conservative result and requires less computational time and storage, especially for dynamic analysis. The typical software used for pipeline installations are OFFPIPE, PipeLay and Orcaflex. The computer program used for this study in Orcaflex 9.6. The software developed by Orcina is a general analysis tool for flexible risers, pipes and umbilical cables subject to gravity, wave and current loads, and external loads and displacement (Orcina 2013).

In this study, the PIP was modelled in 3 methods. The inner pipe is placed inside the carrier pipe to simulate the actual PIP system effect during the installation, as illustrated in Fig. 2(a). The carrier and inner pipelines are independent structures. The second method of modelling is a single pipe (SP) with increasing pipe weight as shown in Fig. 2(b). The pipe density is modified to get a certain submerged weight. The third technique uses an equivalent method by modifying the pipe's thickness. Both the carrier and inner pipe were equated to an equivalent second moment of area. The analysis was done to compare these three options.

For PIP systems, the pipe's submerged weight properties play an important role in the design for installation. The weights considered for the inner pipe and carrier pipe are calculated while the pipe is in the air and water due to the system's layout. A correlation study on carrier pipe dimension ratio against submerged weight of pipe is illustrated in Fig. 3. A dimension ratio (DR) of less than 30 resulted in a positive submerged weight, while DR of 30 or more resulted in negative submerged weight. If the submerged weight is positive, the carrier pipe will not float. If the submerged weight is negative, the carrier pipe will float by itself. It is the inner pipe weight that will cause the carrier pipe to become submerged into seawater. However, only results of the carrier pipe with DR of 25.4 are discussed in this paper since both of the results show similar outcomes.

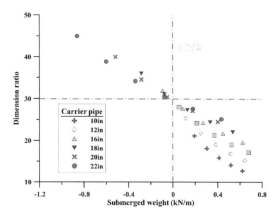

Figure 3. Carrier pipe dimension ratio.

4 DESIGN METHODOLOGY

Lagner's study on suspended pipes demonstrated that a general form of s-lay load conditions can be derived from this equation (Lagner 1984):

$$T_{st} = T_{res} + w.wd - \frac{B.K^2}{2}. \quad (1)$$

where T_{st} = tension at stinger end; T_{res} = horizontal tension on seabed; w = submerged weight; wd = water depth; B = bending stiffness; and K = curvature at the stinger tension. The main parameters that will have an effect on the pipeline's catenary shape at the sagbend region are tension, water depth and submerged weight. Increase in barge tension (at top or residual tension) would reduce the pipeline's curvature in the overbend and sagbend regions.

An initial tension needs to be assumed to perform the pipeline stress calculation. A rough estimate of the required tension can be obtained using a simple method described based on single pipe laying experiences (Herbich 1981). Tension load is directly proportional to water depth. The initial tension equation is:

$$T = 260 \times OD \times \sqrt[3]{wd}. \quad (2)$$

where T = required lay tension in (kN); OD = carrier pipe diameter in (m); and wd = water depth in (m).

In a PIP system, the curvatures on the overbend and sagbend are determined by the properties of both the carrier pipe and inner pipe. For the sliding configuration PIP, the surrounding carrier pipe is connected to the inner pipe through the bulk heads (Fig. 1) at the pull head with only spacers located at certain intervals (Hausner & Dixon 2002). The inner pipe is restrained only at both ends of the pipeline. There will be no tension applied to the inner pipeline during installation. Stress analysis was only conducted for the carrier pipeline. Hence, it is assumed that the inner pipeline will follow the outer pipeline's lay radius.

The theoretical derivation of the resultant weight is fairly straight forward. In a single pipe model, the weight of the pipe is the sum of the carrier pipe's submerged weight and inner pipe's air weight. The resultant weight can be derived from this equation:

$$W_{total} = W_{sub\_carrier} + W_{air\_inner}. \quad (3)$$

Only the carrier pipe displacement is considered for calculating buoyancy force. Therefore, the outer diameter of the equivalent pipe system should remain the same to maintain its hydrodynamic properties. Pipe thickness is modified to obtain similar structural properties of bending and axial stiffness of the PIP system. The equivalent pipe method based on the second moment of area and cross-section area are derived from:

$$I_{eq} = I_{inner} + I_{outer}. \quad (4)$$

$$A_{eq} = A_{inner} + A_{outer}. \quad (5)$$

The second moment of area (Eq. 4) enables one to calculate the pipe's bending stiffness and area. Equation 5 enables one to calculate the pipe's axial stiffness. Many of the details of the principal and calculations of equivalent pipe method were presented in Abeele et al. (2015).

5 RESULTS AND DISCUSSIONS

For comparison of the 3 modelling techniques, stress diagrams are shown in Figs. 4 and 5. In general, the lay configuration gives a good support to the roller load distribution. The graphs show that all three models analysed have very little differences in terms of stress and strain. The SP and PIP models' stress and strain are more or less similar, except that there is a kink near the last roller for the SP model. However, the equivalent model has a similar stress/strain graph shape as the PIP model in all regions, although its load value is very much lower compared to the other 2 models.

In the overbend region, the maximum strain percentage for SP, PIP and equivalent model are 0.116, 0.113 and 0.106 respectively. The strain for the SP model is 2.7% higher, while the equivalent model is 6.2% lower than the expected results. For stress in the sagbend region, both SP and equivalent models' stress are 109.5 MPa (1.9% higher) and 101.4 MPa (9.1% lower) respectively, whereas the PIP model's calculation resulted is 111.6 MPa.

The tension applied on a pipeline is a pull force affecting the tensile stress of the pipeline and the lay curvature in the overbend, last roller and sagbend regions. A suitable lay tension can be obtained by reducing the initial tension applied (292 kN). The lay tension is reduced gradually by 10% reduction at value of 263 kN, 234 kN, 204 kN and 175 kN. Figure 6 shows that the maximum stress on the sagbend region increases from 111.6 MPa to 291.7 MPa due to the reduction of lay tension.

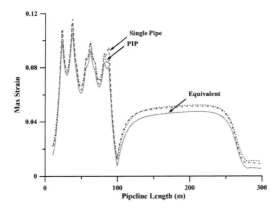

Figure 4. Strain comparison of the 3 models.

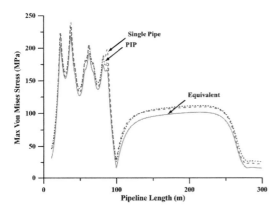

Figure 5. Von Mises stress comparison of the 3 models.

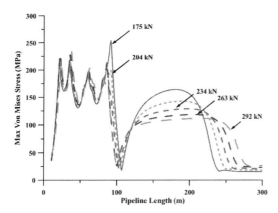

Figure 6. Tension variation for PIP model.

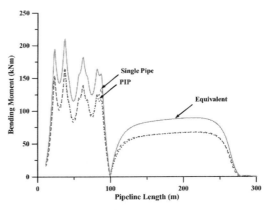

Figure 7. Bending moment comparison of the 3 models.

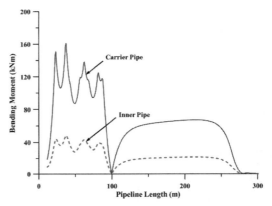

Figure 8. PIP model bending moment.

As the lay curvature increases, more of the pipe at the stinger section is supported by rollers. This improves the pipe's load distribution and indirectly reduces the maximum stress on the ramp section of the overbend region, but increases the lay curvature at the last roller section. Reducing more lay tension will increase the stresses and reaction load on the last roller, which is not always favourable. Some operators' specification require the last roller to be a free load support during static analysis. The effect of increasing the lay tension shows similar results (Al-Kurayshi, 2014).

Figure 7 shows the bending moment comparison results of the 3 models. The graph clearly shows that the maximum bending moment is obtained by the equivalent method. In other words, the equivalent pipe model proves to be very much conservative. This is due to the excessive bending stiffness property which combines both the carrier pipe and inner pipe properties. The lay curvature profile will remain the same for all models.

The last roller or stinger tip and sagbend are considered as load controlled conditions (DNV 2013). Thus, it is important to get accurate bend moment load at both of these regions. The bend moment for the equivalent model is 89 kNm, while for the PIP system is 67 kNm. As shown in Fig. 8, this difference of 22 kNm corresponds to the bend moment of the inner pipe. The reduction of bending moment for the equivalent model can be calculated using this equation:

$$BM_{total} = BM_{eq} - \kappa_{eq}.E.I_{inner} \qquad (6)$$

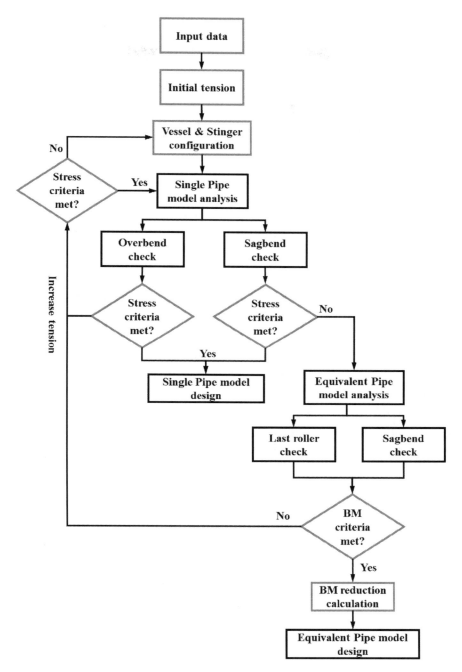

Figure 9. PIP design for installation approach flow diagram.

where κ_{eq} = equivalent model curvature corresponding to the bending moment location.

By applying this load reduction calculation, it would result in more accurate results for the bending moment. As bending moment is reduced, the minimization of installation costs is gained. Overestimation of the lay tension would require a higher capacity of vessels and lay equipment, and would have direct impact on the cost of operations.

Figure 9 shows the proposed PIP installation design approach using both the single pipe and equivalent pipe models.

6 CONCLUSION

In this study, different modelling approaches to simulate a structural response of a Pipe-in-Pipe system were reviewed and compared. Comparison is made on 3 modelling approaches to investigate the load distribution of the PIP system. An approach to obtain quick results for the PIP design installation without modelling the actual carrier and inner pipe interaction is studied.

Results show that the load distribution on the overbend, last roller and sagbend regions are very much comparable with marginal differences. Therefore, the single pipe having the same carrier pipe properties and mass is suitable to determine the stress strain behaviour of the PIP system. Meanwhile, an equivalent pipe having an equivalent section bending stiffness is suitable to determine the bending moment for the corresponding PIP system installation.

The results, presented in this paper, show the methodology for designing the installation of the Pipe-in-Pipe system (sliding configuration). The approach should enable an engineer to quickly and efficiently analyse the pipe laying and determine the integrity of the Pipe-in-Pipe system during installation. The single pipe and equivalent models are both simplified models and can be analysed without having to resort to time-consuming analyses.

ACKNOWLEDGEMENTS

This study was undertaken at the Ocean and Ship Technology (OST) research group under DWT MOR at UTP. This research was supported by the Technology Innovation Program (Grant No.: 10053121) funded by the Ministry of Trade, Industry & Energy (MI) of Korea), the Energy Efficiency & Resource of the Korea Institute of Energy Technology Evaluation and Planning (KETEP), and a grant funded by the Korean Ministry of Knowledge Economy (Grant No.: 2014301002-1870). The authors would also like to thank POSTECH, POSCO, Daewoo E&C, and the Republic of Korea for the great support.

REFERENCES

Abeele, F. V. D., Ville, Q. D., Giagmouris, T., Onya, E. & Njuguna, J. 2015. Finite Element Simulation of Pipe-In-Pipe Systems Installed in Uneven Seabed. *Proceedings, the 34th International Conference on Ocean, Offshore and Arctic Engineering*, St. John's, Canada.

Al-Kurayshi, H. S. 2014. Structural Analysis of Dual Submarine Pipelines During Laying. *International Journal of Scientific and Technology*, 3, 391–391.

Bai, Y. & Bai, Q. 2010. *Subsea engineering handbook*. Elsevier.

Det Norske Veritas 2013. *Submarine pipeline systems*, DNV OS-F101 Standard.

Dhillon, R. S. 2014. Offshore Installation of Technology of Puteri Heated Pipe-In-Pipe System in Malaysia. *Proceedings, the 1st Offshore Technology Conference-Asia*, Kuala Lumpur Malaysia.

Hausner, M. & Dixon, M. 2002. Optimized Design of Pipe-In-Pipe Systems. *Proceedings of Offshore Technology Conference*. Houston, Texas, USA.

Herbich, J. B. 1981. *Offshore pipeline design elements*. M. Dekker Publisher, New York, USA.

Kristoffersen, A. S., Asklund, P. O. & Nystrøm, P. R. 2012. Pipe-In-Pipe Global Buckling and Trawl Design on Uneven Seabed. *Proceedings, The 22nd International Offshore and Polar Engineering Conference*, Rhodes, Greece.

Lagner, C. G. 1984. Relationships for Deepwater Suspended Pipe Spans. *Proceedings, the 3rd International Conference on Offshore Mechanics and Arctic Engineering*, New Orleans, Louisiana, USA.

Offshore magazine 2014. McDermott Achieves Industry First Offshore Asia/Pacific', from *http://www.offshore-mag.com/*.

Orcina 2013. Orcaflex 9.6 software manual.

Method for installation of marine HDPE pipeline

Y.T. Kim
Daewoo Institute of Construction Technology, Daewoo Engineering & Construction Ltd., Republic of Korea

K.S. Park
Steel Business Division, POSCO Center, Seoul, Republic of Korea

H.S. Choi
Graduate School of Engineering Mastership, Pohang University of Science and Technology, Pohang, Republic of Korea

S.Y. Yu & D.K. Kim
Department of Civil and Environmental Engineering, Universiti Teknologi PETRONAS, Seri Iskandar, Perak, Malaysia

ABSTRACT: The seawater intake system of a power plant has the function of delivering cooling water to condenser and auxiliaries. In fact, the HDPE pipeline in the marine environment has been used for seawater intake system due to its advantages, such as immunity to galvanic corrosion, light weight, and flexibility. Hence, it is indeed imminent to employ the HDPE pipeline safely sinking analysis and thus, rigorous study should be carried out. Moreover, the Daewoo Engineering and Construction Ltd. had installed the ultra-diameter of 2500mm HDPE pipeline successfully for Ras Djinet Combined Cycle Power Plant in Algeria. With that, this paper introduces the design and the analysis pertaining to installation of HDPE pipeline.

1 INTRODUCTION

The seawater intake system of a power plant has the function of delivering cooling water (C.W.) to condenser and auxiliaries. The seawater intake system is composed of C.W. intake head structure, C.W. intake pipe, and intake basin structure. In fact, the Daewoo Engineering and Construction (Daewoo E&C) had successfully constructed the intake pipeline system for the project of Ras Djinet Combined Cycle Power Plant in Algeria. The pipeline material for intake involved various resources, such as steel pipe, GRE (Glass Reinforced Epoxy), and HDPE materials. This project adopted HDPE materials after considering a number of benefits in relation to immunity to galvanic corrosion, light weight, and flexibility.

HDPE pipeline has become an attractive alternative in oil and gas industry because of its strength, durability, light weight, and its joint-less construction method (Rinker Pipeline Systems, 2004). Thus, it is suitable for marine application such as subsea pipeline (Idris, 2008). HDPE pipeline needs concrete ballast units to be placed on the seabed which is to withstand buoyancy force and lateral hydrodynamic force. Corrugated HDPE generally has a minimum design service life of 50 years. (Kang et al. 2009). Corrugated HDPE also has a vertical deflection of the central soil prism greater than adjacent backfill soil prisms. It causes a positive arching action and resulting in a vertical arching factor (VAF) less than one. Corrugated HDPE is more sensitively affected by the interface condition where vertical load induced due to shearing stress than the radial pressure (Kang et al. 2009).

In addition, HDPE is also widely used for lining system which is used in pipeline to enhance corrosion resistance as well as to line damaged pipeline. However, polymer liners can fail in service as well. It could be due to buckling collapse induced by external pressure as well as the aging problem. Failure mode due to buckling collapse induced by external pressure is caused by the combined action of two separate factors such as rapid decompression of pipeline and the permeation of oil derived gasses through the liner wall (Rueda et al. 2015). The rapid decompression of pipeline usually happens during service maintenance and inspection shutdowns. Several analytical models were proposed regarding the buckling collapse of pipeline induced by external pressure (Jacobsen 1974, Glock 1997, El-Sawy 2001).

However, it is worth to note that the analytical models are only applied to purely elastic or ideal elastoplastic behavior. They are unable to describe the behavior of highly strain-rate and time-dependent materials as well as the whole deformation process.

To overcome the limitation of these analytical models, an FEM model for the simulation of the buckling collapse of HDPE liners was developed by Rueda et al. (2011). Other than using FEM model to simulate the buckling collapse of polymetric pipeline, there are also several methods proposed by Boyce et al. (1988), and Bergstrom et al. (2004; 2010).

Aging problem is also one of the main factors that cause failure of pipe liners. Numbers of previous studies have been done in this field for different types of polymers in terms of aging temperature effect (Mohd Ishak et al. 2000; Pegoretti and Migliaresi 2002; Chen et al. 2007; Guermazi et al. 2009). The outcome shows that the degradation process of pipeline liner is obviously affected by aging problem.

This study focuses on subsea pipeline installation using HDPE. Song et al. (2015) performed an installation analysis on HDPE pipeline to analyze the structural stability during installation. The purpose of this study is to find a general trend and design consideration throughout the parametric study. To perform installation analysis of the HDPE pipeline, below information is applied. Hence, the study looked into installation analysis and execution for the HDPE pipeline. The sea water intake pipelines were approximately 1.0 km and consisted of 4 lines with 2500 mm OD HDPE pipes.

2 THE DESIGN OF HDPE PIPELINE

The basic design for marine HDPE pipe system is depicted in the following steps.

2.1 Selection of an appropriate pipe diameter

For estimation of the minimum flow diameter, the optimum flow rate has to be determined in order to meet the cooling system criteria. The intake system for the project was designed for the intake of seawater at 162,360 m^3 per hour, while the diameter selected was 2500 mm to meet the limit flow velocity.

2.2 Determination of the pipe wall thickness

The thickness of the wall for the pipelines is based on the selection of diameter procedure and core resistant factor to endure both internal pressure and external loads. Moreover, several criteria need to be satisfied in determining the pipe wall thickness, as given in the following;

- Bursting criteria under internal pressure
- Pipe ring deflection under hydrostatic pressure
- Combined stress under burial load and buckling criteria under installation load.

The wall thickness of SDR 30 and 26 pipes had been confirmed to be sufficient in marines to the wall thickness calculated based on pressure containment.

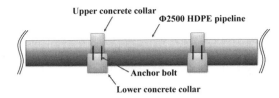

Figure 1. Schematic of HDPE pipeline with concrete collar (Kim, 2015).

2.3 Determination of the required weight and spacing of ballast weight for concrete collar

The submerged weight of pipelines should be calculated after considering both cases of empty and flooded pipes. Empty condition causes pipeline to float in the water surface i.e., pipeline neither sinks nor floats.

If the pipeline sinks, the laid pipeline on the seabed will be exposed to the marine environment, such as current and wave, before backfilling. Thus, 2-year significant wave height and current conditions should be considered in designing the required weighting and spacing for its stability against marine environment due to temporary condition.

Hence, on-bottom stability was performed to determine the stability of the pipe on the seabed. Pipeline resting on the seabed is significantly subjected to the forces due to wave and current loads. If a pipeline is not stable, then it will move under the actions of waves and currents. Therefore, in order to meet the criteria of both on-bottom stability and required submerged weight to sink the pipe, the concrete collar volume was designed at 6.7 m^3 with 6 m of span, which is equivalent to 135 mm of concrete coating thickness along the pipeline.

The span length of the concrete collar was then examined if it did meet the criteria for free-span length after considering static and dynamic conditions. Concrete collars inevitably make gaps between pipe and seabed and the span refers to the length between the concrete collars, as shown in Fig. 1.

As a result, the analysis for stability displayed that a weighting of 40% should offer adequate stability on a pipe laid on flat seabed. Besides, the trench effect would compensate to the force with sheltering effect from the trench estimating at 20%. In addition, a net weighting at 32% of the weight of sea water was displaced by the pipe, where 5.03 t/m was proposed. The pipe with this weighting would resist against wave and current before backfilling (Ian Larsen 2013).

3 INSTALLATION ANALYSIS OF HDPE

An installation analysis for sinking pipeline floating on the surface should be performed by adhering to the basic design of the pipe. This analysis determines the max Von-Mises stress, the bending radius, and the curvature during pipe sinking. Table 1 and Table 2 present

Table 1. HDPE pipeline key design characteristics.

Description		Intake Pipeline A	Intake Pipeline B
Type	–	SDR30	SDR26
Material	–	HDPE	HDPE
Material Grade	–	PE100	PE100
Pipeline Outer Diameter	mm	2500	2500
Service	–	Raw Sea Water	Raw Sea Water
Pressure Regime	–	LP	LP
Wall Thickness	mm	83.3	96.2
Design Pressure	bar(g)	5.5	6.4
Design Temperature	°C	40	40
Operating Temperature	°C	24	24
Minimum Required Strength (MRS)	MPa	10	10

Table 2. High density polyethylene properties for HDPE100.

Property	Unit	Value
Density	kg/m^3	960
Tensile Yield Strength	MPa	23
Elongation at Yield	%	8
Tensile Break Strength	MPa	37
Coefficient of Thermal Expansion	m/m/°C	2.4×10^{-4}
Poisson's Ratio	–	0.4
Minimum Required Strength at 20°C (MRS)	MPa	10.0
Hydrostatic Design Stress at 23°C (HDS)	MPa	7.7

Figure 2. Pre-cast concrete collar.

the key characteristics of the designed HDPE pipeline and material properties.

The designed concrete collar, to give ballast during sinking, was finally constructed, as shown in Fig. 2.

The commercial software Orcaflex 9.6 was used to perform the sinking analysis. Orcaflex is a marine dynamic program developed by Orcina for static and

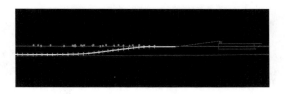

Figure 3. Orcaflex modelling for HDPE pipeline sinking analysis.

dynamic analyses of a wide range of offshore systems, including all types of marine risers (rigid and flexible), global analysis, moorings, installation, and towed systems.

The following results are extracted from this sinking analysis:

- Minimum bending radius (or pipe curvature) along the pipeline
- Maximum von-Mises stress in the pipeline based on the results of the analysis, whereby the following pertinent information required for the sinking installation is obtained:
- Tension capacity required from the tensioning tug and the recommended pulling tensions
- Stress during various stages of flooding/sinking operation.

Nevertheless, as Orcaflex is unable to simulate the pipe content change from empty to flooded condition, the pipe was modelled so that the pipe would initially be flooded during floating out, but additional buoyancy tanks were attached to compensate to the initial flooded water weight inside the pipe. The stripping of the tank one by one from offshore end further simulated the flooding process; the tanks were actually used to sink the pipe to the seabed.

The configurations, as well as the sinking of the offshore and pipeline sections, are illustrated in Fig. 3.

Furthermore, in order to prevent buckling failure, the minimum bending radius was maintained during the pipeline installation. The minimum short-term bending radius is given in the following (AWWA 2013):

- Pipe dimension ratio that ranged between 13.5 and 21 in: 17 times pipe OD
- Pipe dimension ratio larger than 21 in: 20 times pipe OD

This criterion was determined in the pipeline installation analysis. The analysis was performed after considering that the crane was positioned at 60 meters offshore from the inshore end of the pipe to lift the pipe that floated from the water surface. This was to ensure that the initial water flooding began at the inmost end of the inshore pipe.

Moreover, the pulling force at the offshore end was analysed to figure out the effects of pulling force on the bend radius and the von-Mises stress. The results showed that the pulling force for above 50 ton met the criteria for the minimum bending radius and von-Mises stress, as shown Fig. 4. However, the pulling

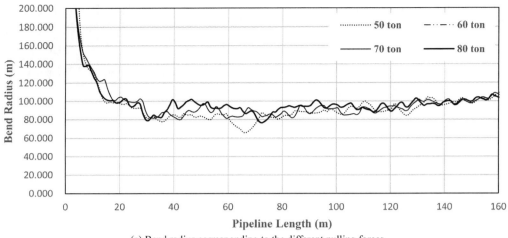

(a) Bend radius corresponding to the different pulling forces

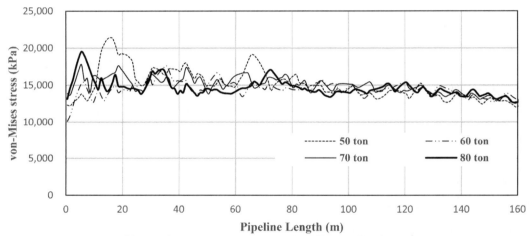

(b) von-Mises stress corresponding to the different pulling forces

Figure 4. Effect of pulling force.

force for above 60 ton offered a margin for safety at 1.5 for the value of 75 m, i.e. 30 times pipe OD.

During installation in the field, the pipe was charged with compressed air at 1.5 bar to reduce the stress in the pipe. The results of analysis for 1.5 bar internal pressure in the pipe showed that the von-Mises stress could be reduced to as low as 20 MPa without larger pulling force, as illustrated in Fig. 5. The reason is that the increased internal overpressure reduced the ovality of the HDPE pipe. The resistance against buckling by applying an internal overpressure further increased the resistance against buckling more efficiently than the loss in resistance by reducing the pulling force from 80 ton to 50 ton. Interestingly, curvature and bend radius were unaffected as much as von-Mises stress.

Other than that, the pulling force of 50 ton was maintained to provide tension for controlling continuous sinking in the field. For that purpose, two water pumps were set in the barge to deliver water into the pipe being charged with compressed air 1.5 bar.

4 CONCLUDING REMARKS

The HDPE pipeline in the marine environment provides numerous advantages, such as immunity to galvanic corrosion, light weight, and flexibility. However, sinking the HDPE pipeline safely into the seabed without accident requires rigorous installation analysis and sensitive control to prevent abrupt change in the pipeline. Hence, some insights gained from the project of Ras Djinet Combined Cycle Power Plant in Algeria are listed in the following.

The weighting of pipe was the most important factor for the pipe to resist any movement against wave and current. The weighting block was generally constructed with the concrete ranging from 25 to 50% of the pipe displacement. Besides, 40% weighting block of the pipe displacement was designed, but it was reduced to 32% after considering the trench effect.

The lifting of the pipe, apart from appropriate distance in the inmost region, prevented the pipe from

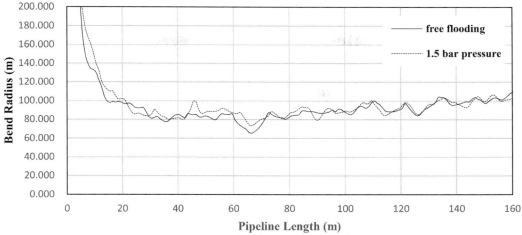

(a) Bend radius in the case of free flooding and 1.5 bar compressed air

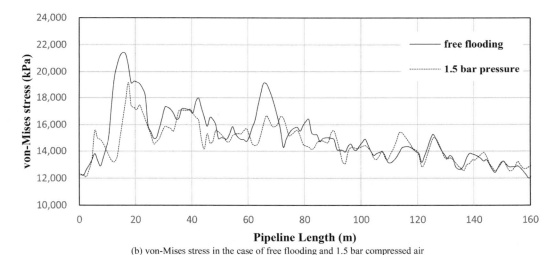

(b) von-Mises stress in the case of free flooding and 1.5 bar compressed air

Figure 5. Effect of internal pressure with pulling force 50 ton.

changing its shape abruptly at the moment of touching the bottom and in making S-shape. Pipe lifting of 60 ton for the 60 m apart from the pipe inmost end was applied for the project.

The pulling force that gave tension to the pipe affected the bend radius and the von-Mises stress in the pipe during installation, specifically the inmost area of the pipeline. Nonetheless, larger pulling force offered a tendency for the pipe to experience lesser von-Mises stress, but internal overpressure could replace the larger pulling force with smaller pulling force, which could save one from charter fee for a barge with larger equipment for pulling force.

ACKNOWLEDGEMENTS

The authors would like to thank Daewoo E&C, POSTECH, and POSCO in the Republic of Korea for the great support. This research was supported by the Technology Innovation Program (Grant No.: 10053121) funded by the Ministry of Trade, Industry & Energy (MI, Korea).

REFERENCES

AWWA. 2006. *M55 PE pipe—design and installation*. Denver, USA: American Water Works Association (AWWA).

Bergström, J. S. & Bischoff, J. E. 2010. An Advanced Thermomechanical Constitutive Model for UHMWPE. *International Journal of Structural Changes in Solids*, 2(1), 31–39.

Bergström, J. S., Rimnac, C. M. and Kurtz, S. M. 2004. An augmented hybrid constitutive model for simulation of unloading and cyclic loading behavior of conventional and highly cross linked UHMWPE. *Biomaterials*, 25(11), 2171–2178.

Chen, Y., Davalos, J. F., Ray, I. & Kim, H. Y. 2007. Accelerated Aging Tests for Evaluations of Durability Performance of FRP Reinforcing Bars for Concrete Structures. *Composite Structures*, 78(1), 101–111.

El-Sawy, K. M. 2001. Inelastic Stability of Tightly Fitted Cylindrical Liners Subjected to External Uniform Pressure. *Thin-Walled Structures*, 39(9), 731–744.

Glock, D. 1997. Behaviour of Liners for Rigid Pipeline Under External Water Pressure and Thermal Expansion. Der Stahlbau, 212–217.

Guermazi, N., Elleuch, K. & Ayedi, H. F. 2009. The Effect of Time and Aging Temperature on Structural And Mechanical Properties of Pipeline Coating. *Material and Design*, 30(6), 2006–2010.

Ian Larsen. 2013. Design report for Ras Djinet seawater intake system.

Idris, K. 2008. Calculation of Concrete Ballast Requirement for Subsea HDPE Pipeline. *Journal Infrastructure and Built Environment*, 4(2), 160–166.

Jacobsen, S. 1974. Buckling of Circular Rings and Cylindrical Tubes Under External Pressure. *Water Power*, 26, 400–407.

Kang, J. S., Han, T. H., Kang, Y. J. & Yoo, C. H. 2009. Short-Term and Long-Term Behaviors of Buried Corrugated High-Density Polyethylene (HDPE) Pipes. *Composites Part B: Engineering*, 40(5), 404–412.

Kim, Y. 2015. Installation and Design of Large Diameter of HDPE Pipeline for Intake, *The 10th Pipeline Technology Conference 2015*, 8–10 June, Berlin, Germany.

Mohd Ishak, Z. A., Ishiaku, U. S. & Karger-Kocsis, J. 2000. Hygrothermal Aging and Fracture Behavior of Short-Glass-Fiber-Reinforced Rubber-Toughened Poly (Butylenes Terephthalate) Composites. *Composite Science and Technology*, 60(6), 803–815.

Pegoretti, A. & Migliaresi, C. 2002. Effect of Hydrothermal Aging on the Thermomechanical Properties of a Composite Dental Prosthetic Material. *Polymer Composites*, 23(3), 342–351.

RPS. 2004. *Rinker Pipeline Systems*. New Orleans, LA, USA.

Rueda, F., Otegui, J. L. & Frontini, P. M. 2011. Numerical Tool to Model Collapse of Polymeric Liners in Pipelines. *Engineering Failure Analysis*, 20, 25–34.

Rueda, F. Torres, J. P., Machado, M., Frontini, P. M. & Otegui, J. L. 2015. External Pressure Induced Buckling Collapse of High Density Polyethylene (HDPE) Liners: FEM Modeling and Predictions. *Thin-Walled Structures*, 96, 56–63.

Song, H. B., Kim, D. K., Choi, H. S. & Park, K. S. 2015. A Study on Structural Stability of HDPE Pipe During Installation. *Journal of Korean Society for Advanced Composite Structures*, 6(1), 59–66.

Dynamic behaviour of tension leg platform in short-crested directional seas

Damilola O. Oyejobi, Mohammed Jameel & Nor H. Ramli Sulong
Department of Civil Engineering, University of Malaya, Kuala Lumpur, Malaysia

ABSTRACT: The representation of the sea condition has great influence on the behaviour of the Tension Leg Platform, (TLP). A real sea condition is a random directional field that varies both in space and time. The aim of this study is to compare responses of the TLP when the ocean surface is simulated as either unidirectional or directional sea. A stochastic sea surface elevation was assumed and simulated using linear wave superposition method for both unidirectional and directional seas. Joint North Sea Wave Project, (JONSWAP) spectrum was used to simulate wave frequency and quantify the energy density. A numerical code was developed in FORTRAN software for solving non-linear uncoupled dynamic analysis. The nonlinearity considered includes relative velocity square drag force in Morison, large displacement, variable submergence and tension fluctuation in the tendon. It was inferred that unidirectional sea simulation overestimates motion responses when compared to directional sea.

1 INTRODUCTION

Tension Leg Platform (TLP) is an excellent station keeping platform due to its high stiffness in vertical direction and compliance in horizontal direction, making it suitable for harsh environmental loads and to avoid exciting wave frequency. Various authors such as (Chandrasekaran and Jain, 2002a), (Kurian et al., 2008), (Ahmad, 1996), (Masciola and Nahon, 2008), (Chen et al., 2013), (Kim et al., 2007), (Chandrasekaran et al., 2004) had either simulated TLP to ocean seas using regular wave or long crested waves. The reported motion responses differ due to the various degrees of nonlinearity considered; different wave simulation method and varied numerical method in their respective studies.

More recently, (Ng et al., 2014) examined responses of classic SPAR to short and long crested wave generation experimentally. It was reported that responses from long-crested simulation were found 35 percent higher than the short-crested sea. Jameel et al. (2016b) carried out dynamic analysis for both intact and a missing tendon TLP in eight different sea states and under different load combinations. In another development, Jameel et al. (2016a) reported effect of sea states with high significant wave heights and periods over low sea states, current and wind loads on the dynamic responses of TLP in random seas. (Liu, 2014) carried out parametric study of TLP under varying incident wave angle, wave height and wave period.

Pascal and Bryden (2011) employed single summation method to generate and measure directional wave spectra in multi-directional wave tank. Their deterministic waves were characterized statistically. Sannasiraj et al. (2001) modelled directional sea by linear superposition of long-crested waves of possible frequencies and directions. This was carried out for analysis of multiple floating structures using diffraction-radiation approach. Hsu et al. (2010) presented solutions for 3D short-crested waves by formulating the problems using lagrangian coordinates Analytical method for estimation of wave-induced current was presented by Myrhaug and Holmedal (2014). It was gathered for both 2D and 3D nonlinear random waves that wave induced current is larger than linear waves and is a function of increase in wave steepness.

There is limited open literature on the motion response of the TLP in directional seas. Since, simulating TLP to unidirectional sea may be uneconomical and its motion response is questionable due to the high motion responses, this paper intends to address the following objectives:

a) To simulate random sea waves from unidirectional and directional seas wave spectrum.
b) To compute TLP motion responses for unidirectional and directional seas
c) To demonstrate effects of different load cases on the TLP response.

2 MATHEMATICAL FORMULATION

The TLP governing equation of motion as represented in Equation 1 was solved in time-domain using numerical method.

$$M\ddot{x} + C\dot{x} + Kx = Q(t, x, \dot{x}) \qquad (1)$$

The matrices [M], [C], [K] were formulated and programmed in Formulation Translation (FORTRAN) software after the approach earlier used by Ahmad (1996) and Jain (1997). In the formulation of equation of motion for uncoupled analysis, tendons are modelled as linear springs and the effects of tendon weight; buoyancy force; current load and damping forces on the tendons were ignored.

2.1 Simulation of random wave and wave kinematics

The ocean surface was assumed to be a random field that is stationary in time and homogeneous in space. Since wave generates free surface motion, superposition of the regular waves with Gaussian random variable was used to create random waves. The wave profile for unidirectional sea (Equation 2) and directional sea (Equation 4) were formulated and coded in FORTRAN software.

$$\eta(x,t) = \sum_{i=1}^{k} A_i \cos(k_i x - w_i t + \emptyset_i) \quad (2)$$

$$A_i = \sqrt{2 S_{\eta\eta}(w_i) \Delta w_i} \quad (3)$$

$$\eta(x,y,t) = \sum_{i=1}^{k} A_i \cos[k_i(x \cos\theta_i + y \sin\theta_i) - w_i t + \emptyset_i] \quad (4)$$

$$A_i = \sqrt{2 S_{\eta\eta}(w_i) D(W,\theta) \Delta w_i \Delta \theta_i} \quad (5)$$

The required wave spectrum, $S_{\eta\eta}(w)$ to compute the energy content and its distribution over the frequency range is given by 5-parameters JONSWAP spectrum in Equation 6, DNV-RP-C205 (2007)

$$S_J(w) = A_\gamma S_{PM}(w) \gamma^{exp\left[-0.5\left\{\frac{w-w_p}{\sigma w_p}\right\}^2\right]} \quad (6)$$

$$S_{PM}(w) = \frac{5}{16} \cdot H_s^2 w_p^4 \cdot w^{-5} \left\{-\frac{5}{4}\left(\frac{w}{w_p}\right)^{-4}\right\} \quad (7)$$

The JONSWAP spectrum used was a modification of Pierson Moskowitz spectrum for developing sea state with peak shape parameter of $\gamma = 3.3$. The directional function, $D(W,\theta)$ for wind sea in Equation 5 was simulated with the approximation that $D(W,\theta) = D(\theta)$, the frequency dependence was neglected.

$$D(\theta) = \frac{\Gamma(1+n/2)}{\sqrt{\pi}\Gamma(1/2+n/2)} \cos^n(\theta - \theta_p) \quad (8)$$

Γ is the Gamma function $|\theta - \theta_p \leq \frac{\pi}{2}|$
θ_p as the main wave direction, DNV-RP-C205 (2007).

The individual wave frequencies and their bandwidths in Equations (2–5) were simulated from the approach of (Chandrasekaran and Jain, 2002b) and (Ahmad, 1996). This follows with calculation of water velocities and accelerations using linearized small amplitude wave theory together with Chakrabarti stretching modification in order to include free surface effect in Equations 9–12.

$$\dot{u}(x,t) = \sum_{i=1}^{k} A_i w_i \cos(k_i x - w_i t + \phi_i) \frac{\cosh k_i z}{\sinh(k_i(d+\eta))} \quad (9)$$

$$\dot{v}(x,t) = \sum_{i=1}^{K} A_i w_i \sin(k_i x - w_i t + \phi_i) \frac{\sinh k_i z}{\sinh(k_i(d+\eta))} \quad (10)$$

$$\ddot{u}(x,t) = \sum_{i=1}^{K} A_i w_i^2 \sin(k_i x - w_i t + \phi_i) \frac{\cosh k_i z}{\sinh(k_i(d+\eta))} \quad (11)$$

$$\ddot{v}(x,t) = -\sum_{i=1}^{K} A_i w_i^2 \cos(k_i x - w_i t + \phi_i) \frac{\sinh k_i z}{\sinh(k_i(d+\eta))} \quad (12)$$

The wave profile time history was simulated for a period of 5000 seconds with 0.2 secs time interval. The wave profile time history and their respective spectrum are shown in Figures 1 and 2 accordingly.

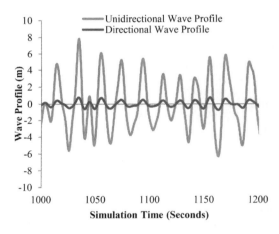

Figure 1. Cross-section of wave profile time history.

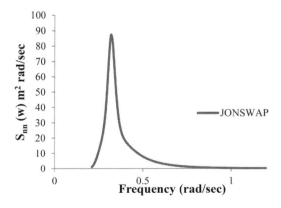

Figure 2. Simulated JONSWAP wave spectrum.

2.2 Simulation of wave, current and wind forces

The modified Morison equation was used to compute drag force that is proportional to the square of the water velocity and inertia force that is proportional to the acceleration, added mass effects was also considered as the third term in Equation 13.

$$f(x,z,t) = 0.5\rho_w C_D D |\dot{u} - \dot{x} + \dot{U}_c|(\dot{u} - \dot{x} + \dot{U}_c) + 0.25\pi D^2 \rho_w C_M \ddot{u} \pm 0.25\pi D^2 [C_M - 1]\rho_w \ddot{x} \quad (13)$$

The parameter $\dot{U}_c$ in the first component of Equation 13 is the wind-drift current velocity and it is taking to be horizontal with the water depth in the form of

$$\dot{U}_c = U_{oW}\left(\frac{z}{d}\right) \quad (14)$$

Where, $\dot{U}_c$ is the current velocity at the level 'z', $z \leq 0$; z is the distance from the Still Water Level (SWL) and positive upwards; U_{oW} is the wind generated current at the SWL; d is the water depth to the SWL.

The mean wind speed and fluctuating wind was simulated from logarithmic profile model, and Simiu and Leigh spectrum according to (DNV-RP-C205, 2007).

3 NUMERICAL STUDY

In order to demonstrate the developed program to calculating the TLP behaviours, International Ship and Offshore Structures Congress (ISSC) TLP that had been widely used for research was adopted, its geometrical and mechanical data can be found in Taylor and Jefferys (1986) and reported in Table 5. The equation of motion was solved by efficient Newmark-Beta numerical method due to its stability and accuracy.

3.1 Results and discussions

In this study, a sea state that corresponds to Gulf of Mexico was considered, the sea state parameters with significant wave height, $H_s = 12$ metres and wave period, $T_p = 14$ seconds were taken from (Jayalekshmi et al., 2010), with angle of wave incidence of zero degree. The current velocity was taken to be 1.05 m/s near the surface and 0.10 at the base of the TLP. Also, wind velocity of 24.38 m/s at reference height of 10 m was used for the projected area of 2579 m². The following load case studies of random wave only; random wave and current, and simultaneous action of wave, current and wind were analysed for uncoupled analysis of TLP.

3.2 Effect of directional short crested wave spectrum on the wave amplitude

The directionality function takes full effect by examining wave profiles as calculated using Equations 2 and 4 respectively. This effect was demonstrated in time history of wave profile shown in Figure 1 and absence of sea direction function made the amplitudes of unidirectional profile to be higher and this invariably affects the water wave kinematics of Equations (9–12) that were used for force computation in Equation 13. The effect of directional short crested wave spectrum undoubtedly predicts the motion responses of the TLP correctly. The sea direction function does not influence the energy density of unidirectional and directional sea of the spectrum due to the fact that energy density distribution over the frequency band was the same. The energy density peak value of JONSWAP spectrum in Figure 2 occurs at ($w_p = 0.3180$ rad/sec and $S_{\eta\eta}(w_p) = 87.27$ m² rad/sec).

3.3 Motion responses of the TLP

In order to capture the effect of directional seas, TLP is simulated to the action of random wave only; random wave and current; and joint action of wave, current and wind.

3.3.1 Surge

The platform oscillation in random wave only was seen to be around the mean position as shown in Figure 3 for unidirectional and directional sea cases. The magnitude of the surge response in unidirectional waves was found higher to directional waves. The maximum, minimum and Root Mean Square (RMS) values ratio of unidirectional seas to directional sea simulation were (1.3: 1); (2.4: 1); and (2.2: 1) respectively. In contrast, when current load was added to random wave, directional sea maximum and RMS motion values were found higher to unidirectional sea. The same trend of increase was observed for combined action of wave, current and wind forces. The statistical response for load combination considered is given in Table 1. This phenomenon can be attributed to the fact that current and wind drag loads increase the motion response along surge direction; lack of wave directionality and the higher wave amplitude suppresses effects of current and wind drag loads. Table 1 shows the statistics of unidirectional and directional sea simulation for the

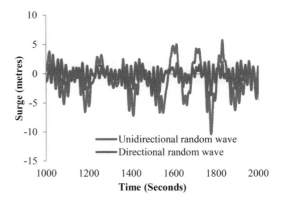

Figure 3. Comparative surge time history for random wave only.

Table 1. Statistics of TLP surge responses.

Surge (M)	A	B	C	D	E	F
Max	5.78	4.55	7.12	10.82	12.28	13.93
Min	−10.38	−4.25	−2.95	−2.84	–	–
Mean	−1.32	−0.13	2.16	3.15	6.74	7.27
RMS	2.77	1.28	2.67	3.77	6.94	7.49

A = Response in unidirectional random wave
B = Response in directional random wave
C = Response in unidirectional random wave and current
D = Response in directional random wave and current
E = Response in unidirectional random wave, current and wind
F = Response in directional random wave, current and wind

Table 2. Statistics of TLP heave responses.

Heave (M)	A	B	C	D	E	F
Max	0.13	0.03	0.02	0.01	–	–
Min	−0.05	−0.03	−0.07	−0.15	−0.19	−0.24
Mean	0.01	–	−0.01	−0.02	−0.06	−0.07
RMS	0.02	0.01	0.01	0.03	0.07	0.08

Table 3. Statistics of TLP pitch responses.

Pitch (0^0)	A	B	C	D	E	F
Max	0.06	0.05	0.05	0.05	0.05	0.04
Mini	−0.05	−0.06	−0.05	−0.07	−0.05	−0.07
Mean	–	–	–	–	–	–
RMS	0.02	0.02	0.02	0.02	0.02	0.02

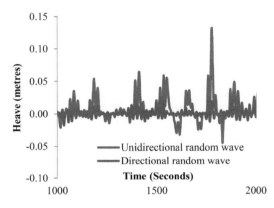

Figure 4. Comparative heave time history for random wave only.

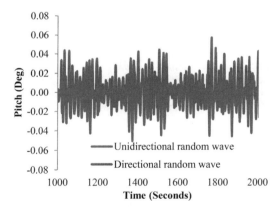

Figure 5. Comparative pitch time history for random wave only.

possible load combinations expected in an ideal ocean environment.

3.3.2 Heave and pitch

Responses from directional sea simulation for heave and pitch were found lower as compared to unidirectional sea, this was shown in Figures 4 and 5 respectively. The ratio of maximum, minimum and RMS values of unidirectional sea to directional sea was calculated as (4.6: 1); (1.49: 1), and (3.3: 1) for load case of random wave only.

There was a gradual decrease in maximum values and increase in minimum values when current, and wind loads were added to random wave, see Table 2. The RMS value of directional sea simulation was found lower in random wave load case and higher under the action of random wave and current and simultaneous occurrence of the three loads. The Pitch maximum and minimum remains almost the same and RMS remains the same in all load cases, Table 3.

This observed trend can be attributed to the influence of sea direction function and drag components of current and wind forces that increases the responses in negative heave direction. Judging by the magnitude of heave response, the value was small due to the high axial stiffness of tendon that restrains the vertical motion.

3.3.3 Tension

The effect of sea state representation was more pronounced in the tension fluctuation. This was due to wider tension fluctuation when the sea state was idealized as unidirectional sea as compared to directional sea simulation. The tension statistical and time history values were represented in Table 4 and Figure 6 respectively. The tension difference between the maximum and minimum values under the action of random wave only; random wave and current, and concurrent action of wave, current and wind were (7.34E7, 2.17E7) Newton, (3.94E7, 1.49E7) Newton, (3.79E7, 1.53E7) Newton for unidirectional and directional seas respectively. Although the RMS for both unidirectional and directional seas and for all load cases were almost the same as seen in the Table 4, the minimum and maximum tension values are needed for design purpose as this will affect the motion response and strengths of

Table 4. Statistics of TLP tension responses ($\times 10^7$).

Tension (N)	A	B	C	D	E	F
Max	9.56	5.57	6.16	4.81	5.95	4.80
Min	2.22	3.38	2.22	3.32	2.16	3.27
Mean	4.12	3.78	3.72	3.72	3.69	3.70
RMS	4.20	3.79	3.77	3.73	3.74	3.70

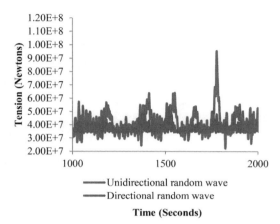

Figure 6. Comparative tension time history for random wave only.

Table 5. Geometrical and Mechanical data, (Taylor and Jefferys (1986)).

Parameters	Values
Column spacing	86.25 m
Column diameter	16.88 m
Pontoon width	7.50 m
Pontoon height	10.50 m
Draft	35.00 m
Displacement	5.346×10^5 kN
Weight	3.973×10^5 kN
Total tether pretension	1.373×10^5 kN
Longitudinal metacentric height	6.0 m
Transverse metacentric height	6.0 m
Platform mass	40.5×10^3 t
Roll mass moment of Inertia	82.37×10^6 t m^2
Pitch mass moment of Inertia	82.37×10^6 t m^2
Yaw mass moment of Inertia	98.07×10^6 t m^2
Vertical position of COG above Keel	38.0 m
Length of Mooring tethers	415.0 m
Vertical stiffness of combined tethers	0.813×10^6 kN/m
Roll and Pitch effective stiffness	1.501×10^9 kNm/rad

the tendon material. The unidirectional sea response when all the loads act together recorded the higher maximum and RMS tension values as compared to the directional sea, this made the fluctuation higher in contrast to directional sea.

4 CONCLUSIONS

The following submissions were made from the dynamic analysis of TLP in directional seas that was carried out:

1. Idealization of ocean surface as unidirectional sea overestimate the wave kinematics used for force computation and this subsequently increased the motion responses.
2. Lack of wave directionality function in long crested seas suppresses the actions of current and wind drag forces when acted with random waves but wave directionality function and drag forces increases motion responses accordingly in short-crested seas.
3. The fluctuation between the maximum and minimum tension values of unidirectional sea simulation was higher as compared to the directional sea.
4. The combined effects of random wave, current and wind load gives the response of an ideal environment for the TLP behaviour.
5. The developed numerical code runs the analysis within few minutes.

ACKNOWLEDGMENTS

This research is funded by the University of Malaya (UM) under the Grant No. RP004E-13AET. It is gratefully acknowledged.

REFERENCES

Ahmad, S. 1996. Stochastic TLP response under long crested random sea. *Computers & Structures*, 61, 975–993.

Chandrasekaran, S. & Jain, A. K. 2002a. Dynamic behaviour of square and triangular offshore tension leg platforms under regular wave loads. *Ocean Engineering*, 29, 279–313.

Chandrasekaran, S. & Jain, A. K. 2002b. Triangular Configuration Tension Leg Platform behaviour under random sea wave loads. *Ocean Engineering*, 29, 1895–1928.

Chandrasekaran, S., Jain, A. K. & Chandak, N. R. 2004. Influence of Hydrodynamic Coefficients in the Response Behavior of Triangular TLPs in Regular Waves. *Ocean Engineering* 31, 2319–2342.

Chen, J. M., Kong, X. Y. & Sun, Y. 2013. Hydrodynamic Responses Analysis for Tension Leg Platform. *In:* Liu, X. H., Zhang, K. F. & Li, M. Z. (eds.) *Manufacturing Process and Equipment, Pts 1–4*. Stafa-Zurich: Trans Tech Publications Ltd.

DNV-RP-C205 2007. Environmental conditions and environmental loads, Recommended Practice. *In:* NO -1322 HOVIK, N. (ed.).

Hsu, H.-C., Chen, Y.-Y. & Wang, C.-F. 2010. Perturbation analysis of short-crested waves in Lagrangian coordinates. *Nonlinear Analysis: Real World Applications*, 11, 1522–1536.

Jain, A. K. 1997. Nonlinear coupled response of offshore tension leg platforms to regular wave forces. *Ocean Engineering*, 24, 577–592.

Jameel, M., Oyejobi, D. O. & Ramli Sulong, N. H. 2016a. Nonlinear Response of Tension Leg Platform subjected to

Wave, Current and Wind Forces. *International Journal of Civil Engineering*, Accepted.

Jameel, M., Oyejobi, D. O., Siddiqui, N. A. & Ramli Sulong, N. H. 2016b. Nonlinear Dynamic Response of Tension Leg Platform under Environmental Loads. *KSCE Journal of Civil Engineering*, Accepted.

Jayalekshmi, R., Sundaravadivelu, R. & Idichandy, V. G. 2010. Dynamic Analysis of Deep Water Tension Leg Platforms Under Random Waves. *Journal of Offshore Mechanics and Arctic Engineering-Transactions of the Asme*, 132, 4.

Kim, C. H., Lee, C. H. & Goo, J. S. 2007. A dynamic response analysis of tension leg platforms including hydrodynamic interaction in regular waves. *Ocean Engineering*, 34, 1680–1689.

Kurian, V. J., Gasim, M. A., Narayanan, S. P. & Kalaikumar, V. Parametric studies subjected to random waves. ICCBT 2008, 2008 Kuala Lumpur, Malaysia. 2130–0222.

Liu, W. M. 2014. Research on Time-domain Dynamic Response of Tension Leg Platform in Regular Wave. *Applied Mechanics, Materials and Manufacturing Iv*, 670–671, 801–804.

Masciola, M. & Nahon, M. 2008. Modeling and Simulation of a Tension Leg Platform. *Proceedings of the Eighteenth (2008) International Offshore and Polar Engineering Conference, Vol 1*, 84–91.

Myrhaug, D. & Holmedal, L. E. 2014. Wave-induced current for long-crested and short-crested random waves. *Ocean Engineering*, 81, 105–110.

Ng, C. Y., Kurian, V. J. & Liew, M. S. 2014. Dynamic responses of classic spar platform: Short crested waves vs long crested waves. *Applied Mechanics and Materials* 567, 235–240.

Pascal, R. & Bryden, I. 2011. Directional spectrum methods for deterministic waves. *Ocean Engineering*, 38, 1382–1396.

Sannasiraj, S., Sundaravadivelu, R. & Sundar, V. 2001. Diffraction–radiation of multiple floating structures in directional waves. *Ocean Engineering*, 28, 201–234.

Taylor, R. E. & Jefferys, E. R. 1986. Variability of Hydrodynamic Load Predictions for a Tension Leg Platform. *Ocean Engineering*, 13, 449–490.

A simplified prediction method for subsea pipeline expansion

J.H. Seo & H.S. Choi
Graduate School of Engineering Mastership, Pohang University of Science and Technology, Pohang, Republic of Korea

K.S. Park
Steel Business Division, POSCO Center, Seoul, Republic of Korea

S.Y. Yu & D.K. Kim
Department of Civil and Environmental Engineering, Universiti Teknologi PETRONAS, Seri Iskandar, Perak, Malaysia

ABSTRACT: In the present study, the estimation method of expansion amount for subsea pipeline is proposed, which could be applied to design a robust structure for the production of oil and gas from the offshore well. This study begins with a general discussion of estimation method for subsea pipeline in terms of previously developed terminologies by literature review. The production fluid from offshore well is normally caused by physical deformation of subsea structures, e.g., expansion and contraction, during the process of transportation due to the effect of high pressure and high temperature. In severe cases, vertical and lateral buckling occurs and it causes the significant negative impact on the structural safety which is associated with on-bottom stability, free-span, structural collapse and many others. In addition, it may give effect on the production rate regarding to flow assurance, wax, hydrate, and many others. Therefore, the applicable diagram with efficient procedure, called standard dimensionless ration (SDR) versus virtual anchor length (L_A) diagram, is proposed for the estimation of subsea pipeline expansion based on applied reliable scenarios. Based on user guideline, the offshore pipeline structural designer could assume reliable amount of subsea pipeline expansion and the obtained results will be useful for the installation, design, and maintenance of subsea pipeline, as well.

NOMENCLATURES

F_t Force due to the temperature
F_p Force due to the pressure
F_v Force due to Poisson contraction
F_f Force due to soil frictional resistance
μ Soil friction coefficient of subsea pipe
W_s Submerged weight of pipe (per unit length)
α Coefficient of thermal expansion
E Young's modulus
A_s Steel section area of pipe
A_i Internal fluid flow area of pipe
T Temperature
P Pressure
N Poisson's ratio
σ_h Hoop stress
ε Strain of pipeline in partially restrained zone
ρ_f Fluid density
OD Pipe outside diameter (without coating)
t Pipe wall thickness
N Residual lay tension
SDR Standard dimension ration (=D/t)
L_a Virtual anchor length

1 INTRODUCTION

For the deepwater development for oil and gas production, it is well recognized that the effect of severe environmental conditions, i.e., high pressure, and high temperature (HPHT) together, have caused various difficulties. Recently, several researches have been performed to design robust offshore and subsea structures against abovementioned several problems. In case of HPHT conditions, it naturally brings thermal expansion phenomenon for structures such as pipelines, subsea systems, topside facilities and many others. Therefore, careful design and its confirmation for offshore oil and gas production structures should be accompanied by engineers. Especially when subsea structures have difficulties to repair and/or replace when unexpected problems occur during the design stage, of course those things will cost a lot compared to onshore or near shore cases.

In order to overcome these HPHT conditions related problems of subsea pipeline, i.e., thermal expansion, global & local buckling, etc., the engineer should consider the amount of thermal expansion. Regarding

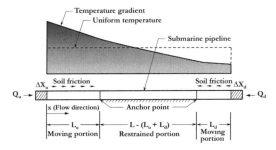

Figure 1. Overview of subsea pipeline expansion (Choi et al. 2008).

this, numerous studies related to expansion of pipeline have been performed in terms of single wall case (Parmer and Ling 1981; AGA 1987; Hobbs et al. 1989; Choi 1995), pipe-in-pipe (PIP) case (Choi 2002; Kershenbaum et al. 1996; Harrison et al. 1997), and many others may also be referred. Especially, Bai and Bai (2014) had widely reviewed and summarized the method for thermal expansion design of subsea pipeline not long ago. Furthermore, new concept for absorbing the pipeline expansion has also been proposed recently by DNV (2013).

The estimation of subsea pipeline expansion is calculated by using complex equations and diverse variables, so there are existing developed equations based on their experience, but it could be simplified for structural design at the initial engineering stage (Pre-FEED and FEED, Front End Engineering Design). In this research, a simple and reliable diagram will be proposed, which is shown as exponential equation, which easily calculates the pipe expansion length by changing some kinds of input values. The insights developed from this study will be useful for engineers to design and install the subsea pipeline.

2 METHOD FOR EXPANSION ANALYSIS

A pipeline is usually installed at the same temperature as the surrounding environment such as seawater temperature. However, during operation, inlet temperature dramatically increases and hence, the temperature of the whole line also changes. At the same time, the corresponding pipeline is pressurized as well.

Normally, the terminologies presented in Fig. 1 are widely used in the expansion analysis of subsea pipeline (Choi et al. 2008). Free expansion analysis case is defined when the pipe end restraining forces are zero and the pipe strain at the end is of course not zero. Anchor point is the fixed points presented in Fig. 1, when the pipe movement due to the expansion is stopped. The portion from the free ends to the anchor points is defined as the unrestrained zone. Lastly, fully restrained zone is the portion between the anchor points.

In general, five types of conditions, i.e., free expansion with uniform temperature, free expansion with temperature gradient, expansion with end restraints,

expansion of pipe-in-pipe system, and lateral deviation, so called snaking, are considered to estimate the expansion of subsea pipeline and it has been widely reviewed by Choi et al. (2008) and Bai and Bai (2014).

2.1 Free expansion with uniform temperature

The pipeline expansion has been analyzed with the assumption of free expansion with uniform temperature. This means that zero restraining forces are assumed at the pipe end but they also consider soil friction along the pipe.

In this case, Q_u and Q_d are zero as shown in Fig. 1 and this analysis is called free expansion problem and also the temperature difference between the product and environment is assumed to be constant along the pipeline. The anchor point in Fig. 1 is defined as the two points where the movement is zero. The anchor length, at each end side, i.e., L_u and L_d, is called as moving portion which is the space from the pipe end to the anchor point presented in Fig. 1 and this two value will be similar due to the constant temperature. Finally, in-between portion of anchor point is defined as fully restrained zone.

Force equilibrium in the pipeline can be expressed as following Eq. (1). The virtual anchor length can be derived once Eq. (1) is solved as shown in Eq. (2) (Choi et al. 2008). The expansion of the pipeline with the free end condition can be calculated by Eq. (3).

$$F_f = F_t + F_p + F_v \quad (1)$$

$$L_a = \frac{1}{\mu W_s}(\alpha E A_s \Delta T + A_i \Delta p - \nu A_s \sigma_h) \quad (2)$$

$$\Delta x = \int_0^{L_a} \varepsilon \cdot dx = \left\{ \alpha \Delta T + \frac{\sigma_h}{2E}(1-2\nu) - \frac{\mu W_s L_a}{2 A_s E} \right\} \cdot L_a \quad (3)$$

More advanced pipeline expansion calculation method, i.e., free expansion with temperature gradient and expansion with temperature gradient can be referred to Choi et al. (2008).

3 APPLIED EXAMPLE

Based on abovementioned procedures 2.1, application studies have been performed by reliable scenarios. Finally, standard dimensionless ratio (SDR) versus virtual anchor length (L_a) diagram is proposed in this section.

3.1 Define the design inputs

3.1.1 Structural characteristics

In general, the outer diameter (OD) of 8 to 14 inch rages are adopted for flowline which is subjected to HPHT conditions. The wall thickness (t) ranges are defined based on the on-bottom stability analysis (DNV-RP-F109) and the long pipe line case ($L_a < L_{total}/2$) under the structural specification and

Table 1. Structural data.

Type	Data
Pipe outer diameter (inch)	8, 10, 12, 14
Pipe wall thickness (mm)	15–28
Density of steel pipe (kg/m^3)	7850
Poisson's ratio	0.3
Young's modulus (GPa)	207
Thermal expansion coefficient (1/°C)	1.171×10^{-5}
Total installed pipe length (km)	100

Note: API-X65 and 70 are considered as structural data.

Table 2. Properties of production fluid and environmental data.

Type	Data
Density of production fluid (kg/m^3)	90
Temperature difference (°C) – 5 cases	30 to 150
Pressure difference (MPa) – 5 cases	10 to 90
Water depth (m)	1000
Seawater density (kg/m^3)	1025
Friction coefficient of soil	0.6

environmental conditions. Structural characteristics are summarized in Table 1 which cover geometric, mechanical, and material properties.

3.1.2 Environmental parameters

The environmental conditions are assumed to be those of deepwater. Those conditions such as properties of seawater, seabed, properties of production fluid, and others should be defined. Pressure and temperature difference means gap between the production fluid and environments, so some kinds of product and environmental properties are omitted: each condition of temperature and pressure. Table 2 shows properties of production fluid and environmental conditions surrounding subsea pipeline.

3.2 Definition of expansion parameters

The parameters causing expansion behaviour of subsea pipeline can be classified by five components, i.e. Poisson effect, thermal effect, pressure effect, soil friction effect, and residual lay tension effect. In this section, these factors are particularized. In the present study, all processes are derived by strain-based approach the total strain, called net strain, can be obtained by the combination of all parameters according to Eq. (4). Each five parameter can be expressed by Eqs. (5.1) to (5.5).

$$\varepsilon_{net} = \varepsilon_E + \varepsilon_v + \varepsilon_T - \varepsilon_f - \varepsilon_r \quad (4)$$

Strain due to end cap effect (Pressure effect) (5.1)

$$\varepsilon_E = \frac{(p_i - p_e) \cdot A_i}{A_s \cdot E} \cong \frac{(p_i - p_e) \cdot OD}{4t \cdot E}$$

Strain due to Poisson effect (5.2)

$$\varepsilon_v = -v \left[\frac{(p_i - p_e) \cdot OD}{2t \cdot E} \right]$$

Strain due to thermal effect (5.3)

$$\varepsilon_T = \alpha \cdot \Delta T$$

Strain due to residual lay tension (5.4)

$$\varepsilon_r = \frac{N}{A \cdot E}$$

Strain due to mobilization of soil friction (5.5)

$$\varepsilon_f = \frac{\mu \cdot W_s \cdot L_A}{A \cdot E}$$

After the calculation of total strain of subsea pipeline, the process of virtual anchor length (L_A) estimation can be performed. Generally, the virtual anchor length can be obtained by following Eq. (6).

$$L_A = \frac{(\varepsilon_E + \varepsilon_v + \varepsilon_T - \varepsilon_r)}{\mu \cdot W_s} \quad (6)$$

Finally, integrating the total strain gives the pipeline expansion length by following Eq. (7),

$$\Delta L = \int_0^{L_a} \varepsilon_{net} \cdot dL \quad (7)$$

3.3 Calculation of virtual anchor length

In this section, calculation of virtual anchor length (L_A) is covered based on Eq. (6). The whole procedure for developing diagram including required input data and output could be summarized as follows:

- **Step 1**: Definition of design input (Structural and Environmental parameters)
- **Step 2**: Definition of expansion parameters (Poisson ratio, thermal, pressure, soil friction, residual laying tension, etc.)
- **Step 3**: Calculation of expansion length
- **Step 4**: Development of SDR versus L_A diagram
- **Step 5**: Investigation of exponential function indices (α and β).

The calculation of virtual anchor length has been performed through the step 1 to 2 presented in sections 3.1 to 3.2. Figures 2(a) to (c) show the obtained SDR versus L_A diagrams. Although five different pressure cases have been adopted, only three represented case results are in this paper. In addition, each case consists of 5 sub-cases which are applied in different temperature conditions. Fig.2 covers general cases to verify the applicability of the suggested five steps above.

The virtual anchor length calculation results for 10 inch diameter is shown in Fig. 9, and from these results, the empirical formula can be developed by following curves as mentioned in Step 5 and Eq. (8). The obtained empirical formula can be defined as SDR versus anchor length diagram. The linear trend could be found between exponential function indices and pipe diameter based on coefficient of determination (R^2).

$$L_A = \alpha \cdot e^{(D/t)\beta} \qquad (8)$$

Where, α and β can be defined as exponential function indices.

3.4 Estimation of pipeline expansion length by developed diagrams

A new method for estimating virtual anchor length of subsea pipeline has been proposed and investigated for its possibility of applicability to the various diameters and wall thicknesses. In this section, guideline is covered for applying SDR and L_a diagram, and this is illustrated in Fig. 3.

When the structural and environmental characteristics have been defined, two types of exponential function indices (α, β) should be selected based on developed chart as shown in Fig. 2. In the previous section, several charts for selecting exponential function indices have been covered in order to verify the proposed method. In addition, the obtained chart for exponential function indices looks good enough to apply the estimation of expansion of subsea pipeline. Nevertheless, it has limitations regarding estimation of each index at the same time.

In order to determine the two types of exponential function indices, new specific relations between pressure & temperature difference are used to get the appropriate equations. Two different specific relations are as below; Eq. (9) and (10).

$$X_\alpha = (8E^3 \cdot \frac{\Delta P}{MPa} + 9.9E^4)^{\left[\frac{1}{0.385 \cdot (\Delta T/°C)+20}\right]} \qquad (9)$$

$$X_\beta = (1.01 \cdot E^{-3} \cdot \frac{\Delta P}{MPa} + 1)^{\left[\frac{1}{\Delta T/°C}\right]} \qquad (10)$$

Finally, the relation between the results of exponential function indices which has considered total 100 test cases and new specific variables are shown in the Figs. 4 and 5. Once the exponential function indices have been determined, empirical formula (Step IV) could be developed by using Eq. (8) which consists of both exponential function indices (α, β), and SDR. Additionally, indices and variables, which are composed by exponential function indices, are directly related to the pipeline OD and this relation is illustrated in the Appendix. Also, exponential function indices are acquired with OD and specific relation ΔP & ΔT. Through the diagram in Figs. 4 and 5, virtual anchor length equation indices can be estimated.

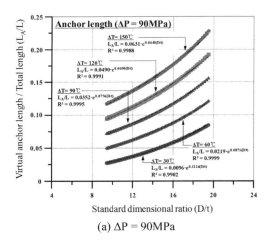

(a) $\Delta P = 90$MPa

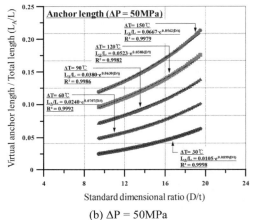

(b) $\Delta P = 50$MPa

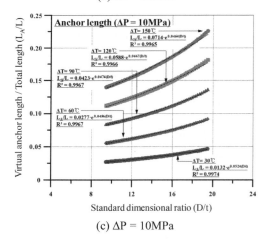

(c) $\Delta P = 10$MPa

Figure 2. Calculation results for the L_a.

4 VERIFICATION OF THE SIMPLIFIED VIRTUAL ANCHOR LENGTH EQUATION

In several cases, diagrams proposed before the steps and simplified exponential equations are verified against the results of conventional calculation by using Eq. (2) and Eq. (3). Expansion length from the new proposed method is determined as exponential function for virtual anchor length and is put in the Eq. (7).

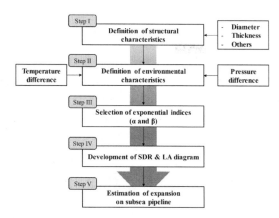

Figure 3. Use of developed diagram.

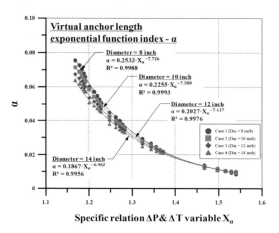

Figure 4. Exponential function index (α).

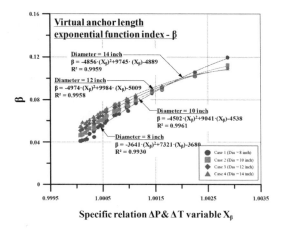

Figure 5. Exponential function index (β).

Selected cases to verify the proposed new method are in Table 3, especially the ones representing compared results are in Fig. 6.

According to Fig. 6, the proposed new method has minor errors when compared with analytical solution.

Table 3. Verification test case data.

Type	Data
Pipeline diameter (inch)	10
Wall thickness (mm)	20, 25, 30
Temperature Difference (°C)	60, 90, 120
Pressure Difference (MPa)	10, 50, 90

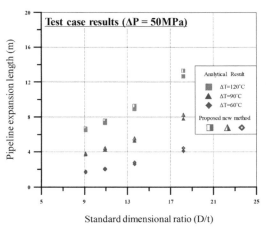

Figure 6. Test case comparison.

Table 4. Test case results and errors.

ΔT (°C)	ΔP (MPa)	D/t	Analytical (m)	New method (m)	Error (%)
60	50	10.92	1.996	2.129	6.638
90	50	10.92	4.246	4.441	4.606
120	50	10.92	7.333	7.593	3.546
90	10	10.92	3.723	3.906	4.919
90	50	10.92	4.246	4.441	4.606
90	90	10.92	4.802	5.010	4.331
90	50	13.65	5.278	5.559	5.326
90	50	10.92	4.246	4.441	4.606
90	50	9.1	3.689	3.840	4.084

To confirm the reason of the errors, the results are summarized by fixing or changing the parameters such as pressure and temperature difference and SDR, and those are in Table. 4.

Regarding Table 4, in certain conditions, that is higher temperature, higher pressure and thicker pipes, the results are more correct.

5 CONCLUDING REMARKS

In this study, useful diagrams have been developed for expansion length of subsea pipeline estimation. The applicability has also been reviewed by applying examples and sensitivity analysis also has been conducted. The purpose of present study is to propose easy, accurate and efficient diagrams for predicting the expansion length of subsea pipeline. In order to draw the expansion length, four representative cases of outer diameter and 25 sub-cases are adopted. When the diameter and difference of temperature & pressure are determined, the exponential function indices (α, β) could be directly calculated. Once the two variables, called exponential indices (α, β) are determined, the expanded length of subsea pipeline is easily obtained by proposing simple exponential equation and the net strain value.

It is concluded that the obtained diagram will be useful for estimating the expansion length of subsea pipeline structure by changing the dimensional inputs under similar conditions. The proposed diagram may also be employed to determine the number of spool or bend systems based on obtained expansion length. It will also be used to develop acceptance criteria for structural buckling behaviour of installed subsea pipeline during design life.

ACKNOWLEDGEMENTS

This research was supported by Korea Evaluation Institute of Industrial Technology (KEIT) grant funded by the Ministry of Trade, Industry & Energy (MI) of Korea (Grant No.: 10053121 and 10051279). In addition, the authors acknowledge for support of Graduate School of Engineering Mastership (GEM), Pohang University of Science and Technology (POSTECH), POSCO and Universiti Teknologi PETRONAS (UTP).

REFERENCES

Choi, H. 1999. Behaviors of high temperature and high pressure marine pipelines. *Journal of Ocean Engineering and Technology*, 2(2): 1–6.

Choi, H., Lee, S. & Chun, E. 2008. A review of the expansion behavior of marine pipelines. *Journal of Ocean Engineering and Technology*, 22 (2): 13–19.

Bai, Q. & Bai, Y. 2014. *Subsea pipeline design, analysis and installation*. Elsevier Science & Technology.

DNV. 2010. On-bottom stability design of submarine pipelines. Recommended Practice. DNV-RP-F109. Oslo: Det Norske Veritas.

DNV. 2012. Submarine pipeline systems. DNV-OS-F101. Oslo: Det Norske Veritas.

Harrison, G.E., Kershenbaum, NY. & Choi, H. 1997. Expansion analysis of subsea Pipe-In-Pipe flowline. *The proceedings of the 7th International Offshore and Polar Engineering Conference*, May 25–30, Honolulu, Hawaii, USA.

Hobbs, R.E. & Liang, F. 1989. Thermal buckling of pipe-lines close to restraints. *The proceedings of the 8th International Conference on Offshore Mechanics and Arctic Engineering*, 4: 121–127.

Palmer, A. & Ling, M.T.S. 1982. Movements of submarine principles close to platforms. *Journal of Energy Resources Technology*, 104(4): 319–324.

APPENDIX

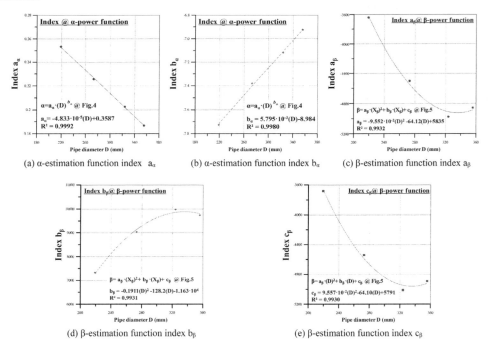

Figure A.1. Index functions at Fig. 4 and Fig. 5.

Tsunami risk assessment: How safe is Malaysia?

D.B. Shahruzzaman & A.M. Hashim
Department of Civil and Environmental Engineering, Universiti Teknologi PETRONAS, Seri Iskandar, Perak, Malaysia

ABSTRACT: The effectiveness of vegetation as coastal protection was evidenced during the Indian Ocean Tsunami 2004. Sufficiency of mangrove forest as natural protection against tsunami along Malaysia's coastline is thus of a great concern. This paper presents the assessment of the overall trend of tsunami occurrence and highlight the level of tsunami exposure for Malaysia. This paper also describes the current status of mangrove forest along west coast of Malaysia in reducing the wave impact. Analysis of global tsunami and earthquake data shows that Malaysia is potentially vulnerable to future tsunami impact. Malaysia had a varying width of mangrove forest distribution which is hoped to protect the coastal area. However, with the increasing trend of threat and decreasing mangrove belt, Malaysia is potentially exposed to such severe wave impacts. Implementation of optimum replanting scheme is inevitable to ensure acceptable level of protection along the coastline of Malaysia.

1 INTRODUCTION

1.1 *Indian Ocean Tsunami*

The occurrence of Indian Ocean Tsunami 2004 (IOT 2004) is beyond the control of mankind and gives impact to both economic and societal factors to affected countries. There were about 13 countries that had been affected by the tsunami, including Indonesia, Sri Lanka, India, Thailand, Somalia, Burma, Maldives, Malaysia, Tanzania, Bangladesh and Kenya (Asmawi & Ibrahim 2013). The incident started when the third largest earthquake (MW 9.1) hit at the epicentre 30 km off the west coast of Northern Sumatra, Indonesia on 26th December 2004 around 07:58:53 local time (00:58:53 GMT) (Koh et al 2012; Asmawi & Ibrahim 2013). The mega thrust earthquake lifted the seabed by 5 m and formed a complex series of waves propagated across the entire Indian Ocean (Cosgrave 2006; Hawkes et al 2007; Athukorala 2012). The wave height exceeded 20 m and contributed to the overall economic losses of approximately USD 9.93 billion, which 75% of damaged attributed from Indonesia, Thailand, Sri Lanka and India. Numerous buildings, mainly along the coastlines were completely destroyed by the impact of the strong wave (Fehr et al 2006). However, less number of damaged had been reported at areas with mangrove buffer as compared to the free-mangrove area (Patel et al 2014).

1.2 *Mangrove forest in Malaysia*

Mangrove trees have unique characteristics to dissipate the wave energy which may result in less impact on the properties at the coastal area (Horstman et al 2014). The effectiveness of vegetation as coastal protection in front of coastal area was evidenced during the Indian Ocean Tsunami on 26th December 2004 (Patel et al 2014). Subsequent from the incident, numerous researches had been undertaken on the capability of mangroves in order to achieve the optimum buffer coastal protection. Although recent technology is able to develop massive breakwaters that are able to protect against 500 to 1000 years return period of the tsunami waves, the cost for such protection can be exorbitantly expensive (Suppasri et al 2012c). It is thus a great concern whether Malaysia has sufficient mangrove forest to protect the coastal properties against tsunami.

2 METHODOLOGY

2.1 *Study area*

This study looks into the global historical tsunami events for the past 50 years. This is followed with the analysis on the trends of potential tsunami generating mechanisms particularly due to earthquake for Malaysia. Overall assessment on the mangrove conditions was then undertaken along the west coast of Peninsular Malaysia with focus on Perak. Perak has the highest contribution of mangrove distribution in Peninsular Malaysia with an area of 43,502 hectares. Matang Mangrove Forest Reserve (MMFR) which is located within Larut Matang district encompasses a total area of approximately 40,535 hectares (97.4%) and was established as a permanent forest reserve in 1906. MMFR is divided into four sub-areas including

North Kuala Sepetang, South Kuala Sepetang, Kuala Trong and Sg Kerang.

2.2 Data analysis

The analysis was carried out in three phases. Firstly, the history of tsunami events occurred globally were reviewed, particularly using available published data. The second part of this paper presents potential tsunami generating events and the possible threats towards Malaysia's coastline. Finally, the varying width of the mangrove forest distribution in Perak was analyzed to deduce the level of natural protection available with reference to the established relationship from recent experimental studies.

3 TSUNAMI EVENT

3.1 Global tsunami event

There were about 15 of tsunami events occurred in the past half-century as illustrate in Table 1. More than 50 % of the tsunamis occurred in the Asean countries (Suppasri et al 2012a). The epicentres of the events were generally situated within the Ring of Fire zone. According to Viroulet et al (2013), the tsunami can be triggered by the earthquakes, volcanic eruption, asteroid impacts and landslides of submarine or sub aerial. However, most of the time, the tsunami events were caused by earthquakes. These earthquakes generated tsunamis were also known as tsunamigenic (Newman et al 2011).

Not all earthquakes will generate tsunami. The tsunamigenic event depends on the earthquake magnitude. Generally, the magnitude below than 6.5 does not trigger tsunami (Whitmore et al 2009). For instance, Ranau, Sabah had experienced a thirty-minute earthquake with MW 6.0 on 5th June 2015. Fortunately, the low magnitude earthquake did not trigger the tsunami event. Nevertheless, there were also a few cases where low magnitude earthquake generated tsunami. However, these earthquakes produced smaller tsunami wave and less destructive to the country (Whitmore et al 2009).

The highest earthquake magnitude of IOT 2004 (MW 9.1; 20 m; Bandar Aceh) did not produce wave height as high as wave height during the tsunami event in 2011 (MW 9.0; 40.5 m; Tohoku) (Fig. 1). The magnitude of the earthquake and the seafloor topography significantly influenced the height of the resulted tsunami (Suppasri et al 2013). According to (Suppasri et al 2012b), the area with large seafloor displacement will induce large amounts of seawater and lead to tsunami generation.

The highest magnitude recorded for the past 50 years was in Sumatra in 2004 (MW 9.1), followed by Tohoku in 2011 (MW 9.0). Although the difference between these two events were only MW 0.1, yet there was a huge difference in the number of the total death toll as shown in Figure 2. Japan ranks

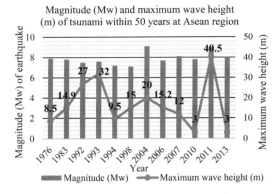

Figure 1. The relationship between magnitude of earthquakes and maximum wave height of the resulted tsunami within 50 years in Asean region.

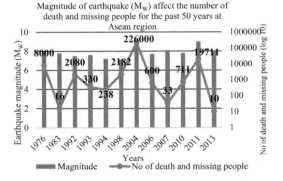

Figure 2. Record of casualties due to earthquake generated tsunami in Asean region for the past 50 years.

as the most affected country by tsunami throughout the world with 48 events up to 2012 (Suppasri et al 2012a). The most tragic tsunami event in Japan was the 3rd November 2011 at Tohoku (Nanto et al 2011). The event was followed by the nuclear explosion from the Fukushima Nuclear Complex, which causes the highest economic losses recorded in Japanese history (Wang & Li 2008; Koyama et al 2012). However, Japan who had experienced more tsunami situation had organized more systematic warning alarm as compared to Indian Ocean countries. The alarm system in Japan which was authorized by Japan Meteorological Agency had introduced a computer-aided simulation system which was used for forecasting of tsunamis (Yamasaki, 2012). There were about 300 seismic sensors placed since 1952 to record the tectonic activities at Japan coast. The residents will be alerted approximately 3–10 minutes earlier prior to the tsunami arrival.

Even though Japan had less number of total death, yet the Tohoku event had contributed to the highest total economic losses recorded in the tsunami history. In Indian Ocean region, the total economic loss was USD 9.93 billion which was 30 times less than the impact of the 2011 tsunami (USD 300 billion). Most countries affected by IOT 2004 had mangrove forest

Table 1. The history of tsunami event occurred within 50 years (*highest value; N/A: Data not available).

Year	Location	Affected Countries	Region	Earthquake Magnitude (M_w)	Max Wave Height (m)	Number Of Losses		References
						Death and Economic (USD)	missing people	
4/2/65	Shemya Island, Alaska	Shemya Island	Pacific Coast	8.7	10.70	10,000	N/A	Wu (1973)
17/8/76	Moro Gulf	Philippines	Celebes Sea	7.9	8.50	N/A	8000	Wiegel & Asce (1980)
12/12/79	Tumaco	Columbia, Ecuador	Pacific Coast	8.0	6.00	N/A	220	Herd et al (1981)
18/5/80	Spirit Lake	Spirit Lake	Pacific Coast	(by Volcanic Eruption)	250	2 B	31	Lander, Whiteside & Lockridge (2003)
26/5/83	Honshu, Japan	Japan, South Korea	Sea Of Japan	7.8	14.93	800 M	16	Lander, Whiteside & Lockridge (2003)
12/12/92	Flores Island, Indonesia	Flores Island	Indian Ocean	7.5	27.00	N/A	2080	Lander, Whiteside & Lockridge (2003)
1/7/93	Hokkaido, Japan	Japan, South Korea, Russia	Sea Of Japan	7.6	32.00	186 M	330	Lander, Whiteside & Lockridge (2003)
2/6/94	Java, Indonesia	Java, Bali	Indian Ocean	7.2	9.50	N/A	238	Lander, Whiteside & Lockridge (2003)
17/7/98	Papua New Guinea	Papua New Guinea, Christchurch, South Island, New Zealand, Japan	Bismarck Sea	7.1	15.00	N/A	2182	Lander, Whiteside & Lockridge (2003)
26/12/04	Sumatra, Indonesia	Indonesia, Sri Lanka, India, Thailand, Somalia, Burma, Maldives, Malaysia, Tanzania, Bangladesh, Kenya	Indian Ocean Coast	*9.1	20.00	9.93 B	*226000	Athukorala (2012)
17/7/06	Pangandaran, Indonesia	Java Island, Indonesia	Indian Ocean Coast	7.7	15.20	44.7 M	600	Mori et al (2001)
2/4/07	Solomon Islands	Solomon Island, Ghizo Island,	Solomon Sea	8.1	12.00	N/A	33	Hagen (2015)
29/9/09	Samoa	Samoa, Tonga, American Samoa, Fiji, New Zealand	South Pacific Coast	8.1	14.00	N/A	192	EERI (June 2010)
27/2/10	Chile	Chile, California, Japan	Pacific Coast	8.8	N/A	N/A	550	Kato et al (2011)
25/10/10	Sumatra, Indonesia	Indonesia	Indian Ocean	7.8	3.00	N/A	711	Newman et al (2011)
11/3/11	Tohoku, Japan	Eastern cost of Japan Coast	Pacific	9.0	*40.50	*300 B	19711	EERI (Nov 2011)
6/2/13	Solomon Islands	Solomon Island, Vanuatu, New Caledona, Japan	South Pacific Coast	8.0	3.00	N/A	10	Bilve et al (2014)
16/9/15	Chile	Chile, Peru, Brazil, Argentina	Pacific Coast	8.3	10.80	N/A	19	EERI (Sept 2015)

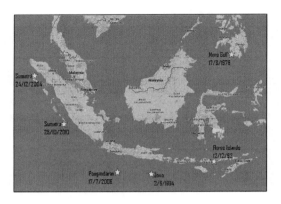

Figure 3. The tectonic setting of earthquake epicenter for past 50 years in Indian Ocean region.

Table 2. The earthquake event at Sabah within 2009–2015 (source: USGS Database).

Date	Location	Magnitude
4th Sept 2009	Kudat	4.5
21st Aug 2010	Lahad Datu	4.2
28th May 2012	Lahad Datu	4.6
1st Feb 2014	Ranau	4.7
24th Oct 2014	Kudat	4.6
5th June 2015	Ranau	6.0
6th June 2015	Ranau	4.5
13th June 2015	Ranau	5.3
23rd June 2015	Ranau	4.5
27th June 2015	Ranau	4.6

as natural protection against tsunami. About 42% of mangrove forest were recorded within Asean region which attributed mostly from Indonesia and Malaysia (Giri et al 2013). The mangrove distribution in Japan was not as extensive as the Indian Ocean region. The mangrove only concentrated at Okinawa Islands and extend northwards on Southern Kyushu (Alsaaideh et al 2013). Thus, the protection provided by mangrove in Japan area was relatively lower as compared to the affected countries in the Indian Ocean region.

3.2 *Threats in Malaysia*

Northwest part of Peninsular Malaysia (Kedah, Pulau Pinang, Perak and Selangor) had been affected by the tsunami in 2004 and at high potential risk to be hit again. Malaysia is surrounded by countries that were directly threatened by the tsunami. Hence, if high magnitude earthquake recurs at nearby critical location, destructive tsunamis reaching Malaysia is also potential. Figure 3 illustrates the relative location of the epicentres and the potential surrounding countries subjected to the generated tsunami for the past 50 years in the Indian Ocean region. Furthermore, the top two catastrophic tsunami events were both occurred in South-East Asia; Indian Ocean Tsunami 2004 and Tohoku Tsunami 2011 (Mazinani et al 2014). The frequent tsunami event that struck the Indian Ocean region caused a concern to the countries especially within the Ring of Fire (Whitmore et al 2009). The high magnitude earthquake may trigger tsunami which can worsen the situation.

The four states (Penang, Kedah, Perak, Selangor and Perlis) states that were hit by the 2004 tsunami are located on the West Coast of Peninsular Malaysia and are exposed to the Straits of Malacca. On the other side, the states located at South China Sea are within the low tsunami risk area (ESCAP, 2009). However, there were few earthquake events occurred in these areas. The latest event was at Ranau, Sabah which had triggered MW 6.0 on 5th June 2015 and lasting for about thirty-minutes. The earthquake was mentioned as the highest magnitude recorded in Malaysian history. However, the earthquake was still low in magnitude to generate tsunami. Based on the USGS database, there were about ten events recorded within a decade (2007–2016) with a scale from MW 4.2 to MW 6.0 out of which 60% of these happened in Ranau, Sabah as shown in Table 2. There were several follow-up earthquake occurrences in Sabah since the largest earthquake magnitude on 5th June 2015. This creates serious concern not only because the frequency of the earthquake, but the magnitude of the earthquake also had a generally increasing trend. The instability of the tectonic plate in the area might lead to another earthquake disaster with higher magnitude.

The above concern warrants serious attention for protection against the tsunami. The mangrove forest in Malaysia was noted as the second largest mangrove distribution in Asia region (Giri et al 2011) and this mainly located in Sabah (61%), Sarawak (21%) and west coast of Peninsular Malaysia. However, during IOT 2004, the tsunami had caused various level of impacts to the properties along to the west coast of Peninsular Malaysia. The states affected by the IOT 2004 and the area of mangrove distribution at the affected states is illustrated in Figure 4. It shows that along the coastline, the protection available was not sufficient especially in the state of Kedah.

3.3 *Assessment of protection level in Malaysia*

This assessment of protection level focusses on the coastline of Perak. Although Perak had the largest distribution of mangrove forest in Peninsular Malaysia, the state was still affected by the previous IOT 2004. The mangrove forest of MMFR are dominated with *Rhizophora* sp. Study by Ibrahim et al (2015) indicated that there were about 83.2% of *Rhizophora* sp in MMFR in 2011 (Fig 5). There are four species found including *Rhizophora* sp., *Avicennia* sp., *Sonneratia* sp., and *Brugueira* sp.

The replanted mangrove trees in MMFR were generally managed in tandem arrangement with spacing of 1.2 m and 1.8 m for *R. apiculata* and *R. mucronata* respectively (Goessens et al 2014). The present width of mangrove forests along the coastline of Perak varies

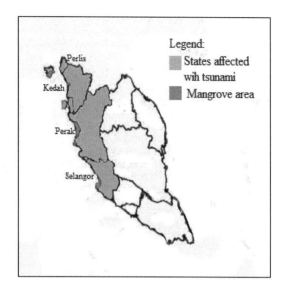

Figure 4. States affected by tsunami with the distribution of mangrove forest.

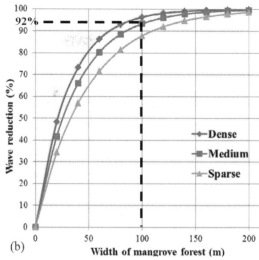

Figure 6. The percentage of wave reduction with respect to density and width of mangrove forest (Hashim & Catherine 2013).

approximately 92% of the incoming wave (Fig 6). Nevertheless, it is crucial to note that the coastline having about 15 m mangrove width is only able to attenuate 25% of wave.

4 CONCLUSION

4.1 Summary

The impact due to the Indian Ocean Tsunami 2004 was considered as one of the worst scenarios among the natural disaster events occurred in world history. Its high magnitude earthquake (MW 9.1) had shocked the world with massive number of casualties ever recorded in any tsunami event. The disaster also had contributed to USD 9.93 billion of total economic losses. Malaysia was one of the countries that have been hit with this catastrophic event which led to 68 numbers of casualties, 6 missing and 4200 displaced people.

Malaysia's experience demonstrated how mangrove has played an important protection buffer. With the increasing trend of threat and decreasing mangrove belt, Malaysia is potentially exposed to such severe wave impacts. Implementation of the optimum replanting scheme is inevitable to ensure an acceptable level of protection along Malaysia's coastline.

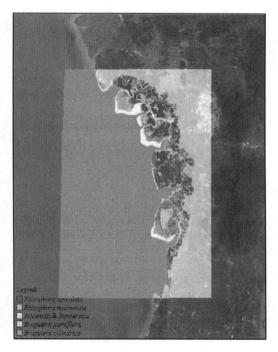

Figure 5. The distribution of mangrove trees at MMFR (Ibrahim et al 2015).

from approximately 15 m to about 500 m. The average width of available mangrove forest at MMFR was observed to be about 100 m and with medium density. Further analysis using the relationship established from recent experimental investigation by Hashim and Catherine (2013) revealed that this 100 m mangrove width is able to potentially dissipate up to

REFERENCES

Asmawi, M. Z., & Ibrahim, A. N. 2013. The impacts of tsunami on the well-being of the affected community in Kuala Muda, Kedah, Malaysia. *Journal of Clean Energy Technologies*, 1(3): 246–250.

Athukorala, P. C. 2012. *Disaster, generosity and recovery: Indian Ocean Tsunami*. Canberra: Australian national University.

Bilve, A., Nogareda, F., Joshua, C., Ross, L., Betcha, C., Durski, K., . . . Nilles, E. 2014. Establishing an early warning alert and response network following the Solomon Islands tsunami in 2013. *Bulletin of the World Health Organ*: 844–848.

Cosgrave, J. 2006. *Tsunami evaluation coalition: Initial findings*. London: ALNAP.

EERI. 2015, September 16. *Chile struck by M8.3 earthquake on September 16, 2015*. Retrieved from Earthquake Engineering Research Institute: https://www.eeri.org/2015/09/chile-struck-by-m8-3-earthquake-on-september-16-2015/

EERI. June 2010, June. *The MW 8.8 Chile earthquake of February 27, 2010*. EERI Special Earthquake Report.

EERI. November 2011. *The Japan Tohoku tsunami of March 11, 2011*. EERI Special Earthquake Report.

ESCAP. 2009. *Tsunami early warning systems in the Indian Ocean and Southeast Asia*. New York: United Nations.

Fehr, I., Grossi, P., Hernandez, S., Krebs, T., McKay, S., Muir-Wood, R., . . . Xie, Y. 2006. *Managing tsunami risk in the aftermath of the 2004 Indian Ocean earthquake and tsunami*. Newark: Risk Management Solutions.

Goessens, A., Satyanarayana, B., Stocken, T. V., Zuniga, M. Q., Mohd Lokman, H., Sulong, I., & Dahdouh-Guebas, F. 2014. Is Matang Mangrove Forest in Malaysia sustainably rejuvenating after more than a century of conservation and harvesting management? *Plos One*, 9(8): 1–14.

Hagen, K. 2015. *The 2007 Solomon Islands earthquake and tsunami: Cascading effects and community recovery. Earthquake Risk and Engineering towards a Resilient World* (pp. 1–9). Cambridge: SEDEC.

Hashim, A. M., & Catherine, S. M. P. 2013. A laboratory study on wave reduction by mangrove forest. *APCBEE Procedia 5*: 27–32.

Hawkes, A. D., Bird, M., Cowie, S., Grundy-Warr, C., Horton, B. P., Hwai, A. T., . . . Aik, L. W. 2007. Sediments deposited by the 2004 Indian Ocean tsunami along Malaysia-Thailand Peninsula. *Marine Geology*:169–190.

Herd, D. G., Youd, L., Meyer, H., C, J. L., Person, W. J., & Mendoza, C. 1981. The Great Tumaco, Colombia earthquake of 12 December 1979. *Science*, 211(4481): 441–445.

Ibrahim, N. A., Mustapha, M. A., Lihan, T., & Mazlan, A. G. 2015. Mapping mangrove changes in the Matang mangrove forest using multi temporal satellite imageries. *Ocean and Coastal Management*, 114: 64–76.

Kato, T., Terada Y, Nishimura, H., Nagai, T., & Koshimura, S. 2011. Tsunami records due to the 2010 Chile earthquake observes by GPS buoys established along the Pacific coast of Japan. *Earth Planets Space*: 63, 5–8.

Koh, H. L., Teh, S. Y., Majid, T. A., Lau, T. L., & Ahmad, F. 2012. Earthquake and tsunami research in USM: The role disaster research nexus. *Pertanika Journal Science and Technology*, 20(1): 151–163.

Koyama, J., Yoshizawa, K., Yomogida, K., & Tsuzuki, M. 2012. Variability of megathrust earthquakes in the world revealed by the 2011 Tohoku-oki earthquake. *Earth Planets Space*: 1189–1198.

Lander, J. F., Whiteside, L. S., & Lockridge, P. A. (2003). Science of tsunami hazards: Tsunami data issue. *The International Journal of the Tsunami Society*, 21(1): 1–88.

Mazinani, I., Hashim, A. M., & Saba, A. 2014. Experimental investigation on tsunami acting on bridges. *International Scholarly and Scientific Research and Innovation*, 8(10): 1043–1046.

Mori, J., Mooney, W. D., Afnimar, Kurniawan, S., Anaya, A. I., & Widiyantoro, S. 2001. The 17 July 2006 tsunami earthquake in West Java, Indonesia. *Seismological Research Letter*: 201–207.

Nanto, D. K., Cooper, W. H., Donnelly, J. M., & Johnson, R. 2011. *Japan's 2011 earthquake and tsunami: Economic effects and implications for the United States*. Congressional Research Service.

Newman, A. V., Hayes, G., Wei, Y., & Convers, J. 2011. The 25 October 2010 Mentawai tsunami earthquake, from real-time discriminants, finite-fault rupture, and tsunami excitation. *Geophysical Research Letters*: 38, 1–7.

Patel, D. M., Patel, V. M., Bhupesh, K., & Patel, K. A. 2014. Performance of mangrove in tsunami resistance. *International Journal of Emerging Technology and Research*, 1(3): 29–32.

Suppasri, A., Fukutani, Y., Abe, Y., & Imamura, F. 2013. *Relationship between earthquake magnitude and tsunami height along the Tohoku coast based on historical tsunami trace database and the 2011 Great East Japan Tsunami*. Report of Tsunami Engineering, 37–49.

Suppasri, A., Futami, T., Tabuchi, S., & Imamura, F. 2012a. Mapping of historical tsunamis in the Indian and Southwest Pacific Oceans. *International Journal of Disaster Risk Reduction*: 62–71.

Suppasri, A., Imamura, F., & Koshimura, S. 2012b. Tsunamigenic ratio of the Pacific Ocean earthquakes and a proposal for a tsunami index. *Natural Hazards and Earth System Sciences*: 175–185.

Suppasri, A., Muhari, A., Ranasinghe, P., Mas, E., Shuto, N., Imamura, F., & Koshimura, S. 2012c. Damage and reconstruction after the 2004 Indian Ocean tsunami and the 2011 Great East Japan tsunami. *Journal of Natural Disaster Science*: 19–39.

Viroulet, S., Cebron, D., Kimmoun, O., & Kharif, C. 2013. Shallow water waves generated by sub aerial solid landslides. *Geophysical Journal International*: 1–16.

Wang, J. F., & Li, L. F. 2008. Improving tsunami warning systems with remote sensing and geographical information system input. *Risk Analysis*: 1–16.

Wiegel, R., & Asce, F. 1980. *Tsunamis along west coast Luzon, Philippines. Proceedings of the 17th Conference on Coastal Engineering 1980* (pp. 652–671). Sydney: Proceedings 1980.

Wu, F. T. 1973. Source mechanism of February 4, 1965, Rat Island Earthquake. Journal of Geophysical Research, 78(26), 6082–6092.rove, A.T. 1980. Geomorphic evolution of the Sahara and the Nile. In M.A.J. Williams & H. Faure (eds), *The Sahara and the Nile*: 21–35. Rotterdam: Balkema.

Wave transmission characteristics of sand containers used as submerged breakwaters

S. Gupta, H.M. Teh & A.M. Hashim
Department of Civil and Environmental Engineering, Universiti Teknologi PETRONAS, Seri Iskandar, Perak, Malaysia

R.B.R. Sulaiman
Forest Management and Ecology Program, Forest Research Institute Malaysia (FRIM), Kepong, Selangor Darul Ehsan, Malaysia

ABSTRACT: Sand containers have been commonly used for wave protection and beach stabilization at coastal areas mainly due to their low construction cost and simplicity in design. In Malaysia, the majority of the breakwaters consisting of several arrrays of sand containers are installed at muddy coasts so as to provide required level of hydrodynamic regimes in its vicinity. Due to unavailability of the established design guidelines of the sand container breakwaters, the geometry of these low crested structures is largely ascertained by engineering judgement and experience of coastal engineers. The present study aims at investigating wave transmission characteristics of the submerged sand container breakwaters of different layers, i.e. single-, double-, triple- and quadruple-layers, with respect to varying water depths, breakwater submergence and wave steepness via physical modelling. The test models were subjected to unidirectional monochromatic waves of different periods and heights using a wave flume facility. Wave transmission characteristics of the respective test models were quantified in terms of wave transmission coefficients. The experimental results revealed that wave transsmission ability of the test models were largely governed by the degree of breakwater submergence and the wave steepness imposed to structures. Wave attenuation ability of sand container breakwater could be further enhanced by increasing the breakwater size and reducing the breakwater submergence.

1 INTRODUCTION

Conventional rubble mound breakwaters are widely used to provide effective wave control for maritime applications. They are also used to mitigate coastal erosion problems and to realign shorelines. Nevertheless, these massive gravity-type structures are not suitable to be built at muddy coasts due to the poor soil condition. The presence of these structures is very likely to pose some problems to the coastal and marine environments, e.g. interruption to water circulation, sediment transport and fish migration. Hence, construction of rubble mound breakwaters must be properly planned and executed so as to minimize impacts to the environment.

There are a number of innovative and environmental friendly breakwaters that have potential applications in muddy coasts, e.g. floating breakwaters (Teh et al. 2012) and pile supported free surface breakwaters (Teh et al. 2013). Geotextile technology is recently introduced to the construction of coastal structures and its importance to the industry is prominent (Shin & Oh 2007). This technology has been adopted in constructing sand containers that are used to form an array of submerged breakwater exist in various configurations and sizes for coastal protection. Each container is made of a geotextile bag filled with sand. The containers are claimed to be more cost effective than the conventional breakwaters (Hornsey et al. 2011). Nonetheless, the life span of these structures is rather short, i.e. 5 years. Depletion on workability of the sand container due to abrasion by sand, ultraviolet light interruption from the sun and vandalism are the key factors contributing to material failures (Harris and Sample 2009).

Unlike emerged breakwaters, submerged breakwaters permit waves and currents past over the crest. They limit the transmission of waves and currents but they are not designed to be the total barriers to these environmental forces. They promote breaking and energy dissipation as waves propagate over the submerged structures. A good number of literature on low-crested or submerged rubble mound breakwaters have been reported by many researchers. Among the innovative submerged breakwaters developed in the past are L-Block (Yuliastuti and Hashim 2011), stepped slope breakwater (Kerpen et al. 2015) and reef balls (Sindhu and Shirlal 2015). These breakwaters provide unique features to promote wave breaking and energy dissipation.

The study on submerged sand container breakwater, particularly on the design optimization, is rather limited. Figure 1 shows the key parameters used to

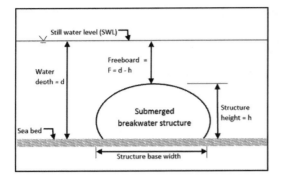

Figure 1. Single layer sand container.

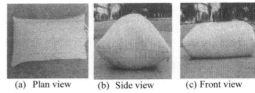

Figure 2. Sand container models.

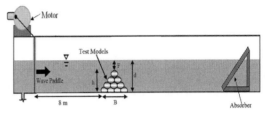

Figure 3. Laboratory setup for quadruple layer breakwater.

describe geometry of a sand container. It is important to note that the sand container normally do not reach the same height as the theoretical diameter but a maximum height of 60% to 70% of the diameter which depends on the filling methods and whether filling is done below or above water level (Koffler et al. 2008). Recio and Oumeraci (2009) studied stability of the geotextile sand containers. They further proposed hydraulic stability formulae for the sand containers that are subjected to the types of failure moods, i.e. sliding and overturning. Teh (2015) developed a submerged breakwater using sand containers and the wave transmission ability was assessed with respect to a range of wave periods using physical modelling. It is not clear that the hydraulic response of the sand container of different layers when subjected to different water depths and submergence levels. Hence, this study is set to investigate wave transmission characteristics of the submerged sand container breakwaters of different layers, i.e. single-, double-, triple- and quadruple-layers, with respect to varying water depths, breakwater submergence and wave steepness via physical modelling.

2 LABORATORY SET-UP

2.1 Test models

In this study, a total of ten units of sand container (sand bag) were scaled down to a factor of 1:12.5 for the present experimental study. The sand bags were made of cotton fabric and each has a plan dimensions of 8 cm in width and 30 cm in length. Upon filling the bag with sand (80–85%), each sand bag has an average weight of 2 kg and has an average cross sectional dimensions of 10 cm in width and 6.5 cm in height, as shown in Figure 2.

2.2 Experimental set-up and test programme

A series of tests were conducted in a 10 m long, 0.30 m wide and 0.45 m high open channel flume at the Hydraulics Laboratory of Universiti Teknologi PETRONAS to evaluate wave transmission ability of the test models. The flume has rigid steel beds and two walls lined with glass panels for the entire length of the flume. The steady monochromatic non-breaking waves of varying steepness were generated by the flap-type wave paddle at one end of the flume. The unwanted reflected waves were reduced by the sloping wave absorber placed at the other end of the flume. The test models were placed at 8 m from the wave paddle as shown in Figure 3. Wave probes were placed in the vicinity of the test model for measurement of water level variations. These wave probes were mounted on two steel rails at the sidewall tops of the flume. The moving-probe method was adopted to obtain both incident and reflected wave heights in front of the test models. The wave probe placed at 1 m leeward of the test models was used to record the transmitted wave heights. Details of the laboratory set-up is presented in Figure 3.

Four layouts of the sand bag, i.e. single layer, double layer, triple layer and quadruple layer, were developed and the wave transmission characteristics were ascertained by the means of physical modeling. Table 1 summarizes the physical properties and test parameters (i.e. water depth, wave height and wave period) of the test models. A total of 85 test runs were conducted in this study.

3 DISCUSSION OF RESULTS

Wave transmission past the submerged breakwaters is numerically quantified in terms of the transmission coefficient $C_T = H_t/H_i$, where H_i and H_t are the incident and transmitted wave heights, respectively. In regular wave tests, H_i and H_t were computed based on the average of a minimum of 10 waves. Small C_T values indicates better efficiency of the breakwater in wave attenuation. The following section presents wave transmission coefficient of the individual test models (i.e. single layer, double layer, triple layer and quadruple layer) of different freeboard-to-water depth ratios,

Table 1. Test parameters.

Test Models			Water Depth, d (cm)	Wave Height, H_i (cm)	Wave Period, T (s)
Configurations	Width, B (cm)	Height, h (cm)			
Single Layer	30	6.5	8,12,16,20	1.0 - 7.5	0.6, 1.0, 1.3, 1.6, 2.0
Double Layer	30	13	15,18,20,22,24	2.20 - 8.73	
Triple Layer	30	19.5	22,24,26,28	4.43 - 10.23	
Quadruple Layer	40	26	26,28,30,32	5.32 - 11.34	

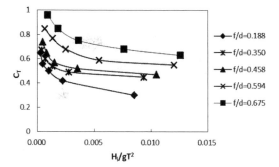

Figure 4. C_T of the single layer breakwater.

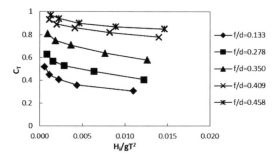

Figure 5. C_T of the double layer breakwater.

f/d plotted against a dimensionless wave steepness parameter, H_i/gT^2, where g and T are acceleration of gravity and wave period, respectively. Note that f/d ranges from 0 (i.e. the breakwater crest is at the still water level) to 0.675 (i.e. the freeboard is 67.5% of the water depth).

3.1 Single layer breakwater

Figure 4 presents the C_T of the single layer breakwater ($h/B = 0.216$) exposed to five relative freeboards f/d of 0.188, 0.350, 0.458, 0.594 and 0.675. The test model was confronted by a wide range of wave steepness H_i/gT^2 ranging from 0.001 to 0.013. It is clear from the figure that the C_T of respective f/d ratios gradually decreases as wave steepness increases. This indicate that wave suppression ability of the single layer breakwater improves when subjected to waves of higher steepness. This is sensible because waves of higher steepness prone to breaking when travelling over the crest of the breakwater. Note that the performance improvement of the breakwater is rather limited as H_i/gT^2 is well beyond 0.0025.

Further, it is also seen from the figure that C_T decreases with the decreasing f/d. This finding deduces that breakwater freeboard is one of the significant variables affecting the hydraulic efficiency of the single layer breakwater. For single layer breakwater, an attainment of 50% wave height reduction can only be achieved as $f/d < 0.35$ and $H_i/gT^2 < 0.001$. A breakwater of $f/d = 0.188$ is capable of attenuating the incident wave height up to 70% at $H_i/gT^2 \approx 0.008$. In short, the single layer sand container with limited freeboard ($f/d < 0.35$) may be an effective submerged breakwater under regular waves.

3.2 Double layers breakwater

Figure 5 demonstrates C_T of the double layer breakwater ($h/B = 0.433$) with $f/d = 0.133$, 0.278, 0.350, 0.409 and 0.458, plotted with respect to H_i/gT^2. Note that the f/d ratios used in Figures 4 and 5 are subjected to different geometrical factor of the breakwater (i.e. h/B); hence, direct comparison of f/d ratios between single and double layer breakwaters is not recommended.

Similar to the single layer breakwater, the C_T of the double layer breakwater slowly decreases with an increase of H_i/gT^2. This implies that wave transmission of the double layer breakwater is restricted by waves of larger steepness. The double layer breakwater offers more interaction area for energy dissipation compared to the single layer breakwater. In addition, the decrease of C_T with f/d is also found to be evident in Figure 5. As the breakwater freeboard decreases (f/d reduces), the wave energy suppression is found to be greatly enhanced. A drop of C_T of about 60% is observed when submergence of the double layer breakwater is reduced from 0.458 to 0.133. At deeper water, breakwaters of higher h/B, e.g. double layer breakwater, and of smaller f/d are recommended for the attainment of satisfactory wave attenuation performance.

3.3 Triple layer breakwater

Figure 6 displays the plots of C_T with $f/d = 0.114$, 0.188, 0.25 and 0.304 for the triple layer breakwater ($h/B = 0.65$). This breakwater is designed to cater

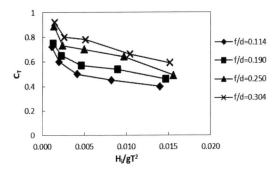

Figure 6. C_T of the triple layer breakwater.

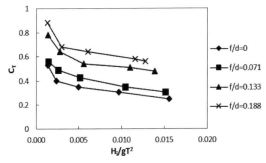

Figure 7. C_T of the quadruple layer breakwater.

for applications in deeper water. The experimental results shown in Figure 6 are somewhat agreeable with those of single and double layer breakwaters as presented in Figure 4 and 5, i.e. a decrease of C_T corresponding with H_i/gT^2 and f/d. For instance, the triple layer breakwater of $f/d = 0.114$ was capable of attenuating the incident wave height of approximately 50% at $H_i/gT^2 = 0.005$. Whereas, the breakwater of $f/d = 0.304$ was only able to reduce 20% of the incident wave height when opposed by waves of similar steepness. These observations leads to a conclusion that wave transmission of the triple layer breakwater is strongly governed by H_i/gT^2 and f/d, and improvement of wave attenuation would be anticipated when the breakwater submergence is further limited.

When compared with the C_T values of single and double layer breakwaters as shown in Figures 4 and 5, those of triple layer breakwater consistently exhibit lower values signifying that the breakwaters of higher h/B do help in increasing the overall wave attenuation performance. Again, this is a sound judgment as breakwater of higher h/B would provide more contact points for wave-structure interactions to take place.

3.4 Quadruple layer breakwater

In the test case of the quadruple layer breakwater, both width and height of the structure increased proportionally, giving h/B of 0.65 (similar to that of the triple layer breakwater). When compared the cross sectional area of the quadruple layer breakwater ($0.5Bh$) with the water column area of the breakwater base (Bd), a new geometrical parameter is $h/2d$ developed. $h/2d$ gives a representation of the breakwater coverage in the water domain. The greater the $h/2d$ values, the greater domain of the water filled with the breakwater. In this case, $h/2d$ of the quadruple layer breakwater ranges from 0.41 to 0.50 compared to the triple layer breakwater $0.35 < h/2d < 0.44$. The breakwater was subjected to smaller range of f/d, i.e. 0, 0.071, 0.133 and 0.188. Note that at $f/d = 0$, breakwater crest is located at the still water level, and the structure is alternatively submerged.

Figure 7 shows the variation of C_T corresponding to different f/d ratios, including $f/d = 0$. At $f/d = 0$, the C_T recorded are consistently lower than those of higher f/d, implying that the quadruple layer breakwater achieves the largest wave attenuation performance when the still water level is located at the crest level. The C_T plot of $f/d = 0$ in Figure 7 shows a value as low as 0.23. At this stage, maximum wave structure-interactions would take place, facilitating significant amount of energy dissipation through wave breaking, frictional loss and energy loss through sound and heat.

Similar to other breakwater layouts, wave transmission ability of the quadruple layer breakwater is also affected by wave steepness. The larger the H_i/gT^2, the smaller the C_T values for all the test cases of f/d. The results also show that increment of breakwater size, i.e. larger $h/2d$ values, helps to improve the wave attenuation ability.

3.5 Summary

For a given water depth, hydraulic performance of the submerged sand container breakwater can be optimized or enhanced by the following approaches:

- Increase the breakwater size – by increasing $h/2d$
- Reduce the breakwater submergence – by decreasing f/d, preferably $f/d = 0$
- Exposed the breakwater to steeper waves

4 CONCLUSION

A series of experiment were conducted to study the wave transmission characteristics of the submerged sand container breakwaters of different layers, i.e. single-, double-, triple- and quadruple-layers, with respect to varying water depths, breakwater submergence and wave steepness. Wave transmission of the breakwater is strongly affected by wave steepness and the relative breakwater submergence. Wave attenuation of all the tested breakwater increased with the decrease of relative breakwater submergence and with the increase of wave steepness. It is recommended to increase the breakwater size (i.e. increasing $h/2d$), reduce the freeboard depth by decreasing f/d and lastly use steeper wave in order to maximize the performance of submerged breakwater using sand bags.

REFERENCES

Harris, L.E. & Sample, J.W., 2009. The Evolution of Multi-Celled Sand-Filled Geosynthetic Systems for Coastal Protection and Surfing Enhancement, 1(1), pp. 1–15.

Hornsey, W.P. et al., 2011. Geotextiles and Geomembranes Geotextile sand container shoreline protection systems: Design and application, 29.

Kerpen, N.B. et al., 2015. Experimental Investigations on Wave Transmission at Submerged Breakwater with Smooth and Stepped Slopes. *Procedia Engineering*, 116(Apac), pp. 713–719. Available at: http://dx.doi.org/10.1016/j.proeng.2015.08.356.

Koffler, a. et al., 2008. Geosynthetics in protection against erosion for river and coastal banks and marine and hydraulic construction. *Journal of Coastal Conservation*, 12(1), pp. 11–17.

Recio, J. & Oumeraci, H., 2009. Process based stability formulae for coastal structures made of geotextile sand containers. *Coastal Engineering*, 56(5–6), pp. 632–658. Available at: http://linkinghub.elsevier.com/retrieve/pii/S037838390900009X.

Shin, E.C. & Oh, Y.I., 2007. Coastal erosion prevention by geotextile tube technology. *Geotextiles and Geomembranes*, 25(4–5), pp. 264–277. Available at: http://linkinghub.elsevier.com/retrieve/pii/S0266114407000295.

Sindhu, S. & Shirlal, K.G., 2015. Prediction of wave transmission characteristics at submerged reef breakwater. *Procedia Engineering*, 116(Apac), pp. 262–268. Available at: http://dx.doi.org/10.1016/j.proeng.2015.08.289.

Teh, H.M. & Mohammed, N.I., 2012, December. Wave interactions with a floating breakwater. In Humanities, Science and Engineering (CHUSER), 2012 IEEE Colloquium on (pp. 84–87). IEEE.

Teh, H.M. 2013. Hydraulic performance of free surface breakwaters: A review. Sains Malaysiana, 42(9), pp. 1301–1310.

Teh, H.M., 2015, October. Wave Transmission over the Low-Crested Sand Container Breakwaters. In Applied Mechanics and Materials (Vol. 802, pp. 57–62). Trans Tech Publications.

Yuliastuti, D.I. & Hashim, A.M., 2011. Wave Transmission on Submerged Rubble Mound Breakwater Using L-Blocks, 6, pp. 243–248.

The need of early response system to HNS accident based on case analysis

J.W. Ryu, J.M. Kim, S.M. Shin & H.K. Park
KAIST Institute of Disaster Studies, Korean Advanced Institute of Science and Technology, Daejeon, Republic of Korea

ABSTRACT: Recently, it has been an important issue that the damage of HNS accident has probability to become a catastrophic disaster. But there is no HNS-specialized response system which enables the successful response in time. So this research analyzes two HNS accidents by identifying the agents, errors, and causes in the accidents. Based on the analysis, the drawback of previous Korean response system to HNS accidents was determined. Finally, it is found that the concept of new platform whose important features are quickness and preciseness is necessary to Korean HNS response system.

1 INTRODUCTION

The OPRC-HNS Protocol defines that HNS (Hazardous and Noxious Substances) is likely to create hazards to human health, to harm living resources and marine life, to damage amenities or to interfere with other legitimate uses of the sea. HNS is defined as a substance other than oil. It has been used widely in industry and other fields.

The marine transportation of HNS in worldwide has been on the rise. In the case of Korea in particular, the traffic growth is about 66% and is 2.5 times of the world average. For this reason, the risk of HNS spill and accident have been increased and there are many dangerous HNS accidents which is passible disaster like the accident of Maritime Maisie.

Even though we can use the response system when oil spill occurs, it is hard to control them with effect once the HNS accident has happened. HNS spill at sea is especially required to optimization of the early response and equipment arrangement (Purnell, 2009). With this consideration, a development of equipment and system is just beginning all over the world. Although research about response system of HNS spill is also needed in Korea, no one has ever researched about this.

Marine accidents are usually caused by human error. To analyze the error, it is required to focus on the agents who make a mistake. Moreover, we need to improve this situation finding the fundamental reason why human error occurred.

Therefore, this research shows analysis of Korean HNS spills based on agent and the problems caused by human or agent error. And in this paper, concept of HNS emergency response platform to solve the problems is suggested.

2 BACKGROUND INFORMATION

There are many accident analysis processes over the world such as Root-cause analysis, Events and Causal Factors Chart, and Sequentially timed events Plotting Procedure(STEP). These accident analysis processes sometimes have limitation to make the case complicated or are hard to find the fundamental causes. So agent based analysis can improve these weaknesses and find the cause of the errors.

2.1 Brief statement of accident cases

2.1.1 Maritime Maisie accident case
Around 2 a.m. on 29th December, the Maritime Maisie vessel which was shipping 39,000 tons of HNS including p-xylene crashed on another vessel G near Busan harbor, resulting in huge fire. Rescue team and patrol ships moved to accident site and started firefighting. But it was very difficult because of toxic gas. The firing vessel moved to Japanese EEZ and keep floating to north-east. The fire was extinguished after 442 hours with the support from Netherlands.

2.1.2 Ulsan Hanyang-Ace accident case
Around 2 p.m. on 11th January, 2015, the Hanyang Ace vessel was loading the mixed acid at Ulsan harbor. And the gas explosion occurred at the container in the vessel. The loaded mixed acid was spilled and toxic gas was generated at midnight because of the second reaction of mixed acid. Citizens near that site were encouraged to evacuate. The response was terminated after reshipment of mixed acid to another safe container.

3 METHODOLOGY

3.1 Agent based accident analysis

Some important facts and processes during the accident should be identified and put chronologically as figure 1 shows. The events such as the state of agents and the occurrence or amplification of damage can be treated as facts. If necessary, facts can have branches. Process between two facts is what happened at that time.

After constructing that diagram by identifying facts, processes, and their context, which parts of processes are performed inappropriately can be analyzed. To do that, what agents did in respective process should be identified. Table 1 shows the organization of above things. If an action of any agents is wrong, the error, how it is wrong, and cause, why the error occurred, should be determined. One important point is that the action of agents means what they did actually during response. Some unperformed duty or task should be ignored on the view of System SafetyII.

With these two steps, the details of accident, actions done by agents, whether actions are suitable to each fact can be determined. What was error and cause also can be determined. Finally, the fundamental cause can be identified so that how to develop previous failed response can be suggested to reduce the damage of an accident.

4 RESULTS AND DISCUSSION

4.1 Results of agent-based analysis of HNS accident

4.1.1 Maritime Maisie accident case

Figure 2 shows three facts and two processes of Maritime Maisie accidents case; the voyage of G and M ships, collisions of two ships, and fire lasted 18 days due to failure of firefighting.

As Table 2 shows, the direct action directly related to accident occurrence was rapid turning to right. It was too late to detect forward vessels so that the vessel M tried to turn right rapidly to avoid collision, failing to detect the backward-right vessel G. So the collision of M and G happened. It was also found that the vessel G didn't keep contacting to vessel M. So, the Error was absence of their communication and corresponding cause was their carelessness and weak regulation.

Originally, the control center in harbor should recognize all vessels near the harbor. But the recognition radius was too short to warn M and G. So it was also found that the policy of control center was insufficient. The environment affected a little because the darkness shortened vessel's sight.

According to Table 3, the directly responding agents moved to accident site in time. But they could not perform their task because of inadequate system and equipment so that the damage of accident amplified very much. The command center on site failed to

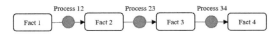

Figure 1. Diagram of agent-based analysis.

Table 1. Identification of actions, errors, and causes in a Process.

Agent	Action	Error	Cause
Agent 1			
Agent 2			
Agent 3			

Figure 2. Facts of Maritime Maisie accident case.

Table 2. Analysis of Process 12 in Maritime Maisie accident case.

Agent	Action	Error	Cause
Vessel M	Rapid turning to right	NOT recognition of near ships	Carelessness and weak regulation
Vessel G		NOT communication with ship M	Carelessness and weak regulation
Harbor control center	Controlling ships	Short controlling radius	Insufficient policy
Environment			Short sight due to darkness

Table 3. Analysis of Process 23 in Maritime Maisie accident case.

Agent	Action	Error	Cause
Vessel M	Self-escape by lifeboat		
Vessel G	Self-firefight and move to harbor		
Command center		Failed to decision making	Inadequate organization Insufficient manual
Prevention team	Immediate move	Absence of appropriate response	Unawareness of appropriate response
Firefighting team	Immediate move and firefight	Absence of appropriate response	Unawareness of appropriate response
Environment	Floating of firing vessel to Japan		

decision making. So it couldn't operate other response teams properly.

The spilled HNS had low ignition point so it was impossible to quench the fire with water. But the prevention team and firefighting team didn't recognize this kind of information. So they could not perform their task because they were unaware of what the appropriate response is. Even if they knew this kind of information, there was no equipment for HNS.

4.1.2 Hanyang Ace accident case

Figure 3 shows the 4 facts of Hanyang Ace accident case; loading chemical, explosion, toxic gas generation, and response termination.

As shown in Table 4, the captain and crew were loading mixed acid to the vessel during Process 12. At that time, they should be very careful because the mixed acid has high reactivity to water. And it was very important to perfectly dry the ballast tank before loading. But the ballast tank was not dried perfectly. Furthermore, the containers on vessel got crack on themselves due to inadequate maintenance. So the mixed acid reacted with water and exploded. It was found that the manual and training program of vessel's company didn't give that information to crews. It is really notable that the company had same accident on same vessel. Even though, the company didn't develop their manual, regulation, and training program. It was one of fundamental cause of this accident.

Figure 3. Facts of Hanyang Ace accident case

After F2, the Table 5 and 6 shows that the escape of crews after explosion was carried rightly. But, the firefighters didn't obey the response process, which they didn't recognize the kind of HNS. Also the port authority was just monitoring the situation passively. So they failed to detect the risk of second explosion and gas generation, leading to being late to evacuate other close vessels and call for Prevention Council. In fact, the Prevention Council should be gathered before F3. It means the authority was organized inadequately to respond HNS accident.

4.2 The drawback of previous Korean HNS response system

From the analysis of two HNS accident case, the drawback of Korean HNS response system can be identified.

At first, and most importantly, there is no system or equipment which is specialized to HNS accident. It is true that equipment for oil spill is still utilized so that the prevention of HNS cannot be performed successfully, amplifying the damage of accident.

Secondly, the tool or manual is essential to support fast and precise decision making. The failure of exact decision making to certain situation leads the damage of accident amplified. There was also some unawareness of agents more than failure to decision making. The capability of agents should be improved with training program specialized to HNS.

Finally, the HNS response system need to be integrated. Previous response system has vertical structure and the function of agents are quite divided so that the recognition and propagation of accident are difficult.

Table 4. Analysis of Process 12 in Hanyang Ace accident case.

Agent	Action	Error	Cause
Captain and crews	Loading mixed acid to vessel	Crack occurred on cargo	Inadequate maintenance of system
		Imperfectly dried ballast tank	Insufficient manual to HNS
Company		Absence of effort to train crew	Insufficient management and training program

Table 5. Analysis of Process 23 in Hanyang Ace accident case.

Agent	Action	Error	Cause
Captain and crew	Escape		
Firefighting team	Firefight and support to escape	NOT identifying of the kind of HNS	NOT obeying the process Insufficient training
Port authority	Passive monitoring	Unawareness of the risk of second accident	Misjudging the situation

Table 6. Analysis of Process 34 in Hanyang Ace accident case.

Agent	Action	Error	Cause
Firefighting team	firefighting and prevention		
Port authority	Call for Prevention Council	Being too late	Inadequate organizations

For the safety of agents, this platform should include HNS-specialized equipment such as chemical protective clothing, telecommunication which is applicable to accident site, and early warning system of offshore. To support the decision making, the platform should include the response manual and strategies of HNS. The prevention equipment for HNS and tools for estimation, analysis, prediction are also necessary. And these technologies and equipment in platform should be operated and managed as an integrated response system.

5 CONCLUSION

The drawback of previous Korean HNS response system was identified through the analysis of Korean HNS accident cases. To improve these drawbacks, the development of new platform, which is a developed system, specialized to HNS is essential. So this research can give the direction to Korean government what kind of HNS disaster response system should be planned to develop.

The important feature of this platform can be considered as the quickness and the preciseness. These features will guarantee the safety of agents, exact decision making, and the integrated disaster response.

In a word, this research suggests the necessity of the development of a new platform, which is HNS-specialized system and is able to result in successful early response to HNS, based on the drawback of previous Korean response system concluded from the analysis of HNS accident. To conclude the detailed components of this platform, in fact, further research should be performed. And some system dynamics models can be constructed to research quantitatively the improvement of resilience related to sustainability by this platform.

ACKNOWLEDGEMENT

This research was supported by a grant [MPSS-CG-2015-01] through the Natural Hazard Mitigation Research Group funded by Ministry of Public Safety and Security of Korean government.

REFERENCES

E.B. Lee, 2013. A study on the development of the response resource model of Hazardous and Noxious Substances based on the risk of marine accidents in Korea, *Journal of Navigation and Port Research*, 36(10): 857–864

H.T. Kim, 2011. A case study of marine accident investigation and analysis with focus on human factor, *Journal of the Ergonomics Society of Korea*, 30(1): 137–150

Korea Coast Guard. 2014. *The overview, risk, and response capability of HNS by marine transportation*, Seoul: Ministry of Public Safety and Security

KAIST. 2015a. *The research of improvement of resilience based on context cognition for integrated disaster management*. Daejeon: KAIST

KAIST. 2015b. *Development of technologies and equipment for HNS outflow disaster response.* Seoul: Ministry of Public Safety and Security

K.H. Kim, 2015. A study on decision making process of OSC and on-scene command system on occurring of disaster at sea, *Journal of the Korean Society of Marine Environment and Safety*, 20(6): 692–703

Construction and project management

Coordination process in construction projects management

W.S. Alaloul, M.S. Liew & N.A. Zawawi
Department of Civil and Environmental Engineering, Universiti Teknologi PETRONAS, Seri Iskandar, Perak, Malaysia

ABSTRACT: Coordination process occupies a central role in construction projects management. This paper investigated the coordination process through selected articles from well-known academic journals. The articles were classified based on coordination process consideration; as a core issue of the study or appeared in the study results as a main factor. The aim of this paper is to review critically the literature for addressing the gap and presenting a research agenda of coordination process. The main findings are: (1) construction industry suffers from poor performance as a result of lack or insufficient coordination; (2) further research should focus on coordination as a core issue; and (3) coordination is one of the critical success factors in construction projects, and affect its performance significantly. Finally, coordination process still in the infancy stage in construction management and need more research to reap its dividends, using artificial intelligent, in order to enhance construction projects performance.

1 INTRODUCTION

Construction industry has suffered widely from fragmentation during the projects life cycle. Construction projects involve a large number of stakeholders e.g. client, designer, consultants, contractor, suppliers, and subcontractors whom have to work on design and construction of unique projects (Neeraj Jha and Misra, 2007, Saram and Ahmed, 2001). Traditionally, many of those participants are not motivated to work together voluntarily. However, the fragmented nature of the construction projects is considered as the cause of the poor performance, where most of the activities are interdependent and required continuous coordination (Hai et al., 2012, Mohd. Noor, 2010). During the last years, there has been fast-tracking trend toward main initiatives to develop the performance of construction projects. New approaches were studied as; lean construction, rethinking, and public privet partnership (Saram and Ahmed, 2001, Hai et al., 2012, Ali and Rahmat, 2009). In addition, coordination was proposed as a creative management procedures and technology (Alaghbari et al., 2007). This paper displays a critical review for coordination process in construction projects. The purpose of this review is to investigate the coordination trends and to identify the key areas, which could affect the construction performance. Limitations of current practices on coordination process and future research guidelines for effective, efficient, and viable coordination were discussed as well (Saram and Ahmed, 2001, Neeraj Jha and Misra, 2007).

2 LITERATURE REVIEW

Construction industry has struggled to overcome the fragmentation and interdependency problem. The central problem of construction projects management is the interdependent relationships, which is adversary in most of the cases. Nevertheless, there is a lack of coordination between the technical and organizational interdependences amongst the construction parties (Hai et al., 2012, Ali and Rahmat, 2009, Abdul-Rahman et al., 2006).

Definition of Coordination. In spite of the large number of research about coordination done in different domains, there is no specific definition of the term coordination. Coordination was understood as the design and operation of organizational processes to bring separate elements together (Hai et al., 2012). However, oxford dictionary definition was "The organization of the different elements of a complex body or activity to enable them to work together effectively" (Neeraj Jha and Misra, 2007). Therefore, the common part of the definition is working together, between project stakeholders or different departments to complete the project effectively. At this point, stakeholders work by way of others objectives for the team endeavors, which are satisfying all interested parties. Based on Malone and Crowston (1990), coordination is "managing dependencies between activities and its multilayered, involving the orchestration of relationships". The management seeks to achieve the efforts integration through its basic functions of planning, organizing, staffing, directing and controlling, therefore considering coordination not as a separate function of management (Ali and Rahmat, 2009).

In projects management, coordination can be expressed as "different actors working on a common project, agree to a common definition of what they are building, share information, and harmonize their activities" (Saram and Ahmed, 2001). The concise oxford dictionary of construction, survey/civil engineering defined coordination as "The act of aligning tasks, activities, and resources so that they are managed effectively" (Hai et al., 2012). For the purposes of this study, coordination can be defined as, the art of managing dependencies between activities and it's multilayered.

In fact, coordination in construction projects has applied not for a long time. It is one of the most influential procedures in construction projects. In construction industry, the active coordination could reduce, predict and curative difficulties occur by design modification lead-time, materials supply chain, work force usage, and equipment availability, to grantee smoother implementation of construction projects. Continuous coordination has been developed as a powerful tool to ensure the achievement of construction work, not only among the different parties but also between the members of the same party as general contractor and his subcontractors or suppliers. The satisfaction status and objectives achievement can be increased by minimizing the rework of defects that is the moral of coordination process (Memon et al., 2012, Abdul-Rahman et al., 2006).

3 METHODOLOGY

This paper presents an in-depth review through literature investigation from famous journals and researches in construction management, to map the improvement of coordination process in construction projects. The search was done using keywords; based on two basic criteria: First, the articles discussed coordination process in construction projects. Second, the article was published during the last 10 years. They also have been accepted as the famous and good quality journals and have been considered to review other subjects in construction management before (Rahman et al., 2013, Memon et al., 2012).

For humanizing the success and efficiency of the in-depth review procedure, the articles were rapidly scanned, coordination and construction projects were the most common words frequented in the search outcomes. Finally, coordination in construction projects were carefully chosen as the three keywords to specify the articles to be considered in this study. The procedures for retrieving the articles were as follows:

1. The headings, keywords, and abstracts were scanned with the selected keywords.
2. Abstracts were evaluated to check whether these articles really fitted with this study aim.

After filtering, 19 articles written on coordination in construction projects were selected and investigated in detail.

4 RESULTS

The detailed review and analysis of the selected 18 article can be classified into two types; First, consider coordination as a target issue form the beginning. Second, where the coordination appeared as a significant factor in the results of the study conclusion. The following sections argues the critical review of coordination process in construction projects based on these two perspectives.

Coordination as a Target, in this group, coordination was the main issue and the core of the study. This group consists of five articles, were discussed in details below:

Based on the assumption that coordination process affect the different approaches of project success in different ways, and dissimilar stakeholders prioritization, a questionnaire survey was conducted by Neeraj Jha and Misra (2007). The coordination activities were ordered for four success approaches; 'Regular monitoring of critical path activities'; 'monitoring the budget on all activities and taking corrective action'; 'application of sound technical practices'; 'implementation of all contractual commitments' are the most important coordination activities consistent to schedule, cost, quality and no dispute measures, respectively. Results showed that coordination process can be addressed by attending to essentially the four broad groups of coordination activities resource handling, planning, team building and contract implementation.

The relationship between coordination process activities in term of important and time consuming is very essential. Saram and Ahmed (2001) arranged array included 64 coordination issues; a questionnaire was developed for construction project managers to indicate the relative importance and time consumed on a 3 point scale (i.e., high, mid or low). Identifying strategic activities and potential delays and ensuring the timeliness of all work were the most important activities. Conducting regular meetings and project reviews and analyzing the project performance, detecting variances and dealing with their effects appeared to be the most time-consuming activities.

Hai et al. (2012) studied the key barriers in construction projects coordination. A significant criticism of poor performance is commonplace in construction industry in comparison with the other industries. On the other hand, coordination process is perceived as the best resolution to this dilemma. Five groups of vital barriers were reviewed including; the nature of construction with four subgroups; complexity and intangibility of project activity, uniqueness and low repetition, temporary and labor intensity. Traditional contractual arrangement with four subgroup; uneven risk allocations, fragmentation, lowest-bid-winner, multi-layer subcontracting. Construction participant with three

Table 1. The first group of articles (coordination as core issue).

Authors	Area of study	Data collection method	Sample
(Neeraj Jha and Misra, 2007)	Ranking and classification of construction coordination activities in Indian projects	Questionnaire survey	Construction professionals
Saram D. & Ahmed S. (2001)	Construction Coordination Activities: What Is Important and What Consumes Time	Questionnaire survey	Construction project managers
Hai et al., (2012)	A Conceptual Study of Key Barriers in Construction Project Coordination	Critical review	Articles from international journals
Noor, (2010)	Challenges in Coordination Process for Tall Building Construction	Questionnaire survey	Consultant firms, developers and government
Ali and Rahmat, (2009)	Methods of Coordination in Managing The Design Process of Refurbishment Projects	Interviews, questionnaire survey	Architects

subgroups; myriad and multi-discipline, adversarial relationships, uncommon objective. Characteristic of organization with two subgroups; temporary organization, and management approach. The study initiated, worthwhile new research subject in improving the dimension of coordination-applied efficiency.

For tall building projects, coordination process is more importune than other building projects. It involves many parties and special construction methods and technologies. Therefore, Mohd. Noor (2010) investigated coordination process challenges and obstacles in tall building construction through questionnaire survey. In the results, the current coordination problems of tall building construction need more investigation. Inefficient or poor coordination process is the major responsible for the projects conflicts and disputes, especially during construction phase. Among the proposed suggestions to improve the coordination process were; more tracking and monitoring for work sequence, implementation of quality management system.

Ali and Rahmat (2009) investigated design process coordination of rehabilitation projects in Malaysia. Rehabilitation projects are more complicated than new projects, because of ambiguity factors. The ambiguity of repair projects is reflected in the complexity of gaining design data. The actors involved must be disposed to make choices in their specialized ranges without coordinating with other actors. Together, causes breakdown of their jobs in the projects, which lead to ambiguity in data used and inadequate design. In this case, higher probability of conflicts during construction are expected. Semi-structured interviews and the postal questionnaire survey were used in data collection. The analysis concluded that, coordination process should be used more by architects to get design data with scheduled and arranged meetings.

Coordination as a Significant Factor, in this group of articles, coordination process appeared as a significant factor in the articles results. This group consists of thirteen articles, the most important four were discussed in detailed below and the other were tabulated in Table 2.

Alaghbari et al. (2007) discussed expressively the factors lead to delay in building construction projects. They found that coordination process is the second most essential factor producing delay in construction projects in their questionnaire survey results. They concluded that, coordination problems not just cause delay, but also affect the cost and the quality issues. In their recommendation, they mentioned to implement deeper investigation of coordination in construction industry, as endeavor to improve the performance.

Rahman et al. (2013) discussed cost overrun problem concern for scholars and construction experts. They explored the consequence of different factors on cost performance in small-scale construction projects. The survey was implemented between client, consultants and contractors. Lack of coordination between parties' with 0.932 was the most important factor. This points out that, for achieving better cost performance in small-scale construction projects, coordination procedures and mechanisms during project life cycle required to be improved.

Yih Chong et al. (2011) argued contract conflicts, which cost construction business millions of dollars every year. Due to poor coordination of construction contracts, disputes and conflicts have appeared frequently. Coordination will provide suitable management system, to supply useful information and data for users. Coordination in contract management become very critical and affect projects success serially, minimize claims and disputes in construction projects.

Memon et al. (2012) carried out data collection in two phases, merging quantitative and qualitative approaches. The findings of investigation discovered that, 92% of respondent said, construction projects suffered from time overrun. In terms of cost performance 89% of respondents agreed that, their projects were in front of the cost overrun. One of the mitigation measures for developing time and cost performance suggested was coordination between the parties. After the critical review and the deep investigation of the selected articles, the coordination process progress in construction become clearer. Coordination process

Table 2. The second group of articles (coordination as significant factor).

Authors	Area of study	Data collection method	Sample
Alaghbari et al., (2007)	Factors causing delay of construction projects	Questionnaire survey	Government, contractors, consultants, developers
Abdul-Rhman, et al., (2006)		Survey	Contractors
Hamzah et al., (2011)		Critical Review	Articles from international journals
Memon et al., (2012)		Questionnaire & interviews	Client, consultant and contractors
Memon and Abdul Rahman (2013)	Factors causing cost overrun of construction projects	Questionnaire survey	Client, consultants and contractors
Memon et al., (2012)		Questionnaire & interviews	Client, consultant and contractors
Chong et al., (2011)	Conflicts with traditional procurement and contract management	Delphi technique	12 experts specialized
(Jaffar et al., 2011)		Critical review	Articles from international journals
Kong and Gray, (2006)		Questionnaire survey	Architect, surveyors, contractors, clients
Memon and Zin (2010)	Resource-driven scheduling and productivity	Questionnaire survey	Contractors
Kadir, et al., (2005)		Questionnaire survey	Contractors, developers and consultants
Al-Tmeemy et al., (2011)	Future criteria, framework and application history	Questionnaire survey	Building contractors
Latiffi et al., (2013)		Critical review	Articles from international journals
Rasli, (2006)		Delphi technique & questionnaire	Consultants

still in the infancy phase in construction projects and need more investigation and research to reap its dividends. In terms of the articles considered the coordination issue as a core, it is seems that coordination need more attention as a core issue, Table 1 below summarized those articles.

The second group of articles, where coordination was one of their results, as a significant factor. Most of those articles explored the causes of unsatisfied performance in construction industry, lack or poor coordination was one of the most critical factors in their results. Table 2, below classified them based on the area of study.

Coordination issue appearance in this large number of articles as main factor affecting construction projects, indicate that, coordination is very important and still need more investigation.

5 CONCLUSIONS

Construction projects involve adversary relationships between parties controlled by contract. Management of those parties without coordination led to poor performance and un-satisfaction status. Early coordination is a serious duty in construction projects management and affects the project success. This research defines coordination in construction projects as working together to achieve effectively and efficiently accomplish of construction project. Coordination in construction projects has emerged and implemented not for a long time as a separate concept. Other industries have been very successful in taking advantage of coordination technologies. In construction industry, coordination are still in the infancy stage. Based on this critical review, construction industry does aware enough about the problem in current coordination practiced. All parties involved in construction project must enhance their effort and gives more emphasis on coordination process. They should make more effort to adopt methodology, which can bear improvements for them. A significant improvement of project performance can be archived through the modern intelligent techniques that might useful in improving coordination process in construction projects.

REFERENCES

Abdul-rahman, H., Berawi, M., Berawi, A., Mohamed, O., Othman, M. & Yahya, I. 2006. Delay mitigation in the Malaysian construction industry. *Journal of construction engineering and management,* 132, 125–133.

Alaghbari, W. E., Razali A. Kadir, M., Salim, A. & Ernawati 2007. The significant factors causing delay of building construction projects in Malaysia. *Engineering, Construction and Architectural Management,* 14, 192–206.

Ali, A. S. & Rahmat, I. 2009. Methods of coordination in managing the design process of refurbishment projects. *Journal of Building Appraisal,* 5, 87–98.

Hai, T. K., Yusof, A. M., Ismail, S. & Wei, L. F. 2012. A Conceptual Study of Key Barriers in Construction Project Coordination. *Journal of Organizational Management Studies,* 2012, 1.

Jaffar, N., Tharim, A. A. & Shuib, M. 2011. Factors of conflict in construction industry: a literature review. *Procedia Engineering,* 20, 193–202.

Malone, T. W. & Crowston, K. What is coordination theory and how can it help design cooperative work systems? Proceedings of the 1990 ACM conference on Computer-supported cooperative work, 1990. ACM, 357–370.

Memon, A. H., Rahman, I. A. & Azis, A. A. A. 2012. Time and cost performance in construction projects in southern and central regions of Peninsular Malaysia. *International Journal of advances in applied sciences,* 1, 45–52.

Mohd. Noor, M. N. 2010. *Challenges in coordination process for tall building construction.* Master Thesis, Universiti Teknologi Malaysia.

Neeraj Jha, K. & Misra, S. 2007. Ranking and classification of construction coordination activities in Indian projects. *Construction Management and Economics,* 25, 409–421.

Rahman, I. A., Memon, A. H. & Karim, A. T. A. 2013. Significant factors causing cost overruns in large construction projects in Malaysia. *Journal of Applied Sciences,* 13, 286.

Saram, D. D. D. & Ahmed, S. M. 2001. Construction coordination activities: What is important and what consumes time. *Journal of Management in Engineering,* 17, 202–213.

Yih Chong, H., Balamuralithara, B. & Choy Chong, S. 2011. Construction contract administration in Malaysia using DFD: a conceptual model. *Industrial Management & Data Systems,* 111, 1449–1464.

Effectiveness of preventive safety management in construction

I. Othman, M. Napiah, M.F. Nuruddin & M.M.A. Klufallah
Department of Civil and Environmental Engineering, Universiti Teknologi PETRONAS, Seri Iskandar, Perak, Malaysia

ABSTRACT: Construction industry has higher risk on threat and danger. As employment increases, the number of fatalities also increases. Objective of this study is to improve the percentages of accidents in building construction. This study also will lead the reader to know how far is the implementation of the construction safety management system has been implemented. The study started with gathering information through literature review whereby factors that might affect the standards of construction industry in general were identified. A study questionnaire was then prepared based on list of factors obtained from literature review and sent out to respondents. After analysing the results of the questionnaire survey, it was found that improvements in many aspects need to be considered and monitored frequently in order to ensure the effectiveness of preventive safety management system was implemented.

1 INTRODUCTION

The construction industry has been ascertained as one of the most hazardous industries worldwide, as assessed by work-related fatality rates and workers' compensation. Due to significant changes that the industrial safety has gone through, safety in this aspect has become a complex phenomenon [1]. Nevertheless, construction personnel who engage in the construction industry face a greater risk of fatality than workers in other industries as a high rate of accident-related incidents registered are from construction sector [2]. This incidents frequently occurred due to the lack of preventive measures [3]. Despite that, firms that demonstrate commitment to well-designed and funded safety system can effectively decrease incident rates [4].

1.1 Objectives

This research is mainly focused on three objective:

- To determine the causes underlying accidents at construction sites and its effect to the stakeholders.
- To investigate the importance and effectiveness of safety management system in construction.
- To propose suggestions in the current safety management system practices for improving its performance.

1.2 Scope of study

This research and review focuses on the effectiveness of preventive safety management implemented in construction project in Malaysia. This research also determine the importance of preventive safety management and rank them according to awareness, efficiency of the implemented system and its significance in construction projects.

2 FACTORS INFLUENCING SAFETY PERFORMANCE IN CONSTRUCTION INDUSTRY

2.1 Organizational safety policy

The most influential factor driving safety performance in the construction industry is the organizational safety policy [6]. Meanwhile, better safety performance requires more detailed safety programs [7].

2.2 Safety training

There is congruent agreement among the respondents of their study that worker training is essential to improve safety performance [8] together with more knowledge and training regarding to cause of accident at work place. A research by Construction Industry Institute (CII) in 1993 recognize that safety training as one of five high-impact zero accident techniques [9].

2.3 Safety meetings

Regular site safety meetings are essential to communicate safety information to all parties. Construction company required daily "tool box" safety briefing each morning for their employees to achieve safety performance [10]. In order to improve safety performance at the project level, the number of formal meetings should be increased at project sites [11].

2.4 Safety equipment

The lack of safety protective measures and personal protective equipment (PPE) contributed to poor safety performance [12]. Construction accidents occur because of the absence of safety equipment [9]. Most workers who suffered injuries were due to not wearing PPE like hardhats when performing their normal jobs at their regular sites [10]. The good performance of various fall protection systems can prevent workers from fall injuries [8].

3 METHODOLOGY

The analysis of survey was done by Relative Importance Index (RII) method by using the following formula:

$$\text{Relative importance/difficulty index} = \frac{\sum w}{AN}$$

Where w is the weighting given to each factor by the respondents, ranging from 1 to 5, A is the highest weight (i.e. 5 in the study) and N is the total number of samples.

At the same time, the analysis also was based on the qualitative measurement or ranking system. Rating for the questionnaire is 1 – Totally Disagree, 2 – Disagree, 3 – Moderately, 4 – Agree, 5 – Totally Agree. The Average Index Formula as follow:

$$\text{Average Index (AI)} = \sum (\text{ß} \times n)/N$$

Where, ß is weighing given to each factor by respondents
  n is the frequency of the respondents
  N is the total number of respondents
  With the rating scale as below [13]

- 1 = Never/Totally Disagree (1.00 < Average Index < 1.50)
- 2 = Rarely/Slightly Disagree (1.50 < Average Index < 2.50)
- 3 = Sometimes/Neutral (2.50 < Average Index < 3.50)
- 4 = Often/Slightly Agree (3.50 < Average Index < 4.50)
- 5 = Very Often/Strongly Agree (4.50 < Average Index < 5.00)

Two types of data collection as below:-

a) Primary Data
  Acquired from literature sources, a list of risk factors was prepared and distributed to the respondents from construction industry.

Table 1. Respondent's Background & Questionnaire Feedback Statistics.

Sectors	Distributed		Returned	
	Building	Oil & Gas	Building	Oil & Gas
Consultant	15	15	11	13
Contractor	40	40	21	28
Developer	20	20	16	15
Government	20	20	18	14
TOTAL	95	95	66	70
Response Rate (%)			70	74

b) Secondary Data
  Some of journals, conference paper and thesis related to the research were reviewed as reference. These materials were obtained from UTP's Information Resource Centre (IRC) and e-Resources. Besides that, the general information from PETRONAS safety management is also obtained for the use of this research.

4 RESULTS AND DISCUSSION

The total of 95 questionnaires were distributed through self-delivery of hardcopy questionnaires to the parties involved. The questionnaire consists of four main sections, General Information, Importance of Safety Management System, Factors Causing Accidents, Important Current Practices of Safety Management System and Its Effectiveness in Construction.

The first part of the discussion was looking into the background of the respondents and the rate of response from the survey. Referring to Table 1, 42% (80 persons) works in construction firms, 21% (40 persons) work for developer or operator, 21% (40 persons) work with government agencies and 16% works in consulting firms. The participation of different stakeholders in a typical project organization was important in this survey as it will be a collective and non-biased response which will further improved the findings. In addition, the sample size of this study is 95 persons per industry and referring to Table 1, the response rate for the building construction was 70% and for the oil and gas industry was 74%. These response rates were acceptable and the data was valid to be used.

The second part of questionnaire – Section B, focuses on Importance of Safety Management System in the industry. It indicated in what way the system could benefit a project or company. Referring to Table 2, for building construction, the top three items were; increase overall safety in construction (4.40), ensure smooth execution of work (4.08) and reduce probability of accident at site (4.08). This index indicated that the general population of respondents in the industries were aware of the importance of safety management system. Awareness was part of behavioral

Table 2. Importance of Safety Management in Industries.

No	Item	Relative Important Index	
		Building	Oil & Gas
1	Smooth execution of work	4.08	4.32
2	Increase overall safety	4.40	4.78
3	Improve business profitability	3.69	4.44
4	Reduce operational cost	3.62	4.22
5	Complete job within schedule	3.85	4.13
6	Increase confident of client	3.77	3.95
7	Increase confident of workers	4.02	4.06
8	Reduce the probability of accident at site	4.08	4.25
9	Enhance knowledge and skill of workers	3.85	4.14

Table 3. Factors Causing Accidents in Building Construction.

No	Item	Sub Item	Average Index
1	Unsafe equipment	Without safety devices	0.95
		Equipment failure	0.92
2	Job site condition	Poor site management	0.98
		Excessive noise	0.91
3	Unique nature of industry	Work at high elevation	1.00
		Transient workforce	0.93
4	Unsafe method	Incorrect procedure	0.92
		Knowledge level	0.95
5	Management	No training provided to the workers	0.97
		Poor inspection program	0.99

Table 4. Important Current Practices of Safety Management System in Building Construction.

No	Item	Sub-Item	Average Index
1	Motor vehicle accident	Road leading to well sites lack of safety features	0.90
		Fatigue due to long driving distance and long working shifts	0.94
2	Contact injuries	Improper storage of equipment	0.95
		Incorrect procedures of using tools	0.92
3	Fire, explosion & hazardous gases	Leakage of hazardous gases causing asphyxiation	0.91
		Presence of highly combustible hydrocarbons	0.92
4	Slip, trip & fall	Frequent need to work at elevation	0.95
		Uneven surfaces	0.94
5	Confined spaces	Unfavorable natural ventilation	0.91
		Not designed for continuous employee occupancy	0.93
6	Management	Practice of reduce cost, increase outputs	0.98
		Non-compliance with safety regulations	0.96

perception for safety improvement in the construction industry which was achievable if all parties in the construction sites change their behavior. In addition, the top three items were actually linked to each other in a sense that to increase overall safety in construction, one must reduce probability of accident at a site. This was actually part of the preventive safety management system whereby risks and hazards were identified beforehand. Once the risks and hazards were eliminated, then overall safety will be increased.

The third part of questionnaire – Section C, focused on factors that cause accidents in industry. Referring to Table 3, the top three factors that cause accidents in building constructions were working at high elevation (1.0), poor inspection program (0.99) and poor site management (0.98). Working at high elevation were the most common cause of fatalities in construction industries, accounting for nearly 29% fatal injuries to workers. Besides that, falls from height was part of the Construction's Fatal Four that causes the highest fatalities. Lack of fall arrest systems and guardrails contributed to this factors. Meanwhile, lack in enforcement of safety was acknowledged to be major cause of accidents. Safety performance was affected by monitoring of safety compliance. Safety inspections were the norm in ensuring safety at construction sites. The results from a semi-structured interview showed that the majority of the participating stakeholders "occasionally" conduct safety inspections. Lack of safety inspection lead to more safety violations which increased the probability of site accident. Safety inspections were ways for project managers and site supervisors to engage with the nature of the safety conditions on site. To effectively enforce safety on sites, the responsible party must be able to monitor work on site more frequently. At the same time, poor site management had also contributed to most construction accident on site. Common problems included fall hazards, such as trailing cables and uneven ground and protruding hazards such as nails and scaffolding components. These were often coupled with a lack of clearly defined walkways and poor housekeeping.

A result of the semi-structured interview had indicated that poor site management or difficult side conditions were inevitable due to dynamic nature of the workplace and work activities that occurred in construction. Poor site management in construction appeared to be culture in the industry.

The fourth part of the questionnaire – Section D, it focused on the important practice of safety management in the respective industry and its effectiveness in preventing accidents. Referring to Table 6 and 7, the top three current practices of safety management in building constructions were provision of safety signboard at work place (4.78), inspection of formwork braces and other supports by designated qualified person (4.20) and usage of proper personal protective equipment (PPE) at construction site (4.15).

5 CONCLUSIONS

Implementation of an efficient preventive safety and health performance was very important as effective safety management system can reduce the probability of accident at site. The preventive action taken will ensure smooth execution of work, hence increase overall safety and profitability. Among the factors that caused accident in construction were working at high elevation, poor inspection program, poor site management, non-compliance with safety regulations and improper storage of equipment. Among the influencing factors determining the success of a safety management system were the personal factor and management factor. However, among the sub-factors were safety awareness and commitment of top management towards safety. Based on research done, improvements in many aspects need to be considered and monitored frequently in order to ensure the effectiveness of safety management in construction industry.

REFERENCES

[1] El-Mashaleh, M.S., Al-Smadi, B. M., Hyari, K. H., & Rababeh, S. M. (2010). Safety management in the Jordanian construction industry. Jordan Journal of Civil Engineering, 4(1).

[2] Hallowell, M. and Gambatese, J. (2009). "Construction Safety Risk Mitigation." J. Constr. Eng. Manage., 135(12), 1316–1323.

[3] Jaselskis, E. J., Anderson, S. D., & Russell, J. S. (1996). Strategies for achieving excellence in construction safety performance. Journal of Construction Engineering and Management, 122(1), 61–70.

[4] Kleiner, B., Smith-Jackson, T., Mills, T., III, O'Brien, M., and Haro, E. (2008). "Design, Development, and Deployment of a Rapid Universal Safety and Health System for Construction." J. Constr. Eng. Manage., 134(4), 273–279.

[5] Kumar, S., & Bansal, V.K. (2013). Construction Safety Knowledge for Practicioners in the Construction Industry. Journal of Frontiers in Construction Engineering, 2(2), 34–42.

[6] Mitropoulos, P. and Namboodiri, M. (2011). "New Method for Measuring the Safety Risk of Construction Activities: Task Demand Assessment." J. Constr. Eng. Manage., 137(1), 30–38.

[7] Mohamed, S. (2002). Safety climate in construction site environments. Journal of construction engineering and management, 128(5), 375–384.

[8] Ning, D. C., Wang, J. P., & Ni, G. D. (2010, August). Analysis of factors affecting safety management in construction projects. In Management and Service Science (MASS), 2010 International Conference on (pp. 1–5). IEEE.

[9] Park, M., Elsafty, N., and Zhu, Z. (2014). "Hardhat-Wearing Detection for Enhancing On-Site Safety of Construction Workers." J. Constr. Eng. Manage., 10.1061/(ASCE)CO.1943-7862.0000974, 04015024.

[10] Sunindijo, R. and Zou, P. (2013). "Conceptualizing Safety Management in Construction Projects." J. Constr. Eng. Manage., 139(9), 1144–1153.

[11] Yilmaz, F. (2015). Monitoring and Analysis of Construction Site Accidents by Using Accidents Analysis Management System in Turkey. Journal of Sustainable Development, 8(2), p57.

[12] Othman, I., Idrus, A., and Napiah, M. (2015). Effectiveness of safety management in oil and gas project, Applied Mechanics and Materials, 815, 429–433.

[13] Majid, M.Z., and McCaffer, R. (1997). Discussion of assessment of work performance of maintenance contractor in Saudi Arabia.

Highway project performance evaluation: A study

Widarto Sutrisno
Sarjanawiyata Tamansiswa University, Indonesia

ABSTRACT: Highway projects are the primary Indonesian issue as a development target in The Medium Term Development Plan. Earned Value Management (EVM), a project control technique which provides a quantitative measure of project performance needed to assurance the target. To evaluate work progress in order to identify schedule slippage and areas of budget overruns used the earned value technique as a proven method. It is involves a crediting (earning) of budget as scheduled work is performed. Budgeted cost of work performed is compared against budgeted cost of work scheduled to assess schedule variances, respectively. The selected site for case study is located in Yogyakarta and found that the project has average schedule variance of 2,53 in 2013, 3,09 in 2014 and 4,40 in 2015 that means the project is ahead of schedule. A SPI of 1,16 in 2013, 1,11 in 2014 and 1,23 in 2015 would tell that the project is progressing ahead of the rate originally planned.

1 INTRODUCTION

Indonesian government in 2015–2019 will build 2.650 km new highway projects and maintenance 46.770 km the existing highway to connect regions in order to cut logistic cost and press high cost economy [1]. One of the project evaluations is a review of project cost/schedule [2]. A review of project cost/schedule controls and a comparison of those processes to industry best practices is an essential project controls tool [3]. Evaluation of schedule performance is crucial and should be quick and accurate [4] because times evaluation as well as the functions of management, with good control of time will secure project implementation time [5]. A progress management as good control of time plays an important role in a construction project [6] and the inhibiting factors for cost control and time control have a high level correlation [7].

Earned Value is a well-known project management tool that uses information on cost, schedule and work performance to establish the current status of the project [8]. The technique helps in comparison of budgeted cost of work to actual cost [9]. Earned Value management needed to be a method of performance measurement. With clearer picture, government can create risk mitigation plans based on schedule and technical progress of the work. Before they become insoluble Earned Value management needed to be an "early warning" project management tool that enables to identify and control problems.

2 EVM IN HIGHWAY PROJECT

The mayor performance indicator for construction organization and projects is a schedule. Monitoring this indicator provides valuable information in terms of "current status" [10]. Earned Value is a project management technique that uses "work in progress" which provides schedule performance measurements to indicate what will happen in the future works. It compares actual accomplishment of scheduled work against an integrated schedule and budget plan. EVM provides government with triggers or early warning signals to enhance the opportunities for project success and allow them to take timely actions in response to indicators of poor performance.

The EVM covers some metrics which should be recognized by government which intends to apply this method. The use of metrics is important in the EVM. Planned Value (PV) and Earned Value (EV) form the basis for performance measurement using Earned Value Management.

2.1 *Project plan metric*

A project plan metric identifies the work to be accomplished. Assessment of this metric is known as Planned Value (PV). The planned value is a numerical reflection of the budgeted work to be performed as on the schedule against the actual progress. The main factor related to project plan is Budget Cost of Work Scheduled (BCWS) which compromises the total planned costs for all tasks to be achieved by a given point in time. The PV is often denoted for the sum of budgets for all work packages scheduled to be accomplished within the actual progress in a given time period.

2.2 *Project accomplishment metric*

The metric which quantifies the accomplishment of work is called as Earned Value (EV). The EV reflects

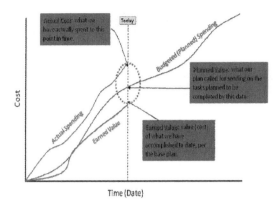

Figure 1. Earned/Planned Value Graph [11].

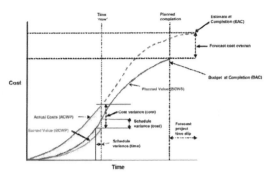

Figure 2. Standard EV elements [11].

the amount of work that has actually been accomplished to date, expressed as the planned value for that work. The EV is often used interchangeably with the Budget Cost of Work Performed (BCWP) as the total planned cost associated with completed work at a given point of time.

2.3 *Performance measurement baseline (PMB)*

For the total program duration, the sum of all work packages Budgeted Cost of Work Scheduled (BCWS) for each time period being calculated. The PMB forms the time-phased budget plan against which project performance is measured.

3 EVM PERFORMANCE ANALYSIS

EVM measures project performance for the current period and the cumulative performance till date given. The important parameters like variances that developed using Planned Value and Earned Value are discussed here.

3.1 *Schedule variance (SV)*

The difference between the work actually performed and the work scheduled. The schedule variance is calculated in terms of the difference value between the work that should have been completed in a given time period and the work actually completed.

3.2 *Schedule performance index (SPI)*

The ratio of work accomplished versus work planned for a specific time period. SPI indicates the amount rate at which the project is progressing.

These variances can be used to answer the key project management questions. Fig 2 is a graphical representation of standard earned value elements. The relationship between those project management questions and the EVM performance measures shown in

Project Management Questions	To Calculate	Formula
[1] How Are We Doing Time-Wise? - Schedule Analysis & Forecasting		
Are we ahead or behind of schedule?	Schedule Variance (SV)	SV= EV-PV
How efficiency are we using time?	Schedule Performance Index (SPI)	SPI= EV/PV

Figure 3. Project management questions and EVM performance measures [11].

SCHEDULE		
SV >0 & SPI > 1	SV =0 & SPI = 1	SV <0 & SPI < 1
Ahead of Schedule	On Schedule	Behind Schedule

Figure 4. Interpretation of basic EVM performance measures [11].

Fig 3. What EVM performance measures indicate about a project in regard to its planned work schedule shown Fig 4.

4 CASE STUDY

The selected site for case study is located in Yogyakarta. Government of Indonesia constructs and maintenance national highway, start in 2013 with 21 project, 2014 with 12 project and 2015 with 15 project. Each year project between 12 month.

From the details of data collected from the government, prepared project cost details which is shown in table 1 respectively. It shows cost for individual project and total cost for the construction of highways.

Cost and schedule details from the collected data, prepared cost vs. duration chart which is also named as S- Curve. This chart shown in fig. 5, give the cumulative planned value of the project for a period of 12 months.

Fig. 6–8 shows the S- Curve for cumulative planned value and earned value of the project for a period of 12 months. This chart shows that the PV line is below the EV line, which means there is good progress in 2013,

Table 1. Project Details.

Project	Budget [Rp.] 2013	2014	2015
P1	13.872.320.849	5.638.110.715	17.495.032.216
P2	18.440.288.821	9.874.095.000	5.134.999.993
P3	15.634.270.423	11.844.844.846	16.986.132.569
P4	10.156.009.223	6.997.799.000	3.754.135.459
P5	23.817.960.745	3.978.158.000	9.693.194.296
P6	37.233.463.402	25.636.277.345	17.331.693.061
P7	28.503.813.432	10.573.042.573	14.352.758.000
P8	10.152.058.578	31.133.727.800	16.178.000.000
P9	9.403.998.254	5.738.039.000	15.451.259.000
P10	11.123.445.192	14.434.752.000	6.895.359.896
P11	4.594.695.047	32.686.677.000	17.177.638.502
P12	3.750.980.000	11.798.453.367	18.980.347.000
P13	4.066.238.719		24.229.203.000
P14	6.008.444.196		9.849.053.259
P15	6.418.937.097		19.876.466.000
P16	7.413.360.700		
P17	9.949.961.473		
P18	8.286.841.000		
P19	16.511.417.758		
P20	4.738.276.000		
P21	114.923.936.000		
Total	365.000.716.908	170.333.976.646	213.385.272.251

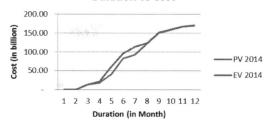

Figure 7. Cumulative Planned Value and Earned Value for the Project in 2014.

Figure 8. Cumulative Planned Value and Earned Value for the Project in 2015.

5 RESULT

From the calculation of various project performance indicators. It is clear that:

1. The project has average schedule variance of 2,53 in 2013, 3,09 in 2014 and 4,40 in 2015 that means the project is ahead of schedule.
2. A SPI of 1,16 in 2013, 1,11 in 2014 and 1,23 in 2015 would tell that the project is progressing ahead of the rate originally planned. SPI indicates the rate at which the project is progressing. SPI > 1 indicates that we are efficient in using time.

6 CONCLUSION

Using this concept to track a real time project is extremely useful and gives a good feel about the performance of the project.

For this study, based on the collected data, schedule of the project and cost for individual project had prepared. S-Curve was drawn which show the relationship between duration and cost of the projects.

Based on the performance we can find out the reasons for variance of schedule and we can predict the future of the projects. Here we found that if work continuous at the current rate in the future the project will take faster than what was originally planned and in 2019 with new target 2.650 km and maintenance 46.770 km will succeed.

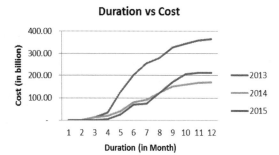

Figure 5. S – Curve.

Figure 6. Cumulative Planned Value and Earned Value for the Project in 2013.

2014 and 2015. From the fig. 6–8 it is cleared that the planned value line is below earned value line, which indicates that the work has been accomplished as per the planned value.

REFERENCES

[1] Information on http://www.bappenas.go.id
[2] Nalewaik, Alexia. Construction Audit: An Essential Project Controls Function, Cost Engineering Vol. 49/No. 10. (2007).
[3] Hasyim, M. H., Unas, S. and Cahyono, N. Matrix of Project Monitoring with Value Concept, Rekayasa Sipil Journal/Volume 4, No.1 (2010).
[4] Vandevelde, Robert. Time Is Up: Assessing Schedule Performance with Earned Value, PM World Today Vol. IX, Issue X. (2007).
[5] Fitrian, M.,A., Acceleration of the Developmnet Project Evaluation Based on Time Schedule Curve S Laboratories Building and Connecting FMIPA Yogyakarta State University, Thesis, 2012.
[6] Chin, S., Yoon, S., Kim, Y,. Jung, Y., Park, S., and Chung, M. A Project Progress Measurement and Management System (2004).
[7] Olawale, Y., and Sun M. "Cost and time control of construction projects: Inhibiting factors and mitigating measures in practice." Construction Management and Economics, 28 (5), (2010) 509–526.
[8] Czarnigowska, Agata. Earned value method as a tool for project control. Budownictwo i Architektura 3. (2008). 15–32.
[9] Verma, A., Pathak, K.K., Dixit, R.K., Earned Value Analysis of Construction Project at Rashtriya Sanskrit Sansthan, Bhopal. International Journal of Innovative Research in Science, Engineering and Technology (An ISO 3297: 2007 Certified Organization) Vol. 3, Issue 4, 2014.
[10] Jung, Y. and Lee, S., Automated progress measurement and management in construction : variables for theory and implementation, Proceeding of the International Conference on Computing in Civil and Building Engineering (ICCBE), Nottingham University Press (2010).
[11] Prasanth, A. and Raja, K.T., Project Performance Evaluation By Earned Value Method. International Journal of Innovative Research in Science, Engineering and Technology (An ISO 3297: 2007 Certified Organization), Volume 3, Special Issue 1, (2014).

Ranking of principal causes of construction waste for Malaysian residential project

Usman Aminu Umar, Nasir Shafiq, Amirhossein Malakahmad, Muhd Fadhil Nuruddin & Ibrahim Umar Salihi
Department of Civil and Environmental Engineering, Universiti Teknologi Petronas, Seri Iskandar, Perak, Malaysia

ABSTRACT: During the last ten decades, many Asian nations around the world have witnessed rapid economic development and social change. The growth of waste generation particularly construction and demolition wastes have attracted considerable attention. The most common causes of waste in construction projects are materials. In Malaysia, the consistent development of the industry offers a strong possibility of greater use of sustainable waste practices, leading towards the nation goals for sustainable development. It is therefore, obvious that the monetary losses from construction material waste can create a great danger to the economic growth and development of a country. Primary and secondary data were gathered using structured questionnaires and interviews. The result indicates that all the primary stakeholders concluded that the leading important factors causing construction waste in Malaysia construction industry are: Site operation; on-site management and planning; materials storage and handling; design and documentation; transportation; and procurement.

1 INTRODUCTION

During the last ten decades, many Asian nations around the world have observed the incredibly rapid economic development and social change, which has tremendously impact urban existence. Six of the world's top ten most populated nations – China, India, Indonesia, Pakistan, Bangladesh, and Japan are located in Asian countries. The region is populated by 3.7 billion people, or around three-fifths of the world's population, about 1.38 billion people are urban residents, and this number makes up about half of the world's urban inhabitants (Shekdar, 2009). Hence, construction waste management will be influenced by the massive amount of construction activity mostly due to highly population and economic development in the region.

The growth of a waste generation, particularly construction and demolition (C&D) wastes have attracted considerable attention. For instance, construction sector generates massive amount waste, equivalent to four times generate in households and over 50% deposited in a landfill (Hwang and Bao Yeo, 2011). Although construction activity has a significant role in developing of cities, it contributes to the degradation of the environment. Some of the negative impacts of this phenomenon include lack of enough area for landfilling of wastes, energy consumption, water usage, dust and gas emission (Lu and Yuan, 2011). As a result, to prevent the waste generation, there is a need to identify out the source of this waste as the factors that lead to the generation of construction waste are many. The most common causes of waste in construction projects are materials. The consequences of materials waste are enormous because materials account for about 50% to 60% construction cost, and they are scarce resources. Lot of materials that get to sites end up as waste through several sources (Formoso et al., 2002). Therefore, the aim of this paper is to identify and rank the most important causes of construction waste materials from Malaysian residential construction project.

Materials wastage on the construction project has emerged as a worm in Malaysian construction sector. The wastes generated at construction areas in physical and non-physical shape. The physical waste from damaged concrete, brick, metals, packaging waste, and so on. While the non-physical wastes are cost overruns and time delays in construction projects (Nagapan et al., 2012). This issue has adversely impacted the general performance of countless projects in Malaysia. (Adewuyi and Otali, 2013), pointed out that wastage in construction businesses has come to stay due to the fact no less than 5% is achieve when making the estimation for any project which is not sufficient. (Teo et al., 2009), noticed that additional construction materials are mostly procured due to material wastage throughout construction. Prior research from numerous nations has proved that waste presents a relatively greater portion of production (Nuruddin et al., 2016). (Tam et al., 2007), in a study in the United Kingdom, stated an extra cost of 15% to construction project cost overruns due to material wastage. While (Oko

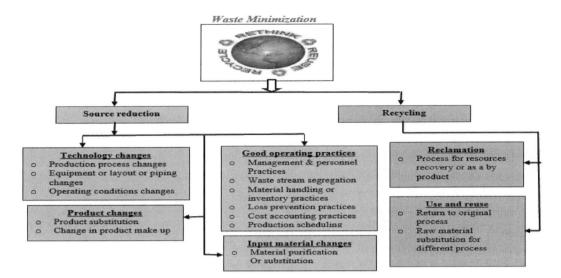

Figure 1. Waste minimization practice.

John and Emmanuel Itodo, 2013) reported that material wastage accounts for between 20–30% project cost overruns.

Many studies in develop nations show that the present of construction waste in the city area tend to grow. Studies in the United States and Europe have shown that substantial amount of waste lies in flow processes of construction (Esin and Cosgun, 2007). Furthermore, research carried out in Sri Lanka also shows that the domestic construction sector labor force is ignorant of flow activities that generate waste and their causes (Senaratne and Wijesiri, 2008). In addition, studies from Nigeria reported waste originate in various phases of construction which can be during planning, estimating or construction phase (Wahab and Lawal, 2011). Other issues based on the Singapore study happen, during design, operational, procurement and material handling factors leading to site waste (Ekanayake and Ofori, 2000). (Alwi et al., 2002), also pointed out construction specialists usually forget to determine or tackle waste in the construction process. It is therefore, obvious that the monetary deficit from construction material waste could cause an incredible risk to the economic growth and development of a country and its environment.

1.1 State of C&D waste in Malaysia

A research by Nasir et al., revealed that 28 percent of municipal solid waste comes from industrial and construction waste in the central and southern regions of Malaysia (Samsudin and Don, 2013). Present publications have mentioned numerous waste minimization alternatives, measures or techniques as shown in Figure 1, (Spies, 2009, Treloar et al., 2003, Abdelhamid, 2014). In Malaysia, there is absolutely no mandatory condition for construction organizations to exercise sustainable resource, waste management, and unlawful dumping continues to be a challenge to the government (Begum et al., 2009). In reaction to that, the Government established an organization known as the Construction Industry Development Board (CIDB); among its goals is to change the sector by enhancing its environmental efficiency.

In promoting to national policy, CIDB has strengthened the sector's dedication to sustainable development and an environmentally accountable sector in the "Construction Industry Master Plan" and is still enlighten the sector's crucial players with sequence training courses, workshops, and awareness raising events (Malaysia, 2007). Moreover, the launch of the Green Building Index (GBI) presents a structure to design and construct sustainable green buildings and increases awareness throughout the industry. In 2014, the Malaysia Productivity Corporation (MPC) revealed a 5.2 percent productivity growth in construction industry (MPC, 2014). This amount shows the vital responsibility the Malaysian construction industry can play in contributing towards the Government's determination towards sustainable development. On the one hand, it supports forecasts that construction waste generation rates will keep growing, positioning raising difficulty on the currently overstretched waste management infrastructure of the nation (MPC, 2014). In general, there are many procedures and voluntary programs supporting sustainable resource and waste management in the Malaysian construction industry; yet the truth remains challenging. The on-going development of the industry offers a possibility for a broader usage of sustainable waste strategies, leading to the country's goals for sustainable development (Papargyropoulou et al., 2011).

2 METHODOLOGY

Text is set in two columns of 9 cm (3.54″) width each Primary and secondary sources of data were gathered

Table 1. Origin of construction waste generated.

S/No	Origin of Construction waste	RII	Rank
1	Site operation	0.733	1st
2	On-site management and planning	0.714	2nd
3	Material storage and handling	0.711	3rd
4	Design and documentation	0.691	4th
5	Transportation	0.674	5th
6	Procurement	0.622	6th
7	External factors	0.568	7th

through structured questionnaires and interviews that have been performed and reviews on scholarly materials on C&D waste. Participants were required to identify which issues they thought to possess major influences on construction waste based on their practical experience. The questionnaire used a 5-point Likert scale starting from 5 to 1, the higher the number, the greater the influence on C&D causes. For the, Sample size, the researcher stratified the accessible population, then adopted the generalized scientific guidelines developed to determine the sample size

Table 2. Sub-categories factors of causes of construction waste.

S/NO	Factors category	RII	Ranking
	Site Operation Related Factors	(1st)	
1	Use of incorrect material, thus requiring replacement	0.760	1st
2	Poor craftsmanship	0.748	2nd
3	Required quantity unclear due to improper planning	0.686	3rd
4	Accident due to negligence	0.684	4th
5	Equipment malfunctioning	0.681	5th
6	Unused materials and products	0.677	6th
7	Time pressure	0.659	7th
	On-site Management and Planning Related Factors	(2nd)	
8	Improper planning for required quantities	0.773	1st
9	Lack of on-site waste management plans	0.751	2nd
10	Lack of supervision	0.751	3rd
11	Lack of on-site material control	0.723	4th
12	Improper planning for required quantities	0.711	5th
	Material Storage and Handling Related Factors	(3rd)	
13	Waste resulting from cutting uneconomical shapes	0.756	1st
14	Damage to materials on site	0.751	2nd
15	Materials supplied in loose form	0.714	3rd
16	Unnecessary inventories on site leading to waste	0.704	4th
17	Poor method of storage on site	0.701	5th
18	Inappropriate site storage space	0.674	6th
19	Manufacturing defects	0.642	7th
	Design and documentation Related Factors	(4th)	
20	Overlapping of design and construction	0.790	1st
21	Design and construction detail errors	0.783	2nd
22	Lack of attention paid to standard sizes	0.783	3rd
23	Design changes	0.780	4th
24	Poor coordination and communication	0.753	5th
25	Designer's unfamiliarity with alternative products	0.736	6th
26	Unclear/unsuitable specification	0.716	7th
27	Design and detailing complexity	0.669	8th
	Transportation Related Factors	(5th)	
28	Damage during transportation	0.726	1st
29	Difficulties for delivery vehicles accessing construction sites	0.630	2nd
30	Inefficient method of unloading	0.600	3rd
31	Insufficient protections during unloading	0.598	4th
32	Procurement Related Factors	(6th)	
33	Ordering errors	0.751	1st
34	Inappropriate methods used for estimation	0.686	2nd
35	Purchased products that do not comply with specification	0.686	3rd
36	Supplier errors	0.667	4th
37	Changes in material prices	0.625	5th
	External factors	(7th)	
38	Weather	0.684	1st
39	Theft	0.610	2nd
40	Vandalism	0.578	3rd

(Kotrlik and Higgins, 2001). The research employed purposive sample to select important informants from all the categories of participants in an attempt to be sure that the appropriate respondents with the relevant experience, recognition and practical knowledge about the various subjects were properly chosen. Equation (1) was applied to obtain the needed sample:

$$[s = (p/P) \times S] \quad (1)$$

Where: S = the sample required for every participating projects, p = is the number of main resource persons in every project, P = the study population and S the total sample size.

A total of 178 participants constituted the sample size. One hundred and seventy-eight (178) questionnaires were given to the selected participants and from all; eighty eight (88) were given back which consists of 49.44%.

The Data analysis was considering the application of Statistic Package for Social Science (SPSS) where the scores allotted to every element from the participants were registered, and thus, the answers from the 88 questionnaires were put through statistical analysis for more observation. The response to each and every factor to entire causes was evaluated, and the position of the elements regarding their criticality as identified by the participants was done by use of Relative Importance Index (RII) which was calculated using equation (2).

$$RII = \frac{\sum W}{A * N} (0 \leq RII \leq 1) \quad (2)$$

Where: W = is the weight given to each factor by the respondents and ranges from 1 to 5, (where "1" is "strongly disagree" and "5" is "strongly agree"), A = the highest weight and, N = is the total number of respondents.

3 RESULTS AND DISCUSSION

The perspectives of respondent of the seven (7) origin and forty (40) sub source of causes of construction waste factors were analyzed based on the relative importance index. The relative importance index and ranks of cause's factors by all the respondents are presented in Table 1 and 2. All the main stakeholders agreed that the top most important factors causing construction waste in Malaysian construction industry were: Site operation; on-site management and planning; materials storage and handling; design and documentation; transportation; procurement and external factors causes by weather, theft and vandalism.

The sub-factors of the causes of materials waste in the Malaysian construction process as identified by the respondents is shown in Table 2. The respondents agree that all the causes generate materials waste in the construction process. However, the respondents have varying level of agreement on these causes as indicated by their different ranks. Their highest level of agreements is on site operation activities because it ranks first in the Table and this could also suggest their order of importance. This is closely followed in descending order as shown in Table 2.

4 CONCLUSION

Building developers and other practitioners provide an essential role to play in minimizing construction waste. This can be achieved by reducing waste through better building design and construction methods, recycling, and re-use of materials can lead to great savings in both natural raw material application associated environmental impacts, as well as the cost to the individual.

ACKNOWLEDGEMENTS

The authors are thankful to the Ministry of Education, Malaysia for providing financial support (Grant No. 0153AB-J13) for this research under the MyRA grant scheme.

REFERENCES

Abdelhamid, M. S. 2014. Assessment of different construction and demolition waste management approaches. *HBRC Journal,* 10, 317–326.

Adewuyi, T. & Otali, M. 2013. Evaluation of Causes of Construction Material Waste: Case of River State, Nigeria. *Ethiopian Journal of Environmental Studies and Management,* 6, 746–753.

Alwi, S., Hampson, K. D. & Mohamed, S. A. 2002. Non Value-Adding Activities in Australian Construction Projects.

Begum, R. A., Siwar, C., Pereira, J. J. & Jaafar, A. H. 2009. Attitude and behavioral factors in waste management in the construction industry of Malaysia. *Resources, Conservation and Recycling,* 53, 321–328.

Ekanayake, L. L. & Ofori, G. 2000. Construction material waste source evaluation. *Proceedings: Strategies for a Sustainable Built Environment, Pretoria,* 23–25.

Esin, T. & Cosgun, N. 2007. A study conducted to reduce construction waste generation in Turkey. *Building and Environment,* 42, 1667–1674.

Formoso, C. T., Soibelman, L., De Cesare, C. & Isatto, E. L. 2002. Material waste in building industry: main causes and prevention. *Journal of construction engineering and management,* 128, 316–325.

Hwang, B.-G. & Bao Yeo, Z. 2011. Perception on benefits of construction waste management in the Singapore construction industry. *Engineering, Construction and Architectural Management,* 18, 394–406.

Kotrlik, J. & Higgins, C. 2001. Organizational research: Determining appropriate sample size in survey research appropriate sample size in survey research. *Information technology, learning, and performance journal,* 19, 43.

Lu, W. & Yuan, H. 2011. A framework for understanding waste management studies in construction. *Waste Management,* 31, 1252–1260.

Malaysia, C. 2007. Construction Industry Master Plan Malaysia 2006–2015. *Kuala Lumpur. Construction Industry Development Board Malaysia.*

M. P. C. 2014. Malaysia Productivity Report. *available at http://www.mpc.gov.my/mpc/images/file/APR%202013% 202014/FullversionAPR20132014.pdf acesss 20 January 2016.*

Nagapan, S., Rahman, I. A. & Asmi, A. 2012. Factors Contributing to Physical and Non-Physical Waste Generation in Construction Industry. *International Journal of Advances in Applied Sciences*, 1, 1–10.

Nuruddin, M. F., Farhan, S. A. & Salihi, I. U. 2016. Application of Structural Building Information Modeling (S-BIM) for Sustainable Buildings Design and Waste Reduction: A Review. *International Journal of Applied Engineering Research*, 11, 1523–1532.

Oko John, A. & Emmanuel Itodo, D. 2013. Professionals' views of material wastage on construction sites and cost overruns. *Organization, Technology & Management in Construction: An International Journal*, 5, 747–757.

Papargyropoulou, E., Preece, C., Padfield, R. & Abdullah, A. A. Sustainable construction waste management in Malaysia: a contractor's perspective. Management and Innovation for a Sustainable Built Environment MISBE 2011, Amsterdam, The Netherlands, June 20–23, 2011, 2011. CIB, Working Commissions W55, W65, W89, W112; ENHR and AESP.

Samsudin, M. D. M. & Don, M. M. 2013. Municipal solid waste management in Malaysia: current practices, challenges and prospects. *Jurnal Teknologi*, 62.

Senaratne, S. & Wijesiri, D. 2008. Lean construction as a strategic option: Testing its suitability and acceptability in Sri Lanka. *Lean Construction Journal*, 4, 34–48.

Shekdar, A. V. 2009. Sustainable solid waste management: an integrated approach for Asian countries. *Waste management*, 29, 1438–1448.

Spies, S. 2009. 3R in construction and demolition waste (CDW)-potentials and constraints. *GTZ–German Technical Cooperation*.

Tam, V. W., Shen, L. & Tam, C. 2007. Assessing the levels of material wastage affected by sub-contracting relationships and projects types with their correlations. *Building and environment*, 42, 1471–1477.

Teo, S., Abdelnaser, O. & Abdul, H. Material wastage in Malaysian construction industry. International conference on Economic and Administration, Faculty of, 2009.

Treloar, G. J., Gupta, H., Love, P. E. & Nguyen, B. 2003. An analysis of factors influencing waste minimisation and use of recycled materials for the construction of residential buildings. *Management of Environmental Quality: An International Journal*, 14, 134–145.

Wahab, A. & Lawal, A. 2011. An evaluation of waste control measures in construction industry in Nigeria. *African Journal of Environmental Science and Technology*, 5, 246–254.

Attributes of coordination process in construction projects

W.S. Alaloul, M.S. Liew & N.A. Zawawi
*Department of Civil and Environmental Engineering, Universiti Teknologi PETRONAS,
Seri Iskandar, Perak, Malaysia*

ABSTRACT: Construction industry is an intricate industry with multi stakeholders in a dynamic environment, which make coordination critical in project success. The aim of this paper is to study coordination attributes in construction projects. Literature review led to identify 29 attributes, which had classified into four groups. A survey was conducted by filling 184 questionnaire. RII and factor analysis was applied to rank and categorize the attributes into clusters. The rank of groups results were; education and gained skills (83.01%), technical (82.26%), personal, (79.50%) and integrity (67.07%). The most important attributes results were; Liaison skill, with being attractive and using all contact approach and communication professionally, RII 92.46%. Then, has a sharp charisma and independence in his decision, RII 90.40%. There is a need for assigning coordinator for each construction project and provide him with communications and transportation facilities, to increase the success probability.

1 INTRODUCTION

Construction industry is considered as a high fragmented, cost and time overruns, and low level of satisfaction. During the construction projects life cycle many parties as contractors, consultants, owner, suppliers, and subcontractor are involved (Xue et al., 2007, Assaf and Al-Hejji, 2006). Management of those dissimilar teams without energetic coordination will lead to poor performance; due to the adversary relationships. Those circumstances make construction industry in a dynamic environment, thus an early coordination is vital in construction projects success. The objective of effective projects management is to ensure the project success. Therefore, continues coordination has been proposed as a powerful tool to ensure the objectives achievement, not only between the different parties, but also between the members of the same party as general contractor, and subcontractor (Cheng et al., 2010). Coordination is unifying, harmonizing and integrating different parties involved in any industry with multiple objectives. Coordination attributes become more critical and must be well described, to facilitate the projects management process. However, attributes of coordination are a much-disputed issue in construction projects (Xue et al., 2007, Enshassi et al., 2009). Consequently, coordination process attributes need more investigation. Thus, the main objective of this paper is, to identify the attributes of construction project coordinator. To achieve this, a postal questionnaire approach was conducted.

In construction industry, active coordination could minimize, predict and remedy problems caused by design modification lead-time, materials availability, workers usage, and equipment availability, consequently the implementation of the construction projects become smoother. Hence, some attributes were recommended for the project coordinator based on the university certification and practical experience (Tatum and Korman, 2000, Assaf and Al-Hejji, 2006).

2 LITERATURE REVIEW

The ability of construction industry to respond effectively to a recovery in complexity of today projects without suffering from capacity constraint or high price inflation is of strategic concern. With current low levels of performance affecting the future productive capacity, the ability of enhancing the construction industry is a key issue. The consequences of a fragmented environment in construction projects contain quite extra transaction costs, augmented requirements for management input and coordination of activities on site, and fewer opportunities to reduce construction waste and defective works (Pheng and Chuan, 2006, Xue et al., 2007).

In construction industry, design variation, materials accessibility, personnel usage and equipment accessibility make projects implementation constrained with several limitations. Active coordination process could minimize and remedy problems caused by industry complication and adversary relationships between parties (Hossain, 2009). Coordination process in construction projects is to harmonize the effort from all parties with the final project objectives. however Coordination can be achieved by the smooth flow of material, information, resources and cash during the project life cycle (Xue et al., 2005, Enshassi et al., 2009).

Several researchers focused on the definition of coordination. Based on Hegazy et al. (2001) coordination is "the integration of several parts into an orderly hole to achieve the project objectives". Coordination can be defined also as, manage dependencies between activities, involving the orchestration, in all levels of project organization (Xue et al., 2005). Wang (2000) and Tatum and Korman (2000) discussed the coordination between successive trades in construction projects. Coordination during the installation of different project components to ensure accessibility for the next component installation, affect significantly on the progress and the project completion. On the other hands, Cheng et al. (2010) discussed coordination in construction sites; the most important aspect was, coordination between every successive task on site, improves the implementation sequence and minimizes the waste. The proposed solution was developed by identified coordination attributes to maximize the net value of project resources (Hegazy et al., 2001, Jha and Iyer, 2006b).

Jha (2005) studied the attributes of coordination process in construction industry. A total of 24 attributes had identified from literatures and the most important ones have established through a questionnaire survey, was conducted among construction professionals. Analysis of responses indicate that, there is a significant relationship between coordination process attributes and project success. Relative important index (RII) was used to choice the highest attributes which affect pointedly in improving coordination process. The analyses indicated that, relationship between client, consultant and contractor; timeliness; technical knowledge of the issue; belief in team live spirit; and coordination for achieving quality are the leading attributes possessed by the successful project coordinators. The responses on the attributes of project coordinators, when analysed through factor analysis suggest the presence of three major categories: team building; contract implementation; and project organization (Jha and Iyer, 2006a, Saram and Ahmed, 2001). In fact, coordination process features in construction projects has applied not for long time. So that, it is one of the most influential procedures in construction projects.

3 METHODOLOGY

As a preliminary stage in this study, relevant literature on coordination process attributes have been combined, which were modified after pilot survey and interviews with professionals. These interviews and pilot survey produced a final list of 29 attributes. Additional information was provided from face-to-face un-structure interviews with selected experts to improve the questionnaire design. A questionnaire was conducted to collect the necessary data. The questionnaire was distributed on 184 of owners, consultants and contractors. Table 1 illustrates the number in each party and their response rate. This technique provides a large number of samples for a meaningful empirical investigation.

Table 1. Sample size and response rate.

Participants	Distributed	Return	Response Rate %
Contractors	70	60	85.7
Owners	100	83	83
Consultants	14	12	85.7
Total	184	188	84.2

All the received and valid data was analyzed via extensively statistical approaches. Univariate and multivariate analysis methods were adapted to identify the vital coordination attributes, and appraise their effect on the projects performance. In order to distinguish those personal characteristics or attributes that are dominant in the successful projects' coordinators RII was calculated. The data was screened for univariate outliers. Three out-of-range values, due to administrative errors, were identified and recoded as missing data. The minimum quantity of responses for factor analysis was satisfied, with the 184 questionnaire. The Kaiser-Meyer-Olkin test of sampling adequacy was 0.72, above the suggested value of 0.6, supporting the presence of all responses in the factor analysis, using both varimax and oblimin rotations of the factor loading matrix (Fang et al., 2004). The three factor solution, which explained 47% of the variance, was preferred because of its previous theoretical support, the 'leveling off' of eigen values on the screen plot after three factors, and the inadequate number of principal loadings and trouble of understanding the fourth factor and succeeding factors. The key goal of factor analysis is to identity simple (items loadings > 0.30 on only one cluster) that are understandable, presumptuous that substances are factorable (Hardcastle et al., 2005).

4 RESULTS

The goal of this study is to recognize the attributes of coordination process in construction projects. The sample was selected to cover the population from various parties of as illustrated in Table 2.

Coordination attributes become even more critical and must be well described with specific in construction, to facilitate the projects management process (Tatum and Korman, 2000). In this part, 29 attributes and characteristics were prioritized and classified. Table 3 below show the coordination attributes Relative Importance Index RII and its groups rank.

The education and gained skills group was the first group with RII of 83.01%. This group represents the academic qualification and gained skills by training courses. The results agreed with Jha and Iyer (2006b) and Xue et al. (2005) who emphasized that, the coordinator education and skills are decisive criterion in project success and the adequacy of coordination process. The second group was technical attributes with RII of 82.26%. Whereas, all previous experiences in the track of the project support the technical attributes. This group of attributes appears from the initial technical feasibility and continues through the sequence of project implementation. The obtained results agreed with Jha and Iyer (2007) who found that, this group has an effective role in disputes and conflicts reduction during the coordination process, where the coordinator familiar with technical details. Personal attributes was the group of the behaviors, conducts and actions occur according to the general thinking and believe. This important group of attributes obtained the third order with RII of 79.50%. The general behaviors and attitude of the coordinator draw the nature of his relation with the others and affect considerably the daily progress and completion of the project. The results agreed with Jha (2005) who emphasized that, this group of attributes illustrate the personal features of the coordinator and reflect the nature of treatment culture in the project. The final group of coordination attributes was integrity group with RII of 67.07%. Since the integrity is an internal feeling, consist from the general context of the project and how the coordinator realizes himself in this project. This important issue in coordination attributes reflects the level of satisfaction because of the project environment. The obtained results agreed with Saram and Ahmed (2001), as those attributes does not obtained a high rank among the coordination attributes.

The first attribute among all coordination attributes was his/her integrity to project, work interest and relationship between parties, with RII of 92.46%. The importance of this duty as the integrity is the internal motivation of the coordinator work. The second attribute was, has sharp charisma and independence in his decision, with RII of 90.40%. Since this coordinator attribute represent the others respect to him and his instructions, it effect on achieving the coordination value in the project completion. The third attribute was, liaison skill, with being attractive and using all contact approaches and communication skills professionally, with RII of 88.11%. Whereas liaison skills are the way to good relations build between construction parties. The fourth attribute was, always meets deadlines for works handing over, with RII of 86.66%. Since this attribute give the trust to coordinator and the others be sure he will do the task professionally (Neeraj Jha and Misra, 2007). Characters are frequently composed of mental, communication, or appropriate knowledge characteristics and they cannot be viewed independent of each other, therefore any change in one will have direct effect on other attributes. However, it can be observed that most of the attributes are correlated with each other attributes. Hence, factor analysis aims

Table 2. Respondent's profiles.

Profile Alternatives	Frequency No.	Percent %
Position in organization		
Project manager	45	29.03
Project coordinator	16	10.03
Site engineer	61	39.30
Office engineer	33	21.30
Years of experience (years)		
From 1 to less than 3	32	20.64
From 3 to less than 5	23	14.83
From 5 to less than 10	44	28.38
More than 11	56	36.16
Degree or education		
B.Sc.	114	73.54
Master	39	25.16
Others	2	1.29

Table 3. RII percent's of coordination attributes and groups.

Group	Attributes	RII	Rank
Education & gained skills (83)	Has liaison skill, with being attractive and using all contact approach and communication skill professionally.	92.46	1st
	Has a civil /architect engineering, planning, certification or a related field.	87.20	5th
	Has computer and modern software's application professionalism	82.86	9th
Technical attributes (82.26)	His/her technical knowledge about the main subject of the work "pavements in roads, pumps in wells".	85.94	7th
	His/her experience in the type of project works or nearest track.	83.31	8th
	Familiar with measuring methods and evaluation criteria's in similar projects/contracts.	82.17	10th
Personal attributes (79.50)	Has sharp charisma and independence in his decision.	90.04	2nd
	Always meets deadlines for works handing over.	87.66	4th
	Has a physiognomy in prediction and planning.	86.86	6th
Integrity attributes (67.07)	His/her integrity to the project, work interest and relationship between parties.	88.11	3rd
	The coordinator not just done his/her work, but he looks for improving and developing the work environment.	81.83	11th
	The coordinator satisfaction on his/her current work from "salary or respect"	80.45	12th

at providing greater insight of relationship among numerous correlated, but seemingly unrelated, variables in terms of a relatively few underlying factor variety. A Principal Axis Factor (PAF) with a Varimax (orthogonal) rotation of 29 Likert scale questions from this arrogance survey questionnaire was conducted on the data (Gerbing and Hamilton, 1996). An inspection of the Kaiser-Meyer Olkin measure of sampling adequacy suggested that the sample was factorable (KMO = 0.698) (Fang et al., 2004, Brown, 2015). All coordination attributes in each group were distributed, based on their loading factor. It is customary to apply rotation in an effort to set of loadings that fit the observations equally well, but more easily interpreted. The goal is to make some of these loadings as large as possible, and the rest as small as possible. All the loading values were in the acceptable range for each attributes to its group. For example, the first group, education and gained skills attributes loading values were all larger than 0.7. (Has liaison skill, with being attractive and using all contact approach and communication skill professionally, 0.974. Has a civil/architect engineering, planning, certification or a related field, 0.878. Has computer and modern software's application professionalism, 0.832).

5 CONCLUSIONS

In construction projects, a project coordinator has to work with many different groups or departments to perform his duties, and for this, he needs to ensure cooperative relationships. To achieve proper relationship he must be good at interpersonal skills and should have good communication skills.

The assigned objectives of this study were achieved through questionnaire survey. After intensive literature review to identify the coordination attributes related to construction projects. The questionnaire was necessary to prioritize coordination attributes based on relative importance. The literature review and interviews led to identify 29 coordination attributes, which classified into four groups using factor analysis. Coordination process in construction industry is a sequential series, so that any failure on any stage will cause the coordination system collapse. Coordination duties and its groups were ranked using RII. Coordination attributes groups order was, education and gained skills attributes group (83.01%), technical attributes group (82.26%), personal attributes group, (79.50%) and integrity attributes group (67.07%). In each group, the twelve attributes obtained more than 80% Coordination process in the Gaza Strip construction projects is a very fecundity context for research. Team building requires conciliatory approach and not the confrontationist approach. A coordinator needs to show concern for other's ego and must have a sound understanding of human psychology. Most importantly, a coordinator must believe in the team spirit. A coordinator must be able to communicate properly, both through verbal and written communication, and he must be proficient in interpersonal skill.

REFERENCES

Cheng, J. C., Law, K. H., Bjornsson, H., Jones, A. & Sriram, R. 2010. A service oriented framework for construction supply chain integration. *Automation in construction*, 19, 245–260.

Enshassi, A., Mohamed, S. & Abushaban, S. 2009. Factors affecting the performance of construction projects in the Gaza Strip. *Journal of Civil engineering and Management*, 15, 269–280.

Hegazy, T., Zaneldin, E. & Grierson, D. 2001. Improving design coordination for building projects. I: Information model. *Journal of Construction Engineering and Management*, 127, 322–329.

Hossain, L. 2009. Communications and coordination in construction projects. *Construction Management and Economics*, 27, 25–39.

Jha, K. Attributes of a project coordinator. 21st Annual ARCOM Conference, 2005. 115–124.

Jha, K. & Iyer, K. 2006a. Critical determinants of project coordination. *International Journal of Project Management*, 24, 314–322.

Jha, K. N. & Iyer, C. K. 2006b. What attributes should a project coordinator possess? *Construction Management and Economics*, 24, 977–988.

Pheng, L. S. & Chuan, Q. T. 2006. Environmental factors and work performance of project managers in the construction industry. *International Journal of Project Management*, 24, 24–37.

Saram, D. D. D. & Ahmed, S. M. 2001. Construction coordination activities: What is important and what consumes time. *Journal of Management in Engineering*, 17, 202–213.

Tatum, C. & Korman, T. 2000. Coordinating building systems: process and knowledge. *Journal of Architectural Engineering*, 6, 116–121.

Wang, Y. 2000. Coordination issues in Chinese large building projects. *Journal of Management in Engineering*, 16, 54–61.

Xue, X., Li, X., Shen, Q. & Wang, Y. 2005. An agent-based framework for supply chain coordination in construction. *Automation in construction*, 14, 413–430.

Xue, X., Wang, Y., Shen, Q. & Yu, X. 2007. Coordination mechanisms for construction supply chain management in the Internet environment. *International Journal of project management*, 25, 150–157.

Structural relationship of success factors for Small Medium Enterprises (SME) contractors in PLS-SEM model

I.A. Rahman, N.I. Rahmat & S. Nagapan
Department of Building and Construction, Faculty of Civil and Environmental Engineering, Universiti Tun Hussein Onn Malaysia

ABSTRACT: This paper presents development of PLS-SEM path model which demonstrates graphical relationships of success factors for SME contractors. The model consisted of 33 success factors which are clustered into 5 independent variables and one dependent variable. Data used for this model were from questionnaire survey amongst 100 SME contractors. The model was constructed and assessed rigorously at measurement and structural components levels. At measurement level, the model has achieved the required threshold values. At structural level, it was found that the model has substantial explaining power to represent groups of success factor to the SME contractors and also the resources related group is having the highest impact value of 0.299 which means giving the strongest influential success factors to SME contractors. This model will benefit the contractors' community in sustaining their success in competitive construction industry.

1 INTRODUCTION

PLS-SEM path modelling uses Partial Least Square (PLS) technique which simultaneously evaluates multiples variables. It is primarily used to develop theories in exploratory research by focusing on explaining the variance in the dependent variables when examining the model. PLS-SEM path model consists of two elements. First element is structural model also called inner model and the second element measurement models also called outer models. The outer models display the relationships between the constructs and indicators variables while the inner model displays the relationships between the constructs. Constructs are variables that are not directly measured while indicators which are also called items or manifest variables are the directly measured proxy variables that contain the raw data (Hair *et al.*, 2014). Manifests are the factors while the Constructs are variables which either independent variables (groups of the factors) or dependent variable (success of SME contractors).

2 MODEL DEVELOPMENT

A hypothetical PLS-SEM model for SME contractors' success factors was developed based on 33 manifests clustered in 5 independent variables and one dependent variable. The manifests are success factors, independent variables are groups of the factors while dependent variable is the success of the contractor which consisted of 2 manifests (success of getting new project and also in completing the

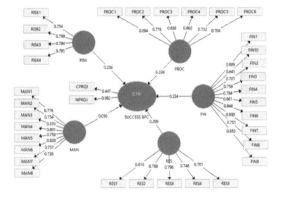

Figure 1. Generated Values after the Simulation Process.

awarded project). In this study, the 33 success factors which were identified through rigorous literature review and verified by experts. These factors are clustered in 5 groups/independent variables. Input data for this model were derived from questionnaire survey involved 100 respondents. This hypothetical model was constructed in SmartPLS software and input data were assigned to the respective items. Then simulation on the model was carried out using PLS algorithm function and the generated results are as in the Figure 1.

Based on these generated values, the model is assessed at two stages which are at the measurement component and also at the structural components of the model. The assessments were carried in accordance to (Hair *et al.*, 2014; Henseler, 2009) suggestions.

Table 1. Convergent Validity of Measurement Model.

Groups/Independent variables	Convergent validity parameters		
	AVE (≥ 0.5)	CR (≥ 0.7)	Alpha (≥ 0.7)
Finance (FIN)	0.544	0.922	0.905
Management (MAN)	0.564	0.911	0.889
Procurement (PROC)	0.533	0.872	0.824
Resources (RES)	0.594	0.879	0.828
Risk (RISK)	0.609	0.862	0.786

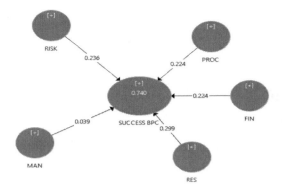

Figure 2. β-values of Structural Model.

3 MEASUREMENT MODEL ASSESSMENT

Assessment of measurement model involves in examining two parts of assessment which are checking model's performance and examining its discriminant validity.

3.1 Model performance assessment

It is carried out by checking individual item reliability and convergent validity concurrently after each iteration. Individual item reliability is to check the correlation of each manifest variable toward the assigned independent variables. According to (Hulland, 1999; Chin, 1998; Hair et al., 2011), individual item reliability is checked based on the loadings of each manifest variable which should be greater than 0.5 which is the threshold value for the individual item reliability and results of the simulation of the model found that all the factors loadings are above 0.5. While in convergent validity, three parameters are used to determine convergent validity which are the Average Variance Extracted (AVE), Composite Reliability (CR) and Cronbach's alpha. The required threshold values for the parameters are Average Variance Extracted (AVE) ≥ 0.5, Composite Reliability (CR) ≥ 0.7 and Cronbach's alpha ≥ 0.7 (Hair et al., 2011; Aibinu et al., 2010; Akter et al., 2011; Ramayah et al., 2011). Thus, for the whole independent variables of the model, the convergent validity values are as in table 1.

Table 1 indicates that all the independent variables are above the threshold values. Then this model does not need any further iterations process. By achieving convergent validity and individual item reliability, it indicates the consistency of all the manifests toward the assigned independent variables.

3.2 Discriminant validity assessment

Discriminant validity of the measurement model is to verify that each independent variable is differing from other independent variables using analysis of Cross Loading and Average Variance Extracted (AVE) (Hulland, 1999; Aibinu et al., 2010; Akter, 2011). Two steps assessment of discriminant validity are Analysis of Cross Loading and Analysis of Average Variance Extracted (AVE). For analysis of cross loading, it was found that the generated values of cross loading from this model for each manifest variable were higher in their relative independent variable. Thus, indicates that the model has achieved partly its discriminant validity. For analysis of Average Variance Extracted (AVE), the diagonal correlation value of the same independent variable is replaced with its value of square root of AVE in which the AVE was determined earlier in Convergent Validity. It was found that the diagonal correlation values of the independent variables are higher than non-diagonal values and this satisfies the discriminant validity of the tested model.

4 STRUCTURAL MODEL ASSESSMENT

Assessment of structural model is carried out by checking the strength of impact path of independent variable to the dependent variable and its explanatory power.

4.1 Impact strength

A structural model is considered acceptable if all the β-value (impact strength of variable toward to the dependent variable) is above 0.1 (Hair et al., 2011; Aibinu et al., 2010). According to Hair et al., (2014) the signage of β-value whether it is positive or negative is not a concern because the impact of path is the absolute value of the β. Results of impact path analysis for this study is as Figure 2.

Figure 2 shows that Risk, Procurement, Finance and Resources groups are having β-values above 0.1 but not Management group where β-value less than 0.1 which indicates that Management group is giving the least impact to the success of the contractors. Resources group with value of 0.299 is considered having the strongest impact on the success of the contractors.

4.2 Assessing explanatory power (R2)

Explanatory Power is to examine the overall ability of the model in representing the impact of independent variables toward the dependent variable. The indicator used to describe the explanatory power is the model's

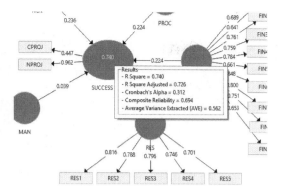

Figure 3. Model's Coefficient of Determination R^2 Value.

coefficient of determination which is R^2 value (Hair et al., 2011; Aibinu et al., 2010; Akter et al., 2011; Memon et al., 2014; Urbach et al., 2010). According to Chin (1998) a model can be considered having substantial, moderate and weak explanatory power if R^2 values are equal or more than 0.67, 0.33 and 0.19 respectively. Figure 3 shows the model's value R^2 is 0.74 which indicates that it has substantial explanatory power.

Based on the outcomes as in Figure 3, it can be concluded that the developed model has substantial explanatory power in representing the impact of the 5 groups of success factors toward the contractor success. In which resources related group is the most influential to the success of the contractors, then followed by Risk, Procurement and Finance. However the management related group is considered insignificant or null and in term of statistic, this incident could be due to the data given by the respondents does not have sufficient power to detect the dependency. This may due to some discrepancies amongst the respondents in giving their opinions during the questionnaire survey.

5 CONCLUSION

This paper has briefly described the development and assessment of PLS-SEM model of SME Contractors' success factors. It has described the assessment processes involved both at measurement and structural components of the model found that all the required parameters are above the required threshold values. The model has also been verified to have an explanatory power (R^2) of 0.740 which indicates it has substantial explanatory power in representing the impact of the 5 groups of success factors toward the success of the contractors. However, only four out of five groups of the model can be accepted because having β-values above 0.1 and the groups are Risk, Procurement, Finance and Resources. While management related group is considered null. This developed model can be shared with the interest parties especially SME contractors in Johor who are the respondents of this study.

ACKNOWLEDGMENT

The authors are grateful to FRGS Grant No 1221, University Tun Hussein Onn Malaysia for supporting this research.

REFERENCES

Hair, J.F, Hilt, G.T.M, Ringle, C.M. & Sarstedt, M. 2014. A Primer on Partial Least Squares Structural Equation Modelling (PLS-SEM), United State of America: SAGE Publications.

Henseler, J. 2009. On Convergence of partial least squares path modelling algorithm, Computational Statistics, 25(1): 107–120.

Hulland. 1999. Use of Partial Least Squares (PLS) in Strategic Management Research: A Review of Four Recent Studies, Strategic Management Journal, 20(2): 195–204.

Chin, W.W. 1998. The Partial Least Squares Approach to Structural Equation Modelling, Modern Metods for Business Research, 295(2): 295–336.

Hair, J.F, Sarstedt, M.C, Ringle, M. & Mena, J. 2011. An Assessment of the Use of Partial Least Squares Structural Equation Modelling in Marketing Research, Journal of the Academy of Marketing Science, 40(3): 414–433.

Aibinu, A. & Al-lawati, A. 2010. Using PLS-SEM Technique to Model Construction Organizations Willingness to Participate in E-bidding, Automation in Construction, 19(6): 714–724.

Akter, S. Ambra, J.D. & Ray, P. 2011. An evaluation of PLS based complex models: the roles of power analysis, predictive relevance and GoF index, Proceedings of the 17th Americas Conference on Information Systems, Detroit, Michigan, 2011, pp. 1–7.

Ramayah, T. Lee, J.W.C. & Chyaw, J.B.C. 2011. Network Collaboration and Performance in the Tourism Sector, Service Business, 5(4): 411–428.

Memon, A.H., Rahman, I.A., Akram, M., Ali, N.M. 2014. Significant factors causing time overrun in construction projects of Peninsular Malaysia, Modern Applied Science, Volume 8, Issue 4, Pages 16–28, Canadian Center of Science and Education.

Urbach, N. and Ahlemann, F. 2010. Structural Equation Modelling in Information Systems Research Using Partial Least Squares, Journal of Information Technology Theory and Application, 11(2): 5–40.

Engineering Challenges for Sustainable Future – Zawawi (Ed.)
© 2016 Taylor & Francis Group, London, ISBN 978-1-138-02978-1

Categorization of Saudi Arabia's construction delay factors using factor analysis technique

I.A. Rahman, N. Al-Emad & S. Nagapan
Faculty of Civil and Environmental Engineering, Universiti Tun Hussein Onn Malaysia, Parit Raja – Batu Pahat, Johor, Malaysia

ABSTRACT: This paper presents categorization of 37 delay factors of Saudi Arabia construction industry. The factors were identified from literature review and verified by construction experts through questionnaire survey involved 100 respondents using 5-points Likert's scale. Collected data from the survey were used to categories the factors into groups using factor analysis function in SPSS software. Nine principal components were generated based on factors loading and these components are meant as groups for the factors. Generated results were evaluated to ensure that factors are assigned to an appropriate group. Evaluation involved relocation of factors to ensure that the factors are logically placed in their respective group. The evaluation resulted in 7 groups only where the groups were given names related to the nature of factors in them. Results from categorization can be used for advance analysis such as structural equation modelling which required groups as latent variables.

1 INTRODUCTION

Construction industry is highly dynamic sector and plays very important role in the development of the country (Memon et al., 2014). According to (Al-Emad & Nagapan, 2015), one of the majors issues engulf the construction industry is delay in completing the project within the specific/agreed duration. Saudi Construction Industry is relatively young (Albogamy et al., 2012) and according to Samargandi et al., (2013) the Saudi Arabia construction sector is the largest and fastest growing market in the Gulf region. Billions of dollars were spend on construction projects in Saudi Arabia with projects in Mecca are no exception. Mostafa & Al-Buzz (2015) Stated that more than $100 billion has been allocated for the construction projects which are being implemented in Mecca including Grand mosque (Al-Haram mosque) in order to improve services being rendered to millions of pilgrims who come for Hajj and Umrah. However, there are many challenges faced by the Saudi Arabian construction industry. One of them is project delay that commonly occurred in most of the projects. Construction industry has a very poor reputation in coping with delays (Yusof et al. 2004). A study conducted by Assaf & Al-Hejji (2006) found that around 70% of all public sector construction projects were delayed due to several factors in construction projects in Saudi Arabia. The issue of delay in the construction projects in Saudi Arabia is not a new problem and has created a negative impact to the industry (Assaf et al., 1995; Al-Khalil & Al-Ghafly, 1999; Assaf and Al-Hejji, 2006; Al-Kharashi & Skitmore, 2009; Albogamy et al., 2012; Mahamid, 2013; Alotaibi at al., 2014; Mahamid, 2014 and Elawi, 2015) to explore the issue as in many previous studies in regard to Kingdom of Saudi Arabia context.

For this study the literature work found 37 causative factors to construction delay faced by Saudi Arabia construction industry. These factors are not being clustered into groups and been used in questionnaire survey regarding their level of significance toward construction delay. The survey was participated by 100 respondents of Saudi Arabia construction community such as clients, consultants and contractors. They were required to rate each factor based on 5-points Likert's scale on the level of significance. Data generated from this survey are used in factor analysis to segregate the factors in their respective groups.

2 FACTOR ANALYSIS AND ITS ASSUMPTIONS

In this paper, factor analysis technique was applied to categorize inter-related factors causing construction delay in Mecca city. Factor analysis approach is the first generation of statistical multivariate analysis which usually used for exploratory work. It is to obtain multivariate interrelationships existing among the major components where it exposes the hidden dimensions that may or may not be apparent directly in the univariate analysis. While carrying out factor analysis, user must understand several important functions in the analysis. One of it is the sampling adequacy test on the data in which Kaiser–Meyer–Olkin (KMO) test

Table 1. KMO and Bartlett's Test Results.

Kaiser-Meyer-Olkin Measure of Sampling Adequacy		0.822
Bartlett's Test of Sphericity	Approx. Chi-Square	2439.901
	df.	666
	Sig.	0.000

and Bartlett's test of sphericity were used to evaluate the satisfactoriness of the survey data for factor analysis. The value of KMO can be varied from 0 to 1, a minimum value of 0.50 was proposed suitable for factor analysis (Williams et al., 2010). According to (Doloi et al., 2012) a value close to 1 illustrates that the pattern of correlations is relatively compact and hence factor analysis should give distinct and reliable results. (Young-Mok et al., 2008), suggested that the overall KMO measure should be greater than 0.80; however, a measure of above 0.60 is tolerable. Besides KMO measure the Bartlett's test of sphericity relates to the significance of the study shows the validity and suitability of collected data. According to (Williams et al., 2010), for factor analysis to be recommended suitable, the Bartlett's test of sphericity must be less than 0.05. KMO test and Bartlett's tests are conducted concurrently while conducting factors analysis in SPSS software. Users need to tick appropriate functions regarding these tests before finally doing factor analysis. Results from these tests are generated together with the factor analysis results.

For this study, the results for KMO test and Bartlett's test of sphericity are 0.822 and 0.000 respectively as summarised in table 1. The results indicate that the data sampling are adequate for factor analysis work.

In factor analysis technique, it requires to decide the types of rotation to be used for reducing the number factors on which the variables under investigation have high loadings. This study used an Equamax rotation for sorting out the construction delay factors into related groups. It is because Equamax rotation is a compromise between Variamax rotation (which attempts to make as many values in each factor column to be as close to zero as possible) with Quartimax rotation (which attempts to make as many values in each row to be as close to zero as possible (Abdi, 2003) and also due to the loadings so that a variable loads high on one factor but low on others. Beside that the technique requires to select an appropriate extraction method to extract the number of the components/factors.

For this study, maximum likelihood method of extraction was adopted which resulted in extracting 9 components (with eigenvalues greater than 1) (Costello & Osborne, 2005). Considering only 9 components, the generated results from analysis process for all these 9 components along together with their inter-related items are tabulated as in table 2.

Generated results in table 2 indicate that each component is consisted number of inter-related items.

However, there are delay factors/items which are not suitable for original extracted component/group theoretically and should be part of other group. Thus, the results require some modifications processes which are discussed below.

3 RELOCATION OF FACTORS

Results presented in table 2 which generated from factor analysis are reviewed to give appropriate names to the groups and also to ensure that the factors are logically placed in their respective group according to literatures. In this process it was noticed that there few factors which are theoretically not suitable in the assigned groups which required some modification process. Thus, the results generated from factor analysis are revised and modified according to the following discussions:

I. **Modification 1**: In component 1, all the factors represented material and machinery issues except one factor i.e. "delay in obtaining permits from municipality". Since obtaining the permit from the municipality is the responsibility of the client. While component 6 was consisting of factors representing the client and consultant related factors. Hence, the factor "delay in obtaining permits from municipality" was placed in component 6 rather than component 1.

II. **Modification 2**: Component 7 consisted of 3 factors in which 1 of the factors i.e. "Delay in approving shop drawings" normally part of the client and consultant duties and not suitable placed in this group. Hence, this factor was merged in component 6 which contained factors related to client and consultant. Similarly, the other 2 factors i.e. "delay in preparation of shop drawings & delay in preparation and submission of materials samples" were representing contractor's scope and they are not appropriate for this group. Thus, these two factors were moved to component 2 which consisting factors related to contractor's site management group.

III. **Modification 3**: Component 9 consisted of 2 factors in which 1 factor i.e. "delay in progress payment" was related to the scope governed by the client. Hence this factor was moved from component 9 to component 6 which containing factors related to client and consultant factors. Also, factor "difficulties in financing project by contractor" in component 9 was related to contractor issues, and since component 2 consisting items related to contractor's site management. Therefore, "difficulties in financing project by contractor" was shifted to component 2.

These modifications resulted in reduction of components/groups from 9 to 7 groups as follows:

1. Material and Machinery Related Factors (MMF) consisting of 3 factors

Table 2. Factor Analysis Loading Results (Rotated Factor Matrix).

Factors Causing Construction Delay	Component								
	1	2	3	4	5	6	7	8	9
Late procurement of materials	.853	.096	.032	.181	.194	.112	.136	.071	.162
Delay in materials delivery	.802	.259	.065	.037	.132	.151	.213	.197	.169
Delay in equipment delivery	.762	.135	-.030	.026	.245	.159	.218	.155	.190
Delay in obtaining permits from Municipality	.477	.276	.238	-.030	.001	.452	.073	.141	.157
Poor qualification of staff assigned to the project	.254	.783	.015	.047	.196	.136	.123	.317	.176
Shortage of professionals in organization	.117	.764	.093	.129	.202	.164	.120	.103	.128
Inadequate Contractor's experience	.203	.582	.046	.158	.215	.147	.116	.084	.198
Inaccurate estimation of project duration during the bidding stage	.160	.532	.208	-.035	.189	.395	.127	.102	.255
Incompetent subcontractors	.312	.388	.220	.064	.183	.280	.133	.227	.065
Mistakes in design documents	.027	.170	.738	.060	.005	.127	.158	.179	.215
Changes in design documents	-.010	.120	.613	.001	.279	.236	.180	-.020	.092
Late in approving of design documents	.102	-.210	.586	.144	.159	.180	.270	.277	.168
Inadequate details provided in drawings	.054	.157	.561	.205	.197	.211	.106	.057	.142
Delays in producing design documents	.112	-.047	.526	.170	.172	.027	.194	.349	.073
Insufficient data collection and survey before design	.144	.123	.433	-.026	.254	.210	.109	.148	-.057
Poor communication between parties	.136	.073	.036	.925	.102	.065	.200	.153	.067
Poor coordination between parties	-.020	.023	.048	.779	.089	.087	.244	.192	.138
Ineffective monitoring and controlling of the project progress	.069	.403	-.029	.451	.373	-.151	.188	.154	.214
Low productivity level of labour	.023	.204	.254	.186	.763	.052	.120	.108	-.079
Unqualified workforce	.187	.251	.145	.212	.755	.075	.135	.255	.081
Shortage of manpower	.358	.057	.025	-.003	.693	.177	.152	.121	.208
Delay in approving of change orders	.165	.111	.354	-.001	.009	.576	.215	-.078	.282
Frauds practice among the parties involved	.323	.309	.086	.090	.110	.518	.201	.282	.040
Inadequate consultant experience	.062	.041	.270	.366	.110	.511	.167	.138	.210
Lack of coordination with authorities	.130	.340	-.049	.169	.221	.474	.129	.113	.066
Slow decision-making process	.102	.065	.205	.144	.118	.364	.175	.116	.283
Difficulty in accessing the site	.201	.131	.253	-.180	.182	.355	.085	.101	.133
Delay in approving shop drawings	.063	.046	.094	.140	.054	.222	.923	-.023	.245
Delay in preparation of shop drawings	.128	.081	.092	.201	.127	.040	.559	.188	.005
Delay in preparation and submission of materials samples	.314	.117	.151	.165	.121	-.041	.422	-.009	.262
Poor contract management	-.011	.087	.039	.396	.222	.021	.143	.615	.146
Unrealistic contract duration	.196	.113	-.002	-.096	.007	.165	.003	.528	.406
Poor site management and supervision	.019	.197	.217	.346	.113	.272	.278	.493	-.113
Improper planning and scheduling of the project	.168	.081	.143	.341	.261	-.246	.231	.484	.268
lack of teamwork	.183	.194	.142	.264	.173	.076	-.050	.441	-.185
Delay in progress payment	.134	.059	.005	.031	-.003	.128	.249	-.074	.795
Difficulties in financing project by contractor	.092	.186	.144	.198	.067	.050	.035	.187	.598

* Extraction Method: maximum likelihood.
** Rotation Method: Equamax with Kaiser Normalization.

2. Contractor's Site Management Related Factor (CSMF) consisting of 8 factors
3. Design and Documentation Related Factor (DDF) consisting of 6 factors
4. Information and Communication Technology Related Factors (ICTF) consisting of 3 factors
5. Labour Management Related Factors (LAB) consisting of 3 factors
6. Client and Consultant Related Factors (CCF) consisting of 9 factors
7. Project Management and Contract Administration Related Factors (PMCAF) consisting of 5 factors

After the modification processes, classification of construction delay factors in the respective groups are as in table 3.

4 CONCLUSION

This paper has demonstrated the application of factor analysis approach for classification of construction delay factors of Saudi Arabian construction industry into several related groups. Based on factor analysis, the 37 factors are categorized into 9 groups. However, this categorization requires logical modification based on prior knowledge of construction discipline. Finally, the factors were rearranged into 7 groups which were given name according to the character of the factors in that group: named as *Material and Machinery Related Factors (MMF) consisting of 3 factors, Contractor's Site Management Related Factor (CSMF) consisting of 8 factors, Design and Documentation Related Factor (DDF) consisting of 6 factors,*

Table 3. Categorization of Delay Factors.

Group	Factors Causing Construction Delay	Factor Code	Factor Loading
Material and Machinery Related Factors (MMF)	Late procurement of materials	MMF1	0.853
	Delay in materials delivery	MMF2	0.802
	Delay in equipment delivery	MMF3	0.762
Contractor's Site Management Related Factor (CSMF)	Poor qualification of staff assigned to the project	CSMF1	0.783
	Shortage of professionals in organization	CSMF2	0.764
	Difficulties in financing project by contractor	CSMF8	0.598
	Inadequate Contractor's experience	CSMF3	0.582
	Delay in preparation of shop drawings	CSMF6	0.559
	Inaccurate estimation of project duration during the bidding stage	CSMF4	0.532
	Delay in preparation and submission of materials samples	CSMF7	0.422
	Incompetent subcontractors	CSMF5	0.388
Design and Documentation Related Factor (DDF)	Mistakes in design documents	DDF1	0.738
	Changes in design documents	DDF2	0.613
	Late in approving of design documents	DDF3	0.586
	Inadequate details provided in drawings	DDF4	0.561
	Delays in producing design documents	DDF5	0.526
	Insufficient data collection and survey before design	DDF6	0.433
Information and Communication Technology Related Factors (ICTF)	Poor communication between parties	ICTF1	0.925
	Poor coordination between parties	ICTF2	0.779
	Ineffective monitoring and controlling of the project progress	ICTF3	0.451
Labour Management Related Factors (LAB)	Low productivity level of labour	LAB1	0.763
	Unqualified workforce	LAB2	0.755
	Shortage of manpower	LAB3	0.693
Client and Consultant Related Factors (CCF)	Delay in approving shop drawings	CCF8	0.923
	Delay in progress payment	CCF9	0.795
	Delay in approving of change orders	CCF1	0.576
	Frauds practice among the parties involved	CCF2	0.518
	Inadequate consultant experience	CCF3	0.511
	Delay in obtaining permits from Municipality	CCF7	0.477
	Lack of coordination with authorities	CCF4	0.474
	Slow decision-making process	CCF5	0.364
	Difficulty in accessing the site	CCF6	0.355
Project Management and Contract Administration Related Factors (PMCAF)	Poor contract management	PMCAF1	0.615
	Unrealistic contract duration	PMCAF2	0.528
	Poor site management and supervision	PMCAF3	0.493
	Improper planning and scheduling of the project	PMCAF4	0.484
	lack of teamwork	PMCAF5	0.441

Information and Communication Technology Related Factors (ICTF) consisting of 3 factors, Labour Management Related Factors (LAB) consisting 3 factors, Client and Consultant Related Factors (CCF) consisting of 9 factors and Project Management and Contract Administration Related Factors (PMCAF) consisting of 5 factors. Once, the factors are categorized into their respective groups, and then following analysis such as structural equation modeling which required latent variables for the factors can be proceed.

ACKNOWLEDGMENT

The authors are grateful to FRGS Grant No 1221, University Tun Hussein Onn Malaysia for supporting this research. Furthermore, the authors are thankful to construction practitioners who are working in Mecca city for their contributions and providing comprehensive and important information.

REFERENCES

Abdi, H. 2003. Factor rotations in factor analyses. In M. Lewis-Beck, A. Bryman., & T. Futing (Eds.), Encyclopedia of social sciences research methods (pp. 978–982). Thousand Oaks, CA: SAGE, 2003.

Albogamy, A., Scott, D., & Dawood. N., Addressing Construction Delays in the Kingdom of Saudi Arabia. Centre for Construction Industry Studies, Teesside Universit: Ph.D Thesis (2012).

Al-Emad, N. & Nagapan, S. 2015. Identification of Delay Factors from Mecca's Construction Experts Perspective.

International Journal of Sustainable Construction Engineering & Technology, Vol 6, No 2, 2015, pp. 1–4.

Al-Khalil, M. & Al-Ghafly, M. A. 1999. Delay in public utility projects in Saudi Arabia. International Journal of Project Management, 17(2), (101–106), 1999.

Al-Kharashi, A. & Skitmore, M. 2009. Causes of delays in Saudi Arabian public sector construction projects. Construction Management and Economics, 27(1), 3–23, 2009.

Alotaibi, N., Sutrisna, M. & Chong, H. 2014. Managing Critical Factors Causing Delays in Public Construction Projects In Kingdom Of Saudi Arabia. Curtin University: Ph.D Thesis (2014).

Assaf, S. A. & Al-Hejji, S. 2006. Causes of delay in large construction projects. International Journal of Project Management, 24 (2006) 349–357.

Assaf, S.A., Khalil, M. & Al-Hazmi, M. 1995. Causes of delay in large building construction projects. Journal of Management and Engineering, 11(2), 45–50, 1995.

Costello, A. B. & Osborne, J. W. 2005. Best Practices in Exploratory Factor Analysis: Four Recommendations for Getting the Most From Your Analysis. Practical Assessment Research & Evaluation, Vol 10, No 7, July 2005.

Doloi, H., Sawhney, A., Iyer, K. C. & Rentala, S.2012. Analysing factors affecting delays in Indian construction projects. International Journal of Project Management. 30(4), pp. 479–489.

Elawi, G. S. A. 2015. Owners' Perspective of Factors Contributing to Project Delay: Case Studies of Road and Bridge Projects in Saudi Arabia. Arizona State University: Master Thesis (2015).

Mahamid, I. 2013. Contributors to Schedule Delays In Public Construction Projects In Saudi Arabia: Owners' Perspective. Journal of Construction Project Management and Innovation Vol. 3 (2): 608–619, 2013.

Mahamid, I. 2014. Micro and macro level of dispute causes in residential building projects: Studies of Saudi Arabia. Journal of King Saud University – Engineering Sciences (2014), Retrieved from http://dx.doi.org/10.1016/j.jksues.2014.03.002

Memon, A.H., Rahman, I.A., Akram, M., Ali, N.M. 2014. Significant factors causing time overrun in construction projects of Peninsular Malaysia. Modern Applied Science, Volume 8, Issue 4, Pp. 16–28, Canadian Center of Science and Education.

Mostafa, M.M. & Al-Buzz, W.W. 2015. Calculation of the Construction Time- Systematic Management of Project Uncertainties in Rush Projects in Makkah. International Journal of Research In Engineering And Technology, Vo. 04, Issue: 05 (2015).

Samargandi, N., Fidrmuc J., & Ghosh. S. Financial development and economic growth in an oil-rich economy: The case of Saudi Arabia. Economics and Finance Working Paper. Brunel University, (2013) p. 13–12.

Sweis, G., Abu-Hammad, A. & Shboul, A. 2007. Delays in Construction Projects: The Case of Jordan. International Journal of Project Management, 26 (6) 665–674, 2007.

Williams, B., Onsman, A. & Brown, T. 2010. Exploratory factor analysis: A five-step guide for novices. Journal of Emergency Primary Health Care (JEPHC), Vol. 8, Issue 3, 2010 – Article 990399, p. 6.

Young-Mok, K., Soo-Yong, K. & Truong-Van, L. 2008. Causes of construction delays of apartment construction projects: Comparative analysis between Vietnam and Korea (2008). 214–226.

Yusof, W.Z.W., Singh, B., Hamid, A.R.A., & Ahmad, W. R. W. 2004.Variation order problem by client during construction, Civil Engineering Research Seminar (SEPKA 2004), FKA & Construction Focus Group, Dewan Alumi UTM, Skudai, 294–308.

The awareness of green building ratings among university students

Gasim Hayder & Osamah Abdulhakim Ahmed
Department of Civil Engineering, College of Engineering, Universiti Tenaga Nasional (UNITEN), Malaysia

Suhir Bin Che Selia
College of Graduate Studies, Universiti Tenaga Nasional (UNITEN), Malaysia

ABSTRACT: A Since the international society paid attention to the issue of global warming, huge efforts was made by governments and environmental institutes to persuade designers and construction firms into adopting green building concept to reduce the greenhouse gas emissions, since the construction industry is one of the main emitters. In Malaysia, the greenhouse gas emission is still rising, although the government has made many steps regarding the matter. This study investigate the awareness of engineering students those who are expected to lead the construction industry in the future regarding the implementation of green buildings in Malaysia. From the survey results, only thirty-three percent of the participants were aware of the existence of a green building assessment tool in Malaysia while only ten percent of them had a positive view of it. This leads to the conclusion that more efforts should be done to educate the students about the importance of implementing Green Building Index (GBI) and its essential role in the future of Malaysian environment, economy and society. Moreover, this study compare between Malaysia and other countries regarding its rating aspects of GBI where it was found that all the aspect are similar with other countries and there are items that exist in other countries, which might be beneficial to be added to Malaysian GBI.

1 INTRODUCTION

Construction activities are among the biggest contributors to greenhouse emission, which has high negative impact on the environment (USGBC, 2016). A building produce greenhouse gas through its life cycle that includes construction, occupancy, renovation, repurposing and deconstruction. United States of America's Environmental Protection Agency (EPA) released data showing that buildings consume 39.8 percent of total energy in the United States of America, 19 percent of total water usage, and contribute by 38 percent to the total carbon dioxide emission (Ali et al. 2012). Furthermore, the building sector is responsible for 72 percent of the electricity in the US (Winston et al. 2013 & Mitton et al. 2011), which contribute to greenhouse gas emission as electricity is mostly generated by fossil fuel which is one of the main sources of carbon dioxide emission. Carbon dioxide emissions from the residential correlate closely with its energy consumption as 72 percent of residential emissions result from electricity (Beering, 2011). As the issue of global warming arise, the effects of greenhouse gas were brought into attention. Therefore, many solutions were proposed to solve the issue including the idea of green buildings, which is defined by United States of America's Environmental Protection Agency as a structure that is environmentally responsible and resource-efficient throughout its life cycle (EPA, 2016).

Although Malaysian government had a plan to reduce its greenhouse gas emission by 40 percent in the 2020 announcement, it kept on rising to reach 200 percent of that of 1990 (Zaid et al. 2014). Taking into consideration the continuous increase in emission, the government took many steps to help spreading green buildings such as releasing the National Green Technology Policy in 2009, the implementing of green initiative in Malaysia, which made a dramatic increase in the number of applications for Green Building Index certificate (GBI) (Bahaudin et al. 2014). GBI is the main Green Building assessment tool that was developed and lunched by Pertubuhan Arkitek Malaysia (PAM) and the Association of Consulting Engineers Malaysia (ACEM) in 2008 (Liang et al. 2014). It rates the characteristics of a building and the degree of sustainability achieved in its life cycle in an attempt to drive the construction market in Malaysia toward a greener future (Mun et al. 2014).

This paper focuses on the awareness of Malaysians regarding the existence of Malaysia's Green Building assessment tool and its effectiveness in Malaysia. In addition, it assesses how well Malaysians perceive Green Building Index, and compares it with other ratings around the world.

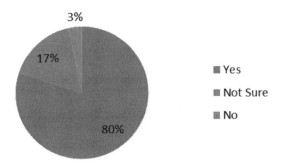

Figure 1. Answers of Question "Do you know that the Rating System is able to improve the quality of Green Building in Malaysia".

Figure 2. Answers of Question "Is Malaysia has its own assessment tool of Green Building".

2 METHODOLOGY

The study was completed by conducting a survey to measure the perception and awareness level regarding green building assessment tool in Malaysia. The method used is collecting the data, which is relevant to the study's objective from websites, articles, and online journals. The study focuses on eight countries, which are Malaysia (GBI), Britain (BREEAM), Egypt (GPRS), Singapore (Green Mark), USA (LEED), UAE (PBRS), Qatar (QSAS), and South Korea (GBCC) that perform the green building rating system for a new construction. In addition, it was analyzed based on the similarities and differences.

3 RESULTS AND DISCUSSION

The survey was conducted to investigate the objectives of the study where most of the surveys were taken by only one local university students. Thirty samples were collected to meet the requirements of the selecting size of Roscoe.

The first question asked was if a green building assessment tool would help improve the green building in Malaysia. The results shows that 80 percent of the participants believed that an assessment tool would help to improve the quality of the green building, where 17 percent of them were not certain. While the 3 percent did not believe that an assessment tool will have any benefit on its quality as shown in Figure 1.

The second question was if Malaysia has its own assessment tool for green building. 30 percent said that Malaysia has its own assessment tool while 40 percent answered the opposite. On the other hand, the rest 27 percent were not sure if it has or not as shown in Figure 2.

The third question asked was if they know the name of the green building assessment tool of Malaysia. Only 70 percent chose the correct answer and the rest chose the wrong answers or they did not give as answer for this question as shown in Figure 3.

The last question was dependent on the previous question, for those who chose the green building

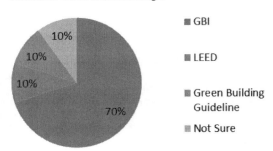

Figure 3. Answers of Question "Do you know the name of the assessment tool of Green Building of Malaysia".

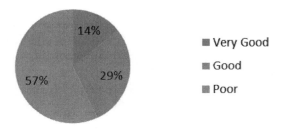

Figure 4. Answers of Question "How you will rate the assessment of Green Building of Malaysia".

assessment tool of Malaysia; they were asked, "How you will rate the assessment of green building of Malaysia". There was 57 percent who rated it as poor and 29 percent rated it as good assessment tool. While the 14 percent rated it as very good assessment tool as shown in Figure 4.

The survey shows that few of the students who answered the survey knows that Malaysia has its own green building assessment tool and few students had enough knowledge about the green building assessment tools. This shows the need at the universities level to educate the students more about the green building assessment tools.

All assessment tools that had studied in this paper shows that there are three rating criteria take high attention which are Water Efficiency, Indoor Environment Quality, and Energy Efficiency which were included in all assessment tools as shown in Table 1. Followed by Site Planning & Management, Innovation,

Table 1. Comparing GBI, GPRS, PBRS, LEED, QSAS, Green Mark, KGBCC, and BREEAM regarding criteria assessment categories.

	GBI	GPRS	PBRS	LEED	QSAS	Green Mark	KGBCC	BREEAM
EE	✓	✓	✓	✓	✓	✓	✓	✓
WE	✓	✓	✓	✓	✓	✓	✓	✓
IEQ	✓	✓	✓	✓	✓	✓	✓	✓
SPM	✓	✓	✓	✓	✓		✓	
I	✓		✓	✓		✓		
MR	✓	✓	✓	✓	✓		✓	✓
EP						✓		
T				✓		✓		
LU							✓	
EEn							✓	
M		✓	✓		✓		✓	
CE					✓			✓

*Energy Efficiency (EE), Water Efficiency (WE), Indoor Environment Quality (IEQ), Site Planning & Management (SPM), Innovation (I), Material & Resources (MR), Environmental Protection (EP), Transport (T), Land Use (LU), Ecological Environment (EEn), Management (M), Culture & Economics (CE).

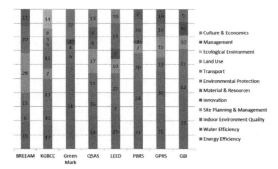

Figure 5. Comparing GBI, GPRS, PBRS, LEED, QSAS, Green Mark, KGBCC, and BREEAM regarding criteria assessment categories.

The comparison shows that the six rating items in GBI are similar with LEED and those in the Middle East, however, LEED and the studied Middle Eastern assessment tools have one more item which is Transport (Attia, 2014, & Ahankoob et al. 2013). In addition, some of the Middle Eastern assessment tools such as UAE and Qatar emphasized on retaining the cultural heritage aspect in the design of new buildings, furthermore, Qatar also implemented an item regarding desertification (Attia, 2014). On the other hand, some rating items in GBI are not included in Singapore and South Korea assessment tools as shown in Figure 5. Generally, the cost of green building is high, therefore, the Malaysian government has countered this rise by implementing the green initiative (Attia, 2014).

and Material & Resources. Most of the rating systems in Middle East, namely Egypt, UAE and Qatar, are a combination of Leadership in Energy and Environmental Design (LEED) and Building Research Establishment Environmental Assessment Methodology (BREEAM) systems. Generally, all systems take into account the climate of the country except the Egyptian rating system, which adopt excessively LEED rating system (Attia, 2014). There were only GBI and Singaporean assessment tools that do not have Transport in their items while only the South Korean assessment tool have a Land Use and Ecological Environment items (Bahaudin et al. 2014, & Bahaudin et al. 2012). Moreover, UAE and Qatar included items in their assessment tools to maintain the cultural heritage of the county in new buildings. All studied assessment tools in the Middle East need more attention to be able to attract developers unlike LEED and BREEAM assessment tools, which are highly established and customized to its countries (Attia, 2014).

4 CONCLUSION

A regardless of the efforts made by the Malaysian government, the level of greenhouse gas emission is still increasing, which means there are still actions need to be taken to guide the construction industry to greener buildings. The results have shown that although 80 percent out of the 30 participants expressed that a rating system for green building would improve the green building construction in Malaysia, only 33 percent of the participants were aware of the fact that Malaysia has its own green building assessment tool while 40 percent of them were not and the rest did not have definitive answer. Furthermore, only 70 percent out of the 33 percent participants recognized its name, where 43 percent of them had a favorable opinion on the quality of the tool. This result is highly unfavorable since students, the future leaders; do not have a clear idea about the Malaysia's green building assessment tool, which might hold back the efforts made to implement a greener construction industry. The concerning parties should step forward by raising awareness among students regarding the challenges the country is facing to inspire a new generation of engineers and leaders capable of taking new challenges to be able to achieve the goals set at the 2020 announcement.

REFERENCES

Ahankoob, A., Morshedi, R., Rad, K.G. 2013. A Comprehensive Comparison between LEED and BCA Green Mark as Green Building Assessment Tools.

Ali, R., Daut, I., Taib, S. 2012. A review on existing and future energy sources for electrical power generation in Malaysia, 4047–4055.

Attia, S. 2014. The Usability of Green Building Rating Systems in Hot Arid Climates.

Bahaudin, A.Y., Elias, E.M., & Saifudin, A.M. 2014. A Comparison of the Green Building's Criteria.

Bahaudin, A.Y., Elias, E.M., & Nadarajan, S. 2012. An Overview of the Green Building's Criteria Non Residential New Construction.

Beering, S.C. 2011. Building a Sustainable Energy Future: U. S. Actions for an Effective Energy Economy Transformation. 1437925782, 9781437925784. DIANE Publishing.

EPA, Environmental Protection Agency. 2016. Green Building: Basic Information. Information on http://archive.epa.gov/greenbuilding/web/html/about.html

Liang, S.Y., Putuhena, F.J., Ling, L.P., Baharun, A. 2014. Towards implementation and achievement of construction and environmental quality in the Malaysian construction industry.

Mitton, M. & Nystuen, C. 2011. Residential Interior Design: A Guide to Planning Spaces. 9781118013038. Wiley.

Mun, T.L., & Baharun, A. 2014. The Development of GBI Malaysia (GBI).

USGBC, Green Building Council 2016. Information on http://www.usgbc.org

Winston, M., Edelbach, R. 2013. Society, Ethics, and Technology. 1133943551, 9781133943556. Cengage Learning.

Zaid, S.M., Myeda, N.E., Mahyuddin, N. and Sulaiman, R., 2014. Lack of energy efficiency legislation in the malaysian building sector contributes to Malaysia's growing GHG emissions. In E3S Web of Conferences (Vol. 3, p. 01029). EDP Sciences.

Factors affecting the embodied carbon footprint potential—Assessment of conventional Malaysian housing habitat

Syed Shujaa Safdar Gardezi, Nasir Shafiq, Muhd. Fadhil Nuruddin & Noor Amila Zawawi
Department of Civil and Environmental Engineering, Universiti Teknologi PETRONAS, Seri Iskandar, Perak, Malaysia

Faris B. Khamidi
Department of Built Environment, University of Reading Malaysia, Menara Kotaraya, Johor Bahru, Johor, Malaysia

ABSTRACT: The development housing consumes natural resources and fossil fuels. Thus not only depleting the resources but also damaging the environment by generating embodied carbon footprint, extent of which is governed by various factors including physical characteristics of built facility. The current study investigates the relationship between these physical factors and embodied CO_2 for conventional housing construction in tropical climate. Five selected single units with different areas of construction have been evaluated. The units were redeveloped in virtual environment by Building Information Modeling (BIM). Life Cycle Assessment (LCA) methodology was adopted to quantify the embodied CO_2 content. The variables of area, weight of materials and carbon footprint, observed a significant positive interrelationship. A 2.5 times increase in area resulted in 50% increase in carbon footprint content which ranged from 19 kg-CO_2/sft to 31 kg-CO_2/sft. Bricks, concrete and steel metals were top three materials with an average share of more than 80%.

1 INTRODUCTION

Housing sector is one of the basic elements in infrastructure development. However, such development has different negative environmental impacts (Gardezi, Shafiq, Zawawi, & Farhan, 2014). The embodied emissions are an integral part of life cycle carbon footprint. The embodied carbon footprint represents the CO_2 emissions resulting from the energy consumed by building materials from the extraction in raw form, manufacture of the final product, including transportation (Rossi, Marique, Glaumann, & Reiter, 2012). However, apart from the simple assessment of individual cases or comparative analysis of different materials, various factors related to physical characteristic of any housing facility also affect the embodied carbon footprint potential (Blengini & Di Carlo, 2010; Ortiz-Rodríguez, Castells, & Sonnemann, 2010). Therefore, the evaluation of different factors impacting the carbon footprint of conventional housing sector is important to promote low-carbon societies. The current study investigates the relationship between these factors and embodied CO_2. The area of housing units and carbon footprint observed a significant positive relationship. Similarly, the quantity of materials also depicted a good positive relation with the emissions and area.

2 LITERATURE REVIEW

Embodied energy or emissions from the housing sector have been a keen field of environmental evaluation around the world. Numerous studies have been conducted to highlight the embodied carbon footprint potential of different housing units. Most of these studies were observed to be limited to assessment (Dahlstrøm, Sørnes, Eriksen, & Hertwich, 2012; Mithraratne & Vale, 2004; Oyarzo & Peuportier, 2014) or comparative analysis (Nasir Shafiq et al.; Peuportier, 2001; Shafiq, Nurrudin, Gardezi, & Kamaruzzaman, 2015). Although, numerous factors affecting the embodied CO_2 emissions have been highlighted previously, table 1, the impact of these factors still requires to be investigated to determine the level of their impact.

3 RESEARCH METHODOLOGY

Process based Partial Life cycle assessment (LCA) with "cradle to site" boundary limitations guided to observe the environmental impact. Simple regression statistical technique was used to observe the relationship between the variables identified from previous

Table 1. Factor affecting the carbon footprint potential highlighted in previous studies.

Researcher	Factors highlighted
Cole (1998)	Climatic variations, fuel mix, transport distances, regional practices
Harris (1999)	Geographic variations, design, construction, location
Ortiz-Rodríguez et al. (2010)	Household size, climate, geography, energy sources
Blengini and Di Carlo (2010)	House size, no. of occupants
Ramesh, Prakash, and Shukla (2010)	Climatic conditions, energy mix, building materials, construction techniques, building features
González and García Navarro (2006)	Quantity of materials
Rajagopalan, Bilec, and Landis (2012)	Geographic locations, construction methods, materials, construction area, data availability, boundary limits, functional unit
Basbagill, Flager, Lepech, and Fischer (2013)	Building type, size, location, geometry
Mateus, Neiva, Bragança, Mendonça, and Macieira (2013)	Weight of construction material
Allouhi et al. (2015)	Human comfort levels, architectural design, geography, climate data

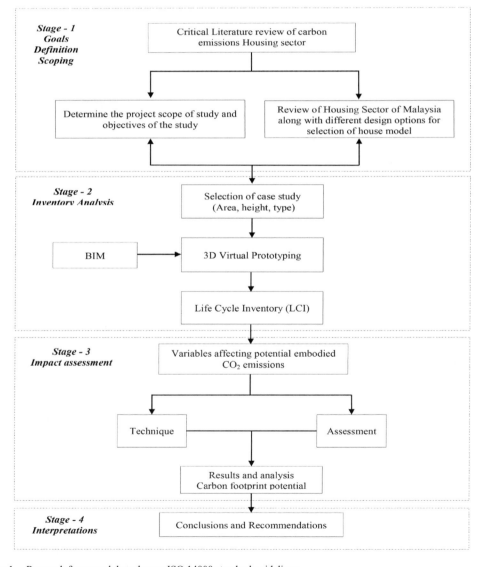

Figure 1. Research framework based upon ISO 14000 standard guidelines.

research works. Figure 1 details the flow of research in a graphical format.

4 SIGNIFICANCE OF THE STUDY

The study has significance in addressing the gap of existing research regarding the impact of physical characteristics of tropical housing facilities in embodied carbon footprint contributions.

5 CASE STUDY

Five houses were selected for the current study. These comprise of detached and semi-detached type construction usually being adopted within Malaysia with a single storey height. The details of housing units have been highlighted in table 2.

6 RESULTS AND DISCUSSIONS

The total embodied carbon footprint is the summation of emissions from the manufacturing process and transporting activity of construction materials, mathematically by eq. (1).

$$\sum CO_{2\ embodied} = \sum CO_{2\ manufacturing} + \sum CO_{2\ tran}. \quad (1)$$

The quantum of emissions depends upon the type of materials, type of fuel and distance for transportation. The study was based on standard transporting equipments locally available in construction works. The study resulted in an average contribution of 50 to 77 tons-CO_2 with a variation of 52%. Figure 2 details the total carbon footprint from each of the housing units graphically.

On average, bricks resulted in 22 ton-CO_2 (34%), concrete 18 ton-CO_2 (28%), steel metals 14 ton-CO_2 (21%), roof tiles 3.5 ton-CO_2 (5.5%), ceramic tiles 2.6 ton-CO_2 (4%) and plaster resulted 2.54 ton-CO_2 (2.5). Among materials, bricks, concrete and steel were the three top emitters. These three materials contributed more than 80% in each of the respective cases. The contribution trend of each of the materials have been elaborated in figure 3 The top emitting materials observed a non-linear whereas the low contributing materials show a linear trend in their contributions. The percentage share of carbon footprint from the transportation of materials to the disposal point remained less than 1% in total content. The contributions ranged from 0.50 tons-CO_2 to 0.90 tons-CO_2. The CO_2 share by different materials have been given in table 3.

Statistical technique of simple regression analysis was adopted to observe relationship between the area and carbon footprint. The results highlighted a positive significant relationship between the two variables ($R^2 > 0.70$), figure 4. An increase in area by 2.5 times highlighted almost 1.5 times increase in carbon footprint.

However, an inverse relationship achieved for emissions per unit area, figure 5. In other words, as the area increased, the per unit contributions observed

Table 2. Details of Housing Models.

Model	Area	Type of House	Type of Structure	Type of Roof
H#01	1625	Semi-Detached	RCC Frame	Roof tiles
H#02	2925	Detached	RCC Frame	RCC Slab
H#03	2800	Detached	RCC Frame	Roof tiles
H#04	4000	Detached	RCC Frame	RCC Slab
H#05	2625	Semi-Detached	RCC Frame	RCC Slab

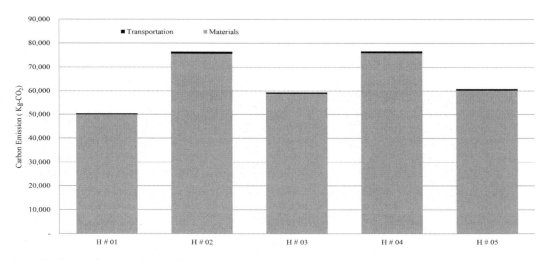

Figure 2. Total carbon footprint contributions.

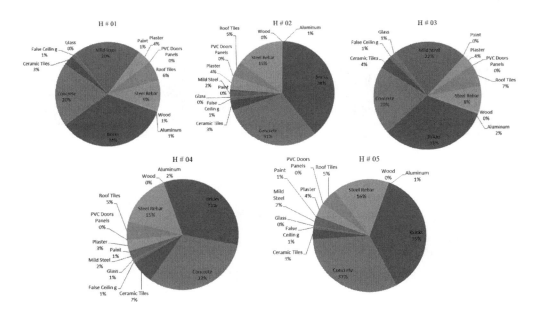

Figure 3. Percentage contribution of materials in housing units.

Table 3. Material wise breakdown of carbon footprint contributions.

Description of Material	Carbon footprint (Kg-CO_2)				
	H # 01	H # 02	H # 03	H # 04	H # 05
Aluminum	396.53	1,090.47	911.33	1,622.62	810.46
Bricks	16,945.71	28,929.65	18,511.32	24,057.93	21,430.37
Concrete	10,053.92	23,316.27	12,153.24	24,410.34	19,436.66
Ceramic Tiles	1,531.20	2,401.22	2,201.73	5,178.46	1,809.11
False Ceiling	594.79	558.57	741.99	495.15	401.62
Glass	138.05	301.97	707.13	332.91	194.89
Mild Steel	10,101.74	1,306.84	12,793.43	1,536.93	1,389.17
Paint	352.21	374.50	262.32	425.56	281.97
Plaster	2,228.27	3,222.41	2,257.14	2,579.02	2,426.23
PVC Doors Panels	47.88	47.88	47.88	47.88	47.88
Roof Tiles	3,131.73	3,642.53	3,890.43	4,077.72	2,915.73
Steel Rebar	4,669.42	11,165.59	4,604.66	11,799.09	9,580.86
Wood	227.57	106.23	312.59	66.10	114.49
Total	50,419.01	76,464.13	59,395.17	76,629.70	60,839.44

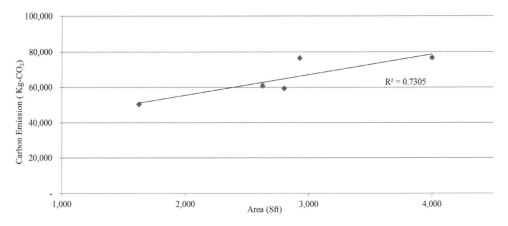

Figure 4. Statistical relationship between dependent and independent variable.

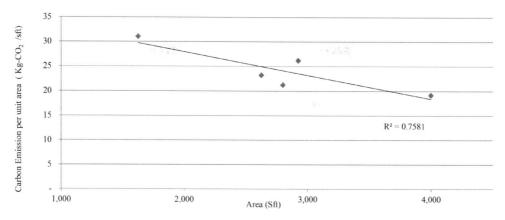

Figure 5. kg-CO2/sft Vs. Area.

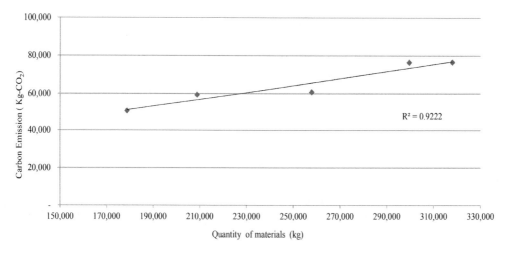

Figure 6. Carbon footprint Vs. Quantity of materials.

a downward trend. As the area increased from 1625 sft to 4000 sft, the per unit area emissions dropped from 31 to 19 kg-CO_2. The relationship between the CO_2 emissions and the material quantity was observed directly proportional to each other with a significant ($R^2 > 0.90$) positive trend, figure 6. Thus also validating higher the quantity of materials, higher is the embodied CO_2. On overall basis, a variation of 80% in quantity of materials resulted in 50% change in carbon footprint content.

The study has also highlighted a good positive relationship between the area and quantity of materials ($R^2 = 0.92$), However, it is necessary to observe the trend of relationship between the area, quantity and emissions on the basis of individual construction materials.

7 CONCLUSIONS

The current study aimed to assess the impact of some physical features of housing units. Single storey conventional houses from Malaysian tropical climate have been studied by adopting partial life cycle assessment with "cradle to site" boundary and BIM. The study concludes that:

a) The studied physical feature possessed significant potential to affect embodied CO_2.
b) An average contribution of 65 tons-CO_2 was achieved with standard deviation of 35%.
c) The transportation share in emissions remained less than 1%.
d) Bricks, concrete and steel metals, the top materials shared more than 80% (average).
e) A positive relation among dependent and independent parameters was observed.
f) With an increase in area from 1625 sft to 4400 sft, the quantity of material increase by 80%, thus resulting a 50% increase in carbon footprint emissions.

ACKNOWLEDGEMENT

This research study is supported by the Ministry of Education (Higher Education Department), Malaysia under MyRA Incentive Grant (0153AB-J11) for Smart Integrated Low Carbon Infrastructure Model Program.

REFERENCES

Allouhi, A., El Fouih, Y., Kousksou, T., Jamil, A., Zeraouli, Y., & Mourad, Y. (2015). Energy consumption and efficiency in buildings: current status and future trends. Journal of Cleaner Production, 109, 118–130. doi: http://dx.doi.org/10.1016/j.jclepro.2015.05.139

Basbagill, J., Flager, F., Lepech, M., & Fischer, M. (2013). Application of life-cycle assessment to early stage building design for reduced embodied environmental impacts. Building and Environment, 60(0), 81–92. doi: http://dx.doi.org/10.1016/j.buildenv.2012.11.009

Blengini, Gian Andrea, & Di Carlo, Tiziana. (2010). The changing role of life cycle phases, subsystems and materials in the LCA of low energy buildings. Energy and Buildings, 42(6), 869–880. doi: http://dx.doi.org/10.1016/j.enbuild.2009.12.009

Cole, Raymond J. (1998). Energy and greenhouse gas emissions associated with the construction of alternative structural systems. Building and Environment, 34(3), 335–348. doi: http://dx.doi.org/10.1016/S0360-1323(98)00020-1

Dahlstrøm, Oddbjørn, Sørnes, Kari, Eriksen, Silje Tveit, & Hertwich, Edgar G. (2012). Life cycle assessment of a single-family residence built to either conventional- or passive house standard. Energy and buildings, 54, 470–479.

Gardezi, Syed Shujaa Safdar, Shafiq, Nasir, Zawawi, Noor Amila Wan Abdullah, & Farhan, Syed Ahmad. (2014). Embodied Carbon Potential of Conventional Construction Materials Used in Typical Malaysian Single Storey Low Cost House Using Building Information Modeling (BIM). Paper presented at the Advanced Materials Research.

González, María Jesús, & García Navarro, Justo. (2006). Assessment of the decrease of CO2 emissions in the construction field through the selection of materials: Practical case study of three houses of low environmental impact. Building and Environment, 41(7), 902–909. doi: http://dx.doi.org/10.1016/j.buildenv.2005.04.006

Harris, D. J. (1999). A quantitative approach to the assessment of the environmental impact of building materials. Building and Environment, 34(6), 751–758. doi: http://dx.doi.org/10.1016/S0360-1323(98)00058-4

Mateus, Ricardo, Neiva, Sara, Bragança, Luís, Mendonça, Paulo, & Macieira, Mónica. (2013). Sustainability assessment of an innovative lightweight building technology for partition walls–Comparison with conventional technologies. Building and Environment, 67, 147–159.

Mithraratne, Nalanie, & Vale, Brenda. (2004). Life cycle analysis model for New Zealand houses. Building and Environment, 39(4), 483–492. doi: http://dx.doi.org/10.1016/j.buildenv.2003.09.008

Nasir Shafiq, Muhd, Nuruddin, Fadhil, Gardezi, Syed Shujaa Safdar, Farhan, Syed Ahmad, Haiyl, A, & Al Rawy, Mohammad. Reduction of Embodied CO2 Emissions from Conventional Single Storey House in Malaysia by Recycled Materials using Building Information Modeling (BIM).

Ortiz-Rodríguez, Oscar, Castells, Francesc, & Sonnemann, Guido. (2010). Life cycle assessment of two dwellings: One in Spain, a developed country, and one in Colombia, a country under development. Science of the total environment, 408(12), 2435–2443.

Oyarzo, Juan, & Peuportier, Bruno. (2014). Life cycle assessment model applied to housing in Chile. Journal of Cleaner Production, 69, 109–116.

Peuportier, B. L. P. (2001). Life cycle assessment applied to the comparative evaluation of single family houses in the French context. Energy and Buildings, 33(5), 443–450. doi: http://dx.doi.org/10.1016/S0378-7788(00)00101-8

Rajagopalan, Neethi, Bilec, Melissa M, & Landis, Amy E. (2012). Life cycle assessment evaluation of green product labeling systems for residential construction. The International Journal of Life Cycle Assessment, 17(6), 753–763.

Ramesh, T., Prakash, Ravi, & Shukla, K. K. (2010). Life cycle energy analysis of buildings: An overview. Energy and Buildings, 42(10), 1592–1600. doi: http://dx.doi.org/10.1016/j.enbuild.2010.05.007

Rossi, Barbara, Marique, Anne-Françoise, Glaumann, Mauritz, & Reiter, Sigrid. (2012). Life-cycle assessment of residential buildings in three different European locations, basic tool. Building and Environment, 51(0), 395–401. doi: http://dx.doi.org/10.1016/j.buildenv.2011.11.017

Shafiq, Nasir, Nurrudin, Muhd Fadhil, Gardezi, Syed Shujaa Safdar, & Kamaruzzaman, Azwan Bin. (2015). Carbon footprint assessment of a typical low rise office building in Malaysia using building information modelling (BIM). International Journal of Sustainable Building Technology and Urban Development, 6(3), 157–172.

Optimization of residential roof design using system dynamics and building information modelling

S.A. Farhan, N. Shafiq, K.A.M. Azizli, F.K. Soon & L.C. Jie
Universiti Teknologi PETRONAS, Perak, Malaysia

ABSTRACT: There are many studies that evaluate the effectiveness of different types of cool roof and roof insulation technologies. However, there is a lack of study that attempts to aid the decision-making on which technology to adopt. Building Information Modelling (BIM) provides stakeholders with a tool to support decision-making, but BIM itself is not able to perform the decision-making. The capability of BIM should be enhanced to enable it to perform decision-making by performing optimization of Building Information models, which can be realized by looking at buildings as dynamic systems that can be optimized. The paper demonstrates an approach to reduce CO_2 emission and improve the thermal comfort level of residential buildings in Malaysia by representing the roof design as a BIM-based dynamic system.

1 INTRODUCTION

Use of air-conditioners (AC) in buildings account for a significant amount of energy use in developed countries, such as 28% in Taiwan (Lai, Huang and Chiou, 2008), 40% in Hong Kong (Lam *et al.*, 2005) and 40–50% in Singapore (National Environmental Agency, 2010). The roof is most exposed to solar radiation among other components of the building envelope and heat flow through the roof system significantly contributes towards the rise in indoor temperature and dependence on AC (Sadineni, Madala and Boehm, 2011). Dependence on AC leads to increased use of electricity and contributes towards CO_2 emission.

Urban and industrial developments continuously emit CO_2 and other greenhouse gases (GHG), cause global warming and contribute significantly to climate change. Concentration of CO_2 in the atmosphere has increased substantially; air temperatures have increased; solar radiation, wind and humidity patterns have been affected; and occurrence of natural disasters related to climate change has risen. It is important that action must be taken to counter the negative environmental impacts of climate change. The residential sector is a major contributor of GHG emission. Population growth, smaller family sizes and higher indoor comfort demand affects the use of heating and cooling systems and increases the GHG emission from this sector even further in the future. In hot climate, heat transfer into residential buildings significantly affects energy consumption and CO_2 emission.

Adoption of cool roof and roof insulation technologies can reduce heat transfer into residential buildings as most of the heat transfer is through the roof. There are many studies that evaluate the effectiveness of different types of cool roof and roof insulation technologies. However, there is a lack of study that attempts to aid the decision-making on which technology to adopt. Building Information Modelling (BIM) provides stakeholders with a tool to support decision-making, but BIM itself is not able to perform the decision-making. The capability of BIM should be enhanced to enable it to perform decision-making by performing optimization of Building Information models. The paper demonstrates an approach to reduce CO_2 emission and improve the thermal comfort level of residential buildings in Malaysia by representing the roof design as a BIM-based dynamic system.

2 LITERATURE REVIEW

2.1 *System dynamics modelling*

There are a lot of researches that adopt system dynamics modeling (SDM) to analyze their respective systems. For example, Elshkaki (2013) utilized SDM to analyze future resources, emissions and losses of platinum, Xiao (2013) developed a system dynamics model to describe the development and change of enterprise cluster complex network resources integratability and Wang *et al.* (2012) adopted SDM for free-water surface constructed wetlands. One of the very few researches that adopted SDM in the green building research area is Zhang *et al.* (2014), which used SDM to assess the sustainability of construction projects. From the review, it has been found that the use of SDM to analyze building systems in research has not been explored in detail. Table 1 presents the review of previous research on SDM.

2.2 Building information modelling

Recently, the field of BIM has gained interest among researchers. Bosché and Guenet (2014) used BIMs with terrestrial laser scanning to automate surface flatness control. Faghihi, Reinschmidt and Kang (2014) adopted the concept of coupling BIM with Genetic Algorithm to develop an application that can automatically derive structurally stable construction schedules. Hiyama et al. (2014) developed a new method to reuse Building Information models of a past project to optimize the default configuration for performance simulations. Wang et al. (2014) integrated BIM with construction process simulations to support project scheduling. Xu, Feng and Li (2014) investigated the factors that influence the adoption of BIM in the Chinese construction industry. Literature findings in this area indicated that most of the research on BIM focused more on project scheduling and model visualization and not so much on the energy, CO_2 emission and cost analyses of energy-efficient building technologies adopted in Building Information models. Table 2 presents the review of previous research on BIM.

2.3 CO_2 emission of roof systems

The CO_2 emissions that result from the adoption of energy-efficient roof systems have been investigated by researchers. Gao et al. (2014) investigated the effect of cool roof on the CO_2 emission of office buildings. Kim, Kim and Lee (2014) presented a method to calculate the life cycle cost that includes the cost of CO_2 emission and used it to analyze the CO_2 emission costs of roof waterproofing methods. Whittinghill et al. (2014) quantified the CO_2 sequestration of green roof systems. Chen (2013) reviewed the strategies to evaluate the performances of green roofs in Taiwan. Liu et al. (2012) measured the drought tolerance and thermal effect of plants for green roof planting. Rowe (2011) reviewed previous research on how green roofs can abate pollution. Li et al. (2010) studied the effect of green roof on ambient CO_2 concentration. Chel, Tiwari and Chandra (2009) estimated the CO_2 emission reduction potential of a skylight mud-house. Mithraratne (2009) evaluates the net CO_2 emission of roof-top wind turbines in urban houses in New Zealand. Table 3 shows that most of the researches are focused on green roofs and that there is a lack of research that analyzes the CO_2 emissions of energy-efficient roof systems that are more affordable and available in the market such as solar-reflective, ventilated and insulated roofs.

Table 1. Review of previous research on SDM.

Research on SDM	Application of SDM	
	Relevant to research on green buildings	Not relevant to research on green buildings
Zhang et al. (2014)	■	
Dukić and Sarić (2013)		■
Elshkaki (2013)		■
Xiao (2013)		■
Li et al. (2012)		■
Wang et al. (2012)		■
Mi et al. (2011)		■
Venkatesan et al. (2011)		■
Xiao, Cao and Zheng (2011)		■
Kiani and Ali Pourfakhraei (2010)		■

Table 2. Review of previous research on BIM.

Research on BIM	Application of BIM			
	Project Scheduling	Model Visualization	Performance Analysis	Others
Bosché and Guenet (2014)			■	
Faghihi, Reinschmidt and Kang (2014)	■			
Hiyama et al. (2014)			■	
Lawrence et al. (2014)			■	
Wang et al. (2014)	■			
Xu, Feng and Li (2014)				■
Akula et al. (2013)			■	
Anil et al. (2013)		■		
Kim, Kim and Son (2013)			■	
Leite et al. (2011)		■		

Table 3. Review of previous research on CO_2 emission of roof systems.

Research on CO_2 Emission of Roof Systems	Energy-Efficient Roof System Adopted			
	Green Roof	Cool Roof	Skylight	Roof-Top Wind Turbine
Gao et al. (2014)		■		
Kim, Kim and Lee (2014)				
Whittinghill et al. (2014)	■			
Chen (2013)	■			
Liu et al. (2012)	■			
Rowe (2011)	■			
Li et al. (2010)	■			
Chel, Tiwari and Chandra (2009)			■	
Mithraratne (2009)				■

3 METHODOLOGY

To develop a system dynamics model, firstly, a roof system is represented as a causal loop diagram, which is a simple map of a system with all constituent components and their interactions. Then, from the causal loop diagram, the stock and flow diagram is developed. The equations that determine the flows are then constructed. Parameters and initial conditions are then estimated. Lastly, the model is then simulated and the results are analyzed. STELLA is employed to execute SDM.

To conduct parametric study on the energy-efficient roof technologies, the baseline Building Information model is developed using Revit. Weather data files for the cities of Kuala Lumpur, Ipoh and Johor Bahru are loaded into the software. From the baseline model, modified models are developed by changing the roof system configuration.

4 RESULTS AND DISCUSSION

Figure 1 presents the stock and flow diagram of the system dynamics model as developed in STELLA.

Adoption of insulation in the roof system is one of the factors that contribute to the improvement of the thermal performance of the roof system and also the entire building envelope. The thermal conductivities and mass densities of several insulation materials are presented in Table 4. The thermal conductivities of rock wool, expanded polystyrene (EPS) and glass foam are presented in Table 5 and illustrated in Figure 2. The table also shows the U-Values and annual cooling loads of each insulation material with thicknesses of 5 mm, 10 mm and 20 mm respectively. It is found that the adoption of insulation material in the roof system results in a significant decrease in the annual cooling load from the initial annual cooling load of the baseline design, which is 41,011.74 kWh. Another finding

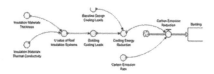

Figure 1. Stock and flow diagrams of the system dynamics model.

Table 4. Thermal conductivities and mass densities of common insulation materials.

Insulation Material	Thermal Conductivity (W/mK)	Mass Density (kg/m³)
Organic/Natural		
Cork	0.045–0.055	80–500
Wool	0.04	20–25
Cotton	0.04	20
Synthetic		
EPS	0.035–0.04	15–30
Polystyrene, XPS	0.035–0.04	25–40
Polyurethane rigid foam	0.025–0.035	30
Inorganic		
Foam glass	0.04–0.0055	10–160
Mineral fibre	0.035–0.05	15–80
Air, motionless	0.0025	1.2
Evacuated panel	0.006	

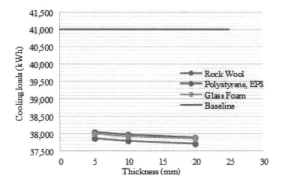

Figure 2. Cooling loads against the thickness of different roof insulation systems.

Table 5. Annual cooling load of Building Information models with different roof insulation systems.

Insulation Material	Thermal Conductivity (W/mK)	Thickness (mm)	U-Value (W/m²K)	Cooling Loads (kWh)
Rock Wool	0.034	5.0	4.84	37,874.82
		10.0	3.55	37,799.90
		20.0	2.32	37,719.54
Expanded Polystyrene	0.035	5.0	4.90	38,061.90
		10.0	3.60	37,985.88
		20.0	2.36	37,901.66
Glass Foam	0.052	5.0	5.55	38,010.11
		10.0	1.81	37,932.18
		20.0	1.34	37,867.77
Baseline Model				41,011.74

Table 6. Annual CO2 emissions reduction of Building Information models with different roof insulation systems.

Insulation Material	K-Value (W/mK)	Thickness (mm)	U-Value (W/m^{2}K)	Cooling Loads Reduction (kWh)	CO$_2$ Emissions Reduction (tonnes)
Rock Wool	0.034	5.0	4.84	3136.92	2.1144
		10.0	3.55	3211.84	2.1649
		20.0	2.32	3292.20	2.2191
Expanded Polystyrene	0.035	5.0	4.90	2949.85	1.9883
		10.0	3.60	3025.87	2.0396
		20.0	2.36	3110.08	2.0963
Glass Foam	0.052	5.0	5.55	3001.63	2.0232
		10.0	1.81	3079.56	2.0758
		20.0	1.34	3143.98	2.1192

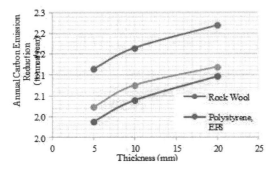

Figure 3. Building annual CO$_2$ emission reduction against the thickness of different roof insulation systems.

is that the rock wool is the best performing insulation material in comparison to EPS and glass foam; however, the difference in performance is not very big as they possess thermal conductivity values that are very close to one another. For instance, the 20-mm thick rock wool insulation results in an annual cooling load of 37,719.54 kWh whereas the EPS and glass foam insulations with the same thicknesses result in annual cooling loads of 37,910.66 kWh and 37,867.77 kWh respectively.

The annual CO$_2$ emission reduction for each roof insulation system configuration is presented Table 6 and illustrated in Figure 3. The highest CO$_2$ emission reduction is obtained from the adoption of the 20-mm thick rock wool, which is 2.22191 tonnes.

5 CONCLUSION

The aim of the paper is to demonstrate an approach to reduce CO$_2$ emission and improve the thermal comfort level of residential buildings in Malaysia by representing the roof design as a BIM-based dynamic system. The paper has highlighted the problem that the present research is attempting to overcome, which is that there is a lack of study that attempts to aid the decision-making on which technology to adopt in the roof system of energy-efficient residential buildings, and that BIM provides stakeholders with a tool to support decision-making, but it itself is not able to perform the decision-making. The present research is suggesting that the capability of BIM should be enhanced to enable it to perform decision-making by performing optimization of Building Information models by coupling BIM with SDM.

ACKNOWLEDGMENT

This work is supported by the Ministry of Education (Higher Education Department) under MyRA Incentive Grant for Smart Integrated Low Carbon Infrastructure Model Program and Universiti Teknologi PETRONAS.

REFERENCES

Akula, M., Lipman, R. R., Franaszek, M., Saidi, K. S., Cheok, G. S. and Kamat V. R. (2013). Real-time drill monitoring and control using building information models augmented with 3D imaging data. *Automation in Construction*, 36, pp. 1–15.

Anil, E. B., Tang, P., Akinci, B. and Huber, D. (2013). Deviation analysis method for the assessment of the quality of the as-is Building Information Models generated from point cloud data. *Automation in Construction*, 35, pp. 507–516.

Bosché, F. and Guenet, E. (2014). Automating surface flatness control using terrestrial laser scanning and building information models. *Automation in Construction*, 44, pp. 212–226.

Chel, A., Tiwari, G. N. and Chandra, A. (2009). A model for estimation of daylight factor for skylight: An experimental validation using pyramid shape skylight over vault roof mud-house in New Delhi (India). *Applied Energy*, 86(11), pp. 2507–2519.

Chen, C.-F. (2013). Performance evaluation and development strategies for green roofs in Taiwan: A review. *Ecological Engineering*, 52, pp. 51–58.

Dukić, S. D. and Sarić, A. T. (2013). A new approach to physics-based reduction of power system dynamic models. *Electric Power Systems Research*, 101, pp. 17–24.

Elshkaki, A. (2013). An analysis of future platinum resources, emissions and waste streams using a system dynamic model of its intentional and non-intentional flows and stocks. *Resources Policy*, 38(3), pp. 241–251.

Faghihi, V., Reinschmidt, K. F. and Kang, J. H. (2014). Construction scheduling using Genetic Algorithm based on Building Information Model. *Expert Systems with Applications*, 41(16), pp. 7565–7578.

Gao, Y., Xiang, R., Su, B., Xu, J. and Tang, X. (2014). Cool roof's effect on the carbon emission of office buildings in chongqing, China. *Energy Education Science and Technology Part A: Energy Science and Research*, 32(4), pp. 2451–2460.

Hiyama, K., Kato, S., Kubota, M. and Zhang, J. (2014). A new method for reusing building information models of past projects to optimize the default configuration for performance simulations. *Energy and Buildings*, 73, pp. 83–91.

Kiani, B. and Ali Pourfakhraei, M. (2010). A system dynamic model for production and consumption policy in Iran oil and gas sector. *Energy Policy*, 38(12), pp. 7764–7774.

Kim, C., Kim, C. and Son, H. (2013). Automated construction progress measurement using a 4D building information model and 3D data. *Automation in Construction*, 31, pp. 75–82.

Kim, S., Kim, G.-H. and Lee, Y.-D. (2014). Sustainability life cycle cost analysis of roof waterproofing methods considering LCCO$_2$. *Sustainability (Switzerland)*, 6(1), pp. 158–174.

Lai, C.-M., Huang, J. Y. and Chiou, J. S. (2008). Optimal spacing for double-skin roofs. *Building and Environment*, 43, pp. 1749–1754.

Lam, J. C., Tsang, C. L., Li, D. H. W. and Cheung, S. O. (2005). Residential building envelope heat gain and cooling energy requirements. *Energy*, 30, pp. 933–951.

Lawrence, M., Pottinger, R., Staub-French, S. and Nepal, M. P. (2014). Creating flexible mappings between Building Information Models and cost information. *Automation in Construction*, 45, pp. 107–118.

Leite, F., Akcamete, A., Akinci, B., Atasoy, G. and Kiziltas, S. (2011). Analysis of modeling effort and impact of different levels of detail in building information models. *Automation in Construction*, 20(5), pp. 601–609.

Li, J.-F., Wai, O. W. H., Li, Y. S., Zhan, J.-M., Ho, Y. A., Li, J. and Lam, E. (2010). Effect of green roof on ambient CO_2 concentration. *Building and Environment*, 45(12), pp. 2644–2651.

Li, C., Lu, G., Ma, P. and Wu, S. (2012). System dynamic model of risk element transmission in project chain. *Przeglad Elektrotechniczny*, 88(7 B), pp. 307–310.

Liu, T.-C., Shyu, G.-S., Fang, W.-T., Liu, S.-Y. and Cheng, B.-Y. (2012). Drought tolerance and thermal effect measurements for plants suitable for extensive green roof planting in humid subtropical climates. *Energy and Buildings*, 47, pp. 180–188.

Mi, L., Yin, G., Sun, M. and Wang, X. (2011). Column-spindle system dynamic model based on dynamic characteristics of joints. *Transactions of the Chinese Society of Agricultural Machinery*, 42(12), pp. 202–207.

Mithraratne, N. (2009). Roof-top wind turbines for microgeneration in urban houses in New Zealand. *Energy and Buildings*, 41(10), pp. 1013–1018.

National Environmental Agency. (2010). *Singapore's Second National Communication: Under the United Nations Framework Convention on Climate Change*, Singapore.

Rowe, D. B. (2011). Green roofs as a means of pollution abatement. *Environmental Pollution*, 159(8–9), pp. 2100–2110.

Sadineni, S. B., Madala, S. and Boehm, R. F. (2011). Passive building energy savings: a review of building envelope components. *Renewable and Sustainable Energy Reviews*, 15, pp. 3617–3631.

Venkatesan, A. K., Ahmad, S., Johnson, W. and Batista, J. R. (2011). Systems dynamic model to forecast salinity load to the Colorado River due to urbanization within the Las Vegas Valley. *Science of the Total Environment*, 409(13), pp. 2616–2625.

Wang, Y.-C., Lin, Y.-P., Huang, C.-W., Chiang, L.-C., Chu, H.-J. and Ou, W.-S. (2012). A system dynamic model and sensitivity analysis for simulating domestic pollution removal in a free-water surface constructed wetland. *Water, Air and Soil Pollution*, 223(5), pp. 2719–2742.

Wang, W.-C., Weng, S.-W., Wang, S.-H. and Chen, C.-Y. (2014). Integrating building information models with construction process simulations for project scheduling support. *Automation in Construction*, 37, pp. 68–80.

Whittinghill, L. J., Rowe, D. B., Schutzki, R. and Cregg, B. M. (2014). Quantifying carbon sequestration of various green roof and ornamental landscape systems. *Landscape and Urban Planning*, 123, pp. 41–48.

Xiao, M., Cao, J. and Zheng, W. X. (2011). Bifurcation with regard to combined interaction parameter in a life energy system dynamic model of two components with multiple delays. *Journal of the Franklin Institute*, 348(9), pp. 2647–2669.

Xiao, B. (2013). System dynamic model of enterprise cluster complex network resources integratability. *Information Technology Journal*, 12(24), pp. 8088–8094.

Xu, H., Feng, J. and Li, S. (2014). Users-orientated evaluation of building information model in the Chinese construction industry. *Automation in Construction*, 39, pp. 32–46.

Zhang, X., Wu, Y., Shen, L. and Skitmore, M. (2014). A prototype system dynamic model for assessing the sustainability of construction projects. *International Journal of Project Management*, 32(1), pp. 66–76.

The development of embodied carbon emission benchmark model for purpose built offices in Malaysia

Mustafa M.A. Klufallah, Muhd Fadhil Nuruddin & Idris Othman
Department of Civil and Environmental Engineering, Universiti Teknologi PETRONAS, Seri Iskandar, Perak, Malaysia

Mohd Faris Khamidi
Department of Built Environment, University of Reading Malaysia, Menara Kotaraya, Johor, Malaysia

ABSTRACT: Over the years, greenhouse gases (GHG) emissions have been increasing in Malaysia. Previously, the total carbon dioxide (CO_2) emission from energy sector was 118,806 kiloton Per capita. Moreover, the buildings construction consume huge amount of natural resources and emitted million tons of carbon emission. Malaysia is one of the top 30 countries for carbon emission in the world. Therefore, this study determines carbon emission from the materials of office buildings and developed a carbon emission benchmark model for Malaysian office buildings. In addition, it is concluded that the office building which falls under 30,000 m² have carbon emission values of ($>$3,000–9,999 tCO_{2eq}), with an average of approximately (0.340 tCO_{2eq}/m^2), the office buildings with an area of ($>$30,000–50,000 m²) have carbon emissions values of (10,000–19,999tCO_{2eq}), at an average value of (0.361 tCO_{2eq}/m^2). While the office buildings with GFA ($>$50,000 m²) have the highest carbon emission values of (20,000–29,000tCO_{2eq}) with an average of (0.373tCO_{2eq}/m^2).

1 INTRODUCTION

The construction of buildings projects has a major impact on the environment and it is considered one of the major consumers of naturally occurring and synthesized resources. Despite the recognition that environmental issues are important to the survival of the construction industry, its activities are major contributors of environmental issues. In addition, changing its conventional practices to incorporate environmental performance as part of its decision making process is gradually slow process in developed and developing countries. However, with increased awareness and knowledge of these impacts, efforts are being made to avoid these adverse effects and to work towards impact mitigation.

Among these are sustainable building materials selection and carbon emissions reduction regulations. These are the most crucial steps and important issues in building design decision-making process.

To mitigate the impact throughout the life cycle of buildings, the construction industry and the related activities are the pressing issues faced by all the stakeholders to promote sustainable buildings [1].

Although the construction industry provides a sustainability measures with wide-range of benefits to the environment and society, it suffers from different kinds of market barriers and assessments measures in developing countries including Malaysia. Nowadays, the management of carbon emission from construction projects is an important competitive advantage in businesses.

This paper introduces a new carbon emission benchmark model based on pre-determined embodied carbon emissions from selected office buildings, which can be used as a starting point to build more robust data set and also provide a useful benchmark opportunity for the practitioner in the absence of any other source of carbon emission assessment model within the Malaysian construction industry which would help at design stage of construction project.

2 BACKGROUND

2.1 Building constructions & climate change

Regardless of whether people argue that global warming is triggered by natural phenomena or by humans or both it is one particular of the most controversial subjects in the scientific world nowadays, and the relationship between GHGs and global temperature has been acknowledge in science since the 19th century. It is showed that the relationship between sea level, global temperature and CO_2 concentrations in the atmosphere is existed over the last 400 million years. Without doubt, the climate change effects on human and natural ecosystems have already existed. Despite current efforts on the part of Malaysian government to curb emissions, Malaysia are ranks one of the top 30 countries in the world that have the largest amount of carbon emission [2].

In fact, more than one third of total energy use and GHGs emissions come from Building construction both in developed and developing countries. Environmental issues are not only linked to technological or economic activities but also associated with cultural and behavioral aspects as well. In terms of economic and technological activities, which are the direct cause of environmentally destructive behavior, individual beliefs and societal norms guide the development of these activities [3–4].

The building construction consumes a great quantity of energy. For instance, in the European Union, buildings through their life cycle consume approximately 50% of the total energy demand and contribute almost about 50% of the carbon emissions to the environment [5].

2.2 Construction & environmental issues

The Construction activities are considered as a major contributor to environmental pollution [6–10], and the impact of construction industry produces undesirable remnants [11–13]. The amount of GHG emissions released in 2004 is found to be increased by 70% compared to those of 1970 [14]. This increase which is caused by human activities is causing a significant year-to-year changes in the amount of CO_2 emission and also it is considered as a fundamental factor in current global issues such as rising of earth temperature, rising of the water level in the seas and oceans, disruption of rainfall, shortage or loss of food sources, reducing of fresh water availability, increasing the intensity and frequency of floods, depletion of the stratospheric ozone layer, late freezing and earlier break-up of ice on rivers and lakes, lengthening of growing seasons, shifts in plants and animal ranges, increasing the probability and extreme climatic events such as hurricanes, droughts, wildfires and other natural disasters.

The most important effect from the concentration of CO_2 emission is the rise in earth temperature, because its impact is the primary reasons for all climate change issues Thus, is it important to control the level of GHG emissions in order to maintain the earth temperature at acceptable level as reported by the Intergovernmental Panel on Climate Change (IPCC-2007, fourth assessment report) and also to reduce these emissions at 50 to 85% by 2050 [15].

A finding of recent studies classifies construction as one of the top three industrial sectors in terms of contribution to GHG emissions in the United States [16]. This sector is responsible for release of more than one third of all GHG and consumption of 40% of the global energy [17]. Furthermore, the applications for cooling, refrigeration, and insulation materials make construction responsible of releasing significant amounts of non-CO_2 GHG emissions such as Halocarbons, CFCs, and HCFCs, and Hydro-Fluorocarbons (HFCs) [18].

Since the construction industry takes such a role in environmental degradation, controlling and reducing GHGs emissions have become an imminent task to be

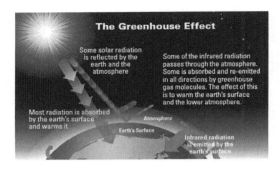

Figure 1. The effects of GHGs on the atmosphere.

managed. The prime minister of Malaysia announced in Copenhagen in 2009 that Malaysia is adopting an indicator of voluntary reduction of up to 40% in terms of emissions intensity of Gross Domestic Product (GDP) by the year 2020 compared to 2005 levels. This announcement is now challenging the financial performance of different economic sectors in Malaysia because they should consider the impact of their investments on the environment. This announcement comes as a response to two facts reported by United Nations Development Program (UNDP) in 2007/2008; Malaysia has classified as number 26 in a top 30 CO_2 emitters in the world and the Malaysia's emission growth rate in the period from 1990 to 2004 is the highest in the world [19].

2.3 Productions of greenhouse gases

As shown in Figure 1, the greenhouse effect is induced when the atmospheric gases trap the ultraviolet rays that come directly from the sun in the earth's atmosphere. The CO_2 is considered as the most significant GHG component by-product in the manufacturing of building materials [20]. However, assessment of CO_2 involves direct and indirect emissions during production and manufacturing of these materials. Direct emissions normally released from consumption of energy by on-site and off-site activities such as construction, prefabrication, transportation and administration. Indirect emission involves energy consumption from fossil fuel during production and manufacturing of building materials in the main processes, upstream and downstream processes like extraction and distribution of raw materials [21].

Furthermore, there are four sources of GHGs emission in the construction of buildings, which are; the manufacture and transportation of building materials, energy consumption of construction equipment, and processing resources, and disposal of construction's waste [5].

On the other hand, statistics show that in Malaysia, the buildings account for about 20% of the production of GHGs that comes in third after transportation 27% and industries 21% [22]. The materials used in buildings, which consist mainly of fossil fuels.

Table 1. Calculation database of the material to CO_2 emission equivalent.

Code of Material	Quantity	Unit	Total Volume (m3)	Density (kg/m3)	Weight of material (Kg/m3)	CO2e Conv. Factor	Carbon Emission (KgCO2e)	Total kgCO2e (Category)
CG30-1	6822.00	m3	6,822.00	2,400.00	16,372,800.00	0.1130	1,850,126.40	
CG30-2	5439.00	m3	5,439.00	2,400.00	13,053,600.00	0.1130	1,475,056.80	
CG30-3	3308.00	m3	3,308.00	2,400.00	7,939,200.00	0.1130	897,129.60	5,549,023.20
CG30-4	34.00	m3	34.00	2,400.00	81,600.00	0.1130	9,220.80	
CG30-5	2945.00	m3	2,945.00	2,400.00	7,068,000.00	0.1130	798,684.00	
CG30-6	1913.00	m3	1,913.00	2,400.00	4,591,200.00	0.1130	518,805.60	

Thus, displace million tonnes of CO_2 emissions during mining and consume more energy [23].

3 RESEARCH METHODOLOGY

In this study, the need for adopting case studies as one of this research strategies arises out of the desire to determine the total amount of embodied carbon emission from building materials of the purpose built offices in peninsular Malaysia. The amount of carbon emission for each material was extracted from the bill of quantities (BQ), which is a document used in tendering in the construction industry in which materials, parts and labors (including costs) are itemized. For carbon emission, the calculation was achieved through the utilization of Inventory of Carbon Emission ICE V2.0. Then, the conversions to tonnes are the density value, where it can relate the volume and weight of each material were calculated as shown in Table 1.

The materials were coded for easy referencing into the calculation database, for example; "concrete grade30" was coded as "CG30" and it was applied to other materials. The method used for the CO_2 equivalent is the boundary of cradle to gate method as the boundary condition are specified compare to cradle to grave. Then, the averages of carbon emission for selected projects were identified and classified accordingly.

4 RESULTS AND DISCUSSION

Figure 2 shows the overall carbon emission contributions of building materials from the office buildings; concrete (38.01%) makes the greatest contribution to embodied carbon which represents the highest amount of emission among all construction materials. The dominance of the building material of concrete could be attributed to its utilization in very large quantity as revealed by an analysis of the material percentage contribution by material mass in the construction of the office buildings.

The second highest carbon emitter after concrete is steel reinforcement which is accounted for about 29.04 percent of an average value of 4,674.66 tCO_{2eq}. The carbon emission factor of steel is considered high compared to various grades of concrete, even though the usage of steel structures is considered at minimum level in building construction in Malaysia but still the steel constitutes the single largest component of whole

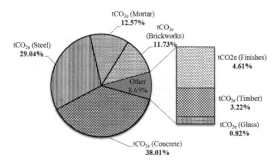

Figure 2. Average CO_{2eq} of office buildings.

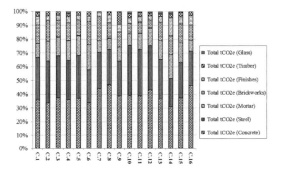

Figure 3. Contribution of overall CO_{2eq} emission by materials.

building emissions, it suggests that the usage of recycled steel is a primary goal to achieve significant and meaningful embodied reductions.

Furthermore, the carbon emission of mortar represents about 12.57% which is the third highest carbon emission value after steel. While brick and block works for internal and external walls, external accounted for 11.73%. Finishes for internal and external wall, floor and ceiling finishes which is about 4.61%, followed by timber for formwork and doors which is about 3.22% and glass for windows, doors, wall partitions and cladding systems accounted for 0.82% which is considered the lowest emission compared to other building elements for office building in Malaysia.

Based on a comparison of building materials as shown in Figure 3, the structural systems of office building accounts for the largest proportion of total emissions at 67.04% comprising the majority of carbon emissions among other building materials.

It can be seen that the concrete is the largest component of the element and is the largest single component in terms of the whole building and ranging from 32-48 percent for all projects. Steel is the next largest element in terms of carbon emission compromising of range between 20 to 33 percent, mortar have a carbon emission value of 10-13 percent, brick/block works ranging between 10–12 percent, finishes represents about 3–8 percent, while timber and glass have the lowest contribution of CO_{2eq} with a value of 3–4 percent and 0.5–1.5 percent respectively.

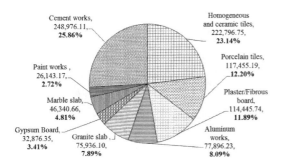

Figure 4. Major CO_{2eq} emission of finishes.

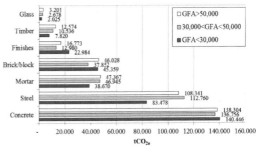

Figure 5. Average CO_{2eq} emission based on GFA.

Buildings Finishes element was ranked in the fifth place in a total average of embodied carbon intensity for the buildings with the majority of emissions embodied in the sub-elements. As can be seen in Figure 4, the highest contributors of embodied carbon emission came from cement works with an average value of 248,976.11 $kgCO_{2eq}$ which is accounted for 25.86% this is due to the great quantities used of tiles works. The second highest quantity of average carbon emission was observed from homogenous and ceramic tiles which have embodied emission of 222.796.75 $kgCO_{2eq}$ and represents 23.14%, this is because most of the purposed built offices in this study, floors and walls were mostly covered with tiles. In addition, porcelain tiles have a great contribution of carbon emission since it presents 12.20% with an average of value 117,455.19 $kgCO_{2eq}$. While ceiling tiles and boards have an average embodied carbon emission of 114,445.74 with 11.89%, followed by aluminum works such as in cladding, ceilings, windows and doors have an embodied emission of 8.09% at an average value of 77.896.23 $kgCO_{2eq}$.

Granite slabs accounted for 7.89% with an average value of 75,936.10 $kgCO_{2eq}$ and for marble slab is 46,340.66 $kgCO_{2eq}$ at 4.81%. The lowest carbon emission was observed from gypsum boards (3.41%), with a value of 32,876.35 $kgCO_{2e}$ and paint works (2.72%) at 26,143.17 $kgCO_{2eq}$. The results revealed that tiles were used widely in office buildings in Malaysia which indicate its higher percentage of embodied carbon emission among other materials for finishes.

In addition, the embodied carbon and embodied energy of tiles which is about (1.14 $kgCO_{2eq}$), it is considered near to that of the steel carbon emission factor which is 1.40 $kgCO_{2eq}$ and even have higher value than other main building materials like various types of concrete grades (0.113–0.151 $kgCO_{2eq}$), brick (0.240 $kgCO_{2eq}$), mortar (0.221 $kgCO_{2eq}$) and glass (0.910 $kgCO_{2eq}$).

The overall average value of GFA and embodied carbon emission ratio as shown in Figure 5 for steel is slightly higher in buildings size of between 30,000–50,000 m^2 than other two size categories, it can be seen the average carbon emission of steel for medium size building is 112.760 tCO_{2eq}/m^2, while in building size above 50,000 m^2 is about 108.341 tCO_{2eq}/m^2 and in small buildings size is approximately 83.478 tCO_{2eq}/m^2. However, the average emission of concrete for small size building is about 140.446 tCO_{2eq}/m^2 which is higher than that of large and medium size office buildings. This is expected because, for example, there is an estimated proportionality between the amount of steel reinforcement and quantity of concrete ratio in a building for each GFA category.

Furthermore, the average carbon emission of concrete is 138.50 tCO_{2eq}/m^2 among the predefined GFA, followed by the average emission of steel reinforcements (101.53 tCO_{2eq}/m^2), which represent the highest in emissions compared to other elements contributions. The average emission of mortar is about (44.33 tCO_{2eq}/m^2), brickworks (43.08 tCO_{2eq}/m^2), finishes (17.56 tCO_{2eq}/m^2), timber (10.31 tCO_{2eq}/m^2) and glass (2.64 tCO_{2eq}/m^2) respectively.

5 EMBODIED CARBON BENCHMARKING MODEL (ECBM)

After extensive extraction and classification of embodied carbon emissions from buildings materials, it was necessary to develop useful carbon emission guidelines which would act as a benchmark to the current buildings practices. Although, benchmarking can be a lengthy and complex process but results in numerous benefits including a greater understanding of how a building portfolio operates, allowing comparisons of buildings, identifying areas of improvement and helping preparation for new legislation. In additions, monitoring the embodied carbon emissions of buildings is a moderately new field of research especially in the Malaysian construction industry and, therefore, there are scarce regulatory standards or academic studies which provide peer-reviewed benchmark values for use at the early project stages in preparation and design especially for the Malaysian office building projects.

Furthermore, measuring carbon emissions relative to GFA (m^2) is widely used and simplest normalized indicator in carbon emissions benchmark practice. This indicator was originally chosen because it has a long history record in management purposes of all types of properties. In order to carry out benchmark assessment based on embodied emissions per GFA

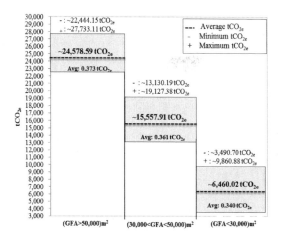

Figure 6. Embodied Carbon Benchmark Model (ECPM).

indicator for various office buildings sizes, the relevant data should be available, relatively accurate, replicable and verifiable.

The average embodied emissions values of office buildings in Figure 6 was established to fill this information gap and provide benchmark figures, which can be used by the quantity surveyors for example when providing a preliminary estimate of cradle-to-gate embodied carbon emissions. The carbon benchmark model can be especially useful during the early stages of a construction project when materials specification has not been drafted.

It was observed that the minimum average embodied emissions value for buildings with a GFA above 50,000 m^2 is 22,444.15 tCO$_{2eq}$ and the maximum embodied emission is around 27,733.11 tCO$_{2eq}$. While GFA of medium size buildings has minimum carbon emissions about 13,130.19 tCO$_{2e}$/m^2 and maximum average emissions is 19,127.38 tCO$_{2eq}$ and buildings with small GFA have a minimum average value of 3,490.70 tCO$_{2eq}$ and maximum average carbon emission of 9,860.88 tCO$_{2eq}$.

Furthermore, the average emissions values were further calculated based on buildings sizes. It was observed that, the office buildings with GFA fall under 30,000 m^2 have an average carbon emission value of 6,460.02 tCO$_{2eq}$ with an average embodied carbon emission of approximately equal to 0.340 tCO$_{2e}$/m^2, whereas the average values of carbon emission of GFA ranged between 30,000–50,000 m^2 have average carbon emission value of approximately 15,557.91 tCO$_{2eq}$ which represents an average embodied emissions value of approximately 0.361 tCO$_{2eq}$/m^2.

While the office buildings with GFA above 50,000 m^2 have an average embodied carbon emission of approximately equal to 24,578.59 tCO$_{2eq}$/m^2.

Unfortunately, there are few comprehensive and peer-reviewed datasets covering embodied carbon emissions associated with Malaysian office buildings. Therefore, embodied carbon determination is relatively new and still unregulated indicator in the Malaysian construction industry. However, the embodied carbon emissions average values presented in Figure 6 introduce a benchmark guideline for the Malaysian construction stakeholders.

The average embodied emissions values of office buildings was established to fill this information gap and provide benchmark figures, which can be used by the quantity surveyors for example when providing a preliminary estimate of cradle-to-gate embodied carbon emissions. The carbon benchmark model can be especially useful during the early stages of a construction project when materials specification has not been drafted.

6 CONCLUSION

The results in this paper represent an absolute estimation of embodied carbon emission for office buildings which highlighted the trend of carbon emission from office buildings in Malaysia. However, the developed ECBM introduce a new benchmark guideline for the Malaysian construction stakeholders, which can be used as a starting point to build a more robust data set and also provide a useful benchmarking opportunity for the construction practitioner which can be used as a starting point to build a more robust data set and also provide a useful benchmark opportunity for the practitioner in the absence of any other source of carbon emission benchmark models within Malaysian construction industry. It is also can be used to assess in the decision-making process in understanding the level of carbon emission for better carbon emission mitigation management.

REFERENCES

[1] Chan, A. "Key performance Indicators for measuring construction success: Benchmarking," An International Journal, Vol. 11(2), pp. 203–221, 2004.

[2] Nation Master Statistic, CO$_2$ Emission by Country. Available at: http:// www.nationmaster.com October, 19, [NOV, 2012].

[3] Barley, S. 1986. Technology as an occasion for structuring: Evidence from observations of CT scanners and the social order of radiology departments. Administrative Science Quarterly 31: 78–108.

[4] David, P. 1985. Clio and the Economics of QWERTY. Economic History 75: 227–332.

[5] Yan H., Shen Q., Fan L C.H., Wang Y., Zhang L. "Green Gas Emission in Building Constrction: A case study of one Peking in Hong Kong" Building and Environment, vol. 45, pp. 1–7, 2010.

[6] Chan, A. "Key Performance Indicators for Measuring Construction Success," Benchmarking: An International Journal, vol. 11(2), pp. 203–221, 2004.

[7] Ding. "Developing a multicriteria approach for the measurement of sustainable performance," Building Research & Information, vol. 33(1), pp. 3–16, 2005.

[8] Yahya, K., & Boussabaine, A. H. "Eco-costing of construction waste", Management of Environmental Quality: An International Journal, vol. 17(1), pp. 6–19, 2006.

[9] Yao, H., Shen, L.-Y., Hao, J., & Yam, C.-h.M. "A Fuzzy-Analysis-Based Method For Measuring Contractor' Environmental Performance", Management of Environmental Quality, An International Journal, vol. 18(4), pp. 442–458. 2007.

[10] Zimmermann, M., Althaus, H.J., & Haas, A. "Benchmarks for sustainable construction: A contribution to develop a standard," Energy and Buildings, vol. 37(11), pp. 1147–1157, 2005.

[11] Augenbroe, G., & Pearce, A. R. "Sustainable Construction in the United States of America A perspective to the year 2010," 1998.

[12] Nelms, C. E., Russell, A. D., & Lence, B. J. "Assessing the performance of sustainable echnologies: a framework and its application," Building Research & Information, vol. 35(3), pp. 237–251, 2007.

[13] San-Jose, J. T., Losada, R., Cuadrado, J., Garrucho, I. "Approach to the quantification of the sustainable value in industrial buildings," Building and Environment, vol. 42, pp. 916–3923, 2007.

[14] I. P. O. C. C. IPCC, "CLIMATE CHANGE 2001: SYNTHESIS REPORT," Valencia, Spain, 12–17 November 2007.

[15] A. A. M. Ali, A. Hagishima, M. Abdel-Kader, and H. Hammad, "Vernacular and Modern Building: Estimating the CO2 emissions from the building materials in Egypt," presented at the Building Simulation Cairo 2013 – Towards Sustainable & Green Life, Cairo.

[16] E. P. Agency, "Quantifying GreenHouseGas Emissions from key industrial sectors in the United States," 2008.

[17] U. N. E. Programme, "Buildings and Climate Change, Summary for Decision-Makers," UNEP, France, 2009.

[18] M. Levine, D. Ürge-Vorsatz, K. Blok, L. Geng, D. Harvey, S. Lang, *et al.*, *Residential and commercial buildings. Climate change 2007; Mitigation. Contribution of Working Group III to the Fourth Assessment Report of the IPCC*, 2007.

[19] K. Watkins, Human Development Report 2007/2008: fighting climate change. Palgrave Macmillan, Houndmills, Basingstoke, Hampshire RG21 6XS and 175 Fifth Avenue, New York, NY 10010: Palgrave Macmillan, 2007.

[20] S. Liu, R. Tao, and C. M. Tam, "Optimizing cost and CO_2 emission for construction projects using particle swarm optimization," *Habitat International*, vol. 37, pp. 155–162, 2013.

[21] IPCC, "Climate Change: The Physical Science Basis, Geneva: Intergovernmental Panel on Climate Change," 2007. America, Washington, DC, Aug. 2008, pp. 6.

[22] M. H. A. Samad, A. M. A. Rahman, and F. Ibrahim, "Green Performance Ratings for Malaysian Buildings With Particular Reference to Hotels," in Proc. of International Conf. on Environmental Research and Technology, Penang, Malaysia, 2008, pp. 313–317.

[23] H. Van, "Minerals and Metals: Cement," Dept. Geology., US Government., Washington, D.C, 1999, vol. 1, pp. 16.1–16.13.

Environmental and water resource engineering

Influence of environmental factor on the performance of *Pistia Stratiotes* in a surface sequencing baffled steep flow constructed wetland

S.R.M. Kutty, E.H. Ezechi & A.T.A. Nagum
Department of Civil and Environmental Engineering, Universiti Teknologi PETRONAS, Seri Iskandar, Perak, Malaysia

ABSTRACT: The performance of *Pistia Stratiotes* for organic matter and ammoniacal nitrogen removal was monitored in a surface sequencing baffled steep flow constructed wetland reactor at various intensity of sunlight. The reactor has 5 compartments of various sizes (C1–C5). *Pistia Stratiotes* were planted in all the compartments of the reactor. Influent wastewater was applied to compartment 1 (C1) at a hydraulic loading rate (HLR) of 10.2 m^3/m.d^{-1} and hydraulic retention time (HRT) of 6 hours. The average organic loading rate (OLR) and ammonia loading rate (ALR) were in the range 6–8 g COD/m^3.m.d^{-1} and 0.15–0.25 g NH$_4^+$-N/m^3.m.d^{-1} respectively. Reactor temperature was daily monitored alongside sample collection at 9 AM, 12 PM, 3 PM and 7 PM respectively for 7 days. On day 4, (3 PM at 35.9°C), NH$_4^+$-N removal reached 94%. COD removal reached 50–55% between 9 AM and 12 PM but fluctuated between 5–14% at 7 PM from day 1 to day 7. This study demonstrates that temperature can influence the performance of *Pistia Stratiotes*.

1 INTRODUCTION

Various industrial and agricultural activities produce high volumes of wastewater containing nitrogenous compounds that can have a negative impact on the environment if not properly removed (Lee et al., 2009, Ezechi et al., 2016, Ezechi et al., 2014). Due to the inherent economic cost associated with the conventional wastewater treatment infrastructures, the need for the development of alternative, low cost, small scale and innovative treatment technology has arisen (Werker et al., 2002).

The use of natural self adaptive living systems within microbial and plant based unit processes for wastewater treatment has become increasingly feasible. These ecologically engineered systems are eco-friendly, low cost and are naturally powered by solar radiation (Werker et al., 2002). Constructed wetlands (CWs) are ecologically engineered low cost systems which utilize natural processes and their associated microbial assemblages in wastewater treatment (Kouki et al., 2009). CWs could be surface flow (SF), sub-surface flow (SSF) or tidal flow (TF). It is a potential wastewater treatment method which has shown satisfying capacities to meet different requirements for contaminant and pathogen removal (Werker et al., 2002). CWs are generally economical due to the utilization of natural elements in the treatment process. It eliminates energy cost, high construction cost, maintenance cost, carbon source, aeration and plant operation (Lee et al., 2009). However, in some regions, the performance of CWs are influenced by climatic conditions (Werker et al., 2002).

Macrophytes are important natural aquatic plants which have high capacity for nutrient and contaminant removal from wastewater (Vitória et al., 2015). Macrophytes could be floating or submerged in the CW. Floating CW provides adequate treatment tolerance to wide fluctuations in water depth. In floating CWs, the emergent macrophytes grow in a floating mat on the surface of the water. Biofilm growth and suspended particle entrapment occur at the plant roots hanging beneath the floating mat. These plants obtain their nutrition from the water column which result in nutrient and element uptake into biomass (Tanner and Headley, 2011).

However, as a natural system, its performance could be affected by seasonal changes and vegetative cycles (Rousseau et al., 2008). Several environmental factors affect the performance of constructed wetlands. Prominent among these factors is temperature. The majority of the studies on CWs has focused on its performance during active plant growth. However, very little study has focused on the influence of temperature on CWs on a day to day approach. The role of temperature and its associated mechanisms in CWs have continued to elicit diverse interest from wastewater treatment researchers and regulators. For instance, temperature can offset both oxygen transfer and pH in CWs (Stein and Hook, 2005, Reid and Mosley, 2016). The seasonal conditions and variations in CWs temperature can influence a vast majority of both physical and biological activities within the system

(Kadlec, 1999). Several phenomenon such as community selection, resultant biogeochemical nutrient cycling, plant and microbial activities will be sensitive to seasonal and temperature variations (Werker et al., 2002).

The objective of this study is to monitor the influence of temperature on the performance of *Pistia Stratioites* for the removal of COD and NH_4^+-N in a five stage surface sequencing baffled steep flow constructed wetland reactor.

2 METHODOLOGY

2.1 Reactor design

The reactor was fabricated using concrete and installed in the effluent line of the sewage treatment plant in Universiti Teknologi PETRONAS (UTP). The reactor consists of 5 compartments (C1–C5) with varying depths and width. Compartment 1 (C1) received the influent sample. The dimension of C1 is 25 cm depth, 48 cm length and 47 cm width. The depth of C2–C5 was sequentially increased by 5 cm (30–45 cm) respectively. Two vertical baffles were installed in C1 to prevent short circuiting. In C2 and C4, a baffle was installed in the middle of the compartments. A 12 cm gap connects wastewater from C1 to C2, C3 and C4. Three 2 cm circular holes were made at the wall of C4 to connect wastewater to C5. A 2 cm circular hole was made between 39–41 cm height of C5 as effluent outlet. *Pistia Stratiotes* were planted on the surface of all the reactor compartments (C1–C5).

2.2 Reactor operation

The reactor was installed at the effluent line of the sewage treatment plant in Universiti Teknologi PETRONAS (UTP) under a sun-permeable transparent shield that allows the penetration of sunlight but not of rainfall. As the influent receiving compartment, C1 dimension was used to calculate the hydraulic loading rate (HLR), organic loading rate (OLR), ammonia loading rate (ALR) and hydraulic retention time (HRT). The influent sample was pumped into C1 at a flowrate of 225L/d using a Masterflex Peristaltic Pump. The influent flowrate corresponded to a HLR of $10.2\,m^3/m.d^{-1}$ at a HRT of 6 hours. Experiments were planned to simulate different weather conditions. Samples were daily collected in the morning (9 AM), afternoon (12 PM and 3 PM) and evening (7 PM). The average OLR and ALR for a seven days period of the experiment at 9 AM were $7\,g\,COD/m^3.m.d^{-1}$ and $0.15\,g\,NH_4^+$-$N/m^3.m.d^{-1}$. The average OLR and ALR for 12 PM were $8.62\,g\,COD/m^3.m.d^{-1}$ and $0.17\,g\,NH_4^+$-$N/m^3.m.d^{-1}$. For 3 PM, the average OLR and ALR was $8\,g\,COD/m^3.m.d^{-1}$ and $0.2\,g\,NH_4^+$-$N/m^3.m.d^{-1}$ whereas for 7 PM, the average OLR and ALR was $6\,g\,COD/m^3.m.d^{-1}$ and $0.25\,g\,NH_4^+$-$N/m^3.m.d^{-1}$ respectively. Samples were daily collected at 9 AM, 12 PM, 3 PM and 7 PM for analysis.

Temperature was measured in the reactor at the sampling time using a thermometer. COD and NH_4^+-N were measured using the Hach method. pH of the samples were measured using a pH meter (Sension^{TM}). All sample analysis were triplicated.

3 RESULT INTERPRETATION

The experiments were conducted at a fixed HLR of $10.2\,m^3/m.d$-1. The average OLR for the duration of the study were about $7\,g\,COD/m^3.m.d^{-1}$, $8.6\,g\,COD/m^3.m.d^{-1}$, $8\,g\,COD/m^3.m.d^{-1}$ and $6\,g\,COD/m^3.m.d^{-1}$ for 9 AM, 12 PM, 3 PM and 7 PM sampling time respectively. The average ALR were $0.15\,g\,NH_4^+$-$N/m^3.m.d^{-1}$, $0.17\,g\,NH_4^+$-$N/m^3.m.d^{-1}$, $0.2\,g\,NH_4^+$-$N/m^3.m.d^{-1}$ and $0.25\,g\,NH_4^+$-$N.m^3/m.d^{-1}$ for 9 AM, 12 PM, 3 PM and 7 PM sampling time respectively. The time course profile for temperature variation at different sampling time and the decrease of organic matter and ammoniacal nitrogen is shown Fig. 1. All negative results (points) were eliminated. The reactor temperature increased from 9 AM to 7 PM in some days and from 9 Am to 3 PM in others. The sampling time at 9 AM represents the morning when the sunlight intensity is low and on the rise. The 12 PM and 3 PM sampling are done when the sunlight intensity is high. At 7 PM, the sunlight intensity takes a downtrend. The four sampling period have a maximum temperature difference of 7.1°C. At 9 AM, the temperature was near linear from day 1 to day 3 (29.8°C) but increased on day 4 (31.8°C).

From day 5 to day 7, the temperature decreased and was near linear. At 12 PM, the temperature was linear from day 1 to day 2 (31°C), decreased to about 30.1°C on day 3 and increased to 32.7°C on day 4. From day 5 to day 7, the temperature fluctuated but was higher than 30.2°C. At 3 PM, the temperature was linear from day 1 to day 3 (32.1°C) and increased to 35.9°C on day 4. From day 5 to day 7, the temperature dropped to about 32.4°C. At 7 PM, the temperature was lowest from day 1 to 2 (28.9°C) but increased on day 3 to day 4 (35.9°C). Henceforth, the temperature at 7 PM decreased to about 30.9°C on day 7. A maximum increase of temperature for all the sampling time was noticed on day 4. The high temperature noticed between day 3 to day 7 at 7 PM could be a result of heat accumulation in the reactor.

3.1 Discussion

It is well known that plants require several elements for their growth and maintenence, among which include nitrogenous compounds and carbon (Wang et al., 2015). Several complex physical, chemical and biological processes occur in a CWs (Lee et al., 2009). Pries (Council and Pries, 1994) reported that cold temperature is a major perceived problem of CWs in Canada. Although Psychrotrophic bacteria can adapt and acclimate to colder temperatures, contaminant and pollutant removal in CWs involves several complex factors which are sensitive to climate (Madigan

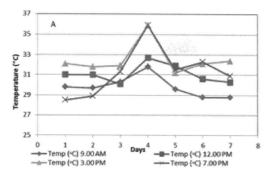

Figure 1a. Temperature variation at various sampling times.

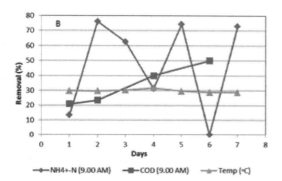

Figure 1b. COD and NH_4^+-N removal at 9 AM.

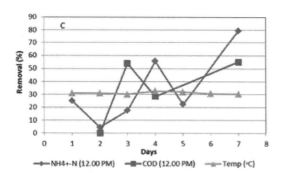

Figure 1c. COD and NH_4^+-N removal at 12 PM.

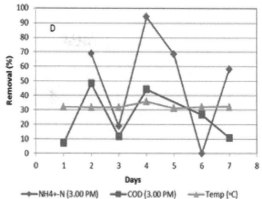

Figure 1d. COD and NH_4^+-N removal at 3 PM.

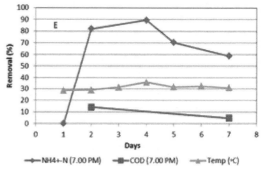

Figure 1e. COD and NH_4^+-N removal at 7 PM.
(Note: negative points were deleted).

et al., 1997, Kadlec, 1999). The reactor temperature corresponded to the mean temperature in July, 2015. NH_4^+-N and COD removal was high throughout the experiment at 12 and 3 PM. Maximum NH_4^+-N (94%) and COD (55%) removal were achieved at 3 PM and 12 PM respectively. At 9 AM and 7 PM, pollutant uptake was low. The latitudinal distribution of submerged freshwater macrophyte could be affected by temperature depending on the specific specie differences and the basic life cycle of the macrophyte (Barko et al., 1982). The seasonal conditions and variations in CWs temperature can influence a vast majority of both physical and biological activities within the system (Kadlec, 1999). Several phenomenon such as community selection, resultant biogeochemical nutrient cycling, plant and microbial activities will be sensitive to seasonal and temperature variations (Werker et al., 2002). The principle of nitrogen removal in CWs is mainly microbial interaction, sedimentation, chemical adsorption, plant uptake, nitrification and denitrification. The uptake and storage of nutrient by macrophytes depends on the concentration of nutrients in their tissue. Therefore, suitable macrophytes used in CWs should have fast growth rate and a high tissue nutrient content (Lee et al., 2009). Temperature, oxygen and pH are inter-related in CWs. Oxygen is required for the oxidation of ammonia. Insufficient oxygen transports through the aerenchyma and roots of macrophytes during early morning (9 AM) and late evening (7 PM) could result to anaerobic conditions which would inhibit nitrifying activity (Hammer and Knight, 1994). The permeation of sunlight into the bottom of the CW basin at 12 PM and 3 PM could prompts a faster rate of microbial growth and photosynthetic reaction. Thus, any loss of nitrification at 9 AM and 7 PM could be attributed to cold temperature and its associated parameters such as plant dormancy and insufficient oxygen transport. However, the influence of temperature on CWs may vary with the type of macrophyte. It is reported that *Phragmites* perform optimally during summer whereas *Typha spp.*

has an all round optimum performance irrespective of season (Hatano et al., 1991). Reid and Mosley (Reid and Mosley, 2016) demonstrated that photosynthesis of macrophytes can be affected by light/dark phenomenon. The authors showed a diurnal increase in the pH of mesocosms maintained in the light and a decrease or a continued fall in the dark. Thus, light influenced the pH values and the pH values influence nitrification/denitrification. The high NH_4^+-N removal in this study indicated an adequate biological nitrification/denitrification and plant assimilation/uptake mechanisms in the CW. Although maximum COD removal occurred at high temperature, COD removal at 9 AM was also high. This indicated that carbonaceous compound removal may not entirely be temperature dependent.

It is therefore evident that water temperature in CWs influence the removal rate of pollutants for the biological reactions that are temperature dependent. On a seasonal basis, low temperature may induce dormant vegetation and reduced reaction rate which could affect both physical and biological activities of aquatic microbes. The biological activities in CWs could therefore be enhanced through influent wastewater temperatures and bed insulation. It is obvious that the processes of ammonification, nitrification and denitrification are all temperature dependent. Thus, the rate of nitrogen removal by the microbial biofilm at the root of the macrophytes could be temperature dependent. Nitrification rate become inhibited below 10°C and complete failure below 6°C (Werker et al., 2002).

4 CONCLUSIONS

The influence of temperature on the performance of *Pistia Stratiotes* in a surface sequencing baffled steep flow constructed wetland reactor was monitored for the removal organic matter and ammoniacal nitrogen at different sampling time. The results obtained show that increase in the reactor temperature favored ammoniacal nitrogen removal but did not show much effect on COD removal. The maximum reactor temperature was noticed on day 4 (35.9°C) and maximum ammoniacal nitrogen removal of 94% was achieved. COD removal also reached 55% throughout the study. Maximum removal of organic matter and ammoniacal nitrogen was more favorable to 12 PM and 3 PM. At 7 PM, COD removal was poor. This study, therefore, shows that high reactor temperature could improve plant photosynthesis, nutrient uptake/assimilation and temperature dependent biological activites in a CW.

ACKNOWLEDGMENT

The authors are grateful to Universiti Teknologi PETRONAS (UTP) for the support.

REFERENCES

Barko, J., Hardin, D. & Matthews, M. 1982. Growth and morphology of submersed freshwater macrophytes in relation to light and temperature. *Canadian Journal of Botany*, 60, 877–887.

Council, N. A. W. C. & PRIES, J. H. 1994. *Wastewater and stormwater applications of wetlands in Canada*. Sustaining Wetlands Issue Paper Series, No. 1994-1. Ottawa, Ont., North America Wetlands Conservation Council.

Ezechi, E. H., Kutty, S. R., Isa, M. H., Malakahmad, A., Ude, C. M., Menyechi, E. J. & Olisa, E. 2016. Nutrient Removal from Wastewater by Integrated Attached Growth Bioreactor. *Research Journal of Environmental Toxicology*, 10, 28–38.

Ezechi, E. H., Kutty, S. R. B. M., Isa, M. H. & Rahim, A. F. A. Treatment of Wastewater Using an Integrated Submerged Attached Growth System. Applied Mechanics and Materials, 2014. Trans Tech Publ, 167–171.

Hammer, D. A. & Knight, R. L. 1994. Designing constructed wetlands for nitrogen removal. *Water Science & Technology*, 29, 15–27.

Hatano, K., Trettin, C., House, C. & Wollum, A. 1991. Microbial populations and decomposition activity in three subsurface flow constructed wetlands. Constructed wetlands for water quality improvement; 541–548.

Kadlec, R. H. 1999. Chemical, physical and biological cycles in treatment wetlands. *Water Science and Technology*, 40, 37–44.

Kouki, S., M'hiri, F., Saidi, N., Belaïd, S. & Hassen, A. 2009. Performances of a constructed wetland treating domestic wastewaters during a macrophytes life cycle. *Desalination*, 246, 452–467.

Lee, C. G., Fletcher, T. D. & Sun, G. 2009. Nitrogen removal in constructed wetland systems. *Engineering in Life Sciences*, 9, 11–22.

Madigan, M. T., Martinko, J. M., Parker, J. & Brock, T. D. 1997. *Biology of Microorganisms*, prentice hall Upper Saddle River, NJ.

Reid, R. & Mosley, L. 2016. Comparative contributions of solution geochemistry, microbial metabolism and aquatic photosynthesis to the development of high pH in ephemeral wetlands in South East Australia. *Science of The Total Environment*, 542, 334–343.

Rousseau, D., Lesage, E., Story, A., Vanrolleghem, P. A. & De Pauw, N. 2008. Constructed wetlands for water reclamation. *Desalination*, 218, 181–189.

Stein, O. R. & Hook, P. B. 2005. Temperature, plants, and oxygen: how does season affect constructed wetland performance? *Journal of Environmental Science and Health*, 40, 1331–1342.

Tanner, C. C. & Headley, T. R. 2011. Components of floating emergent macrophyte treatment wetlands influencing removal of stormwater pollutants. *Ecological Engineering*, 37, 474–486.

Vitória, A. P., Da Silva Santos, J. L., Salomão, M. S. M. B., De Oliveira Vieira, T., Da Cunha, M., Pireda, S. F. & Rabelo, G. R. 2015. Influence of ecologic type, seasonality, and origin of macrophyte in metal accumulation, anatomy and ecophysiology of Eichhornia crassipes and Eichhornia azurea. *Aquatic Botany*, 125, 9–16.

Wang, Z., Xia, C., Yu, D. & Wu, Z. 2015. Low-temperature induced leaf elements accumulation in aquatic macrophytes across Tibetan Plateau. *Ecological Engineering*, 75, 1–8.

Werker, A., Dougherty, J., Mchenry, J. & Van Loon, W. 2002. Treatment variability for wetland wastewater treatment design in cold climates. *Ecological Engineering*, 19, 1–11.

Influence of organic loading rate on a submerged two-stage integrated bioreactor without external carbon addition

E.H. Ezechi, S.R.M. Kutty, M.H. Isa, A. Malakahmad & E. Olisa
Department of Civil and Environmental Engineering, Universiti Teknologi PETRONAS, Seri Iskandar, Perak, Malaysia

ABSTRACT: The influence of organic loading rate (OLR) on a two stage integrated bioreactor was investigated in this study. The bioreactor consisted of two biological zones (aerobic and anoxic), a clarifier after the biological zones and an effluent zone, operated as a unit compartmental post-anoxic system. The aerobic zone was integrated into the anoxic zone and the anoxic zone was integrated into the clarifier. A 10 layer hollow rectangular and circular acrylic made submerged materials were installed in the aerobic and anoxic zones, respectively. A synthetic wastewater simulating a low and medium strength domestic wastewater with biochemical oxygen demand (BOD) concentration of 115 and 260 mg/L, corresponding to two different OLR of 0.33 kg $BOD_5/m^3/m^2 \cdot d^{-1}$ (phase 1) and 0.75 kg $BOD_5/m^3/m^2 \cdot d^{-1}$ (phase 2), was used at a fixed HRT of 12 days. The performance of the bioreactor was monitored by the BOD_5 removal. Results show that BOD_5 removal was stable at 97.3% with increasing OLR from 0.33 kg $BOD_5/m^3/m^2 \cdot d^{-1}$ in phase 1 to 0.75 kg $BOD_5/m^3/m^2 \cdot d^{-1}$ in phase 2. The BOD_5 removal rate increased from 0.314 kg $BOD_5/m^3/m^2 \cdot d^{-1}$ in phase 1 to 0.73 kg $BOD_5/m^3/m^2 \cdot d^{-1}$ in phase 2. However, the residual effluent concentration increased from 3 to 7 mg/L with increasing OLR. The aerobic zone was more prominent for BOD_5 removal throughout the study. NH_4^+-N and NO_3-N was also monitored and was found to reach 98% and 73% removal in phase 2. This study, therefore shows that OLR did not cause any decrease of the bioreactor performance at the operating conditions investigated.

1 INTRODUCTION

The discharge of wastewater with high pollutant concentration can cause significant environmental deterioration (Ramos et al., 2007, Ezechi et al., 2015). This can provoke unwanted phenomenons such as eutrophication as a result of the enrichment of water bodies with nutrients and blue baby syndrome, caused by long term consumption of nitrate-rich water in infants (Ramos et al., 2007, Ezechi et al., 2014). In addition, it has been reported that about 25% of all water body impairment are caused by nutrients (USEPA, 2007). Thus, it is necessary to ensure that effluent concentration of nutrients does not cause environmental deterioration and health problems.

The treatment of wastewater by biological method have been widely practiced with different configurations such as the bardenpho system (Mostafa M. Emaraa, 2014) and the modified Ludzack Ettinger (Park et al., 2012). However, the conventional biological treatment process such as the Ludzack Ettinger and Bardenpho systems have been associated with various challenges such as unstable effluent quality (Wang et al., 2005, Eddy, 2004) and large space requirement (Hamoda et al., 1996). This can lead to the discharge of wastewater with high nutrient concentration into water bodies, which can promote eutrophic and health related problems. Therefore, there is a dire need to address the challenges of the conventional process.

Submerged attached growth treatment plants were developed to overcome the challenges of the conventional process. It is more compact and its efficiency is independent of the sludge separation characteristics (Jahren et al., 2002). Several attached growth materials such as gravel (Tziotzios et al., 2005) and Polyurethane foam blocks (Sarti et al., 2001) have been successfully utilized for wastewater treatment. Their advantages include rapid biofilm formation, flexibility with pollutant load, small space requirement and modular construction (Osorio and Hontoria, 2001). These advantages are mainly derived from the capacity of the submerged material to retain high concentration of biomass (Nicolella and Rovatti, 1997). Thus, several phenomenons such as solid filtration, carbonaceous removal and nitrification could be achieved in a single unit (Mann and Stephenson, 1997).

The unit integration of various bioreactor zones has been investigated for the treatment of simulated domestic wastewater (Ezechi et al., 2016). Integrated bioreactors have demonstrated high treatment capabilities and have advantages such as smaller footprints, ease of operation and cost reduction (Chan et al., 2012b).

The objective of this study was therefore, to investigate the influence of organic loading rate on the performance of a submerged integrated bioreactor at fixed a HRT and HLR.

2 METHODOLOGY

2.1 Wastewater prepration

This study utilized a synthetic wastewater, developed by dissolving appropriate amount of purina alpo substrate in tap water according to the experimental plan at a fixed HRT of 12 days (influent flowrate of 15 L/d). Ammonium chloride and sodium carbonate were also added to provide nitrogen and alkalinity for pH stability between 7-8. The BOD_5 and NH_4^+-N concentration in phase 1 was 115 mg/L (V_{OLR} 0.33 $BOD_5/m^3.m^2.d^{-1}$) and 15 mg/L while in phase 2, the concentration was raised to 260 mg/L (V_{OLR} 0.75 $BOD_5/m^3/m^2.d^{-1}$) and 26 mg/L respectively which is within the range of the typical low and medium strength domestic wastewater (Eddy, 2004).

2.2 Bioreactor setup and operation

The bioreactor has a total liquid volume of about 180 L. It consist of the aerobic (10 L), anoxic (20 L), clarifier (150 L) and effluent zone, respectively. It was set up in a post-anoxic pattern. The aerobic zone was inserted into an un-aerated zone of a near equal size. The aerobic zone has four openings at the bottom for the flow of wastewater into the unaerated zone. The unaerated zone was used as a baffle system for phase separation and flow of wastewater into the anoxic zone. Ceramic plate diffusers were placed at the bottom of the aerobic zone to provide fine air bubbles in upflow pattern, 40 mm from the unaerated zone. The un-aerated zone has a semi-permeable filter at the topmost part, 200 mm from the bottom for the flow of wastewater into the anoxic zone in the downflow pattern. A 10 layer rectangular hollow submerged media were installed into the aerobic zone and occupy about 70% of the reactor volume. The aerobic zone was then integrated into the anoxic zone. In the anoxic zone, a 10 layer circular hollow submerged media was installed and occupied about 50% of the reactor volume (space left after the installation of the aerobic zone). The anoxic zone was then integrated into the clarifier. The influent feed was applied to the aerobic zone in a downflow pattern. The wastewater then passes into the unaerated zone and flows through the edge of its semi-permeable filter into the anoxic zone in downflow pattern while retaining the biomass. Through this process, the effluent of the aerobic zone becomes the influent of the anoxic zone. Samples then flow through the bottom hole of the anoxic zone into the secondary clarifier. Mixed liquor recirculation (MLR) was conducted into the aerobic zone at 180% to the influent flowrate. Single samples were collected from the exit point of the aerobic, anoxic and effluent zones every two days and were triplicated during analysis.

2.3 Bioreactor start-up

During bioreactor start-up, seed sludge was collected from the sewage treatment plant in Universiti Teknologi PETRONAS (UTP). Influent wastewater was applied at a low flowrate to initiate biofilm formation on the media. After about 2–3 weeks, sufficient biofilm was formed on the media. The experiments were then initiated at a fixed hydraulic loading rate (HLR) of $3.5\,m^3.m^2.d^{-1}$ (Influent flowrate of 15 L/d) and HRT of 12 days. No external carbon source was added to the anoxic zone as it utilized the carbonaceous substrate from the aerobic zone.

2.4 Analytical techniques

BOD_5 was measured according to the standard method number 5210 B (Eaton, 2005). Ammonia and nitrate concentration was measured using the colorimetric method (Nessler and cadmium reduction methods, respectively). pH, dissolved oxygen and temperature was daily monitored concurrently using a pH meter (Sension^{TM}), DO meter (YSI 550 A) and a thermometer (Thermolyne P/N MEX–147 IMM 76 MM MCT.

3 RESULT INTERPRETATION

The hydraulic loading rate (HLR) and the hydraulic retention time (HRT) were fixed at $3.5\,m^3/m^2.d^{-1}$ and 12 days respectively. The volumetric organic loading rate (V_{OLR}) was varied by increasing the average influent BOD_5 concentration from 115 mg/L (V_{OLR} of 0.33 kg $BOD_5/m^3.m^2.d^{-1}$) in phase 1 to 260 mg/L (V_{OLR} of 0.75 kg $BOD_5/m^3.m^2.d^{-1}$) in phase 2. The time course profile for BOD_5 removal from each zone of the bioreactor is presented in Fig. 1. In phase 1, the experiment was initiated with a V_{OLR} of 0.33 kg $BOD_5/m^3.m^2.d^{-1}$ with a C/N ratio of 8:1. In the aerobic zone, BOD_5 concentration decreased from day 1 to day 18 and reached a value of about 11 mg/L (0.026 kg $BOD_5/m^3.m^2.d^{-1}$). From day 20 to day 26, the BOD_5 concentration attained steady state with a final value of about 9 mg/L (0.023 kg $BOD_5/m^3.m^2.d^{-1}$). In the anoxic zone, a similar trend from day 1 to day 18 and from day 20 to day 26 was noticed with a final BOD_5 concentration of about 8 mg/L (0.022 kg $BOD_5/m^3.m^2.d^{-1}$). BOD_5 concentration in the effluent zone was about 3 mg/L (0.08 kg $BOD_5/m^3.m^2.d^{-1}$) on day 26. A total BOD_5 removal of about 97.3% (0.314 kg $BOD_5/m^3.m^2.d^{-1}$) was achieved in this phase with a residual effluent BOD_5 concentration of 3 mg/L (0.08 kg $BOD_5/m^3.m^2.d^{-1}$). After steady state attainment in phase 1, phase 2 (day 27) was initiated by increasing the BOD_5 concentration to 260 mg/L (V_{OLR} 0.75 kg $BOD_5/m^3/m^2.d^{-1}$) at the same conditions of HRT and HLR. On day 28, effluent BOD_5 concentration increased in each zone. BOD_5 concentration reached 44 mg/L (0.1 kg $BOD_5/m^3.m^2.d^{-1}$), 36 mg/L (0.09 kg $BOD_5/m^3.m^2.d^{-1}$) and 13 mg/L (0.04 kg $BOD_5/m^3.m^2.d^{-1}$) for the aerobic, anoxic and effluent

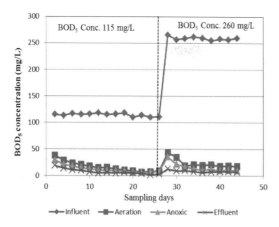

Figure 1. BOD_5 concentration in each zone.

zones, respectively. However, from day 30 to day 36, BOD_5 concentration decreased and reached 20 mg/L (0.032 kg $BOD_5/m^3.m^2.d^{-1}$) in the aerobic zone, 11 mg/L (0.03 kg $BOD_5/m^3.m^2.d^{-1}$) in the anoxic zone and 6.5 mg/L (0.02 kg $BOD_5/m^3.m^2.d^{-1}$) in the effluent zone. Between day 38 to day 44, BOD_5 concentration fluctuated and attained a steady state in the aerobic, anoxic and net effluent zones with a concentration of about 18 mg/L (0.029 kg $BOD_5/m^3.m^2.d^{-1}$), 10 mg/L (0.027 kg $BOD_5/m^3.m^2.d^{-1}$) and 7 mg/L (0.018 kg $BOD_5/m^3.m^2.d^{-1}$), respectively. A total BOD_5 removal of about 97.3% was achieved. However, the removal rate and residual effluent BOD_5 concentration increased to 0.73 kg $BOD_5/m^3.m^2.d^{-1}$ and 7 mg/L, respectively.

4 DISCUSSION

A notable BOD_5 removal trend was observed in this study. It was noticed that BOD_5 removal was stable but removal rate and residual BOD_5 effluent concentration increased with increasing V_{OLR}. In phase 1 (V_{OLR} 0.33 kg $BOD_5/m^3.m^2.d^{-1}$), a total BOD_5 removal rate of about 0.314 kg $BOD_5/m^3.m^2.d^{-1}$ was achieved with a residual effluent concentration of 3 mg/L (0.08 kg $BOD_5/m^3.m^2.d^{-1}$). In phase 2 (V_{OLR} 0.75 kg $BOD_5/m^3.m^2.d^{-1}$), BOD_5 removal and residual effluent concentration increased to 0.73 kg $BOD_5/m^3.m^2.d^{-1}$ and 7 mg/L (0.02 kg $BOD_5/m^3.m^2.d^{-1}$) respectively. The aerobic zone of the bioreactor was more prominent for BOD_5 removal. In phase 1, the aerobic zone achieved a total removal of 92.2% (0.3 kg $BOD_5/m^3.m^2.d^{-1}$) while the anoxic zone accounted for about 5.1% (0.0013 kg $BOD_5/m^3.m^2.d^{-1}$). In phase 2, BOD_5 removal in both the aerobic and anoxic zones were 93.1% (0.72 kg $BOD_5/m^3.m^2.d^{-1}$) and 3.05% (0.0016 kg $BOD_5/m^3.m^2.d^{-1}$), respectively. Several mechanism could involve in the substrate removal. The high BOD_5 removal in the bioreactor could be due to high biofilm formation, exogenous consumption of substrate, dynamic balance in the ecological structure of microbial consortium and re-attachment of biomass on the surface of the media (Hamoda et al., 1996, Liu et al., 2008). The uptrend in BOD_5 concentration at the start of phase 2 (day 28) and downtrend on day 30, could be attributed to endogenous respiration and the self regulatory capacity of microbial community to re-stabilize, re-acclimate and re-adjust to higher V_{OLR} (Hamoda et al., 1996, Chan et al., 2012a). The increase in BOD_5 removal at higher V_{OLR} and in the aerobic zone was due to higher substrate concentration feed applied to the zone.

(Hamoda et al., 1996) noted that COD removal increased with increasing organic loading and decreasing hydraulic loading in their study of biological nitrification kinetics in a fixed film reactor with four compartments. A COD removal of 42 g m^{-2} day^{-1} was achieved at a maximum organic loading rate of 50 g m^{-2} day^{-1} and COD removal was remarkedly prominent (80%) in the first compartment receiving the influent feed. Chan et al., (Chan et al., 2012a) demonstrated that COD removal increased with increasing OLR in their study of pre-anaerobic/aerobic integrated bioreactor for palm oil mill effluent (POME) treatment. COD removal was 87.2% at OLR of 3.0 g COD/L but increased to 88.9% and 92.1% when OLR was increased to 5.4 g COD/L and 10.5 g COD/L) respectively. The authors noted that 92.6% of the total COD removal was achieved by the first compartment (anaerobic) and 7.1% by the second compartment (aerobic). Pozo and Diez (Del Pozo and Diez, 2005) also demonstrated that COD removal was more prominent in the aeration zone in an integrated anaerobic-aerobic fixed film reactor for the treatment of slaughterhouse wastewater. COD removal of 96% and 2.6% were achieved in the aeration and anoxic zones, respectively. The high removal of BOD_5 in the aerobic zone and at increased V_{OLR} was a result of higher microbial activity at higher organic loading rate.

NH_4^+-N and NO_3-N were also monitored during this study and presented in Fig. 2. In both phase 1 and 2 (NH_4^+-N concentration of 15 mg/L and 26 mg/L), a steady state residual effluent concentration of 0.3 and 0.6 mg/L were obtained. The steady state effluent NO_3-N concentration in phases 1 and 2 were 0.2 and 0.55 mg/L, respectively.

In addition, NO_3-N accumulation was not noticed in the aerobic zone as NO_3-N removal reached 85% and 73% in phase 1 and 2 while NH_4^+-N removal was 98%, respectively. This implies that the bioreactor has a good oxygen mass transfer capacity.

5 CONCLUSIONS

The performance of a two stage (aerobic and anoxic) integrated bioreactor was evaluated in this study for the removal of BOD_5 from simulated synthetic low and medium strength domestic wastewater. Experiments were conducted at a fixed HRT and HLR while the V_{OLR} was varied. The results clearly show that

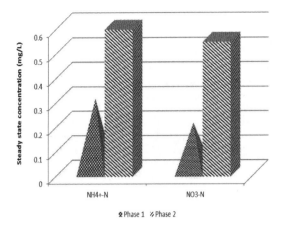

Figure 2. NH_4^+-N and NO_3-N removal at steady state.

the bioreactor has the capacity to effectively remove carbonaceous compounds from wastewater. BOD_5 removal and residual effluent concentration increased with increasing V_{OLR}. The aerobic zone of the bioreactor was more prominent for BOD_5 removal due to the application of the influent feed in this zone. At the start of phase 2, BOD concentration in each zone increased due to the increased V_{OLR}. However, after 4 days (days 26 to 30), the biofilm was observed to stabilize and BOD concentrations decreased to levels similar to what was observed in phase 1. NH_4^+-N and NO_3-N removal of 98% and 73% was achieved in phase 2. The increase of the BOD removal rate with increasing V_{OLR} indicates that the maximum loading rate was not attained. This study, therefore, shows that the integrated bioreactor possess high carbonaceous substrate removal capacity at the V_{OLR} investigated and could tolerate higher V_{OLR}, HLR and shorter HRT.

ACKNOWLEDGMENT

This study was supported by Universiti Teknologi PETRONAS (UTP) through the Graduate Asistantship Scheme (GA). The authors are therefore grateful to UTP.

REFERENCES

Chan, Y. J., Chong, M. F. & Law, C. L. 2012a. An integrated anaerobic–aerobic bioreactor (IAAB) for the treatment of palm oil mill effluent (POME): Start-up and steady state performance. *Process Biochemistry*, 47, 485–495.

Chan, Y. J., Chong, M. F. & Law, C. L. 2012b. Start-up, steady state performance and kinetic evaluation of a thermophilic integrated anaerobic–aerobic bioreactor (IAAB). *Bioresource Technology*, 125, 145–157.

Del Pozo, R. & Diez, V. 2005. Integrated anaerobic–aerobic fixed-film reactor for slaughterhouse wastewater treatment. *Water Research*, 39, 1114–1122.

Eaton, A. D. 2005. *Standard methods for the examination of water and wastewater,* Washington, D.C., APHA-AWWA-WEF.

Eddy, M. 2004. *Wastewater Engineering, Treatment and Reuse, fourth edition.*

Ezechi, E. H., Kutty, S., Isa, M. H., Malakahmad, A. & Ibrahim, S. U. 2015. Chemical oxygen demand removal from wastewater by integrated bioreactor. *J. Environ. Sci. Technol,* 8, 238–243.

Ezechi, E. H., Kutty, S. R., Isa, M. H., Malakahmad, A., Ude, C. M., Menyechi, E. J. & Olisa, E. 2016. Nutrient Removal from Wastewater by Integrated Attached Growth Bioreactor. *Research Journal of Environmental Toxicology,* 10, 28–38.

Ezechi, E. H., Kutty, S. R. B. M., Isa, M. H. & Rahim, A. F. A. Treatment of Wastewater Using an Integrated Submerged Attached Growth System. Applied Mechanics and Materials, 2014. Trans Tech Publ, 167–171.

Hamoda, M., Zeidan, M. & Al-Haddad, A. 1996. Biological nitrification kinetics in a fixed-film reactor. *Bioresource Technology,* 58, 41–48.

Jahren, S. J., Rintala, J. A. & Ødegaard, H. 2002. Aerobic moving bed biofilm reactor treating thermomechanical pulping whitewater under thermophilic conditions. *Water Research,* 36, 1067–1075.

Liu, F., Zhao, C.-C., Zhao, D.-F. & Liu, G.-H. 2008. Tertiary treatment of textile wastewater with combined media biological aerated filter (CMBAF) at different hydraulic loadings and dissolved oxygen concentrations. *Journal of Hazardous Materials,* 160, 161–167.

Mann, A. & Stephenson, T. 1997. Modelling biological aerated filters for wastewater treatment. *Water Research,* 31, 2443–2448.

Mostafa M. Emaraa, F. A. A., Farouk M. Abd El-Azizc and Ahmed M.A. Abd El-Razek 2014. Biological Aspects of the Wastewater Treatment Plant "Mahala Marhoom" in Egypt and Modified with Bardenpho Processes. *Nature and Science* 12, 41–51.

Nicolella, C. & Rovatti, M. 1997. Biomass concentration in fluidized bed biological reactors. *Water Research,* 31, 936–940.

Osorio, F. & Hontoria, E. 2001. Optimization of bed material height in a submerged biological aerated filter. *Journal of Environmental Engineering,* 127, 974–978.

Park, C., Lee, J. W., Moon, S., Park, K. Y. & Jutidamrongphan, W. 2012. Advanced treatment of wastewater from food waste disposer in modified Ludzack-Ettinger type membrane bioreactor. *Environmental Engineering Research,* 17, 59–63.

Ramos, A., Gomez, M., Hontoria, E. & Gonzalez-Lopez, J. 2007. Biological nitrogen and phenol removal from saline industrial wastewater by submerged fixed-film reactor. *Journal of Hazardous Materials,* 142, 175–183.

Sarti, A., Vieira, L. G. T., Foresti, E. & Zaiat, M. 2001. Influence of the liquid-phase mass transfer on the performance of a packed-bed bioreactor for wastewater treatment. *Bioresource Technology,* 78, 231–238.

Tziotzios, G., Teliou, M., Kaltsouni, V., Lyberatos, G. & Vayenas, D. 2005. Biological phenol removal using suspended growth and packed bed reactors. *Biochemical Engineering Journal,* 26, 65–71.

USEPA. 2007. National Section 303(d) List Fact Sheet. Available: Information Online at www.iaspub.epa.gov/waters/national_rept.control.

Wang, Y., Huang, X. & Yuan, Q. 2005. Nitrogen and carbon removals from food processing wastewater by an anoxic/aerobic membrane bioreactor. *Process Biochemistry,* 40, 1733–1739.

Performance of a bench scale anaerobic baffled reactor operated at ambient temperatures

U.A. Abubakar, M.M. Muhammad, F.B. Ibrahim, M.A. Ajibike & A. Ismail
Ahmadu Bello University, Zaria, Nigeria

ABSTRACT: The potential of the anaerobic baffled reactor (ABR) for efficient wastewater treatment was evaluated. A laboratory bench set-up, which consisted of two ABR, R1 at 37°C and R2 at ambient temperatures each comprised of 6 – compartments, was operated for 8 months, with each reactor receiving the same organic loading rate of 1.25 kg chemical oxygen demand (COD) per m^3.day. 40% of the volumes of the 2nd–6th compartments of each of the two reactors were inoculated using anaerobic biomass. The reactors were operated for 45 days before determination of steady state by stability of pH. R1 showed higher performance efficiency than R2 due to the high COD removal in the 1st compartment, 54% of influent loading for R1 against 25% for R2, and low concentrations of volatile fatty acids. Thus, low treatment performance was observed at ambient temperature compared to the one observed at stable mesophilic temperature.

1 INTRODUCTION

The anaerobic baffled reactor (ABR) has the capability of achieving enhanced contact between wastewater and anaerobic biomass through the influence of a series of vertical baffles which break the reactor volume into several compartments and force the wastewater to flow through any entrapped biomass (Baloch & Akunna 2003; Barber & Stuckey 1999). Consequently, the proper establishment of appropriate microbial biomass in the ABR, normally during start-up periods before steady state performance is achieved, is considered critical before high organic load removal can be achieved (Bodkhe 2009; Liu et al. 2010). Zhu et al. (2015) identified the required long start-up periods and the absence of a clear definition of the relationship between organic loading rates (OLRs), hydraulic retention times (HRTs) and chemical oxygen demand (COD) removal efficiencies as the key limitations towards the full scale application of the ABR as a conventional wastewater treatment system. Feng et al. (2015) reported a start-up period of 65 days for an ABR reactor operated at ambient temperature (22.0°C–24.8°C) with average influent COD of 444 mg/L. The average effluent COD during the start-up period was 323 mg/L, with a range of 1.66 to 60.05% removal efficiency which improved to 66.4% during the steady state performance period which was achieved after approximately 130 days. The observed average effluent COD during steady state operation was 71 mg/L for influent COD ranging from 100 mg/L to 250 mg/L (Feng et al. 2015). Bodkhe (2009) reported a biological acclimatization period of 90 days before COD removal efficiency was steady at 97% for experiments without initial inoculation of biomass. Boopathy & Tilche (1991) observed four stages during the operation of a hybrid anaerobic baffled reactor, with the first stage lasting for 40 days with low organic loading rate before suitable biomass was established in the reactor. This study aims to advance the understanding of ambient temperature operation of the ABR, and the subsequent development of high efficiency operational conditions and configurations of the ABR for efficient wastewater treatment.

2 MATERIALS AND METHODS

2.1 Laboratory bench set-up

The laboratory bench set-up adopted consisted of two bench reactors (Fig. 1), receiving the same feed at a set organic loading rate using Masterflex variable speed peristaltic pumps (Cole-Parmer UK).

The bench reactors design was based on an initial reactor design from Baloch et al. (2008) and Shanmugam & Akunna (2008), comprising five equal compartments in series with a total effective treatment volume of 8.75 litres (1.75 litres for each compartment). Each compartment was divided into equal up-comer and down-comer volumes using baffles (Fig. 2), and in order to influence solid retention and enhance hydrolysis, the model design adopted provided for an additional 1st compartment that is larger than the subsequent compartments as recommended by Boopathy & Sievers (1991), and therefore the effective treatment volume was increased to 17.5 litres.

Materials used for the fabrication of the bench reactors were 10 mm thick acrylic sheets and glue, where the sheets were formed to sections and then glued

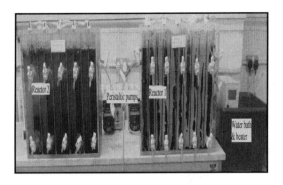

Figure 1. Laboratory bench set-up. The major items in the set-up include: 2 variable peristaltic pumps; Reactor 1 set at 37 ± 3°C; Reactor 2 at ambient temperature and a water bath and heater with circulating pump for temperature control.

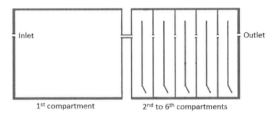

Figure 2. A schematic representation of the bench reactors design.

together to ensure water and air tightness. 12 mm diameter holes were threaded at various locations to provide for sample collection from the six compartments and also to provide for biogas outlets from the top of the compartments. The 6 sampling points for liquid samples are located in the centre of the up-comer sections, 20 mm below the outlet of each compartment.

2.2 Bench experiment

The two ABR bench models were operated simultaneously from March to October, 2013. The bench experiment was initiated with introduction of anaerobic biomass and synthetic feed into two bench reactors (R1 and R2). For each of the two bench reactors, 40% of the volumes (3.5 L) of the 2nd – 6th compartments were inoculated with anaerobically digested sludge sourced from Hatton wastewater treatment plant, while the 1st compartments were not inoculated. The inoculum was initially conditioned and degassed by incubating at 37°C for 3 days (72 hours) prior to inoculation.

2.3 Organic loading

The feed used for this study was based on a synthetic feed with characteristics from Shanmugam & Akunna (2008), Gopala-Krishna et al. (2009) and Ghaniyari-Benis et al. (2009). The components of the feed were: 2.8 g/L pure cane molasses, 75 mg/L Ammonium bicarbonate (NH_4HCO_3), 400 mg/L Potassium dihydrogen phosphate (KH_2PO_4), 1 mg/L Magnesium sulphate (Epsom salts) ($MgSO_4$), 1 mg/L Iron (III) chloride (Ferric chloride) ($FeCl_3$), 1 mg/L Calcium chloride ($CaCl_2$), 1 mg/L Potassium chloride (KCl), 0.2 mg/L Cobalt (II) chloride ($CoCl_2$), 0.2 mg/L Nickel chloride ($NiCl_2$) and 2 g/L Sodium bicarbonate ($NaHCO_3$). The feed was prepared daily and loading for the two reactors was from the same feed tank with an organic loading rate (OLR) of 1.25 kg COD/m^3.day.

2.4 Steady state determination

After the initial inoculation and start-up, the system was operated for 45 days before analysis was performed due to the reported start-up periods in literature (Barber & Stuckey 1999; Bodkhe 2009; Nachaiyasit & Stuckey 1997). Collection of liquid samples from the bench reactors was carried out using clean 25 mL plastic vials through the sampling points provided. Steady state was determined by monitoring variation of pH values in the compartments of the two bench reactors after operation for five times the duration of the design HRT. If the variation of the pH in each compartment remained within a range of ±0.2 over a three days period, the system was considered to have achieved steady state performance.

2.5 pH analysis

The pH of the samples was determined using a SenSION3 pH probe and meter (Hach Company, Loveland Colorado U.S.A). The pH meter is programmed with a pH slope determined through measurements of standard pH solutions, and the pH of any subsequent solution is determined based on direct comparison with the standard pH slope. Calibration of the pH probe was carried out before evaluation of the samples using standard 4.00, 7.00 and 10.00 pH buffer solutions supplied with the pH probe. According to Heirholtzer (2013), the accuracy (closeness of agreement between a test result and a reference value) of this method is ±0.2% for the pH probe and meter as reported by the manufacturer.

2.6 VFA concentrations determination

VFA concentrations, expressed as Acetic acid (mg/L HOAC) within the range of 27–2800 mg/L, were determined by spectrophotometry with the Ferric hydroxamate method for determination of carboxylic esters (Hierholtzer et al. 2013), also known as the Montgomery method, using a DR 5000 Hach Lange spectrophotometer (Hach Lange, Salford Manchester, UK). The analysis, defined as Method 8196 in the DR 5000 user manual (Hach Company 2005), was performed in triplicates for each sample, and the average of the three measurements was adopted as the VFA concentration for the sample. This method has a reported precision (closeness of agreement between results from several independent tests under standard conditions) of 4.1 (Hierholtzer 2013).

2.7 COD concentrations determination

Analysis for COD concentrations (mg/L) was carried out using a DR 5000 spectrophotometer (Hach Lange, Salford Manchester, UK). The concentrations of COD (mg/L) in collected samples were determined using colorimetric determination with the Hach-Lange DR 5000 spectrophotometer Method 8000 (Hach Company 2005). Only one COD measurement was obtained for each sample, and Heirholtzer (2013) reported the accuracy and precision of this method of analysis as 6.5 and 2.7%, respectively.

2.8 Determination of solids concentrations

Total solids concentrations were determined according to standard methods (American Public Health Association 1998), by drying the samples in an oven at 105°C over 24 hours, while the volatile solids concentrations were determined by igniting the dried samples in a furnace at 550°C for two hours. Duplicate measurements were performed for each sample, and the average TS and VS concentrations were adopted.

3 RESULTS AND DISCUSSION

The characteristics of the inoculum after incubation were: 7.02 pH, 90 mg/L volatile fatty acids (VFA), 25.21 g/L total solids (TS) and 14.29 g/L volatile solids (VS). Similarly, the characteristics of the influent feed were: 2479.50 mg/L mean COD (range was 2150.00–2727.00 mg/L); 98.50 mg/L mean VFA (range was 64.00–148.00 mg/L) and 7.42 mean pH (range was 6.91–7.86).

Monitored pH in the effluents of the compartments of the two bench reactors indicated that achieving stable pH values at 48 hours HRT required a longer time period for R2 compared to R1, indicating a longer acclimatization period for R2 to the organic loading rate of 1.25 kg COD/m^3.day compared to the period taken by R1. The time taken for stable pH values to be observed at 48 hours HRT and 1.25 kg COD/m^3.day ORL was approximately 70 days for R1 and 110 days for R2. The removal of COD loading by R1 was 88% of the influent feed on experimental Day 67, while by Day 105 of the ABR bench experiment, a 73% influent COD removal was observed for R2. Apart from the quicker acclimatization of the biomass to the organic loading in R1 than in R2, there was also eventually a higher COD removal achieved in R1 compared to R2 at steady state. Figure 1 shows the effluent COD values for the two reactors at pseudo-steady state during the period stable pH values were observed in R1 and R2 at 48 hours HRT.

The difference in the results of the 1st compartments from the two reactors (Fig. 3), which were initially not inoculated with any biomass, may be due to the temperature differences because both compartments were receiving the same feed at the same flow rates. From Figure 3, a 54% influent COD removal was observed

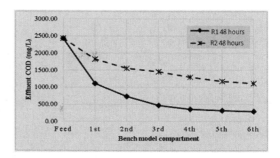

Figure 3. Effluent COD (mg/L) for each compartment in R1 and R2 during operation at 48 hours HRT with 1.25 kg COD/m^3.day OLR.

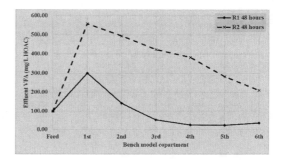

Figure 4. Volatile fatty acids concentrations (mg/L HOAC) in the effluents of compartments of R1 and R2.

for the 1st compartment of R1, against a 25% influent COD removal for the 1st compartment of R2. The COD removal in the 1st compartments appear critical to the overall performance of the reactors; if the 1st compartments are disregarded, the observed COD removal from the last five compartments of R1 was 34% of the influent COD, while R2 showed a removal of 30% of the influent COD in the last five compartments (Fig. 3). Figure 4 presents the observed volatile fatty acids concentrations in the effluents from the compartments of the bench reactors.

From Figure 4, the VFA profiles shown were for 48 hours HRT, corresponding to 1.25 kg COD/m^3.day OLR, and the effluent VFA concentrations for R2 were higher than effluent VFA concentrations observed for R1. For the VFA profiles for the two reactors, Figure 4, the 1st compartments are again the critical sections of the reactor operation, where R1 was able to maintain approximately 300 mg/L VFA concentrations, while R2 maintained VFA concentrations above 500 mg/L in its 1st compartment. For R1, once the biomass acclimatized, the readily biodegradable substrates were mostly removed before the 4th compartment (Fig. 3), indicating that four compartments may be adequate to achieve high COD removal at 37°C. While for R2, there appears to be no substantial change in VFA concentrations between consecutive compartments, and the change was relatively constant between the compartments.

The pH values recorded were between 5.90 and 7.30, indicating a pH range suitable for methane producing organisms, however, for R2 there were considerable VFA concentrations observed in the final effluents of the reactor, presented as effluent from the 6th compartment in Figure 4, indicating inefficient depletion of the VFA concentrations by methanogens. This suggests that zones of microbial dominance were developed in R1, while in R2 there were no definite zones of dominance established. The critical aspect of the operations at ambient temperature, R2, is the savings in energy consumption by avoiding heating requirements, however the results presented in Figures 3 and 4 indicate a loss in treatment performance in R2 when compared to a reactor operated at a stable mesophilic temperature (R1). According to Yaun et al. (2011), change in microbial populations as a result of changes in operational conditions, for example temperature, can lead to shifts in the anaerobic digestion pathway and consequently the nature of the intermediate products.

During sludge digestion experiments under thermophilic conditions with a continuous flow reactor, Aitken et al. (2005) observed that the most abundant VFA in the effluent was propionate. In experiments with an anaerobic membrane bioreactor, Yuzir et al. (2011) reported detection of propionate with a HRT of 1 day, however propionate was not detected when HRTs longer than 1 day were examined (3, 7 and 17 days). Yuzir et al. (2011) observed an increase in butyric acid as a percentage of VFA concentrations after a decrease in pH, where potentially the low pH conditions induced a stressful environment for the microorganisms causing a change in pathway and led to increase in butyric acid production. Consequently the low organic loading removal rates observed at ambient temperature may be due to the predominance of a different form of intermediate acid which is not readily converted to methane compared to the volatile fatty acids composition at 37°C. Therefore, apart from the observed low substrates reduction, Figure 3, at ambient temperature, another limiting factor for efficient ambient temperature anaerobic digestion may be the transformation of the substrates to volatile acids that can be readily converted to methane.

Changes in COD concentrations for successive compartments at ambient temperature were not substantial, and available volatile fatty acids were not substantially removed. The operational areas for consideration with respect to improvement of the performance efficiency of the ABR for ambient temperature treatment of wastewater are the process stability, system hydraulics and final effluent quality. Future research on anaerobic treatment of wastewater at ambient temperature should focus on identified process limitations that are stopping the achievement of performance efficiency. Determination of the size of the 1st compartment relative to the other compartments is also very important for system efficiency and hydrodynamics at ambient temperature. An advantage of the reactor operated at 37°C over the ambient temperature reactor was the observed COD removal in the 1st compartment and the stability of VFA concentrations in ranges that could not inhibit methanogens.

4 CONCLUSIONS

Energy efficiency in domestic wastewater treatment can be achieved when energy consumption is minimized and energy recovery is maximised. Wastewater treatment systems are expected to become energy efficient if heating is avoided and the treatment processes occur at ambient temperature. Identified process challenges at ambient temperature are the high concentrations of organic loads in the effluent, and the low conversion of substrates to final products of the anaerobic process. Low treatment performance was observed at ambient temperature compared to the performance observed at a stable mesophilic temperature.

ACKNOWLEDGEMENT

The authors gratefully acknowledge the financial support of the Petroleum Technology Development Fund (PTDF) of Nigeria and Ahmadu Bello University, Zaria, Nigeria. The authors also acknowledge the support of the staff of Abertay University, Dundee, United Kingdom, where the bench experiment was carried out.

REFERENCES

Aitken, M. D., Walters, G. W., Crunk, P. L., Willis, J. L., Farrell, J. B., Schafer, P. L., Arnett, C. & Turner, B. G. 2005. Laboratory evaluation of thermophilic-anaerobic digestion to produce Class A biosolids. 1. Stabilization performance of a continuous-flow reactor at low residence time. *Water environment research.* 77(7): pp. 3019–3027.

American Public Health Association. 1998. *Standard methods for the examination of water and wastewater.* 20th ed. Washington, DC: American Public Health Association (APHA) – American Water Works Association (AWWA) – Water Environment Federation (WEF).

Baloch, M.I., Akunna, J.C., Kierans, M. & Collier, P.J. 2008. Structural analysis of anaerobic granules in a phase separated reactor by electron microscopy. *Bio-resource Technology.* 99(5): pp. 922–929.

Baloch, M.I. & Akunna, J.C. 2003. Effect of rapid hydraulic shock loads on the performance of granular bed baffled reactor. *Environmental Technology.* 24(3): pp. 361–368.

Barber, W. P. & Stuckey, D. C. 1999. The use of the anaerobic baffled reactor (ABR) for wastewater treatment: a review. *Water Research.* 33(7): pp. 1559–1578.

Bodkhe, S.Y. 2009. A modified anaerobic baffled reactor for municipal wastewater treatment. *Journal of Environmental Management.* 90(8): pp. 2488–2493.

Boopathy, R. & Sievers D. M. 1991. Performance of a modified anaerobic baffled reactor to treat swine waste. *Transactions of the ASAE (USA).* 34(6): pp. 2573–2578.

Boopathy, R. & Tilche A. 1991. Anaerobic digestion of high strength molasses wastewater using a hybrid anaerobic baffled reactor. *Water research.* 25(7): pp. 785–790.

Feng, J., Wang, Y., Ji, X., Yuan, D., & Li, H. 2015. Performance and bio-particle growth of anaerobic baffled reactor (ABR) fed with low-strength domestic sewage. *Frontiers of environmental science and engineering*. 9(2): pp. 352–364.

Ghaniyari-Benis, S., Borja, R., Monemian, S. A., & Goodarzi, V. 2009. Anaerobic treatment of synthetic medium-strength wastewater using a multistage biofilm reactor. *Bioresource technology*. 100(5): pp. 1740–1745.

Gopala-Krishna, G. V. T., Kumar, P., & Kumar, P. 2009. Treatment of low-strength soluble wastewater using an anaerobic baffled reactor (ABR). *Journal of environmental management*. 90(1): pp. 166–176.

Hach Company. 2005. *DR5000 Spectrophotometer procedures manual: Catalog number DOC082.98.00670*. 2nd ed. Germany.

Hierholtzer, A. 2013. *Investigating factors affecting the anaerobic digestion of seaweed: modelling and experimental approaches*. [PhD dissertation]. Abertay University, Dundee, United Kingdom. Available online at: https://repository.abertay.ac.uk/jspui/bitstream/handle/10373/2025/Heirholtzer_InvestigatingFactorsAffectingthe%20AnaerobicDigestionofSeweed_Redacted_2013.pdf;jsessionid=3C1A64153FB6FB04DE0D340EA9D038EB?sequence=2.

Hierholtzer, A., Chatellard, L., Kierans, M., Akunna, J. C.,& Collier, P. J. 2013. The impact and mode of action of phenolic compounds extracted from brown seaweed on mixed anaerobic microbial cultures. *Journal of applied microbiology*. 114(4): pp. 964–973.

Liu, R., Tian, Q., & Chen, J. 2010. The developments of anaerobic baffled reactor for wastewater treatment: A review. *African Journal of Biotechnology*. 9(11): pp. 1535–1542.

Nachaiyasit, S. & Stuckey, D.C. 1997. Effect of low temperatures on the performance of an anaerobic baffled reactor (ABR). *Journal of Chemical Technology and Biotechnology*. 69(2): pp. 276–284.

Shanmugam, A. S. & Akunna, J. C. 2008. Comparing the performance of UASB and GRABBR treating low strength wastewaters. *Water Science and Technology*. 58(1): pp. 225–232.

Yuan, Q., Sparling, R. & Oleszkiewicz, J.A. 2011. VFA generation from waste activated sludge: effect of temperature and mixing. *Chemosphere*. 82(4): pp.603–607.

Yuzir, A., Chelliapan, S. & Sallis, P. J. 2011. Influence of step increases in hydraulic retention time on (RS)-MCPP degradation using an anaerobic membrane bioreactor. *Bioresource technology*. 102(20): 9456–9461.

Zhu, G., Zou, R., Jha, A. K., Huang, X., Liu, L., & Liu, C. 2015. Recent developments and future perspectives of anaerobic baffled bioreactor for wastewater treatment and energy recovery. *Critical reviews in environmental science and technology*. 45(12): pp. 1243–1276.

Use of low frequency ultrasound for solids solubilization in Palm Oil Mill Effluent

Lai-Peng Wong & Mohamed Hasnain Isa
Department of Civil and Environmental Engineering, Universiti Teknologi PETRONAS, Seri Iskandar, Perak, Malaysia

Mohammed J.K. Bashir
Department of Environmental Engineering, Faculty of Engineering and Green Technology, Universiti Tunku Abdul Rahman, Kampar, Malaysia

ABSTRACT: Palm Oil Mill Effluent (POME) is a thick suspension that contains high strength of Chemical Oxygen Demand (COD) of 48,000 mg/L, and Biochemical Oxygen Demand (BOD_5) of 22,000 mg/L. In the present study, effects of low frequency ultrasound on solids solubilization in POME were investigated. Different ultrasonic densities (1.0–3.5 W/mL) were applied for 10 s. The Soluble COD (SCOD)/Total COD (TCOD) ratio and biodegradability of soluble organic matter increased after sonication; indicating an increase in disintegration in POME. The maximum SCOD/TCOD ratio reached almost 50% while particle size reduced by 25% for sample sonicated at 1.25 W/mL for 10 s. The BOD_5/SCOD ratio also increased after sonication suggesting the biodegradability of the soluble organic material increased during the treatment. While sonication density exhibited the most significant role in cavitation bubble behaviours, particle disruption could be optimized for energy input by sonicating at the optimum sonication density range.

1 INTRODUCTION

POME is an acidic brownish colloidal suspension, characterized by high organic content and high temperature. Among biological processes, anaerobic wastewater treatment system has been employed by most palm oil mills in Malaysia due to the high organic content of POME (Poh et al., 2010). However, the major concern in the application of anaerobic treatment is the high concentration of particulate organic matter in POME that limits the hydrolysis stage of anaerobic treatment (Vavilin et al., 2008). POME contains complex organic and phenolic compounds that can be difficult to degrade biologically (Wu et al., 2010). Therefore, the performance of anaerobic treatment can be enhanced by introduction of a pretreatment stage. Combination of several pretreatments methods such as electrocoagulation (Agustin et al., 2008) and membrane filtration (Wu et al., 2007) have been reported. However, the cost implications of these methods are very prohibitive and may not completely solve the problem (Cesaro and Belgiorno, 2014). In anaerobic treatment processes, hydrolysis of complex organic substances is usually the rate-limiting step. A high fraction of particulate organic matter could be hydrolyzed with the mechanical attack mechanisms of ultrasound (Tiehm et al., 2001).

Ultrasonication treatment has been used as an alternative pretreatment to accelerate the hydrolysis rate in anaerobic digestion for sludge management (Pilli et al., 2011). Cavitation bubbles formed during sonication, grow and violently collapse when they reach a critical size and produce intense local heating and high pressure on liquid-gas interface, turbulence and high shearing phenomena in the liquid phase (Bougrier et al., 2005, Wang et al., 2007). Thus, ultrasonic disintegration is a powerful method for breaking up of floc structure to extract intracellular material. Due to the disintegration process, organic compounds are transferred from the solids into the aqueous phase resulting in enhanced biodegradability. Therefore, ultrasound disintegration could be a promising method to enhance anaerobic digestion rates. In the case of disintegrated activated sludge several researchers reported that low frequency ultrasound pretreatment could decreased sludge particle size (Show et al., 2007) and increase soluble COD (Tiehm et al., 2001). Besides, up to two-fold higher production of biogas was achieved in the subsequent anaerobic digestion process. (Salsabil et al., 2010).

Considering composition of POME which consists of recalcitrant soluble compounds, solids and long chain fatty acids, ultrasonication could theoretically convert these substances to simpler compounds. Studies of applying ultrasonic pretreatment before anaerobic process for POME are still limited. Saifuddin and his group (Saifuddin, 2009) have applied 1–30 min ultrasonication pretreatment for POME, and the best result achieved was 29% SCOD/TCOD after 30 min ultrasonication.

Table 1. Characteristics of raw POME.

Parameters	Mean value	Units
Temperature	80	°C
pH	5.07	–
Biochemical Oxygen Demand (BOD_5)	22,000	mg/L
Chemical Oxygen Demand (COD)	48,450	mg/L
Soluble COD (SCOD)	14,850	mg/L
Total Solids (TS)	43.89	g/L
Total Suspended Solids (TSS)	20.72	g/L
Particle Size, Volume weighted mean	68.82	μm

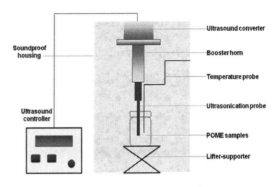

Figure 1. Schematic diagram of ultrasonic processor.

The current study aims to examine the effects of low frequency ultrasound on solids solubilization of POME. The system was operated at a low frequency of 20 kHz with varying density of 1.0–3.5 W/mL for 10 s. The contribution of ultrasonication system to COD solubilization and particle size profile were evaluated.

2 MATERIALS AND METHODS

2.1 Characterization of Palm Oil Mill Effluent (POME)

POME was obtained from a local palm oil mill in Perak, Malaysia. It was stored at 4°C in a closed plastic container until use. The characteristics of raw POME are given in Table 1. All analyses were carried out in triplicate according to the procedures given in Standard Methods (APHA, 2005).

2.2 Ultrasonic treatment

A Cole-Parmer 500-Watt Ultrasonic processor was used for ultrasonication of POME as shows in Figure 1. The apparatus is equipped with a 13 mm titanium transducer with an operating frequency of 20 kHz and supplied power of 500 W. For each experiment, sample in batches of 50–100 mL were ultrasonicated without temperature adjustment. Different ultrasonic densities (1.0, 1.25, 1.5, 1.75, 2.0, 2.5, 3.0, 3.5 W/mL) were applied for 10 s. The temperature probe was submerged into the beaker along with the ultrasonicated probe to record the temperature before and after ultrasonication.

2.3 Analytical methods

The variations of COD, BOD, TSS content and temperature in both raw samples and ultrasonicated samples were monitored to assess the treatment performance. All analyses were carried out according to the procedures given in the Standard Methods (APHA, 2005). Particle size distributions of the samples were measured by Malvern Mastersizer Hydro2000MU to further evaluate the effects of ultrasonication on solubilization of organic matter in POME.

3 RESULTS AND DISCUSSION

3.1 Effect of low frequency ultrasound on TCOD and SCOD

In anaerobic treatment processes, hydrolysis of complex organic substances is usually the rate-limiting step. A high fraction of particulate organic matter could be hydrolyzed with the mechanical attack mechanisms of ultrasound (Tiehm et al., 2001). Thus, hydrolysis step of anaerobic treatment may be accelerated by sonication.

TCOD is the organic matter in both the solid and liquid state of the POME sample while Soluble COD (SCOD) represents the organic matter only in the liquid portion. As expected the TCOD remained almost constant during the ultrasound application. The increase in SCOD/TCOD ratio denotes the release of the organic materials from solid to liquid state after ultrasonication treatment. Therefore, the variation of SCOD/TCOD ratio under different ultrasonic density was studied to evaluate the degree of POME solids solubilization by ultrasonication treatment. The maximum SCOD/TCOD ratio reached about 51% for sample ultrasonicated at density of 1.25 W/mL as shows in Figure 2. The SCOD/TCOD ratio decreased for higher ultrasonication density as represented by the particle size distribution profile in this study. The SCOD increase was mainly attributed to the break-up of suspended solids leading to the release of intracellular materials. Higher ultrasonication density might re-agglomerate the particle which could block the further release of organic matter from the particle (Prozorov et al., 2004).

3.2 Effect of low frequency ultrasound on solubilization of organic matter

The increase in BOD_5/SCOD ratio in Figure 2 suggested the enhancement of biodegradability of POME by sonication pretreatment (Saifuddin, 2009). The BOD_5/SCOD ratio of the raw POME was 0.839; the

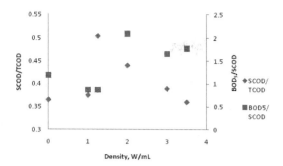

Figure 2. Effect of ultrasonic treatment on SCOD/TCOD and BOD$_5$/SCOD ratio at different sonication density.

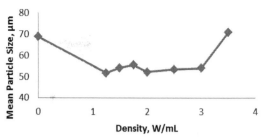

Figure 3. Mean particle size at different ultrasonication density.

ratio increased with the ultrasonication density. The BOD$_5$/SCOD ratio was in the range of 1.65–2.10 for samples ultrasonicated at density of 1.5–3.0 W/mL.

Different ultrasound mechanisms for the transformation of compounds have been reported. These included sono-chemical (Hua and Hoffmann, 1997) and sono-physical effects (Oz and Uzun, 2015). However, it has also been reported that the sono-chemical effects are more pronounce at frequencies above 100 kHz (Sangave et al., 2007). In our study which used 20 kHz ultrasound treatment, sono-physical effects may play a major role in the disintegration of coarse particle to soluble form. The sono-physical mechanisms induced the hydrodynamic shear force in aqueous phase due to rapid collapse of microbubbles formed during cavitation. The implosion of microbubbles affect the rate of mass transfer and reaction, hence affect the ratio of BOD$_5$/SCOD.

3.3 Effect of low frequency ultrasound on particle size profile

Particle size reduction is one of the most significant effects shown after ultrasonication treatment. The particle size decreased as soon as the ultrasonication was employed as shown in Figure 3. The reduction of particle size reflected that particles were broken up and free substances were released, which offers better conditions for the subsequent digestion (Show et al., 2007). As showed in Figure 3, the mean particle size decreased by as much as 25% after 10 s of ultrasonication with density of 1.25 W/mL. Figure 3 depicts that ultrasonication for 10 s could reduce the particle size most effectively within the density range of 1.25–3.0 W/mL. The reduction was in the range of 19%–25%. After this initial density, further ultrasonication density has little contribution to particle size reduction but lead to coarsening of secondary agglomerates and re-agglomeration (Prozorov et al., 2004).

Table 2 shows the particle size changes for different ultrasonication densitiy. Particle size distribution shown as dp10, dp50 and dp90 indicates that 10%, 50% and 90% of particles in volume having a diameter lower or equal to the values shown in the respective column in Table 2.

Table 2. Particle size profile for different ultrasonication density.

Sonication density, W/mL	Particle Size, μm			
	Mean	dp10	dp50	dp90
0	68.82	11.51	51.52	122.30
1.25	51.76	11.41	47.23	95.81
1.50	54.18	10.68	47.84	97.43
1.75	55.54	10.33	47.95	98.14
2.00	52.29	9.90	48.26	96.22
2.50	53.44	8.42	49.50	98.95
3.00	54.12	8.76	49.58	100.93
3.50	71.15	9.99	55.05	133.07

Large particles in liquid suspension are exposed to surface erosion or particle size reduction when subjected to ultrasonication (Jeganathan et al., 2007). The reduction in particle size generally eases the hydrolysis of solids and enhances degradation of the organic fraction in the solid phase (Jiang et al., 2014). During bubbles collapse, intense shock waves are generated that propagate through the liquid at velocities above the speed of sound (Suslick, 1990). The particle size increased for sample ultrasonicated at 3.5 W/mL. This could be due to the re-agglomeration of particles in POME. The particle size has a very strong effect on the outcome of the ultrasonic irradiation where qualitatively agglomeration would occur for particles in the range of 100 nm–100 μm (Prozorov et al., 2004).

4 RECOMMENDATION

The primary aim of ultrasonication is to increase the organic solubilization therefore enhance the biogas production at lower hydraulic retention time in an anaerobic digester. The biomethane emission rate is high from POME thus there is indeed a huge potential in utilizing POME to produce renewable energy and generate high commercial return. Oz and colleagues (Oz and Yarimtepe, 2014) reported that 40% more biogas was obtained from an anaerobic batch reactor feed with ultrasonically pretreated leachate compared to unsonicated leachate reactor. Show and colleagues (Show et al., 2011) has calculated energy recovery from anaerobic process for POME treatment with COD concentration of 58,000 mg/L. They estimated a project activity of biogas recovery with an investment of about USD 2.5 millions with a return period of less than 2 years. Thus, the addition of readily biodegradable organic matter after ultrasonication in POME could significantly increase biogas production during anaerobic treatment process. The payback period for investment on capital and operating cost for ultrasonicated-anaerobic reactor could be shortened. However, further study is needed to be carried out to verify this hypothesis.

5 CONCLUSION

This study investigated the effects of ultrasound treatment of POME on solid solubilization by examining the COD solubilization and particle size profile. Experimental results showed that low frequency ultrasound treatment could improve the characteristics of POME by the reduction of particle size and increase of SCOD under optimum ultrasonication conditions. The overall results indicated that ultrasound treatment could be influenced by ultrasonication density. The higher the ultrasonication power employed, the more particles are ruptured and the more completely the structure is deteriorated. However, the ultrasound performance would decline if the ultrasonication density exceeded the optimal range. In this study, the optimum ultrasonication density range suggested is 1.25 W/mL to 2.5 W/mL for 10 s ultrasonication treatment of POME.

REFERENCES

Agustin, M.B., Sengpracha, W.P. & Phutdhawong, W. 2008. Electrocoagulation of palm oil mill effluent. *International Journal of Environmental Research and Public Health*, 5, 177–180.

Bougrier, C., Carrere, H. & Delgenes, J.P. 2005. Solubilisation of waste-activated sludge by ultrasonic treatment. *Chemical Engineering Journal*, 106, 163–169.

Cesaro, A. & Belgiorno, V. 2014. Pretreatment methods to improve anaerobic biodegradability of organic municipal solid waste fractions. *Chemical Engineering Journal*, 240, 24–37.

Hua, I. & Hoffmann, M.R. 1997. Optimization of Ultrasonic Irradiation as an Advanced Oxidation Technology. *Environmental Science and Technology*, 31, 2237–2243.

Jeganathan, J., Nakhala, G. & Bassi, A. 2007. Hydrolytic pretreatment of oily wastewater by immobilized lipase. *Journal of Hazardous Materials*, 145, 127–135.

Jiang, J., Gong, C., Wang, J., Tian, S. & Zhang, Y. 2014. Effects of ultrasound pre-treatment on the amount of dissolved organic matter extracted from food waste. *Bioresource Technology*, 155, 266–271.

Oz, N.A. & Uzun, A.C. 2015. Ultrasound pretreatment for enhanced biogas production from olive mill wastewater. *Ultrasonics Sonochemistry*, 22, 565–572.

Oz, N.A. & YARIMTEPE, C.C. 2014. Ultrasound assisted biogas production from landfill leachate. *Waste Management*, 34, 1165–1170.

Pilli, S., Bhunia, P., Yan, S., Leblanc, R.J., Tyagi, R.D. & Surampalli, R.Y. 2011. Ultrasonic pretreatment of sludge: A review. *Ultrasonics Sonochemistry*, 18, 1–18.

Poh, P.E., Yong, W.J. & Chong, M.F. 2010. Palm oil mill effluent (POME) characteristic in high crop season and the applicability of high-rate anaerobic bioreactors for the treatment of pome. *Industrial and Engineering Chemistry Research*, 49, 11732–11740.

Prozorov, T., Prozorov, R. & Suslick, K.S. 2004. High velocity interparticle collicions driven by ultrasound. *Journal of the American Oil Chemists' Society*, 126, 13890–13891.

Saifuddin, N., & Fazlili, S.A. 2009. Effect of microwave and ultrasonic pretreatments on biogas production from anaerobic digestion of palm oil mill effleunt. *American J. of Engineering and Applied Sciences*, 2, 139–146.

Salsabil, M.R., Laurent, J., Casellas, M. & Dagot, C. 2010. Techno-economic evaluation of thermal treatment, ozonation and sonication for the reduction of wastewater biomass volume before aerobic or anaerobic digestion. *Journal of Hazardous Materials*, 174, 323–333.

Sangave, P.C., Gogate, P.R. & Pandit, A.B. 2007. Ultrasound and ozone assisted biological degradation of thermally pretreated and anaerobically pretreated distillery wastewater. *Chemosphere*, 68, 42–50.

Show, K.Y., Mao, T. & Lee, D.J. 2007. Optimisation of sludge disruption by sonication. *Water Research*, 41, 4741–4747.

Show, K.Y., Ng, C.A., Faiza, A.R., Wong, L.P. & Wong, L.Y. 2011. Calculation of energy recovery and greenhouse gas emission reduction from palm oil mill effluent treatment by an anaerobic granular-sludge process. *Water Sci. Technol*, 64, 2439–2444.

Suslick, K.S. 1990. Interparticle collisions driven by ultrasound. *Science*, 247, 1439–1445.

Tiehm, A., Nickel, K., Zellhorn, M. & Neis, U. 2001. Ultrasonic waste activated sludge disintegration for improving anaerobic stabilization. *Water Research*, 35, 2003–2009.

Vavilin, V.A., Fernandez, B., Palatsi, J. & Flotats, X. 2008. Hydrolysis kinetics in anaerobic degradation of particulate organic material: An overview. *Waste Management*, 28, 939–951.

Wang, J.X., Huang, Q.D., Huang, F.H., Wang, J.W. & Huang, Q.J. 2007. Lipase-catalyzed Production of Biodiesel from High Acid Value Waste Oil Using Ultrasonic Assistant. *Chinese Journal of Biotechnology*, 23, 1121–1128.

Wu, T.Y., Mohammad, A.W., Jahim, J.M. & Anuar, N. 2010. Pollution control technologies for the treatment of palm oil mill effluent (POME) through end-of-pipe processes. *Journal of Environmental Management*, 91, 1467–1490.

Wu, T.Y., Mohammad, A.W., Jahim, J.M. & Anuar, N. 2007. Palm oil mill effluent (POME) treatment and bioresources recovery using ultrafiltration membrane: Effect of pressure on membrane fouling. *Biochemical Engineering Journal*, 35, 309–317.

The performance investigation of three glass cover solar stills using different basin absorbents

Khamaruzaman Wan Yusof, Ali Riahi, Nasiman Sapari, Mohamed Hasnain Isa,
Noor Atieya Munni Zahari, Emmanuel Olisa & Khouna Mohamed Khouna
*Department of Civil and Environmental Engineering, Universiti Teknologi PETRONAS,
Seri Iskandar, Perak, Malaysia*

Balbir Singh Mahinder Singh
Department of Fundamental and Applied Sciences, Universiti Teknologi PETRONAS, Perak, Malaysia

ABSTRACT: Solar stills are used to produce drinking water as low cost technologies. This work aims at investigating the productivity of three passive double sloped solar stills fabricated with similar shapes and glass cover materials, but having different basin materials of stainless steel (CSS), black paint (SBP) and black soil (SBS), respectively. Each solar still had triangular slope cover, with dimensions; length = 60 cm and width = 50 cm. All had bottom basin with dimensions, length = 50 cm, depth = 8 cm and width = 30 cm. Experimental outputs indicated that SBP had 22.88% and 72.61% higher yield than SBS and CSS throughout the experiment. The solar stills utilized in this work can be used to produce potable water in the areas having lack of fresh water accessibility.

1 INTRODUCTION

Generally, solar stills are classified into two broad categories which include; active and passive solar stills (Malik et al. 1982). The freshwater generated from the use of solar stills, meet the WHO standard for drinking water (Jasrotia et al. 2012, Ahsan et al. 2014, Syuhada et al. 2013, Riahi et al. 2015, Riahi et al. 2016). This technology can be applicable in rural, remote, coastal and remote regions where potable water is lacking. Previous researchers have conducted experiments to investigate the performance of solar stills applying various materials for heat storage in basin in order to enhance the productivity of passive solar stills with glass cover. The performance of symmetrical greenhouse type solar still with glass as cover and with the use of black painted stainless steel basin was studied in Jordan (Al-Hayek, I. & Badran, O.O. 2004). The water production of 4.5 L/m^2 was obtained during 11 hours of the experimental work. A double slope single basin glass cover and black painted Galvanized Iron steel basin was fabricated in Saudi Arabia (Al Garni 2012). Productivity reached to 2.5 L/m^2 at the end of experiment. A single slope single basin solar still with glass and black painted stainless steel as cover and basin materials respectively has been fabricated in Malaysia (Sapari et al. 2014). The maximum productivity of 2.26 L/m^2 has been obtained. A glass cover pyramid shape solar still having black plate basin was constructed in Jordan (Taamneh, Y. & Taamneh, M.M. 2012). The peak yield of 2.485 L/m^2 acheived at the end of experiment. In a comprehensive study in Jordan (Abu-Hijleh, B. & Rababa'h, H.M. 2003), the performance of the application of black coal, black sponge cubes, yellow sponge cubes and black steel as heat absorbent materials in basin of single slope solar still with glass cover was investigated. The results indicated improvement in productivity of 2.00 L/m^2 by the use of yellow sponge cubes when compared with the other absorbents. The use of suspended aluminium and galvanized iron floating plates as absorbents in the basins of two single slope solar stills was investigated and compared with the conventional solar still in Mehsana, India (Panchal, H. & Shah, P.K. 2012). The productivity increased significantly to 3.703 L/m^2 using suspended aluminium floating plates in the basin of solar still as compared to the solar still using galvanized iron floating plates and conventional solar still with the yield of 3.10 and 2.54 L/m^2 respectively. The objective of this work is to evaluate the performance of three double slope solar stills with glass covers having different types of materials for heat absorbents (e.g. one layer of black paint in the stainless steel basin, use of only stainless steel basin and addition of 2 cm depth of black soil to the solar still basin, respectively). The novelty of this study has to do with the various heat absorbents applied in the basins of the glass cover double slope solar still.

2 METHODOLOGY

Figures 1(a) and (b) shows the sketch and photograph of the experimental setup of three glass cover solar stills, respectively. Three types of single basin double slope glass cover passive solar stills were fabricated

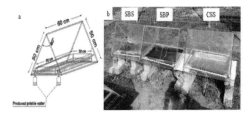

Figure 1. a) Sketch of solar still and b) Photograph of experimental set-up.

with similar triangular shapes. Each solar still had a stainless steel trough as basin having length 50 cm, width 30 cm and depth 8 cm. The basins of the three solar stills were placed inside their frames in order to reduce the basin plate heat losses to the atmosphere. Glass, glass and stainless steel were used to construct the frame, double slope cover and basin of all the three solar stills. The solar stills investigated are:

Solar still 1 (CSS)—solar still using double slope glass as cover and stainless steel trough as basin
Solar still 2 (SBS)—solar still using double slope glass cover and stainless steel trough with an additional 2 cm depth of black soil in its basin trough
Solar still 3 (SBP) — solar still using glass cover and stainless steel trough with an additional layer of black paint in its basin trough

The first solar still (CSS) was tested using only stainless steel basin, while SBS had a layer of 2 cm black soil in the basin and SBP was used with a basin having a layer of black paint. Experiment was conducted in Universiti Teknologi PETRONAS (UTP) campus to analyse the performance of the three different heat absorbents in the basin of solar stills:

Solar stills 1, 2 and 3 were exposed to similar solar irradiance in an open field (Fig. 1b).

The water temperature in the various basins; ambient temperature and inner cover temperature of solar stills were measured at time intervals of 1 hour using digital multimeter. Solar irradiance within the UTP campus was measured using a pyranometer. Lake water in UTP (4.5 L) was collected and filled in each solar still basin for the experimental work. Water vapor condensed on the glass covers of solar stills. Condensed water vapor was collected using four aluminum sheets collectors attached to two funnels at the bottom of the each cover of solar stills which were used to drain and collect condensed water vapor from inner surface cover of the still as shown in Figure 1 which was measured using measuring cylinder.

3 RESULTS AND DISCUSSION

3.1 *Solar irradiance effect on the temperatures of water, inner cover and hourly yield of three solar stills*

Figure 2 shows the solar irradiance variations (I_s) from 9 am to 5 pm on a typical day on the 12th of August

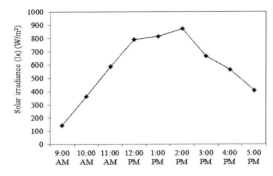

Figure 2. Solar irradiance variations from 9 am to 5 pm on 12th August 2015.

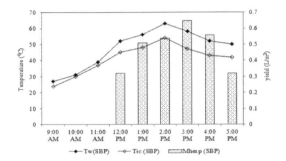

Figure 3. Water temperature variations and tem-perature of inner cover of solar still against the yield of SBP from 9 am to 5 pm on 12th August 2015.

2015. Figures 3, 4 and 5 present the variations of water temperature (T_w) and temperature of inner cover (T_{ic}) of SBP, SBS and CSS, respectively, against hourly yield (M_{hexp}) of these solar stills on 12th August 2015.

Increase and decrease in temperature corresponded to increase and decrease in hourly solar irradiance (Figs. 2–5). It is observed that when solar irradiance attained a maximum value from 144.30 W/m² at 9:00 am to 870.78 W/m² at 2:00 pm, the highest T_w (SBP), T_w (SBS) and T_w (CSS) were recorded at 2:00 pm with the values of 63, 59 and 53°C, respectively, while the corresponding maximum inner cover temperatures T_{ic} (SBP), T_{ic} (SBS) and T_{ic} (CSS) were 54, 49 and 48°C. It was observed that the maximum hourly experimental water production of 0.65, 0.54 and 0.42 L/m² for SBP, SBS and CSS occurred at 3.00 pm respectively; while the solar irradiance and water temperature decreased at 2.00 pm to 664.13 W/m² and 58, 55 and 51°C at 3.00 pm for SBP, SBS and CSS respectively (Figs. 2–5). Such trend has been reported in some previous studies too (Ahsan et al. 2014, Riahi et al. 2015, Riahi et al., 2016, El-Sebaii, A.A. 2004, Zurigat, Y.H. & Abu-Arabi, M.K. 2004, El-Bahi, A. & Inan, D. 1999, Delyannis, A.A. & Piperoglou, E. 1965, Akash et al. 2000, Tiwari, G.N. 1991, Moustafa et al. 1979, Al-Hayek, I. & Badran, O.O. 2004, Kamal, W.A. 1988).

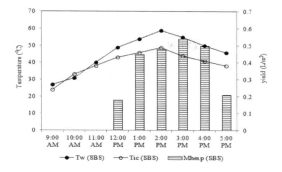

Figure 4. Water temperature variations and tempera-ture of inner cover of solar still against the yield of SBS from 9 am to 5 pm on 12th August 2015.

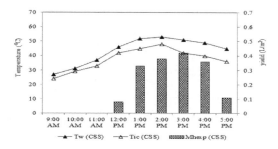

Figure 5. Water temperature variations and temperature of inner cover of solar still against the yield of CSS from 9 am to 5 pm on 12th August 2015.

3.2 Cumulative water production

The variations in cumulative water production (M_{cexp}) for the three solar stills in a typical day (12th August 2015) were presented in Figure 6. The least cumulative water yield throughout the day was obtained from CSS, while the highest (M_{cexp}) was from SBP, indicating that the adopted configuration for SBP was more effective than the rest. The overall quantity of water produced from CSS, SBS and SBP were 1.68, 2.36 and 2.90 L/m², respectively in that typical day. In this work, maximum yield of SBP reached to the amount of 2.90 L/m² using black paint in the stainless steel basin, while the stepped solar still using fin, pebble and sponge together in the basin produced 1.8 L/m² in India (Velmurugan et al. 2008). The cumulative yield of SBP (2.90 L/m²) showed higher water production than the yields of 2.50, 1.55, 2.26, 1.43, 1.47, 2.10, 2.36, 2.50 and 2.00 L/m² for solar stills in Saudi Arabia (Al-Garni 2012), Malaysia (Ahsan et al. 2014, Sapari et al. 2014, Riahi et al. 2013a, b, Riahi et al. 2014a, b), Jordan (Taamneh, Y. & Taamneh M.M. 2012 and Abu-Hijleh, B. & Rababa'h, H.M. 2003), respectively.

3.3 Water quality analysis of lake water and produced water using three solar stills

Table 1 shows the average of important water quality parameters of lake water, the water produced by CSS, SBP and SBS and WHO standards for drinking water.

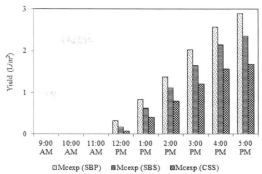

Figure 6. Values of cumulative yield of SBP, SBS and CSS from 9 am to 5 pm on 12th August 2015.

Table 1. Performance assessment of solar stills used for treatment of the lake water in UTP campus.

Parameters	Lake water (average)	Produced water (average)	WHO standards (WHO 2011)
pH	6.95	6.50	6.5–8.0
(TDS) (mg/l)	280	30	<600
Salinity (ppt)	0.1	0	<0.25
Iron (mg/l)	3.2	0.01	<0.3
Turbidity (NTU)	9.30	0.98	<5

The values of pH, Total dissolved solids (TDS), Salinity, Iron and Turbidity reduced from 6.95, 280 mg/l, 0.1 ppt, 3.2 mg/l, 9.30 NTU before experiment to 6.50, 30 mg/l, 0 ppt, 0.01 mg/l, 0.98 NTU for solar stills after the experiment respectively, which showed that the water produced from three solar stills are within the acceptable range of WHO drinking water standards (WHO 2011) in this work.

4 CONCLUSION

Solar energy can be effectively recovered and converted to heat energy for the production of potable water using the three solar stills studied i.e. with similar cover materials of glass, but having different cover materials of black paint (SBP), black soil (SBS) and only stainless steel (CSS). In this work, the experimental investigation of these solar stills were done and showed that SBP produced higher productivity of 2.90 L/m² as compared to the SBS and CSS with the yield of 2.36 and 1.68 L/m² respectively. Therefore, SBP can be suggested to use in rural, coastal and remote areas for potable water production.

REFERENCES

Abu-Hijleh B. & Rababa'h, H.M. 2003. Experimental study of a solar still with sponge cubes in basin. *Energy Convers. Manage.* 44: 1411–1418.

Ahsan, A., Imteaz, M., Thomas, U.A., Azmi, M., Rahman, A., Nik Daud, N.N. 2014. Parameters affecting the performance of a low cost solar still, *Appl. Energy* 114: 924–930.

Akash, B.A., Mohsen, M.S., Nayfeh, W. 2000. Experimental study of the basin type solar still under local climate conditions. *Energy Convers. Manag.* 41: 883.

Al-Garni AZ. 2012. Productivity enhancement of solar still using water heater and cooling fan. *J. sol. energy eng., Transactions of the ASME* 134: 031006.

Al-Hayek, I. & Badran, O.O. 2004. The effect of using different designs of solar stills on water distillation. *Desalination* 169: 121–127.

Delyannis, A.A., Piperoglou, E. 1965. Solar distillation in Greece. *First International Symposium of Water Desalination, Washington, D.C. USA* 2: 627–633.

El-Bahi, A., Inan, D. 1999. A solar still with minimum inclination, coupled to an outside condenser. *Desalination* 123: 79–83.

El-Sebaii, A.A. 2004. Effect of Wind Speed on Active and Passive Solar Stills. *Energy Convers. Manag.* 45: 1187–1204.

Jasrotia, S., Kansal, A., Kishore, V.V.N. 2012. Application of solar energy for water supply and sanitation in arsenic affected rural areas: a study for Kaudikasa village, India. *J. Clean. Prod.* 37: 389–393.

Kamal, W.A. 1988. A theoretical and experimental study of the basin-type solar still under the Arabian Gulf climatic conditions. *Solar and Wind* 5: 147–157.

Malik, M.A.S., Tiwari, G.N., Kumar, A., Sodha, M.S. 1982. *Solar Distillation*. Oxford: Pergaman Press.

Moustafa, S.M.A., Brusewitz, G.H., Farmer, D.M. 1979. Direct use of solar energy for water desalination. *Solar Energy* 22: 141–148.

Panchal, H. & Shah, P.K. 2012. Investigation on solar stills having floating plates. *Int. J. Energy Environ. Eng.* 3: 1–5.

Riahi, A., Wan, Y.K., Mahinder Singh, B.S., Isa, M.H., Olisa, E., Zahari, N.A.M. 2016. Sustainable potable water production using a solar still with photovoltaic modules-AC heater, *Desal. Water Treat.* 57: 14929–14944.

Riahi, A., Yusof, K.W., Isa M.H., Mahinder Singh, B.S., Sapari, N.B. 2014a. Solar stills productivity with different arrangements of PV-DC heater and sand layer in still basin: A comparative investigation, *Res. J. Appl. Sci. Eng. Tech.* 8: 1363–1372.

Riahi, A., Yusof, K.W., Isa, M.H., Singh, B.S.M., Malakahmad, A., Sapari, N.B. 2014b. Experimental investigation on the performance of four types of solar stills in Malaysia. *Appl. Mech. Mater.* 567: 56–61.

Riahi, A., Yusof, K.W., Mahinder Singh, B.S., Olisa, E., Sapari, NB., Isa, M.H. 2015. The perfomance investigation of triangular solar stills having different heat storage materials. *Int. J. Energy Environ. Eng.* 6: 385–391.

Riahi, A., Yusof, K.W., Sapari, N.B., Malakahmad, A., Hashim, A.M., Singh, B.S.M. 2013a. Potable water production by using triangular solar distillation systems in Malaysia. *Proc. IEEE Conf. Clean Energy Technol.* 6775679: 473–477.

Riahi, A., Yusof, K.W., Sapari, N.B., Singh, B.S.M., Hashim, A.M. 2013b. Novel configurations of solar distillation system for potable water production. *IOP Conf. Ser. Earth Environ. Sci.* 16: 012135.

Sapari N, Ahmadan NAM, Riahi A, Orji KU. 2014. The performance of trapezoidal glass cover solar still during monsoon period of tropical environment, *Appl. Mech. Mater.* 567: 161–166.

Syuhada, N., Ahsan, A., Thomas, U.A., Imteaz, M., Ghazali, A.H. 2013. A low cost solar still for pure water production. *J. Food Agric. Environ.* 11: 990–994.

Taamneh, Y., Taamneh, M.M. 2012. Performance of pyramid-shaped solar still: Experimental study. *Desalination* 291: 65–68.

Tiwari, G.N. 1991. Feasibility study of solar distillation plants in South Pacific countries. *Desalination* 82: 233–241.

Velmurugan, V., Deenadayalan, C.K., Vinod, H., Srithar, K. 2008. Desalination of effluent using fin type solar still. *Energy* 33: 1719–1727.

World Health Organization (WHO), 2011. "Guidelines for drinking-water quality 4th ed. Geneva, Switzerland". Accessed at: http://whqlibdoc.who.int/publications/2011/9789241548151_eng.pdf?ua=1 on 10 March 2013

Zurigat, Y.H., Abu-Arabi, M.K. 2004. Modelling and Performance Analysis of a Regenerative Solar Desalination Unit. *Appl. Therm. Eng.* 24: 1061–1072.

Artificial neural network approach for modeling of Cd (II) adsorption from aqueous solution by incinerated rice husk carbon

T. Khan, M.H. Isa & M.R. Mustafa
Department of Civil and Environmental Engineering, Universiti Teknologi PETRONAS, Seri Iskandar, Perak, Malaysia

ABSTRACT: This study examined implementation of Artificial Neural Network (ANN) for the prediction of Cd (II) adsorption from aqueous solution by Incinerated Rice Husk Carbon (IRHC). Batch adsorption tests showed that extent of Cd (II) adsorption depended on initial concentration, contact time and pH. Equilibrium adsorption was achieved in 60 min, while maximum Cd (II) adsorption occurred at pH 5. The Levenberg-Marquardt algorithm (LM) training algorithm was found to be the best among 8 backpropagation (BP) algorithms tested; lowest Mean Square Error (MSE) of 20.99 and highest R^2 was 0.96. Langmuir constants $Q°$ and b were 40 and 0.04, and Freundlich constants K_f and $1/n$ were 2.15 and 0.69, respectively. Adsorption capacity of IRHC was compared with other adsorbents and activated carbons reported in the literature. Being a low-cost carbon, IRHC has potential to be used for the adsorption of Cd (II) from aqueous solution and wastewater in developing countries.

1 INTRODUCTION

The discharge of heavy metals to the environment is a matter of global concern, especially in developing countries. Heavy metal pollution is mainly due to various human activities and rapid industrialization. Heavy metals are non-biodegradable and easily get bioaccumulated and bioconcentrated in living tissues (Kumar & Bandyopadhyay 2006). Cadmium (Cd (II)) is a non-essential element and is one of the most toxic heavy metals in the environment. The main sources of water contamination by Cd (II) ions are effluents from metallurgical and chemical industries, ceramics, electrogalvanization and textile industries (Zhao et al. 2014, Hizal & Apak 2006). Cd (II) contamination causes serious cases of acute toxicity and diseases such as lung cancer, kidney failure, Itai-itai disease, renal damage, emphysema, hypertension and testicular atrophy (El-Shafey 2007, Balkaya & Cesur 2008). The protective and efficient removal of Cd (II) contaminated wastewater is a challenging issue that needs to be addressed because alternative low-cost removal techniques are not readily available. The main objective currently is to implement suitable methods and to develop appropriate techniques either to prevent metal contamination or to reduce it to acceptable levels (Totlani et al. 2012). Many techniques such as precipitation, ion exchange, membrane processes, and different electrolytic methods are practiced for removal and recovery of metals from wastewater. Various researchers suggest that the adsorption of metal ion onto insoluble compounds used as adsorbents is more effective than the above mentioned techniques.

Activated carbon, charcoal, and resin are used as adsorbents, but due to their high price, alternative low-cost adsorbents are sought (Chand et al. 2014). A low-cost adsorbent is waste material from industry or agriculture and requires little processing (Bailey et al. 1999).

Rice husk is an agricultural waste. It consists of cellulose (32.23%), hemicelluloses (21.34%), lignin (21.44%) and mineral ash (15.05%) (Rahman et al. 1997) with high percentage of silica (96.34%) in the mineral ash (Rahman & Ismail 1993). Expectantly, rice husk-based adsorbents would be effective in adsorbing Cd (II) from water. However, the rice husk needs to be modified or treated before being applied for adsorption (Chakraborty et al. 2011). Chemical or thermal treatment reduces cellulose, hemicelluloses and lignin crystalinity, leading to an increase of specific area for adsorption (Daffala et al. 2010).

Recently, artificial neural network (ANN) has drawn attention of researchers as an alternative tool to determine complex relationships among variables. Based on the potential of ANN to describe precisely non linear relationships, it has been applied in many areas of environmental engineering (El-Shafey 2005). However, only limited amount of work has been devoted to the use of ANN for the prediction of adsorption of heavy metals from aqueous solution. Hence, this study focuses on the application of ANN to predict the adsorption of Cd (II) by incinerated rice husk carbon (IRHC) from aqueous solution. The main objective of the study was to develop a low-cost adsorbent for inexpensive removal of Cd (II). The efficiency of the IRHC to remove Cd (II) from aqueous solution was tested and the adsorption mechanism was investigated. Cadmium (II) adsorption capacity by the IRHC was compared with commercial activated carbon and other low-cost adsorbents.

2 MATERIALS AND METHODS

2.1 Preparation of IRHC

Raw rice husk was collected from a rice mill located in Perak (Malaysia). The collected raw rice husk was washed with tap water followed by washing with distilled water several times in order to remove dust. The washed rice husk was dried in an oven at 105°C for 24 h. The washed and dried rice husk was then incinerated in a muffle furnace at 300°C for 4 h. The resulting IRHC was ground to a finer size of 212–500 μm and used in various batch adsorption tests.

2.2 Batch adsorption studies

Unless specified, batch adsorption tests were conducted by shaking 100 mL of Cd (II) solution of desired concentration with 0.2 g of IRHC in a conical flask at room temperature (22°C), using an orbital shaker at 150 rpm. After a predetermined contact time, the flask was removed from the orbital shaker and the supernatant was filtered through 0.45 μm membrane filter and analyzed for remaining Cd (II) concentration using atomic absorption spectrophotometer. The effects of pH (1–8), contact time (5–180 min) and Cd(II) concentration (20–80 mg/L) were also determined. The pH of the solution was adjusted by 0.1 N NaOH or 0.1 N HCl. Adsorption isotherms were studied based on batch equilibrium test using the optimum contact time and pH.

2.3 Selection of backpropagation (BP) training algorithm

To determine the best training algorithm that can accurately predict the adsorption of Cd (II) based on 40 experimental sets of batch study onto IRHC, eight BP training algorithms were tested. The optimization for all BP algorithms was done by varying the number of neurons in the range of 4–40. It was found that Levenberg–Marquardt backpropagation algorithm (LMA) yielded the lowest mean square error (MSE) of 20.99 and highest R^2 of 0.96 among the eight BP algorithms tested. Thus, LMA was selected as the most suitable training algorithm for the prediction of Cd (II) adsorption.

3 RESULTS AND DISCUSSION

3.1 Effect of pH

It is known that the removal of metal ions from aqueous solution is dependent on pH. Therefore, to know the effect of pH on the adsorption of Cd (II) ion, the

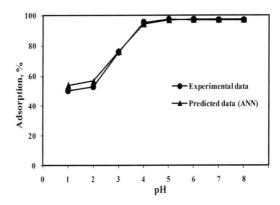

Figure 1. Effect of pH on adsorption of Cd (II) on IRHC.

pH of the solution was varied in the range of 1–8 for a 20 mg/L Cd (II) solution. The results are shown in Fig. 1. It is noticed that the adsorption of Cd (II) increased with an increase in pH. Maximum adsorption of 97.28% occurred at pH 5. The pH of the aqueous solution of Cd (II) affects its uptake on the IRHC, and the uptake increases at higher pH values. This can be explained because of proton-competitive sorption reactions. It can be said that at lower pH, H^+ ions compete with Cd (II) ions for the surface binding sites of the adsorbent. When the pH was increased, the competing effect of H^+ ions decreased and the positively charged Cd^{2+} and $Cd(OH)^+$ ions hook up the free binding sites. Hence, the metal uptake was increased on the surface of the adsorbent with the increase in pH (Balkaya & Cesur 2008). A pH of 5 has also been found optimal for the adsorption of Cd (II) using chemically modified apple pomace (Chand et al. 2014) and sugarcane baggase (Homagai et al. 2010).

A comparison of the ANN model predictions and the experimental data as function of pH is depicted in Fig. 1. The difference between the ANN model predictions and experimental data were found to be less than 4%. Thus, predicted values generated by ANN model are in good agreement with the experimental data.

3.2 Effect of contact time and initial concentration

To investigate equilibrium adsorption time, Cd(II) adsorption by IRHC at four different initial concentrations was determined as a function of contact time. The effect of contact time on Cd (II) adsorption is shown in Fig. 2. The extent of Cd (II) adsorption increased as the initial Cd (II) concentration decreased and as the contact time increased. The adsorption rate of Cd (II) was rapid in the first 15 min. Equilibrium was achieved at 60 min and thereafter remained constant. A similar contact time of 60 min has also been reported for the adsorption of Cd (II) using chemically modified apple pomace (Chand et al. 2014). Based on these results, a contact time of 60 min was taken as optimal contact time for all the following tests.

The experimental data and the data predicted by the ANN are shown in Fig. 2. It can be seen that the ANN

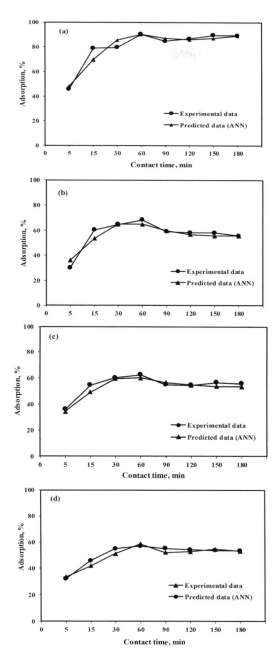

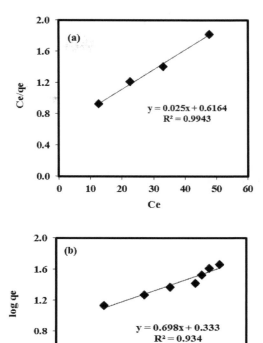

Figure 3. Linearized (a) Langmir and (b) Freundlich isotherm for Cd (II) adsorption by IRHC.

Figure 2. Effect of contact time and initial Cd (II) concentration on adsorption of Cd (II) on IRHC ((a) 20 mg/L, (b) 40 mg/L, (c) 60 mg/L and (d) 80 mg/L).

can satisfactorily predict adsorption of Cd (II) in terms of initial concentration and contact time.

3.3 Adsorption isotherms

In adsorption in a solid-liquid system, the distribution ratio of the solute between the liquid and the solid phase is a measure of the position of equilibrium. The preferred form of depicting this distribution is to express the quantity q_e as a function of C_e at a fixed temperature, the quantity q_e being the amount of solute adsorbed per unit weight of the solid adsorbent, and C_e the concentration of solute remaining in the solution at equilibrium. An expression of this type is termed an adsorption isotherm (Weber 1972). The Langmuir adsorption isotherm is

$$q_e = \frac{Q°bC_e}{1+bC_e} \quad (1)$$

where, $Q°$ is the amount of solute adsorbed per unit weight of adsorbent in forming a monolayer on the surface (monolayer adsorption capacity) and b is a constant related to the energy of adsorption.

The Freundlich adsorption isotherm is

$$q_e = K_f C_e^{1/n} \quad (2)$$

where, K_f is the Freundlich constant (adsorption capacity) and $1/n$ represents the adsorption intensity or surface heterogeneity.

Adsorption isotherm for Cd (II) adsorption by the IRHC was fitted to the linear form of the Langmuir $(C_e/q_e = 1/bQ° + C_e/Q°)$ (Fig. 3(a)) and Freundlich

Table 1. Comparison of adsorption capacity ($Q°$) of different agricultural-based adsorbents and activated carbon for Cd (II).

Adsorbent	Adsorption capacity (mg/g)	Reference
Apple pomace	4.45	(Chand et al. 2014)
Peanut hulls	5.96	(Brown et al. 2000)
Pinus pinaster bark	8.0	(Vasconcelos & Beca 1993)
Spent grain	17.3	(Low et al. 2005)
Olive stone waste	6.88	(Fiol et al. 2006)
Powdered activated carbon	3.7	(Zacaria et al. 2002)
Granular activated carbon	3.7	(Zacaria et al. 2002)
Corncobs	8.89	(Zacaria et al. 2002)
Sugar beet pulp	17.2	(Orhan & Buyukgungor 1993)
Exhausted Coffe	1.48	(Orhan & Buyukgungor 1993)
IRHC	40	This study

($log\ q_e = log\ K_f + 1/n\ log\ C_e$) (Fig. 3 (b)) adsorption isotherm. Langmuir constants $Q°$ and b were 40 and 0.04, and Freundlich constants K_f and $1/n$ were 2.15 and 0.69, respectively. The Langmuir isotherm model gave better fit to the experimental data than Freundlich isotherm model ($R^2 = 0.9943$ compared to 0.934).

3.4 *Comparison of Cd (II) adsorption by different adsorbents*

The adsorption capacity of the IRHC for Cd (II) is compared with that of other adsorbents reported in literature (Table 1). IRHC exhibited higher adsorption capacity for Cd (II) (40 mg/g) compared to that of other adsorbents (4.45–17.3 mg/g), powder activated carbon (3.7 mg/g) and granular activated carbon (3.7 mg/g).

4 CONCLUSIONS

Incinerated rice husk carbon was found effective in the adsorption of Cd (II) from aqueous solution and maximum adsorption occurred in 60 min at pH 5. Langmuir isotherm model fitted better than Freundlich isotherm model to the experimental data. Langmuir constants $Q°$ and b were 40 and 0.04, and Freundlich constants K_f and $1/n$ were 2.15 and 0.69, respectively. IRHC exhibited higher adsorption capacity for Cd (II) (40 mg/g) compared to that of other adsorbents (4.45–17.3 mg/g), powder activated carbon (3.7 mg/g) and granular activated carbon (3.7 mg/g). ANN results showed that neural network modeling could effectively simulate and predict the adsorption of Cd (II) from aqueous solution. Therefore, the experimental output for the respective operational parameters adsorption of Cd (II) by IRHC are well explained using the prediction by ANN. Being a low-cost agricultural by-product, rice husk is a good source for the preparation of rice husk carbon that can be employed as an effective adsorbent for the removal of Cd (II) and potentially for other heavy metals, from aqueous solution and wastewater in developing countries.

REFERENCES

Bailey, S.E., Olin, T.J., Bricka, R.M. & Adrian, D.D. 1999. A review of potentially low-cost sorbent for heavy metals. *Water Res.* 33: 2469–2479.

Balkaya, N. & Cesur, H. 2008. Adsorption of cadmium from aqueous solution by phosphogypsum. *Chem. Eng. J.* 140: 247–254.

Brown, P., Jefcoat, I.A., Parrish, D. Gill, S. & Graham, E. 2000. Evaluation of the adsorptive capacity of peanut hull pellets for heavy metals in solution. *Adv. Environ. Res.* 4: 19–29.

Chakraborty, S., Chowdhury, S. & Saha, P.D. 2011. Adsorption of Crystal Violet from Aqueous Solution onto NaOH-modified Rice Husk. *Carbohydr. Polym.* 86: 1533–1541.

Chand, P., Shil, A.K., Sharma, M. & Pakade, Y.B. 2014. Improved adsorption of cadmium ions from aqueous solution using chemically modified apple pomace: Mechanism, kinetics, and thermodynamics. *Int. Biodeterior. Biodegrad.* 90: 8–16.

Daffala, S.B., Mukhtar, H.& Shaharun, M.S. 2010. Characterization of Adsorbents Developed from Rice Husk: Effect of Surface Functional Group on Phenol Adsorption. *J. Appl. Sci.* 10: 1060–1067.

El-Shafey, E.I. 2005. Behaviour of reduction–sorption of chromium (VI) from an aqueous solution on a modified sorbent from rice husk. *Water Air Soil Poll.* 163: 81–102.

El-Shafey, E.I. 2007. Sorption of Cd(II) and Se(IV) from aqueous solution using modified rice husk. *J. Hazrd. Materi.* 147: 546–555.

Hizal, J. & Apak, R. 2006. Modeling of cadmium(II) adsorption on kaolinite-based clays in the absence and presence of humic acid. *Appl. Clay Sci.* 32: 232–244.

Homagai, P.L., Ghimire, K.N. & Inoue, K. 2010. Adsorption behavior of toxic metals onto chemically modified sugarcane bagasse. *Bioresour. Technol.* 101: 2067–2069.

Kumar, U. & Bandyopadhyay, M. 2006. Sorption of cadmium from aqueous solution using pretreated rice husk. *Bioresour. Technol.* 97: 104–109.

Low, K.S., Lee, C.K. & Lie, S.C. 2000. Sorption of cadmium and lead from aqueous solution by spent grain. *Process Biochem.* 36: 59–64.

N. Fiol, I. Villaescusa, M. Martınez, N. Miralles, J. Poch, J. Serarols, Sorption of Pb(II), Ni(II), Cu(II) and Cd(II) from aqueous solution by olive stone waste, Sep. Purif.Technol. 50 (2006) 132–140.

Orhan, Y. & Buyukgungor, H. 1993. The removal of heavy metals by using agricultural wastes. *Water Sci. Technol.* 28: 247–255.

Rahman, I.A. & Ismail, J. 1993. Preparation and Characterization of a Spherical Gel from a Low-Cost Material. *J. Mater. Chem.* 3: 931–934.

Rahman, I.A., Ismail, J. & Osman, H. 1997. Effect of Nitric Acid Digestion on Organic Materials and Silica in Rice Husk. *J. Mater. Chem.* 7: 1505–1509.

Totlani, K., Mehta, R. & Mandavgane, S.A. 2012. Comparative study of adsorption of Ni (II) on RHA and carbon embedded silica obtained from RHA. *Chem. Eng. J.* 181–182: 376–386.

Vasconcelos, L.A.T.D & Beca, C.G.G. 1993. Adsorption *equilibria* between pine bark and several ions in aqueous solution Cd(II), Cr(III) and Hg(II). *Eur. Water Pollut. Control.* 3: 29–39.

Weber, W.J. Jr., Adsorption. In: W.J Weber, Jr. (ed.), 1972. *Physiochemical processes for water quality control,* Wiley-Interscience, New York.

Zacaria, R., Gerente, C., Andres, Y.& Cloirec, P.L. 2002. Adsorption of several metal ions onto low-cost biosorbent: kinetic and equilibrium studies. *Env. Sci. Technol.* 36: 2067–2073.

Zhao, X. Jiang, T. & Du, B. 2014. Effect of organic matter and calcium carbonate on behaviors of cadmium adsorption–desorption on/from purple paddy soils. *Chemosphere* 99: 41–48.

Engineering Challenges for Sustainable Future – Zawawi (Ed.)
© 2016 Taylor & Francis Group, London, ISBN 978-1-138-02978-1

Implementation of attached growth system in Malaysia: An overview

Gasim Hayder, Nur Farah Syazana Bt. Mohamad Fu'ad & Puniyarasen A./L. Perumulselum
Department of Civil Engineering, College of Engineering, Universiti Tenaga Nasional (UNITEN), Malaysia

ABSTRACT: Attached growth system to treat wastewater has been in practice for many years. It has been the main concern as it is using media such as slag, rock, plastic and other materials that can be used as a physical bed on which the biofilm will develop. A numerous research has been done in Malaysia by using attached growth system. However there are advantage and disadvantage of using attached growth system. Therefore, this paper presents the overview on how far the attached growth system has been implementing in terms of authorities involved and type of attached growth treatment process that has been used in Malaysia.

1 INTRODUCTION

Biological treatment is the important part of the sewage treatment process which is the process involves removal of the remaining dissolved and non-settle able organic material in the wastewater by living bacteria (SPAN, 2009). It can be classified as suspended growth, attached growth and hybrid. Attached growth process is using media such as slag, rock, plastic and other materials that can be used as a physical bed for microorganism attached to. However there are advantages and disadvantages of using attached growth system.

The media used in attached growth process are usually lighter, easy to carry and install into mechanical plant, and cheaper. According to (Abdulgader et al. 2007), they proved that attached growth process performed better as it has achieved highest COD removal at high organic loading rates compare to conventional suspended growth process in treating high strength food processing wastewater. The material of media is usually cheap and easy to install in the treatment plant. Free floating media used in attached growth process can avoid biomass recycling as the media is providing large surface areas for biomass growth at a longer time (Quan et al. 2012). Moreover, mechanical plant with media shows better process performance than mechanical plant without the media in terms of COD and TSS removal and also sludge settling properties during power failure (Aygun et al. 2014). Attached growth process has been proved to have more energy efficient compare to suspended growth (Steven et al. 2012).

Despite from all the advantages of attached growth process, there are also some limitations has been reported. Common problem with biofilm is the proliferation of slime bacteria due to high organic loading which caused clogged film (Khaled et al. 2014). Poor control and less knowledge in handling the mechanical plant also can cause failure for the whole system (Ray, 2010). Slow water flows will inhibit the removal of excess microorganisms (Steven et al. 2011). The consequences of media clogging are excess head loss, short circuiting and frequently of backwashing is increasing (Wang et al. 2005). Media clogging also caused malodour problem due to anaerobic condition (Khaled et al. 2014). With the information gathered on advantages and limitations of attached growth process, this paper presents the overview on how far the attached growth process has been implemented in Malaysia.

2 RELATED AUTHORITIES IN MALAYSIA

In Malaysia, sewerage management has undergone several of challenging to treat wastewater. The proliferation of small plants which is always for less than 5000 population equivalent (PE) in new developments by developers has increased due to lack of investment to build large sewerage treatment plant (Din, 2011). As the sewerage infrastructure need to be developed by developer, there was a need for proper guidelines to control the sewerage issues in terms of designing, operation and maintenance of treatment plant. Therefore, establishment of enforcement law and sewerage services by certain agencies will help to improve the development of sewerage infrastructure in the country.

The main agencies responsible to manage sewerage wastewater treatment in Malaysia are the National Water Services Commission (SPAN), Indah Water Konsortium (IWK) and the Department of Environment (DOE) which under observation and support by Ministry of Energy, Green Technology & Water (KeTTHA), Ministry of Finance (MoF) and Ministry of Natural Resources & Environment as in Figure 1. These agencies were set up to provide services according to the needs of sewerage wastewater treatment in Malaysia.

Figure 1. Governance Structure for Sewerage Services (Din, 2011).

2.1 *The National Water Services Commission (SPAN)*

SPAN has been established in 2008 to regulate the sewerage services industry (Din, 2011). The function of SPAN also to implement and enforce sewerage services law and to recommend improvements to sewerage services laws. It is important to ensure the productivity of the sewerage services industry and the monitoring of operators compliance with specified standards, contractual duties and applicable laws and rules (SPAN, 2016).

Malaysian sewerage industry guideline of 2009 by SPAN has been set up to guide all the developer throughout the country in order to build a proper sewerage treatment plant. The type of biological treatment processes of attached growth system of rotating biological contactor (RBC), trickling filter (TF) and hybrid or combination multistage design system are optional for use in Malaysia (SPAN, 2009).

2.1.1 *Recommended design requirements*
Generally, SPAN has developed the guideline for design parameters for developer such as IWK. Design parameters for RBCs plants in Section 5 shows that to build the RBCs, the most important thing is the media used for attached growth process (SPAN, 2009).

In Section 5, it can be seen from design parameters for trickling filters plants, types of low rate, intermediate rate and high rate are depends on the organic loading, hydraulic loading, and sludge yields for influent of wastewater (SPAN, 2009). The media such as PVC, slag, rock or other material which is can be random or standard arrangement are acceptable to use with minimum depth of 1.5 m.

A new technology has been explored which is a hybrid system. It is a combination of suspended and attached growth system. For now, hybrid system may be considered if the design criteria to be implemented have a verified performance and result (SPAN, 2009).

2.1.2 *Package sewage treatment plant*
It can be used in minimum land area and for small population. It consists of prefabricated biological treatment system and limited for population equivalent (PE) between the ranges of 150–5000 only. Therefore, it is only applicable to Class 1 and Class 2 of sewerage treatment plant (STP) as defined in Section 4 (SPAN, 2009).

Prefabricated biological treatment must be packaged in terms of layout, piping, arrangement of the tanks and the biological processes. Each model of the prefabricated system must have fixed dimension of each tank as all items cannot be changed once approved. Package treatment plant is developed for long term. Therefore the minimum life span of prefabricated tank and other structure must be not less than 50 years (SPAN, 2009). According to Appendix A3 in (SPAN, 2009), it shows a major biological treatment process used for STP. The performance of biological process of attached growth system such as BOD_5 removal will be compared with the effluent standard that have been regulated by DOE depending on type of wastewater and treatment plant used.

2.2 *The department of environment*

It is an agency who regulates the effluent standard of wastewater control pollution and environment. The effluent standard is under Environmental Quality Act 1974 and its regulations such as the Environmental Quality (Sewage) Regulations 2009 and Environmental Quality (Industrial Effluent) Regulations 2009. Environmental Quality Act 1974 has set a standard that should meet Standard A if the effluent is discharged upstream water supply intake. While, if the effluent is discharged downstream, it supposed to meet Standard B.

2.2.1 *Environmental quality (sewage) regulations 2009*
The regulation is for discharged sewage from any housing or commercial development. The regulation is consists of standard effluent that must be met for new treatment system, existing system which is approved before January 1999 and also standard effluent for existing system which is approved after January 1999 (DOE, 2009a).

BOD_5 and COD for new sewage treatment system are 20 and 120 mg/L for Standard A and 50 and 200 mg/L for Standard B. This also applied for existing sewage treatment system which is approved before

January 1999. However, for new sewage treatment system, there are additional parameter of nitrate-nitrogen and phosphorus that must be tested and comply the standard. DOE has revised the discharged limit of ammonia-nitrogen in sewage effluent from 50 mg/L to 5 mg/L in 2009 due to the implications of severe water body pollution mainly from sewerage effluent (Ezerie et al. 2016). Meanwhile, acceptable conditions for sewage discharge of existing sewage treatment system which is approved after January 1999 is according to the type of sewage treatment plant system as defined in (DOE, 2009a). Generally, it can be seen that the effluent standard for BOD_5 and COD removal have been set up higher than the other two regulation of acceptable condition. For BOD_5 and COD of standard A and B are between 60–200 mg/L and 180–360 mg/L, respectively (DOE, 2009a).

2.2.2 Environmental quality (industrial effluent) regulations 2009

The regulation is for any premises which discharge industrial or mixed effluent. Industrial effluent is any waste in a form of liquid or wastewater which is generate from manufacturing process. Meanwhile, mixed effluent is a form of liquid or wastewater which is consisted of industrial effluent and sewage. The acceptable condition for discharge of industrial or mixed effluent of Standard A and B are explained in (DOE, 2009b).

2.3 Indah Water Konsortium (IWK)

IWK has been given responsibilities to provide sewerage services almost for entire Malaysia since 1993. The services include operation, maintenance, and development which are upgrading, rehabilitation, and expansion of sewerage infrastructure (JSC, 2011). According to IWK (IWK, 2016a), attached growth treatment plant systems which always used in urban areas in Malaysia are high rate trickling filters, rotating biological contactors and submerged biological contactors. Most of attached growth treatment plant system uses an aerobic condition. Rotating biological contactors and trickling filters also suitable for small communities but it is usually prefabricated or also known as package plant before delivery to site.

2.3.1 Type of attached growth wastewater treatment process in Malaysia

They can be categorized into RBC, TF, submerged biological contactors (SBC), bio-soil system (BS), bio-filter system (BF), bio-drum system (BD), moving bed bioreactor (MBBR), and SJTU Wastewater Treatment System (SJTU-WTS) (IWK, 2016a, IWK, 2016b, Wahyu, 2010, Fei, 2007, Edes, 2016, & UPUM, 2015).

Table 1 lists the type of treatment process which involved attached growth system until January 2016 under monitoring and services by IWK. The highest attached growth treatment system is BF and the lowest is BS. RBC is a horizontal shaft mounted across

Table 1. Types of treatment process (IWK, 2016b).

No	Attached Growth Treatment Process	Total
1	Rotating Biological Contactors (RBC)	29
2	Trickling Filter (TF)	29
3	Bio-Soil (BS)	5
4	Bio-Filter (BF)	90
5	Bio-Drum (BD)	24
	TOTAL	177

the tank of a contoured bottom tank with series of corrugated discs. Excess biomass growth on the disks can be removed due to the disks arrangement (Najafpour et al. 2012). The disks media is usually made from of high quality plastic (IWK, 2016a). TF consists of highly permeable bed media made from rocks or plastic. It can be classified based on hydraulic or organic loading rates as low rate, intermediate rate, high rate, and super high rate and roughing. High rate filters release higher organic loading due to re-circulation of filter effluent (IWK, 2016a). The disks media are 80%–100% submerged in the Submerged Biological Contactor (SBC) as it is a modified version of the conventional RBC. Forced air is introduced in order to supply sufficient air at the bottom of the treatment tank (IWK, 2016a). A bio-soil is a constructed of a multi soil layering system. The treatment system could demonstrate the gradually high pollutant removal efficiency without affected by different of water quality and fluctuation of weather temperature. BS is a low cost and simple structure, yet has a very easy maintenance (Wahyu, 2010). The system normally has two tanks, where one of it is for aeration and the other one is for settling sludge (Ambu et al. 2003). BF is comprises of pump station to lift wastewater flow to bio-filter media tower, and into a clarifier and sludge digester. The process is as the same as BS (Wahyu, 2010). IWK claimed that common problems faced by BF are odour that produced from bio-filter tower due to less of sufficient oxygen transfer and complications in sustaining the clarifiers (Wahyu, 2010). BD system is similar to RBC (Fei, 2007) and usually available in a packaged plant (Husham et al. 2012). BD is commonly consists of a wire mesh cylinder and is partially or completely filled or it is put at the lower part of the drum with randomly media (Šíma et al. 2012 & Domínguez et al. 2001). The drum will rotate the media to be in contact with the air of the upper part of tank, allowing sufficient oxygen transfer (Domínguez et al. 2001). However, some of BDs were reported not running smoothly as the system always collapse and not meet the effluent discharge of standard B (Fei, 2007). It was claimed by IWK that the BD also have high electric bill charges compare to other type of system (Fei, 2007). There are several advanced MBBR technologies using many types of patented media. Its effluent can meet stringent water quality standard for discharge and reuse. The system gives BOD effluent less than 20 mg/L and TSS than 1 mg/L (with membrane technology) (Edes, 2016). SJTU is a modular

Table 2. Types of treatment plant (Kadir, 2010).

No	Type of STP	Total	Percentage (%)
1	Communal Septic Tank	3632	36.4
2	Imhoff Tank	745	7.5
3	Oxidation Ponds	425	4.3
4	Aerated Lagoon	163	1.6
5	Network Pump Station	773	7.8
6	Mechanical Plants with Media	215	2.2
7	Mechanical Plant without Media	4019	40.3
	TOTAL	9972	100

hybrid process which designed to meet the effluent discharge of Standard A (UPUM, 2015). Integrated Fixed Film Activated Sludge (IFAS) is a hybrid process taking place in a single aeration or clarifier tank (Mittal, 2001), (Ishak et al. 2012). It integrates several compartments into one single unit to reduce necessity of large space requirement. The aerobic compartment oxidizes ammonia whereas the anoxic compartment reduces nitrate. The presence of an attached growth media into the integrated bioreactor offers more surface area for microorganism's growth (Ezerie et al. 2016). In MBBR, the media in the reactor is maintained in suspension by aeration or mechanical mixing (Ishak et al. 2012).

2.3.2 Implementation of attached growth system in Malaysia

Table 2 shows that mechanical plant without media is the highest treatment plant that has been used in Malaysia year 2010 under IWK's operation and maintenance. There are 5185 of treatment process that applied suspended growth system in Malaysia which is based on updated data until January 2016 (IWK, 2016b). It is higher than treatment process with attached growth system as shows in Table 1. Most of public sewerage treatment plants are mechanical plants operating using mechanical equipment which will accelerate the sewage breakdown (IWK, 2016a).

Obviously, treatment process with attached growth system is still not widely used in Malaysia. Data from IWK is relevant to use as IWK give services of operation and maintenance of treatment plant for almost the entire part of Malaysia. They also provide technical expertise for remaining un-serviced area (Kadir, 2010). Treatment process without media has popularity in Malaysia maybe due to long term used. Mechanical system has been introduced in Malaysia since 1980s where oxidation ponds started transformed to aerated lagoon system. At the late 1980s and 1990, fully usage of mechanical system in the form of biological filters and activated sludge system was accelerated (Candiah, 2004). Mechanical system plant provides full secondary treatment which performed better than other systems that comply with DOE's effluent standards (Haniffa et al, 2005).

Perhaps attached growth system is limited in Malaysia because of the drawbacks of the systems which cause poor performance of treatment plant. Even though the effluent are still achieved the quality standard by DOE, it often faced mechanical failure that cause excessive growth of bacteria on the media (Ashikin, 2010) resulting in media clogging. However, several research has been done in Malaysia shows the capability in using attached growth system that fulfill acceptable condition for discharge.

3 CONCLUSION

A numerous researched has been done in Malaysia by using attached growth system. Even though, most of them proved that attached growth system perform well in organic removal rate under low and high organic loading, less sludge production, and offer energy efficiencies; but the implementation of attached growth treatment system under IWK are still not widely used. It is possible due to limitation of attached growth system and media clogging that can be affected by several issues such as high organic loading, less oxygen transfer, and others. Attached growth system also caused odour problem due to clogging media. Therefore, if improvement to overcome the limitation of attached growth system has been done, it might increase the implementation.

REFERENCES

Abdulgader, M.E., Yu, Q.J., Williams, P., & Zinatizadeh, A.A.L. 2007. A review of the performance of aerobic bioreactors for treatment of food processing wastewater, the International Conference on Environmental Management, Engineering, Planning and Economics 1131–1136.

Ambu, S., Krishnasamy, M. & Singh, B. 2003. The incidence of parasites in a Survey of various types of Wastewater Treatment Ponds and their run-off, Environmental Health Focus Vol 1 (10) 46–48.

Ashikin, N. 2010. Effect of domestic sewage treatment plant effluent on river water quality, Degree of Civil Engineering, Universiti Teknologi Malaysia.

Aygun, A., Nasa, B., Berktaya, A., & Ates, H. 2014. Application of sequencing batch biofilm reactor for treatment of sewage wastewater treatment: effect of power failure, Desalination and Water Treatment.

Candiah, R.G. 2004. Sewerage industry in Malaysia – The way forward, Jurutera.

Din, A.K.M. 2011. Sewerage management in Malaysia, 2011. Information on http://www.mlit.go.jp/common/0001355 15.pdf

DOE, Department of Environmental, 2009b. Environmental quality act 1974, Environmental Quality (Industrial Effluent) Regulations.

DOE, Department of Environmental. 2009a, Environmental quality act 1974, Environmental Quality (Sewage) Regulations.

Domínguez, A., Rivela, I., Couto, S.R., Sanromán, M.A'. 2001. Design of a new rotating drum bioreactor for lignolytic enzyme production by Phanerochaete chrysosporium grown on an inert support, Process Biochemistry 37 549–554.

Edes Technology Sdn Bhd. 2016. Edes company profile, Kuala Lumpur. Information on http://www.edes.com.my/Attachment/edescompanyprofile.pdf.

Ezerie, H.E., Shamsul, R.B.M.K., Isa, M.H., Malakahmad, A., Clement, M.C., Ezerie, J.M. & Olisa E., 2016. Nutrient removal from wastewater by integrated attached growth bioreactor, Research Journal of Environmental Toxicology 10 (1) 28–38.

Fei, K.Y. 2007. Performance of biodrum in Malaysia, degree of Master of Engineering (Wastewater), Universiti Teknologi Malaysia,UTM . Unpublished thesis.

Haniffa, H. & Aminuddin, M.B. 2005. Sewage treatment trends in Malaysia, The Ingeniur. Information on http://www.bem.org.my/publication/march-may2005/ENVIRO (IWK3).pdf.

Husham, T.I., Qiang, H., Al-Rekabi, W.S. & Qiqi, Y. 2012. Improvements in biofilm processes for wastewater treatment, Pakistan Journal of Nutrition 11 (8) 610–636.

Ishak, S., Malakahmad, A. & Isa, M.H. 2012. Refinery wastewater biological treatment: A short review, Journal of Scientific & Industrial Research Vol 71, 251–256.

IWK, Indah Water Konsortium. 2016a. Information on https://www.iwk.com.my/do-you-know/sewage-treatment-plant.

IWK, Indah Water Konsortium. 2016b. Nationwide statistics data, Planning Unit, IWK, Kuala Lumpur. Unpublished.

JSC, Country sanitation assessment in Malaysia report, 2011. Information on http://www.jsanic.org/publications/Country_Survey_Reports/Malaysia/JSC_Malaysia_Sanitation_Assessment_Report.pdf.

Kadir, A. 2010. Towards sustainable sewerage development – Sharing Malaysia's Experiences for Replication, 6th Ministerial Conference on Environment and Development in Asia and the Pacific (MCED-6). Information on http://www.unescap.org/sites/default/files/9.%20MCKadir-Malaysia.pdf.

Khaled, S., Azni, I., Rozita, O. & Hamdan, M.Y. 2014. Review on biofilm processes for wastewater treatment, Life Science Journal, 11(11) 1–13.

Mittal, A. 2011. Biological wastewater treatment, Water Today.

Najafpour, G.D. & Sadeghpour, M.H. 2012. Evaluation and characterization of biological processes: Aerobic cersus Anaerobic processes, Linnaeus ECO-TECH.

National Water Services Commission (SPAN), Malaysia sewerage industry guidelines, vol 4-Sewage Treatment Plants, 3rd Ed., 2009.

Quan, F., Yuxiao, W., Tianmin, W., Hao, Z., Libing, C., Chong, Z., Hongzhang, C., Xiuqin, K., & Hui, X.X. 2012. Effects of packing rates of cubic-shaped polyurethane foam carriers on the microbial community and the removal of organics and nitrogen in moving bed biofilm reactors, Bioresource Technology 117 201–207.

Ray, R. 2010. Effects of reduced aeration in a biological aerated filter, Degree of Master of Applied Science, University of Windsor.

Šíma, J., Pocedič, J., Roubíčková, T. & Hasal, P. 2012. Rotating drum biological contactor and its application for textile dyes decolorization, Procedia Engineering 42 1579–1586.

SPAN, The National Water Services Commission, 2016. Information on http://www.span.gov.my/index.php/en/about-us/vision-mision.

Steven, E., & Williams, P. E. 2011. Reconsidering rotating biological contactors as an option for municipal wastewater treatment, Williams & Works, Inc. Information on http://williams-works.com/images/pdf/ReconsideringRotatingBiologicalContactors.pdf.

Steven, E., & Williams, P.E. 2012. Energy usage comparison between activated sludge treatment and rotating biological contactor treatment of municipal wastewater. Information on http://williams-works.com/articles/RBC%20v%20AS%20energy%20comparison.pdf.

UPUM, Unit Perundingan Universiti Malaya. 2015. SJTU wastewater treatment system, Ronser Bio-Tech Berhad and Solano Advanced Materials Sdn Bhd.

Wahyu, N. 2010. Factors influencing to the selection of sewage treatment plant, degree of Master of Science (Construction Management), Universiti Teknologi Malaysia, UTM.

Wang, R.C., Wen, X.H., & Qian, Y. 2005. Influence of carrier concentration on the performance and microbial characteristics of a suspended carrier biofilm reactor, Process Biochemistry 40 2992–3001.

Comparative study of polymeric adsorbent for copper removal from industrial effluents

Shuaib M. Laghari & Mohamed Hasnain Isa
Department of Civil and Environmental Engineering, Universiti Teknologi PETRONAS, Seri Iskandar, Perak, Malaysia

Zeenat M. Ali
Department of Chemical Engineering, Mehran UET, Jamshoro, Pakistan

Abdul Jabbar Laghari
IARSCS/Hightech CRLabs, University of Sindh, Jamshoro, Pakistan

ABSTRACT: Chitosan and carboxymethyl chitosan was prepared on laboratory scale and used as adsorbent for copper removal from cottage industry waste effluent. The chitosan was extracted by chemical method from Indian prawn (*Fennerpenaeus Indicus*) and CM-chitosan was extracted by alkylation method. The prepared polymeric adsorbent was characterized by fourier transforms spectroscopy, electro X-ray and thermogravimetric analysis. Further their effectiveness for copper removal was analyzed by column elution method. In this regard the adsorbents packed in two separate glass columns and sample elution was collected and analyzed. The Ethylene diaminetetracetic acid (EDTA) was used as desorption agent and it removed copper upto 99.6% in both columns. The results showed that chitosan filled column removed 47.1–52.1% copper and CM-chitosan filled column removed 58.8–69.2% copper. It was concluded that CM-chitosan filled column is more effective for copper removal.

1 INTRODUCTION

Industrialization is the main cause of pollution and contamination in water. Many industrial and other activities cause copper contamination in water bodies like production of industrial catalysts, electrodes, copper wire, plumbing pipe and rods, radiators, heat exchangers, fertilizer, coating in cathode ray tubes and batteries etc. The untreated industrial effluent discharge, pipe corrosion, mine or agricultural runoffs are also the causes of copper contamination.

Copper availability is 50 ppm in Earth's crust, as native copper or in mineral form such as copper carbonates, copper sulfides and copper(II)oxide (Lenntech.com, 2016).

Copper is an essential micro-nutrient for living organisms but its excess accumulation could cause a number of health problems. Copper content of 1.3 mg/L is allowable limit in drinking water (Davis and Bennet, 1985). Copper levels in surface water range from 0.5–1,000 ppb and in seawater <1–5 ppb (Gotoha et al., 2004). Excess quantity of copper alters order, smell, and color of water from transparent to bluish green and the taste from tasteless to bitter metallic. It creates disturbance in digestive system, and may cause vomiting, diarrhea, kidney problems, anemia and liver failure.

Chitosan is a unique biodegradable linear polymer and adsorbent composed of anhydroglucosamine. The reactive groups in chitosan chain are amino, hydroxyl and secondary hydroxyl groups. Chitosan has high tendency to adsorb metal ions under different conditions (Wan Ngah et al., 2004). The chemical modification in chitosan chain increased its sorption and selectivity capabilities. Maximum +ve charge density was observed in solution at pH <6.5.

In this study the chitosan was extracted from Indian prawn (*Fennerpenaeus Indicus*) and characterized. The chemical modification through direct alkylation was done to alter the polymeric chain of chitosan. The modified chitosan derivative "Carboxymethyl Chitosan" was formed. Copper removal was examined for comparative means from chitosan and its derivative treatment of water solutions.

2 METHODS

2.1 Chitosan extraction

The Chitosan was extracted from crustacean shell named Indian prawn or white shrimp (*Fenneropenaeus Indicus*) by chemical method. The extraction method comprised of deacetylation, demineralization, deproteinzation and decolorization (Ali et al., 2013a).

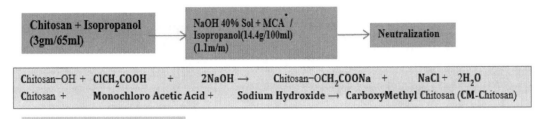

Figure 1. Preparation of Carboxymethyl chitosan.

2.2 Carboxymethyl chitosan

The carboxymethyl chitosan was prepared by direct alkylation process from extracted chitosan of Fenneropenaeus Indicus (Ali et al., 2013b) as shown in Figure 1.

2.3 Characterization of chitosan and carboxy methyl chitosan

The Chitosan and carboxymethyl chitosan was characterized by fourier transform spectroscopy, electro X-ray and thermogravimetric analysis.

2.4 Preparation of standard copper solution

Standard aqueous copper solution was prepared from sulfate pentahydrate (Merck). A 1000 ppm stock solution was prepared and further diluted from 1 mg/L to 10 mg/L. The flame ionization atomic absorption spectroscopy (Varian Spectr AA-20) with acetylene–air flame and standard burner head was used for determination of calibration curve.

2.5 Preparation of columns for copper removal

The chitosan and carboxymethyl chitosan was ground well up to particle size 100 μm mesh screen. A 10 gram of each adsorbent was filled in quick fit glass column (15X0.9 cm). The bottom of column was sealed with glass wool to prevent the adsorbent losses. The adsorbent in the column was packed compactly and clamped in iron stand using pronged clamp. When the rubber bulb tied on top of columns, was pressed, it enhanced the volumetric flow rate of the solution. The eluted sample was collected in beaker, pH optimized and analyzed through atomic absorption spectroscopy (Figure 2).

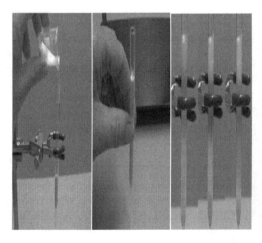

Figure 2. The preparation of column for copper removal.

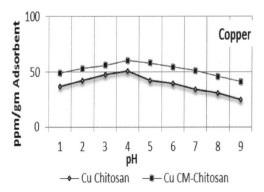

Figure 3. The pH optimization of Chitosan filled column and CM- Chitosan filled column.

Table 1. Copper removal and Desorption efficiency in Chitosan filled column and CM-Chitosan filled column.

		S-1	S-2	S-3	S-4	S-5
Chitosan Filled column	Cu Removal Efficiency (%)	51.1 ±0.5	47.1 ±1.2	49.3 ±0.9	50.1 ±1.2	52.1 ±2.4
	Cu desorption Efficiency (%)	**98.3**	**99.9**	**98.67**	**97.56**	**98.91**
CM-Chitosan Filled column	Cu Removal Efficiency (%)	67.6 ±3.1	58.8 ±3.2	68.4 ±1.9	68.2 ±2.2	69.2 ±3.1
	Cu desorption Efficiency (%)	**99.0**	**97.9**	**97.87**	**99.66**	**97.91**

2.6 Copper desorption

Ethylene diaminetetracetic acid (EDTA) was used as desorption agent for copper removal from column. 10 ml of 0.1M EDTA was eluted after samples run and analyzed on atomic absorption spectroscopy (Varian Spectr AA-20).

3 RESULTS AND DISCUSSION

3.1 pH optimization

Copper adsorption and chelation was pH dependent. The cationic and anionic behavior of chitosan and carboxymethyl chitosan were due to variation in pH and in acidic medium the cationic metal exchange behavior was observed. The van der Waals forces were active in acidic medium and took part in copper chelation and adsorption, whereas; alkaline medium promoted copper binding to polymeric chain through active sites. Chitosan molecule is bonded via amino and hydroxyl groups, while in carboxymethyl chitosan the O and N atoms of chitosan molecules were the additional sites for binding. Hence carboxymethyl chitosan was found more effective for copper removal. The pH was optimized experimentally by running and testing with acetate buffer and phosphate buffer. The data showed that pH 4 was most suitable for copper removal (Figure 3).

Copper removal by chitosan filled column and CM-chitosan filled column was evaluated by running five samples, taken from waste effluent discharge of cottage industries in Hyderabad city. The initial copper concentration in samples was found to be 132, 117, 112, 127, and 218.1 mg/L. The samples were run smoothly, and after elution, collection, and analyzation, the copper removal percentage was calculated (Table 1).

The copper removal efficiency was the ratio of copper in elution versus copper in terms of percentage. The net result of copper removal percentage was 51.1, 47.1, 49.3, 50.1 and 52.1% respectively. On the other hand, CM-chitosan filled column exhibited copper removal percentage of 67.6, 58.8, 68.4, 68.2 and 69.2% respectively. The CM-chitosan was more efficient because more binding sites were available for adsorption due to secondary hydroxyl groups.

The adsorbed copper on the filled columns was desorbed using 0.1M EDTA solution, and it was calculated as under:

$$Cu_{desorbed} = (S_{Cu} - E_{Cu})/S_{Cu} * 100 \qquad (1)$$

where

S_{Cu} & E_{Cu} = Copper concentration in sample and elution respectively.

Calculated percentage of desorbed copper in the columns was 97.56-99.6% respectively.

4 CONCLUSION

It was concluded that chitosan filled column removed copper between the range of 47.1–51.1% and the CM-chitosan column removed copper 58.8–69.2% at optimum pH. Ethylene diaminetetracetic acid (EDTA) desorbed the copper in both column up to 99.6%.

REFERENCES

Ali, Z. M., Laghari, A. J., Ansari, A. K. & Khuhawar, M. Y. 2013a. Extraction and Characterization of Chitosan from Indian Prawn (Fenneropenaeus Indicus) and its Applications on Waste Water Treatment of Local Ghee Industry. *International Organization of Scientific Research Journal of Engineering (IOSRJEN)* 3, 28–37.

Ali, Z. M., Laghari, A. J., Ansari, A. K. & Khuhawar, M. Y. 2013b. Synthesis and Characterization of Carboxymethyl Chitosan and its Effect on Turbidity Removal of River Water. *International Organization of Scientific Research Journal of Applied Chemistry (IOSRJAC)*, 5, 72–79.

Davis, J. Q. & Bennet, V. 1985. Human Erythrocyte Clathrin and Clathrin-uncoating Protein. *The Journal of Biological Chemistry*, 260, 14850–14856.

Gotoha, T., Matsushima, K. & Kikuchi, K.-I. 2004. Preparation of alginate–chitosan hybrid gel beadsand adsorption of divalent metal ions. *Chemosphere* 55, 135–140.

Lenntech.com. 2016. *Copper – Cu, Chemical properties of copper – Health effects of copper – Environmental effects of copper* [Online]. Available: http://www.lenntech.com/periodic/elements/cu.htm [Accessed 14 March 2016].

Wan Ngah, W. S., Kamari, A. & Koay, Y. J. 2004. Equilibrium and kinetics studies of adsorption of copper (II) on chitosan and chitosan/PVA beads. *International Journal of Biological Macromolecules*, 34, 155–161.

Analysis of mould growth contamination in library building

S. Ngah Abdul Wahab & N.I. Mohammed
Faculty of Engineering, Universiti Teknologi PETRONAS, Seri Iskandar, Perak

M.F. Khamidi
*School of the Built Environment, University of Reading Malaysia (UORM),
Persiaran Graduan, Educity Kota Ilmu, Iskandar Puteri (Nusajaya), Johor*

Z. Mohd Noor
Faculty of Applied Sciences, Universiti Teknologi Mara, Shah Alam, Selangor

M. Ismail
School of Housing, Building and Planning, Universiti Sains Malaysia, USM, Penang

ABSTRACT: The purpose of this paper is to analyse the mould growth and yeast contamination in libraries building. The mould growth causes identified during an investigation and visual inspection of Library PTAR building. The mould contamination level had been measured using 3M Petrifilm Rapid Yeast and Mould Count Plate. The objective of the measurement is to detect environmental concentration for indoor library environment using swab contact monitoring procedures and air sampling procedures. The result recorded yeast and moulds colonies on the 3M Petrifilm with variably pigmented colonies with certain characters. The yeast colonies recorded are small, tan colonies with defined edges and no foci while the mould colonies are with diffuse edges and centre foci. They are large, crowd, sporulate and overlap each other on the plate. The result confirmed high contamination count using swab contact, and low contamination recorded using air sampling procedure.

1 INTRODUCTION

Moulds growth can appear in different colours and patterns. Three patterns of growth have been identified in the recent past, and these may be in the form of moisture stains, radiate from spots or bloom around the damaged material (Aibinu et al. 2009). Moulds are microscopic in size and their spores travel through the air. The potential for growth and survivals on surfaces is influenced by the micro environmental conditions of both the material and the indoor surface (Adan Olaf CG & Robert 2011). The factors contributed such as moisture or high humidity, nutrition, radiation and light, oxygen, pH level, temperature and particularly the water con-tent of the material or substrate (Armstrong & Liaw 2003; Nianping et al. 2011).

Moulds growth on library building materials and components may range from small areas covering a few centimetres to a very large extent. Also, moulds have the potential to cause degradation in virtually all building materials they colonies (Nielsen et al. 2004; Hoang et al. 2010). Numerous studies have documented that Sick Building Syndrome (SBS) are the causes that came from airborne bacteria and fungi found in the indoor environment of the building (Crook & Burton 2010; Goldstein n.d.; Aibinu et al. 2009). Moreover, research also showed that SBS and certain diseases such as humidifier fever or asthma are often associated with high airborne microbial concentration exposure (Shoemaker & House 2006). Among the effects that had attracted the most attention is biological growth such as moulds with various species of Cladosporium, Aspergillus and other fungi types. This mould species can be found in both indoor and outdoor environments of library buildings and it is completely linked to most building deterioration (Hollis 2005). The growing awareness of the consequential health effect has, in turn, resulted in several litigation against researchers, designers, facilities manager and building owners. Therefore, the study of mould growth in the library building is important for the preservation of library collections, material and archives as well as users health consequences.

2 THE METHODOLOGY

The objective of the study is to know the relation of mould growth and contamination condition in the libraries building. Airborne mould spores are of concern because they can be directly inhaled. Area samples are used to measure the concentration of mould in an environment. Sampling can also be used as a determination of clearance in an area cleaned of mould infestation. The study was carried out in a library building (35 years old) with a central air conditioning system and open spaces reading areas located in Shah Alam, Malaysia. The investigation works encompass visual inspection of walls, floors, and ceilings, air samples, surface samples, and, where necessary, inspection of interiors of bookshelves, walls, ceilings, air conditioning and mechanical ventilation systems, and other building elements. These process identified mould growth area and possible causes that relate to its growth. The 3M Petrifilm Rapid Yeast and Mould Count plate are a sample ready culture medium system which contains nutrients supplemented with antibiotics, a cold water soluble gelling agent and an indicator system that facilitates yeast and mould enumeration (App et al. 2012).

To ensure the accurate result and measurement of environment microbial contamination, the procedure shall comply with the steps and recommendation by the 3M Petrifilm manufacturer. The procedure that to follow including hydration procedure, inoculation procedure, incubation, interpretation guidelines and storage. Table 1, shows the requirement and standards that need to comply with the sampling and data collection process.

Figure 1. The 3M petrifilm air sampling method.

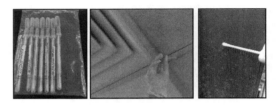

Figure 2. Surface sampling method.

2.1 Sampling method

The measurement of environmental mould and yeast contamination for the indoor library are using air sampling procedures and swab contact monitoring procedures. The sampling is similar with traditional agar method but it is easy to count Yeast and Mould Colonies. The Mould Count Plate method is faster and easier. The 3M Petrifilm air sampling method that applied to detect environment mould and yeast contamination in the library followed the method recommended by 3 M Petrifim manufacturer to ensure the accurate result and finding. Two level open-spaces library and two sampling area were selected for air sampling method.

It is located under air duct distribution from the fresh air intake. The air was sampled for microbiological analysis in the centre of each open-space library, at 1.5m from the floor (on the table level) and was collected after 15 minutes exposed to air. Three unit of 3M petrifilm plate were placed on the table for each sampling location. The Petrifilm Plate was incubated and enumerated as directed in product in-struction. Fig. 1, shows the 3M petrifilm air sam-pling method that placed on the table and exposed time been recorded.

The swab contact monitoring procedures can determine whether a visible stain found in the library has resulted from mould growth or other species. The 3M Quick Swab Method used to identify and measure some colonies on the plate from the area sampled. The swabbing method and 1mL inoculation procedure followed the guideline and manufacturer recommendation. From the visual inspection, a visible stain and sign of mould growth were recorded and mark as sampling location for swabbing procedures. In Library PTAR, six areas or elements had been identified for surface swabbing including on the books, window frame, air conditioning outlet, carpet area, ceiling board and on timber column where the symptoms and stain of possible mould growth visibly identified. The three units of 3M Quick Swabs were used for each sampling location and elements. The process of swabbing method at the sampling location and mould count is shown in Fig. 2 below.

2.2 Mould count

Mould and yeast colonies can be count on Petrifilm Yeast and Mould Count plates. Yeast and mould can be differentiating based on typical characteristics such as sizes, colours, edges and focus area of the colony. Time and temperature are important to ensure the growth of the mould. The manufacturer recommended to ensure optimum growth; the plates should be incubated at 20°C to 25°C (room temperature) and to check plates for growth at both three and five days (Boyle et al. 2014). The results of yeast and mould would be read and count at 48 hours and to enhance interpretation of these moulds additional 12 hours of incubation time is required.

3 RESULT AND DISCUSSION

3.1 Air sampling result

The result reported in Table 2 showed that the total yeast and mould count from all air sampling location at Library PTAR was low (range between 0 to 3 CFU per 1 mL).

Table 1. The 3M petrifilm rapid yeast and mould count procedures.

No	Procedure	Standards/Guideline
1	Hydration	For Yeast and mould count using Air or Direct Contact hydrate plates with 1 ml of appropriate sterile diluent. Allow hydrated plates to remain closed for a minimum of 1 hour before use.
2	Inoculation	Use appropriate sterile diluents such as Butterfield's phosphate buffer (ISO55411) or distilled water. Use Flat Spreader (6425) and refer procedure as recommended by the manufacturer.
3	Incubation	Incubate 3M Petrifilm Rapid Yeast and Mould Count Plate at 25–28°C for 48 ± hours in a horizontal position with the clear side up in stacks not more than 40.
4	Interpretation	Read yeast and mould results at 48 hours. Certain slower growing yeast and moulds may appear faint at 48 hours. To enhance interpretation of these moulds allow for an additional 12 hours of incubation time.
5	Storage	Seal by folding the end of the pouch over and applying adhesive tape. To prevent exposure to moisture, do not refrigerate opened pouches. Store resealed pouches in a cool, dry place (20–25°C/<60% RH) for no longer than four weeks.

Table 2. Total yeast and mould count from air sampling.

Level/Area	Date Cultured	Sample No	Env/Element	Total Yeast & Mould
Level 4 (A)	18/12/15-20/12/15	Env 1	Env	$3/60cm^2$
Level 4 (A)	18/12/15-20/12/15	Env 2	Env	$3/60cm^2$
Level 4 (A)	18/12/15-20/12/15	Env 3	Env	$1/60cm^2$
Level 4 (B)	18/12/15-20/12/15	Env 1	Env	0
Level 4 (B)	18/12/15-20/12/15	Env 2	Env	0
Level 4 (B)	18/12/15-20/12/15	Env 3	Env	0
Level 5 (A)	18/12/15-20/12/15	Env 1	Env	$2/60cm^2$
Level 5 (A)	18/12/15-20/12/15	Env 2	Env	$2/60cm^2$
Level 5 (A)	18/12/15-20/12/15	Env 3	Env	$1/60cm^2$
Level 5 (B)	18/12/15-20/12/15	Env 1	Env	$1/60cm^2$
Level 5 (B)	18/12/15-20/12/15	Env 2	Env	$2/60cm^2$
Level 5 (B)	18/12/15-20/12/15	Env 3	Env	0

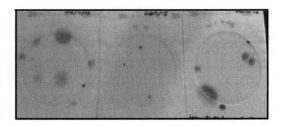

Figure 3. Yeast and Mould contamination appear from air sampling.

Figure 3 shows the yeast and mould appear on the Yeast and Mould Count Plate (YM). The yeasts appear as small, regularly-shaped, and aqua green colonies while mould appear as larger, variable-coloured colonies with diffuse edges and a central focal point.

3.2 *Quick swab result*

All samples collected from library PTAR demonstrated that the sampling was contaminated with yeast and mould. From the previous work by authors, confirmed that the most abundant mould was found to be *Aspergillus sp.* followed by *Penicillium sp.*, *Stachybotrys sp.* and *Rhizoctonia Solani*. The result reported in Table 3 showed the total yeast and mould count from all sampling location at Library PTAR was high. The findings indicated that the mould identified earlier have a consistent reading when the total yeast and mould count recorded very high concentration and even on carpet surfaces total yeast and mould too numerous to count (TNTC).

There are a few pinpoint spots on the YM Count Plate for book 1 and book 3 and very little recorded for book 2. The YM Count Plate from the surface of window frame recorded range from 60 to too numerous to count (TNTC). Figure 4 shows the yeast and moulds appear on the Yeast and Mould Count Plate (YM) from the surface of books and window frame sampling location.

The YM Count Plate recorded mould and yeast too numerous to count (TNTC) from carpet and ceiling board sampling location. Figure 5 shows the yeast and moulds appear on the Yeast and Mould Count Plate (YM) from the surface of carpet and timber sampling location.

The mould appears as large colonies and has diffuse edges and centre foci with variable coloured colonies were recorded from the sample taken from the air conditioning outlet. It was recorded very high (250 CFU per 1 Ml) at AC outlet 3, where the sign and stain of mould visibly appear at this point compared to AC outlet 1. The YM Count Plate recorded consistently high at all timber columns sampling location (range between

Table 3. Total yeast and mould count from surface sampling.

Level/Area	Date Cultured	Sample No	Env/Element	Total Yeast & Mould
Level 4 (A)	22/12/15-25/12/15	BOOK 1	Book	42
Level 4 (A)	22/12/15-25/12/15	BOOK 2	Book	2
Level 4 (A)	22/12/15-25/12/15	BOOK 3	Book	35
Level 4 (B)	22/12/15-25/12/15	WF 1	Window Frame	TNTC
Level 4 (B)	22/12/15-25/12/15	WF 2	Window Frame	104
Level 4 (B)	22/12/15-25/12/15	WF 3	Window Frame	60
Level 4 (B)	22/12/15-25/12/15	AC 1	Air Cond Outlet	26
Level 4 (B)	22/12/15-25/12/15	AC 2	Air Cond Outlet	86
Level 4 (B)	22/12/15-25/12/15	AC 3	Air Cond Outlet	250
Level 5 (A)	22/12/15-25/12/15	CPT 1	Carpet	TNTC
Level 5 (A)	22/12/15-25/12/15	CPT 2	Carpet	TNTC
Level 5 (A)	22/12/15-25/12/15	CPT 3	Carpet	TNTC
Level 5 (A)	22/12/15-25/12/15	CL 1	Ceiling Board	TNTC
Level 5 (A)	22/12/15-25/12/15	CL 2	Ceiling Board	TNTC
Level 5 (A)	22/12/15-25/12/15	CL 3	Ceiling Board	TNTC
Level 5 (B)	22/12/15-25/12/15	COL 1	Timber Column	300 / TNTC
Level 5 (B)	22/12/15-25/12/15	COL 2	Timber Column	204 / TNTC
Level 5 (B)	22/12/15-25/12/15	COL 3	Timber Column	300 / TNTC

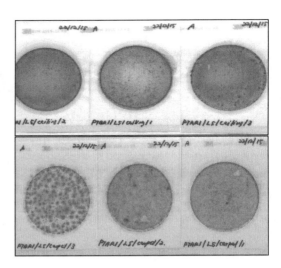

Figure 5. Mould and Yeast contamination appear from the samples of Ceiling and Carpet.

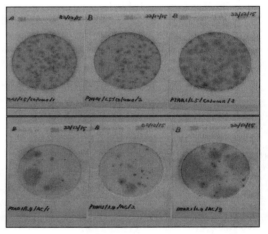

Figure 6. Mould and Yeast contamination appear from the samples of Timber Column and AC Outlet.

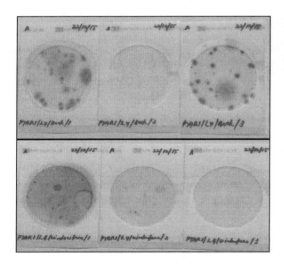

Figure 4. Mould and Yeast contamination appear from the sampling of Books and Window Frame.

204 to 300 CFU per 1 mL and TNTC). Fig. 6 shows the yeast and mould appear on the Yeast and Mould Count Plate (YM) from the surface of AC outlet and timber column sampling location.

4 CONCLUSION

In this paper, methods of mould count using 3M Petrifilm Yeast and Mould Count Plate had been conducted. The area of contribution of this research work is the count colonies of mould growth and the acquired image from air sampling and swabbing method on the surfaces that had been identified earlier as the potential area of mould growth. The total count of mould and yeast base on the sampling area determined the yeast and mould contamination of the library environment. The most abundant mould was found to be *Aspergillus sp. Penicillium sp.*, *Stachybotrys sp.* and *Rhizoctonia Solani*. The sample size and particular characteristics of each library will result in the mould growth pattern and finding. The research expected to identify the best solution to protect a collection that will depend on various factors such as type and number of visitors, characteristics of the library building and level of maintenance work provide.

REFERENCES

Adan Olaf CG & Robert, S.A., 2011. Introduction Of Fundamentals of Mold Growth in Indoor Environments and Strategies for Healthy Living. In *Fundamentals of Mold growth in Indoor Environments and Strategies for Healthy Living*. Netherlands: Wageningen Academic Publishers, p. 302.

Aibinu, A.M. et al., 2009. Assessment of Mould Growth on Building Materials using Spatial and Frequency Domain Analysis Techniques. *IJCSNS International Journal of Computer Science and Network Security*, 9(7), pp. 154–167.

App, B.M.B., Tafe, R. & Biological, S., 2012. A Brief Introduction to Microbiology and the Use of 3M™ Petrifilm Plates™.

Armstrong, S. & Liaw, J., 2003. The Fundamentals of Fungi, (April), pp. 14–19.

Boyle, M. et al., 2014. Method Comparison of the 3M™ Petrifilm™ Rapid Yeast and Mold Count Plate Method for the Enumeration of Yeast and Mold., pp. 1–16.

Crook, B. & Burton, N.C., 2010. Indoor moulds, Sick Building Syndrome and building related illness. *Fungal Biology Reviews*, 24(3–4), pp. 106–113.

Goldstein, W.E., Development, Directions in Construction Practice, and Summary Recommendations, pp.209–217.

Hoang, C.P. et al., 2010. Resistance of Green Building Materials to Fungal Growth. *International Biodeterioration & Biodegradation*, 64(2), pp. 104–113.

Hollis, M., 2005. *Surveying Buildings 5th Edition* 5th ed., RICS.

Nianping, L. et al., 2011. Indoor Mildew Pollution in Building and Control Strategies. *2011 Third International Conference on Measuring Technology and Mechatronics Automation*, pp. 394–397.

Nielsen, K.F. et al., 2004. Mould growth on building materials under low water activities. Influence of humidity and temperature on fungal growth and secondary metabolism. *International Biodeterioration & Biodegradation*, 54(4), pp. 325–336.

Shoemaker, R.C. & House, D.E., 2006. Sick building syndrome (SBS) and exposure to water-damaged buildings: Time series study, clinical trial and mechanisms. *Neurotoxicology and Teratology*, 28(5), pp. 573–588.

Petroleum sludge thermal treatment and use in cement replacement – A solution towards sustainability

E.N. Pakpahan
Division on Hazardous & Toxic Substances Registration & Notification, Deputy Minister for Hazardous & Toxic Substances, Hazardous & Toxic Wastes and Garbage Management – Ministry of Environment, Republic of Indonesia

N. Shafiq, M.H. Isa, S.R.M. Kutty & M.R. Mustafa
Department of Civil and Environmental Engineering, Universiti Teknologi PETRONAS, Seri Iskandar, Perak, Malaysia

ABSTRACT: Petroleum sludge contains polycyclic aromatic hydrocarbons (toxic and carcinogenic-mutagenic compounds) and heavy metals. It needs to be suitably treated and disposed in an environmental friendly fashion. This paper reports the effectiveness of additives ($Ca(OH)_2$ and $NaHCO_3$) and excess air in thermal treatment of petroleum sludge cake and the suitability of the residual of Petroleum Sludge Ash (PSA) as a Cement Replacement Material (CRM). Polycyclic Aromatic Hydrocarbons (PAH) concentrations in samples were determined using GC-MS. Leaching of heavy metals from petroleum sludge cake and PSA was determined following TCLP. Heavy metals concentrations were determined using AAS. Mortar mixes were prepared with different amounts of cement replacement by PSA and their compressive strengths were determined. Additives and excess air were found to improve sludge treatment. The strength development of mortar cubes showed that PSA has the potential to be used as CRM. The studied treatment method provides a solution that also enhances sustainability.

1 INTRODUCTION

Petroleum sludge is one of the major wastes produced by the petroleum refining industry. Typically, the sludge contains 10–30% hydrocarbons, 5–20% solids and 50–85% water (Ward et al., 2003). In some places, the sludge is simply dumped at unused land. A local refinery in Malaysia produces more than 3,200 m^3 of sludge per year. It reduces to 40 to 75 m^3 of sludge cake after oil recovery. The oil sludge, if unused, is considered as hazardous waste and is included in listing of the Resource Conservation and Recovery Act (RCRA). As a part of waste oil, the sludge is also regulated as code A4060 in Annex VIII List A of the Basel Convention on the Control of Transboundary Movements of Hazardous Wastes and Their Disposal (1989). In Malaysia, petroleum sludge falls under the Environmental Quality Act (Schedule Waste) Regulation for control of its storage, transportation and disposal.

Because of the hazardous characteristics of petroleum sludge and its potential impact on the environment, its safe disposal poses several issues and challenges. Petroleum sludge contains about 550 mg/kg total polycyclic aromatic hydrocarbon (PAH) which are toxic and carcinogenic-mutagenic (Bojes and Pope, 2007). The United States Environmental Protection Agency (USEPA) has listed 16 PAH as priority pollutants: Naphthalene (Nap; 2-rings, $C_{10}H_8$), Acenaphthylene (Acy; 3-rings, $C_{10}H_8$), Acenaphthene (Ace; 3-rings, $C_{10}H_{10}$), Fluorene (Flu; 3-rings, $C_{13}H_{10}$), Phenanthrene (Phe; 3-rings, $C_{13}H_{10}$), Anthracene (Ant; 3-rings, $C_{13}H_{10}$), Fluoroanthene (Fla; 4-rings, $C_{16}H_{10}$), Pyrene (Pyr; 4-rings, $C_{16}H_{10}$), Benzo[a]Anthracene (BaA; 4-rings, $C_{18}H_{12}$), Chrysene (Chr; 4-rings, $C_{18}H_{12}$), Benzo[b] Fluoranthene (BbF; 5-rings, $C_{20}H_{12}$), Benzo[k]Fluoranthrene (BkF; 5-rings, $C_{20}H_{12}$), Benzo[a]Pyrene (BaP; 5-rings, $C_{20}H_{12}$), Indeno[1,2,3-cd]Pyrene (Ind; 6-rings, $C_{22}H_{12}$), Dibenzo[a,h]Anthracene (DbA; 6-rings, $C_{22}H_{14}$), and Benzo[g,h,i]Perylene (BPer; 6-rings, $C_{22}H_{12}$).

Thermal incineration is one of the common methods of treatment of such sludge (Chang et al., 2000). The technique is capable of degrading of all priority PAH, but the high energy cost associated with its high operating temperature of 820 to 1600°C is a concern (Pakpahan et al., 2012). Also, ash produced from sludge incineration poses problems, challenges and issues regarding public health protection and environmental safety. Prevention of heavy metals leaching from the ash into the environment is also a major concern. Solidification and stabilization (S/S) techniques are usually applied to prevent or minimize the release of heavy metals into the environment. Solidification refers to the technique of ash encapsulation, forming a solid material, and does not necessarily involve a chemical reaction between the heavy metals and the solidifying additives, while stabilization refers to chemical conversion of the heavy metals into less soluble, mobile or toxic forms (Patel and Pandey, 2009). S/S is well suited for inorganic (heavy metals) immobilization.

Ordinary Portland cement, OPC is a widely used construction material; in 2010, its global annual production was estimated at 2.2 billion tons, which was about 50% more than the production made in 1999. Continuous increase in the rate of production of cement poses serious concerns of greenhouse gases emission; approximately 1 ton of CO_2 is emitted for every ton of cement produced (Meman et al., 2011). Several waste materials, such as industrial byproducts and processed agricultural wastes, containing amorphous silica have been introduced for partial replacement of cement in the construction industry. These are commonly called cement replacement materials (CRM). Silica fume and fly ash have been widely used in producing high performance concrete. Various studies have reported that 30–40% partial replacement of cement showed significant improvement in compressive strength and long-term durability of concrete as compared to concrete made of 100% OPC (Shafiq et al., 2007; Shafiq, 2011).

This paper reports the effectiveness additives ($Ca(OH)_2$ and $NaHCO_3$) and excess air in thermal treatment of petroleum sludge cake and the suitability of the residual petroleum sludge ash (PSA) as a cement replacement material.

2 MATERIALS AND METHODS

2.1 Petroleum sludge cake

Petroleum sludge cake samples were collected from the sludge treatment plant of a local Malaysian petroleum refinery. The samples were kept in dark glass bottles and stored at 4°C in a cold room. They were placed in a desiccator and brought to room temperature prior to use.

2.2 Thermal treatment of petroleum sludge cake

Petroleum sludge cake (500 g) was thermally treated in a rotary-drum electric heater (Pakpahan et al., 2013; 2015) at a temperature of 650°C using additives $Ca(OH)_2$ (1 mole) + $NaHCO_3$ (1 mole) (1:1) with and without 30% excess air addition. The heater was an assembly of two concentric cylinders; 0.3 m ID of outer cylinder, 0.2 m OD of inner cylinder, and 0.6 m effective length of cylinders. The space between the cylinders served as the heating chamber. The outer cylinder rotated at 20 rpm during operation. A 9 kW ceramic electric heater, housed in the inner cylinder, was used to provide heat for thermal treatment and a blower was used to provide air.

2.3 PAH analytical procedure

Sonication extraction (SW-846 Method 3550B) using acetonitrile as solvent was performed to extract PAH

Table 1. Compositions of mortar mixes.

Code	PSA:OPC ratio	OPC g	PSA g	Sand g	Water G
M0 (control)	0 : 100	55.5	0	222.2	22.2
M5	5 : 95	52.7	2.7	222.2	22.2
M10	10 : 90	50.0	5.5	222.2	22.2
M20	20 : 80	44.4	11.1	222.2	22.2
M30	30 : 70	38.8	16.6	222.2	22.2

from PSA (Evagelopoulos et al., 2010; Chantara et al., 2010). Stock solution of a PAH mixture standard (Restek Corp) was used to prepare standard solutions of 17 concentrations from 0.001 to 9.0 ppm for calibration. PAH in the samples were determined using a 6890N GC, 5973N MS with 7683 auto-injector (Agilent Technologies).

2.4 Heavy metals leaching from raw sludge and PSA

Triplicate tests of the Toxicity Characteristic Leaching Procedure (TCLP) were conducted for petroleum sludge cake and PSA (100 g samples) prior to use as cement replacement according to the USEPA SW-864, 1996. The eight target metals in the test were Chromium (Cr), Cadmium (Cd), Lead (Pb), Arsenic (As), Barium (Ba), Silver (Ag), Selenium (Se), and Mercury (Hg). Concentrations of the heavy metals were determined using Atomic Absorption Spectroscopy (AAS).

2.5 Mortar mixes

Compositions of mortar mixes used in this study are presented in Table 1. OPC, PSA and siliceous sand were the main constituents of the mortar mixes. After thermal treatment of petroleum sludge cake, the produced ash was ground in an LA Abrasion Machine for 2000 cycles and then passed through sieve # 100 for use in mortar mixes.

2.6 Characterization of Petroleum Sludge Ash (PSA)

Table 2 shows the chemical composition of OPC and PSA. PSA contained about 13.3% of calcium oxide (CaO). According to ASTM specifications, the ash can be classified as Class-C i.e. such type of ash possesses cementing properties, which could be verified by the compressive strength test results.

SEM images of cement and PSA as shown in Figure 1 reveal that both cement and PSA possessed identical morphological characteristics and grain size. The ash due to grinding at 2000 cycles and passing through sieve #100 showed slightly higher surface area and low pore volume than the cement as given in Table 3. Chindaprasirt et al. (2005) reported similar findings.

The XRD results of cement and PSA samples show that the PSA samples are composed of hexagonal

Table 2. Chemical composition of OPC and PSA.

Chemical constituents	Composition %	
	OPC	PSA
Na_2O	–	7.53
K_2O	0.41	1.93
MgO	1.72	2.22
Al_2O_3	4.39	12.2
SiO_2	18.99	47.28
Fe_2O_3	3.09	6.48
SO_3	5.27	3.61
CaO	65.02	13.30
P_2O_5	0.77	1.93

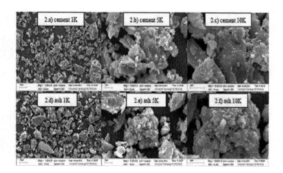

Figure 1. Surface morphology of OPC and PSA, (1kX, 5kX, 10kX magnification).

Table 3. Surface area and porosity of OPC and PSA.

Constituents	Surface area m^2/g	Pore volume cm^3/g	Pore size Å
OPC	0.0116	0.5702	0.9489
PSA	0.0119	0.2698	0.8747

crystalline structure; its mineral phases consist of quartz, syn-SiO_2 and carlinite, syn-Tl_2S. Therefore, based on chemical composition, morphological character and crystalline structure, PSA is a potential CRM for green concrete.

2.7 Preparation, casting, curing and testing of mortar samples

Mortar mixes were prepared using a Hobart mixer, following relevant ASTM specifications. Following the mixing of mortar, 50 mm cubes were cast using standard steel molds. After casting, the samples were covered with black thick plastic sheet and left for 24 hours. Subsequently, all cubes were removed from the moulds and immersed into a water-bath at room temperature for curing. The cubes were taken out for compressive strength test (at the age of 3, 7, 14, 28, 56, and 90 days) according to ASTM C-109 using a universal hydraulic testing machine.

Table 4. Priority PAH in raw and treated petroleum sludge.

PAH	Raw sludge µg/g	0% excess air µ g/g	30% excess air µ g/g
2–3 rings	1.05×10^5	19500	838
4–6 rings	4.77×10^3	2690	336
Σ PAH	1.10×10^5	22190	1170

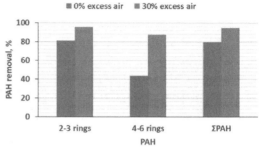

Figure 2. PAH removal from petroleum sludge.

3 RESULTS AND DISCUSSION

3.1 PAH in thermally treated petroleum sludge cake

Table 4 shows concentrations of the priority PAH in raw and treated petroleum sludge. The total concentration of priority PAH (ΣPAH) in the ash was 22,190 µg/g when no excess air was supplied and only 1,170 µg/g in the presence of 30% excess air. With reference to the concentration of ΣPAH in raw sludge, ΣPAH removal was 80% in the absence of excess air and increased to 95% with 30% excess air (Figure 2). Thus, excess air supply was shown to be beneficial to thermal degradation of PAH. A lack of excess air could not provide adequate interaction between oxygen and radicals, and resulted in inefficient degradation of PAH.

3.2 TCLP for petroleum sludge cake and PSA

The mean concentrations of the targeted heavy metals in the raw sludge were below the standard limit stipulated by the USEPA. Cr, Cd, and Pb were not detectable (<0.01 mg/L) and the other heavy metals were below the limit: As 0.032 mg/L, Ba 1.301 mg/L, Ag 0.204 mg/L, Se 0.030 mg/L, and Hg 0.056 mg/L.

The test of heavy metals leaching from PSA for use as partial cement replacement yielded lower concentrations compared to the sludge. Cr, Cd, and Pb were not detectable (<0.01 mg/L), whereas As 0.027 mg/L, Ba 0.674 mg/L, Ag 0.189 mg/L, Se 0.020 mg/L, and Hg 0.005 mg/L were far below the limits. Hence, use of PSA as CRM will not pose a heavy metal leaching issue.

3.3 Compressive strength and strength activity index

Compressive strength of different mortar mixes was determined at the age of 3, 7, 28, 56, and 90 days, the

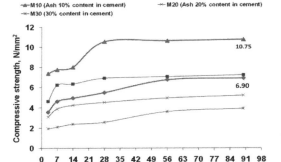

Figure 3. Compressive strength of mortar cubes.

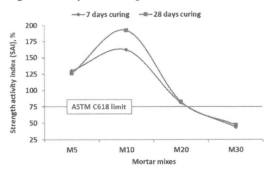

Figure 4. Strength activity index (SAI) of mortar at 7 and 28 days curing.

Table 5. Compressive strength gain/loss of PSA mixes with respect to the mix M0.

	M5	M10	M20	M30
Compressive strength gain (CSG) N/mm^2	1.356	4.550	−0.772	−2.036

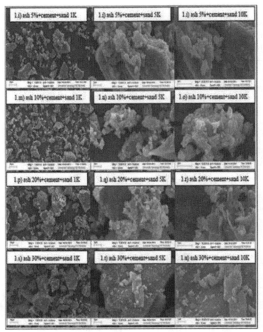

Figure 5. SEM images of mortars samples (1kX, 5kX, 10 kX magnification).

results are shown in Figure 3. As a general trend at the age of 7 days, all mixes achieved approximately 75% to 80% of the 28 days strength. Beyond 7 days until the age of 28 days, strength was developed at a slower rate. Whereas from 28 days to 90 days, very little increase in strength was observed. By comparing the effects of PSA content; mortar mix M10 with 10% PSA exhibited highest strength at every age as compared to the other mixes. At the age of 3 days, it achieved 7.4 N/mm^2, which was double the strength of the control mix M0 (3.6 N/mm^2). Similarly, at 28 days, M10 achieved 10.6 N/mm^2 as compared to 5.5 N/mm^2 that was achieved by mortar mix, M0 at the same age.

Strength activity index, SAI (Mx/M0) is defined as the ratio of the strength of sample with different CRM content, Mx, to the strength of the sample with 100% cement, M0. SAI is a useful parameter that usually defines the effectiveness of a particular type of cement replacement material; whether the material possesses pozzolanic or cementing characteristics. According to ASTM C 618 standard, SAI must be higher than 75% at the age of 7 and 28 days in order to justify the use of ash as partial replacement of cement (Celik et al., 2008). Figure 4 shows that mix M10 exhibited highest value of SAI as determined at the age of 7 days and 28 days. The mix with 30% PSA, M30 showed the SAI value less than the value as specified by ASTM as qualification for CRM. Referring to compressive strength gained (CSG) as shown in Table 5, only mixes M5 and M10 met the criteria of cement replacement material, whereas mixes M20 and M30 failed to satisfy the criteria.

3.4 *Microstructure of mortar samples using SEM*

Figure 5 shows the SEM micrographs of all mortar samples obtained at different magnification level. From the images of mortar mix, M10, traces of agglomerate can be seen, which are slightly smaller than the ash particles. This agglomeration may have occurred due to reaction of Na$_2$O with other material. During thermal treatment of petroleum sludge, some additives were added, which could have refined the pore structure and improved the bond strength of mortar mixtures, which is significantly reflected in the compressive strength results of the mix M10. It is noted that smaller particles size and irregular shape reduced the void space of mixture, which consequently minimized the deformation behavior. Similarly, due to morphological character of PSA, it filled the pores of the mixture and offered strong interlocking within the mortar constituents.

Table 6. Chemical composition of mortar obtained using XRF technique.

Element oxides	Composition %				
	M0	M5	M10	M20	M30
K_2O	0.394	0.446	0.528	0.608	0.709
MgO	1.32	1.22	1.39	1.31	1.1
Al_2O_3	5.20	5.76	6.12	6.28	7.26
SiO_2	27.58	34.1	35.73	36.17	37.2
Fe_2O_3	2.840	2.468	2.992	3.184	3.126
SO_3	3.95	3.84	3.77	3.89	3.77
CaO	57.17	48.09	51.88	46.83	43.15
P_2O_5	1.09	1.51	1.62	1.21	2.57

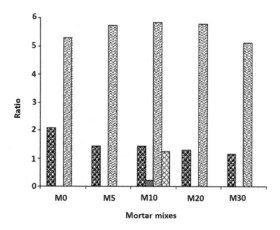

Figure 6. Ratio of CaO/SiO_2 of constituents in mortar mixtures.

Karim et al. (2011) and Fernandez et al. (2010) studied mineralogical characteristics of pozzolanic reaction and found that microstructure and amorphous mineralogical conditions significantly govern further formation of calcium silicate hydrate, C-S-H. C-S-H formation is also governed by the CaO/SiO_2 ratio, lower value of CaO/SiO_2 causes early-formation of C-S-H gel with rounded-shape and flaky appearance in a wide size range, whereas high CaO/SiO_2 ratio promotes well-formed C-S-H due to intergrowth of thin layers.

Concerning the ash prepared from petroleum sludge cake heating process, Lin et al. (2010) mentioned that the agglomeration is gradually formed by accumulation of the liquid-phase eutectics, which corresponds to the reduced quality of fluidization during thermal treatment. Therefore, stickiness or glutinous (gel phase) formation in the mortar mixes is a part of agglomeration mechanisms During thermal treatment of sludge, $Ca(OH)_2$ together with $NaHCO_3$ was used as additive, which appeared in the hydrated phase of the mortar samples as given in Table 6. Figure 6 shows that the M0 samples have negligible values of Na_2O/SiO_2 and Na_2O/Al_2O_3, and lower SiO_2/Al_2O_3, whereas mix M10 showed significant values of Na_2O/SiO_2, SiO_2/Al_2O_3 and Na_2O/Al_2O_3; hence mix M10 has shown highest compressive strength.

4 CONCLUSIONS

Thermal treatment of petroleum sludge cake with additives and excess air was shown to provide an effective solution to an environmental problem in a way that would also enhance sustainability. Based on the conducted study, the following conclusions can be made:

– Thermal treatment of petroleum sludge cake with additives and excess air was beneficial; PAH removal was over 99%.
– Leaching of heavy metals from PSA was low and within limits.
– PSA has the potential to be used as CRM when the replacement content is kept below 20%.
– PSA content of 10% showed twice the compressive strength as that of the control mix (100% cement). It also showed well refined microstructure of the mortar sample.
– PSA increased the strength development until 28 days of curing and beyond that the strength was found almost constant. Therefore it can be characterized as cementing material rather than pozzolanic.

ACKNOWLEDGEMENT

This work was supported by the Prototype Research Grant Scheme (PRGS), Ministry of Education (MOE), Malaysia, under cost centre 0153AB-I34. The authors are also thankful to Mr. Zaaba bin Mohammad, Mr. Mohd Khairul Anuar bin Jamaluddin, Ms. Yusyawati binti Yahaya, and Mr. Imtias bin Amir B. Bahauddin from the Department of Civil and Environmental Engineering, Universiti Teknologi PETRONAS, Malaysia for their assistance.

REFERENCES

Bojes, H.K. & Pope, P.G. 2007. Characterization of EPA's 16 priority pollutant polycyclic aromatic hydrocarbons (PAHs) in tank bottom solids and associated contaminated soils at oil exploration and production sites in Texas. *Regulatory Toxicology and Pharmacology* 47: 288–295.

Celik, O., Damci, E. & Piskin, S. 2008. Characterization of fly ash and its effects on the compressive strength properties of portland cement. *Indian Journal of Engineering and Material Sciences* 15: 433–440.

Chang, C.Y., Shie, C.Y., Je, L., Lin, J.P., Wu, C.H., Lee, D.J. & Chang, C.F. 2000. Major products obtained from the pyrolysis of oil sludge. *Energy & Fuels* 14: 1176–1183.

Chantara, S., Wangkarn, S., Sangchan, W. & Rayanakorn, M. 2010. Spatial and temporal variations of ambient PM10-bound polycyclic aromatic hydrocarbons in Chiang Mai and Lamphun Provinces, Thailand. *Desalination and Water Treatment* 19: 17–25.

Chindaprasirt, P., Jaturapitakkul, C. & Sinsiri, T. 2005. Effect of fly ash fineness on compressive strength and pore size of blended cement paste. *Cement & Concrete Composites* 27: 425–428.

Evagelopoulos, V., Albanis, T.A., Asvesta, A. & Zoras, S. 2010. Polycyclic aromatic hydrocarbons (PAHs) in fine and coarse particles. Global NEST Journal 12: 63–70.

Fernández, R., Nebreda, B., de la Villa, R.V., García, R. & Frías, M. 2010. Mineralogical and chemical evolution of hydrated phases in the pozzolanic reaction of calcined paper sludge. *Cement & Concrete Composites* 32: 775–782.

Karim, M.R., Zain, M.F.M., Jamil, M., Lai, F.C. & Islam, M.N. 2011. Strength development of mortar and concrete containing fly ash: A review. *International Journal of the Physical Sciences* 6(17): 4137–4153.

Lin, C.L., Tsai, M.C. & Chang, C.H. 2010. The effects of agglomeration/defluidization on emission of heavy metals for various fluidized parameters in fluidized-bed incineration. *Fuel Processing Technology* 91: 52–61.

Meman, F.A., Nuruddin, M.F. & Shafiq, N. 2011. Compressive strength and workability characteristics of low-calcium fly ash-based self-compacting geopolymer concrete. *International Journal of Civil and Environmental Engineering* 3(2).

Pakpahan, E.N., Isa, M.H., Kutty, S.R.M., Chantara, S., Wiriya, W. & Faye, I. 2012. Comparison of polycyclic aromatic hydrocarbons emission from thermal treatment of petroleum sludge cake in the presence of different additives. Journal of Scientific and Industrial Research 71(6): 430–436.

Pakpahan, E.N., Isa, M.H., Kutty, S.R.M., Chantara, S. & Wiriya, W. 2013. Polycyclic aromatic hydrocarbons removal from petroleum sludge cake using thermal treatment with additives. Environmenatl Technology 34(3): 407–416.

Pakpahan, E.N., Isa, M.H., Kutty, S.R.M., Chantara, S., Wiriya, W. & Farooqi, I.H. 2015. Effect of excess air on polycyclic aromatic hydrocarbons removal from petroleum sludge using thermal treatment with additives. *Journal of Scientific and Industrial Research* 74(4): 245–249.

Patel, H. & Pandey, S. 2009. Exploring the reuse potential of chemical sludge from textile wastewater treatment plants in India - A hazardous waste. *American Journal of Environmental Sciences* 5(1): 106–110.

Shafiq, N., Nuruddin, M.F. & Kamaruddin, I. 2007. Comparison of engineering and durability properties of fly ash blended cement concrete made in UK and Malaysia. *Advances in Advanced Ceramics* 106(6): 314–318.

Shafiq, N. 2011. Degree of hydration and compressive strength of conditioned samples made of normal and blended cement systems. *KSCE Journal of Civil Engineering* 15(7): 1253–1257.

Ward, O., Singh, A., & Van Hamme, J. 2003. Accelerated biodegradation of petroleum hydrocarbon waste. *Journal of Industrial Microbiology and Biotechnology* 30: 260–270.

Residential landscape minimize indoor temperature in tropical climate: Myth or reality

N. Kamarulzaman, N.A. Zawawi & N.I. Mohammed
Department of Civil & Environmental Engineering, University Technology, Malaysia

ABSTRACT: In a tropical climate, outdoor environment is clearly warmer than indoor due to higher air temperatures. Since the indoor environment is influenced by its surroundings, this situation indirectly contributes to the discomfort condition. Thus, the dependence on mechanical ventilation and energy consumption will increase. Plants are viewed to be useful elements that relieve a higher air temperature in an urban area. Strategically placed plants around a building could decrease the energy consumption in buildings by reducing the adverse impact of some climate elements. However, it is difficult to estimate the quantitative relation between plants, thermal environment and energy. Overall, this paper aims to analyze the relationships between the amount of plants and climatic parameters in a residential area. The different landscape density around two Semi-Detached houses were study in term of building configuration, landscape design and climatic parameters such as air temperature, relative humidity and wind velocity in this paper.

1 INTRODUCTION

In general, the issue of increasing of heat and energy consumption in a building is due to the increasing of outdoor temperature, and it relates to environmental problems such as global warming and heat island effects as discussed before. Memon, R. A et al.,(2008) stated that there are three reasons of Urban Heat Island which such as i) Anthropogenic heat released from vehicles, power plants, air conditioners and other heat from electrical appliances, ii) Enormous amounts of solar radiation due to less vegetation from rapid urbanization. iii) High unevenness structure is another problem of urban areas, which decreases the convective heat removal (Memon, R. A., Leung, D. Y. C., & Chunho, 2008).

International Energy Agency (2009) stated that by 2030 average air temperature will rise by 2°C represent a critical limit (Yeo, Saito, & Said, 2014). Ismail, A (2008) has mentioned that when the outdoor air temperature increases, the building will experience indoor discomfort conditions. The process of heat transfer through building fenestration such as windows and absorbed through the building envelope (walls and roofs) is the primary source of heat flow into buildings. According to Akbari and Taha (1992), climatic factors that affect outdoor thermal comfort are (i) surface and air temperature; (ii) relative humidity; (iii) solar radiation; and (iv) wind velocity. Therefore, the interaction of these climate elements and its relation to local climate should be fully understood to define the appropriate approach to improving the urban thermal environment (Thani, Mohamad, & Idilfitri, 2012).

Technically, there are many ways to improve the energy efficiency of the built environment. Improvements in building envelope technologies, such as wall, floor, and roof insulation, high-performance windows and doors, and air infiltration, have a priority role in producing a comfortable interior. However, to improve energy efficiency using strategies mentioned might require an extra financial cost. Therefore, the simplest and inexpensive method is by greening strategies in urban, especially in residential areas to improve the local air temperature.

In Malaysia scenarios, the effect of surrounding landscaping on the thermal and energy performance of tropical domestic building has not been widely recognized or quantified. Therefore, this paper aims to achieve two objectives i) To analyze the relations between the amount of plants in the residential area and local air temperature ii) To identify typical landscape design around a single family residential area.

2 MYTH OR REALITY

Vegetation is usually regarded to be excellent materials that relieve a heat island in an urban area. However, it's difficult to estimate the quantitative relation between plants and the local air environment. A number of plants are thought to be a primary factor that has a connection to ambient air temperature, then the shape of plans and the species of plants should be considered (Tsutsumi, Ishii, & Katayama, 2003).

McPherson, Herrington et al. (1988) stated that landscaping can have a predominantly direct effect on local climate and linked to building thermal performance and one of the simplest strategies to reduce solar heat gain. Strategically placed vegetation around a building has recognized as a one of cooling method. It decreases the amount of solar entering the building envelope through shading, moderating temperatures, evapotranspiration processes, and by controlling the wind direction to keep the building cool. With appropriate amount, species, and placement of vegetation can slow heat build-up on a hot summer day (Misni, 2012).

Expert researcher in landscaping also agrees that plant (trees, shrubs and grasses) can give significant effect to the local environment directly. By producing shading surfaces and channeling the wind, and by evapotranspiration of water (Akbari, H., Davis, S., Dorsano, S., Huang, J., & Winnet, 1992; DOE, 1993; Misni, 2013). Shading is the most cost-effective way to diminish solar heat gain. Vegetation planted nearby the house will decrease the surrounding air temperatures and provide the best evaporative cooling (Misni, 2013).

3 FIELDWORK MEASUREMENT

The field measurement had implemented in the residential area of Perak Tengah District. There are Four measurement points, which set in Two Semi-Detached Houses located at Taman Iskandar Perdana, Seri Iskandar Perak, Malaysia (4°N latitude and 100°E Longitude). The weather recording was carried out in one day during daytime from 10.30 to 16.30hours according to owner's permission and availability of the instrument. The fieldwork measurement was to analyze the relations between the differences amount of plants in two residential area and local air temperature.

The measurement points were located at several position in these sites, for example, one measurement point (MP) for indoors had recorded in the living room due to the constraint of access. Meanwhile, three measurement points for the outdoor climate were taken around this site. The distance of each measurement point is approximately 3 m from the building envelope and 1m height from the ground. The climatic data recorded were not influenced by shadows or reflected solar radiation.

The field study measured meteorological parameters such as dry and wet bulb temperature, relative humidity and wind velocity at 10-minute intervals time using Two sets of portable weather station devices, model DeltaLog10 version 0.1.5.29 which set indoor and outdoor. One set of a compass and measuring tapes used to identify the sun orientation and measure the house configurations. A set of layout plans is also used to sketch and record the house arrangements and landscape design on the site.

3.1 Comparison of two case study

In this paper, the study house was termed as House A and House B according to the numbers of tree either

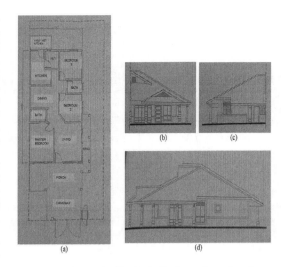

Figure 1. Figure (a), (b), (c) & (d) shows floor plan, front, rear and side elevation of study houses.

single or group species, canopy size, and their height. The selection of the case study is according to four criteria such as:

i) Ages of buildings
ii) Size of green area
iii) Standard of construction
iv) Similar location but different in their landscape density, size, and orientation.

Overall, both houses configuration was similar where the largest building facade is east-west oriented. Both of Semi-D houses had a similar size of a built-up area approximately $90.8\,m^2$ or $7543\,mm \times 12049\,mm$. A similar conventional tropical design of architecture, building materials, and construction methods had been used. The building consists of a living room, dining, kitchen, three bedrooms with two bathrooms and one store room *(refer Figure 1)*.

The primary structure was a reinforced concrete pad foundation; all had brick walls and a pitched aluminum- framed roof covered by the similar colour of concrete tiles. Ceiling heights for the two houses were around 3m height and finished with fibrous, cement plaster, and asbestos free fiber cement ceiling sheets. All of the building openings constructed using glass with metal window frames for all casement windows and the sliding doors. Climatic parameters and landscape design around two case study houses were surveyed to measured their significant effect on local air temperature. The results discussed in section 3.11 & 3.12.

3.1.1 *Analysis of climatic parameter*
From the data recorded, the air temperatures inside and outside of the densely landscaped house (House A) were lower than those in the sparsely landscaped house (House B). The average outdoor temperature for House A and B is 32.5°C and 33.2°C respectively. Meanwhile, the average indoor temperature level shows a

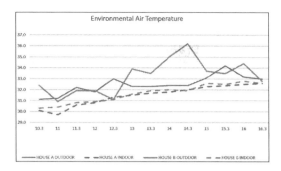

Figure 2. Indoor and outdoor air temperature results for House A and House B.

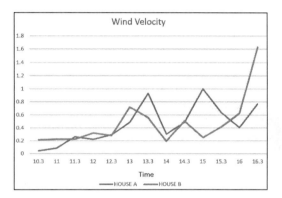

Figure 3. Wind velocity for the House A and House B.

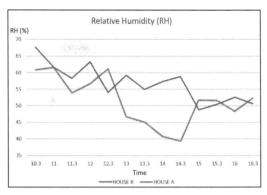

Figure 4. Relative Humidity for the House A and House B.

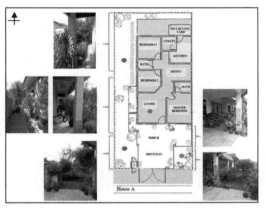

Figure 5. The plan shows the location of measurement points and landscape design for House A.

small differences rate at 0.1°C. At 10.30 hours the indoor temperature in the House, A was 30.1°C while the House B was 30.3°C as shown in *Figure 2*.

The different landscaping amount and species resulted in only minor differences in this temperature level. Both houses reached their lowest indoor temperatures at 11.00 hours; 30.4°C at the House B and 29.7°C at the House A. The outdoor temperature for House B continues fluctuated until reaching the highest point of 36.2°C at 14.30 hours. The same pattern occurred at the House A, which reached its highest outdoor temperature of approximately 34.2°C at 15.30 hours.

Overall, the outdoor temperature of the House A was as much as 3.8°C lower than the House B during the peak time of the day. In contrast, the indoor temperature at the House A and House B had a similar pattern and reading at all times. The apparent difference in temperature readings for the outdoor not much influence the ambient indoor temperatures might be due to activity in the house during the data recorded. The two houses use natural ventilation during the day.

Figure 3 shows the air velocity pattern in m per second (m/s) at 1.5 m heights in two study houses. Outdoor wind velocity around both houses gradually rose in the morning, starting at 10.30 hours; 0.04 m/s at the House A and 0.21 m/s at House B until they reached the highest point at 0.93m/s and 0.56 m/s respectively at 13.30 hours. Overall, an average data of wind velocity demonstrates that House B reached a higher rate for indoor and outdoor compared to House A. Less in cross ventilation due to enclosed space (*windows and door closed*) causes in decreasing of indoor wind velocity in both houses. The outdoor wind velocity then continues in the fluctuating pattern at throughout the day for both houses.

Analysis of relative humidity (RH) shows that during the morning, the ambient RH readings at the House A is higher around 6.8% difference at 10.30 hours until the RH pattern slowly fell to the lowers point at 39.4% at 15.00 hours. In the middle of the day, the relative humidity levels of the House B were dramatically dropped around 21.7% starting at 12.30 hours to 14.30 hours. The graph also shows a big difference of RH between two houses at 14.30 hours around 19.4% (shown in *Figure 4*). However, all two house have a slightly similar in declining pattern in relative humidity.

3.1.2 *Analysis of landscape design*

The landscape surrounding of the two houses used a combination of trees, shrubs, ground cover and turf.

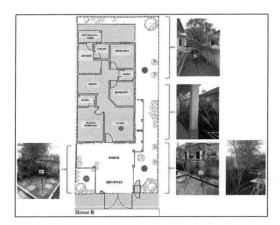

Figure 6. Location of measurement points and landscape design for House B.

Table 1. List of trees, shrubs and ground cover at Densely Landscape House (House A).

No	Species	Azimuth[1]	Distance to wall[2]	Trees, shrubs & ground cover Index / typology / biomass structure		
				Area[3]	Amount of leaves[4]	Size of leaves[5]	Height[6]
1	Calathea lutea	3*	3	2	2	3	2
2	Bambusa ventricosa	3*	3	2	3	1	2
3	Calathea lutea	3*	2	2	2	2	2
4	Yucca aloifolia	3*	3	2	1	1	1
6	Asplenium nidus	3*	1	2	2	3	2
8	Calathea lutea	3*	2	2	2	2	2
9	Yucca aloifolia	3*	3	1	2	1	2
10	Rhoeo spathacea	3*	1	2	2	1	1
11	Anthurium sp.	3*	1	2	3	2	1
12	Ophiopogon jaburan variegata	3	3	2	3	2	1
13	Heliconia sp.	3	3	1	2	2	2
14	Bambusa ventricosa	3	3	3	3	1	2
15	Hymenocallis littoralis	3	3	2	3	2	2
16	Yucca aloifolia	3	3	1	1	1	1
17	Bambusa sp	3	2	2	3	1	3
18	Yucca aloifolia	3	2	2	2	1	3
19	Heliconia sp.	3	1	2	3	2	3
20	Anthurium sp.	3	1	2	3	3	1
21	Yellow palm	3	1	1	2	2	2
22	Hymenocallis littoralis	4	1	3	3	2	1
23	Calathea altissima	4	1	2	3	2	3
24	Yucca aloifolia	4	1	2	2	2	2
25	Yucca aloifolia	4	2	2	3	2	3
26	Aracene sp	4	2	2	2	3	2
27	Calathea lutea	4	2	2	2	2	2
28	Euphoria malaiense	1	2	4	3	2	3
29	Musa spp.	1	2	4	3	3	4

Table 2. List of trees, shrubs and ground cover at Sparsely Landscape House (House B).

No	Species	Azimuth[1]	Distance to wall[2]	Trees, shrubs & ground cover Index / typology / biomass structure			
				Area[3]	Amount of leaves[4]	Size of leaves[5]	Height[6]
1	Bambusa ventricosa	3*	3	2	3	1	4
2	Ixora javanica	3*	3	2	2	1	1
3	Costus Woodsonii	3*	3	2	2	1	1
4	Plumeria rubra	3*	2	2	1	2	4
5	Bambusa ventricosa	3	3	2	3	1	4
6	Hymenocallis littoralis	3	3	1	3	2	1
7	Costus Woodsonii	3	3	3	3	1	1
8	Calathea altissima	3	3	3	3	2	2
9	Plumeria rubra	3	2	1	1	2	4
10	Ixora javanica	3	2	1	1	1	1
11	Costus Woodsonii	3	2	1	1	1	1
12	Podocarpus	2	2	1	2	1	3
13	Podocarpus	2	2	1	2	1	2
14	Ixora javanica	2	2	2	3	1	2
15	Calathea lutea	2	2	4	3	3	3
16	Plumeria rubra	2	3	4	3	2	4
17	Mangifera indica Linn.	1	3	5	3	2	4
18	Musa spp.	1	2	3	3	3	4
19	Pandanus amaryllifolious	1	2	1	1	1	1

1= North(right), 1*= North(left), 2=East, 3=South(right), 3*= South(left), 4=West
1= proximity to walls, 2=below 3m, 3= 3-6m
1= below 1m², 2=1-4m², 3=5-9m², 4=10-14m², 5=15-19m²
1= rare, 2=medium, 3=dense
1= small, 2=medium, 3=large
1= below 1m, 2=1-2m, 3=2-3m, 4= 3m an above

as for beautiful view, edible, to create shade and to yield a lovely, comfortable and healthy environment (*refer figure 6*). Both houses surrounded by almost similar species of tree such as *Calathea lutea, Crinum asiaticum, Bambusa sp.* and all landscape garden covered with turf. They also used different amounts of trees and shrubs, species, and configurations of the plant (*refer Table 1&2*). Edible plants planted in the backyard.

4 CONCLUSION

This paper attempts to answer the issue either myth or reality regarding the ability of residential landscape in reducing indoor air temperature. A quantitative finding on three climatic parameters and landscape design in two Semi-Detached houses in the tropical region were presented. Study of the temperatures, relative humidity and wind velocity of both houses of the outdoor and indoor facades revealed an interesting aspect of the passive cooling potential of landscaping.

Overall, the results show that outdoor temperature of the House A was 3.8°C lower than the House B and the percentage of Relative Humidity for House A were much higher about 19.4% during the afternoon. Meanwhile, an average data of wind velocity shows that House B reached a higher rate for indoor and outdoor compared to House A.

Strategic arrangement and design, sufficient numbers, and sizes of vegetation can provide the most efficient means of reducing outdoor air temperature around a residential building. The combination of

The study of all landscape elements was done by observation and interviewed directly to the owner of the houses. Landscape elements for each building were drawn not to scale in layout plan, and the photo of the landscape was captured to make a comparison as shown in *Figure 5&6*.

Both of the study house had a medium sized area for landscaping and situated on the left and right side of the building. A small area of landscape garden in rectangular shape approximately 2.5 m × 2.5 m located at the front side between the Semi-Detached houses. Meanwhile, the other side of the landscape area is wider approximately 3 m width × 21 m length measured from the front side to the backyard. For the House A, the largest area of landscape garden situated to the left while the for House B is to the right of the building.

The owners personally inspired the tropical landscape design around the gardens. Each house had been planted with tropical plants for many reasons such shade trees, shrubs, vines, and groundcovers was predicted found to be the most efficient landscape strategy. This preliminary study apparently demonstrates the best technique to create a favorable environment for appropriate choice of plants, quantity, species and arrangement of landscaping for a tropical residential landscape.

REFERENCES

Akbari, H., Davis, S., Dorsano, S., Huang, J., & Winnet, S. (1992). *Cooling our communities, a guidebook on tree planting and light-coloured surfacing Washington.* Lawrence Berkerly Laboratory.

Asmat Ismail, M. H. A. S. A. A. M. A. R. (2008). Using Green Roof Concept As A Passive Design Technology To Minimise The Impact Of Global Warming.. *2nd International Conference On Built Environment In Developing Countries (ICBEDC 2008).*

DOE. (1993). Tomorrow's Energy Today for Cities and Countries: Cooling Our Cities U.S. *Department of Energy*, 1–6.

Memon, R. A., Leung, D. Y. C., & Chunho, L. I. U. (2008). A Review On The Generation, Determination And Mitigation Of Urban Heat Island., 20, 120–128.

Misni, A. (2012). *The Effects of Surrounding Vegetation, Building Construction and Human Factors on the Thermal Perfomance of Housing in a Tropical Environment.*

Misni, A. (2013). Modifying the Outdoor Temperature around Single-Family Residences: The Influence of Landscaping. *Procedia - Social and Behavioral Sciences*, *105*, 664–673. http://doi.org/10.1016/j.sbspro.2013.11.069

Thani, S. K. S. O., Mohamad, N. H. N., & Idilfitri, S. (2012). Modification of Urban Temperature in Hot-Humid Climate Through Landscape Design Approach: A Review. *Procedia - Social and Behavioral Sciences*, *68*, 439–450. http://doi.org/10.1016/j.sbspro.2012.12.240

Tsutsumi, J. G., Ishii, A., & Katayama, T. (2003). QUANTITY OF PLANTS AND ITS EFFECT ON LOCAL AIR TEMPERATURE.

Yeo, O. T. S., Saito, K., & Said, I. (2014). Quantitative Study of Green Area for Climate Sensitive Terraced Housing Area Design in Malaysia. *IOP Conference Series: Earth and Environmental Science*, *18*, 012101. http://doi.org/10.1088/1755-1315/18/1/012101

Enhancement of waste activated sludge disintegration and dewaterability by H_2O_2 oxidation

G.C. Heng, K.W. Chen & M.H. Isa
Department of Civil and Environmental Engineering, Universiti Teknologi PETRONAS, Seri Iskandar, Perak, Malaysia

ABSTRACT: This study investigates the effect of hydrogen peroxide (H_2O_2) dose, pH and reaction time on the disintegration of Waste Activated Sludge (WAS) by H_2O_2 oxidation alone and H_2O_2 oxidation with acid/alkaline hydrolysis. H_2O_2 oxidation with alkaline hydrolysis enhanced the sludge disintegration and dewaterability. Under the optimum operating conditions (pH 11, 1250 g H_2O_2/ kg TS and reaction time 45 min), H_2O_2 oxidation aided with alkaline hydrolysis resulted in good removal of solids (TS 28.1%, VS 25.3%, MLSS 24.6% and MLVSS 21.4%), CST reduction 28.4% and sCOD concentration 8863 mg/L. The current method was shown to be an adequate pretreatment for further treatment of the waste activated sludge by biological digestion.

1 INTRODUCTION

Waste activated sludge (WAS) is generated during biological treatment of the wastewater and the cost associated with various type of sludge handling methods can cost up to 50% of the total wastewater treatment plants operating cost [1, 2]. It contains mainly microbial cells that are complex polymeric of 59–88% (w/v) of organic matter, which is decomposable. The excess activated sludge need to be treated before disposed of. However, the cost for sludge treatment is highly dependent on the volume and water content of the produced sludge. Conventional sludge disposal methods such as incineration, disposal in landfills and land application are facing increasing pressure from environmental authorities and from the public domain. Therefore, sludge treatments and pretreatments for subsequent biological process were studied in the literature, such as microwave irradiation, ultrasonication, ozonation, high pressure homogenizer method and chemical pretreatment with acid or alkali [3].

Hydrogen peroxide (H_2O_2) is a strong oxidant that has been used in industrial applications and in water treatment processes. When catalyzed in water, hydrogen peroxide may generate a wide variety of free radicals and other reactive species that are capable of transforming or decomposing organic chemicals. However, H_2O_2 alone is not effective for high concentrations of certain refractory contaminants because of low rates of reaction at reasonable H_2O_2 concentrations. Improvements can be achieved by using transition metal salts (e.g. iron salts) or ozone and UV light that can activate H_2O_2 to form hydroxyl radicals, which are strong oxidants. Hydrogen peroxide has a number of advantages, including the ability to use it in a wide range of temperatures and pH values, high selectivity of oxidation of various wastewater impurities, good solubility in water solutions, high stability of commodity solutions during storage, simplicity of hardware design of water purification processes and ecological compatibility [4]. Besides, H_2O_2 yields no noxious or polluting by-products which are only water and oxygen.

Alkali treatment is effective in sludge solubilization, with in order of efficacy being (NaOH > KOH > Mg(OH)$_2$) [5]. It is normally combined with thermal treatment [6, 7]. Dogan and Sanin [8] found an improvement on the dewaterability (measured by CST) by about 22% after anaerobic digestion of pretreated sludge (pH of 12, 160°C microwave) compared to anaerobic digestion of untreated waste activated sludge. Kim et al. [9] studied on alkaline pretreatment to enhance the efficiency of hydrogen peroxide oxidation of sludge by evaluations of total solids concentration and particle size distribution. The authors found that total solids removal was 49% and the median diameter of sludge particle decreased from 34.5 μm to 10.8 μm under the conditions of pH 11 and 1800 g H_2O_2/kg TS dose.

In this study, hydrogen peroxide oxidation was performed under different conditions (H_2O_2 alone, H_2O_2 with acid/alkaline hydrolysis) to study the effect of hydrogen peroxide dose, pH and reaction time for sludge disintegration.

2 MATERIALS AND METHODS

2.1 Sludge sampling

The waste activated sludge (WAS) was obtained at the return pipe from secondary clarifier sludge hopper of a municipal wastewater treatment plant in

Table 1. Sludge characteristics.

Parameter	Unit	Value (mean)
pH	–	5.81
Total solids	mg/L	29268
Volatile solids	mg/L	19871
Mixed liquor suspended solids	mg/L	24585
Mixed liquor volatile suspended solids	mg/L	18146
Soluble chemical oxygen demand	mg/L	1127.2
Capillary suction time	s	109

Universiti Teknologi PETRONAS. The sludge samples were thickened to the required solids concentration and stored in the cold room prior to use. The characteristics (range from triplicate samples) of the raw samples are presented in Table 1.

2.2 Experimental procedures

The hydrogen peroxide oxidation process was performed in a 500-mL Pyrex reactor using 250 mL of the activated sludge. Hydrogen peroxide (H_2O_2) was added according to the selected dosage of H_2O_2 per kilogram of total solids (g H_2O_2/kg TS), varied from 500 to 1750 g. The effect of pH was also studied by adjusting the pH using sulphuric acid and sodium hydroxide to pH 2–12. The mixture was stirred continuously to ensure complete homogeneity during reaction. Aliquots were taken at selected reaction time for the measurements of solids (total solids, volatile solids, mixed liquor suspended solids and mixed liquor volatile suspended solids), soluble chemical oxygen demand (sCOD) and capillary suction time (CST).

2.3 Analytical measurements

pH measurement was performed using a pH meter (HACH sension 4) and a pH probe (HACH platinum series pH electrode model 51910, HACH company, USA). Solids analysis was made according to Standard Methods [10]. sCOD was measured by the Reactor Digestion HACH Method No. 8000 [11]. CST was measured using Triton type 319 Multi-CST (Triton Electronics Ltd.).

3 RESULTS AND DISCUSSION

3.1 Effect of hydrogen peroxide dosage

Figure 1 shows different hydrogen peroxide dosages (500–1750 g H_2O_2/kg TS) were carried out at unadjusted pH (5.81) of sludge sample and reaction time 30 min. The solids removal (TS, VS, MLSS and MLVSS) increased with the increasing of hydrogen peroxide throughout the dosages range. The maximum

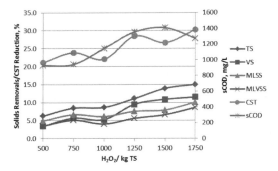

Figure 1. Effects of hydrogen peroxide oxidation (g H_2O_2/kg TS) on solids removal, CST reduction and sCOD concentration.

solids removal (TS, VS, MLSS, MLVSS), CST reduction and concentration of sCOD were 15%, 11.5%, 10.1%, 8.6%, 30.3% and 1408 mg/L, respectively. The results show that the solids removal efficiencies and sCOD concentration were relatively low, indicating that sludge solubilization was not effective by H_2O_2 oxidation alone.

3.2 Effect of pH

Figure 2 shows the effect of hydrogen peroxide oxidation on the removal of solids, reduction of CST and concentration of sCOD by 750 g H_2O_2/kg TS and reaction time 30 min at different pH, ranging from 2 to 12. Relatively higher removal of solids and CST were observed at acidic and alkaline pH than at unadjusted pH of the sludge sample. The solids removal and CST reduction were TS 8.5%, VS 5.7%, MLSS 6.6%, MLVSS 5.1% and CST reduction 23.9% by H_2O_2 oxidation at 750 g H_2O_2/kg TS (Fig. 1). When the alkaline hydrolysis (pH 11) was applied prior to H_2O_2 oxidation, the solids removal and CST reduction were enhanced to TS 20%, VS 14.6%, MLSS 15.1%, MLVSS 13.1% and CST reduction 30.3%. Sludge hydrolysis can be expressed by the changes of sCOD concentrations [12]. It was increased from 941 mg/L to 4372 mg/L, which indicated that more and more particulate organics in WAS became soluble substrates. The sCOD concentration was significantly higher than those at near unadjusted pH or acidic pH. When acid or alkali were added to the sludge samples, the cell loses its viability and can not maintain an appropriate turgor pressure and disrupts. Saponification of the lipids in the cell walls occurs and lead to solubilization of the cell membrane. Disrupton of the sludge cells leads to a leakage of intracellular material out of the cell [6]. According to Battisa [13], hydrolysis occurs slowly and crystallization generates pieces that are more resistant to acid hydrolysis. There are three phases attributed to endogenous attack of amorphous cellulose by acid followed by exogenous acid attack of the ends of crystalline cellulose, and finally simultaneous endogenous/exogenous hydrolysis of the remaining cellulose to break down the cells

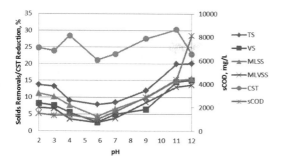

Figure 2. Effects of pH at 750 g H2O2/kg TS and reaction time 30 min on solids removal, CST reduction and sCOD concentration.

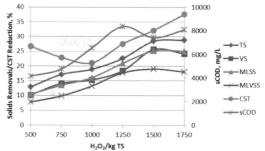

Figure 3. Effects of hydrogen peroxide oxidation (g H2O2/kg TS) at pH 11 and reaction time 30 min on solids removal, CST reduction and sCOD concentration.

by acid [14]. WAS hydrolysis rate was accelerated under alkaline conditions, which has also been observed by other researchers [15, 16]. Katsiris et al. [17] stated that the bacterial surfaces were negatively charged as the pH of sludge samples increased. This creates high electrostatic repulsion which causes desorption of some part of extracellular polymeric substances (EPS). Hence, it explains alkaline range has better effect on sludge solubilization compared to acidic range and unadjusted pH of sludge. Kim et al. [9] proved that the pretreatment of alkaline hydrolysis applied prior to H_2O_2 oxidation could enhanced the total solids removal from 33% to 49% at pH 11 and ~1800 g H_2O_2/kg TS dose. However, at extremely high pH, the maximum sCOD concentration was increased to 8121 mg/L at pH 12 but reduced the CST reduction. One possible reason was due to the soluble EPS had strong water binding capability and hence deteriorate the sludge dewaterabilty [18]. Based on the results, alkaline hydrolysis was proved performed better than acidic hydrolysis and pH 11 was chosen to obtain the optimum dosage of H_2O_2.

3.3 Effect of hydrogen peroxide oxidation and alkaline hydrolysis

The same hydrogen peroxide dosages range (500–1750 g H_2O_2/kg TS) at pH 11 and reaction time 30 min were applied to investigate the effect of hydrogen peroxide oxidation with alkaline hydrolysis (Fig. 3). The solids removal increased significantly with the increasing of hydrogen peroxide from 500 to 1250 g H_2O_2/kg TS and has adverse effect beyond 1500 g H_2O_2/kg TS. The maximum solids removal were achieved at 1500 g H_2O_2/kg TS – TS 28.4%, VS 25.7%, MLSS 25.1, MLVSS 19%. CST reduction was increased with the increasing of hydrogen peroxide throughout the range while sCOD increased from 4128 mg/L to 8355 mg/L at 500 and 1250 g H_2O_2/kg TS, respectively and reduced to 7408 mg/L and 8071 mg/L at 1500 and 1750 mg/L, respectively. Increased H_2O_2 dose produced more hydroxyl radicals leading to higher substrate degradation [19]. Further increase of H_2O_2 dose did not

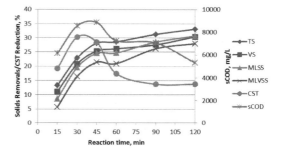

Figure 4. Effects of reaction time at pH 11 and 1250 g H_2O_2/kg TS on solids removal, CST reduction and sCOD concentration.

improve the removal efficiency. This was due to scavenging of •OH radicals (Reaction 1) [20]. This leads to the production of hydroperoxyl radical, a species with much weaker oxidizing power compared to hydroxyl radical [21]. Ksibi et al. [4] reported that hydrogen peroxide can effectively improve the biodegradability of sludge by degrading the intermediates such as short chain carbon carboxylic acids produced from a ramified aliphatic chain. Hydrogen peroxide dose of 1250 g H_2O_2/kg TS was chosen to obtain the optimum reaction time, as good solids removal, CST reduction and the highest concentration of sCOD were achieved.

$$\cdot OH + H_2O_2 \longrightarrow HO\cdot_2 + H_2O \qquad (1)$$

3.4 Effect of reaction time

Figure 4 shows the effect of reaction time on hydrogen peroxide oxidation at pH 11 and 1250 g H_2O_2/kg TS. Solids removal were increased with the increasing of reaction time 15–120 min from 13.4, 11.2, 8.6 and 5.7% to 33.1, 30.4, 30.7 and 28% on TS, VS, MLSS and MLVSS, respectively. The solids removal increased significantly at the first 45 min. CST reduction was improved in the first 30 min to 30.3% and then decreased from 45 min to 120 min. The maximum sCOD concentration, 8863 mg/L was achieved after 45 min and decreased to 5313 mg/L at 120 min of oxidation. Mineralization was occurred at the beginning of oxidation (15–45 min) and then solubilization took place and hence sCOD start decreasing. High sCOD can possibly be applied to anaerobic digestion to produce biogas. Thus, 45 min of reaction time was chosen as the most effective operating condition.

4 CONCLUSIONS

When 750 g H_2O_2/kg TS was applied to the activated sludge (H_2O_2 oxidation), the solids removal were TS 8.5%, VS 5.7, MLSS 6.6%, MLVSS 5.1% and CST reduction 23.9%. Whereas when alkaline hydrolysis (pH 11) was applied prior to H_2O_2 oxidation by adding 750 g H_2O_2/kg TS, the solids removal were enhanced to TS 20%, VS 14.6%, MLSS 15.1, MLVSS 13.1% and CST reduction to 30.3%. Furthermore, sCOD was increased from 941 mg/L to 4372 mg/L. Therefore, H_2O_2 oxidation with alkaline hydrolysis was proved to enhance the sludge solubilization compared to oxidation by H_2O_2 alone. Acid hydrolysis was not as effective as alkaline hydrolysis due to the difficulty of three phases cells breaking down and crystallization generates pieces that are more resistant to acid hydrolysis. However, at extremely high pH of alkaline (pH 12), the dewaterabilty of sludge can be deteriorated. Under the optimum operating conditions (pH 11, 1250 g H_2O_2/kg TS and reaction time 45 min), H_2O_2 oxidation with alkaline hydrolysis resulted in good removal of solids (TS 28.1%, VS 25.3%, MLSS 24.6% and MLVSS 21.4%), CST reduction 28.4% and sCOD concentration 8863 mg/L. H_2O_2 oxidation with alkaline hydrolysis was proved to be able to enhance the sludge disintegration and dewaterability and it appears to be an adequate pretreatment for further treatment of the waste activated sludge by biological digestion.

REFERENCES

[1] M. Tokumura, M. Sekine, M. Yoshinari, H. T. Znad, and Y. Kawase, "Photo-Fenton process for excess sludge disintegration," *Process Biochemistry*, vol. 42, pp. 627–633, 2007.

[2] S. Pilli, T. T. More, S. Yan, R. D. Tyagi, and R. Y. Surampalli, "Fenton pre-treatment of secondary sludge to enhance anaerobic digestion: Energy balance and greenhouse gas emissions," *Chemical Engineering Journal*, vol. 283, pp. 285–292, 2015.

[3] V. K. Tyagi and S.-L. Lo, "Application of physicochemical pretreatment methods to enhance the sludge disintegration and subsequent anaerobic digestion: an up to date review," *Reviews in Environmental Science and Biotechnology*, vol. 10, pp. 215–242, 2011.

[4] M. Ksibi, "Chemical oxidation with hydrogen peroxide for domestic wastewater treatment" *Chemical Engineering Journal*, vol. 119, pp. 161–165, 2006.

[5] J. Kim, C. Park, T. H. Kim, M. Lee, S. Kim, S. W. Kim, et al., "Effects of various pretreatments for enhanced anaerobic digestion with waste activated sludge," *Journal of Bioscience and Bioengineering*, vol. 95, pp. 271–275, 2003.

[6] E. Neyens, J. Baeyens, and C. Creemers, "Alkaline thermal sludge hydrolysis," *Journal of Hazardous Materials*, vol. B97, pp. 295–314, 2003.

[7] S. Jomaa, A. Shanableh, W. Khalil, and B. Trebilco, "Hydrothermal decomposition and oxidation of the organic component of municipal and industrial waste products," *Advances in Environmental Research*, vol. 7, pp. 647–653, 2003.

[8] I. Dogan and F. D. Sanin, "Alkaline solubilization and microwave irradiation as a combined sludge disintegration and minimization method," *Water Research*, vol. 43, pp. 2139–2148, 2009.

[9] T.-H. Kim, S.-R. Lee, Y.-K. Nam, J. Yang, C. Park, and M. Lee, "Disintegration of excess activated sludge by hydrogen peroxide oxidation," *Desalination*, vol. 246, pp. 275–284, 2009.

[10] APHA, *Standard Methods for the Examination of Water and Wastewater*. Washington D. C., USA: 21st ed. American Public Health Association, 2005.

[11] HACH, *Water analysis handbook* Loveland, CO, USA.: 4th ed. Hach Company, 2003.

[12] K. Andreasen, Petersen, G., Thomsen, H., Strube, R., "Reduction of nutrient emission by sludge hydrolysis," *Water Sci. Technol.*, vol. 35, pp. 79–85, 1997.

[13] O. A. Battista, "Hydrolysis and crystallization of cellulose," *Industrial & Engineering Chemistry*, vol. 42, pp. 502–507, 1950.

[14] J. Bouchard, Abatzoglou, N., Chornet, E., and Overend, R.P., "Characterization of depolymerized cellulosic residues: Part 1: Residues obtained by acid hydrolysis processes," *Wood Science and Technology*, vol. 23, pp. 343–355, 1989.

[15] A. G. Vlyssides, Karlis, P.K., "Thermal-alkaline solubilization of waste activated sludge as a pre-treatment stage for anaerobic digestion," *Bioresource Technol.*, vol. 91, pp. 201–206, 2004.

[16] Y. Chen, S. Jiang, H. Yuan, Q. Zhou, and G. Gu, "Hydrolysis and acidification of waste activated sludge at different pHs," *Water Research*, vol. 41, pp. 683–689, 2007.

[17] N. Katsiris and A. Kouzeli-Katsiri, "Bound Water Content of Biological Sludges in Relation to Filtration and Dewatering," *Water Research*, vol. 21, pp. 1319–1327, 1987.

[18] H. Yuan, N. Zhu, and F. Song, "Dewaterability characteristics of sludge conditioned with surfactants pretreatment by electrolysis," *Bioresource Technology* vol. 102, pp. 2308–2315, 2011.

[19] Y. Deng and J. D. Englehardt, "Treatment of landfill leachate by the Fenton process." *Water Research*, vol. 40, pp. 3683–3694, 2006.

[20] R. Andreozzi, M. Canterino, R. Marotta, and N. Paxeus, "Antibiotic removal from wastewaters: The ozonation of amoxicillin," *Journal of Hazardous Materials*, vol. 122, pp. 243–250, 2005.

[21] W.-P. Ting, M.-C. Lu, and Y.-H. Huang, "Kinetics of 2,6-dimethylaniline degradation by electro-Fenton process," *Journal of Hazardous Materials*, vol. 161, pp. 1484–1490, 2008.

Engineering Challenges for Sustainable Future – Zawawi (Ed.)
© 2016 Taylor & Francis Group, London, ISBN 978-1-138-02978-1

Application of integrated bioreactor system (i-SGBR) for simultaneous treatment of wastewater and excess sludge degradation

S.R.M. Kutty, N. Aminu, M.H. Isa & I.U. Salihi
Department of Civil and Environmental Engineering, Universiti Teknologi PETRONAS, Seri Iskandar, Perak, Malaysia

ABSTRACT: An integrated suspended growth bioreactor system (*i-SGBR*) was designed to treat wastewater. It has anoxic chamber *(ANX-C)* (40 L) for denitrification, extended aeration chamber *(EA-C)* (125 L) for combined carbon removal and nitrification, aerobic digester chamber *(AD-C)* (75 L) sub-system to degrade sludge, and clarifier chamber *(CLR)* (100 L) to settle the biomass. Evaluation of the treatment performance for organics (chemical oxygen demand, COD), total suspended solids (TSS), and total nitrogen (TN) between the system *COD* volumetric loading operated between 0.6–0.9 kgCOD/m^3. d, resulted in an average effluent *COD*, total nitrogen *(TN)*, and *TSS* of 41 mg/L, 14 mg/L and 21 mg/L, respectively. The corresponding average influent concentrations for *COD, TN* and *TSS* were 1,019 mg/L, 84 mg/L and 504 mg/L, respectively. The *AD* could achieve 12% daily mixed liquor volatile solids *(MLVSS)* reduction. The average daily sludge wasting (Q_w) based on *MLSS* was 1.07 L/d.

1 INTRODUCTION

1.1 Background

Rapid urbanization and industrialization have amplified environmental problems such as water pollution and competition for land (Keng et al., 2013). According to the Environmental Quality Report (EQR, 2013), there were a total of 1,475,444 water pollution sources identified in Malaysia. These sources include pollution from 4,595 manufacturing industries, 10,336 sewage treatment plants, 1,262,185 individual septic tanks *(IST)*, 3,629 communal septic tanks *(CST)*, 602 animal farms, 508 agro-based industries, 879 wet markets and 192,710 food services establishments ((DOE), 2013). These *IST* and *CST*, otherwise classified as decentralized systems *(DS)*, have long been implicated in being the major source of nitrogen inputs to surface and ground waters (Oakley et al., 2010).

Conversely, the *DS* were reported to be ineffective in appropriately decreasing the nutrient strength of wastewater (Gill et al., 2009). The *DS* reduce nitrogen loading only by 10–20%, which is grossly inadequate (Keeney, 1986, Siegrist and Jenssen, 1989, Lamb et al., 1990).

Due to the high cost of centralized systems *(CS)*, the *DS* systems are no longer perceived to be a short-term measure, in the subsequent planning to replace them with the *CS* collection and treatment systems (Ujang and Buckley, 2002, Massoud et al., 2009).

The discharge of these partially treated wastewater into the surface water (rivers, lakes, coastal areas), and groundwater may generate various problems, ranging from elevated five days biochemical oxygen demand *(BOD$_5$)*, significant nutrients impacts into the aquatic environment, high *TSS*, ecosystem disruption, and health exposures due to potential pathogens (Oakley et al., 2010, Ling et al., 2010).

The production of excess sludge from activated sludge process *(ASP)* is an additional problem that presents one of the serious challenges encountered in the aerobic wastewater treatment plants *(WWTP)*. The handling of excess sludge could account for 25–65% of the plant's total related operational cost (Liu, 2003). These large quantities of sludge produced, need to be treated and reused, or disposed. However the safety of public health and the environment must be guaranteed (Fytili and Zabaniotou, 2008).

The realistic and cost-effective technologies are desirable, particularly considering the older plants, with large footprint (Lundquist et al., 2010). In Malaysia, effluent discharge, irrespective of sewage or industrial, must conform to standards A or B, where standard A (120 mg COD/L) is applicable to water catchment areas, while standard B (200 mg COD/L) is applicable to the non-water catchment areas (DOE, 2009). Therefore, to practically attain these strict limits, application of BNR technology will be suitable (Oakley et al., 2010), in order to meet the set limits by the DoE Malaysia (DOE, 2009).

2 MATERIALS AND METHOD

2.1 Working process of the i-SGBR and sampling process

The bioreactor used for this pilot study is illustrated in Figure 1. It was designed and operated as a continuous flow, complete mix and suspended growth system (Tchobanoglous et al., 2003). The

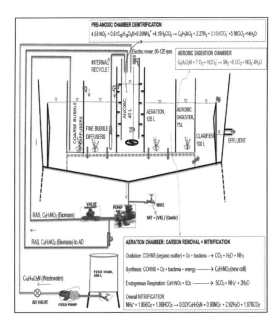

Figure 1. The *i – SGBR* structure and biochemistry.

wastewater in the reservoir tank (200 L) was fed into the *ANX-C*, receiving the wastewater as raw carbon (electron donor) source for denitrification (Metcalf & Eddy, 2003). The nitrate produced in the *EA-C* and *AD-C* are transferred through the internal recycle (*IR*) pump, into the *ANX-C*, with nitrate as terminal electron acceptor (Baeza et al., 2004). The *AD-C* is an separate, intermittently fed from the return activated sludge *(RAS)*, with effluent released after required *HRT* to the *EA-C*. The recycle of thickened *RAS* from the clarifier *(CLAR)* was done through periodic flow of biomass, to maintain its concentration in the bioreactor (Ilies and Mavinic, 2001). The three distinctive processes occur in the *EA-C*, namely; oxidation, synthesis and endogenous respiration (Metcalf & Eddy, 2003). The nitrification process results from the ammonia nitrogen produced, which is eventually oxidized to nitrate, and nitrate reduced to nitrogen gas in the *ANX-C* (Katsou et al., 2014).

Due to the biomass yield, the *MLSS* control was achieved through daily wasting and *AD*. The sludge wasting is necessary to get rid of non-biodegradable portion of inert matter within the system (Yang and Chen, 1977, Al-Malack, 2006). During this study, as illustrated in Figure 1, samples were obtained from the influent (A), the exit of *ANX-C* (C) (corresponding to influent *EA-C*), exit of the *EA-C* (B) (corresponding to *IR*), *EAD* (D), *IAD* (E) (corresponding to *RAS*), and *CLR* effluent (F), respectively.

2.2 Aerobic digestion in the i-SGBR

AD can be considered as an extension of *ASP* under endogenous metabolism conditions (Grady Jr et al., 2012). Conversely, there was no raw sewage fed to the *AD-C*. The degradation process continues where the biodegradable organic matter is hydrolyzed and transformed to soluble organic matter, with discharge of nutrients. This is followed by the conversion of biodegradable soluble organic matter to carbon dioxide, water, and active biomass by the aid of heterotrophic bacteria. Subsequently, the active biomass goes through process of decay, resulting in generation of carbon dioxide and water, with additional inactive biomass as cell debris (Foladori et al., 2010). Though, the non-biodegradable portion of the particulate matter in the influent will remain as is. Hence, it would be directly part of the digested solids. *AD* results in the destruction of both *MLVSS* and *MLSS* (Benefield and Randall, 1978, Zupančič and Roš, 2008). The sludge degradation was evaluated on this basis (Zupančič and Roš, 2008), based on Equation 1 below:

$$D_{MLVSS} = (IAD_{MLVSS} - EAD_{MLVSS})/(IAD_{MLVSS}) \quad (1)$$

Where, D_{MLVSS} is the degraded *MLVSS* or *MLSS*, IAD_{MLVSS} and EAD_{MLVSS} are the influent and effluent *MLVSS* or *MLSS*, respectively.

The concept of mass balance was adopted (Koers, 1979) to evaluate the *MLVSS* degradation in the *AD-C*. The mass of daily wasting rate was determined (Al-Malack, 2006), and compared with digested sludge. The daily sludge wasting for each day was made based on the difference from *AD-C* efficiency.

The Q_w was the variable used to maintain the *SRT* during the operation at each organic loading, according to the Equation 2 below:

$$\theta c = ((VX)/(X_r Q_w)) \quad (2)$$

Where, θc (d) is the *SRT*, X (mg/L) is the *EA-C MLSS*, V (L) is the volume of the *EA-C*, X_r (mg/L) is the *MLSS* concentration in the *CLR* underflow, and Q_w is the daily sludge wasting (L/d).

2.3 Analytical methods

The analysis was done in accordance with the procedure in standard methods (APHA, 1998), with succeeding individual procedures described. The BOD_5 was measured once a week to validate the *COD* measurements. The sampling was conducted three times per week, with each sample collected and analyzed in triplicates.

The *MLSS*, *MLVSS* and *TSS* were carried out by gravimetric method, using digital analytical balance (MAX 200G), oven (103–105°C, muffle furnace (ashing at 550°C, vacuum suction pump, whatman filter paper 934 AH, 47 mm, cat 1827-047, whatman filter paper no. 1, 47 mm, cat 1001047. The colorimetric measurements were performed, using spectrophotometer DR 3800, 10/20 mL cuvettes, digestion vessel DRB 200, were used to analyze total phosphorus *(TP)*, by acid persulphate digestion, method 8190, the total chemical oxygen demand *(TCOD)* was carried out using reactor digestion method 8000, the ammonia nitrogen *(AMN)* was analyzed using nessler

Table 1. Experimental design and operation.

Activity period, (days)	164-278 (phase IV)				
Objective	Operate EA HRT 20-30hrs, SRT 20-40 days				
	Operate 24 HRT AD				
	Monitor removal of organics, nitrogen and TSS on increased organic loading.				
	Monitor O & M parameters				
Flow rate, I_Q (L/d)	Stage 1	Stage 2	Stage 3	Stage 4	Stage 5
	100	110	120	135	150
Hydraulic retention time Θ, HRT (days)					
Anoxic, Θ_{ANX-C}	0.40	0.36	0.33	0.30	0.27
Aeration, Θ_{EA-C}	1.25	1.14	1.04	0.96	0.83
Aerobic digester, Θ_{AD-C}	1.000				
Clarifier, Θ_{CLAR}	1.25	1.14	1.04	0.96	0.83
Solids retention time Θc, SRT (days)	40	35	30	25	20
MLSS in **AE-C**, MLSS, mg/L (after biomass build up) = **5,000**					
Internal recycle flow, $I_Q *R$, (L/d)	1200	1320	1440	1620	1800
Internal recycle ratio, IR = R					
Return activated sludge flow, RAS = 2* I_Q, (L/d)	200	220	240	270	300
RAS flow EAD after 24 hrs (once)	75	75	75	75	75
RAS vol. (L) to be pumped from under flow (2Q – AD_Q)	125	145	165	195	225
Automation of pump for biomass RAS to AT					
Biomass from AD	EAD	EAD	EAD	EAD	EAD	EAD
8am	11pm	2pm	5pm	8pm	11pm	2am	5am	8am
Dissolved oxygen, DO (mg/L)					
Anoxic, DO_{ANX-C}	< 0.2 mg/L (60-120 rpm mixer)				
Aeration, DO_{EA-C}	4 – 6 mg/L (Fine bubble diffusers)				
Aerobic digester, DO_{AD-C}	15 – 18 mg/L (Coarse bubble diffusers)				
pH in aeration and AD-C	7.0 – 7.8 (Controlled)				
Temperature in AT, °C	25 – 33 (Mesophilic range)				
Alkalinity as NaHCO$_3$, g/d	15.9	17.5	19.1	21.4	23.7

Notes: Pumps are automated. RAS is made 8 times in every 24 hrs (2:59 hrs pump idle, 1 minute active). RAS pump capacity ranges between 10 – 50 Lpm, with flow meter and adjustable control valve. Volume from AD occupies 2 time slots (8am – 2pm), while volume from RAS occupy 6 time slots (2pm – 8am). The IR - made 12 times every 24hrs, with pump capacity ranging between 40 – 200 Lpm

Table 2. Influent wastewater characteristics.

Parameter	Max.	Min.	Mean±SD
Undiluted COD (mg/L)	1593	994	1275 ± 187
Diluted COD (mg/L)	1096	952	1019 ± 28.8
BOD$_5$ (mg/L)	661	483	591 ± 53
TSS (mg/L)	562	461	504 ± 20
TP (mg/L)	60.8	38.0	51.1 ± 5.68
AMN (mg/L)	57.0	48.2	52.6 ± 2.2
Nitrate-nitrogen (mg/L)	0.6	0.1	0.4 ± 0.2
TN (mg/L)	92.3	78.7	84.0 ± 3.5
TKN (mg/L)	91.1	79.1	84.7 ± 3.25
pH	4.2	6.7	

method 8038, the nitrate was analyzed using cadmium reduction method 8039, the total kjeldahl nitrogen *(TKN)* was done using nitrogen simplified *TKN (s-TKN)* method 10242, the biochemical oxygen demand *(BOD$_5$)* was carried out using membrane electrode method, with 5 days incubation at 20°C, 300 mL *BOD$_5$* bottles, nutrient buffers, *DO* (dissolved oxygen) meter YSI 5100, and distilled/aerated water. Other parameters such as *DO*, *pH*, and temperature, were measured in situ, using HACH pH sensor +, and probe, YSI 550 A.

2.4 Experimental design and operational conditions

The experimental design for the procedure was according to Table 1. The removal of organics, nitrogen and *TSS*, and the operation were discussed in literature (Raj and Anjaneyulu, 2005, RAJ D et al., 2004, Liu et al., 2013, Yang and Chen, 1977, Zupančič and Roš, 2008, Ding et al., 2013).

2.5 Features of the wastewater composition and seed sludge

The characteristics of the wastewater used for the experiment are presented in Table 2. The industry produces daily wastewater, ranging between 80–150 m^3/d. This stage of the experiment follows the phase IV of the program, where the bio-kinetics study was carried-out (not reported here). The reactor biomass build up and stability were achieved during the previous phases. Earlier to this stage, volume of 250 L activated sludge was used for the seeding of the bioreactor; 150 L from the plant's clarifier *RAS* storage, and 100 L from the *EA* tank (100 L). The sludge was allowed to settle down for 30 minutes, before emptying into the *ANX-C*, *EA-C* and *AD-C* of the bioreactor, as explained (Mehrabadi and Zinatizadeh, 2014). The *MLSS* and sludge volume index *(SVI)* for the seeding were determined accordingly (Mahiroglu et al., 2009), both in the EA tank and the *RAS*, as 3,188 mg/L and 250 mL/g, and 15,800 mg/L and 850 mL/g, respectively. The *pH* of the sludge samples tested were in the range of 6.2–6.8. The influent wastewater was deficient in nitrogen. Thus, stabilized in order to accomplish requirements for the carbon, nitrogen and phosphorus *(C: N: P)*, and trace elements (Liu et al., 2013, Baeza et al., 2004, Davies, 2005, Van Loosdrecht et al., 1997). Irrespective of the influent *AMN* concentration, 30 mg/L as *AMN* (NH$_4$Cl) was added.

The wastewater was diluted to 1000 mg/L during this phase, to ensure steady and consistent organic loading for the kinetic analysis. The initial *COD* of

Table 3. Effluent wastewater characteristics.

Parameter	Max.	Min.	Mean ± SD	Removal eff. (%)
COD (mg/L)	89	19	41.0 ± 12.5	92.0–98.0
BOD_5 (mg/L)	43.8	12.9	20.7 ± 7.5	93.3–97.3
TSS (mg/L)	42	13	21.0 ± 5.96	92.5–96.9
TP (mg/L)	57.5	31.8	45.2 ± 6.08	5.4–16.3
AMN (mg/L)	5.5	1.9	3.6 ± 0.9	90.4–96.1
Nitrate (mg/L)	16.1	8.0	10.3 ± 1.6	Denitrification% = 99
TN (mg/L)	21.7	9.0	14.1 ± 2.8	76.4–88.6
TKN (mg/L)	5.7	2.3	3.9 ± 1.03	93.7–97.1
pH	7.2	7.8		

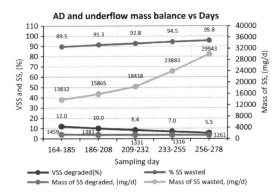

Figure 2. The MLSS digested mass and wasted mass profile.

sample was determined during each time according to Equation 3 below:

$$C_1 V_1 = C_2 V_2 \qquad (3)$$

Where, C_1 (mg/L) is the concentration of the raw wastewater before dilution, V_1 (L) is the quantity required for dilution (unknown), C_2 is the target concentration of 1,000 mg/L, and V_2 is the working volume, which is from 100–150 L operated for the five stages in the phase IV. Wastewater reserve above 100 L was always reserved for 2 days at 4°C before being disposed.

3 RESULTS AND DISCUSSION

3.1 Removal of COD, TSS, nitrogen and phosphorus

The average C : N : P ratio of 100:8:3 was operated with system TCOD volumetric loading between 0.6–0.9 kg $COD/m^3 . d$. The steady state was assumed to be reached during each phase of the organic loading, when the standard deviations of the mean efficiencies for the COD removal, nitrification and denitrification were less than 10% in variability (Kim and Novak, 2011). Table 3 presents the summary of the effluent wastewater parameters, during the period. The average effluent COD, total nitrogen (TN), and TSS of 41 ± 12.5 mg/L, 14.1 ± 2.8 mg/L and 21 ± 5.9 mg/L were obtained, respectively. The corresponding average influent concentrations were 1019 ± 28.8 mg/L, 84 ± 3.5 mg/L and 504 ± 20 mg/L, respectively. The raw wastewater concentration averages 1275 ± 187 mg/L (with equivalent average volume of 99.14 ± 17.6 L), and 25.4 ± 16.4 L of dilution tap water for the volumes between 100–150 L. The COD/BOD_5 ratio obtained at average of 1.7 was usually low, but could likely suggest high biodegradability of the wastewater, and purely organic in nature. Low ratio is possible (El-Kamah et al., 2010).

The DoE Malaysia limits for COD, (120 and 200 mg/L as standards A and B) were not violated. The bioreactor achieved maximum TSS removal of 95%, with average effluent value of 21 ± 5.96 mg/L, with an average influent concentration of 562 mg/L (50 and 100 mg/L allowed as standard A and B, respectively). Yet, there was no violation as evident from the minimum and maximum values.

Total nitrogen (TN) comprises both organic and inorganic (ammonium, nitrate and nitrite) forms (Downing and Nerenberg, 2008). The average influent AMN was 52.62 ± 2.2 mg/L, with its effluent having an average value of 3.62 ± 0.9 mg/L, and minimum of 1.87 mg/L. The effluent AMN did not violate the regulatory limits (as 10 and 20 mg/L for standards A and B, respectively). The entire nitrate concentration (ammonia oxidized from AD, EA-C, and influent) was 99% denitrified, with effluent average of 10.3 ± 1.6 mg/L, minimum of 8.0 mg/L, and maximum of 16.1 mg/L (20 and 50 mg/L effluent are allowed for the discharge for standard A and B into rivers, respectively). The denitrification (%) was evaluated according to (Fu et al., 2009), and 99% was achieved. The i-SGBR has not provided significant treatment for TP, due to the TP removal requires anaerobic condition (Baeza et al., 2004, Fu et al., 2009). The effluent pH ranges between 7.2 to 7.8, which was favorable for disposal.

3.2 The underflow and AD mass balance

Referring to Figure 2, it can be observed that the sludge degradation efficiency declines, with the increase in the organic loading, which resulted in the biomass build up, and solids accumulation in the sludge blanket. Thus, increased the solids concentration fed into the AD-C. The SRT for the five successive stages from 40 to 20 days could also influence the rate of biomass production in the bioreactor. The highest (12.0%) and lowest (5.5%) MLVSS degradation were achieved, when the IAD_{MLVSS} were at an average of 10,577 mg/L in stage 1, and 15,100 mg/L at stage 5, respectively. The DO supplied into the AD was between 15–18 mg/L. However, the actual evaluation of the mass balance was based on MLSS. 89.5% (13,832 mg/d) and 1,439 mg/d (1.0 L/d) were the mass wasted and degraded in stage 1. In contrast, during stage 5, 95.8% (29,943 mg/d) and 1,261 mg/d (1.5 L/d) were the wasted and digested mass, respectively.

When the ASP system operates at sufficiently extended aeration period, with long SRT, the excess sludge production is significantly reduced due to endogenous metabolism (Foladori et al., 2010). *AD* has have been studied and outcomes comparable with the present work (Zupančič and Roš, 2008, Grady Jr et al., 2012, Tyagi and Lo, 2012, Bernard and Gray, 2000, Eikum et al., 1974).

4 CONCLUSIONS

The law prohibits industries from discharge into centralized sewers, thus their individual mandates to ensure provision of suitable *DC* treatment options. The removal of *COD*, *TN* and *TSS* in the *i-SGBR* has achieved satisfactory performance for disharge into rivers. The biological *AD* process in *i-SGBR* was slower, compared to mechanical or other chemical processes. However, the *AD* served to further enhance aerobic treatment of the wastewater, and sludge stabilization within a small footprint. The *i-SGBR* operated at high organic loading between 0.6–0.9 kg COD/m^3.d, which likely facilitated the augmented sludge production.

ACKNOWLEDGEMENT

The authors would like to thank the funding body, prototype research grant scheme (*PRGS, 0153AB-I40*) for the financial support, and the esteemed University Teknologi PETRONAS for providing an enabling environment to conduct this research.

REFERENCES

(DOE), D. O. E. 2013. Environment Quality Report, 2013. Department of Environment, Ministry of Natural Resources and Environment, Malaysia, Kuala Lumpur.

Al-Malack, M. H. 2006. Determination of biokinetic coefficients of an immersed membrane bioreactor. *Journal of Membrane Science*, 271, 47–58.

Apha, A., Wef, 1998. Standard Methods for the Examination of Water and Wastewater (20th edition), American Public Health Association/American Water Works Association/Water Environment Federation, Washington DC.

Baeza, J., Gabriel, D. & Lafuente, J. 2004. Effect of internal recycle on the nitrogen removal efficiency of an anaerobic/anoxic/oxic (A 2/O) wastewater treatment plant (WWTP). *Process Biochemistry*, 39, 1615–1624.

Benefield, L. D. & Randall, C. W. 1978. Design relationships for aerobic digestion. *Journal (Water Pollution Control Federation)*, 518–523.

Bernard, S. & Gray, N. 2000. Aerobic digestion of pharmaceutical and domestic wastewater sludges at ambient temperature. *Water Research*, 34, 725–734.

Davies, P. S. 2005. The biological basis of wastewater treatment.

Ding, A., Qu, F., Liang, H., Ma, J., Han, Z., Yu, H., Guo, S. & Li, G. 2013. A novel integrated vertical membrane bioreactor (IVMBR) for removal of nitrogen from synthetic wastewater/domestic sewage. *Chemical engineering journal*, 223, 908–914.

DOE 2009. Environmental Quality (Sewage) Regulations 2009, Environmental Quality Act 1974, Environmental Quality (Control of Pollution from Solid Waste Transfer Station and Landfill) Regulations 2009. *In:* ENVIRONMENT, D. O. (ed.). Malaysia.

Downing, L. S. & Nerenberg, R. 2008. Total nitrogen removal in a hybrid, membrane-aerated activated sludge process. *water research*, 42, 3697–3708.

Eikum, A. S., Carlson, D. & Paulsrud, B. 1974. Aerobic stabilization of primary and mixed primary-chemical (alum) sludge. *Water Research*, 8, 927–935.

El-Kamah, H., Tawfik, A., Mahmoud, M. & Abdel-Halim, H. 2010. Treatment of high strength wastewater from fruit juice industry using integrated anaerobic/aerobic system. *Desalination*, 253, 158–163.

Foladori, P., Andreottola, G. & Ziglio, G. 2010. *Sludge reduction technologies in wastewater treatment plants*, IWA publishing.

Fu, Z., Yang, F., Zhou, F. & Xue, Y. 2009. Control of COD/N ratio for nutrient removal in a modified membrane bioreactor (MBR) treating high strength wastewater. *Bioresource technology*, 100, 136–141.

Fytili, D. & Zabaniotou, A. 2008. Utilization of sewage sludge in EU application of old and new methods—a review. *Renewable and Sustainable Energy Reviews*, 12, 116–140.

Gill, L., O'Luanaigh, N., Johnston, P., Misstear, B. & O'Suilleabhain, C. 2009. Nutrient loading on subsoils from on-site wastewater effluent, comparing septic tank and secondary treatment systems. *Water Research*, 43, 2739–2749.

Grady Jr, C. L., Daigger, G. T., Love, N. G. & Filipe, C. D. 2012. *Biological wastewater treatment*, CRC Press.

Ilies, P. & Mavinic, D. 2001. Biological nitrification and denitrification of a simulated high ammonia landfill leachate using 4-stage Bardenpho systems: system startup and acclimation. *Canadian Journal of Civil Engineering*, 28, 85–97.

Katsou, E., Frison, N., Malamis, S. & Fatone, F. 2014. Controlled Sewage Sludge Alkaline Fermentation to Produce Vola-tile Fatty Acids to be Used for Biological Nutrients Removal in WWTPs. *Journal of Water Sustainability*, 4, 1–11.

Keeney, D. R. 1986. Sources of nitrate to ground water. CRC Crit. Rev. Environ. Control. 16, 257–304.

Keng, P.-S., Lee, S.-L., HA, S.-T., Hung, Y.-T. & Ong, S.-T. 2013. Cheap materials to clean heavy metal polluted waters. *Green Materials for Energy, Products and Depollution*. Springer.

Kim, J. & Novak, J. T. 2011. Combined anaerobic/aerobic digestion: effect of aerobic retention time on nitrogen and solids removal. *Water Environment Research*, 83, 802–806.

Koers, D. A. 1979. *Studies of the control and operation of the aerobic digestion process applied to waste activated sludges at low temperatures.* University of British Columbia.

Lamb, B., Gold, A., Loomis, G. & Mckiel, C. 1990. Nitrogen removal for on-site sewage disposal: a recirculating sand filter/rock tank design. *Transactions of the ASAE*, 33, 525–0531.

Ling, T.-Y., Siew, T.-F. & Lee, N. 2010. Quantifying pollutants from household wastewater in Kuching, Malaysia. *World Applied Sciences Journal*, 8, 449–456.

Liu, G., Xu, X., Zhu, L., Xing, S. & Chen, J. 2013. Biological nutrient removal in a continuous anaerobic–aerobic–anoxic process treating synthetic domestic wastewater. *Chemical Engineering Journal*, 225, 223–229.

Liu, Y. 2003. Chemically reduced excess sludge production in the activated sludge process. *Chemosphere*, 50, 1–7.

Lundquist, T. J., Woertz, I. C., Quinn, N. & Benemann, J. R. 2010. A realistic technology and engineering assessment of algae biofuel production. *Energy Biosciences Institute*, 1.

Mahiroglu, A., Tarlan-Yel, E. & Sevimli, M. F. 2009. Treatment of combined acid mine drainage (AMD)—Flotation circuit effluents from copper mine via Fenton's process. *Journal of hazardous materials*, 166, 782–787.

Massoud, M. A., Tarhini, A. & Nasr, J. A. 2009. Decentralized approaches to wastewater treatment and management: applicability in developing countries. *Journal of environmental management*, 90, 652–659.

Mehrabadi, Z. S. & Zinatizadeh, A. 2014. Performance of a compartmentalized activated sludge (CAS) system treating a synthetic antibiotics industrial wastewater (SAW). *Journal of Water Process Engineering*, 3, 26–33.

Metcalf & EDDY 2003. *Wastewater Engineering, Treatment and Reuse*, Mc Graw Hill.

Oakley, S. M., Gold, A. J. & Oczkowski, A. J. 2010. Nitrogen control through decentralized wastewater treatment: Process performance and alternative management strategies. *Ecological Engineering*, 36, 1520–1531.

Raj D, S. S., Chary, N. S., Bindu, V. H., Reddy, M. & Anjaneyulu, Y. 2004. Aerobic oxidation of common effluent treatment plant wastewaters and sludge characterization studies. *International journal of environmental studies*, 61, 99–111.

Raj, D. S. S. & Anjaneyulu, Y. 2005. Evaluation of biokinetic parameters for pharmaceutical wastewaters using aerobic oxidation integrated with chemical treatment. *Process Biochemistry*, 40, 165–175.

Siegrist, R. & Jenssen, P. 1989. Nitrogen removal during wastewater infiltration as affected by design and environmental factors. *Proceedings of the 6th Northwest on-site wastewater treatment short course. University of Washington, Seattle*, 304–318.

Tchobanoglous, G., Stensel, H. D., Tsuchihashi, R. & Burton, F. L. 2003. *Wastewater Engineering: Treatment and Resource Recovery*, 5th ed. McGraw-Hill.

Tyagi, V. K. & Lo, S.-L. 2012. Enhancement in mesophilic aerobic digestion of waste activated sludge by chemically assisted thermal pretreatment method. *Bioresource Technology*, 119, 105–113.

Ujang, Z. & Buckley, C. 2002. Water and wastewater in developing countries: present reality and strategy for the future. *Water Sci Technol*, 46, 1–9.

Van Loosdrecht, M., Smolders, G., Kuba, T. & Heijnen, J. 1997. Metabolism of micro-organisms responsible for enhanced biological phosphorus removal from wastewater, Use of dynamic enrichment cultures. *Antonie van Leeuwenhoek*, 71, 109–116.

Yang, P. & Chen, Y. 1977. Operational characteristics and biological kinetic constants of extended aeration process. *Journal (Water Pollution Control Federation)*, 678–688.

Zupančlč, G. D. & Roš, M. 2008. Aerobic and two-stage anaerobic–aerobic sludge digestion with pure oxygen and air aeration. *Bioresource technology*, 99, 100–109.

Hydraulic assessment of grassed swale as bioengineered channel

M.M. Muhammad, K.W. Yusof & M.R. Mustafa
Department of Civil and Environmental Engineering, Universiti Teknologi PETRONAS, Seri Iskandar, Perak, Malaysia

A.Ab. Ghani
River Engineering and Urban Drainage Research Centre, Universiti Sains Malaysia, Pinang, Malaysia

ABSTRACT: This paper evaluates the suitability of a grassed swale for channel stabilization against erosion, as an alternative to concrete drainage systems. To achieve this objective, hourly flow measurements due to different rainfall events were made along 10 m length of the grassed swale. The data collected were cross sections, grass geometry, flow depth and velocity. Rating curves, velocity distributions, stage and velocity – Manning's resistance relationships were developed, respectively, with correlation coefficients, R^2, mostly above 70%. Also, the centerline velocity of grassed swale was determined to range from 0.05 to 0.76 m/s. The shear stresses fall between 3.47 to 6.59 N/m^2 both on bed and bank of the swale. Comparing the velocity and shear stresses values with standard charts, it was found that all values are within the allowable limit. Thus, the relationships developed would be suitable for designing an erosion resistant grassed swale based on Malaysia climate conditions and alike.

1 INTRODUCTION

A grassed swale may be defined as a shallow and bio-engineered open channel with trapezoidal, parabolic or non-uniform shape. Swales are usually vegetated with flood and erosion resistant plants. The design of grassed swales promotes the conveyance of storm water at a steady and regulated rate. It also acts as a filter medium in removing pollutants and improving water quality, where it can detain storm water for several hours or days (SuDS, 2015, DID, 2012). Also, from the economic point of view, vegetated drainage systems are less expensive compared to concrete drainage systems which in several circumstances are being gradual eroded or failed by water actions.

Swales generally have two different types of flows that can be distinguished as surface and sub-surface flows. The surface flow occurs due to flow – vegetation interactions with mostly grass being planted along the wetted perimeter of the swale channel. While for the case of sub-surface flow, flow occurs underneath the grass swale through geo-synthetic polypropylene modules in form of mesh, which are normally filled with river sands. It is important to note that the surface flow can only take place when the sub-surface modules are saturated through infiltration of run-off, after several rainfall events.

The hydraulic characteristics of grass swales can be analyzed using the flow resistance such as Darcy-Weisbach friction (f), Chezy's resistance (C) and Manning's roughness (n) factors which have been used in a wide range of hydraulic and hydrologic analyses. However, the most common used equation in determining flow resistance in the hydraulic industry is Manning's equation (Kirby et al., 2005). The roughness components in open channels (swales) are conceptually divided into three parts via: form roughness, soil grain roughness and vegetative roughness (Temple, 1999). In most vegetated open channels, vegetative roughness characterized by vegetation density, vegetation height and type of vegetation dominates the flow resistance of the channel.

Investigation of flow characteristics in grass swale have been conducted by numerous researchers. However, the researchers failed to evaluate the swales as an erosion resistant in addition to the hydraulic analysis. Also, the authors recommended that further assessment of grassed swales with different vegetation types, vegetation density, and geometrical cross sectional channel is required for developing a more reliable model in predicting vegetative resistance (Ahmad et al., 2011).

Recently, Department of Irrigation and Drainage (DID) Malaysia has developed a manual termed Urban Storm Water Management Manual (DID, 2012) that will help in designing drainage systems. However, the guidelines in MSMA did not reflect the exact climatic conditions of Malaysia, the manual used foreign standards which are different from the climate of Malaysia. Based on this fact, further studies are required in order to verify the guidelines in MSMA with respect to Malaysia climatic conditions.

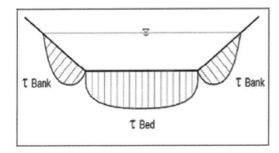

Figure 1. Distribution of shear stress on streambed and banks.

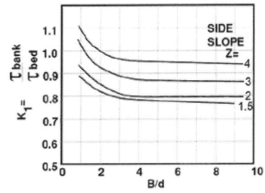

Figure 2. Coefficient K_1 vs. side slope z, and width/depth B/d Ratio.

Therefore, the present study has attempted in solving the above issues through field experiments in a grassed swale. Through this, the effectiveness of grass swale for channel stability was examined. And comparisons were made from our findings with the standard charts provided in MSMA, in order to assess the suitability of adapting the guidelines in MSMA within the environment of Malaysia.

2 METHODOLOGY

2.1 Modelling the vegetated flows through the swale

The usual method for estimating energy losses in hydraulic modeling is through the use of popular equation by Manning.

$$Q = \frac{1}{n} A R^{2/3} S_0^{1/2} \quad (1)$$

Where: A = cross-sectional area of flow (m2), R = Hydraulic radius in (m), S = Bed slope of the channel, and n = Manning's coefficient.

2.2 Evaluation of channel stability in the swale

The resistance to erosion may vary on the nature of the biomaterial and the location of the channel. Lane's diagram given in Figure 1, shows theoretical distribution of shear stress on trapezoidal channel section. This means biomaterials of greater shear resistance are required lower on the bank, while a lighter-duty treatment may be sufficient near the top of the bank.

Shear stress is an essential parameter in channel rehabilitation, because all materials, whether manufactured or natural, must be able to withstand the expected shear stress at the design discharge (Saldi-Caromile, 2004). Thus, for maximum shear stress on the bed the following expression is used:

$$\tau_{bed} = \gamma RS = \gamma (A/P) S \quad (3)$$

$$\tau_{bed} = 9806 \, RS \quad (4)$$

where, τ_{bed} = maximum bed shear stress (N/m^2), γ = the specific weight of water = 9806 N/m^3,

R = hydraulic radius in m, A = flow cross-sectional area (m^2), P = wetted perimeter (m) and S = energy slope = 0.002.

Also, multiplying the maximum bed shear stress, τ_{bed}, by a coefficient or factor, K_1, given in Figure 2. The value of K_1, varies based on channel side slope (z) and the ratio of bottom width (B) to depth (d). This approach is used for a relatively straight channel reach. For maximum shear stress on the bank, τ_{bank}, equation (5) can be applied:

$$\tau_{bank} = K_1 \tau_{bed} \quad (5)$$

It should be noted that K_1 is estimated by knowing the aspect ratio = B/d, that is, the ratio of width (B) to the flow depth (d), and the side slope z of the channel. For this project z = 1:2.

Table 1 presents velocity range for various channel boundaries conditions which has been provided in MSMA 2nd Edition. And to account for the limiting values of the shear stresses as well as erosion limits based on flow duration, Figs. 3 and 4 may be used as a guide to assess the suitability of channels in erosion control. These were also provided by the guidelines of MSMA 2nd Edition.

2.3 Study area

Figure 5 shows the locations of swales at USM Engineering Campus, Nibong Tebal, Pineng, Malaysia. There are 3 types of swale system in USM namely, swale Type A, swale Type B, and swale Type C, based on their sizes and capacities. The swale Type A consists of a single sub-surface module, swale Type B (which was considered in the present study) has two numbers of single sub-surface modules and swale Type C contains 3 numbers of single modules respectively (Zakaria et al., 2003).

Also, *Axonopus Compressus* commonly known as *Cow grass* was used within the channel bed of the swale Type B. Figure 6 shows the pictorial representation of the grass. The technical details of swale components was well illustrated by Ghani et al (Ab. Ghani et al., 2004).

Table 1. Stability of channel linings for given velocity limits (Fischenich, 2001).

Lining	Velocity limits				
	0 – 0.61 m/s	0.61 – 1.22 m/s	1.22 – 1.83 m/s	1.83 – 2.44 m/s	> 2.44 m/s
Sandy Soils					
Firm Loam					
Mixed Gravel and Cobbles					
Average Turf					
Degradable RECPs Stabilizing Bioengineering					
Good Turf					
Permanent RECPs Armoring Bioengineering CCMs & Gabions					
Riprap					
Concrete					
Key :					
	Appropriate				
	Use Caution				
	Not Appropriate				

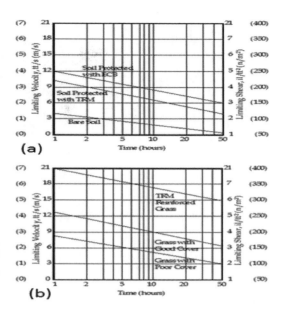

(a)

(b)

Figure 4. Limiting value for velocity and shear for (a) TRM, ECB and/or Soil; (b) Grass and/or TRM (Sprague, 1999).

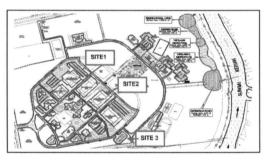

Figure 5. Study area of grassed swale.

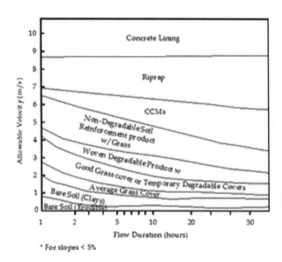

* For slopes < 5%

Figure 3. Erosion limits as a function of flow duration (Fishenich and Allen, 2000).

Figure 6. Sample of *cow grass*.

2.4 Measurement of physical quantities

Field measurements were obtained by selecting a control section of 10 m length along the channel in swale Type B on a bed slope of 1:500. The data collected were cross section, slope, flow depth, velocity, grass height and flow measurements were carried out using current meter at 1 hour intervals due to different rainfall events from September to October 2015 as shown in Fig. 7. In this study the grass height was maintained at 150 mm. This is because a minimum grass heights of 75 mm and maximum of 150 mm should be sustained within grass swale [1]. And the swale was regularly clean, to ensure that it is free of debris and excessive silt [11].

Figure 7. Flow scenarios when the grass is Submerged (A) and Unsubmerged (B).

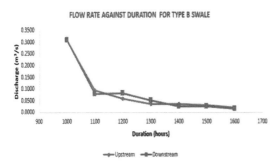

Figure 8. Discharge variation in upstream and downstream on 9/9/15.

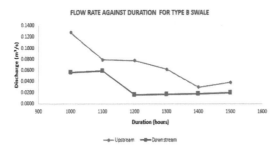

Figure 9. Discharge variation in upstream and downstream on 17/9/15.

3 RESULTS AND DISCUSSION

3.1 Flow parameters relationship

From Figs. 8 and 9 the discharge was plotted against time for different rainfall events. Generally, the discharge decreases over time, with the upstream discharge being higher than the downstream discharge. This has agreed with the statement by several researchers that vegetation reduces the magnitude of flow and obstructing water to reach the downstream smoothly (Guscio et al., 1965, Muhammad et al., 2015).

Figure 10 shows the rating curve of grass swale Type B. The curves indicate that strong relationship exist between discharge, Q and flow area, A, as the correlation coefficients, R^2, are all above 70%. This

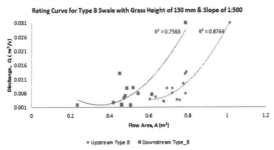

Figure 10. Rating curves for Type B swale.

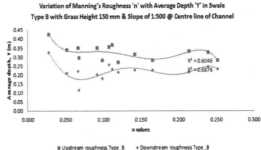

Figure 11. n–y relationship.

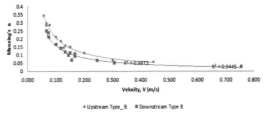

Figure 12. n–V relationship.

means the upstream accommodates larger flow area compared to the downstream.

From Fig. 11, the results reveal that Manning's n at upstream was high compared to downstream. This can be explained based on the greater roughness exerted by the grass in obstructing the discharge from the upstream. Thus, Manning's n – values decrease with increase in the flow depth, y. This agrees with the findings of other researchers (Chow, 1959, García Díaz, 2005, Chen et al., 2009), where smaller Manning's n was observed for greater flow depth. However, Manning's n was observed to be constant in both upstream and downstream when the flow depth was very high and the grass was completely submerged.

Similarly, Figs. 12 and 13, show the variations of Manning's n with Velocity, V and Discharge, Q, respectively. It was observed that, n, declines as both, V and Q, increased which is similar to the findings of Ahmad et al (Ahmad et al., 2011).

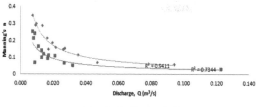

Figure 13. n–Q relationship.

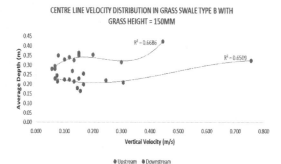

Figure 14. Centre line velocity distribution in swale Type B.

Table 2. Upstream Shear Stresses for 9th and 17th September, 2015.

Depth, y	Area, A	Wetted Perimeter, P	Hydraulic Radius, R	Bed Shear Stress, τ_b	Bank Shear Stress, τ_s
(m)	(m²)	(m)	(m)	(N/m3)	(N/m3)
Date of observation: 09/09/15					
0.45	1.013	3.136	0.323	6.332	5.066
0.37	0.777	2.797	0.278	5.448	4.304
0.35	0.788	3.063	0.257	5.042	4.009
0.37	0.759	2.708	0.280	5.492	4.339
0.33	0.743	2.606	0.285	5.588	4.415
0.30	0.675	2.610	0.259	5.072	3.982
Average Values				5.496	4.353
Date of observation: 17/09/15					
0.36	0.785	2.967	0.265	5.188	4.124
0.36	0.720	2.633	0.273	5.363	4.291
0.33	0.693	2.798	0.248	4.858	3.862
0.34	0.719	2.825	0.255	4.992	3.969
0.33	0.677	2.671	0.253	4.968	3.949
0.30	0.605	2.488	0.243	4.764	3.788
Average Values				5.022	4.796

Table 3. Downstream Shear Stresses for 9th and 17th September, 2015.

Depth, y (m)	Area, A	Wetted Perimeter, P (m)	Hydraulic Radius, R	Bed Shear Stress, τ_b	Bank Shear Stress, τ_s
(m)	(m²)	(m)	(m)	(N/m³)	(N/m³)
Date of observation: 09/09/15					
0.35	0.788	3.062	0.257	5.044	3.985
0.25	0.488	2.471	0.197	3.870	3.057
0.26	0.520	2.537	0.205	4.020	3.176
0.24	0.480	2.531	0.190	3.719	2.919
0.13	0.234	2.143	0.109	2.141	1.681
0.21	0.420	2.522	0.167	3.266	2.564
Average Values				3.677	3.476
Date of observation: 17/09/15					
0.30	0.615	2.648	0.232	4.556	3.622
0.28	0.546	2.450	0.223	4.371	3.475
0.26	0.507	2.451	0.207	4.056	3.205
0.26	0.507	2.457	0.206	4.048	3.198
0.24	0.461	2.382	0.193	3.795	2.998
0.23	0.460	2.521	0.182	3.579	2.810
Average Values				4.067	3.861

3.2 *Velocity profiles*

Figure 14 shows the trend pattern of the average vertical velocity profiles at the centerline of grass swale Type B considering both upstream and downstream flow conditions. Comparing the velocity ranges of the upstream and downstream of Fig. 14 with values in Table 1 provided in MSMA 2nd edition, it follows that all velocities fall within the allowable standards and hence, the channel will be resistive to erosion.

3.3 *Shear stresses*

Tables 2 and 3 show the variations of average bed and bank shear stresses with the flow depth, y, respectively. As the depth decreases the shear stresses decreases for both cases of rainfall events. Also, comparing the shear stresses calculated in Tables 2 and 3 with charts in Figures 3 and 4 as provided in MSMA 2nd edition, it shows that the shear stress values are within the allowable range and hence the biomaterials are stable against erosion.

4 CONCLUSION

The results obtained from this study shows that the grassed swale is suitable to be used in urban drainage systems to convey water in a non-erosive manner. This is obvious as velocity values were found to be within the allowable standards, and appropriate for designing a stable bio-channel, with minor cautions especially when the velocity is very large. Thus, the curves developed can be used to enhance the estimation of roughness and discharge of grassed channels in guidelines provided by the MSMA.

ACKNOWLEDGMENT

We would like to thank the support from Ministry of Education under HiCOE's niche area Sustainable Urban Stormwater Management, Universiti Sains Malaysia, USM (Grant No. 311.PREDAC.4403901). The first author would also like to acknowledge Universiti Teknologi PETRONAS for the fellowship scheme to undertake the postgraduate study.

REFERENCES

AB. Ghani, A., Zakaria, N. A., Abdullah, R., Yusof, M. F., Mohd Sidek, L., Kassim, A. H. & Ainan, A. 2004. Bio-ecological drainage system (Bioecods): concept, design and construction. *Internariond Conference on Hydro-Science and Engineering.* Brisbane, Australia.

Ahmad, N. A., AB. Ghani, A. & Zakaria, N. A. 2011. Hydraulic characteristic for flow in swales. *3rd International Conference on Managing Rivers in the 21st Century: Sustainable Solutions for Global Crisis of Flooding, Pollution and Water Scarcity.* Rivers, Penang, Malaysia.

Chen, Y.-C., Kao, S.-P., Lin, J.-Y. & Yang, H.-C. 2009. Retardance coefficient of vegetated channels estimated by the Froude number. *Ecological Engineering*, 35, 1027–1035.

Chow, V. T. 1959. *Open channel hydraulics*, New York, McGraw-Hill.

DID 2012. Urban storm water management manual for Malaysia. Department of Irrigation & Drainage (DID) Malaysia.

Fischenich, C. 2001. Stability thresholds for stream restoration materials. DTIC Document.

Fishenich, C. & Allen, H. 2000. Stream Management, Water Operations Technical Support Program. *US Army Corps of Engineers, Environmental Laboratory. Waterways Experiment Station. Vicksburg, MS.*

García Díaz, R. 2005. Analysis of Manning coefficient for small-depth flows on vegetated beds. *Hydrological Processes*, 19, 3221–3233.

Guscio, F. J., Bartley, T. R. & Beck, A. N. 1965. Water resources problems generated by obnoxious plants. *Journal of the Waterways and Harbors Division*, 91, 47–62.

Kirby, J. T., Durrans, S. R., Pitt, R. & Johnson, P. D. 2005. Hydraulic resistance in grass swales designed for small flow conveyance. *Journal of Hydraulic Engineering*, 131, 65–68.

Muhammad, M. M., Yusuf, K. W., Mustafa, M. R. & AB. Ghani, A. Vegetated Open Channel Flow for Urban Stormwater Management: A Review. 36 th IAHR World Congress, 2015 The Hague, the Netherlands.

Saldi-Caromile, K., Bates, K., Skidmore, P., Barenti, J., and Pineo, D. 2004. Stream habitat restoration guidelines: Final draft. Washington Department of Fish and Wildlife and Ecology and the U.S. Fish and Wildlife Service, Olympia, WA.

Sprague, C. 1999. Green Engineering: Design principles and applications using rolled erosion control products. Part Two. *Civil Engineering News*, 11.

SUDS 2015. Sustainable Drainage Manual: Guidance on Design of Swales. Construction Industry Research and Information Association. *In:* (CIRIA) (ed.) C753 ed.

Temple, D. 1999. Flow resistance of grass-lined channel banks. *Applied engineering in agriculture.*

Zakaria, N. A., AB. Ghani, A., Abdullah, R., Mohd Sidek, L. & Anita, A. 2003. Bio-ecological drainage system (BIOECODS) for water quantity and quality control. *International Journal of River Basin Management*, 1, 237–251.

Sulfide precipitation as treatment for iron rich groundwater

Nasiman Sapari, Hisyam Jusoh, Emmanuel Olisa & Ezerie Henry Ezechi
*Department of Civil and Environmental Engineering, Universiti Teknologi PETRONAS,
Seri Iskandar, Perak, Malaysia*

ABSTRACT: Iron concentrations above the permissible limit of 0.3 mg/L in groundwater makes it unacceptable for drinking. Conventionally, iron removal from groundwater involves aeration and rapid filtration. This paper describes iron removal under anaerobic conditions by batch experiment. The operation involved purging 500 ml of groundwater sample with hydrogen sulfide for about 360 minutes. About 83% removal of iron concentration for 1 mg/L of iron solution was attained. For higher initial iron concentrations of 3.55 and 5.01 mg/L, the removal efficiency reduced to 82% and 75%, respectively. At the end of the experiment, the average residual sulfide concentration in groundwater was 25 μg/L, which is lower than 50 μg/L, the standard for drinking and the Eh level was less than −272 mV.

1 INTRODUCTION

The use of groundwater as a source of water supply for drinking is a common practice everywhere in the world and it is usually a cleaner form of water than surface water, especially as typically it is void of microorganisms that cause diseases. Nonetheless, groundwater is known to be contaminated by organic and/or inorganic pollutants, which is a source of utmost concern to the public across the globe [1]. Groundwater makes up the underground part of the 'water cycle' which makes up about 40% of fresh water in the world [2]. In areas where there is short supply of surface water, groundwater can effectively be utilized [3]. Groundwater quality is largely dependent on the parent materials, especially the contents of fluoride, hardness, iron, sulfide, total dissolved solids and pH [4]. During groundwater treatment Iron is removed. Nonetheless, removal of iron from groundwater is normally incomplete as particles of iron made up of of Fe (III) oxides and hydroxides in water supplies results in diverse aesthetic and operational challenges such as bad taste, water colouration, deposition and suspension in the supply systems. More so, dissolved iron (Fe (II)) can engender post-treatment flocculation resulting in additional iron particles [5]. Iron in groundwater is found in various forms such as dissolved iron (II), inorganic complexes, organic complexes, colloidal and suspended iron. Dissolved iron minerals in groundwater are essential for good health but can have adverse health effect if consumed above the permissible limit [6]. It is therefore essential to remove iron from groundwater before consumption. Oxidation and filtration is a frequently used method for iron removal in groundwater. This method was first used in 1874 by the Germans to treat groundwater [7]. More so, iron can also be removed by anaerobic process in the absence of oxygen. Usually, dissolved oxygen (DO) level less than 1.0 mg/l is considered anaerobic [8] In anaerobic groundwater, iron may be found in high concentrations [9] and can be removed by precipitation as sulfide under anaerobic condition [10] The operation involves purging the system with pure nitrogen gas to expel traces of oxygen present in it [11] Alkalay et al. [12] reported that iron can be removed from groundwater by precipitation of hydrogen sulfide (H_2S) followed by rapid filtration. Introducing sulfides, basically in the form of H_2S, at pH of 8.5 was found to be effective in iron removal when it was combined with aqueous bi-sulfide ion to form precipitate as solid sulfide [13] especially as precipitation is not possible in acid solution due to insufficiency of sulfide ion concentration (S^{2-}) to surpass products of iron sulfide solubility. However, the addition of sodium acetate solution increases the availability of sulfide ion, which allows the occurrence of black iron sulfide precipitation [14]. In his report, indicated that under anaerobic condition, the reaction between sulfide and metals generates precipitates that are insoluble hence making separation very easy by filtration. The reaction between iron and sulfide generates a black iron-sulfide (FeS) precipitate [15].

$$H_2S + Fe^{2+} \rightarrow 4FeS_{(S)} + 2H^+ \quad (1)$$

$$Fe^{2+} + HS^- \rightarrow FeS_{(S)} + H^+ \quad (2)$$

The formation of hydrogen ions in the reaction, results in a drop in pH hence buffering is needed. Sodium acetate has been applied as a buffer by previous researchers [14]. A study carried out by [16] showed nearly 100% removal of iron and manganese in groundwater. In another study by [17] about 69% of iron removal from groundwater was achieved under anaerobic condition in tube wells [18] Recorded about

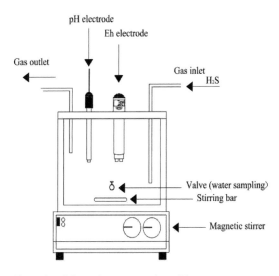

Figure 1. Schematic representation of the reactor.

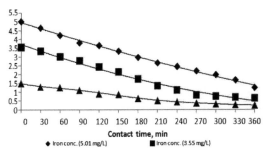

Figure 2. Effect of optimum pH, 7 on iron removal from groundwater samples.

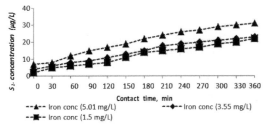

Figure 3. Effect of contact time on initial sulfide concentration at pH 7.

82% removal of iron from synthesized water containing 1.95 mg/L of iron for a contact duration of 14 days. Increased redox potential (Eh) has been found enhance removal efficiency of iron to about 99% in another study [19]. Redox potential around 0 mV to −100 mV under anaerobic condition favours iron removal from water [20] while maintaining a constant pH of 7 with Eh value of −218 mV [21]. Anaerobes have been reported to grow well at Eh level between −200 to −350 mV at pH of 7 [22]. This paper reports the main findings on iron removal from groundwater by sulfide precipitation under anaerobic condition.

2 MATERIALS AND METHOD

Groundwater samples were collected from three different monitoring wells within the Universiti Teknologi PETRONAS (UTP) campus. Hach spectrophotometer (DR 2400) was used to determine the initial total iron concentration in the three samples using the phenanthroline method (APHA 1980) and were found to be 1.5, 3.55 and 5.01 mg/L, respectively. The experiment was carried out at room temperature of about 25°C.

A 3 L stainless steel reactor was used in conducting the experiment, Fig. 1. It was furnished with pH and Eh probes to constantly monitor hydrogen ion activities and the redox potential. The reactor was filled with groundwater to its full capacity, thirty grams of sodium acetate was added to the samples and stirred with a magnetic stirrer. Oxygen was expelled from the reactor by purging it with nitrogen gas about 30 minutes, followed by H_2S purging for another 6 hours. Water samples were collected at intervals of 30 minutes for 6 hours and iron concentration was measured. Total sulfide was determined concurrently using methylene blue method.

3 RESULTS AND DISCUSSION

3.1 Effect of pH on Iron Removal from groundwater

The pH for this study was varied between 1 and 8 and was found to be optimum at pH 7. This is similar to findings reported elsewhere [24]. Figure 2 below shows the effect of optimum pH of 7 for iron removal from the groundwater. For over a 6 hour contact time, iron removal varied between the three groundwater samples with various iron concentrations.

Iron was removed from groundwater samples with initial iron concentration of 1.5, 3.55 and 5.01 mg/L. The final iron concentration after the treatment were 0.25, 0.65 and 1.24 mg/L which represents 83%, 82% and 75%, respectively. It is observed from the obtained result that there was higher removal of iron from the groundwater at lower concentration than in higher concentrations. This may be as a result of the inability of the sulfide to precipitate large amount of iron concentrations in the groundwater.

3.2 Effect of Contact Time on Initial Sulfide Concentration at Optimum pH

The unconsumed sulfide concentration in the groundwater is shown in figure 3. Sulfide concentration were found to be within the permissible standard level in drinking water. Reaction contact time was 6 hours for initial iron concentration of 5.01, 3.55 and 1.5 mg/L. Sulfide levels increased from 7 µg/L to 31 µg/L, 4 µg/L to 23 µg/L and from 2 µg/L to 22 µg/L for all the three samples, respectively. Influence of contact time was observed in all three samples.

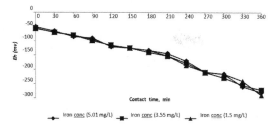

Figure 4. Effect redox potential on initial concentration of iron at pH 7.

As contact time increased a concurrent increase in sulfide level was observed. However, the lowest iron concentration gave the highest sulfide level after 6 hours. Nonetheless, all the samples had sulfide levels that are less than the standard allowable limit of 50 μg/L [23].

3.3 Effect of Redox Potential on Initial Iron Concentration at Optimum pH

Figure 4 shows the effect of redox potential (Eh) on the initial concentration of iron at pH 7. Eh gradually decreased from an initial average value of $-50\,mV$ to $-300\,mV$ after six hours of contact time. The Eh reduced to $-80\,mV$ during the period of purging the reactor with nitrogen gas and this was within the first 30 minutes of contact time. The least Eh values throughout the experiment for groundwater with iron concentration of 1.5, 3.55 and 5.01 mg/L were -290, -280 and $-272\,mV$, respectively. These results were in agreement with the requirements in sulfide precipitation process. These conditions should be constantly maintained at Eh value from -200 to $-400\,mV$ for iron to be effectively removed [25]. The decrease of Eh value is due to the starting sulfate reduction in the groundwater which results to the dissolution of iron to form iron sulfides [26].

4 CONCLUSION

This study has shown that iron concentration in groundwater up to 1.5 mg/L can be removed up to 83%. After the treatment, the final iron concentration was found to have reduced from 1.5 mg/L to 0.25 mg/L. The optimum pH was found to be between 7 and 8. However, at higher initial concentrations of 3.55 mg/L and 5.01 mg/L the removal efficiency dropped to 82% and 75%, respectively. Residual sulfide amount in the water were less than the permissible limit, with an average of 25 μg/L after the experiment. The Eh level reduced until $-290\,mV$.

ACKNOWLEDGEMENTS

The authors would like to thank the management and governing body of the Universiti Teknologi PETRONAS (UTP), who sponsored this research through the provision of the Graduate Assistantship (GA) Scheme.

REFERENCES

[1] Schwarzenbach, R.P., Escher, B.I., Fenner, K., Hofstetter, T.B., Johnson, C.A., Von Gunten, U and Wehrli, B. The challenge of micropollutants in aquatic systems, Science, 313 (2006) 1072.

[2] Sampat, P. 2000. Groundwater shock. *World Watch*, 13, 16–24.

[3] Mohammed, T. & Ghazali, A. 2007. Evaluation of Yield and Groundwater Quality for Selected Wells in Malaysia. *Pertanika Journal of Sciences and Technology, Vol. 17 (1)*, 33–42.

[4] Murukesan, V. K., Timothy, S. and Christina, F. 2006. Qualitative Examination of Groundwater From Yap and Some of its Neighboring Islands. *Technical Report: Water and Environmental Research Institute of the Western Pasific University of Guam*, 115, 10–12.

[5] Teunissen, K., Abrahamse, A., Leijssen, H., Rietveld, L., van Dijk, H,. 2008. Removal of both dissolved and particulate iron from groundwater. Drink. Water Eng. Sci. Discuss., 1, 87–115.

[6] Faust, S.D. and Aly, G.M. (1998) *Chemistry of Water Treatment* Second Edition, Ann Arbor Press Inc., USA.

[7] O'Connor, J. T. 1971. *Iron and manganese. In: Water quality and Treatment – A handbook of public water supplies*, New York, McGraw Hill Book Company.

[8] Castle II R.J, J. N.-H. 2002. Case Studies: Aerobic vs Anaerobic Pretreatment of Groundwater. *Groundwater, Avsapof*.

[9] Wolthoorn, A., Temminghoff, E. & Van Riemsdijk, W. 2004. Effect of synthetic iron colloids on the microbiological NH4+ removal process during groundwater purification. *Water research*, 38, 1884–1892.

[10] Rott, U. 1990. *Protection of groundwater quality by biochemical treatment in the aquifer*, IAHS Publ.

[11] John, D. 2003. Chemical Equilibrium Diagrams For Ground-Water Systems.

[12] Alkalay, D., Guerrero, L., Lema, J. M., Mendez, R. & Chamy, R. 1998. Review: Anaerobic treatment of municipal sanitary landfill leachates: the problem of refractory and toxic components. *World Journal of Microbiology and Biotechnology*, 14, 309–320.

[13] Cohen, R. R. H. 2006. Use of microbes for cost reduction of metal removal from metals and mining industry waste streams. *Journal of Cleaner Production*, 14, 1146–1157.

[14] Vogel 1979. V*ogel's : Textbook of Macro and semimicro qualitative inorganic analysis revised by G. Shevla*.

[15] Gonçalves, M. M. M., Da Costa, A. C. A., Leite, S. G. F. & Sant'anna Jr, G. L. 2007. Heavy metal removal from synthetic wastewaters in an anaerobic

bioreactor using stillage from ethanol distilleries as a carbon source. *Chemosphere,* 69, 1815–1820.
[16] Willow, M. & Cohen, R. 2003. Technical Reports-Bioremediation and Biodegradation-pH, Dissolved Oxygen, and Adsorption Effects on Metal Removal in Anaerobic Bioreactors. *Journal of Environmental Quality,* 32, 1212–1221.
[17] Grindstaff, M. & Office, U. S. E. P. A. T. I. 1998. *Bioremediation of chlorinated solvent contaminated groundwater*, Citeseer.
[18] Saunders, J. A., Lee, M. K., Shamsudduha, M., Dhakal, P., Uddin, A., ChowdurY, M. T. & Ahmed, K. M. 2008. Geochemistry and mineralogy of arsenic in (natural) anaerobic groundwaters. *Applied Geochemistry,* 23, 3205–3214.
[19] Jong, T. & Parry, D. L. 2003. Removal of sulfate and heavy metals by sulfate reducing bacteria in short-term bench scale upflow anaerobic packed bed reactor runs. *Water research,* 37, 3379–3389.
[20] Peterson. H, R. P., R. Neapetung, O. Sortehaug 2006. Integrated Biological Filtration and Reverse Osmosis treatment of cold poor quality groundwater on the North American prairies *Safe Drinking Water Foundation,IWA Publishing,* 5, 424–432.
[21] Wang, L., Zhou, Q. & LI, F. T. 2006. Avoiding propionic acid accumulation in the anaerobic process for biohydrogen production. *Biomass and Bioenergy,* 30, 177–182.
[22] Morris, J. G. 1975. The physiology of obligate anaerobiosis. *Adv. Microb. Physiol.,* 12, 169–246.
[23] WHO 1996. Guidelines for drinking-water quality, 2nd ed. Vol. 2. Health criteria and other supporting information. World Health Organization, Geneva, 1996.
[24] Ghosh, D., Solanki, H., Purkait, M. K., (2008). Removal of Fe(II) from tap water by electrocoagulation technique, *Journal of Hazardous Materials,* (155) 135–143.
[25] J. Cao, G. Zhang, Z. Mao, Z. Fang, and C. Yang, "Precipitation of valuable metals from bioleaching solution by biogenic sulfides," *Minerals Engineering,* vol. 22, pp. 289–295, 2009.
[26] A. Haaning Nielsen, P. Lens, J. Vollertsen, and T. Hvitved-Jacobsen, "Sulfide–iron interactions in domestic wastewater from a gravity sewer," *Water Research,* vol. 39, pp. 2747–2755, 2005.

Climate change impact on water resources of Bakun hydroelectric plant in Sarawak, Malaysia

M.R. Mustafa
Department of Civil and Environmental Engineering, Universiti Teknologi PETRONAS, Seri Iskandar, Perak, Malaysia

M. Hussain
Hydro Department, Sarawak Energy Berhad, The Isthmus, Kuching, Malaysia

K.W. Yusof
Department of Civil and Environmental Engineering, Universiti Teknologi PETRONAS, Seri Iskandar, Perak, Malaysia

ABSTRACT: River inflow works as fuel for hydroelectric plants (HEPs) and it drive the energy output from HEPs. This study is carried out to predict changes in Balui river inflow due to potential climate changes. A hydrological model was developed to simulate river inflows at Bakun HEP. Model was calibrated using observed river inflows and then further used for long term river flow generation for future period of 2011–2070 by projected CanESM2 future rainfall. It is noted, under CanESM2 RCPs, low flow (Q_{95}) at Bakun HEP would increase by 35% to 42%, median flow (Q_{50}) would increase by 6% to 10% and high flow (Q_5) would decrease by 9% to 11%. From Q_{30} to Q_{100}, the inflow to the Bakun Dam would significantly improve especially under RCP 8.5, which would result improvement in Bakun HEP future reservoir operation.

1 INTRODUCTION

Water resources play important role in hydropower industry as it works as a fuel for the power stations. Hydropower plants are relatively cheaper sources of energy and contribute significantly to ensure the stability of national grid system. Hydropower plant also support the grid system to supply power during peaking off-peaking, daily and seasonal demands, it integrate with the other intermittent sources of energy i.e. solar, thermal and wind power plants. The impact of global warming on hydropower energy is difficult to assess as is interconnect with hydrology, engineering, economy and politics (Gaudard et al., 2013). Currently, hydropower is the main source of renewable and sustainable energy producing more than 16 percent of electricity in all over the World. Global hydropower production increased by 175 percent during 1973–2011 (Cole et al., 2014).

Climate change and anthropogenic effects are considered as the two major drivers of changes in stream flow (Piao et al., 2007). Climate change, results, increase in temperature and changes in precipitation intensities and patterns, as well as evapotranspiration changes, significantly impact on regional hydrological processes (Labat et al., 2004). In this study, we assessed the possible impacts of climate change on the water resources of upper reaches of Rajang River basin contributing to Bakun HEP in Sarawak modeled under three Representative Concentration Pathways (RCPs) of CanESM2.

2 STUDY AREA AND DATA DISCRIPTION

2.1 Study area

Bakun Hydroelectric Plant (HEP) located in central Sarawak, on Balui River, a major tributary of Rajang River Basin. As part of the project, Bakun Dam is the second tallest concrete-faced rockfill dam in the world. It has installed capacity of 2,400 megawatts (MW), which is the highest along the other Hydroelectric Plants in Malaysia. The catchment area for the Bakun HEP is 14,750 km^2 and it experiences a wet and humid tropical climate throughout the year. Three main rivers contribute to the Bakun HEP are Murum River, Linau River and Balui River.

2.2 Data description

The historical daily rainfall data of fourteen rainfall recording stations (as shown in Figure 1) was utilized in this study. The data was obtained from the Department of Irrigation and Drainage; Sarawak (DID) for the period of 1981–2010. Two types of the daily predictors required for this study were obtained from the Canadian Climate Data and Scenarios website (http://ccds-dscc.ec.gc.ca/?page=dst-sdi); (a) 26

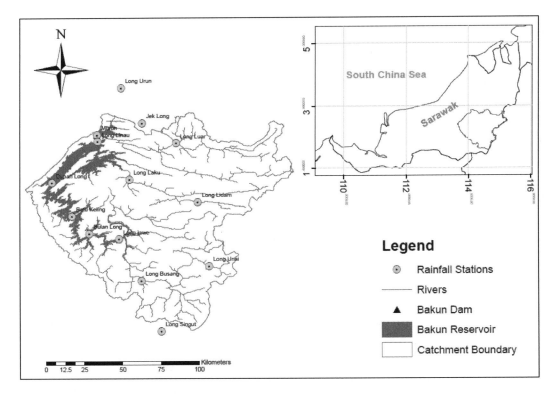

Figure 1. Rainfall Stations in the Bakun HEP Catchment.

predictors of the National Center of Environmental Prediction (NCEP) for the period of 1961–2005; and (b) the 26 predictors of CanESM2 for historical period and for three future scenarios; RCP 2.6, RCP 4.5 and RCP 8.5 for the period of 2006–2070. These data sets for predictors were especially processed for the SDSM model.

3 METHODOLOGY

3.1 Hydrological modeling using HEC-HMS

The Hydrological Modeling system (HEC-HMS) is a rainfall-runoff modeling tool, being used for large river basins to small urban areas to model the long term river inflow and various intensity floods. The model was developed by the U.S. Army Corps of Engineers (USACE) at the Hydrologic Engineering Center (HEC). A complete river basin model consists of a basin model, a meteorological model, control specification, and input time series i.e precipitation, discharge, temperature and evaporation (Center, 2015). A hydrological model was developed for Bakun HEP catchment and was calibrated with the observed inflow for the period of 2001–2010.

During the study, the basin model was developed using the Clark unit hydrograph method for transforming direct runoff, Lag for channel routing, and the constant monthly for base flow method. The meteorological model was established using the Thiessen polygon gauge weight method for precipitation calculation. The Thiessen polygons were created and their weights were calculated by HEC-GeoHMs in accordance with the precipitation gauges.

During the model calibration period, the simulated inflow was compared with the observed inflow using the statistical measures i.e coefficient of determination (R2), percent deviation (D), and Nash-Sutcliffe efficiency (E). The value of R2 describes how good the variations in the observed dataset are picked by the simulated dataset, the value of D indicates the average percentage deviation between observed and simulated data sets, and the value of E indicates how well observed date plot fits with the simulated data plot. Following equations were used to calculate the performance parameters – R2, D, and E;

$$R^2 = \frac{\Sigma(Q_{obs}-\overline{Q_{obs}})\times (Q_{sim}-\overline{Q_{sim}})}{\sqrt{\Sigma(Q_{obs}-\overline{Q_{obs}})^2\times (Q_{sim}-\overline{Q_{sim}})^2}} \quad (1)$$

$$D(\%) = 100 \times \frac{\Sigma(Q_{sim}-Q_{obs})}{\Sigma Q_{obs}} \quad (2)$$

$$E = 1 - \frac{\Sigma(Q_{sim}-Q_{obs})^2}{\Sigma(Q_{obs}-\overline{Q_{obs}})^2} \quad (3)$$

The E value lies between 0 and 1, positive value of E closer to 1 describes model is calibrated well while a negative value of E closer to 0 shows the model is not acceptable. If E value is greater than 0.75, it is considered as a good model, and if it is in the range of 0.36 and 0.75, it is considered satisfactory (Van Liew and Garbrecht, 2003, Mahmood et al., 2016)

3.2 Projected changes in river flow

After the calibration, model was run using historical rainfall for the baseline period of 1981–2010 and using projected rainfall for three RCPs (RCP 2.6, RCP 4.5 and RCP 8.5) for the future period of 2011–2070. The simulated data were divided into two periods: the 2011–2040 and 2041–2070 and were compared with the baseline period of (1981–2010) to assess the potential changes in river inflow in the future. Different flow indicators i.e low flow (Q95), median flow (Q50), and high flow (Q5) were computed for the two future periods and the results were compared with the baseline period to assess the climate change impact on the inflow in the Balui river basin. Q95 is the inflow value available upto 95 percent of the time duration, Q50 is the inflow value available 50 percent of the time and Q5 is the inflow available during the 5 percent of the time duration. All these flow parameters derived from the flow duration curves.

To analyze river inflow pattern at a hydroelectric plant, flow duration curves are the primary tools used to address the frequently asked questions i.e. how often specific amount of inflow occur and what would be the magnitude of lowest and peak inflow. Flow duration curves describes the percentage of times that inflow is most likely equal or exceed a specified inflow value. Probability of inflow is calculated using the equation (4) and then plotted against the inflow to construct the flow duration curve;

$$P(\%) = 100 \times \frac{M}{(n+1)} \quad (4)$$

P is flow probability (% of time), M is the rank of events, and N is the total number of events in a specified time period. During this study, a time series of daily inflows was used to develop the flow duration curves for the base period (1981–2010) and for the two future periods of (2011–2040 and 2041–2070).

4 RESULTS AND DISCUSSION

4.1 Model calibration

During the model calibration, simulated monthly inflows were compared with the observed monthly inflow using R2, D, and E. The value of R2 close to 1, value of D close to 0% and the value of E close to 1 implies good calibration. During this study, the coefficient of determination (R2) was noted as 0.80, percent deviation (D) as 9.35%, and Nash-Sutcliffe efficiency (E) as 0.72. These results are satisfactory and complement well with some previous studies such as (Meenu et al., 2013) in India, (Verma et al., 2010) in India, (Yimer et al., 2009) in Ethiopia, (García et al., 2008) in Spain and (Mahmood et al., 2016) in Pakistan. All these studies also used HEC-HMS to simulate stream flow for climate change studies, with E ranging from 0.48–0.83 and R2 from 0.63–0.84.

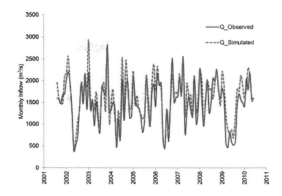

Figure 2. Comparison of observed and simulated monthly inflows at Bakun Dam during 2001–2010.

4.2 Potential changes in river inflows at Bakun Dam

By forcing the historical and predicted rainfall in calibrated model, river inflows at Bakun Dam have been simulated for historical period and for future period upto 2070 under CanESM2 RCPs. Q95, Q50 and Q5 for the baseline period were compared with future simulated period. It was noted, under CanESM2 RCPs, Q95 at Bakun HEP would increase by 35% to 42%, Q50 would increase by 6% to 10% and Q5 would decrease by 9% to 11%. From Q30 to Q100, the inflow to the Bakun Dam would significantly improve especially under RCP 8.5, which would help to improve Bakun reservoir operation in longer term.

It is concluded in (IPCC, 2007) that future tropical cyclones (hurricanes and typhoons) will likely to be more intense, with higher peak of wind speeds and extreme precipitation resulting from increasing trend of tropical sea surface temperature. The increase in the future potential inflow yield for Bakun HEP is associated with potential increase in rainfall yield over the catchment. It is expected that in future due to increased Greenhouse Gas Emission and Global warming, the sea surface temperature would increase significantly and would cause significant increase in monsoon rainfalls in the South East Asian region.

Simulated peak flow under each RCP for 2040s and 2070s were also analyzed and compared with the peak flow during the baseline period. It was noted that the peak flow would increase by 13% under RCP2.6 and decrease by 3% and 13% for RCP4.5 and RCP8.5 respectively during 2040s. During 2070s, it was noted that the peak flow would increase by 16%, 18% and 8% under RCP2.6, RCP4.5 and RCP8.5 respectively as shown in Figure 6 below.

5 CONCLUSIONS

It is noted that the rainfall in the upper reaches of Rajang River Basin would improve seasonal distribution, which would help to improve the flow duration across the year as compared to the baseline period. The Bakun HEP is planned to supply firm power to the

Table 1. Future changes in inflow at Bakun HEP with respect to the baseline period of (1981–2010).

Flow Probability (%)	1981–2010 (m3/s)	2011–2040	2041–2070
RCP 2.6			
Q5	3768	−10	−10
Q50	1232	6	8
Q95	237	35	39
RCP4.5			
Q5	3768	−9	−11
Q50	1232	7	6
Q95	237	35	37
RCP8.5			
Q5	3768	−10	−11
Q50	1232	10	6
Q95	237	42	38

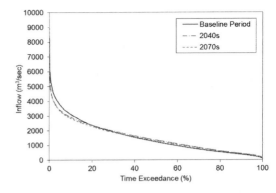

Figure 5. Flow duration curves under RCP 8.5.

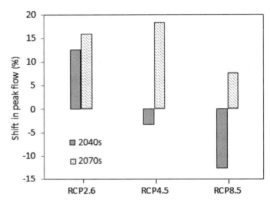

Figure 6. Shift in peak flow under RCPs with comparison to the baseline period of 1981–2010.

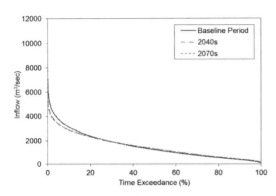

Figure 3. Flow duration curves under RCP 2.6.

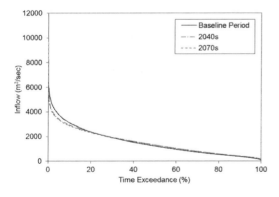

Figure 4. Flow duration curves under RCP 4.5.

Sarawak grid and with improved flow duration, Bakun HEP would be able to improve its plant utilization factor.

6 LIMITATIONS OF THE STUDY

In this study, climate change impact on the river inflows to Bakun HEP was assessed by using CanESM2, although it is recommended to use more than one GCM to cover the uncertainties related to GCM outputs.

Murum HEP is located in the Bakun HEP catchment, has recently been commissioned in December, 2014. It covers eighteen percent of Bakun HEP catchment and will also help to improve the flow duration at Bakun HEP with regulated discharge through Murum power station. The impact of Murum HEP operation over the Bakun HEP water resources has not been assessed in the current study.

ACKNOWLEDGMENT

This study was carried out under a climate change research project sponsored by Sarawak Energy Berhad, Malaysia. The authors would like to express thanks to Department of Irrigation and Drainage, Sarawak (DID) for providing the historical rainfall data for Rajang River Basin.

REFERENCES

Center, H. E. 2015. Hydrologic Modeling System HEC-HMS. *Institute for Water Resources: Davis, CA, USA.*

Cole, M. A., Elliott, R. J. R. & Strobl, E. 2014. Climate Change, Hydro-Dependency, and the African Dam Boom. *World Development,* 60, 84–98.

García, A., Sainz, A., Revilla, J. A., Álvarez, C., Juanes, J. A. & Puente, A. 2008. Surface water resources assessment in scarcely gauged basins in the north of Spain. *Journal of Hydrology,* 356, 312–326.

Gaudard, L., Romerio, F., Dalla Valle, F., Gorret, R., Maran, S., Ravazzani, G., Stoffel, M. & Volonterio, M. 2013. Climate change impacts on hydropower in the Swiss and Italian Alps. *Sci Total Environ*.

IPCC 2007. Summary for Policymakers. In: Climate Change 2007: The Physical Science Basis. Contribution of Working Group I to the Fourth Assessment Report of the Intergovernmental Panel on Climate Change [Solomon, S., D. Qin, M. Manning, Z. Chen, M. Marquis, K.B. Averyt, M. Tignor and H.L. Miller (eds.)]. Cambridge University Press, Cambridge, United Kingdom and New York, NY, USA.

Labat, D., Goddéris, Y., Probst, J. L. & Guyot, J. L. 2004. Evidence for global runoff increase related to climate warming. *Advances in Water Resources,* 27, 631–642.

Mahmood, R., Jia, S. & Babel, M. 2016. Potential Impacts of Climate Change on Water Resources in the Kunhar River Basin, Pakistan. *Water,* 8, 23.

Meenu, R., Rehana, S. & Mujumdar, P. P. 2013. Assessment of hydrologic impacts of climate change in Tunga-Bhadra river basin, India with HEC-HMS and SDSM. *Hydrological Processes,* 27, 1572–1589.

Piao, S., Friedlingstein, P., Ciais, P., de Noblet-Ducoudre, N., Labat, D. & Zaehle, S. 2007. Changes in climate and land use have a larger direct impact than rising CO2 on global river runoff trends. *Proc Natl Acad Sci USA,* 104, 15242–7.

Van Liew, M. W. & Garbrecht, J. 2003. Hydrologic simulation of the Little Washita River experimental watershed using SWAT. *Journal of the American Water Resources Association,* 39, 413–426.

Verma, A. K., Jha, M. K. & Mahana, R. K. 2010. Evaluation of HEC-HMS and WEPP for simulating watershed runoff using remote sensing and geographical information system. *Paddy and Water Environment,* 8, 131–144.

Yimer, G., Jonoski, A. & Griensven, A. V. 2009. Hydrological Response of a Catchment to Climate Change in the Upper Beles River Basin, Upper Blue Nile, Ethiopia. *Nile Basin Water Eng. Sci. Mag.,* 2, 49–59.

Investigation of the influence of particle size on the migration of DNAPL in unsaturated sand

M.Y.D. Alazaiza & S.K. Ngien
Faculty of Civil Engineering and Earth Resources, Universiti Malaysia Pahang, Lebuhraya Tun Razak, Kuantan, Pahang, Malaysia

M.M. Bob
Department of Civil Engineering, College of Engineering, Taibah University, Madinah City, Kingdom of Saudi Arabia

S.A. Kamaruddin
UTM Razak School of Engineering and Advanced Technology, Universiti Teknologi Malaysia

ABSTRACT: Four experiments were conducted to investigate the migration of dense non-aqueous phase liquid (DNAPL) in unsaturated porous media using Light Reflection Method (LRM). The porous media was natural sand collected from a river and segregated into different sizes through sieving. Three different sizes of the sand were used in the first three experiments while the fourth experiment used a mixture of these sands. The sands were packed separately in rectangular acrylic columns and then DNAPL was injected from the top of the column. The migration of DNAPL, modeled by tetrachloroethylene (PCE), was observed using a digital camera connected to a laptop and controlled using special software. The images were captured according to a predetermined time interval set into the software. The results show a significant difference in the migration of PCE through these sands. The migration of PCE in Experiment 3 was much faster than the migration in the other experiments. This is most likely due to the large pores in the sand samples. Moreover, in the experiment using mixture sand, it was observed that the migration was uneven and relatively slower than other experiments. LRM provides a non-intrusive and non-destructive tool for studying fluid flow for which rapid changes in fluid flow in the entire flow domain is difficult to measure using conventional techniques.

1 INTRODUCTION

Non-aqueous phase liquids (NAPLs) have a low solubility in water and occur in the subsurface as a separate phase and are considered as one of the most spread hazardous chemical (Newell et al., 1995). Theses NAPLs can be resulted from leakage of underground storage tanks and huge pipelines. NAPLs are classified into two types based on its density: the first type is light non-aqueous phase liquid (LNAPL) which has density less than water and the second type which is termed dense non-aqueous phase liquid (DNAPL) is denser than water. LNAPL include many hydrocarbon fuel components, for instance, toluene, benzene, xylenes (BTEX) and ethyl benzene. Tetrachloroethylene (PCE) and trichloroethylene (TCE) are examples of DNAPL materials (Newell et al., 1995). NAPL migration in the subsurface system depends on its relative density. When LNAPL enter the subsurface system, it will pass through unsaturated soil and migrate downward due to gravity and float on the surface of water table (Morris et al., 2003) resulting in deterioration of groundwater quality. On the other hand, DNAPL will pass through unsaturated zone and continue its downward migration under the effect of gravity until it reaches the saturated zone (Soga et al., 2004).

In laboratory scale experiments, non-destructive imaging techniques have gained more attention during the last decades which make the characterization and understanding of the multiphase system more accurate (Agaoglu et al., 2015). In the last few decades, there was a quantum leap in the using of image analysis techniques to investigate and measure multiphase fluid contents in laboratory experiments (Oostrom et al., 2007). Accordingly, LRM technique based on image analysis is one of the most important and promising techniques. LRM was used by several researchers to investigate NAPL infiltration using digital cameras under controlled lighting conditions (Kechavarzi et al., 2000). LRM is considered a cheap technique and requires only limited equipment [7, 8]. This paper qualitatively analyze the migration of PCE in three different sizes of natural sand, whereas the forth experiment used a mixture of these sizes using LRM technique. The migration of PCE was monitored using digital camera and capturing of the photos was carried out according to specific time intervals.

Table 1. Chemical and Physical Properties of Sand.

Properties	Parameter	Value	Unit
Chemical Properties	Mn	77.3	ppm*
	Fe	2406.8	ppm
	Cu	1.4	ppm
	Zn	0.9	ppm
	Cd	> 0.5	ppb*
	Pb	11.3	ppm
	Si	1.3	ppm
Physical Properties	Density	2.63	g/cm^3
	Porosity	0.42	
	D_{50}	0.33	
	C_u	2.23	

* ppm is part per million, ppb is part per billion.

Table 2. PCE Characteristics.

Term	Value
Chemical formula	C_2Cl_4
Density	1.62
Viscosity	0.89
Molar mass	165.82
Melting point	19
Boiling point	121

2 MATERIALS AND METHODS

2.1 Experimental setup

A 1-D rectangular column made of acrylic was used in our experiments. The acrylic material of the column provided a clear view of the medium within. The dimensions of the column were 30 cm height × 10 cm wide × 5 cm depth. The column was packed separately with the sand and every experiment used a new column. A digital camera Nikon D7100 (Nikon SDN, BHD, Malaysia) was fixed on the same position during the experiments at a distance of 1.5 m from the column. The camera captured images that has 24 mega pixels (6000 × 4000 pixels), and a 12 bit dynamic range that results in 4096 grey levels. The camera was connected to a laptop and the capturing of the images was controlled by (Control My Nikon 5 software).

2.2 Materials selection and preparation

Four experiments were conducted to achieve the goal of this research. Natural sand used for this study was collected from a river in Kuantan, Malaysia. Table 1 shows the main chemical and physical properties of the sand.

The first three experiments were used three different sizes retained on sieve 600 μm, 425 μm and 1.18 mm while the forth experiment was packed using a mixture of these sand sizes. These experiments were referred to as Experiment 1, 2, 3, and 4 respectively. The methodology of the four experiments was similar. However, the sole difference was the size of the sand used in the experiments. The sand was washed with water and dried in the oven for 48 hours at 45° C. After drying, the sand was packed by pouring 1 cm layers of sand into the column, using a spatula to mix each layer with the previous one, and then tapped the outer frame of the model using a plastic hammer to achieve a dense pack. Each experiment used 50 ml PCE. Table 2 shows the main properties of PCE. Since PCE is colorless, 0.1 g/L of Oil-Red-O was used to dye the DNAPL for better visualization. Kechavarzi et al. (Kechavarzi et al., 2000) has stated that mixing a small amount of dye is enough. This concentration is deemed to be sufficient to facilitate good visual observation of the DNAPL movement through the acrylic wall of the column. The dye was weighted and then the PCE was added and mixed with the dye using spatula. PCE was injected using injection syringe needle from the top of the column. The needle was pushed to penetrate the soil to about 1 cm below the top of the model before PCE was released to allow the movement of it downward directly and avoid the movement of PCE on the top of the soil. After the injection of PCE, the digital camera starts to capture images automatically every 10 seconds. Consequently, all images were processed using Image-Pro Premier 9.1 software (Media Cybernetics Inc., Silver Spring, MD).

3 RESULTS

3.1 Experimental 1

As mentioned before, in this experiment the sand retained on sieve 600 μm was used. The migration of the PCE was relatively fast in the first minute, and then it was slowly declined and the change in the behavior and movement of PCE was very small. It was also observed that some of the PCE was migrated faster that some parts as shown in Figure 1. This behavior can be due to the uneven compaction of the sand. The PCE was reached the bottom of the column after 4 minutes. The observation of the PCE movement was continued until the change of the PCE behavior stopped. The overall required time for PCE migration was about 124 minutes.

3.2 Experimental 2

In this experiment the sand retained on sieve 425 μm was used. A significant difference in the behavior of PCE migration was observed. The migration of the PCE was more rapid compared with Experiment 1. This behavior can be explained due to the larger pores volume between sand particles. The PCE reached approximately the half of the model in the first 30 seconds, and reached the bottom of the column after 60 seconds which is around one-fourth of the time needed in Experiment 1. The shape of PCE diffusion was more regular compared with Experiment 1 as shown in Figure. 2. Furthermore, it is obvious that no significant difference in the PCE behaviour after the fourth minutes as shown in Figure. 2,e and Figure. 2,f.

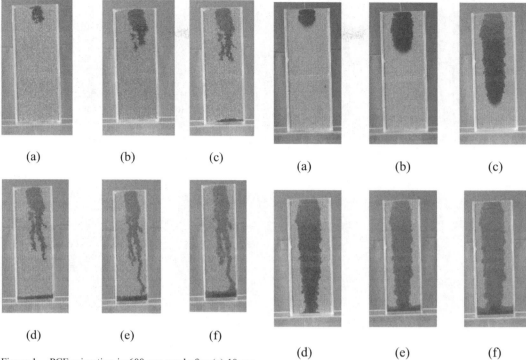

Figure 1. PCE migration in 600 μm sand after (a) 10 sec (b) 20 sec (c) 30 sec (d) 60 sec (e) 240 sec (f) 124 min.

Figure 2. PCE migration in 425 μm sand after (a) 10 sec (b) 20 sec (c) 30 sec (d) 60 sec (e) 240 sec (f) 124 min.

3.3 *Experimental 3*

In this experiment the sand retained on sieve 1.18 mm was used. As expected, the migration of PCE was dramatically rapid compared with the Experiments 1 and 2 where the PCE reached the bottom of the column only after 30 seconds. This is due to the fact that the pore volume is larger than that of experiment 1 and 2 which lead to larger permeability that increased the velocity of fluid movement. The observation was continued until 124 minutes similar to the previous experiments. No significant difference was observed in the behavior of PCE after the second minutes as shown in Figure 3.

3.4 *Experimental 4*

A significant variation in the behaviour of the PCE was observed as shown in Figure 4. At the first minute, the migration was very rapid, and then it was slowly declined. PCE was reached the bottom of the column after 22 minute. This marked variation was due to the different pore size distribution which decreased the volume of voids and lead to less permeability. Moreover, uneven movement of PCE was observed due to the uneven compaction of sand layers.

4 CONCLUSION

Four laboratory experiments using LRM technique were conducted to investigate the migration of the DNAPL modeled by PCE in natural sand with different

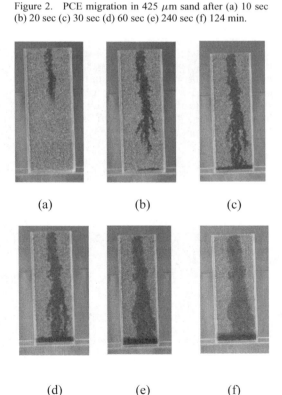

Figure 3. PCE migration in 1.18 mm sand after (a) 10 sec (b) 20 sec (c) 30 sec (d) 60 sec (e) 240 sec (f) 124 min

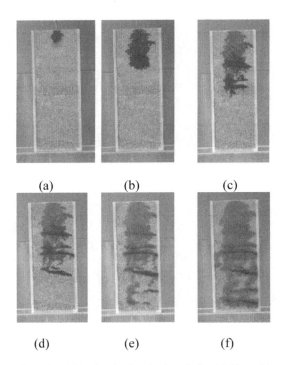

Figure 4. PCE migration in mixed sand after (a) 10 sec (b) 20 sec (c) 30 sec (d) 60 sec (e) 240 sec (f) 124 min

sizes. The results show a significant difference in the migration movement of PCE through these sands. The migration of PCE in Experiment 3 was much faster than the migration other experiments. This is most likely due to the large pores in the sand samples. Moreover, in the experiment using mixture sand, it was observed that the migration was uneven and relatively slower than other experiments. LRM provides a non-intrusive and non-destructive tool for studying fluid flow for which rapid changes in fluid flow in the entire flow domain is difficult to measure using conventional techniques.

ACKNOWLEDGMENT

This research was funded by the Fundamental Research Grant Scheme (Project RDU 130139) led by the second author. Also, the authors are grateful for Universiti Malaysia Pahang (UMP)'s support to the first author.

REFERENCES

Agaoglu, B., Copty, N. K., Scheytt, T. and Hinkelmann, R. 2015. Interphase mass transfer between fluids in subsurface formations: A review. *Advances in Water Resources*, vol.79, pp. 162–194.

Kechavarzi, C., Soga, K. and Wiart, P. 2000. Multispectral image analysis method to determine dynamic fluid saturation distribution in two-dimensional three-fluid phase flow laboratory experiments. *Journal of Contaminant Hydrology*, vol. 46, pp. 265–293.

Morris, B. L., Lawrence, A. R., Chilton, P., Adams, B., Calow, R. C. and Klinck, B. A. 2003. Groundwater and its susceptibility to degradation: A global assessment of the problem and options for management. Early warning and assessment report series , RS. 03-3. United Nations Environment Programme, Nairobi, Kenya.

Newell, C. J., Acree, S. D., Ross, R. and Huling, S. 1995. Ground water issue: Light nonaqueous phase liquids. U.S. Environmental Protection Agency. Groundwater Services, Inc., Houston, TX (United States), EPA/540/S-95/500.

Oostrom, M., Dane, J. and Wietsma, T. W. 2007. A review of multidimensional, multifluid, intermediate-scale experiments: Flow behavior, saturation imaging, and tracer detection and quantification. Vadose Zone J, vol.6, pp. 610–637.

Soga, K., Page, J. and Illangasekare, T. 2004 A review of NAPL source zone remediation efficiency and the mass flux approach. J Hazard Mater, vol.110, pp. 13–27.

Comparison of support vector machines kernel functions for pore-water pressure modeling

N.M. Babangida, K.W. Yusof, M.R. Mustafa & M.H. Isa
Department of Civil and Environmental Engineering, Universiti Teknologi PETRONAS, Seri Iskandar, Perak, Malaysia

ABSTRACT: Modeling pore-water pressure (PWP) responses to rainfall is an important part of monitoring hydrological behavior of hill slope. Of recent, soft computing techniques had been used to model these responses. Using support vector regression (SVR) these responses can be modeled with very good accuracy. However, selection of appropriate kernel for such modeling is a necessity. Using PWP and rainfall data from an instrumented slope, four kernel function (linear, sigmoid, polynomial and radial basis function) were used to develop four Models to predict PWP. Input features were selected using a wrapper algorithm, and the SVR meta-parameters were calibrated using k-fold cross validation and grid search. The radial basis function (RBF) was found to be the most suitable for modeling PWP responses, due to its competitive results and less complexity in implementation.

1 INTRODUCTION

Porewater pressure (PWP) or matric suction (in the negative sense), has been central to the study of rainfall induced slope failure. Its excessive rise can trigger failure in slopes, especially those with unsaturated soil zone. Rainfall infiltration is known to increase PWP. Although it is difficult to tell how much rainfall becomes infiltration, there exist a good relationship between rise in PWP and rainfall infiltration (Huang et al. 2012). For a failure prone, hill slope, it is vital to monitor PWP responses during and after rainfall. As PWP levels could rise to some undesirable levels, and possibly cause failure. A rise in PWP can result in among other things a decrease in soil shear strength parameters (Yeh & Lee 2013, Lee et al. 2009). Because of this, negative PWP are desirable.

For effective evaluation of hydrological behavior of slopes, Damiano & Mercogliano (2013) intimated that in-situ monitoring of some parameters, amongst which is PWP should be conducted. PWP measurements is typically conducted on site, via a filed instrumentation program. A simple instrumentation may entail the installation of tensiometers to monitor the PWP and a raingauge to monitor the onsite rainfall. This endeavor is not only time consuming but, labor and resource expensive, and an alternative means could be quite helpful.

Of recent PWP models have appeared in the literatures of Mustafa et al. (2013) and Mustafa et al. (2012). All these models are artificial neural network (ANN) based models. ANN is known to model such complex systems, and it did with remarkable accuracy in the case of PWP. Despite the success enjoyed by ANN, another machine learning algorithm, the support vector machines (SVM), has been making progress, and it is proved to have at least the same accuracy as ANN, or even outperform ANN in many situations (Burges 1998).

SVM is a kernel based algorithm, and because of this, the choice of kernel is a vital choice one has to make when implementing SVM. The choice of kernel function depends on the type of problem to be modelled. Radial basis function (RBF) is the most widely used kernel functions. In several applications, including soil related studies, RBF kernels is deemed as the default kernel, and used as such. Often without proper justification, other than that, it is the most common. Some of such studies, include studies of Samui & Sitharam (2011) where they used SVM to predict liquefaction susceptibility of soils. The study of Lamorski et al. (2008) and Twarakavi et al. (2009) in which they developed a SVM model of pedo-transfer function for approximating soil moisture retention curve using soil hydraulic properties. Studies of Zhao (2008) and Samui et al. (2013) whom used SVM to predict slope reliability index from factor of safety (FS) and FS respectively, using data that included soil shear strength parameters. In spite of the success recorded in these models, using RBF because it is the most common kernel is not enough a justification. A good literature showing the best kernel in the study area (where available should suffice). Better yet, is to evaluate some few kernels and compare their performance, with regard to the system being modeled. Therefore this study will evaluate and compare the performance metrics of some basic kernels in modeling the non-linear complex responses of pore-water pressure to rainfall. This is with a view to selecting the best kernel for modeling such system.

2 SUPPORT VECTOR MACHINES THEORY

Support vector machines was originally developed by Vapnik (1995) for classification, and then later extended to regression problems, and called support vector regression (SVR). SVM is firmly developed on the frame work of structural risk minimization (SRM) and theory of the vapnik-chevenonkis (VC) bounds. This is to say that SVM preforms SRM instead of the conventional empirical risk minimization (ERM). And by doing so, it creates a model with minimized VC dimension. In its simples form, SVM seek a linear separating hyper plane that creates the widest of margins between classes of data. If the problem is non-linearly separable, the original input space is mapped to higher dimensional feature space using an appropriate kernel function. There after SVM determines an optimal linear separating plane in this higher dimension feature space. For detailed explanation about SVM, one is referred to the literature of Kecman (2001).

In the mapping to higher dimensions, SVM can use any of several kernel functions. Kernel functions provide a simple avenue, which leads from non-linearity to linearity for algorithms, by simple expression of the dot product of the input. Some of the basic kernels are linear kernel, polynomial kernel, radial basis function (RBF) kernel and sigmoid kernel, and they are respectively written in Equations 1–4.

$$k(\mathbf{x},\mathbf{y}) = \mathbf{x}^T\mathbf{y} + c \qquad (1)$$

$$k(\mathbf{x},\mathbf{y}) = \left(-\gamma\mathbf{x}^T\mathbf{y} + r\right)^d \qquad (2)$$

$$k(\mathbf{x},\mathbf{y}) = \exp\left(-\gamma\|\mathbf{x}-\mathbf{y}\|^2\right) \qquad (3)$$

$$k(\mathbf{x},\mathbf{y}) = \tanh\left(\gamma\mathbf{x}^T\mathbf{y} + r\right) \qquad (4)$$

Where $\mathbf{x}$ and $\mathbf{y}$ are input vectors, γ, c, r and d are parameters that need to be tuned for the respective kernel.

3 METHOD

3.1 Data source

Data of PWP and rainfall was collected from an instrumented slope at Universiti Teknologi PETRONAS, Malaysia. The data used here has a resolution of 30 minutes and was collected at a depth of 0.6m below the surface of the slope crest.

3.2 Input feature selection

The first step towards a successful model development is finding the right combination of input features. The literature of Mustafa et al. (2012) has shown that five antecedent PWP, present and two antecedent rainfall are necessary for a good PWP model. A broad selection involving five antecedent PWP, present and five antecedent rainfall was made, making a total of 11 input features. The selection was trimmed down using the wrapper technique of input feature selection.

The wrapper algorithm entails the use of a search algorithm and an evaluator algorithm. The search algorithm searches within the given input feature space, each time selecting a certain set of features base on some given search criteria and then pass it to the evaluator. The evaluating algorithm uses the given set of features to conduct some classification and determine the accuracy base on a given measure, say root mean squared error (RMSE). The result is used as the objective function of the search engine and is thus passed back to it. The search algorithm takes the result and evaluate which sub features need to be discarded or not. And thus create another subset of features. The process continues until a given stopping criteria is met. Evolutionary algorithm (EA) with tournament selection, single point crossover, bit flip mutation and generational replacement with elitism was used as the search algorithm. Support vector regression was used as the evaluator, with RMSE as measure of accuracy. This implementation was carried out using Library SVM in Waikato environment for knowledge analysis (WEKA) developed by El-Manzalawy & Honavar (2005).

It is worth noting that EA searches randomly through the input features search space. It can return different subsets every time (with RMSE within just about the same range). Therefore, there could be not only one feasible results or subset of features. This is echoed by Hussain et al. (2011) where they obtained less than 0.2% difference in the accuracy, between sigmoid kernel models (as evaluator algorithm) evaluated with different feature subsets that were obtained using genetic algorithm as the search algorithm. A difference of 0.6% for RBF kernel model and 0.3% for polynomial kernel model were obtained as well. Therefore, in this study, several runs were conducted to obtain different sets of features for each kernel type. Common input features from all kernels were selected to build set of features in Equation 5, which yielded similar accuracy with other obtained feature subsets, and was thus used for the model development.

$$U_t = f\left(U_{t-1}, r_{t-2}, r_{t-3}\right) \qquad (5)$$

Where U = PWP; r = rainfall; t = time index. Lag intervals are of the order of 30 minutes.

4 MODEL DEVELOPMENT

Once the input features has been decided, implementing the SVR model for any kernel is more or less the same procedure as explained below. However, some few necessary steps are vital to the model development. First there is the need to scale the data to some appropriate scale, so as to avoid large values dominating over smaller values. All data was thus scaled between 0 and 1. There is also the need to decide on the measure of accuracy one needs to use to determine model performance. Low values of RMSE (Eqn. 6) and mean absolute error (MAE) are desirable. They show how close the model predictions and the observed values are. In contrast a high value close to unity is desired when coefficient of determination (R^2) shown in Equation 8 is used. These show the strength of relationship between, the model predicted and observe values. The use of Akaike information criterion (AIC) given in Equation 7, provides a tradeoff between model complexity and its fitness. It however, does not provide much on the model quality. A low value is desirable. Models with many fitting parameters are punished for complexity.

All kernels have some parameters that have to be calibrated. Therefore in model implementation, the kernel parameters along with two other SVM parameters C (is a regularization parameter) and ε (represents the width of the corridor that is minimized by SVM) were tuned. This was done by an exhaustive search of the parameter space using a grid search. A grid search with k-fold cross validation is a slow technique of tuning SVM parameters. However, if searched within the right space, it is certain to yield the optimum set of parameters. For each kernel Model, once the optimum set of parameters was obtained. They were used to train and then test the model.

$$RMSE = \left[\frac{1}{n}\sum_i^n (\hat{u}_i - u_i)^2\right]^{\frac{1}{2}} \quad (6)$$

$$AIC = 2k + n\ln\left[\frac{1}{n}\sum_i^n (\hat{u}_i - u_i)^2\right] \quad (7)$$

$$R^2 = \frac{\sum(u_i - \bar{u}_i)^2 - \sum(\hat{u}_i - u_i)^2}{\sum(u_i - \bar{u}_i)^2} \quad (8)$$

Where u and $u=$ observed and predicted values of PWP respectively; $\bar{u}=$ mean of observations; $n=$ number of observations; $k=$ number of model parameters.

5 RESULTS AND DISCUSSION

The use of high resolution and immediate antecedent condition data, i.e. one lag data of PWP from the model feature selection algorithm has given advantage to the Models, particularly the linear kernel Model. It is important to note that with high resolution data there is increased strong linear correlation between one lag antecedent PWP and the present PWP, and because of this, the linear kernel Model has achieved very good result, just like the other kernels (the presented results are with scaled data). This has made it increasingly difficult to evaluate the best performing Model. All the models showed very similar calibration and test accuracy, as shown in Table 1, however, the linear kernel Model performed slightly less than the others at calibration level.

Table 1. Comparison of performance metrics between all the four kernel functions, in calibration and in testing.

Kernel Type	Calibration		Test		
	R^2	RMSE (kPa)	R^2	RMSE (kPa)	AIC
Poly.	0.9604	0.0215	0.9684	0.0228	−1239
RBF	0.9616	0.0211	0.9690	0.0225	−1251
Sig.	0.9612	0.0212	0.9690	0.0225	−1248
Lin.	0.9592	0.0218	0.9694	0.0222	−1253

Table 2. Excerpts of prediction results. With Linear kernel always showing lagged predictions.

Observed (kPa)	Linear (kPa)	Sigmoid (kPa)	Polynomial (kPa)	RBF (kPa)
−11.5	−11.6	−11.5	−11.5	−11.5
−11.5	−11.5	−11.4	−11.4	−11.4
−11.4	−11.5	−11.4	−11.4	−11.4
−11.3	−11.4	−11.3	−11.3	−11.3
−11.5	−11.3	−11.2	−11.2	−11.2
−11.5	−11.5	−11.4	−11.4	−11.4

On further scrutiny of the test results, it was realized that, fast changing trends are not easily captured by all the Models. The Linear kernel Model particularly can only make predictions in a lagged manner. It is difficult to show such lagged predictions in a graphical manner due to large number of test sets, and thus will require high magnification of the prediction plots. However, a section of the prediction results is captured in Table 2. Where the linear kernel model is always predicting the previous observed value (one lag). In a way it behaves like a bad autoregressive model, which completely retains the one lag value, without incorporating any error term. Addition of more antecedent records could possibly reduce this effect. Therefore, albeit the linear kernel Model provide competitive accuracy measures, its predicted values are not reliable. As for the other three kernels, the use of AIC was the basis for best selection. The RBF and Sigmoid kernel Models have similar measures in the test results, with slight edge by the RBF in the calibration, indicating comparatively better learning. The AIC showed RBF, having the lowest of value.

The event based comparison of the predicted and observed, was shown in the scatter plot shown in Figure 1. All Models showed similar performance, with most of the points clustered in some what a straight line. This indicates good correlation between the observed and the predicted. However few other points fell quite away from the cluster, indicating either outliers or some level of miss-estimation.

Base on the results of the accuracy measure, the RBF is the better choice, particularly because of the ease of implementation, when number of parameters to be tuned is considered. Its calibration measures and test measures are as good as any.

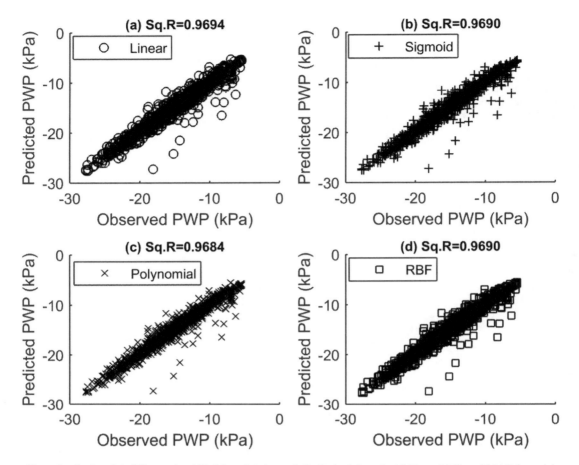

Figure 1. Scatter plot of Observed and Model predicted records for the basic kernels. (a) Linear (b) Sigmoid (c) Polynomial (d) Radial basis function

6 CONCLUSIONS

The models presented here are more of complimentary models, they rely heavily on data from the previous conditions in the system. Using antecedent records to make a one lead time prediction. The results from all four different kernel Models are competitive. The radial basis function kernel is adjudge to be the best for modeling PWP in hill slope study. This is not only because it has the best results in validation and testing, but it has relatively fewer parameters to calibrate. Parameter calibration in SVM is not an easy task, yet it is an essential stage of the model development. The more parameters to calibrate, the more difficult and time consuming it becomes. This makes it a good choice. Although the linear kernel Model has even fewer parameters. It lacks the capability to make good predictions and instead is relegated to predictions that constantly retain lagged records (one lag).

ACKNOWLEDGEMENT

This study is financed by Ministry of Education Malaysia under fundamental research grant scheme (FRGS) of cost center no 0153AB-I61. The first author is thankful to UTP for study sponsorship, under the graduate assistantship scheme.

REFERENCES

Burges, C.J.C., 1998. A tutorial on support vector machines for pattern recognition. *Data Mining and Knowledge Discovery*, 2: 121–167.

Damiano, E. & Mercogliano, P., 2013. Potential Effects of Climate Change on Slope Stability in Unsaturated Pyroclastic Soils. In C. Margottini, P. Canuti, & K. Sassa, eds. *Landslide Science and Practice SE – 2*. Springer Berlin Heidelberg : 15–25.

EL-Manzalawy, Y. & Honavar, V., 2005. WLSVM: Integrating LibSVM into Weka Environment, 2005.

Huang, A.B., Lee, J.T., Ho, Y.T., Chiu, Y.F. & Cheng, S.Y., 2012. Stability monitoring of rainfall-induced deep landslides through pore pressure profile measurements. *Soils and Foundations*, 52(4): 737–747.

Hussain, M., Wajid, S.K., Elzaart, A. & Berbar, M., 2011. A Comparison of SVM Kernel Functions for Breast Cancer Detection. In *2011 Eighth International Conference Computer Graphics, Imaging and Visualization*. Singapore: IEEE, 145–150.

Kecman, V., 2001. *Learning and Soft Computing*, London, England: MIT press.

Lamorski, K., Stawi, C. & Walczak, R.T., 2008. Using Support Vector Machines to Develop Pedotransfer Functions for Water Retention of Soils in Poland. *Soil Science Society of America Journal*, 72(5): 1243–1247.

Lee, L.M., Gofar, N. & Rahardjo, H., 2009. A simple model for preliminary evaluation of rainfall-induced slope instability. *Engineering Geology*, 108(3–4): 272–285.

Mustafa, M.R., Rezaur, R.B., Saiedi,S., Rahardjo, H. & Isa, M.H., 2013. Evaluation of MLP-ANN Training Algorithms for Modeling Soil Pore-Water Pressure Responses to Rainfall. *Journal of Hydrologic Engineering*, 18(1): 50–57.

Mustafa, M.R., Rezaur, R.B., Rahardjo, H. & Isa, M.H., 2012. Prediction of pore-water pressure using radial basis function neural network. *Engineering Geology*, 135–136: 40–47.

Samui, P., Lansivaara, T. & Bhatt, M.R., 2013. Least Square Support Vector Machine Applied to Slope Reliability Analysis. *Geotechnical and Geological Engineering*, 31(4): 1329–1334.

Samui, P. & Sitharam, T.G., 2011. Machine learning modelling for predicting soil liquefaction susceptibility. *Natural Hazards and Earth System Science*, 11(1): 1–9.

Twarakavi, N.K.C., Šimůnek, J. & Schaap, M.G., 2009. Development of Pedotransfer Functions for Estimation of Soil Hydraulic Parameters using Support Vector Machines. *Soil Science Society of America Journal*, 73(5): 1443.

Vapnik, V.N., 1995. *The Nature of Statistical Learning Theory* M. Jordan & S. L. Lauritzen, eds., New York, New York, USA: Springer.

Yeh, H.F. & Lee, C.H., 2013. Soil water balance model for precipitation-induced shallow landslides. *Environmental Earth Sciences*, 70(6): 2691–2701.

Zhao, H., 2008. Slope reliability analysis using a support vector machine. *Computers and Geotechnics*, 35(3): 459–467.

Toxicological studies of Perak River water using biological assay

T.S. Abd Manan & A. Malakahmad
Department of Civil and Environmental Engineering, Universiti Teknologi PETRONAS, Seri Iskandar, Perak, Malaysia

S. Sivapalan
Department of Management and Humanities, Universiti Teknologi PETRONAS, Seri Iskandar, Perak, Malaysia

ABSTRACT: Perak River, the second longest river in Peninsular Malaysia supplies water mainly for domestic, agricultural and industrial purposes, whilst contributing to the state's economy along the river. However, the increasing demand for water has strained anthropogenic environmental pressures on the river basin itself. This paper presents the findings of a toxicological study conducted to evaluate pollution levels in the Perak River using *Allium cepa* (AC) assay, an excellent biological indicator of pollution that is frequently used in environmental monitoring studies. The water samples, both treated and raw, were taken from 3 water treatment plants (WTPs), namely Parit (P), Kampung Senin (KS) and Teluk Kepayang (TK). ACs were grown in these water samples for a 96-hour period of exposure at 22°C of incubation temperature. The presence of cytotoxic and genotoxic substances in the water samples collected from Perak River were confirmed based on root morphology deformation and cytogenetic alterations observed in the root meristem cells of the ACs.

1 INTRODUCTION

Rivers are natural chains of networking that link all land ecosystems within the river basin (Scholes et al. 2008). However, this unified system is very susceptible to anthropogenic impacts leading to river pollution and degradation of ecosystems, whilst affecting inhabitants, including the public itself (Scholes et al. 2008, Sayeda et al. 2010, Hayzoun et al. 2015). Anthropogenic impacts as such inadmissible management, knowledge illiteracy and high population growth rate co-existing with rapid urbanization, massive industrialization, and agriculture contribute to river pollution.

To date, the mortality rates from cancer are continuously rising (Kitchin 1999). More than 90% of the carcinogenic loading comes from domestic areas which make rivers a potential water-borne transmission media of various health threats (Scholes et al. 2008, Sayeda et al. 2010, Hayzoun et al. 2015, Masao 2013, Jain 2001, Beatriz et al. 2011). These areas can achieve loading values greater than industrial zones with standard loading over a discharge rate of more than 10^9 L/d (Scholes et al. 2008, Sayeda et al. 2010, Hayzoun et al. 2015, Masao 2013, Jain 2001, Beatriz et al. 2011). Water borne cancer cases were first reported in the United States of America in the early 1970s. The States was severely contaminated with synthetic organic chemicals that lead to a high incidence of bladder cancer. The Safe Drinking Water Act (SDWA) was legislated by the US Congress in 1974 to overcome this problem (Jaffe 2000, Malakahmad 2013, Abd-Manan 2013). In India, cancer deaths were 52 and 30 per 100,000 populations per year in two areas namely Talwandi Sabo and Chamkaur Sahib. The water was highly contaminated with heavy metals and pesticides (Thakur 2008). In developing nations, 80% of diseases are water-related.

Water borne pollutants are introduced to water bodies through point and non-point sources. The point sources are any identifiable origins such as domestic and industrial effluents (Santschi et al. 2001), petroleum leakage and chemical spills (Anas-Ahmad et al. 2014). Non-point sources originate from diffused sources such as atmospheric deposition (Knighton 1999), sediments deposit and eroding stream banks due to gravel and sand mining (Feller 2009) or forestry and construction (Motelay-Massei et al. 2006), urban run-off (Santschi et al. 2001, Malakahmad 2013], residential, industrial (Abd-Manan 2013), agricultural (Guéguen et al 2004, Santschi et al. 2001) and petroleum refineries (Anas-Ahmad et al. 2014). Water-borne pollutants such as heavy metals ion, polychlorinated biphenyls (PCB), polycyclic aromatic hydrocarbons (PAHs), N-nitroso compounds, mycotoxins and Heterocyclic Amines (HCAs) are carcinogenic upon continuous exposure and ingestion under certain concentration (Department of Environment 2007).

Chromosomal aberrations (CA) can thus be detected by employing various biological test systems such as plant cells, microorganisms, and mammalian cells. These bio-assays serve as a bio-marker for the presence of carcinogens in water. As cancer is often linked to DNA damage, they are also highly reliable in the estimation of the toxicity level or carcinogenic potential of a compound. Plant bioassays have several strong points over microbial and mammalian systems. Plants and mammals have a similar response to mutagens. Plant bioassays also have similarity in the chromosomal morphology of mammals (Radić et al. 2010, Malakahmad et al. 2015).

The *Allium cepa* (AC) test (2n = 16) or onion test was first introduced by Levar in 1938 (Fenech 2008). AC is a locally commercialized crop which makes its conveniently accessible. Although the test is time consuming, it is cost effective and highly reliable. The chromosomes are large in size and easily observed compared to other plants. This makes it a very convenient test system to measure a variety of morphological and cytogenetic alterations such as micronuclei and chromosomal aberrations that can serve as toxicity indicators of environmental pollutants in water bodies. Prior research studies utilizing the Allium test have monitored the effects of a mixture of pollutants including heavy metals (Kitchin 1999, Wescoat 2003, Clayson 2000, Kwon & Choi 2014), herbicides (Nieves 2013), pesticides (Kanzari et al. 2014), and silk dyeing industry (Mohamad et al. 2001), coal (Jamhari et al. 2014), electromagnetic fields (EMF) (Clayson 2000) and petroleum hydrocarbons (Boonyatumanond et al. 2007).

The Perak River is the second longest river in Peninsular Malaysia. It mainly supplies water for domestic use, agricultural activities and industrial operations along the river (Department of Environment 2007).

The AC assay was presented in this study as a biological indicator in the environmental monitoring of water pollution levels of the Perak River, particularly from 3 WTPs, i.e. Parit (P), Kampung Senin (KS) and Teluk Kepayang (TK). The aim of this research was two-fold, i.e. to evaluate the possible cyto- and genotoxic effects of raw and treated water samples from the aforementioned WTPs.

2 MUTATION

Mutations are the indications of presence of any genotoxic compounds.

2.1 *Chromosomal aberrations*

Genotoxicity is the quality of being toxic to the cells' genetic materials. In genotoxicological studies, scientists most commonly refer to mutation as chromosomal aberration (CA). They measure the genotoxicity effects of a compound/solvent based on CA observed via microscopy observation and analyses. The microscopy analyses consist of identification of different types of CA, Anaphase/Metaphase (AM) and Micro-nucleus (MN) analyses (Nieves 2013).

Plants that are commonly used for the genotoxicty studies were AC (onion), *Tradescantia* sp. (flower), *Crepis capillaris* (flower) and *Vicia faba* (broad beans) (Table 1).

Table 1. Average length of root growth and its morphogenetic deformations observed.

Water	Location	Mean ± sd	Growth (%)	H	S	C	T
NC	DW	4.5 ± 0.2	100.0				
	P	9.3 ± 1.8	206.7	✓	✓		
Tr	KS	8.2 ± 0.2	182.2	✓	✓		
	TK	6.8 ± 0.9	151.1	✓	✓		
	P	6.5 ± 1.1	144.4	✓		✓	✓
Rw	KS	7.3 ± 0.6	162.2	✓	✓	✓	
	TK	5.8 ± 1.3	128.9	✓	✓		
PC	H_2O_2	0.5 ± 1.3	–				

There are numerous genotoxicity studies using the AC. The AC test has been widely exploited to other plants with the same concept and method of measurement. AM, MI and percentage of root length are reported in the test. Water samples tested are commonly from sewage effluent (Boonyatumanond et al. 2007), industrial effluent (Jamhari et al. 2014), pharmaceutical effluent (Wishart et al. 2008), industrial wastewater (Rinaldi 2005), surface water (Mossa et al. 1997) as well as wastewater sludge (Feller 209).

Zeng (1999) studied the pollution levels of three rivers passing through Fuzhou city using Tradescantia sp. with MN frequencies from stamen hair. A survey by Grant and Owens (1998) assessed chromosome damage induced by chemicals and environmental pollutants (Sang & Li 2004) using Crepis capillaris via AM and MN from stamen hair.

The genotoxicity of municipal landfill leachate was carried out by Sang and Li (2004) via the Vicia faba bioassay using AM, MI and MN from stamen hair. The results showed that leachate may pose a genotoxic risk to bio-organisms (Michael & Craig 2010).

2.2 *Clastogenic and Aneugenic effects*

Genotoxic effects can be classified into clastogenic (CE) and aneugenic (AE). CE effects on the structural chromosomes such as chromosome losses (CL), chromosome bridges (BD) and breaks (BR) causing sections of the chromosome to be deleted, added or rearranged. Moreover, cells that are supposed to be exterminated are not removed. These can lead to carcinogenesis. It will also cause teratogenic effects by increasing the frequency of abnormal germ cells in paternal males leading to impaired development in fetus. Binucleated cells (or multipolarity) (BC) are commonly found in cancer cells (Radiæ et al. 2010).

Meanwhile, AE effects on numerical chromosomes such as chromosome delays (or laggards) (CD), adherence (or stickiness) (AD) and C-Metaphase (CM). AE effects change the number of chromosomes. The numbers might be extra or less, and prompt diseases of genetic disorder origin (Radiæ et al. 2010) such as congenital cerebral palsy, down syndrome, alzheimer's disease, cystic fibrosis, breast cancer, spinal muscular strophy and many more (Weinberg 2007).

3 METHODOLOGY

3.1 Sampling locations

The Perak River basin under study consists of 3 WTPs: P (4.4973440°, 100.9224010°), KS (4.3825730°, 100.9022410°) and TK (4.3157490°, 100.8802380°). This river basin acts as the principal water source for Pangkor, Lumut, Manjung, Seri Iskandar, Tronoh, Siputeh and Pengkalan, Ipoh. There are many villages as well as massive agricultural land along the river. The river is aligned with two major highways: Bota Kiri (intermediate highway for northern and southern Perak) and Bota Kanan (Ipoh-Lumut highway). Treated (Tr) and raw water (Rw) samples were collected from the WTPs for further analysis on toxicology studies using AC.

3.2 Test System

Equal size bulbs (25–30 mm in diameter, average weight 20 g) of the commercial sweet onion (AC, 2n = 16) were used as biological assay for these toxicology studies: genotoxicity and cytotoxicity test. The procedure was previously described by Fiskesjo, 1985 and Rank and Nielsen, 1998 (Wishart et al. 2008, Rinaldi et al. 2005, Mossa & Mclean 1997). The bulbs were purchased from the local supermarket. The outer scales of the bulbs were peeled and the dried brownish bottom plate was carefully removed leaving a fresh wet root primordia left intact. Bulbs were set up in distilled water for 24 h to let the homogenously adventitious roots to grow in an incubator at 20–22°C in darkness. Prior to the germination of adventitious roots of 0.5–1.0 cm, bulbs were transferred to the real water samples for 96 hours. The experimental set up had three replicates for each water sample. The water samples were changed for every 24 hours. The morphology of the root growth was observed.

The root tips were cut after 96 hours. These root tips were immediately fixed in absolute ethanol:glacial acetic acids (3:1 v/v) for 48 hours. The solvents were discarded and the roots were rinsed with 75% of ethanol twice. The roots were soaked in 75% of ethanol before it was kept in 4°C. The slide preparation followed the procedure described by Rank and Nielsen (1998). In order to evaluate cell damages, 5 slides per water sample were prepared. Approximately 200 cells were analyzed per slide, giving 1000 cells per water sample.

3.3 Equation

The Mitotic Index (MI) (Motelay-Massei et al. 2006) calculation is based on mitotic phases (prophase, metaphase, anaphase and telophase) over total number of cells (Equation 1).

$$MI = \left(\frac{P + M + A + T}{Total\ Number\ of\ Cells} \right) \quad (1)$$

4 RESULTS AND DISCUSSION

4.1 Root morphology

In this paper, morphogenetic root deformations observed in AC may be subjected to chemical stresses obtained from Tr and Rw samples collected from P, KS and TK. Distilled water (DW) was used as negative control (NC) and 0.3 mM of hydrogen peroxide (H_2O_2) as positive control (PC).

The root length was measured for each controls and water samples. The length was measured from the end tip of RAM area towards the root primordia plate of the AC. The average length measurement was done on roots with common length. The longest and shortest lengths of root length were disqualified.

After 96-hours of exposure, the root growth in NC gives 4.5 ± 0.2 cm in root length. PC gives 0.5 ± 1.3 cm in root length. The overall growth of root length for all replicates was more than PC. Therefore, the growth baseline percentage was made based on positive control.

Root growth from Tr samples of all WTPs showed an average length equal to 9.3 ± 1.8 cm (P), 8.2 ± 0.2 cm (KS) and 6.8 ± 0.9 cm (TK). These roots had shown average root growth percentages of more than NC (4.5 ± 0.2 cm, 100%). In ascending order, the results were 151.1% (TK), 182.2% (KS) and 206.7% (P).

Meanwhile, root growth from Rw samples showed average length equals to 6.5 ± 1.1 cm (P), 7.3 ± 0.6 cm (KS) and 5.8 ± 1.3 cm (TK). Rw samples had shown pattern equivalent to average root growth percentages of Tr water samples. The lowest percentages were 128.9% (TK), followed by 144.4% (P) and 162.2% (KS).

All replicates' for Tr water samples had abnormal lateral root growth such as spiral and hooked shape (Figure 1). All replicates' for Rw samples from P, KS and TK shown abnormal lateral root growth such as spiral and hooked shape similar to treated water samples. Replicates for Rw water samples from KS were hooked, spiraled, cranky and turgid roots.

To date, hooked and spiral shaped observed in AC is an indication of the presence of heavy metals (Wescoat &White 2003) and phenol (Knighton 1999) in water.

Further indication on root morphogenetic responses towards any environmental exposures have yet been reported. However, many published literature were mainly focusing on MI and CA for the detailing's on molecular biology and genetic materials (Kitchin 1999, Knighton 1999).

Herbicide activity on trees can be expressed by various visual symptoms depending upon the nature of the chemical used. While some herbicides produce rather distinct symptoms, others result in diseases to plants that can resemble damages from an assortment of causes. Following is a description of the more common symptoms produced by herbicide contact with trees along with similar mimicking diseases (Feucht 1988).

Table 1 shows average root length and types of morphogenetic deformations observed in root growth

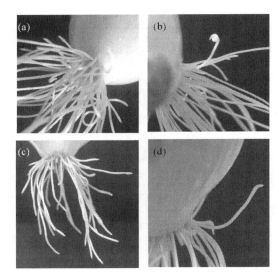

Figure 1. The presence of morphogenetic deformation observed: (a) Hooked; (b) Spiraled; (c) Cranky; and (d) Turgid.

from water samples tested. Figure 1 shows types of morphogenetic deformation observed such as hooked, spiraled, cranky and turgid. Further investigations were discussed in sub-sections that follow based on cytotoxicity and genotoxicity analysis.

4.2 Cytotoxicity

Cytotoxicity is the quality of being toxic to cells. The MI has been used to assess the cytotoxicity of environmental pollutants. It is a parameter that allows estimation of the frequency of cellular division. Cytotoxicity was recorded as the MI from a total of 1000 cells of each water sample. The cytotoxicity of an agent can be determined by the increase or decrease in the MI (Leme & Morales 2009).

MIs for Tr were found higher than the NC(11.3%). These were 28.1 (P), 41.0 (KS) and 58.0 (TK). It showed an increase in cell division (Table 2).

MI for Rw from P (3.2) was found lesser than PC (5.1). MI for Rw from KS (11.0) was slightly different from negative control (11.3). MI for raw water samples from TK (46.0) was found higher than PC (11.3).

According to (Firbas 2010), MIs of samples higher than PC can be harmful to the cell leading to disordered cell proliferation and even to the formation of tumor tissues.

MIs lesser than PC is an indication of the presence of cytotoxic substance in the water sample which causes inhibition of mitotic activities. Reduction in the mitotic activity could be due to inhibition of DNA synthesis or a blocking in the G2 phase of the cell cycle, preventing the cell from entering mitosis (Leme & Morales 2009, Rank & Nielsen 1993, Fiskesjö 1981, Firbas 2010).

The cytotoxic effects of Tr and Rw samples were confirmed as shown by a significant decrease and increase in the MIs in comparison to both types of control (PC and NC).

Table 2. The MI and the percentage of mitosis stages in the AC root meristem treated with water samples from P, KS and TK.

Types of Water	Location	MI	Mitotic Phases (%)			
			P	M	A	T
NC	DW	11.3	54.5	18.9	15.2	11.4
Tr	P	28.1	44.7	34.0	11.3	10.0
	KS	41.0	63.4	19.5	12.2	4.9
	TK	58.0	58.0	69.0	13.8	8.6
R	P	3.2	87.5	0	12.5	0
	KS	11.0	45.5	36.4	18.2	0
	TK	46.0	65.2	23.9	8.7	2.2
PC	H_2O_2	5.1	55.5	14.6	6.7	23.2

*NC: Negative control, Tr: Treated water, Rw: Raw water, PC: Positive control, P: Prophase, M: Metaphase, A: Anaphase, T: Telophase

4.3 Genotoxicity

Genotoxicity is the quality of being toxic to cells' genetic materials, DNA. It will cause CE and AE effects as discussed in the earlier sections. The genotoxic potential was determined according to the observation and quantification of any CA in the meristematic cells of all water samples. At a high level of genotoxicity, maximum of 100 aberrated cells will be observed from 200 metaphase cells (Firbas 2010).

The percentages of CA for Tr as compared to Rw were P (35.1:26.8), KS (40:55.7) and TK (45.6:70.7) (Table 3).

Tr from P, KS and TK has both CE and AE effects. This can be observed from the percentages of CA effects such as AD (38:47:0), CD (2:15:11), CM (9:0:68), CL (15:0:10), BD (0:0:22), BR (3:21:0), and BC (284:317:331). Rw also has both effects.

CE consisted of AD (0:29:54), CM (0:36:0), CL (35:0:121), BD (0:0:42) BR (6:0:0) and BC (428:482:490). AE were mini cell (3:0:0) and lobulated nuclei (0:10:0).

5 CONCLUSION

The inclusion of toxicology studies using bioassays is of great importance to allow understanding of the impacts and consequences of genotoxic substances on living organisms. The morphogenetic deformations observed were likely due to the presence of heavy metals, phenoxy compounds or herbicides. The cytotoxic effects of Tr and Rw were confirmed as shown by a significant decrease and increase in the MIs in comparison to both controls (PC and NC). The genotoxicity effects showed that Tr and Rw water samples have both CE and AE effects. Appallingly, BC type of CA was the highest in all water samples. This indicates high potential of canker in plants or carcinogenicity towards other

Table 3. Frequencies and percentages of CA observed in meristematic cells of AC treated with water samples from P, KS and TK.

CA	Treated			Raw		
	P	KS	TK	P	KS	TK
AD	38	47	0	0	29	54
CD	2	15	11	0	0	0
CM	9	0	68	0	36	0
CL	15	0	10	35	0	121
BD	0	0	22	0	0	42
BR	3	21	0	6	0	0
BC	284	317	331	428	482	490
MC	0	0	14	3	0	0
LC	0	0	0	0	10	0
Total CA	351	400	456	268	557	707
CA (%)	35.1	40	45.6	26.8	55.7	70.7

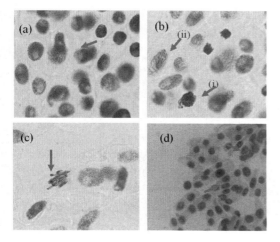

Figure 2. Types of CA under 2000X of magnification; (a) Lobulated nuclei, (b) i. Metaphase with AD and ii. Vagrant chromosomes in Metaphase (CD) (c) Mini cell in metaphase with AD (d) BC.

bio-organisms. This research has confirmed the presence of cytotoxic and genotoxic substance that has CE and AE properties in both Tr and Rr samples collected from P, KS and TK that comes from the same river basin.

ACKNOWLEDGEMENT

The authors thankfully the financial support from Ministry of Higher Education (ERGS 15-8200-136), MyBrain15, Centre for graduate Studies and Civil & Environmental Engineering Department, Universiti Teknologi Petronas for the continuous support.

REFERENCES

Anas Ahmad J., Mazrura S., Mohd Talib L., Lok Meng C., Hock Seng T., Md Firoz K., Norhayati M. T. 2014. Concentration and source identification of Polycyclic Aromatic Hydrocarbons (PAHs) in PM10 of urban, industrial and semi-urban in Malaysia. *Atmos. Environ.* 86:16–27.

Chakraborty R., Mukherjee A. K., Mukherjee A. 2009. Evaluation of genotoxicity of coal fly ash in Allium cepa root cells by combining comet assay with the Allium test. *Environ. Monit. Assess.* 15:351–357.

Chauhan L. K. S., Gupta S. K. 2005. Combined cytogenetic and ultratructural effects of substituted urea herbicides and synthetic pyrethroid insecticide on the root meristem cells. *Pest. Biochem. Physiol.* 82:27–35.

Chauhan L. K. S., Saxena P. N., Gupta S. K. 1999. Cytogenetic effects of Cypermethrin and Fenvalerate on the root meristem *Allium cepa*. *Environ Ept Bot.* 42, 181–189.

David Knighton A. 1999. The gravel-sand transition in a disturbed catchment. *Geomorphology*. 27:325–341.

Department of Environment. Ministry of Natural Resources and Environment Malaysia. Malaysia Environmental Quality Report. ISSN 0127-6433.

Feller M. C. 2009. Deforestation and nutrient loading to fresh waters. Reference Module. In Earth Systems and Envi ronmental Sciences: Encyclopedia of Inland Waters. 29–44.

Fenech M. 2002. Biomarkers of genetic damage for cancer epidemiology. *Toxicology* 181–182:411–416.

Feucht J. R. 1988. Herbicide injuries to trees-symptoms and solutions. Journal of Arboriculture 14(9):215–219.

Fiskesjö G. 1981 Allium test on copper in drinking water. Vatten 37:231–240.

Fiskesjö G. 1988. The Allium test- an alternative in environ mental studies: the relative toxicity of metal ions. *Mutat. Res.* 197:243–260. fs297/en/

Gonzalez P. S., Maglionc G. A., Giordana M., Paisio C. E., Talano M. A., Agostinin E. 2015. Evaluation of Phenol deoxification by *Brassica napus* hairy roots using *Allium cepa* test. ESPR ISSN 0944-1344.

Green P. B. 1962. Mechanism for plant cellular morphogenesis. Science 138:1404–1405.

H. F. Kraybill 1978. Carcinogenesis induced by trace contam inants in potable water. A symposium on carcinogens held by Blue Cross and Blue Shield of Greater New York at the New York Academy of Medicine May 25, 1977. 54 (4), April 413-427 (1978) Website: http://europepmc.org/backend/ptpmcrender.fcgi?accid=PMC1807517&blobtype=pdf

Hayzoun H., Garnier C., Durrieu G., Lenoble V., Le Poupon., Angeletti B., Ouammou A., Mounier S. 2015. Organic carbon and major and trace element dynamic and fate in a large river subjected to poorly regulated urban and industrial pressures (Sebou River, Morocco). *Sci. Total Envi ron* 502 (1):296–308

Hye-Ok K., Sung-Deuk C. 2014. Polycyclic Aromatic Hydrocarbons (PAHs) in soils from a multi-industrial city, South Korea. Sci. Total Environ. 470–471, 1494–1501.

Inceer H., Beyazoglu O., Ergul H. A. 2000. Cytogenetic effects of wastes of copper mine on root tip cell of *Allium cepa* L. Pakistan J. Biol. Sci. 3, 376–377.

Kanzari F., Syakti A.D., Asia L., Malleret L., Piram A., Mille G., Doumenq P. 2014. Distributions and sources of persistent organic pollutants (aliphatic Hydrocarbons, PAHs, PCBs and pesticides) in surface sediments of an industrialized urban River (Huveaune) France. *Sci. Total Environ.* 478:141–151.

Koivusalo M., Jaakkola J. J. K., Varliainen T., Hakulinen T., Karjalainen S., Pukkala E., Tuomisto J. 1994. Drinking water mutagenicity and gastrointestinal and urinary tract cancers: an ecological study in Finland. *Am. J. Public Health* 84:1223–1228.

Leme D. M., Morales M. A. 2009. *Allium cepa* test in environmental monitoring: a review on its application. *Mutat. Res.* 682, 71–81.

Malakahmad A., Abd Manan T. S., Sivapalan S. 2015. Detection Methods of Carcinogens in Estuaries: A Review. International Journal of Sustainable Development and Planning 10/2015; 10(5):601–619.

Mohamad Pauzi Z., Tomoaki O., Hideshige T. 2001. Polycyclic Aromatic Hydrocarbons (PAHs) and Hopanes in stranded tar-balls on the coasts of Peninsular Malaysia: applications of biomarkers for identifying sources of oil pollution. *Mar. Pollut. Bull.* 42(12):1357–1366.

Motelay-Massei A., Garban B., Tiphagne-larcher K., Chevreuil M., Ollivon D. 2006. Mass balance for Polycyclic Aromatic Hydrocarbons in the urban watershed of Le Havre (France): transport and fate of PAHs from the atmosphere to the outlet. *Water Res.* 40(10):1995–2006.

Nakielski J, Hejnowicz Z. 2003. The description of growth of plant organs: a continuous approach based on the growth tensor. In: Nation J, Trofimova I, Rand JD, Sulis W, editors. Formal description of developing systems. Netherlands: Kluwer Academic Publishers; 119–136.

Odeigah P. G. C., Nurudeen O., Amund O. O.1997. Genotoxicity of oil field wastewater in Nigeria. *Hereditas.* 126: 161–167.

R. L. Jaffe (2000). Drinking Water Toxicity in New York City Reservoir and Tap Water Samples. Report to the New York City Council, January, Website: http://docplayer.net/9138085-Drinking-water-toxicity-in-new-york-city-reservoir-and-tap-water-samples.html

Radić S., Stipaničev D., Vujčić V., Rajčić M. M., Širac S., Pevalek-Kozlina B. 2010. The evaluation of surface and wastewater genotoxicity using the *Allium cepa* test. Science of the Total Environment 408:1228–1233.

Rank J. & Nielsen M. H. 1998. Genotoxicity testing of wastewater sludge using the *Allium cepa* anaphase-telophase chromosome aberration assay. *Mutat. Res. Gen. Tox. En.* 418(2–3):113–119.

Rank J., Nielsen M. H. 1993. A Modified Allium test as a tool in the screening of the genotoxicity of complex mixtures, Hereditas 18:49–53.

Sayeda M. A., Shawky Z. S., Mohammed F., Mo hammed M., Nabil A. H. 2010.The influence of agro-industrial effluents on river nile pollution. *JARE* 2(1): 85–95.

Scholes L., Faulkner H., Tapsell S., Downward S. 2008. Urban rivers as pollutant sinks and sources: a public health concern for recreational river users? *Water Air Soil Pollute Focus* (8): 543–553.

Sehgal R., Roy S., Kumar D.V.L. 2006. Evaluation of cytotoxic potential of latex of Calotropis procera and Podophyllotoxin in *Allium cepa* root model. *Biocell*, 30(1):9-13.

Sudhakar R., Ninge Gowda K. N., Venu G. 2001. Mitotic abnormalities induced by silk dyeing industry effluents in the cells of *Allium cepa*. *Cytologia* 66:235–239.

Thakur J. S., Rao B. T., Rajwanshi A., Parwana H. K., Kumar R. 2008. Epidemiological study of high cancer among rural agricultural community of Punjab in North ern India. *Int. J. Environ. Res. Public Health*. 5(5):399–407

Tkalec M., Malaric K., Pavlica M., Pevalek Kozlina B., Vidakovic Cifrek Z. 2009. Effects of radiofrequency electromagnetic fields on seed germination and root meristematic cells of *Allium cepa* L. *Mutat. Res.* 672:76–81.

Ukaegbu M. C., Odeigah P. G. C. 2009. The genotoxic effect of sewage effluent on *Allium Cepa*. *The Report and Opinion* 1(6):36–41.

World Health Organization (WHO). Fact Sheet: Cancer. Website:http://www.who.int/mediacentre/factsheets/

Predicting CMIP5 monthly precipitation over Kuching using multilayer perceptron neural network

M. Hussain
Hydro Department, Sarawak Energy Berhad, The Isthmus, Kuching, Malaysia

K.W. Yusof & M.R. Mustafa
Department of Civil and Environmental Engineering, Universiti Teknologi PETRONAS, Seri Iskandar, Perak, Malaysia

ABSTRACT: In this study, four General Circulation Models (GCMs) from Coupled Model Intercomparison Project Phase 5 (CMIP5) were applied to predict monthly precipitation over Kuching, Sarawak. A feed forward neural network technique was pursued using the Levenberg-Marquardt method to train and predict monthly precipitation. HadGEM2-AO and MIROC5 performed better than BCC-CSM1.1 and CSIRO-Mk3.6.0 when compared by correlation coefficient and root mean square error. Overall HadGEM2-AO performed better than all GCMs when compared for the monthly precipitation prediction. All models underestimated monthly precipitation during the December to February and overestimated monthly precipitation during March to May. Except HadGEM2-AO, all other models were unable to predict monthly precipitation during Jun to November. However, HadGEM2-AO was able to predict monthly precipitation more realistically in the historical run for all months.

1 INTRODUCTION

Floods caused by heavy rains are regular natural disaster in Malaysia and it happens almost every year during the monsoon season. Kuching, the capital city of Sarawak state of Malaysia is one of the most vulnerable cities to these flood events. Climate change is one of the reasons for frequent flooding in Malaysia during last few years. Global warming caused by greenhouse gas emission is real threat to the future climate.

General Circulation Models (GCMs) are widely used to predict the impact of climate change on, for instance, the regional precipitation trend (Campozano et al., 2016). The resolution of these models is not suitable for climate change impact assessment on a catchment or city unless downscaled to the area of interest. There are few techniques commonly used for climate change downscaling i.e statistical downscaling (SD) and dynamic downscaling (DD). Statistical downscaling is easier and faster than dynamic downscaling as it is simpler method. Using SD methods, global-scale climate variables are linked with local-scale variables, and this is done by producing some statistical/empirical relationships Wetterhall et al. (2006). Many statistical downscaling models have been developed with various level of success in various regions based on local climate condition.

Statistical downscaling can be roughly grouped into four categories, weather typing method, stochastic weather generators, resampling methods and regression methods. A regression method constructs a linear or non-linear empirical function between local-scale variables and large-scale GCM variables, and is preferred among statistical downscaling methods because it is easy to implement (Chen et al., 2010).

Artificial Neural Network (ANN) is an emerging modeling technique which has the ability of self-adaptation, pattern recognition and capturing non-linear complex behavior between the input and output parameters (Mustafa et al., 2012). A number of studies have been conducted on precipitation prediction using ANN. (Campozano et al., 2016) noted ANN performed better when compared with Support Vector Machine (SVP) and Statistical Downscaling Model (SDSM). (Mekanik et al., 2013) used Levenberg–Marquardt (LM) algorithm in multilayer Perceptron (MLP) Neural Network and noted ANN models generally showed lower errors and are more reliable for prediction purposes when compared with Multy regression methods.

The objectives of this study are; a) to develop the ANN model for four CMIP5 GCMs data sets using Levenberg–Marquardt (LM) algorithm in multilayer Perceptron (MLP) Neural Network, b) performance evaluation of all GCMs for precipitation prediction over Kuching and c) to compare the predicted historical monthly precipitation over Kuching to recommend the most suitable GCMs for future precipitation projection over Kuching.

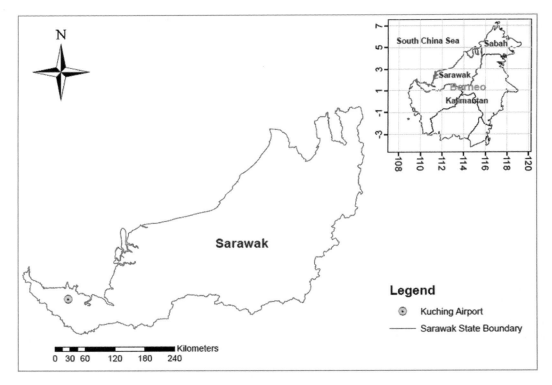

Figure 1. Location of rainfall station at Kuching Airport in Sarawak State of Malaysia.

2 STUDY AREA AND DATA DISCRIPTION

2.1 Study area

The study site Kuching, has coordinate of Longitude 110.35E, Latitude 1.50N, is the capital and most populous city of Sarawak state of Malaysia. Kuching has a tropical rainforest climate. The city is situated on the Sarawak River at the southwest tip of the state of Sarawak on the island of Borneo.

2.2 Data description

The historical daily rainfall data of Kuching Airport was used in this study. The location of rainfall station is shown in Figure 1. The data was obtained from the Department of Irrigation and Drainage (DID), Sarawak for the period of 1971–2010.

In this study, precipitation output from four CMIP5 models (as listed in Table 1) were used to predict the historical monthly precipitation at Kuching. Ensemble member (r1i1p1) run for each of these models has been downloaded and used similar to (Kitoh et al., 2013, Sharmila et al., 2015). More details about this experimental design of CMIP5 are available in (Taylor et al., 2012). Output of all of selected models is freely available at World Data Center for Climate Hamburg (DKRZ) maintained by IPCC Data Distribution Center at; http://www.ipcc-data.org/sim/gcm_monthly/AR5/Reference-Archive.html.

3 METHODOLOGY

3.1 ANN model development

An artificial neural network (ANN) is composed of several interconnected layers of processing units (the neurons) that transform inputs into outputs. The inputs at the neurons are multiplied by weights and then inserted into an activation function. ANNs are characterized by their topology, and probably the most widely known neural network is the multilayer perceptron (MLP). It consists of multiple layers of adaptive weights with full connectivity between inputs and hidden units and between hidden units and outputs. MLP is feed-forward artificial neural network mapping sets of input data onto a set of appropriate outputs.

Neural network toolbox (nntool) of Matlab was used in this study, and optimization of the neural network was pursued using the Levenberg-Marquardt method, minimizing the mean square error. The performance of a total of four ANNs was tested, respectively, a model considering either one or two intermediate neural layers, or a linear or sigmoidal transfer function in the neurons. For the input layer all networks had a single neuron and for the network with one hidden layer 100 neurons were used.

3.2 Input data selection

Selection of input data is very important in the development of an appropriate network. Appropriate ANN

Table 1. General Circulation Models used in present study.

Cmip5 model	Modeling group
BCC-CSM1.1	Beijing Climate Centre, China Meteorological Administration, China
MIROC5	Atmosphere and Ocean Research Institute (The University of Tokyo), National Institute for Environmental Studies, and Japan Agency for Marine-Earth Science and Technology, Japan
CSIRO-Mk3.6.0	Commonwealth Scientific and Industrial Research Organization in collaboration with the Queensland Climate Change Centre of Excellence, Australia
HadGEM2-AO	National Institute of Meteorological Research, Korea Meteorological Administration, South Korea

Table 2. Performance of GCMs over the Kuching during training and testing period.

Model	R	RMSE	RS
Training (1971–1995)			
BCC-CSM1.1	0.43	190	0.69
MIROC5	0.64	156	0.69
HadGEM2-AO	0.67	151	0.69
CSIRO-Mk3.6.0	0.43	184	0.57
Testing (2001–2005)			
BCC-CSM1.1	0.36	207	0.54
MIROC5	0.53	187	0.71
HadGEM2-AO	0.66	164	0.56
CSIRO-Mk3.6.0	0.37	201	0.44

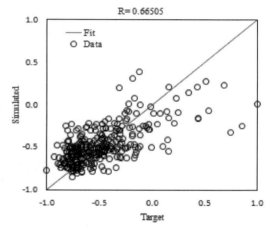

Figure 2. Performance of MLP neural network during training period for HadGEM2-AO historical monthly precipitation.

modeling highly depends on input data selection which represents the non-linearity and complexity of the data during training the network .().(Rojas, 1996). In the present study, about 72% of the data (January 1971 to December, 1995) have been selected for training, 14% data (January, 1996 to December, 2000) for validation and 14% data (January, 2001- to December, 2005) for testing the MLP neural network model.

3.3 Data normalization

It is necessary to normalize the raw data to ensure fast convergence and minimization of global error during network training. Input and target data was using the following equation;

$$N_p = a \frac{(x_p - x_{min})}{(x_{max} - x_{min})} + b \quad (1)$$

where, a and b are the scaling factors; N_p is the normalized value and x_p is the original value while x_{min} and x_{max} are the minimum and maximum values in the data respectively. In the current study, the value of a and b were assigned as 2 and −1 respectively to normalized the data between −1 to +1.

3.4 Model performance

Performance of all GCMs were evaluated by determining the coefficient of correlation (R), root mean square error (RMSE) and ratio of standard deviation (RS) between simulated and observed monthly precipitation. These parameters are defined as in equations below:

$$R = \frac{\sum_{i=1}^{n}(Obs_i - \overline{Obs}) \cdot (pred_i - \overline{pred})}{\sqrt{\sum_{i=1}^{n}(Obs_i - \overline{Obs})^2 \cdot \sum_{i=1}^{n}(pred_i - \overline{pred})^2}} \quad (2)$$

$$RMSE = \sqrt{\frac{\sum_{i=1}^{n}(X_{obs,i} - X_{pred,i})^2}{n}} \quad (3)$$

In which, obs = observed data value; pred = predicted data value; obs = mean observed data value and pred = predicted mean data. The closer R value to 1 and RMSE value to 0, the predictions are better. To build confidence with the analysis's performance, mean rainfall including wet and dry spell lengths is compared graphically with the observed data. These graphical comparisons are able to identify pattern and variations captured by all models (Hassan et al., 2014). RS is ratio of standard deviation of predicted data to observed data.

4 RESULTS AND DISCUSSION

All of four GCMs used in this study have difference different spatial resolution; all models were regridded to 1.0° × 1.0° grid using bilinear interpolation to avoid the possible effect of special resolution on the evaluation. After this, the monthly precipitation data for the

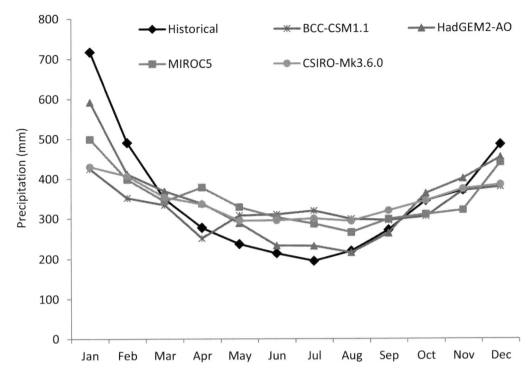

Figure 3. Comparison of historical monthly precipitation with predicated monthly precipitation over Kuching using various GCMs (1971–2005).

grid which cover the Kuching was extracted for further analysis.

ANN models developed using each GCMs precipitation as input and observed monthly precipitation as target were trained by hit and trial method to achieve higher coefficient of correlation between targeted and output data. Simulated output for training and test period was compared with observed historical monthly precipitation over Kuching. The HadGEM2-A0, with correlation coefficient 0.69 and RMSE 154, have better performance compared to the other three models. The performance of HadGEM2-AO using ANN approach is better than (Hussain et al., 2015) applied to Limbang (Sarawak) using HadCM3 model in SDSM and (Hassan et al., 2014) applied to Peninsular Malaysia applying HadCM3 using SDSM and LARS-WG.

HadGEM2-AO and MIROC5 performed better than BCC-CSM1.1 and CSIRO-Mk3.6.0 when compared as shown in Table 2 below. Overall HadGEM2-AO performed better than MIROC when compared for the monthly precipitation prediction as shown in Figure 3. The performance of HadGEM2-AO for the calibration and testing period is shown in Figure 2 and Figure 4 respectively.

5 CONCLUSIONS

It is noted that MLP neural network performed well for the performance evaluation of various GCMs for

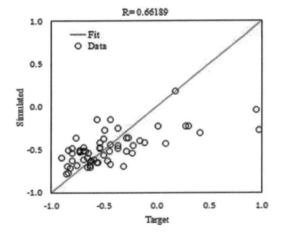

Figure 4. Performance of MLP neural network during test period for HadGEM2-AO historical monthly precipitation.

precipitation downscaling. It relatively performed well when compared with some recently studies done by (Hussain et al., 2015) and (Hassan et al., 2014) in Malaysia. Among the four GCMs from CMIP5 project, HadGEM2-AO performed relatively better for precipitation prediction over Kuching. And it can be used for historical monthly precipitation prediction and future projection over Kuching. However, it is recommended to extend this analysis to other participating GCMs under CMIP5 project to identify the most suitable GCMs for the precipitation prediction over this region.

ACKNOWLEDGMENT

This study was sponsored by Sarawak Energy Berhad, Malaysia. The authors would like to express thanks to Department of Irrigation and Drainage, Sarawak (DID) for providing the historical rainfall data for Kuching Airport Meteorological Station.

REFERENCES

Campozano, L., Tenelanda, D., Sanchez, E., Samaniego, E. & Feyen, J. 2016. Comparison of Statistical Downscaling Methods for Monthly Total Precipitation: Case Study for the Paute River Basin in Southern Ecuador. *Advances in Meteorology,* 2016, 1–13.

Chen, S. T., Yu, P. S. & Tang, Y. H. 2010. Statistical downscaling of daily precipitation using support vector machines and multivariate analysis. *Journal of Hydrology,* 385, 13–22.

Hassan, Z., Shamsudin, S. & Harun, S. 2014. Application of SDSM and LARS-WG for simulating and downscaling of rainfall and temperature. *Theoretical and Applied Climatology,* 116, 243–257.

Hussain, M., Yusof, K. W., Mustafa, M. R. & Afshar, N. R. 2015. Application of statistical downscaling model (SDSM) for long term prediction of rainfall in Sarawak, Malaysia. *WIT Transactions on Ecology and The Environment,* 196.

Kitoh, A., Endo, H., Krishna Kumar, K., Cavalcanti, I. F. A., Goswami, P. & Zhou, T. 2013. Monsoons in a changing world: A regional perspective in a global context. *Journal of Geophysical Research: Atmospheres,* 118, 3053–3065.

Mekanik, F., Imteaz, M. A., Gato-Trinidad, S. & Elmahdi, A. 2013. Multiple regression and Artificial Neural Network for long-term rainfall forecasting using large scale climate modes. *Journal of Hydrology,* 503, 11–21.

Mustafa, M. R., Rezaur, R. B., Saiedi, S. & Isa, M. H. 2012. River Suspended Sediment Prediction Using Various Multilayer Perceptron Neural Network Training Algorithms—A Case Study in Malaysia. *Water Resources Management,* 26, 1879–1897.

Rojas, R. 1996. Neural networks: A systematic introduction. . *Springer Verlag, Berlin,* 151–184.

Sharmila, S., Joseph, S., Sahai, A. K., Abhilash, S. & Chattopadhyay, R. 2015. Future projection of Indian summer monsoon variability under climate change scenario: An assessment from CMIP5 climate models. *Global and Planetary Change,* 124, 62–78.

Taylor, K. E., Stouffer, R. J. & Meehl, G. A. 2012. An Overview of CMIP5 and the Experiment Design. *Bulletin of the American Meteorological Society,* 93, 485–498.

Wetterhall, F., Bárdossy, A., Chen, D., Halldin, S. & Xu, C.-Y. 2006. Daily precipitation-downscaling techniques in three Chinese regions. *Water Resources Research,* 42, n/a-n/a.

Drought analysis of Cheongmicheon in Korea based on various drought indices

K.J. Won, S.H. Kim & E.S. Chung
Seoul National University of Science and Technology, Seoul, South Korea

S.U. Kim
Kangwon National University, Kangwon, South Korea

M.W. Son
Chungnam National University, Chungnam, South Korea

ABSTRACT: This study assessed drought of Cheongmicheon watershed of Korea from 1985 to 2015 according to the duration. Because drought is a sequence phenomenon, we calculated various drought indices such as meteorological, agricultural, and hydrological drought index. Standardized Precipitation Index (SPI) and Standardized Precipitation Evapotranspiration (SPEI) were used as meteorological drought index. Palmer Drought Severity Index (PDSI) and Streamflow Drought Index (SDI) calculated by results of Soil and Water Assessment Tool (SWAT) were applied as agricultural and hydrological drought index. As a result, 2015 and 2014 have the most vulnerable in all drought indices of average of extreme and averaged drought. Variation of drought showed different tendency based on analysis of frequency. Also, the extreme and averaged drought have high correlation between drought indices excluding between PDSIs. However, each drought index showed different occurrence year and severity of drought. Therefore, drought indices with various characteristics were used to analysis drought.

1 INTRODUCTION

Climate change have affected human, natural, industrial and agricultural in various ways as flood and drought etc. Among that, drought is a natural hazard which happen for a long period combined with various cause. In general drought is categorized as three groups: meteorological, agricultural, hydrological and socioeconomic drought index [Wilhite & Glantz (1985), Correia et al. 1991 and Tate & Gustard (2008)]. Drought is a sequence procedure associated with various impacts. So, many drought indices was used to quantify drought [Zarch et al. 2015, Hong et al. 2015 & Jiang et al. 2015]. However, it is useful and necessary to consider more than one index to examine the sensitivity and accuracy of indices, the correlation between them and explore how well they confirm each other in the context of a specific research or management objective [Morid et al. 2006, Smakhtin & Hughes (2007) and Banimahd & Khalili (2013)]. Therefore, this study compared the drought of Cheongmicheon from 1985 to 2015 according to duration 90 days, 180 days and 270 days and calculated correlation coefficient and frequency analysis of drought indices to estimate variation. Standardized Precipitation Index (SPI) and Standardized Precipitation Evapotranspiration Index (SPEI) were used as meteorological drought index. Palmer Drought Severity Index (PDSI) and Streamflow Drought Index (SDI) which were agricultural and hydrological drought index were calculated using soil moisture and runoff from Soil and Water Assessment Tool (SWAT) model.

2 METHODOLOGY

2.1 Model configuration

SWAT model was developed to predict the impact of land management practices on water sediment and agricultural chemical yields in large complex watersheds with varying soils, land use and management conditions over long periods of time. The hydrological component of SWAT is based on the following water balance equation [Ghaffari et al. 2010]:

$$SW_t = SW_0 + \sum_{i=1}^{t}(R_i - Q_{surf} - ET_i - W_{seep} - Q_{gw}) \quad (1)$$

where SW_t is the final soil water content (mm), SW_0 is the initial soil water content on day i (mm), t is the time (days), R_i is the amount of precipitation on day i (mm), Q_{surf} is the amount of surface runoff on day i (mm), ET_i is the amount of evapotranspiration on day i (mm), W_{seep} is the amount of water entering the vadose zone from the soil profile on day i (mm), and Q_{gw} is the amount of return flow on day i (mm).

2.2 Model optimization

SWAT-CUP model was used for optimization of SWAT's parameters according to runoff and soil moisture. This model included 5 algorithms consisted by Sequential Uncertainty Fitting ver. 2 (SUFI-2), Generalized Likelihood Uncertainty Estimation (GLUE), Parameter Solution (ParaSol), Particle Swarm Optimization (PSO) and Markov Chain Carlo (MCMC). The 18 model parameters were optimized using SUFI-2 algorithm based on min and max values suggested by Absolute SWAT Values in SWAT-CUP model.

2.3 Drought index

This study used meteorological, agricultural and hydrological drought index. SPI is a general meteorological drought index that is only based on precipitation. SPEI was calculated using precipitation and evapotranspiration. Computation of SPEI is based on monthly difference between precipitation and evapotranspiration. It was recommended to overcome the disadvantages of SPI on drought assessment. The PDSI by using a water balance model between moisture supply and demand for a two-layer soil is a tool to determine spatial and temporal variation in soil moisture stresses. Its calculation includes determination of the Z index, the monthly soil moisture indicator, which reflects the deviation of actual precipitation, and soil moisture supply from the expected values for the normal condition known as climatically appropriate for the existing condition [Karamouz et al. 2013]. SDI is calculated using cumulative streamflow volumes from different time periods.

3 RESULTS

3.1 Model configuration and optimization

SWAT was used to calculate soil moisture and runoff necessary to apply agricultural and hydrological drought index. The data to construct SWAT model were classified with two aspects, following Figure 1. One is the spatial data such as Digital Elevation Model (DEM, grid size 30 m), 1:25,000 land use and soil map. Another is the meteorological data such as daily precipitation (mm/day), max and min temperature (°), solar radiation (MJ/m^{3}day), wind speed (m/s) and humidity (%). This study used Sequential Uncertainty Fitting ver. 2 (SUFI-2) algorithm to optimize SWAT parameter about soil moisture and runoff. Two model performances were applied to quantify how well the data of optimized parameters. As a result, monthly and seasonal discharge variation are fitted well as shown in Figure 2.

3.2 Drought assessment

As shown in Figure 3, this study quantified and estimated the worst drought according to duration 90 day, 180 day and 270 day of various drought indices such as SPI, SPEI, PDSI and SDI. As a result, the most extreme drought showed −4.27 and −2.91 with duration 90 day of SPI and SDI. Drought severity of SPEI and PDSI has a high value in duration 180 day and 270 day different with results of those.

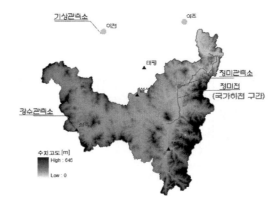

(a) DEM

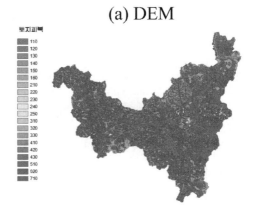

(b) Land use

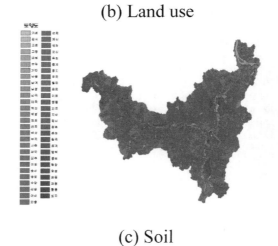

(c) Soil

Figure 1. Spatial input data of study area.

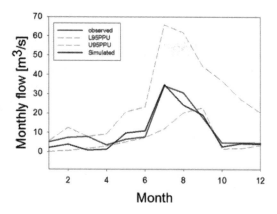

Figure 2. Result of optimization.

Table 1. Calculation of averaged drought of extreme drought severity.

Rank	SPI		SPEI	
	Year	Drought severity	Year	Drought severity
1	2001	−2.743	2015	−2.400
2	2015	−2.728	2014	−2.304
3	2014	−2.233	2001	−1.876
4	2000	−2.177	1988	−1.862
5	1988	−1.957	1994	−1.792

Rank	SDI		PDSI	
	Year	Drought severity	Year	Drought severity
1	2015	−2.547	2015	−8.219
2	2014	−2.307	2014	−6.186
3	2001	−1.889	2002	−5.742
4	2000	−1.479	2001	−5.036
5	1988	−1.388	1994	−4.635

3.3 Drought comparison

This study averaged extreme and average drought based on drought severity of drought indices according to durations, following Table 1 and Table 2. 2015 has the most vulnerable in SPEI, SDI, and PDSI in average of extreme drought. In average drought, 2014 showed the highest drought severity in SPI, SPEI and SDI. 2015 has the most vulnerable in overall average of extreme and average drought.

Frequency analysis was accomplished according to duration of drought indices. Kolmogorov-Simirnov method which was one of goodness of fit test was applied to select probability distribution. Results for 5% significance level based on calculated drought indices. Generalized Extreme Value (GEV) distribution was better than the other distributions in goodness of fit test. Drought indices showed different trend according to increase duration. Frequency based on

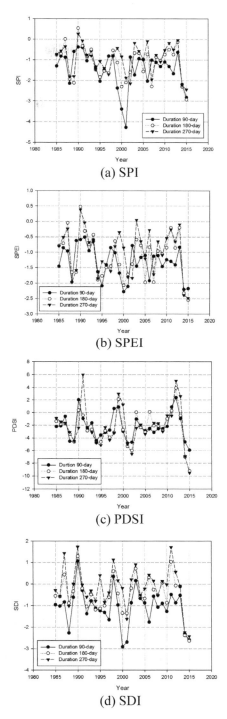

Figure 3. Calculation results of drought indices according to durations.

results of Table 1 was calculated as shown in Table 3. As a result, SPI and SDI have high possibility which was occurred extreme drought in duration 90 day. SPEI and PDSI showed low frequency in duration 180 day

Table 2. Calculation of averaged drought of average drought severity.

Rank	SPI		SPEI	
	Year	Drought severity	Year	Drought severity
1	2014	−1.190	2014	−1.190
2	2015	−1.179	2015	−1.179
3	2001	−0.891	2001	−0.891
4	1988	−0.677	1988	−0.677
5	1994	−0.665	1994	−0.665

Rank	SDI		PDSI	
	Year	Drought severity	Year	Drought severity
1	2014	−1.357	2014	−1.357
2	2015	−1.260	2015	−1.260
3	2001	−0.996	2001	−0.996
4	1988	−0.676	1988	−0.676
5	1994	−0.579	1994	−0.579

and 270 day as well as those of results of calculating drought severity.

This study calculated correlation coefficient between extreme and average drought. These have relatively high correlation between drought indices. SPI and SPEI showed a high correlation opposite to those of SPI and PDSI.

4 CONCLUSION

This study assessed drought of Cheongmicheon watershed using drought indices according to the duration 90 day, 180 day and 270 day. Meteorological drought index based on SPI and SPEI was calculated and agricultural and hydrological which were PDSI and SDI were applied using results of SWAT simulation. As a result, drought indices showed different severity and occurrence year of extreme drought. 2015 has the most vulnerable in overall average of extreme and average drought. Drought indices showed different trend according to increase in duration. In addition, extreme

Table 3. Results of frequency analysis for drought indices.

Rank	Duration	SPI Frequency	SPEI Frequency	SDI Frequency	PDSI Frequency
1	90-day	14.5	23.7	19.1	356.3
	180-day	24.2	19.2	46.7	39.9
	270-day	37.4	20.2	33.2	30.6
2	90-day	14.3	17.7	14.5	49.6
	180-day	23.6	14.7	34.1	13.3
	270-day	36.5	17.0	26.6	11.9
3	90-day	9.8	4.8	9.0	32.2
	180-day	10.2	4.5	19.7	10.5
	270-day	15.1	7.8	18.0	9.6
4	90-day	9.3	4.6	5.6	16.3
	180-day	9.2	4.3	11.5	7.2
	270-day	13.7	7.6	12.3	6.9
5	90-day	7.9	3.7	5.1	11.0
	180-day	6.4	3.6	10.2	5.8
	270-day	9.3	6.7	11.3	5.8

Table 4. Correlation coefficient between extreme drought.

	SPI3	SPI6	SPI9	SPEI3	SPEI6	SPEI9	SDI3	SDI6	SDI9	PDSI3	PDSI6	PDSI9
SPI3	-	0.591	0.344	0.888	0.570	0.389	0.754	0.521	0.452	0.358	0.258	0.198
SPI6		-	0.877	0.596	0.965	0.885	0.715	0.754	0.672	0.786	0.695	0.532
SPI9			-	0.347	0.828	0.978	0.566	0.679	0.679	0.808	0.821	0.698
SPEI3				-	0.630	0.449	0.724	0.560	0.447	0.394	0.286	0.274
SPEI6					-	0.869	0.699	0.790	0.692	0.755	0.654	0.498
SPEI9						-	0.618	0.733	0.715	0.813	0.819	0.690
SDI3							-	0.799	0.737	0.664	0.501	0.306
SDI6								-	0.917	0.737	0.681	0.411
SDI9									-	0.752	0.656	0.416
PDSI3										-	0.856	0.693
PDSI6											-	0.850
PDSI9												-

Table 5. Correlation coefficient between averaged drought.

	SPI3	SPI6	SPI9	SPEI3	SPEI6	SPEI9	SDI3	SDI6	SDI9	PDSI3	PDSI6	PDSI9
SPI3	-	0.900	0.754	0.986	0.884	0.752	0.984	0.928	0.812	0.827	0.629	0.299
SPI6		-	0.930	0.901	0.993	0.928	0.906	0.941	0.940	0.904	0.781	0.468
SPI9			-	0.770	0.922	0.993	0.771	0.828	0.866	0.935	0.930	0.737
SPEI3				-	0.901	0.780	0.978	0.922	0.807	0.843	0.652	0.338
SPEI6					-	0.933	0.897	0.936	0.932	0.892	0.770	0.465
SPEI9						-	0.774	0.835	0.865	0.922	0.915	0.725
SDI3							-	0.946	0.852	0.812	0.622	0.297
SDI6								-	0.927	0.816	0.670	0.336
SDI9									-	0.793	0.690	0.369
PDSI3										-	0.931	0.715
PDSI6											-	0.884
PDSI9												-

and averaged drought have relatively high correlation between drought indices excluding between PDSIs. However, each drought index showed different severity and occurrence year of extreme and averaged drought. It is useful to examine the sensitivity and accuracy of drought indices. Therefore, this study indicated that drought indices with various characteristics were used to analyze and quantify drought condition.

REFERENCES

Banimahd, S.A. & Khalili, D. 2013. Factors influencing Markov chains predictability characteristics, utilizing SPI, RDI, EDI and SPEI drought indices in different climatic zones. *Water. Resour. Manag* 27: 3911–3928.

Correia, F.N., Santos, M.A. & Rodrigues, R.P. 1991. Reliability in regional drought studies. *Water Resources Engineering Risk Assessment*: 63–72.

Ghaffari, G., Keesstra, S., Ghodousi, J. & Ahmadi, H. 2010. SWAT-simulated hydrological impact of land-use change in the Zanjanrood basin, northwest Iran. *Hydrol. Proc.* 24: 892–903.

Hong, X., Guo, S., Zhou, Y. & Xiong, L. 2014. Uncertainties in assessing hydrological drought using streamflow drought index for the upper Yangtze River basin. *Stoch Environ Res Risk Assess* 29(4): 1235–1247.

Jiang, R., Xie, J., He, H. & Luo, J. 2015. Use of four drought indices for evaluating drought characteristics under climate change in Shaanxi, China: 1951–2012. *Natural Hazards* 75(3): 2885–2903.

Karamouz, M., Nazif, S. & Falahi, M. 2013. *Hydrology and hydroclimatology: Principles and applications*. New York: CRC Press.

Morid, M., Smakhtin, V.U. & Moghadasi, M. 2006. Comparison of seven meteorological indices for drought monitoring in Iran. *Int. J. Climatol* 26: 971–985.

Smakhtin, V.U. & Hughes, D.A. 2007. Automated estimation and analyses of meteorological drought characteristics from monthly rainfall data. *Environ. Modell. Softw* 22: 880–890.

Tate, E.L. & Gustard, A. 2008. Drought characterization and monitoring in regions of Greece. *European Water*: 29–39.

Wilhite, D.A. & Glantz, M.H. 1985. Understanding: the drought phenomenon: The role of definitions. *Water. Int.* 10: 111–120.

Zarch, M.A.A., Sivakumar, B. & Sharma, A. 2015. Droughts in a warming climate: A global assessment of standardized precipitation index (SPI) and reconnaissance drought index (RDI). *J. Hydrol.* 526: 183–195.

Introduction of decision support system for design of LID based on SWMM5.1: A case study in Korea

J.Y. Song, E.S. Chung & S.H. Kim
Department of Civil Engineering, Seoul National University of Science and Technology, Seoul, Korea

S.-H. Lee
Department of Civil Engineering, Pukyong National University, Pusan, Korea

ABSTRACT: Climate change, urbanization, and land use change influences the urban environmental compositions and eventually occurs water related problems such as flood, drought, and water quality deterioration. Recently, Low Impact Development (LID) techniques, which can be effective alternatives to river rehabilitation, have been introduced and EPA SWMM has also developed an innovative LID control module that involves with infrastructure practices. However, the LID control module should be repeatedly simulated to develop the detailed design concept. Therefore, Water Management Analysis Module (WMAM), an effective decision support system (DSS) based on EPA SWMM5.1 for LID designing and planning was developed. In this study, the developed WMAM is applied to a case study in Korea to show the effectiveness and usefulness of it. WMAM can simulate multiple-scenarios of water management facilities as well as design and planning parameters, simultaneously, in a selected sub-catchment. By using its sensitivity analysis, it finds several powerful parameters among various LID design parameters. Moreover, it generates and simulates multiple scenarios of the watershed management facility's design and planning parameters, and it provides the most effective scenario for LID design and planning parameters in the hydrologic aspect. It was found that this DSS provides a better detailed design concept among various plausible LID options for water cycle rehabilitation.

1 INTRODUCTION

Due to climate change, urbanization, and land use change, the hydrological cycle has been severely distorted. Therefore, sustainable development is important for the rehabilitation of river environment and sustainable infrastructures including the natural water cycle should be included to the urban watershed planning and design. Low impact development (LID) is a term used to describe a land planning and engineering design approach to manage stormwater runoff.

Recently, EPA SWMM, which is a broadly used dynamic rainfall-runoff model for simulations in urban areas, has been updated to model the hydrologic performance of specific types of LID controls that are very useful (Rossman, 2010). However, developing a SWMM LID control and usage editor need time and effort to enter accurate values to find a prospective result for each parameter related to the water management design and planning. Moreover, decision support systems (DSSs) are need for integrated water resources management in South Korea.

Therefore, the objective of this study is to develop a DSS based on EPA' SWMM for the performance evaluation system for urban watershed management in a changing environment and apply the developed program to an urbanized area in South Korea.

2 WATER MANAGEMENT ANALYSIS MODULE

Water Management Analysis Module (WMAM) is a DSS for urban watershed planning and management based on EPA's SWMM5.1 (Fig. 1) (Song and Chung, 2015). WMAM uses sensitivity analysis on all design parameters related to the SWMM LID controls and derives the effective parameters which influence the hydrologic result. It also generates and simulates multi-scenario analysis of all plausible water management design and planning parameters to derive the best management scenario in hydrologic aspect. Fig. 2 shows 6 types of LID infrastructures such as bio-retention cell, rain garden, green roof, rain barrel, permeable pavement, and infiltration trench which can be simulated by WMAM and the procedure of WMAM.

Fig. 3 shows the procedure of WMAM. In step 1 and 2, users can load SWMM input files and select the study subcatchment and watershed management facility. In step 3 and 4, a simple sensitivity analysis will be done for design parameters. In step 5 to 7, a multi-scenario analysis of design parameters will be conducted and the superior scenarios will be listed in order of priority. Finally, in step 8 to 10, a multi-scenario analysis for planning parameters will

Figure 1. Main screen of WMAM.

Figure 2. Types of LID considered in WMAM.

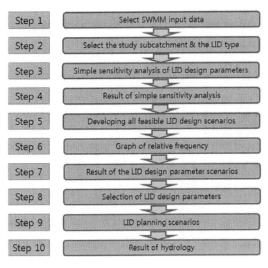

Figure 3. 10 step procedure of WMAM.

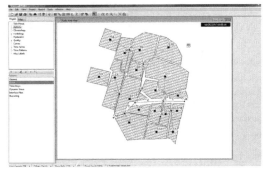

Figure 4. SWMM model map of study area.

be conducted based on a selected design scenario, and the effective scenarios will be shown in order of priority.

3 METHODOLOGY & STUDY AREA

In this study, the effectiveness of WMAM will be shown by comparing the results of the SWMM before constructing LID, after constructing LID, and by using WMAM. Infiltration trench, one of the LID structures, will be modeled in a selected subcatchment area. Peak runoff and infiltration loss will be compared by each case. Since LID structures are designed to maintain runoff rate and duration by infiltrating, storing, filtering and so on, it is more effective when peak runoff gets lower and infiltration loss gets higher.

Fig. 4 shows the SWMM modeling map of the study area based on a university in South Korea. The study area lies between latitudes 37 N, and longitudes 127 E. The total area is 508,690 m² while 92.7% of the area is covered by building sites. Subcatchment S5, out of 18 subcatchments, was selected as a target area for the construction of infiltration trench since the percentage of imperviousness was over 80% and the area was more likely to be developed. A daily time-series rainfall data from 17 Aug 2014 to 26 Aug 2014 of Seoul observatory (Korea) was used as a rainfall data for the SWMM modeling. A small event of rainfall was used because it is more effective for the LID (Gironas, et al., 2009).

At first, a simulation of SWMM before construction of infiltration trench was prepared. Secondly, infiltration trench was designed and planned at subcatchment S5 by using SWMM LID control and usage editor (Fig. 5). Design parameters such as storage height, surface roughness and planning parameters such as area of each unit, number of unit were estimated and entered by the field circumstances. Finally, WMAM was used to analyze various design and planning scenarios for the constructed infiltration trench at subcatchment S5. From the 4th step of WMAM, flow coefficient, seepage rate, thickness, and clogging factor were shown to be sensitive design parameters which influence the modeling result (fig. 6, up). 90 combinations of design parameters (Fig. 6, down) and 72 combinations of planning parameters were generated and simulated by WMAM.

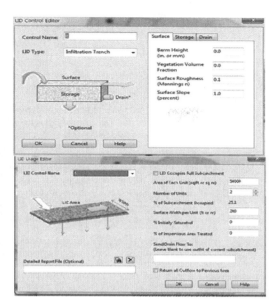

Figure 5. Design and planning of infiltration trench using SWMM LID control(up)/usage(down) editor.

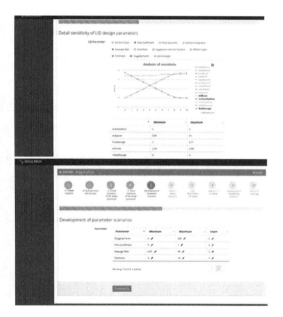

Figure 6. Sensitivity anlaysis (up) and multi-scenario analysis (down) of WMAM.

4 RESULTS

Fig. 7 and Fig. 8 shows the results of peak flow and infiltration loss of the SWMM model. For the comparison of three different cases of results, 1) a basic SWMM file with none-LID management, 2) a SWMM file with *Infiltration Trench* designed and planned in a single subcatchment area, and 3) WMAM were simulated.

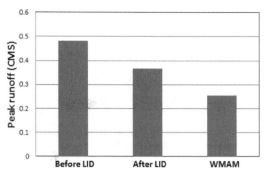

Figure 7. Comparison of peak runoff before/after LID and WMAM.

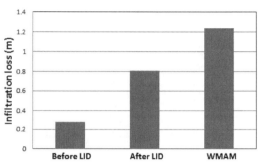

Figure 8. Comparison of infiltration loss before/after LID and WMAM.

The peak runoff when there was no infiltration trench in subcatchment S5 showed the highest value, 0.48 CMS. Peak flow after planning infiltration trench and using WMAM were followed by 0.37 and 0.26, respectively. The result of WMAM showed the highest value of infiltration loss 1.22 m. The SWMM result after planning several infiltration trenches and before planning them followed by the value of 0.80 m and 0.27 m, respectively.

LIDs are known as green infrastructure that considers sustainable water management by storing, delaying runoffs, and eventually prohibits water related problems like flood by reducing the direct flow and water deficit by storing in urbanized area. According to the compared results from Fig. 7 and Fig. 8, *peak runoff* was reduced and *infiltration loss* was increased when WMAM was used. Therefore, it is found to be obvious that using WMAM can be effective in designing and planning LID structures.

5 CONCLUSIONS

By the lack of DSS for urban watershed management and IWRM, we developed a DSS, WMAM, for urban watershed planning and management based on EPA's SWMM5.1. WMAM generates and simulates multi-scenario analysis for LID design and planning parameters and develops the best water management design and planning scenarios in hydrology. By using WMAM, users can:

- Seek sensitive parameters for LID which influences the hydrological results from SWMM (4th step)
- Derive the best scenario for LID design parameter (7th step)
- Reach to an effective watershed management planning (10th step).

Also, as shown is the case studies, using WMAM was more effective than just using the SWMM LID control and usage editor to design and plan a green infrastructure. Since the developed DSS is a web-based, user-friendly tool, it is easy to learn and use. Furthermore, it can simulate a large number of LID design and planning scenarios in fast speed so that users can save their time and effort. Still, WMAM has some weakness of post-process tool, lack of high-performance servers, and several errors to be improved. However, further works on WMAM will be on progress and it is sure to be useful for the users and will contribute for the sustainable development of urban water management and IWRM.

REFERENCES

L.A. Rossman, 2010, *Storm Water Management Model User's Manual Version 5.0*. National Risk Management Research Laboratory, Environmental Protection Agency, Cincinnati, Ohio.

J.Y. Song, E.-S. Chung, 2015, Decision Support System Framework for Design of Low Impact Development Based on SWMM5.1 Using Multi-scenario Analysis. International Conference on Water Management Modeling, Toronto, Canada.

J. Gironas, L.A. Roesner, and J. Davis, 2009, *Storm Water Management Model Applications Manual*. National Risk Management Research Laboratory, Environmental Protection Agency, Cincinnati, Ohio.

Geotechnical engineering and geoinformatics

Engineering Challenges for Sustainable Future – Zawawi (Ed.)
© 2016 Taylor & Francis Group, London, ISBN 978-1-138-02978-1

Effects of tunnel face distance on surface settlement

A. Marto & H. Sohaei
Faculty of Civil Engineering, Universiti Teknologi Malaysia, Johor, Malaysia

M. Hajihassani
Department of Mining, Urmia University, Urmia, Iran

A.M. Makhtar
Faculty of Civil Engineering, Universiti Teknologi Malaysia, Johor, Malaysia

ABSTRACT: Underground activities such as tunnel excavation may cause soil movements around the excavation area and subsequently the ground surface. Several tunnel construction methods have been developed during the last decades and the most significant effort is to reduce the ground surface settlement. New Austrian Tunnelling Method (NATM) is one of the well-known methods which is widely used in the construction of tunnels and metro stations. This paper discusses the effects of excavation sequence and heading distance on the surface settlement induced by tunnelling in Side Gallery (SG) method, as an alternative to NATM, in Karaj Metro Tunnel (KMT) project, Iran. The Abaqus software was used to simulate the Finite Element modelling of both methods. The results showed that the removal of top head in NATM for KMT project caused 80% of the total surface settlement, but it was only 60% in SG method due to the existence of middle liner.

1 INTRODUCTION

Tunnel is a significant underground structure which could help to reduce the congestion at the urban cities when used as transportation purposes. Nowadays, the demand for tunnel excavation has been extensively increased in terms of rail or road tunnel system. Since then, tunnels were progressively utilized in urban transportation system due to the expansion of the cities as the result of economic and population growth. Meanwhile the tunnel construction has been practiced for more than a century, the design has improved gradually based on the analytical solutions proposed by engineers and the advantages of computer technology. Various numerical analyses have been applied to simulate the tunnel construction in different cases. Tunnel excavation affects the in-situ stresses distribution, which sometimes leads to ground deformations. In several cases, damage occurred to the overburden structures and properties (Sohaei et al., 2011). The New Austrian Tunnelling Method (NATM) was introduced by Rabcewicz (1964) and later by Muller (1978) in Austria. Using the NATM tunnelling method has been known to be cost-saving and it has control over the settlement. This method utilizes a thin and flexible shell of shotcrete as a temporary support system in the excavated exposed areas through the soft soil (Rabcewicz, 1964; Muller, 1978; Sohaei et al., 2011; Marto et al., 2013). A number of researchers have been investigating the effects of heading distance on NATM tunnelling operation, such as Farias et al. (2004), Karakus, and Fowell (2009), Masin (2009), Ng et al. (2004) and Yoo (2009). However, limited work has been published regarding the surface settlement induced by heading distance in terms of different partial face excavation on NATM tunneling method. The NATM has been utilized in a large tunnel dimensions since the complete face excavation in the soft soils caused the large settlement and subsequently damage to the surface structures. Reduction of the face excavation into several cross-sections due to NATM tunneling method causes reduction of the stress distribution and yield zone around the tunnel. The cross-section is excavated in each stage of the excavation sequence. One of the important features in NATM tunnelling is excavation sequence design, which depends on the soil condition, tunnel geometry and tunnelling requirements which may reduce the yield zone, and consequently, may minimize the displacement. However, heading distance has also been shown to cause major settlement due to the inability of the soil to restraint and its response to the support system offered (Marto et al., 2013).

A study was conducted to determine the effect of excavation sequence on ground movements due to NATM method of tunneling. Line No.2 of Karaj Metro Tunnel (KMT) project in Iran was used as the case study. In order to determine the optimum excavation sequence for KMT, another method of excavation sequence was investigated. The Side Gallery (SG) cross sections with different multi-face excavation and specific heading distance was introduced. The

Table 1. Geotechnical properties of soil layers (Tunnel Rod, 2011).

Soil Layer	Thickness (m)	Unit Weight (kN/m^3)	Young's Modulus (MPa)	Poisson's Ratio	Friction Angle (°)	Cohesion (kN/m^2)	Permeability (m/s)
Top soil	1.5	17	40	0.4	20	50	1×10^{-9}
L1-1	8	18.6	40	0.4	20	50	1×10^{-9}
L2-1	13	21.1	50	0.35	32	35	1×10^{-7}
L2-2	19	21.4	80	0.3	36	25	2.5×10^{-4}

methods were analyzed and the optimum excavation sequence was determined.

2 CASE STUDY

A new subway system was constructed in Karaj; a large city of Iran, located 40 km to the west of the capital city, Tehran. To address the long-standing traffic congestion problem in this city, the new subway system was constructed. Tunnel construction by Tunnel Rod Company was a part of 27 km of the Line No.2 KMT between Kamal-Shahr and Malard, in the northwestern and south of the Karaj, respectively. The line composed of a single tunnel that was excavated using NATM technique. Based on the observations during excavations it has been found that the subsurface soils are composed of inorganic clay with sandy clay at the top layer, followed by sandy clay (L1-1) and silty sand with over-layer clayey (L2-1) and silty gravel (L2-2) with the properties as shown in Table 1.

The tunnel depth, which is the distance between the crown and the ground surface, was approximately 7 to 14 m, representing a shallow tunnel. The average tunnel depth was taken as 12.5 m. Settlement markers were placed approximately at 25–100 m intervals along the tunnel alignment. The NATM construction technique had been selected for KMT based on geotechnical analysis and economic studies. Tunnelling was designed into two sections of top head and bottom head. The top head was excavated in one step of 1.2 m long upon reaching the 120 m excavation distance from the bench. The bottom head was excavated in two steps, in which the bench was excavated first and then followed by the corner. Finally, subsequent to the installation of a waterproof membrane, completed excavation areas were supported using concrete tunnel liner.

Figure 1 shows the ground profile in the case study area. According to Figure 1a, settlement markers were placed approximately at 25–100 m intervals along the tunnel alignment (line N). This amount was reduced in critical areas such as the portal of stations in order to provide a more accurate inspection. For each transverse section (line M) there are five (5) monitoring points installed at the center, right, and left sides of the tunnel. The point markers have a distance of 7.5 m from each other as shown in Figure 1a. The monitoring points were made of steel rod and grouted about 100 cm to the ground in order to isolate from asphalt movement.

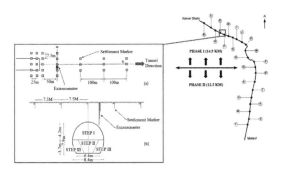

Figure 1. Cross section of the tunnel and the locations of settlement markers at Line No.2 of KMT project.

3 NUMERICAL ANALYSIS

The numerical analysis was conducted using Abaqus software. The analysis was divided into three stages; Stage I started from initial stage consisted of the horizontal stress, which represents the coefficient of lateral earth pressure and vertical geostatic stresses which represents active pressure in initial soil condition. Once the setting of the whole initial condition is completed, the next task involved checking the results of the model in terms of displacement, in which it has to be near to zero. Stage II was geostatic step, which was used for equilibrium between boundary conditions and gravity load in every element of the soil. Stage III was the tunnelling processes. Numerical simulation of the tunneling involved with the Abaqus software was modeled based on two main parts; general and tunnel construction parts. The general part is to draw the geometry dimensions, boundary conditions, meshing and material properties application. The tunnel construction part consists of excavation progress and the installation of shotcrete shell liner. The tunnel is assumed to be excavated in the homogeneous material with top soil and three layers of soils. The thickness of the top soil, layers L1-1, L2-1 and L2-2 were 1.5 m, 8 m, 13 m, and 19 m, respectively. The tunnel was constructed through layer L2-1 which was approximately 12.5 m from the crown below the ground surface.

In terms of the tunnel excavation process, the simulation of the model excavation made used of the de-activation of the elements set. Thus, the software calculated tension of re-distribution from the removing part. Adding a cover on the tunnel wall was conducted with re-activating the coverage elements. Since

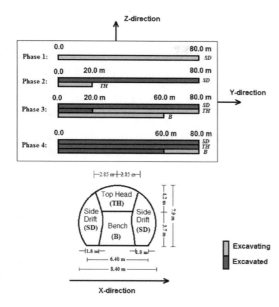

Figure 2. Tunnel profile and excavation sequence of Side Gallery (SG) method.

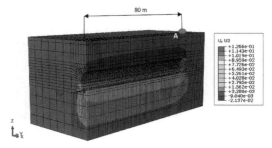

Figure 3. Monitoring point on the tunnel axis at ground surface of the SG model.

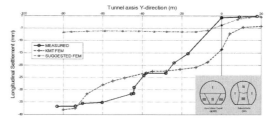

Figure 4. Longitudinal surface settlement for measured and FE predicted of KMT and Side Gallery model.

the model-change has been used to simulate the tunnelling model, therefore, the mesh size was chosen for one step of the excavation. The current KMT was excavated using NATM technique and the tunnel was supported with shotcrete. The KMT tunnel width in the horseshoe-shape was 8.4 m wide and 6.4 m bottom wide. In simulating the Side Gallery (SG) tunnel construction method using FE analysis, the tunnel dimensions used were similar with the actual method (NATM) in advancing the KMT. The shotcrete thickness at 30 cm on the tunnel curve and the thickness of 15 cm for the central shotcrete had been considered as the supporting system. The SG process was simulated by excavating the side drift first, with 1.80 m wide in the narrow openings and 1.20 m into the tunnel face (Y-direction) for each excavation step. After excavation of each 1.2 m, the exposed area was supported as a next step.

After completing the side drift excavation up to 80 m distance, 4.2 m height of the top head was excavated up to 20 m using the same excavation rate as KMT. Each excavation step was 1.2 m in Y-direction. In the next period of the excavation, it started with the bench excavation and the removal of middle lining at the same time. This step of the excavation is carried out together with the excavation of the top head up to 80 m. The process continued until the whole area of 80 m tunnel was excavated. Figure 2 represents the excavation process in the Side Gallery model of the excavation.

4 ANALYSIS AND RESULTS

With the aim of determining the optimal face distance between top heading and bench for SG model, a monitoring point was selected at chainage 80 m (Point A in Figure 3). The optimal face distance is the amount of distance that has the minimum effect on the surface settlement. To find this distance, the cumulative surface settlement at monitoring Point A was recorded when the procedure of tunnel excavation was started.

Figure 4 shows the comparison of longitudinal surface settlement between the measured data, KMT (using NATM) and SG models. The results from the finite element modelling have shown reasonable matching with the measured settlements data for the current KMT excavation model. The results show that the longitudinal surface settlement for recommended SG model is significantly less than the measured settlement and the values obtained from simulating the KMT excavation model using NATM.

Figure 5 shows the cumulative surface settlement at different steps (each step was 1.2 m distance). Surface settlement took place at chainage 80 m when the side drifts reached step 53 (Chainage 64 m). With side drifts advancing, the surface settlement increased rapidly up to step 70, when the side drifts reached to monitoring Point A. Between steps 70 to 130 the settlement was steady state due to the top head and bench faces were far away from monitoring Point A. Surface settlement increased significantly between steps 130 (chainage 56 m) to 140 (chainage 80 m), when the faces of top head and bench reached to the monitoring Point A. The results show that when the top head and bench faces was 24 m (18.6 m + 5.4 m) away from side drifts faces, the surface settlement increased sharply. Therefore, the optimal face distance for this SG method is 24 m.

From Figure 5 it can be seen that excavating the side drifts resulted with 1.5 mm settlement while the settlement occurred was 2.2 mm after excavating the top head and bench. The results show that from the total of

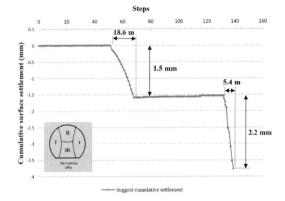

Figure 5. Cumulative surface settlement on monitoring Point A for Side Gallery method.

3.7 mm settlement, about 60 % of the total settlement occurred as the result of excavating the top head and the bench. Marto *et al.* (2013) reported that tunnelling using NATM for KMT caused 80% total surface settlement upon excavating the top head. Hence, the middle lining in SG method shows significant effect in controlling the soil movement around the tunnel which leads to a major reduction of the surface settlement.

5 CONCLUSION

Three-dimensional (3D) simulation using Abaqus software was conducted to simulate finite element analysis of the ground surface settlement induced by tunneling. The Side Gallery (SG) tunneling method was used to determine the effects of the excavation sequence and heading distance on the surface settlement. Using the Karaj Metro Tunnel (KMT) project as the case study, it can be concluded that the top head excavation caused major influence to the surface settlement. Through SG method of tunnelling, the top head excavation only caused about 60% of the total surface settlement compared to 80% through using NATM. This indicated that the middle lining of SG method has a significant effect in controlling the soil movement around the tunnel. On the other hand, the optimal face distance for SG method was found to be 24 m, which was less than the distance obtained using NATM (35 m). It could be concluded that the optimal face distance, the longitudinal settlement and the excavation time were reduced when excavating the tunnel using the Side Gallery method compared to the New Austrian Tunnelling method at the Karaj Metro Tunnel project in Iran.

ACKNOWLEDGEMENT

The authors gratefully acknowledge the financial support given by the Ministry of Higher Education (MOHE), Malaysia and the Universiti Teknologi Malaysia (UTM) through a research project under the Research University Grant (RUG) scheme, Vote No. Q.J130000.2522.06H60. The cooperation given by all parties in undertaking this research is also greatly acknowledged, in particular the researchers from Soft Soil Engineering Research Group (SSRG) of Innovative Engineering Research Alliance / Faculty of Civil Engineering, UTM.

REFERENCES

Farias, M.M., Junior, A.H.M., and Assis, A.P. 2004. Displacement Control in Tunnels Excavated by the NATM: 3-D Numerical Simulations, Tunnelling and Underground Space Technology, 19, 283–293.

Karakus, M., and Fowell, R.J. 2009. Back analysis for tunnelling induct ground movement and stress redistribution, Tunnelling and Underground Space Technology, 514–524.

Marto, A., Sohaei, H., Hajihassani, M. and Namazi, E. 2013. Prediction the Effect of Excavation Sequence on Ground Movements During NATM Tunnelling Through Finite Element Method. New Developments in Structural Engineering and Construction, pp. 736–740.

Masin, D. 2009. 3D Modelling of an NATM Tunnel in High K0 Clay using Two Different Constitutive Models. Geotechnical and Geoenvironmental Engineering ASCE, 1326–1335.

Muller, L. 1978. The Reasons for Unsuccessful Applications of the New Austrian Tunnelling Method, Tunnelling under Difficult Conditions, International Tunnel Symposium, Tokyo, Pergamum. 67–72.

Ng, C.W.W., Lee, K.M., and Tang, D.K.W. 2004. Three-Dimensional Numerical Investigations of New Austrian Tunnelling Method (NATM) Twin Tunnel Interactions. Canadian Geotechnical Journal, 41, 523–539.

Rabcewicz, L. 1964. The New Austrian Tunnelling Method. Part one. Water Power (November), 453–457. Part two. Water Power, 511–515.

Sohaei, H., Hajihassani M., Marto A. and Karimi Shahrbabaki M. 2011. Influence of Construction Stages on Surface Settlement in NATM Tunnelling. Proceeding of the Euro Asia Civil Engineering Forum (EACEF), Indonesia.

Tunnel Rod Construction Consulting Engineers Inc. Instrumentation Report of Line No.2 of Karaj Urban Railway, *Technical Report. 2011.*

Yoo, C. 2009. Performance of Multi-Faced Tunnelling, A 3D Numerical Investigation. Tunnelling and Underground Space Technology, 24, 562–573.

Characterization of Pb and Cd contaminated sandy soil by dielectric means

H.M. Al-Mattarneh
Civil Engineering Department, Najran University, Najran, Saudi Arabia

R.M.A. Ismail, M.F. Nuruddin & N. Shafiq
Department of Civil and Environmental Engineering, Universiti Teknologi PETRONAS, Seri Iskandar, Perak, Malaysia

M.A. Dahim
Civil Engineering Department, King Khalid University, Abha, Saudi Arabia

ABSTRACT: Soils normally have low levels of heavy metals. However, in industrial, agricultural and municipal wastes areas heavy metals concentrations may be higher. Extreme amounts of heavy metals can be harmful to both human and to the fauna and flora. This study was transacted to investigate the level and type of heavy metals such as lead (Pb) and cadmium (Cd) in the saturated sandy soil. In this paper, the permittivity of contaminated soil by lead and cadmium was measured at the frequency 1 kHz–1000 kHz using LCR meter. The results indicate that both dielectric constant and loss factor of heavy metal contaminated soil decrease with increasing frequency. Dielectric constant and loss factor of soil could be used to determine type and level of heavy metal in soil. Complex permittivity of soil using parallel plate capacitor is a promising tool for evaluation of soil pollution by heavy metals.

1 INTRODUCTION

Soil is paramount in some way to all living things; it is a fundamental element of life on our planet. Our survival based on a nourishment of soil as well as that of all organisms. Soils normally have low levels of heavy metals. However, in industrial agricultural and municipal wastes areas heavy metals concentrations may be higher (Shukurov et al. 2005). Extreme amounts of heavy metals can be dangerous to both human and to the fauna and flora (Ahmad & Shuhaimi 2010, Mathews & Fisher 2009). The hazardous degrees of the heavy metal rely on their consistency in soil materials. One of the major sources which participate to increase the levels of heavy metals in soil, seawater and freshwater are oil pollution (Onojake & Frank 2013).

Soil can be easily contaminated by many sources of contamination such as solid waste, industrial wastes, and agricultural substances (fertilizers, pesticides) which compose a major source of heavy metals, corrosion of metal like fences and roofs can also be as heavy metals source (Alloway 1995).

It is known that man behavior has raised the heavy metals amount in the environment by many ways; include industrial production and agricultural production almost all of these materials contain heavy metals. Lead and Cadmium and other heavy metals have negative effects on marine life which directly effects on the seafood (Borg & Johanson 1989).

Soil pollution with heavy metals nowadays became one of the environmental concerns because it causes a financial problem to the landowners. So, the environmental assessment for the soil which contaminated with heavy metals has received high attention in environmental studies (Osakwe et al. 2003).

A good strategy for soil sampling, the appropriate survey and samples analytical methods and the accurate interpretation of the sampling results are required (Wong & Xiangdong 2003).

Several electromagnetic method was used for characterization of soil material (Bhat et al. 2007). These methods include ground penetration radar (Redman & Annan 1992), electrical resistivity (Schwartz et al. 2008), resonant cavity (Johri & Roberts 1990), time domain reflectometer (Rao & Singh 2011) and parallel plate (Johri & Roberts 1990, Mittelbach et al. 2012). Several applications of electromagnetic methods for soil science were investigated such as soil moisture content (Al-Mattarneh et al. 2013), soil density (Dobson et al. 1985), and petroleum contamination (Al-Mattarneh et al. 2014). Parallel plate capacitor has advantages over other methods like low cost, simple, safe and fast (Al-Mattarneh 2014). This paper investigates the use of parallel plate capacitor for assessment of soil contaminated with heavy metals.

2 BASIC THEORY

The response of materials to alternating electric fields is characterized by a complex permittivity (ε^*). Usually, the complex permittivity is separated into real and imaginary parts, which is done by Equation 1.

$$\varepsilon^* = \varepsilon' - j\varepsilon'' \quad (1)$$

Where, ε' is the real part of the permittivity called dielectric constant, which is related to the stored energy within the material and ε'' is the imaginary part of the permittivity called dielectric loss factor, which is related to the dissipation (or loss) of energy within the material.

If the material forms a part of an electronic circuit using parallel plate capacitor, the admittance of the material can be measured. The admittance is a complex number or quantity has real component (Y') called conductance (G) and an imaginary component (Y") called susceptance (B) as given in Equation 2.

$$Y = Y' + jY'' = G + jB \quad (2)$$

Using LCR meter, the admittance (Y) could be measured if the parallel plate capacitor is filled by dielectric material such as soil. The dielectric properties of the soil material could be related to the measured admittance of the material using the following Equations.

$$C_o = \frac{\varepsilon_o A}{d} \quad (3)$$

$$\tan\delta = \frac{Y'}{Y''} \quad (4)$$

$$\varepsilon' = \frac{Y''}{\omega C_o} \quad (5)$$

$$\varepsilon'' = \varepsilon' \tan\delta \quad (6)$$

Where C_o is the capacitance when the capacitor filled by vacuum, ε_o is the permittivity of air, A is the area of the plate electrode, d is the distance between the two electrodes and ω is the angular frequency.

3 DIELECTRIC MEASUREMENT SYSTEM

Copper square plates of 80 × 80 × 2 mm were attached to two adverse flat of the dielectric cell. The test cell was of internal dimensions 80 × 80 × 40 mm is shown in Figure 1. The dielectric cell connected with LCR meter by the fixture to obtain the admittance measurements at frequency range 1 kHz–1000 kHz. The complex permittivity such as dielectric constants and loss factors for all soil samples were calculated using the equations presented in the previous section.

4 MATERIAL AND PROPERTIES

In this study, two heavy metals Pb and Cd were to investigate the effect of heavy metal concentrations on the dielectric properties of saturated sandy soil. The basic properties of these two heavy metals are given in Table 1. For each heavy metal, five samples were

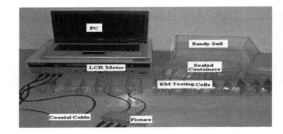

Figure 1. Dielectric measurement system.

Table 1. Basic properties of Lead and Cadmium.

Basic Properties	Cadmium (Cd)	Lead (Pb)
Density (g/cm^3)	8.65	11.34
Electrical resistivity (nΩm)	72.7	208
Atomic number	48	82
Standard atomic weight	112.414	207.2
Crystal structure	hexagonal close-Packed	face-center cubic

Table 2. Samples of Lead-contaminated soil.

Sample code	Lead (mg/l)	Weight (gram)			
		Soil	Water	Pb	Air
Pb0WS	0.0	500	126.26	0.0000	0.00
Pb1WS	2.5	500	114.72	0.0025	0.00
Pb2WS	5.0	500	103.18	0.0050	0.00
Pb3WS	7.5	500	91.64	0.0075	0.00
Pb4WS	10.	500	80.10	0.0100	0.00

Table 3. Samples of Cadmium-contaminated soil.

Sample code	Cadmium (mg/l)	Weight (gram)			
		Soil	Water	Cd	Air
Cd0WS	0.0	500	126.26	0.0000	0.00
Cd1WS	2.5	500	114.72	0.0025	0.00
Cd2WS	5.0	500	103.18	0.0050	0.00
Cd3WS	7.5	500	91.64	0.0075	0.00
Cd4WS	10.	500	80.10	0.0100	0.00

prepared at different heavy metal contents. Table 2 and Table 3 show the samples details for Pb and Cd respectively. All the samples were prepared from the sandy soil with porosity N = 0.4. The sand was classified according to unified classification system as well graded sand (SW). Sandy soil was dried by putting it in the oven for 24 hours at 105°C then sieved through the sieve number 10. Different water amount was added to the soil samples and the measured quantities of Pb and Cd were added, after that the samples were placed in the electromagnetic cell for testing.

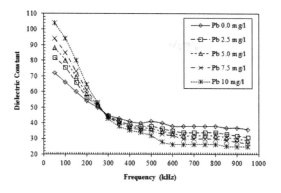

Figure 2. Dielectric constant of Pb contaminated soil versus frequencies.

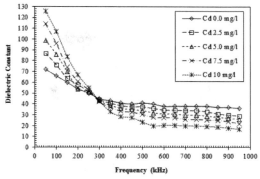

Figure 4. Dielectric constant of Cd contaminated soil versus frequencies.

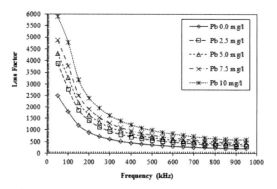

Figure 3. Loss factor of Pb contaminated soil versus frequencies.

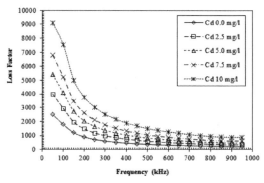

Figure 5. Loss factor of Cd contaminated soil versus frequencies.

5 RESULT AND DISCUSSION

The effect of Pb and Cd in saturated sandy soil using dielectric method was investigated. The results of the permittivity of Pb-contaminated soil at different frequencies are shown in Figures 2–3. Both dielectric constant and loss factor of Pb-contaminated soil decrease with increasing frequency. This trend may attribute to the low current conductance at high frequency and lower atomic polarization. At a frequency below 300 kHz the dielectric constant increase with increasing Pb content whilst at a frequency more than 300 kHz the contrary trend is observed. Loss factor of soil increase with increasing Pb content overall frequency this is a result of the higher conductivity of heavy metal.

The results of the permittivity of Cd-contaminated soil at different frequencies are shown in Figures 4–5. Both dielectric constant and loss factor of Cd-contaminated soil decrease with increasing frequency. This trend may attribute to the low current conductance at high frequency and lower atomic polarization. At a frequency below 250 kHz, the dielectric constant increased with increasing Cd content whilst at a frequency more than 300 kHz the contrary trend is observed. Loss factor of soil increase with increasing Cd content overall frequency this is a result of the higher conductivity of heavy metal.

The result of loss factor of soil sample with lead is very low in comparison with the result of soil contaminated with cadmium. This result may attribute to the low resistivity and high conductivity of cadmium in compare with lead. The electrical resistivity of cadmium is 72.7 nΩm while the resistivity of lead is 208 nΩm. The results also indicate that at a frequency around 260 kHz the trend of the dielectric constant of lead samples is changed while this frequency is shifted to 280 kHz for cadmium samples. This may attribute to the higher density of lead (11.34 g/cm^3) compared with the density of cadmium (8.65 g/cm^3) which may change the resonant frequency of the two heavy metals.

6 CONCLUSIONS

The dielectric permittivity of contaminated soil by Pb and Cd were calculated at the frequency from 1 kHz to 1000 kHz using LCR meters. The results show that both dielectric constant and the loss factor decrease with increasing the operating frequency. A change of trend of dielectric constant versus frequency was observed at 260 kHz for lead contamination and 280 kHz for cadmium contamination. These results can be referred to the ionic polarization, electrical resistivity and density of Pb and Cd. Parallel plate capacitor seems to be a useful tool to determine complex permittivity of soil samples and determine the level and type of heavy metal in soil. Calibration and validation procedures for the new sensor for on-site dielectric pollution sensor are ongoing. A prototype of the new pollution device will be developed.

ACKNOWLEDGEMENTS

The authors would like to thank the Deanship of Research at Najran University, Kingdom of Saudi Arabia for financially supporting of this research under the grant of development of onsite soil pollution sensor.

REFERENCES

Ahmad, A. K. & Shuhaimi, O. M. 2010. Heavy Metal Concentrations in Sediments and Fishes from Lake Chini, Pahang, Malaysia. *Journal of Biological Sciences* 10(2): 93–100.

Alloway, B. J. 1995. Soil Processes and the behavior of Metals. In: *Heavy Metals in Soil*. Alloway, B. J. (ed.) Blackie Academic & Professional, London.

Al-Mattarneh, H. 2014. Enhancement of Parallel Plate Sensor for Electromagnetic Characterization of Material, *European Journal of Scientific Research* 120(3): 348–359.

Al-Mattarneh, H. Alwadie, A. Dahim, M. Ismail R. & Malkawi, A. 2013. Dielectric Spectra and Modeling for Determination Soil Moisture Content, *The 29th Annual International Conference on soil Sediment, Water and energy*, University of Massachusetts, Amherst, MA, USA, October 21–24.

Al-Mattarneh, H. Dahim, M. Ismail R. & Nuruddin, M. F. 2014. Determination of Soil Polluted with Kerosene Using Electromagnetic Cell, *Applied Mechanics and Materials* 567: 183–188.

Bhat, A. Rao, B. & Singh, D. 2007. A generalized relationship for estimating dielectric constant of soils, *Journal ASTM International* 4(7): 1–17.

Borg, H. & Johanson, K. 1989. Metal Fluxes to Swedish Forest Lakes. Water, *Air and Soil Pollution* 47: 427–440.

Dobson, M. C. Ulaby, F. T. Hallikainen M. T. & El-Rayes, M. A. 1985. Microwave dielectric behavior of wet soil. Part II: Dielectric mixing models. *IEEE Trans Geosci. Remote Sensing* 23(1): 35–46.

Johri, G. K. & Roberts, J. A. 1990. Study of the dielectric response of water using a resonant microwave cavity as a probe, *J. Phys. Chem.* 94(19): 7386–7391.

Mathews, T. & Fisher, N. S. 2009. Dominance of dietary intake of metals in marine elasmobranch and teleost fish. *Science of the total environment* 407(18): 5156–5161.

Mittelbach, H. Lehner, I. & Seneviratne, S. 2012. Comparison of four soil moisture sensor types under field conditions in Switzerland, *Journal Hydrology* 430–431: 39–49.

Onojake, M. C. & Frank, O. 2013. Assessment of heavy metals in a soil contaminated by oil spill: a case study in Nigeria. *Chemistry and Ecology* 29(3): 246–254.

Osakwe, S. Otuya, B. & Adaikpoh E. 2003. Determination of Pb, Cu, Ni, Fe, and Hg in the soils of Okpai Delta State, Nigeria. *Nigerian Journal of Science and Environment* 3: 45–49.

Rao, B. and Singh, D. 2011. Moisture content determination by TDR and capacitance techniques: a comparative study, *International Journal Earth Science Engineering* 4(6): 132–137.

Redman, J. D. & Annan, A. P. 1992. Dielectric permittivity in a sandy aquifer following the controlled release of a DNAPL, *Fourth International Conference on Ground Penetration Radar*, Rovaniemi, Finland, June 8–13, 191–195.

Schwartz, B. Schreiber, M. & Yan, T. 2008. Quantifying field-scale soil moisture using electrical resistivity imaging, *Journal Hydrology* 36(2): 234–246.

Shukuroy, N. Pen-Mouratov, S. & Steinberger, Y. 2005. The impact of the Almalyk Industrial Complex on soil chemical and biological properties, *Environmental Pollution* 136(2): 331–340.

Wong, C. S. C. & Xiangdong, L. 2003. Analysis of Heavy Metal Contaminated Soils. Pract. Periodical of Haz., *Toxic, and Radioactive Waste Management* 7(1): 12–18.

Microbially induced cementation to improve the strength of residual soil

Murtala Umar & Khairul Anuar Kassim
*Department of Geotechnics and Transportation Engineering, Faculty of Civil Engineering,
Universiti Tecknologi Malaysia, Johor Malaysia*

Zaharah Ibrahim
*Department of Biosciences and Health Sciences, Faculty of Biosciences and Medical Engineering,
Universiti Tecknologi Malaysia, Johor Malaysia*

ABSTRACT: Microbial carbonate precipitation as a soil improvement technique has the prospect of becoming a substitute to traditional soil improvement in terms of performance and environmental sustainability. This study examines the carbonate biomineralization process mediated by the urease active strain of *Sporosarcina pasteurii* (ATCC® 11859™) in tropical residual soil. The effects of bacteria concentrations, curing conditions and treatment durations were the major parameters evaluated. Bacteria concentrations of 1×10^5 cfu/ml and 1×10^6 cfu/ml and 0.5 M concentration of cementation reagents were used for the study. The soil specimens were cured at atmospheric temperature and 40°C and treated for 24, 36, 48 and 60 hours durations. Shear strength and calcite precipitated at each treatment were evaluated. The results indicated a general increase in strength with increase in curing temperature; an indication of temperature influence in bacterial activity. Meanwhile, higher concentration of bacteria of 1×10^6 cfu/ml resulted in higher strength and calcite content.

1 INTRODUCTION

In many civil engineering works, soil encountered as foundation or construction material often do not satisfy the expected requirements in terms of mechanical and engineering properties. As such, buildings and other civil infrastructures founded on these loose, weak or soft sediments may require some preventive measures to avoid structural damage Van Paassen, 2011). Hence, some of the measures taken by geotechnical engineers include different soil improvement techniques to modify the engineering properties of the soil in order to serve an intended purpose. Meanwhile, according to (Gunaratne, 2013) many techniques of soil improvement are practically utilized, most of which are grouped into three categories. These include compaction, reinforcement and fixation. Fixation involves binding the soil particles to improve their strength and decrease compressibility; jet and permeation grouting falls under this category.

Though, most of these techniques have proved successful in improving the engineering properties of soil, application of some of these methods usually require high amounts of energy, costs, have limitations with regards to treatment range and require materials which have considerable impact on the environment (Karol, 2003). Chemical grouting has commonly been used as a soil improvement technique due to its economic benefits. It is usually achieved with a variety of additives including cement, lime, asphalt, sodium silicate, lignin, urethane and resins. Though many of these additives were found to be successful in improving the engineering properties of the soils, they may often contaminate the soil and groundwater (DeJong et al., 2006, Karol, 2003). These approaches create environmental concerns over their field application and are increasingly under the scrutiny of public policy and opinion; in fact all chemical grouts except sodium silicate are toxic and/or hazardous (DeJong et al., 2010, Karol, 2003). Hence, the need for new and sustainable methods of soil improvement is inevitable.

Bio-mediated method of soil improvement in general refers to the biochemical reaction that take place within a soil mass to produce calcite precipitate to modify the engineering properties of the soil (DeJong et al., 2010). The technique takes the advantage of biological process, technically known as microbially induced calcite precipitation (MICP) to produce calcite in the soil. The calcite generated is responsible for cementing and clogging the soils, and hence improve the strength and reduce the hydraulic conductivity of the soils. MICP can be a viable alternative technique that improve soil supporting new and existing structures and in many geotechnical engineering applications such as liquefiable sand deposits, slope stabilization, and subgrade reinforcement (Cheng et al., 2013, DeJong et al., 2006). The MICP technique has been a subject of research in recent years and interesting findings from the studies conducted have proved its viability and environmental sustainability (De Muynck et al., 2010, Ivanov and Chu, 2008).

Table 1. Microorganisms whose urease activity are not repressed by NH_4^+ (Whiffin, 2004).

Microorganisms	High activity	Not repressed by NH_4^+	Not pathogenic or GM
Sporosarcina pasteurii	yes	yes	yes
Proteus vulgaris	unknown	yes	moderately
Proteus mirabilis	unknown	yes	no
Helicobacter pylori	yes	yes	no
Ureplasmas(Mocllicutes)	yes	yes	no

Table 2. Index properties of the soil.

Properties	Description
Gravel (%)	27
Sand (%)	19
Silt (%)	40
Clay (%)	14
Liquid limit (%)	58
Plastic limit (%)	44
Specific gravity	2.60
MDD (Mg/m^3)	1.402
OMC (%)	26.5
Classification(BSCS)	MIG
UCS (kPa)	32.3
pH	7.3

The microbial method generally involves three stages:

1. Hydrolysis of urea by urease producing bacteria to form ammonium and carbonate ions.
2. The carbonate ions then react with the dissolve calcium from the supplied calcium chloride to form calcium carbonate crystals.
3. The formed calcium carbonate crystals bind the soil particles together to improve the strength and reduce hydraulic conductivity. The chemical reactions are presented in Equations 1 and 2.

$$H_2N\text{-}CO\text{-}NH_2 + 2H_2O \rightarrow 2NH_4^+ + CO_3^{2-} \quad (1)$$

$$Ca^{2+} + CO_3^{2-} \rightarrow CaCO_3\downarrow \quad (2)$$

Therefore, microorganisms whose urease activity are not repressed by high ammonium content are preferred in biomediated soil improvement; since high concentrations of urea are hydrolyzed in the process (Whiffin, 2004). Table 1 shows some urease positive bacteria in which their urease activities are not repressed by high ammonium contents.

Hence, all microorganisms that are found to be good for biomineralization applications because of their urease activity; they must also be safe for the environment during and after the treatment process. Therefore, urease – producing bacteria for biomediated applications should not be pathogenic, genetically modified or contains any transferable elements that may augment the pathogenicity of environmental strains. For this study Sporosarcina pasteurii (ATCC® 11859™) which is a gram-positive, spore forming urease active microorganism was considered as the source of urease. These bacteria can precipitate calcium carbonate in the presence of dissolved calcium and urea (Fujita et al., 2008, Mitchell and Ferris, 2005, Mitchell and Ferris, 2006). Sporosarcina pasteurii are commonly available in soil, sewage and urinal incrustations (Kersters and Vancanneyt, 2005).

2 EXPERIMENTAL PROCEDURES

2.1 Residual soil tested

The soil sample used for this study is a tropical residual soil that contains all range of particle sizes possibly from sedimentary rock origin. The soil was classified as Gravelly silt of intermediate plasticity (MIG) based on British Soil Classification System (BSCS) and was collected from a site at Universiti Tecknologi Malaysia UTM, Johor Campus. Table 2 present the Index properties of the soil.

2.2 Bacteria cultivation and cementation reagents

The urease active strain of Sporosarcina pasteurii (ATCC® 11859™) was used in this study. The strain was cultivated in a yeast extract–based medium of 20 g yeast extract, 10 g ammonium sulphate in a 1 Litre 0.3 M Tris buffer solution of pH 9.0. After 48 hour incubation at 30°C, the culture was harvested and stored at 4°C prior to use. The bacterium was grown to its exponential growth phase and the desired concentrations of 1×10^5 cfu/ml and 1×10^6 cfu/ml were obtained by plate count technique and used in the study. The cementation reagents consist of 3 g nutrient broth and 0.5 M concentrations of $CaCl_2$ and urea.

2.3 Sample preparations and curing conditions

To prepare the soil specimens, a liquid medium containing the microorganism (Sporosarcina pasteurii) was mixed with the residual soil. The quantity of the medium was determined to correspond to the optimum moisture content of the soil of 26.5%. The soil specimens were cured prior to treatment at atmospheric temperature (23 to 32°C) and 40°C. The cured soil specimens were then compacted into the prefabricated steel mould to a maximum dry density of 1.402 Mg/m^3. The soil samples were placed between two clean gravel layers to serve as filter layers to avoid turbulent inflow and clogging at the inlet. The specimens were treated for 24, 36, 48 and 60 hours durations. After treatment and strength determinations, the calcite contents of each treated sample were determined using gravimetric analysis of acidified samples.

3 EXPERIMENTAL RESULTS

3.1 Bacteria concentrations and strength improvement

Bacteria cells concentration is one of the major factors that influence MICP process, the more the cells

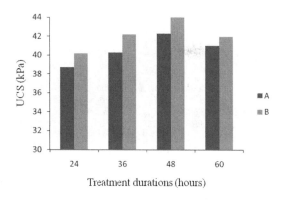

Figure 1. Relationship between bacteria concentrations (A: 1×10^5 cfu/ml, B: 1×10^6 cfu/ml) and Unconfined Compressive Strength (UCS) at atmospheric temperatures.

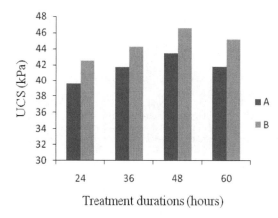

Figure 2. Relationship between bacteria concentrations (A: 1×10^5 cfu/ml, B: 1×10^6 cfu/ml) and Unconfined Compressive Strength (UCS) at 40°C.

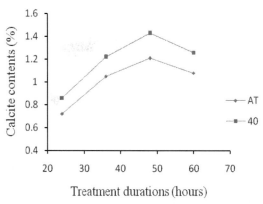

Figure 3. Relationship between calcite contents and treatment durations.

concentrations the more calcites are likely to be precipitated (Okwadha and Li, 2010). Figure 1 and 2 present correlation between the bacteria cell concentrations and strength improvement for atmospheric temperature and 40°C curing conditions. The result shows a general increase in strength as the bacteria concentrations increases for all the treatment durations up to 48 hours. This is because as the concentration increases more active bacteria cells would be available to hydrolyze urea into ammonium and carbonates ions for the precipitation of calcites. Meanwhile, at 48 hours duration a strength improvement of 31 and 36% relative to the untreated samples were recorded for 1×10^5 cfu/ml and 1×10^6 cfu/ml concentrations at atmospheric temperature respectively. Hence, at 40°C a similar improvement of 35 and 44% were recorded accordingly. Considering the 48 hours treatment duration that produces higher strength improvement of 40%; it can be deduced that MICP was most effective within 48 hours as also revealed by (Soon et al., 2014).

3.2 Relationship between calcite contents and treatment durations

Temperature effect is one of the environmental factors that are reported to greatly influence microbial calcite precipitation process (Okwadha and Li, 2010). Hence, the effects of curing temperature on the calcite precipitation and the subsequent increase in the strength of the residual soil have been evaluated. Figure 3 presents the relationship between calcite contents and treatment durations in relation to the curing conditions (temperatures) for 1×10^6 cfu/ml bacteria concentrations. The result shows a general increase in calcite contents as the treatment durations increases up to 48 hours, after which the calcite contents decreases indicating a decline in the bacterial urease activity. This may be attributed to the pH value of 8.5 attained at 60 hours treatment durations; since the bacteria used in the study is reported to have its optimum urease activity at pH above 8.0 (Stocks-Fischer et al., 1999). The effects of temperature on the calcite formations can also be seen from Figure 3. The calcites formed at 40°C are higher than those obtained at atmospheric temperatures (23 to 32°C); this also corresponds with the strength improvement reported in section 3.1. These findings are in conformity with the findings of Whiffin (Whiffin, 2004) which revealed that urease activity increases proportionally with increase in temperature up to 70°C.

3.3 Variation of pH with the treatment durations

Figure 4 presents the pH variations over the treatment durations of 60 hours. The results indicated a continuous increase in the pH for all the treatment durations. The first measured pH was 7.5, slightly above the pH of the untreated soil; which is an indication of very little ammonium ions productions at the initial stage of the treatment. The urea hydrolysis induced by the bacteria results in the production of ammonium ions that subsequently increased the pH of the soil environment.

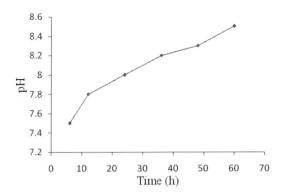

Figure 4. pH variations with treatment durations.

4 CONCLUSIONS

General increase in strength with the increase in curing temperature up to 48 hours treatment durations was observed. This revealed that increase in temperature facilitates the bacterial activity for urea hydrolysis to precipitates more calcite for strength improvement. The longer the treatment duration the more the calcite contents increase (up to 48 hours) causing more binding effects between the soils particles; thereby increasing the strength. At 60 hours treatment duration the calcite content decreases due the decline in the bacterial activity as the pH reached 8.5. Hence, 48 hours treatment duration that produces up to 44% strength improvement relative to untreated samples was found to be the most effective. Therefore, higher concentrations of the bacteria of 1×10^6 cfu/ml resulted in higher strength improvement when compared with strength produced at 1×10^5 cfu/ml concentrations.

ACKNOWLEDGMENT

This research was financially sponsored by the Ministry of Education Malaysia through Universiti Tecknologi Malaysia UTM, under Fundamental Research Grant Scheme (FRGS) (4F628).

REFERENCES

Cheng, L., Cord-Ruwisch, R. & Shahin, M. A. 2013. Cementation of sand soil by microbially induced calcite precipitation at various degrees of saturation. *Canadian Geotechnical Journal*, 50, 81–90.

De Muynck, W., De Belie, N. & Verstraete, W. 2010. Microbial carbonate precipitation in construction materials: a review. *Ecological Engineering*, 36, 118–136.

Dejong, J. T., Fritzges, M. B. & Nüsslein, K. 2006. Microbially induced cementation to control sand response to undrained shear. *Journal of Geotechnical and Geoenvironmental Engineering*, 132, 1381–1392.

Dejong, J. T., Mortensen, B. M., Martinez, B. C. & Nelson, D. C. 2010. Bio-mediated soil improvement. *Ecological Engineering*, 36, 197–210.

Fujita, Y., Taylor, J. L., Gresham, T. L., Delwiche, M. E., Colwell, F. S., Mcling, T. L., Petzke, L. M. & Smith, R. W. 2008. Stimulation of microbial urea hydrolysis in groundwater to enhance calcite precipitation. *Environmental science & technology*, 42, 3025–3032.

Gunaratne, M. 2013. *The foundation engineering handbook*, CRC Press.

Ivanov, V. & Chu, J. 2008. Applications of microorganisms to geotechnical engineering for bioclogging and biocementation of soil in situ. *Reviews in Environmental Science and Bio/Technology*, 7, 139–153.

Karol, R. H. 2003. *Chemical Grouting And Soil Stabilization, Revised And Expanded*, CRC Press.

Kersters, K. & Vancanneyt, M. 2005. Bergey's manual of systematic bacteriology.

Mitchell, A. C. & Ferris, F. G. 2005. The coprecipitation of Sr into calcite precipitates induced by bacterial ureolysis in artificial groundwater: temperature and kinetic dependence. *Geochimica et Cosmochimica Acta*, 69, 4199–4210.

Mitchell, A. C. & Ferris, F. G. 2006. The influence of Bacillus pasteurii on the nucleation and growth of calcium carbonate. *Geomicrobiology Journal*, 23, 213–226.

Okwadha, G. & Li, J. 2010. Optimum conditions for microbial carbonate precipitation. *Chemosphere*, 81, 1143–1148.

Soon, N. W., Lee, L. M., Khun, T. C. & Ling, H. S. 2014. Factors Affecting Improvement in Engineering Properties of Residual Soil through Microbial-Induced Calcite Precipitation. *Journal of Geotechnical and Geoenvironmental Engineering*, 140.

Stocks-Fischer, S., Galinat, J. K. & Bang, S. S. 1999. Microbiological precipitation of CaCO 3. *Soil Biology and Biochemistry*, 31, 1563–1571.

Van Paassen, L. Bio-mediated ground improvement: from laboratory experiment to pilot applications. Geo-Frontiers 2011@ sAdvances in Geotechnical Engineering, 2011. ASCE, 4099–4108.

Whiffin, V. S. 2004. *Microbial CaCO3 precipitation for the production of biocement*. Murdoch University.

The behaviour of electrical resistivity correlated with converted SPT-N results from seismic survey

S.B. Syed Osman, H. Jusoh & K. Chai
Universiti Teknologi PETRONAS, Seri Iskandar, Perak, Malaysia

ABSTRACT: Conventional soil investigation method to derive strength parameters incorporates borehole sampling which is reliable but is invasive, time consuming, and costly. Alternatively, in this study, seismic surface wave method and electrical resistivity survey were utilized to estimate the soil strength parameters. Some drilling work along the seismic and electrical resistivity lines were carried out where samples were then brought to the laboratory and tested for the related index properties. All the data were then analyzed and correlated. The correlation study shows the relationship between inverted SPT-N (using OYO SeisImager and Pickwin software) with field electrical resistivity produced a regression value, R^2 of 0.4696. The comparison of both field and converted SPT-N yield a result of R^2 value of approximately 0.84. Finally, the relationship of moisture content (MC) with field resistivity and plasticity index with field resistivity resulted in regression values of $R^2 = 0.433$ and $R^2 = 0.3156$ respectively.

1 LITERATURE REVIEW

1.1 Introduction

One of the main purposes of SI is to acquire geotechnical profile of soil for construction purpose. SI involves the determination of ground water level, depth of bedrock, drainage situation as well as the soil stratification. Thus, by identifying these parameters will allow efficient foundation design with respect to the soil bearing capacity and factor of safety (FOS) of the soil.

Some of the conventional SI techniques are soil boring, vane shear test and cone penetration test. However, these methods are destructive to soil, at the same time are expensive and tedious. Moreover, it requires high density of samples in order to have reliable information for high variation of soil properties. An alternate non-destructive and rapid in-situ investigation technique such as electrical resistivity and seismic survey are utilized to acquire soil strength parameters because these methods are non-destructive, rapid and economical compared to the conventional methods. However, these geophysical methods do have their own limitations and weaknesses and could never match the conventional methods in terms of their accuracy and reliability. This study primarily looks into the correlation between electrical resistivity and SPT-N values converted from surface wave velocity and then comparison was made with the actual SPT-N obtained from field SPT test.

1.2 Electrical resistivity

Studies show that electrical resistivity has the potential to calculate factor of safety (FOS) or bearing capacity

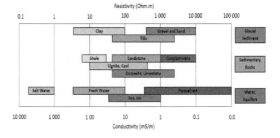

Figure 1. Typical resistivity and conductivity values in various earth materials.

using correlated parameters such as cohesion, internal angle of friction and unit weight with electrical resistivity values (Siddiqui & Syed 2012). In general, different types of soil possess different range of resistivity values as portrayed in Figure 1. However, apart from moisture content, there are a few other parameters which controls resistivity such as salinity, cation exchange capacity etc. (Siddiqui & Syed 2012).

This study was conducted by implementing the Wenner configuration method as shown in Figure 2 involving 4 electrodes where an electrical current (I) was injected into the soil to measure the soil resistivity (Rhoades 1976). Subsequently, the potential difference (V) can be measured, followed by the determination of resistance (R).

The measurement of resistivity (ρ) can be calculated by using the formula:

$$\rho = 2\pi \qquad (1)$$

where R: resistance, Ω
a: distance between each electrode

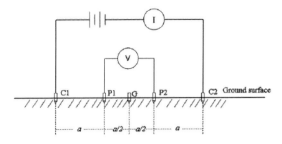

Figure 2. Wenner Array for VES Survey.

This method can be applied to both 1D and 2D resistivity survey methods. Both methods will produce similar results except that 2D resistivity survey will produce the overall two dimension soil stratigraphy with respect to the soil resistivity.

The study of correlation between SPT-N with resistivity behaviour is uncommon as compared to with seismic behaviour and have only been studied for the past decade. Siddiqui and Syed (2012) has achieved moderate coefficient value between SPT-N and electrical resistivity values by implementing the similar method applied in this research.

Several studies have been conducted to obtain the correlation between electrical resistivity and soil parameters. Cosenza et al. (2006) performed a 2D electrical resistivity survey using Wenner electrode configuration to correlate resistivity with cone penetration test (CPT) values. However, there was no strong correlation between resistivity and CPT values and further investigation was recommended to be carried out for better correlation. On the other hand, by applying electrical resistivity tomography method, Cosenza et al. (2006) has acquired a strong correlation between resistivity and moisture content for silty clay sample, with an empirical relationship of $\rho = 1.187w-2.444$. Meanwhile, Celano et al. (2011) has obtained the strongest correlation using pole-dipole on calcic soil.

Furthermore, Abu-Hassanein (1996) proved a curvilinear correlation of plasticity index (PI) and electrical resistivity of clay. In addition, Abu-Hassanein (1996) also came out with the conclusion that high plasticity soil possess lower electrical resistivity values.

1.3 Seismic surface wave

Seismic survey portrays the soil profile by determining the shear wave velocity as the function of depth. When the seismic rays are produced and intruded into the soil, it will be split into two media with distinctive acoustic impedance where it will either be reflected or partially refracted into the lower medium (Roe 1953). In this study, the surface wave, specifically Rayleigh waves was utilized. Rayleigh wave exhibits lower frequency velocity but high amplitude (Sheriff 1991). At a specific mode, surface waves with greater wavelength can reach deeper into the earth than surface waves with shorter wavelength. The surface wave passing through each of the earth materials is mainly controlled by its elastic properties (Babuska & Cara 1991). Hence, different materials will possess different properties and distinctive range of seismic velocity at distinctive wavelength.

Andy & Rosli (2012) previously applied seismic refraction method in a tropical environment and achieved a strong correlation between seismic wave velocity with SPT-N. Meanwhile, in this study, seismic surface wave method were used and subsequently with the aid of OYO SeisImager software, the seismic data obtained were converted into SPT-N value based on correlation by Imai & Tonouchi (1982).

2 METHODOLOGY

2.1 Field work

The fieldworks were comprised of soil boring, field standard penetration test (SPT), vertical electrical sounding survey (VES), 2D seismic (surface wave) survey and details are given below.

2.1.1 Soil boring

The borehole samples were acquired by drilling boreholes at designated locations by using petrol operated percussion drilling set (CobraTT, Atlas Copco) with 1 meter core sampler. Boreholes were drilled at similar area where field resistivity and seismic survey were conducted. The maximum depth of boreholes can reach up to 3 meter.

The undisturbed samples were then preserved in a capped plastic cylinder to minimise loss of moisture content. The samples were brought back to UTP laboratory to acquire both index and engineering properties by conducting laboratory experiments, namely moisture content, direct shear test, particle size distribution test, and bulk and dry density test. The samples were labelled with respect to its location and depth.

2.1.2 Field standard penetration test (SPT)

SPT test was carried out at UTP location where the field resistivity and seismic survey were conducted for comparison purpose. The test procedure was performed in accordance to British Standard BS EN ISO 22476-3. The borehole report from the contractor indicate the SPT values at different depth, which were eventually compared with the SPT values acquired from the seismic surface wave survey.

2.1.3 Vertical electrical sounding survey (VES)

This method requires components such as electrodes, power source (DC power supply), voltmeter, insulated wires and measuring tape.

Figure 3 above shows electrical current of 20 mA was injected through electrodes C & D whereas voltage reading was taken in between electrodes X & Y. The test were conducted at the same location as the borehole samples were drilled.

2.2 2D Seismic (surface wave) survey

Surface wave technique was used to produce the seismic imaging. Also, this survey was conducted at the

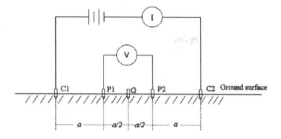

Figure 3. Four electrode Wenner configuration for VES survey.

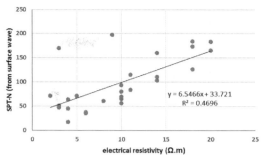

Figure 4. Converted SPT-N vs field electrical resistivity.

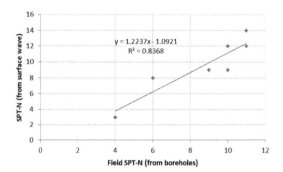

Figure 5. Converted SPT-N vs Field SPT-N.

location where VES and soil boring were performed. The equipment required were; seismograph set; 2 set of 12 channels seismic cables; 24 units of geophones; fully charge car battery; remote cable and trigger switch; hammer and steel plates; measuring tape.

The geophones were clipped on the seismic takeout cable at a fixed distance away from each geophones. The geophones were triggered by hammer (trigger switch attached) hitting on a steel plate. The seismograph unit recorded the trigger level and data saved in a dedicated file in the computer in .sg2 file.

The data were then processed in three phases until seismic image with soil profile were generated. By using the pickwin (surface wave) software, the details such as source interval and source receiver were analysed and verified by the wave equation program before the seismic images were presented in geo plot program. Finally, the seismic image with surface velocity and SPT-N value were generated.

2.3 Laboratory work

Once the samples were transported to the laboratory, soil characterization tests were carried out on the soil samples.

3 RESULTS AND DISCUSSION

The correlation studies consisted of four plotted graphs, involving SPT-N from field SPT test, SPT-N converted from surface wave velocity, field resistivity, moisture content and plasticity index. This correlation study was done based on field and laboratory results on soil samples from selected locations in Ulu Pudu, Selayang, Shah Alam, Parit and in Universiti Teknologi PETRONAS (UTP).

3.1 SPT-N converted from seismic surface wave versus field electrical resistivity

Figure 4 shows a relationship of SPT-N $= 6.5466\rho + 33.721$ with a correlation coefficient, R^2 of 0.4696, obtained from this research for converted SPT-N and electrical resistivity which could be considered moderate. The SPT-N values used were converted from wave velocities by means of Oyo McSEIS seismogragh and SeisImager software instead of the conventional field SPT-N from boreholes. Since among the objective of this research is to determine the suitability and accuracy of Oyo McSEIS seismogragh and SeisImager software in determining SPT-N values in heterogeneous soil in tropical environment like Malaysia, field SPT was also conducted at a location in UTP in order to compare the results obtained from converted SPT-N values.

It should be noted here that the data input into the Seisimager software produces surface wave velocities which were then converted into SPT-N values in which the software was developed based on robust correlation performed by Imai and Tonouchi ($R^2 = 0.87$). The comparison of both field and converted SPT-N results is shown in Figure 5 where it depicts a high R^2 value of approximately 0.84. This shows fairly strong relationship between both the SPT-N methods. Some of the factors which gave rise to errors in the results obtained could be due to the surrounding noise captured in the McSEIS instrument through the geophones during field survey. Furthermore, the manual hammering process to trigger seismic wave can also affects the quality of the waves received by the geophones. Error from the software especially during inversion process could also contribute largely on the inaccuracies. Nevertheless, the moderate correlation shows vast potential in the application of using seismic survey in estimating the SPT-N value for heterogeneous soil.

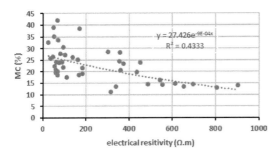

Figure 6. Moisture content vs field electrical resistivity.

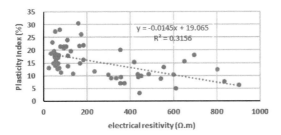

Figure 7. Plasticity index vs field electrical resistivity.

3.2 *Moisture content VS field resistivity*

The relationship of moisture content and field resistivity is formulated to be MC (%) $= 27.42^{-9E-04\rho}$ with correlation coefficient, R^2 of 0.4333 as depicted in Figure 6. Differing from Cosenza et al. (2006), this research showed exponential relationship between moisture content with field resistivity instead of a linear relationship. Nevertheless, the study showed that the moisture content decreases as the resistivity increases. The trend implied that the moisture content in the soil act as a medium for the current flow. Hence, the decrement in moisture content would amplify the resistance for the current flow. This research obtained a relatively lower coefficient of correlation than previous authors. The main reason would be mainly due to the various difference in the characteristic of the previous soil samples being studied such as soil type, mineralogy, salinity etc.

3.3 *Plasticity index VS field resistivity*

In Figure 7, the relationship of plasticity index and field resistivity is given as PI (%) $= -0.0145\ \rho + 19.065$ with correlation coefficient, R^2 of 0.3156. The plasticity index decreases with increasing electrical resistivity. The downward trend has also been studied by Abu-Hassanein (1996) stating that soil with higher plasticity index will have less resistivity values. Soil that possess high plasticity index has higher tendency to have more clay composition whereas the lower range tends to be more silty and zero plasticity index indicates non-plastic soil, commonly sand. The trend is also verified by the range of resistivity as shown in Figure 1 as the resistivity of clayey soil is relatively lower compared to silty and sandy material.

4 CONCLUSION

The objective of this study is to estimate the relationship between soil electrical resistivity and SPT-N values from seismic survey to support the applications of geophysical methods in determining soil strength parameters. Additional relationships between electrical resistivity versus moisture content and plasticity index are included to look into their effects on electrical resistivity. The overall regression obtained for SPT-N inversed from seismic survey versus electrical resistivity is 0.4696 and for SPT-N from seismic versus actual SPT-N is 0.8368. In conclusion, the study has proven the potentials of geophysical methods in estimating soil strength parameters for both checking and design purposes.

ACKNOWLEDGEMENT

The author would like to acknowledge the management of Universiti Teknologi PETRONAS (UTP) for providing sufficient facilities to perform the experiment in this research.

REFERENCES

Abu-Hassanien, Z. S. 1996. Electrical Resistivity of Compacted Clay, Geotechnical Engineering, vol. 120, pp 45–1457.

Andy, B. & Rosli, S. 2012. Correlation of Seismic P-Wave Velocities with Engineering Parameters (N Value and Rock Quality) for Tropical Environmental Study, International Journal of Geosciences, vol. 2012.

Babuska, V. & Cara, M. 1991. Seismic Anisotropy in the Earth, Kluwer Academic Publisher.

Celano, G. & Palese, A. M. & Ciucci, A. & Martorella, E. & Vignozzi, N. & Xiloyannis, C. 2011. Evaluation of Soil Water Content in Tilled and Cover-cropped Olive Orchards by the Geoelectrical Technique, Geoderma, vol. 163, pp. 163–170.

Cosenza, P. & Marmet, E. & Rejiba, F. & Jun, C. Y. &, Tabbagh, A. & Charlery, Y. 2006. Correlations between Geotechnical and Electrical Data: A Case Study at Garchy in France, Journal of Applied Geophysics, 60 (3-4): 165–178.

Imai, T. & Tonouchi, K. 1982. Correlation of N-value with S-wave Velocity and Shear Modulus, Proc. 2nd European Symposium on Penetration Testing, 67–72.

Roe, F. W. 1953. The Geology and Mineral Resources of the Neighbourhood of Kuala Selangor and Rasa, Selangor, Federation of Malaya, with an Account of the Geology of Batu Arang Coal-Field., Selangor, Geological Survey Department Federation of Malaya.

Rhoades, J. D. 1976. Measuring, Mapping and Monitoring Field Salinity and Water Table Depths with Soil Resistance Measurements, FAO Soils Bulletin.

Sheriff, R. E. 1991. Encyclopaedic Dictionary of Exploration Geophysics, 3rd ed. Society of Exploration Geophysicists.

Siddiqui, F. I. and Syed, B. S. O. 2012. Use of Vertical Electrical Sounding (VES) method as an Alternative to Standard Penetration Test (SPT).

Development of a new distributed optical fibre sensor as borehole extensometer

H. Mohamad
Department of Civil and Environmental Engineering, Universiti Teknologi PETRONAS, Seri Iskandar, Perak, Malaysia

ABSTRACT: Borehole extensometer is an example of geotechnical instrumentation used to measure subsurface movements such as settlement below an embankment or ground heaving resulting from basement excavation. The extensometer instrument records deformation along the axis of the borehole with certain number of measurement intervals and resolution. In order to increase the resolution and accuracy of such data, a novel technique of measuring vertical displacement continuously inside a borehole is proposed using distributed optical fibre strain sensor. This article presents initial development of novel fibre-optic borehole extensometer on the basis of Brillouin Optical Time Domain Analysis (BOTDA). A series of laboratory tests were performed to validate the sensor performance using steel pipe tubes containing cement grout embedded with optical fibre strain cables and were subjected to uniaxial loadings. Results from calibrated sensing cables embedded in the cement grout were very much similar to the electrical based strain gauges mounted externally on the steel pipe.

1 INTRODUCTION

Distributed fibre-optic (FO) strain sensors are offering new possibilities in the field of geotechnical monitoring. By integrating a single FO cable into soil or structure, an unprecedented amount of accurate, spatially resolved data could be obtained (Mohamad and Tee, 2015). Current commercially available technology allows for strain measurements in the microstrain ($\mu\varepsilon$) range (0.0001%) with a spatial resolution of 1m along a 30 km long fibre.

In this article, the author describes recent novel geotechnical FO technology applications for subsurface ground movement detection. The emphasis is to calibrate and validate the FO cable sensing performance based on various instrumentation layouts in the laboratory and practicality for field deployment.

Ground movement monitoring using distributed Brillouin sensors have been proposed in many case studies; for instances to investigate landslides, settlements, cavities, embankments, and dams. In addition, such monitoring scheme is used to enhance safety of large, long infrastructure components such as pipelines, roads, tunnels, and railways. A wide range of FO monitoring systems can be found in the literature, depending on the sensor deployment methods. Some of these applications are noted below:

- *single cable embedded in soil* – Lanticq et al. (2009), Belli et al. (2009)
- *cable fixed on plate/ grid formation* – Kluth et al. (2006), Shi et al. (2007)
- *cable integrated with geotextile* – Koelewijn, (2009), Belli et al. (2009)
- *cable fixed with rod* – Shi et al. (2007)
- *cable fixed on aluminium strip* – Ohno et al. (2001)

Accordingly, many of these methods have contributed to the advancement of ground movement monitoring by fiber-optic sensors. However, several open questions remain. The most straightforward sensor integration strategy of simply soil-embedding a cable works only if the soil strain is accurately transferred into cable strain, rather than the soil just flowing around the cable. The drawback is that there is a maximum shear stress that can be transferred to the soil-cable interface after which interface slippage occurs (progressive failure) and the strain measurement ceases to correlate to the soil strain. Investigations of the progressive failure between soil and FO sensors are not discussed in the literature. However, the understanding of the progressive failure is necessary to conclude on the accuracy of the sensor system and the interpretation of the measured strain distribution (Iten, 2011).

In addition, many of the installation process described previously comprised of optical cables that were laid horizontally in the ground. No research so far has been reported for the case of measuring vertical displacement profiles in the form of borehole extensometer. Vertical extensometer has the advantage of providing direct measurements of subsurface ground displacement pattern such as ground heaving and settlement profile below a foundation. To date, FO sensor incorporating fibre Bragg gratings (FBG) technology has recently been implemented as borehole extensometer with quite a success (Mohamad, 2014). However, the FBG sensor is not a fully distributed system but rather a multiplexed sensor, which means

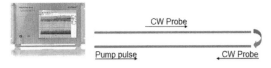

Figure 1. Principle measurement of a BOTDA system.

that the measurement is limited to the number of gratings in the fibre (normally not more than 20). Brillouin based technology such as BOTDA therefore has the desirable advantage to be incorporated as a distributed borehole extensometer particularly for deep borehole measurement. This article describes the works associated with the development of BOTDA borehole extensometer.

1.1 Brillouin Optical Time Domain Analysis (BOTDA)

In this study, a commercially available Brillouin Optical Time Domain Analysis (BOTDA) interrogator (Omnisens) was used to determine the strain distribution along the whole depth of a borehole or in the case of laboratory experiment, the strain distribution of a steel pipe casing casted with cement grout.

The BOTDA sensor uses two different light sources, launched from two ends of an optical circuit. The system utilizes the backward stimulated Brillouin scattering (SBS); i.e., the pumping pulse light launched at one end of the fibre and propagates in the fibre, while the continuous wave (CW) light is launched at the opposite end of the fibre and propagates in the opposite direction (Figure 1). In this configuration, the pump pulse generates backward Brillouin gain whereas the CW light interacts (amplifies) with the pump pulse light to create stimulated Brillouin scattering. The Brillouin frequency shift in the singlemode fibre is proportional to the change in the strain or temperature of that scattering location. By resolving this frequency shifts and the propagation time, a full strain profile can subsequently be obtained. One particular advantage of BOTDA over other type of distributed strain sensing system such as BOTDR (Brillouin Optical Time Domain Reflectometry) is that the technique produces strong signal, which can reduce averaging times (faster acquisition time) and longer measurement distances capabilities (for up to 50 km).

1.2 Optical cable

Figure 2 shows an example of a strain sensing fibre. This 7.5 mm diameter optical cable is specifically designed for embedment in cast-in-situ concrete piles and stiff soils. It consists of a single core singlemode fibre reinforced with steel wires and polyethylene cable jacket. The external plastic coating and the inner glass core are fixed together so that the strain applied externally (from the concrete) is fully transferred from the coating to the inner core.

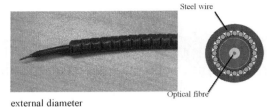

Figure 2. Brugg optical cable with 7.5 mm.

2 BOREHOLE EXTENSOMETER

There are basically two types of borehole extensometer:

- Non-accumulative ground subsidence profile monitoring
- Accumulative ground subsidence profile monitoring

A multi-point magnetic borehole extensometer system is an example of non-accumulative ground subsidence profile monitoring (Smith & Burland, 1976). The system consisted of a series of magnetic rings, axially magnetized which acted as markers embedded at various depths in the ground and a reed switch that served as a sensor. The subsidence profile is obtained by locating the position of the magnets, that is, the reed switch sensor was lowered down the central plastic guide tube with a metallic tape attached to the sensor head.

In the accumulative mode, a sensor unit is used to measure relative ground subsidence within a given depth range, in reference to its neighboring sensor unit. The vertical displacement at a given depth is an accumulation of ground subsidence/heaving from below that level. While the absolute ground displacement, in reference to a fixed datum may be substantial, the differential values for a given depth range or strains are usually small. Often, it is this small strain related readings that are of great interest (Burland, 1989). From this point of view, an accumulative ground subsidence-monitoring scheme would be more desirable.

2.1 Proposed methods

The installation of optical cables into a borehole should follow several criteria before it can be configured into a vertical extensometer: (i) The cables should be pre-tensioned inside the borehole, normally by fixing the bottom point as a reference datum. (ii) A fully grouted borehole is a commonly used method in geotechnical instrumentation. Hence, an appropriate cement-grout design mixture suitably for surrounding ground is applied to the fibre inside the borehole.

In BOTDA technology, a returning a cable is required when making the measurements (the system requires access to both ends of the optical cable). A loose section of the cable is created and protected at the bottom of the borehole to allow joining between two different cables and separation between strained cable

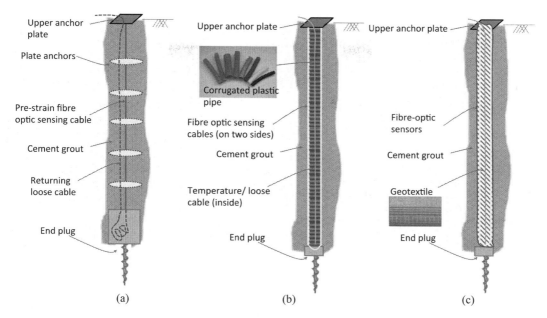

Figure 3. Proposed installation methods (a) cables point fixed with plate anchors, (b) cables with corrugated plastic tube, (c) geotextile.

and the returning loose cable. A strain and a temperature sensing cable should be installed together (side-by-side) to enable temperature compensated strain measurements.

Based on these configurations, several methods of installing fibre-optic cables inside a borehole are suggested and are shown in Figure 3. The proposed methods were based on the ease of access on the available materials, and the simplest but effective way of instrumenting the cables inside the borehole. Figure 3 shows three strategies in which optical cables can be configured into a borehole extensometer.

(i) Optical cables point fixed with plate anchors
(ii) Optical cables with flexible corrugated plastic tube
(iii) Optical cables integrated with geotextile

In the first method (optical cable fixed with plate anchors), a special type of cable from Brugg is used. This cable is chosen considering that the optical cable has been tested and calibrated (Hauswirth et al. 2010). In addition, the cable may be packaged with the their standardize anchorage system, known as V3 anchors.

The second method of utilizing a corrugated plastic tube (or flexible electrical conduit) is an attractive option because the ribbed surface of plastic tube has more contact surface and therefore potentially providing better adhesion between the grout and plastic tube and the fibre attached onto the tube. Moreover, the materials used for this technique are very cheap. For this study, an eight-ribbon Fujikura cable is attached externally and internally inside the plastic tube (Mohamad et al., 2009). As shown in Figure 3b, the fibre fixed externally on the tube is pre-tensioned and spot-glued whereas the inner fibre inside the tube is loosely laid inside. A special type of glue (melted glue sticks) that is flexible when cured (so that it follows the movement shape of the tube) is used.

The geotextile integrated with optical cables manufactured by Tencate, (GeoDetect) may be configured as a vertical extensometer. The experiment shall investigate the bonding performance between the woven fabric of geotextile and the cement grout.

3 EXPERIMENTAL PROGRAM

3.1 Methodology

In order to investigate the accuracy of the proposed measurement systems, a series of uniaxial loading on the specimens consisting of optical cables embedded inside a tube enclosure with cement-grout was established. The validity of the test depends on assumption that full bonding would take place between optical cable, grout mix, and the pipe and hence the shear stress or external strain applied from the two ends of tube is fully transferred to the sensing cable. That is, the criteria for selection were evaluated in terms of bonding performance, as well as robustness (i.e. whether cables survive the field installation process).

The experiment was performed at Magnel Laboratory for Concrete Research, University of Ghent. In order to match the appropriate working load and geometrical configuration of the machine, mild grade steel pipes ($E = 210\,000$ MPa) with 60 mm external diameter and 3 mm thickness were used. The experimental setup is further discussed below.

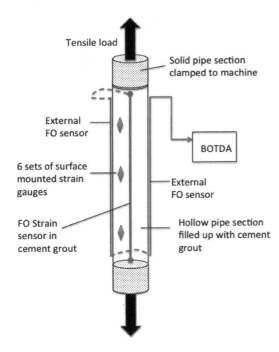

Figure 4. Instrumentation layout of FO sensors in steel pipe.

3.2 Preparation of steel tube casing

The steel tubes of 60 mm in diameter were cut into several segments of 2.3 m long. The length was determined to be long enough to allow correct BOTDA strain samplings at the center distance of the tube. The thickness of the steel pipe is 3 mm; chosen to match an axial stiffness that is appropriate for the loading range of the machine in the Magnel Laboratory. For the specific machine used in this laboratory, the steel tube is clamped at the two ends. Hence the ends of the pipe must be strong enough to withstand the clamping forces. To strengthen the ends, two segments of steel cylinders (30 cm long and 50 mm in diameter) are welded together; the bottom section to be welded before pouring the grout mixture whereas the top end is welded after the cables have been pre-tensioned during the process of grout hardening.

The installation of optical cables inside the steel tube was challenging, partly because the cables have to be pre-tensioned inside the cement-grout, and running the optical cables in and out of the steel tube at a very tight space. The cables were fixed to the bottom end of the tube by gluing them on a steel plate that was welded to the steel cylinder as shown in Figure 4.

Slight adjustments of attachment method were made for the three types of cables. In the case of Brugg cable, the technique of coupling ground anchors along the cable is not possible due to the small diameter tube size. In addition, due to the space limitation, the stiff optical cable was spliced to a smaller cable (wiring cable) at the bottom of the tube so that the returning cable can be pulled out from a tight hole easily. For the geotextile sensor, the width of the panel fabric was cut down to fit the size the tube. A splice was made to connect two fibres of the geotextile at the bottom of the tube.

3.3 Cement-bentonite grout mix

The general rule for grouting any kind of instrument in a borehole is to mimic the strength and deformation characteristics of the surrounding soil rather than the permeability. However, while it is feasible to match strengths, it is unfeasible with the same mix design to match the deformation modulus of cement-bentonite to that of a clay for example. The practical thing to do is to approximate the strength and minimize the area of the grouted annulus. In this way the grout column would only contribute a weak force in the situation where it might be an issue (Mikkelsen, 2002).

Grout mixes should be controlled by weight and proportioned to give the desired strength of the set grout. The conversion factors to be used in this study are 2.5 for water, 1.0 for cement and 0.3 for bentonite, which is in accordance to the mix design for "medium" to "hard soils" (Dunnicliff, 1988). The 28-day compressive strength of this mix is about 50 psi, similar to very stiff to hard clay and elasticity modulus of about 10,000 psi (Mikkelsen, 2002).

In order to keep field procedures simple the emphasis should be on controlling the water-cement ratio. This was accomplished by mixing the cement with the water first. When water and cement are mixed first, the water-cement ratio stays fixed and the strength/modulus of the set grout is more predictable. If bentonite slurry is mixed first, the water-cement ratio cannot be controlled because the addition of cement must stop when the slurry thickens to a consistency that is still pumpable.

A trimmie pipe was used during the pouring of cement-grout mix to ensure the mixture entered the pipe from the bottom first and no air was entrapped. Then, the optical cables were pre-tensioned and left for two weeks to allow the mixture to cure and hardened. Next, the top end of the steel tube was sealed (welded). Additional sensors were also attached on the tube surface for the loading test.

It must be emphasised that for each procedure, i.e. from gluing cables onto the bottom plate, welding, and grout curing, the sensors are checked for the survivability (e.g. using optical fault detector or OTDR). Strain measurements from each tube during the curing stage were recorded. The data indicated some degree of shrinkage stress formed during the early stage of curing based on the minor reduction of pre-tension (tensile strain) in the optical fibre.

4 TENSILE TUBE TESTS

In this study, the tensile tube tests were performed after the cement-grout have been casted at about 28 days. As mentioned previously, the loading tests relied on the interactions between the external tube, the grout, and the optical fibre. Theoretically if full bonding exists between the three interfaces, the strain applied on the steel tube would be similar in the cement-grout and optical cables. In order to measure the strains of the steel tube, strain gauges and distributed optical fibre sensors were installed on the surface of the tube as shown in Figure 4. The optical ribbon fibres (Fujikura) were mounted on two sides of each steel tube whereas the meta-foiled strain gauges were installed on four sides at the middle distance of the tube. In order to minimise of the acquisition time of BOTDA readings at each loading stage, the various optical cables instrumented on the steel tube were connected together to form one continuous cable (see Figure 4) so that a single reading can be taken with one optical channel. In this experiment, the measurement processing time was between two and five minutes.

Since the objective of the test is to compare the strain or displacement between the external strain gauges and optical fibre sensor within the cement-grout, the loading test in terms of displacement control mode was conducted. For this experiment, the typical loading step was 0.25 mm. In the displacement controlled loading, movement of the end plates is held constant while awaiting BOTDA analyzer to perform its strain measurements of several minutes. During this moment, the load stress measured by the machine's transducer may vary in order to maintain the same output value of the displacement transducer.

As shown in the subsequent figures of the test results in Section 4.1, the maximum pipe extension was 5 mm (or approximately 2300 microstrain) whereas compressive loading was limited to the steel compression of 2 mm. Given the axial stiffness of the steel tube, the applied load for the extension and compression was not more than 150 kN. The aim of the experiment was not to excessively deform the tube (i.e. plastically); hence only a minimum loading range was applied.

5 EXPERIMENTAL RESULTS

Figure 5 shows the results from the first experiment consisting of Brugg cable. There are four plots shown; the Brugg optical cable sensor, averaged readings of the strain gauges, averaged readings of Fujikura ribbon cables, and the theoretical strain calculated from the displacement transducers of the machine. In general, all measurements matched each other remarkably well. The surface mounted sensors consisted of optical fibre and electrical based strain gauges were exactly similar. The strain sensors inside the grout and the surface mounted sensors showed exactly the same strain increments after each applied displacement (i.e. increment of 0.25 mm).

It is interesting to note that there was a slight strain deviation between the theoretical (applied strain) and instrumented values at the point after the release of the extension loading to about 1.3 mm displacement. This was because at this point, the machine has changed from a positive load (tension) to a negative (compression) load. In other words, there was a permanent

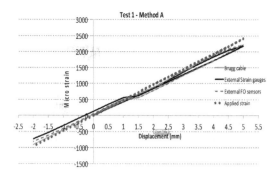

Figure 5. Strain-displacement relationships of Brugg cable and other sensors.

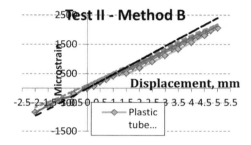

Figure 6. Strain-displacement relationships of Brugg cable and other sensors.

deformation of 1.3 mm when the load has returned to zero and a further compressive load was needed to bring the tube to the original length.

The results of Fujikura 8-ribbon fibres cable installed with corrugated plastic tube are presented in Figure 6. The data showed good agreement between all measurement systems. Surface mounted Fujikura cable recorded the closest values to the theoretical applied strain but the integrated plastic pipe fibre measured a slight lower strain response in particular during the unloading phase. The differences however were very small between three sensors (plastic-tube fibre, strain gauge, and fujikura external fibre), which was, about 6 % strain different when the compression was 2 mm.

Figure 7 presents the results of GeoDetect sensor embedded in cement-grout. Comparison between the geotextile strain response and external sensors indicate that the geotextile has the worst linear response. The electrical strain gauges data matched by the Fujikura surface mounted fibres very well. However, GeoDetect strain data indicate some sort of data instability; at first, the strain was very high (during elongation of 1 mm to 3 mm), but recorded roughly similar strains towards the final phase of tensile loading. Afterwards the strain in the geotextile did not recover as much as expected (when compared with theoretical strain), during the release of tensile load and application of compressive load. For example, the final compressive strain recorded by GeoDetect at −2 mm is

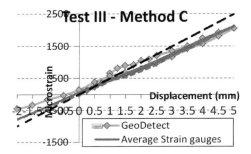

Figure 7. Performance of geotextile sensor (Geodetect) embedded in cement-grout and comparison with other strain results.

−453 microstrain whereas both external strain sensors recorded compressive strain of −787 microstrain.

6 CONCLUSION

This study successfully investigated the performance of optical fibre cables embedded in cement-grout using various installation techniques. Based on the experimental evidence, the cement-grout within the tube was able to move as according to the deformation of the tube, which suggests full adhesion exists between the interfaces of steel, grout, and optical cables. In general, the optical cable strain sensors were able to measure tension and compression of the cement-grout quite accurately and matched the strain data results of surface mounted sensors.

Between the three installation methods, the Brugg cable performed the most satisfactory (the data matches well with the external sensors), followed by the Fujikura corrugated plastic tube sensor. Unfortunately the GeoDetect (geotextile) strain sensor recorded the least satisfactory results with some strain measurements differ 100 microstrain more than the external strain gauges.

ACKNOWLEDGEMENT

The author is grateful to Geotechnical Division of Ministry of Public Works, Belgium for inviting and giving the opportunity to work with the members of the group. In particular, ir Gauthier van Alboom (Head of Geotechnics Division) and fellow engineers; ir Koenraad Haelterman, ir Leen de Vos, ir Leen Vincke, ir Jan Couck, Elke Declercq and Eva Goeminne for providing assistance to conduct the research and monitoring the fieldworks.

REFERENCES

Belli, R., Glisic, B., Inaudi, D. & Gebreselassie, B. 2009. Smart Textiles for SHM of Geostructures and Buildings, *4th International Conference on Structural Health Monitoring on Intelligent Infrastructure* (SHMII-4), Zurich, Switzerland.

Burland J.B, 1989, 9th Laurits-Bjerrum-Memorial-Lecture – Small Is Beautiful – The Stiffness Of Soils At Small Strains, *Canadian Geotechnical Journal*, Vol: 26, Pages: 499–516.

Dunnicliff, J. (1988) *Geotechnical Instrumentation for Monitoring Field Performance*, Wiley-Interscience, New York.

Hauswirth, D., Iten, M., Richli, R. & Puzrin, A. M. (2010). "Fibre optic cable and micro-anchor pullout tests in sand", *Physical Modeling in Geotechnics*, ICPMG 2010, Zurich, Switzerland.

Iten, M. (2011). Novel applications of distributed fiber-optic sensing in geotechnical engineering. *PhD Thesis*, Nr. 19632, ETH Zurich, Switzerland.

Koelewijn, A. (2009). "Deltares GeoAcademy course on Geotechnical Instrumentation for Field Monitoring", New Measurement Technology for Dike Monitoring, 21–23 April, GIFN.

Lanticq, V., Bourgeois, E., Magnien, P., Dieleman, L., Vinceslas, G. Sang, A. and Delepine-Lesoille, S. 2009. Soil-embedded optical fiber sensing cable interrogated by Brillouin optical time-domain reflectometry (B-OTDR) and optical frequency-domain reflectometry (OFDR) for embedded cavity detection and sinkhole warning system, *Meas. Sci. Technol.*, 20(3).

Mikkelsen P.E. (2002) Cement-bentonite grout backfill for borehole instruments, *Geotechnical News*, Dec. issue, pp. 38–42.

Mohamad, H., Soga, K. & Bennett, P.J. (2009) Fibre optic installation techniques for pile instrumentation, *Proceedings of the 17th International Conference on Soil Mechanics and Geotechnical Engineering*, IOS Press, 1873–1876.

Mohamad, H. (2014) *Industrial Training Report – Departement Mobiliteit en Openbare Werken*, Deputy Vice Chancellor (A) Office, UTM.

Mohamad, H. & Tee, B.P. (2015) Instrumented Pile Load Testing with Distributed Optical Fibre Strain Sensor, *Jurnal Teknologi*, 77(11), pp. 1–7.

Ohno, H., Naruse, H., Kihara, M. & Shimada, A. (2001). "Industrial Applications of BOTDR Optical Fiber Strain Sensor", *Optical Fiber Technology*, vol. 7, Academic Press, pp. 45–64.

Shi, B., Sui, H., Zhang, D., Wang, B., Wie, G. & Piao, C. (2007). "Distributed monitoring of the slope engineering", *2nd International Workshop on Opto-electronic Sensor-based Monitoring in Geo-engineering*, Nanjing, China, pp. 81–86.

Smith P.D.K. and Burland JB, 1976, Performance Of A High Precision Multi-Point Borehole Extensometer In Soft Rock, *Canadian Geotechnical Journal*, Vol: 13, Pages: 172–176.

Pile set up investigation for improvement on pile capacity

Indra Sati Hamonangan Harahap
Universiti Teknologi PETRONAS, Malaysia

Look Chee Fai
ExxonMobil Exploration and Production Malaysia Inc.

ABSTRACT: Driven pile capacity often increases over time and this is known as pile set-up. This study is to investigate the pile set-up based on the data obtained from the expansion project in United Arab Emirate. A significant number of piles at the project fall below the required pile capacity. Instead of performing pile remedial work, the effect of pile set-up is then quantified. To evaluate pile set-up, selected piles are re-stroked and evaluated using Case Pile Wave Analysis Program (CAPWAP). The elapsed time for re-strike varied from 4 to 71 days and the mobilized capacity increased varying from 65 to 280 tons. The results are then compared with established correlation and the highest percentage difference is 25%. Overall pile set-up is estimated from the correlation and 40% increment is adopted. The results show that the percentage of failed piles after incorporating set-up reduced from 35.9% to 10.3%.

1 INTRODUCTION

Piles are columnar elements in a foundation which have the function of transferring load from the superstructure through weak compressible strata or through water, onto stiffer or more compact and less compressible soils or onto rocks' (Tomlinson and Woodward, 2008). Pile is widely used in foundation design and its material and installation are often costly in construction project. Studies found out that driven pile capacity increase over time and this phenomenon is named as pile set-up. The phenomenon of pile set-up of driven piles have been studied and published, for example, Kormuka et al. (2003), Tan et al. (2004), Erbland and McGillivray (2004), Hadjuk (2006) etc. Taking into account and incorporating the time effects of pile set-up into design can reduce pile length, pile cross section and thus reduce the cost of pile materials, installation times and construction difficulties. However, the practice of incorporating pile set-up during design phase is very seldom in the real case of construction project due to some uncertainties and because the rate and magnitude of setup is a function of a combination of a number of factors which included the types of pile (steel, wood, concrete), the pile cross section and length of the pile and types of soil (cohesive, non-cohesive, and mixed) (Kormuka et al, 2003). Besides that, Hadjuk (2006) concluded that set-up for driven pile is attributed to soil aging and dissipation of excess pore pressure only contributes a minor degree to overall set-up (see Axelsson, 2000). Therefore, it is important to investigate and study the interrelation of all these factors that affecting the set-up of pile capacity before it can be applied into future foundation design.

Piling always makes up a huge portion of works in construction projects. Over designed pile will added extra cost to the project while under designed pile will required pile remedial works which is time consuming. Taking into consider pile set-up effect helps to optimize pile capacity by minimizing the construction cost and difficulties. The set-up of driven pile can be done by comparing the pile capacity at end of initial drive and the re-strike pile capacity after a period of time. Pile load test such as static load test and dynamic testing involved cost and is time consuming. It is useful to carry out pile load test for certain number of piles and then predict the overall set-up for remaining piles in a project. A case study of the expansion project in United Arab Emirate was conducted to study the pile set-up, which involved 230 piles to be driven. During pile driving, a significant number of piles failed to reach the required pile capacity at end of initial drive as shown by the PDA field results. The pile remedial work such as pile extension and tremie concrete required extra cost and will caused delay for the project. Rather than to carry out pile remedial work, the pile set-up was investigated. Driven piles were test by PDA and then carried out CAPWAP analysis to determine the pile capacity after re-strike. Ten piles were chosen to carry out the re-strike test. The PDA and CAPWAP analysis data at end of initial drive and re-strike were studied to estimate the pile set-up. The set-up dependence on time will be determined and compared with the pile set-up correlation established by past researches.

Table 1. Empirical formula for predicting pile capacity with time.

Author(s)	Equation		Comments
Huang (1988)	$Q_t = Q_{EOD} + 0.236[1 + \log(t)(Q_{max} - Q_{EOD})]$	...(Eqn. 1)	Q_t = pile capacity at time t, days Q_{EOD} = pile capacity at EOD Q_{max} = maximum pile capacity
Svinkin (1996)	$Q_t = 1.4 Q_{EOD} t^{0.1}$ $Q_t = 1.025 Q_{EOD} t^{0.1}$	...(Eqn. 2) ...(Eqn. 3)	Upper bound Lower bound
Guang-Yu (1988)	$Q_{14} = (0.375 S_t + 1) Q_{EOD}$	...(Eqn. 4)	Q_{14} = pile capacity at 14 days S_t = sensitivity of soil
Skov and Denver (1988)	$Q_t = Q_0 [A \log(t/t_0) + 1]$	...(Eqn. 5)	t_0 A Sand 0.5 0.2 Clay 1.0 0.6
Svinkin and Skov (2000)	$R_u(t)/R_{EOD} - 1 = B[\log(t) + 1]$	...(Eqn. 6)	B similar to A

2 BASIC THEORY AND METHODS

Pile capacity is generally refer the shaft (skin friction) and the toe resistance (end bearing capacity) of a pile. Pile set-up is defined as the increase of pile capacity with time due to the following mechanisms. (Kormuka et al. 2003):

- Logarithmically nonlinear rate of excess pore water pressure dissipation – the rate of dissipation of excess pore water pressure is not constant with respect to the log of time as the effect of highly disturbed state of the soil.
- Logarithmically linear rate of excess pore water pressure dissipation – the rate of excess pore water pressure dissipation is constant with respect to the log of time at some time after driving.
- Aging – time dependent change in soil properties at a constant effective stress. Aging effects increase the soil's shear modulus, stiffness and dilatancy and reduce the soil's compressibility.

In the same report, Kormuka et al. (2003) provided a summary of the empirical formulas to predict pile capacities over time developed in the past researches (Table 1).

The loading tests chosen for this project is dynamic loading test which using PDA and CAPWAP analysis. Fellenius (1999) suggested that dynamic loading test is better than static load test in term of cost and time saving. It is suitable to evaluate re-strike pile capacity because the equipment set-up is not as complicated as static load test and STATNAMIC load test.

The data collected from this project is the piling data of 117 driven piles. This including the length of pile, the type of pile, the PDA and CAPWAP analysis results at end of initial drive and piling date. For the purpose to investigate the pile set-up, 10 re stroked piles data are analysed.

From the data collected, the pile set-up will be determined by using the dynamic load test result. The capacities increased of driven piles are determined by comparing the CAPWAP analysis result at the end of initial drive and re-strike. The increment is studied based on the driven pile information. Next, the effect of pile set-up will be investigated from the capacity increased over time. An analysis is carried out to study the capacity increased depending upon the amount of time allowed for set-up and the improvement on pile capacity. The set-up effect is verified with the existing set-up correlation developed by the past researches. Finally, an overall set-up is estimated to incorporate into all driven piles. The improvement on pile capacity due to set-up is analysed.

3 RESULT AND DISCUSSIONS

During the pile driving, a significant number of piles failed to achieve the required capacity. Based on the data given, before considering the effect of pile set-up, 42 out of 117 piles (35.9%) are failed to meet the load requirement. Most of the piles are driven up to design penetration level.

For the purpose to investigate pile set-up, 10 piles are selected to carry out re-strike and the result of CAPWAP analysis from field PDA data at the end of initial drive and during re-strike are shown in Table 2. Initially, all the 10 piles are failed to meet the required capacity. After the re-strike test, the result shows that all the piles achieved the design capacity after taking into account of set-up. This shows that pile set-up is significant.

The pile capacity is included the skin friction, end bearing and mobilized capacity. The mobilized capacity is the sum of skin friction and end bearing capacity. After the re-strike, we can observe that the increment in skin friction, end bearing and mobilized capacity are different for each pile. Besides that, the rate of set-up also varied depending upon the time elapsed for set-up. Figure 1, 2 and 3 show the skin friction, end bearing and mobilized capacity in-creased over time for all the re-strike piles.

The results show that there is increment in pile capacity after set-up. To verify, the pile capacity in-crease over time is compared using established correlation, with the sole purpose to quantify the set-up in term of the mobilized capacity which is the sum of skin friction and end bearing.

Table 2. CAPWAP Analysis Result from Field PDA (End of initial strike and re-strike pile).

Pile	CAPWAP Result (EOID)				CAPWAP Result (Restrike)			
	Skin Friction (tons)	End Bearing (tons)	Mobilized Capacity (tons)	Test date	Skin Friction (tons)	End Bearing (tons)	Mobilized Capacity (tons)	Test date
GP-02	190	255	445	13/03/08	429	213	642	18/04/08
TS-01	200	180	380	03/03/08	390	210	600	05/04/08
TS-02	260	170	430	03/03/08	470	240	710	05/04/08
TS-08	243	159	402	02/03/08	486	184	670	16/04/08
TS-31	216	217	433	31/01/08	295	310	605	06/04/08
TS-33	387	321	708	25/01/08	390	550	940	05/04/08
TS-34	365	370	735	–	215	265	480	–
TS-38	515	215	730	25/01/08	610	250	860	29/01/08
TS-42	200	240	440	29/01/08	305	270	575	09/04/08
TS-43	60	434	494	30/01/08	245	506	751	09/04/08

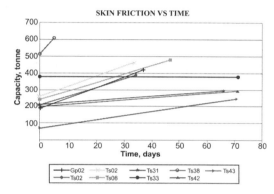

Figure 1. Skin friction with time.

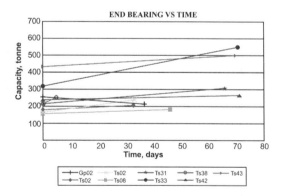

Figure 2. Increase of End Bearing with Time.

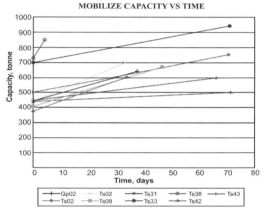

Figure 3. Increase of Mobilized Pile Capacity with Time.

In this study, set-up is verified using the empirical correlation presented by Skov and Denver (1988). It is by far the most popular relationship presented to quantify and predict the pile capacity, which models set-up as linear with respect to the log of time as,

$$Q_t = Q_0[A \log(t/t_0) + 1] \quad (5)$$

where Q_t = axial capacity at time t after driving; Q_0 = axial capacity at time t_0; A = a constant, depending on soil type; t_0 = an empirical value measured in days, and $t_0 = 0.5$, $A = 0.3$ for sand, $t_0 = 1$, $A = 0.6$ for clay

This correlation is used to predict the pile capacity based on the initial pile capacity and time elapsed for set-up. It is suitable to use this relationship because the data obtained from the project fulfil the parameters required to compute the predict capacity. Based on the data available, the soil type of this project is categorized as sand.

This correlation is used to predict the pile capacity based on the initial pile capacity and time elapsed for set-up also named as theoretical capacity. It is suitable to use this relationship because the data obtained from the project fulfil the parameters required to compute the predict capacity. Based on the data available, the soil type of this project is categorized as sand.

This correlation is used to predict the pile capaci-ty based on the initial pile capacity and time elapsed for set-up also named as theoretical capacity. It is suitable to use this relationship because the data ob-tained from the project fulfil the parameters required to compute the predict capacity. Based on the data available, the soil type of this project is categorized as sand.

Table 3. EOID capacity, re-strike capacity and predicted capacity.

Pile	Time lapsed (days)	EOID Capacity (tons)	Theoretical Capacity (tons)	Actual Capacity, re-strike (tons)	% Difference
GP-02	37	445	611	642	5
TS-01	33	380	518	600	14
TS-02	33	430	586	710	17
TS-08	46	402	560	670	16
TS-32	66	433	617	605	-2
TS-33	71	708	1013	940	-8
TS-34		735		480	
TS-38	4	730	862	860	0
TS-42	71	440	629	505	-25
TS-43	70	494	706	751	6

Table 4. EOID Capacity, Re-strike capacity and predicted capacity.

Pile	Time Elapsed (days)	Capacity increased			Capacity increased		
		Skin Friction (tons)	End Bearing (tons)	Mobilize Capacity (tons)	Skin Friction (%)	End Bearing (%)	Mobilize Capacity (%)
GP-02	37	239	−42	197	126	−16	44
TS-01	33	190	30	220	95	17	58
TS-02	33	210	70	280	81	41	65
TS-08	46	243	25	268	100	16	67
TS-32	66	79	93	172	37	43	40
TS-33	71	3	229	232	1	71	33
TS-34		−150	−105	−255	−41	−28	−35
TS-38	4	95	35	130	18	16	18
TS-42	71	105	30	65	53	13	15
TS-43	70	185	72	257	308	17	52

This correlation is used to predict the pile capacity based on the initial pile capacity and time elapsed for set-up also named as theoretical capacity. It is suitable to use this relationship because the data obtained from the project fulfil the parameters required to compute the predict capacity. Based on the data available, the soil type of this project is categorized as sand.

Table 3 shows the end of initial driving (EOID) capacity, re-strike capacity and theoretical capacity. The percentage difference of actual capacity and theoretical capacity is also computed. The result shows that the highest percentage of difference is 25%. The percentage of difference is considered reasonable because in the field of pile set-up study, there is no fix empirical relationship can quantify all set-up. Set-up is a function of a combination of a number of factors which can vary from one case to another.

Table 4 shows the capacity increased of driven piles and the time elapsed between end of initial drive and re-strike. Mobilized capacity increase varied from 65 tons to 280 tons depending upon the amount of time (4 days to 71 days) allowed for pile set and the soil type at the pile location. The percentage of increment is ranged between 15% and 67% in mobilized capacity. The set-up is reasonable because the case history presented by Erbland and McGillivray (2004) indicated that the capacity increase varied from 20 tons to 200 tons.

Besides that, it is unusual to notice that pile TS-34 shows decrement in capacity after re-strike. This is caused by overdrive during re-strike. Rather than re-strike, pile TS-34 is re drive due to some unwanted circumstances. The soil around the pile is disturbed and thus decreases the pile capacity. If the extreme value

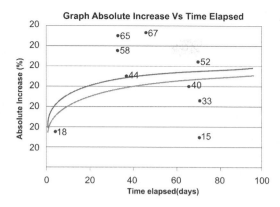

Figure 4. Absolute Increase vs. Time Lapsed.

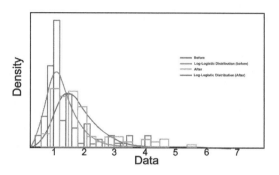

Figure 5. Log-Logistic Distribution Plot Before and After 40% Increment.

(negative value) is not taken into account, the average improvement of mobilized capacity can be assumed as 43%.

Graph Absolute Increase versus Time Elapsed is plotted as shown in Figure 4 to estimate average pile set-up. Based on the available data, 2 lines are fitted to the re-strike result: Log and Power. Both fitted lines showed that the average increment is lies between 40–50% as the time increases. This means the total improvement which can be incorporated is conservatively taken as 40% on top of the PDA predicted mobilized capacity.

Incorporating the 40% increment of mobilized capacity is significant for all the driven piles. The predict pile capacity shows that the total piles failed to meet the capacity are reduced from 42 to 12 piles. The percentage of failure is decrease from 35.9% to 10.3%. Further on, the result shows that 6 out of the 12 failed piles already achieve more than 95% of the required capacity. If these were taking into accounts, the percentage of failure would be less than 10%.

Statistical analysis is carried out to identify the probability of failure pile. The actual and predict pile capacity are divided with the ultimate load to produce a safety factor (SF). This safety factor determines whether the pile meet or not meet the load requirement. If SF > 1, the pile fine; if SF < 1, the pile fails.

Figure 5 shows the distribution of pile before and after incorporating the 40% set-up increment. From

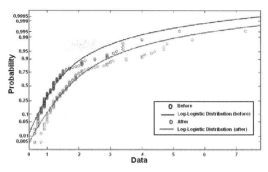

Figure 6. Cumulative Probability Plot.

this, the density of succeed and fail piles can be determined. The figure also shows the probability plot from the distribution data. The type of curve suit the data most is Log-Logistic Distribution. This distribution is the probability distribution of a random variable whose logarithm has a logistic distribution, which is used as a parametric model for events whose rate increases initially and decreases later, as shown by pile set-up behaviour.

The results proved that set-up is taking place on the driven piles. Incorporating set-up helps to im-prove the pile capacity and reduce the construction difficulties. The failed piles achieve the required capacity eventually due to pile set-up.

Then, a cumulative probability (Figure 6) is plotted to investigate the probability of pile will fail before and after the set-up. The plot shows that before the set-up, the cumulative probability less than 1 is 0.3 while after the set-up is only 0.1. This means that the percentage for a pile will fail before taking into account set-up is about 30%. After incorporating the set-up effect, the pile capacity is improve and thus the percentage of failure is about 10% which is very close to the manual computation result.

4 CONCLUSIONS

The objective of this study is to investigate set-up on pile capacity over time. This study was conducted based on the driven piles data of the expansion project of Berth 5 & 6 in United Arab Emirate. Pile set-up was determined by comparing the pile capacity at the end of initial drive and re-strike.

The objective of this study is to investigate set-up on pile capacity over time. This study was conducted based on the driven piles data of the expansion project of Berth 5 & 6 in United Arab Emirate. Pile set-up was determined by comparing the pile capacity at the end of initial drive and re-strike.

Piles subjected to re-strike after its installation (4 days to 71 days) showed that mobilized capacity increase varied from 65 tons to 280 tons. Taking average 40% increment as set-up and incorporated into pile capacity, the failed piles reduce from 42 numbers (35.9%) to 12 numbers (10.3%). Statistical analysis showed that the probability of pile failure before the set-up was 0.3 while after the set-up was only 0.1, a significant reduction of the pile probability of failure.

REFERENCES

Axelsson, G., 2000. "Long-term Setup of Driven Piles in Sand", Doctoral Thesis, Swedish Royal Institute of Technology, Stockholm.

Erbland, P. J. and McGillivray R.T. (2004). Effects of Pile Setup on Pile Design and Construction a Case History. Journal of Geotechnical Engineering, 142: 66–76.

Fellenius, B.H. (1999). Using the Pile Driving Analyzer. Pile Driving Contractors Association, PDCA, Annual Meeting, San Diego.

Guang,-Yu Z. (1988). Wave Equation Applications for Piles in Soft Ground. *In* Proceeding of the 3rd International Conference on the Application of Stress-Wave Theory to Piles, Ottawa, Ontario, Canada. *Edited by* B. H. Fellenius. pp. 831–836.

Hadjuk, E.L. (2006). Full Scale Field Testing Examination of Pile Capacity Gain with Time. PhD Thesis, University of Massachusetts. Lowell.

Huang, S. (1988). Application of Dynamic Measurement on Long H-Pile Driven into Soft Ground. *In* Proceeding of the 3rd International Conference on the Application of Stress-Wave Theory to Piles, Ottawa, Ontario, Canada. *Edited by* B. H. Fellenius. pp. 635–643.

Kormuka, V.E., Wagner, A.B. and Edil, T.B. (2003). Estimating Soil/Pile Set-Up, Wisconsin Highway Research Program #0092-00-14.

Skov, R. and Denver, H. (1988). Time-Dependence of Bearing Capacity of Piles. *In* Proceedings 3rd International Conference on Application of Stress-Waves to Piles. pp 1–10.

Svinkin, M.R. (1996). Setup and Relaxation in Glacial Sand – Discussion. Journal of Geotechnical Engineering. 122(4):217–225.

Svinkin, M.R., Skov, R. (2000). Set-Up Effect of Cohesive Soils in Pile Capacity. *In* Proceedings of the 6th International conference on Application of Stress Waves to Piles, Sao Paulo, Brazil Balkema.

Tan, S.L., Cuthbertson, J. and Kimmerling, R.E. (2004). Prediction of Pile Set-Up in Non-Cohesive Soil. Journal of Geotechnical Engineering, 142:50–65

Tirrant, P.L. ed. (1992). Design Guides for Offshore Structures. Technip, Paris.

Tomlinson, M.J. and Woodward, J.C (2008). Pile Design and Construction Practice. 5th Edition, Taylor & Francis, New York, N.Y.

Whittle, A.J., and Sutabutr, T. (2005). Parameters for Average Gulf Clay and Prediction of Pile Set-Up in the Gulf of Mexico. Journal of Geotechnical Engineering, 169: 440–458.

Feasibility study of P-wave monitoring of selective bioplugging caused by biofilms

Dong-Hwa Noh & Tae-Hyuk Kwon
Department of Civil and Environmental Engineering, Korea Advanced Institute of Science and Technology (KAIST), Daejeon, Korea

ABSTRACT: Selective plugging of high permeability zones using bacterial biofilms is considered as a promising technique to enhance oil recovery (i.e., microbially enhanced oil recovery, or MEOR). We explored the feasibility of using P-wave to monitor the bacterial biofilm accumulation and resulting bioclogging in unconsolidated sediments. A lab-scale experiment was conducted, in which we stimulated our model bacteria Shewanella oneidensis MR-1 to produce biofilms in fully saturated sands and monitored the responses of permeability and P-wave signals. During biofilm accumulation in the sand, we observed the permeability reduction and P-wave amplitude reduction, which provides the unique experimental results for assessing the feasibility of using the response of P-wave amplitudes for bacterial selective plugging for MEOR.

1 INSTRUCTIONS

Bioplugging is a process used to fill the pore spaces of porous medium with biofilms, biopolymers, or biomass. Most bacteria living in the subsurface can produce insoluble materials as by-products, leading to a significant reduction in permeability, which alters the hydrodynamic system of subsurface environments. Some species of bacteria can move using flagella or via advective flows by pressure gradients because their size ranges approximately 0.5–3 μm. Therefore, the plugging effect can easily spread throughout the surrounding environment (Mitchell & Santamarina 2005). Application of bioplugging has been attempted for various practices, including the permeability reduction of dams and dykes, prevention of soil erosion, and hydraulic barriers at soil pollution sites (Ivannov & Chu 2008). Selective plugging using bacterial activities has also received many attention as a way to enhance hydrocarbon production, i.e., referred to as microbially enhanced oil recovery (MEOR) (Gray et al. 2008) This mechanism has been investigated to increase daily oil production at a production well or to reduce water cut, which is the ratio of the produced water volume to the produced total liquid volume. On the other hand, identifying spatial distribution of bacteria and detecting the stimulated area for bioplugging in a reservoir is important for successful implementation of the MEOR technique (Kwon & Ajo-Franklin 2013).

This study therefore investigated the feasibility of monitoring bioplugging caused by bacterial biofilm accumulation in uncemented soils using P-wave responses. The model bacteria, *Shewanella oneidensis* MR-1, were cultured and stimulated to produce biofilms in a fully saturated sand. During biofilm formation, the changes in permeability and P-wave response at an ultrasonic frequency range were monitored. Accordingly, the responses of velocity and attenuation were analyzed and the correlation between permeability and P-wave responses were discussed.

2 MATERIALS AND METHODS

2.1 *Model bacteria*

Shewanella oneidensis MR-1 was used as a model bacterium. *S. oneidensis* is a gram-negative, facultative anaerobe, isolated from a sediment of Lake Oneida, and it can reduce poisonous heavy metal ions such as Fe (III) and Mn (IV) (Thormann et al. 2004a). *S. oneidensis* is rod shaped, 2–3 μm in length and 0.4–0.7 μm in diameter, and can swim with the aid of a single polar flagellum (Thormann et al. 2004b). In this study, a defined growth medium was used to stimulate the growth of the model bacteria *S. oneidensis* MR-1 and the production of bacterial biofilms under a consistent environment, as shown in Table 1. Phosphate buffers were used to maintain the initial growth condition at pH 7.0. The doubling time of the model bacteria, which is the time required for cells to double in cell density, was measured as ∼3.2 hr/generation at 17.5°C by using the optical density (OD) at a wavelength of 660 nm.

2.2 *Column setup*

A column was designed to measure permeability and P-wave responses (velocity and amplitude) during biofilm formation in uncemented soils, as shown in Figure 1. This setup was developed by modifying the column used in our previous research

Table 1. Composition of the defined growth medium for *Shewanella Oneidensis* MR-1.

Compound	Concentration
Tryptic soy broth	30 g/L
Sodium fumarate dibasic	8 g/L
1 M Monobasic KH_2PO_4	41 mL/L
1 M Dibasic K_2HPO_4	59 mL/L

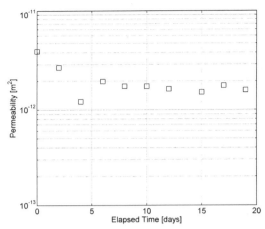

Figure 2. A change in permeability with time during biofilm accumulation.

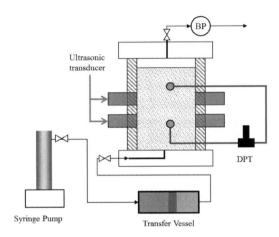

Figure 1. The column setup.

(Kwon & Ajo-Franklin 2013). The acrylic transparent column (internal volume, 538.51 cm^3; internal diameter, 70 mm; height, 140 mm) was instrumented with one differential pressure transducer (PX409, OMEGA), two pairs of immersion-type ultrasonic transducers (I3-0108-S, OLYMPUS). These ultrasonic transducers had a central frequency of 1 MHz. An ultrasonic pulse generator (DPR 300, JSR) was used in a transmission mode to apply a high-voltage excitation to the source ultrasonic transducers. A digital oscilloscope (NI PXI-5105, National Instruments) was used to acquire the propagated and received P-waves. A transparent transfer vessel was connected to a syringe pump (500HP, ISCO Teledyne) and used to inject the fresh growth media into the column. A back-pressure regulator was connected to the outlet of the column to maintain pore pressure and avoid carbon dioxide bubble formation.

2.3 Procedure and measurement

A uniformly graded sand (Ottawa F110; mean particle diameter = 120 μm) was used as a host soil medium. The sand was autoclaved prior to packing and inoculation of *S. oneidensis*. An inoculum was prepared by culturing the model bacteria aerobically in the fresh growth medium. The sand was wet-packed with the inoculum by hand tamping in the column. Porosity of the prepared sand-pack was 0.38. The vertical effective stress of ∼20 kPa was applied using a spring placed at the top of the sand-pack. The baseline permeability was measured by injecting a sterile growth medium at a flow rate of 2 ml/min. During this injection process, the fluid pressure was elevated to and kept at 100 kPa using a back-pressure regulator. Thereafter, the refilling process, where the pore fluid was replaced with a fresh growth medium of ∼500 mL using a syringe pump, was repeated every two days. During these refilling processes, the differential pressure was recorded to monitor a change in the Permeability. The P-wave responses were acquired every hour during biofilm accumulation. This column experiment was conducted for 21 days under an ambient temperature of 24°C for bacteria biofilm accumulation.

3 RESULTS AND ANALYSIS

3.1 *Permeability reduction*

The differential pressure responses (ΔP) were obtained by injecting a growth medium at a flow rate of 2 ml/min (Q). The dynamic viscosity (μ) of the injected fluid was assumed to be the same with that of pure water at 20°C. The permeability (K) was estimated by using ΔP, Q, and column dimension (A: cross-sectional area, L: distance between differential pressure ports) and Darcy's law:

$$K = \frac{Q \cdot L \cdot \mu}{A \cdot \Delta P} \qquad (1)$$

Figure 2 shows the variation in permeability with time. The baseline (initial) permeability was 4.1×10^{-12} m^2. The permeability reduction was significant during the first ∼ 8 days because of the the accumulation of biofilm produced by S. oneidensis. Thereafter, the permeability was fairly converged to 1.6×10^{-12} m^2 at Day 19.

3.2 *P-wave responses*

Acquired P-wave responses were analyzed to determine the velocity, amplitude, and attenuation at a

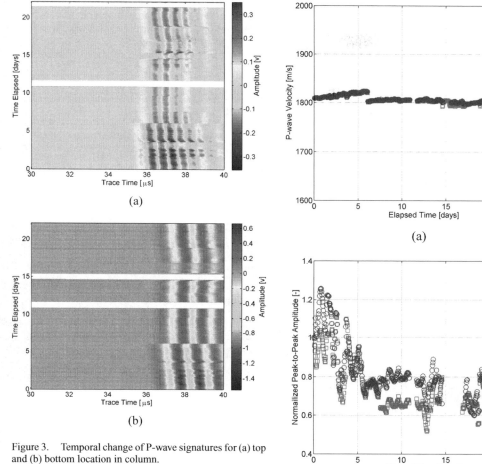

Figure 3. Temporal change of P-wave signatures for (a) top and (b) bottom location in column.

frequency band of 600–1000 kHz. Before the sand packing, the sensor-to-sensor distance was calibrated using deionized water. The continuous P-wave signatures during the experiment were shown in Figure 3.

For wave analysis, the one cycle of all P-waveforms was extracted using rectangle window and then P-wave arrival time was obtained by cross correlation. Figures 3a and 3b depict the changes in P-wave velocity (V_p) and normalized peak-to peak amplitude by Day 0 data. The P-wave velocity was observed to be consistent varying less than ∼10 m/s in the both pairs. As can be seen, the normalized peak-to peak amplitude were reduced by ∼40% in the both pairs. Variations in the P-wave attenuation was estimated and compared to the permeability change during biofilm accumulation. The attenuation change depends on the physical properties of porous media and the fluid saturation conditions along with biofilm formation in porous medium (Toksöz et al. 1979) The spectral ratio technique was chosen to estimate P-wave attenuation at a frequency band of 600–1000 kHz (Toksöz et al. 1979, Sears & Bonner 1981, Frampong et al. 2005). It appeared that the normalized attenuation (Q_P^{-1}/Q_{P0}^{-1}) increased by ∼5% at the both pairs (Figure 5).

Figure 4. Change in (a) p-wave velocity and (b) peak-to-peak amplitude relative to Day 0 by ultrasonic transducer pairs.

It is because the pore space of sand was filled with biofilm produced by *S. oneidensis*. Accordingly, we examined the correlation between the P-wave attenuation increase and the permeability reduction, which occurred with the accumulation of biofilm. Figure 6 shows that the normalized attenuation (Q_P^{-1}/Q_{P0}^{-1}) can be possibly correlated with normalized permeability (K/K_0).

4 CONCLUDING REMARKS

This study explored the feasibility of seismic monitoring of selective plugging caused by biofilm formation in porous media. The permeability of the sand was reduced form 4.1×10^{-12} m^2 to 1.2×10^{-12} m^2 due to plugging caused by biofilm accumulation. P-wave velocity shows almost a constant value of ∼1800 m/s.

P-wave amplitude was notably reduced by ∼40% between 0 and 4 days. Normalized P-wave attenuation slightly increased during accumulation of biofilms. Our experiment result provides the hint of possibility that biofilm-induced plugging can be seismically monitored at this laboratory scale.

ACKNOWLEDGEMENTS

This research was supported by Basic Science Research Program through the National Research Foundation of Korea (NRF) funded by the Ministry of Science, ICT & Future Planning (No.2014R1A1 A1003419).

REFERENCES

Frampong, P., Butt, S. & Donald, A. 2005. Frequency dependent spectral ratio technique for Q estimate, *paper presented at Rainbow in the Earth –2nd International Workshop, 17-18 Agust*, Lawrence Berkely National Laboratory, Berkely, USA.

Gray, M., Yeung, A., Foght, J. & Yarranton, H. W. 2008. Potential microbial enhanced oil recovery processes: a critical analysis, paper presented at SPE Annual Technical Conference and Exhibition, Society of Petroleum Engineers, 21-24 September, Denver, Colorado, USA.

Ivanov, V. & Chu, J. 2008. Applications of microorganisms to geotechnical engineering for bioclogging and biocementation of soil in situ, *Environmental Science and Bio/Technology* 7(2): 139–153.

Kwon, T., & Ajo-Franklin, J. 2013. High-frequency seismic response during permeability reduction due to biopolymer clogging in unconsolidated porous media, *Geophysics* 78(6): EN117-EN127

Mitchell, J. K. & Santamarina, J. C. 2005. Biological considerations in geotechnical engineering, *Journal of Geotechnical and Geoenvironmental Engineering* 131(10): 1222–1233.

Sears, F. M. & Bonner, B. P. 1981. Ultrasonic attenuation measurement by spectral ratios utilizing signal processing techniques, *IEEE Transactions Geoscience and Remote Sensing* 19(2): 95–99

Thormann, K. M., Saville, R . M., Shukla, S., Pelletier, D. A., & A. M. Spormann 2004a. Initial phases of biofilm formation in Shewanella oneidensis MR-1, *Journal of Bacteriology* 186(23), 8096–8104.

Thormann, K. M., Saville, R. M., Shukla, S., & Spormann, A. M. 2004b. Induction of rapid detachment in shewanella oneidensis MR-1 biofims, *Journal of Bacteriology* 187(3): 1014–1021.

Toksöz, M., Johnston, D., & Timur, A. 1979. Attenuation of seismic waves in dry and saturated rocks: I. Laboratory measurements, *Geophysics* 44(4): 681–690, doi:10.1190/1.1440969.

Prioritizing the criteria for urban green space using AHP-multiple criteria decision model

Rabi'ah Ahmad
Department of Civil Engineering, Ungku Omar Polytechnic, Ipoh, Perak, Malaysia

Abdul Nassir Matori
Department of Civil and Environmental Engineering, Universiti Teknologi PETRONAS, Seri Iskandar, Perak, Malaysia

ABSTRACT: This paper presents an Analytic Hierarchy Process (AHP) multiple criteria decision model in prioritizing the criteria (feature attributes) related to urban green space planning. Two prominent recreational parks in Ipoh city was chosen as a case study in prioritizing the criteria pertaining to the needs of urban green space. The decision making procedure uses twenty expert decision makers' opinion through AHP pairwise comparison questionnaire. Nine criteria (desirable feature attributes) that influence urban green space is selected from literature study that satisfy the needs of the public and the environment. An unrestricted, non-commercial AHP Excel Template tool were used in the decision making process. The result has shown that safety criteria have been identified as the most important indicator, followed by maintenance, accessibility and connectivity. Other criteria such as visual pattern, recreation facilities and amenities are identically important features that contribute to the sustainable indicator of urban green space development.

1 INTRODUCTION

1.1 Urban green space

Green space such as community parks in urban areas should be properly located, designed, furnished and maintained in achieving a sustainable environment, lifestyles and patterns. Urban green space features such as size, design, facilities, water elements, location and accessibility are the main decisive factors that determine the satisfaction level of the users or community.

In the preceding research studies by (Rabi'ah and Nassir 2015), qualitative assessment on the satisfaction level of park visitors towards the available features of green space in the local recreational parks was accomplished. The result has shown that the variables in measuring the physical and natural characters of the parks suggested that the diversity of feature attributes of urban green space supports the social interaction of park users for all age groups. The diversity of urban green spaces attributes requires some form of prioritization where the features related to urban green space can be rank according to the public along with planner's perception. Therefore, this research has proposed a documented procedure of quality measure to comprehend the design of green space features that can blend the social attributes and green space properties. The objective is to prioritize the criterion that determines the effectiveness of present urban green space features. This is accomplished through a series of interview sessions with experts in green space planning (decision makers). Their relative preference of criteria that affects the design of green space feature attributes was analyzed using AHP pair wise comparison questionnaire and modelled as priority values.

The preference weight (priority values) will be the sustainable indicator in prioritizing the selection of green space feature attributes that are "desirable" to the needs of the public and the environment of urban residential living areas. The AHP method possesses the variables that aid in decision making towards the sustainable indicator on urban green space. This method has been popular with respect to the assessment of public parks site suitability as published in (Maruani & Amit-Chen 2007; Chandio et al. 2013; Shen et al. 2012; Malczewski 2015) and urban green space design (Lo et al. 2003).

1.2 The Analytical Hierarchy Process (AHP)

AHP is considered the most appropriate mathematical method for multi-criteria decision making (Saaty 1980) and (Saaty 1982). It captures both qualitative and quantitative aspects of a decision and provides a powerful yet simple way of weighting selection criteria thus reducing bias in the decision-making process. The term 'analytical' indicates the problem is simplified and analyzed accordingly into its preferred purpose

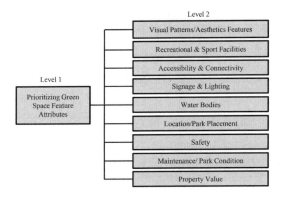

Figure 1. AHP Hierarchy tree definition.

Table 1. AHP Standard value judgment table for expert use (Saaty 1980).

Definition	Intensity
Importance	1
Equal important	2
Equal to moderate important	3
Moderate important	4
Moderate to strong important	5
Strong important	6
Strong to very strong important	7
Very strong important	8
Very to extremely strong important	9

and aim. While 'hierarchy' indicates that a hierarchical tree of the primary elements is formed and decomposed. 'Process' indicates the data judging process to reach the final results by obtaining final weightings scores that belong to the elements contributed. AHP combines the group judgments when several people are involved in the decision making process (Saaty 1980).

2 RESEARCH METHODS

2.1 Structuring the decision problem

The first step in AHP process is to structure the decision problem in a hierarchy as shown or depicted in Figure1. The overall goal of the decision, i.e. prioritizing feature attributes of urban green space, is assigned the top level (Level 1) of the hierarchy. The next level (Level 2) consists of criteria or factors relevant to the goal. Nine criteria (desirable feature attributes) that has influence on urban green space satisfactory level was identified as according to the needs of the public and the environment. These nine criteria were determined from literature summary, the nature and conditions of the study area viz. visual pattern/aesthetics features, recreational & sport facilities, accessibility & connectivity, signage & lighting, water bodies, location, safety, maintenance, and property value (Lo et al. 2003; McCormack et al 2010; Kaczynski et al 2008).

2.2 Pairwise comparison

The second step will be a comparison of criteria (alternatives). They are compared in pairs (pairwise) with respect to the objective of level 1. For the relative comparison, experts express their likelihood comparisons either in verbal, graphical, numerical or in questionnaire models using the fundamental scales in Table 1. This scale is later translated into a quantifiable scale.

The mathematical procedure of AHP follows the simple steps of (Saaty 1980):

1. Establishing a pairwise comparison matrix of multiple criteria evaluation based on expert preferences.

2. Deriving the preference weights by normalization or eigenvector method from the original pairwise comparison matrix by:-
   - Computing the sum of each column matrix.
   - Dividing each entry value in the matrix by the total respective column (sum).
   - Computing the average across rows to get the relative weights.

3. The final step is measuring the consistency ratio (CR) of pairwise comparison. It will evaluate the reliability of the estimated weights. Value of CR equal to 0.1 will be accepted. Otherwise the comparison matrix made by expert should to be revised.

2.3 AHP – Pairwise questionnaire

The pairwise questionnaires have been distributed between the period of June and July 2015. The respondents are mainly experts (i.e. People who indirectly and directly involve in green space planning) from the Institution of Higher learning (IHL), senior officers from Ipoh Local Municipality (MBI) and the Perak State Planning Agency. The initial task in questionnaire distribution is the preliminary explanation (by official appointments) to ensure that the respondent understood what information is needed and how the questionnaire work. There were 20 numbers of respondents (questionnaires collected) in the AHP decision-making exercise. The recommended number mentioned in the literatures varies from five to thirteen (5–13) respondents (Rogers et al. 2001; Balana et al. 2010; Mendoza & Prabhu 1999).

The questionnaire allows decision makers to make pairwise comparisons between the elements. It measures the relative importance between two criteria in a subsequent manner based on the continuous AHP 9-point scale (as in Table 1). The scale one (1) given by an expert gives indicates that his preference between two elements is of 'equal importance' and the scale nine (9) indicates 'extremely important' (Saaty 1980). The questionnaire requires the decision makers to make pairwise comparison between criteria of urban green space feature attributes.

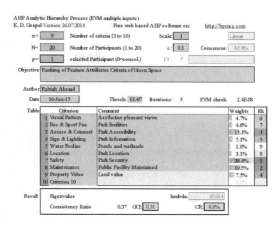

Figure 2. Sample pairwise comparison of one correspondent.

2.4 AHP- Multiple criteria weightings

Business Performance Management Singapore (BPMSG) AHP Excel® Template version 2013 by Klaus (Goepel 2013) provides a non-commercial free web based AHP solution, as a supporting tool for decision making processes. The AHP excel template derives ratio scales from paired comparisons of criteria, and allows for some small inconsistencies in judgments. Inputs can be actual measurements, but also subjective opinions. As a result, priorities (weightings) and a consistency ratio will be calculated. Mathematically the method is based on the solution of an Eigenvalue problem (Goepel 2013). The results of the pair-wise comparisons are arranged in a matrix. The first (dominant) normalized right Eigenvector of the matrix gives the ratio scale (weighting), the Eigenvalue determines the consistency ratio.

The workbook consists of 20 input worksheets for pair-wise comparisons, a sheet for the consolidation of all judgments, a summary sheet to display the result, a sheet with reference tables (random index, limits for geometric consistency index GCI, judgment scales) and a sheet for solving the eigenvalue problem when using the eigenvector method. A sample input-output summary of one participant (total 20 participants) is shown in Figure 2. Once completed, the total weight for each criterion will helps in prioritizing the green feature attributes derived from the decision makers' preferences.

3 RESULTS

3.1 The final weighted criteria

Figure 3 shows the final weighted criteria of 20 decision makers derived by geometric mean from the eigenvector method (EVM). Consequently, Table 3 summarizes the AHP survey results in the form of individual score, the geometric mean, consistency ratio and ranking of the 9 selected criteria. The detail characteristics to describe the importance of the ranked criteria are presented in Table 3.

3.2 Discussions

The result has shown that safety criteria has been identified as the most preferred and was unanimously ranked highest by expert decision makers in relation to recreational park planning in Ipoh city. Safety features refer to people's perceptions and feelings in recreational park issues. This indicates that park activity correlates with sense of safety. This indicator has supported the previous study (Rabi'ah and Nassir 2015),

Figure 3. Final weightage and ranking of criteria.

Table 2. Summary of preference weights (in %).

Decision Makers	Criteria								
	Visual Pattern	Rec & Sport Facilities	Access & Connectivity	Sign & Lighting	Water Bodies	Location	Safety	Maintenance	Property Value	Consistency Ratio (%)
1	5	5	15	5	2	3	39	20	7	7
2	7	16	12	6	7	13	12	16	10	10
3	3	6	21	12	8	12	9	25	4	8
4	17	3	11	5	9	12	8	31	4	10
5	12	7	6	6	7	7	23	24	8	8
6	6	5	2	3	21	28	22	3	9	8
7	21	11	4	3	4	14	13	22	7	10
8	3	6	8	4	2	17	31	27	2	10
9	4	25	17	4	3	36	2	5	2	8
10	9	11	11	14	3	5	17	23	7	9
11	17	21	9	16	3	11	16	4	3	9
12	20	25	17	6	3	15	8	5	2	5
13	6	1	5	8	1	5	34	34	6	12
14	2	3	7	17	5	9	45	9	4	9
15	12	5	11	10	7	8	20	18	10	4
16	15	13	13	4	3	12	9	23	8	6
17	3	5	4	2	28	18	25	11	4	7
18	12	26	5	5	11	5	17	12	7	4
19	3	4	3	9	2	5	33	25	16	9
20	16	12	19	4	4	16	12	14	3	8
Geometric Mean	4.7	4.6	15.3	5.1	1.8	3.1	38.4	19.5	7.5	
Rank	6	7	3	5	9	8	1	2	4	

Table 3. The summarized characteristics of the Ranked Criteria.

Ranked Criteria	Characteristics of Criteria
1. Safety	Criteria that determine the area gives the sense of safety and security with presence of appropriate staff, fencing, presence of lighting, visibility of surrounding houses or roads as part of safety measure. Attractive and safe environment will attract more users and make people stay longer doing active leisure [19]. Survey results from park visitors and decision makers shows that safety criteria need to be upgrade by placing guards and provide safety regulation signage.
2. Maintenance/park condition	Criteria that determine park are well maintained to ensure areas are attractive for the enjoyment of visitors. This includes the maintenance of public toilet, playing area and equipment, sport facilities, benches and other features in park [9]. The maintenance of facilities and others features in the study area parks need to be increased through regular inspection such as children playground and other exercise equipment, trees that need pruning and track acts of vandalism.
3. Accessibility & Connectivity	Criteria that consider location of park function as a focal point within the surrounding areas and are accessible to visitors with good public transport links, adequate parking facilities, no difficult physical barriers such as slopes and cambers.
4. Property Value	Criteria that consider the greenery and aesthetically of the park gives attraction to both residents and investors. It gives a positive impact for parks which provide variety of activities and increase financial returns for land developers [21]. The location of the parks has enabled various businesses to flourish as well as the development of several new housing projects.
5. Signage & Lighting	Criteria that determine park provides good signing, information center/maps, with directional within the area. Park also provides good and sufficient lighting for long stay after dark. This will promotes greater safety and improve emergency access.
6. Visual Patterns/ Aesthetics Features	Criteria that determine aesthetics and pleasant views, scenic beauty that influences visitors and physical activity. These include park size, layout design, landscaping and balance of sun, which gives visual attractiveness and scenic beauty to the surrounding area [20]. Both parks in the study area have remarkable value in terms of environment which attracts many visitors.
7. Recreational & Sport facilities	Criteria that determine the park site provides variety of facilities for physical activities that includes sports field, courts, cycling and exercise/jogging. Park also provides amenities which influence park users that is seating and table, shelters, rest area, toilets, change rooms, litter bins, food services, cultural facilities, fun facilities (playground for children) [9],[10]. Both parks also caters the needs of disabled and elderly people but the facilities provided need to be upgrade to sustain as a popular spot among the locals.
8. Location/park placement	Criteria that take into account the visibility of park, located close to major public transport routes to draw visitors and entrance are easily identifiable and accessible [9]. Both study area parks are located within the urban area and easily accessible by the local thus encouraging social contact in a positive environment.
9. Water Bodies	Criteria that consider the existence of water bodies such as ponds that enhances the design integrity of the environment, gives the sense of naturalness, harmony and mood of relaxation [6]. The presence of water elements that is artificial lake in both parks enhances the visual landscape and character of the place.

where the safety parameter shows that half of the respondents (park users) have that safety feeling in parks. Most of the visitors would stay 1 to 2 hours in parks to perform their regular activities and enjoying surrounding nature. It was suggested that safety level has to be upgraded in the parks through the placement of security guards, more lighting and having fencing around the parks. Hence, enhancing security and enforcement levels in the park has been the top priority issues considered by expert decision makers.

Subsequently, maintenance/park condition which has intangible aspect has been chosen as second priority. This contributes towards community well-being, enjoyment which meets park users need. Poor maintenance and condition can discourage park usage and will negatively affects aesthetics, perceptions of safety, functionality and the overall perception of park quality (McCormack et al. 2010). As mentioned in (Powell et al. 2003), unsafe or poorly maintained parks may discourage use even when they are located within easy walking distance of home. In contrast, accessibility and connectivity is regarded as important as it forms a network of routes which "influence how pleasant it can be to move from one area to another, how much daylight, landscape and beauty we can enjoy" (Rogers 1999). The findings of other attributes imply that visual pattern, recreation facilities and amenities are also important features contribute to the sustainable indicator of urban green space. Visual pattern indicates that people living and working in a busy urban area treasure urban spaces with plants, sunlight and wind. Parks containing a variety of features and amenities may support a wider range of users (Kaczynski et al. 2008; Giles-Corti et al. 2005).

4 CONCLUSIONS

The AHP-decision making process utilizing the pairwise comparison questionnaire is a very structured technique. Decision makers are able to determine their best preferences among multiple criteria with respect to prioritizing feature attributes of urban green space. The participation of experts and officials in the prioritization of criteria has provides comprehensive and rational framework for structuring decision problem in urban green planning. AHP has been proved to be a flexible and practical tool to select features in urban green space.

REFERENCES

Abdul M., Mariapan, M., Mohd, M. and Aziz A. 2011, Assessing the Needs for Quality Neighbourhood Parks. *Australian Journal of Basic and Applied Sciences*, 5(10); 743–753.

Balana, B., Mathijs, E. and Muys, B. 2010. Assessing the Sustainability of Forest Management: an Application of Multi-Criteria Decision Analysis to Community Forests in Northern Ethiopia. *Journal of Environmental Management*, 91(6); 1294–304.

Chandio, I., Matori, M., WanYusof, K., Talpur, M. and Balogun, A. 2013. GIS-based analytic hierarchy process as a multicriteria decision analysis instrument: a review. *Arabian Journal of Geosciences*, 6(8); 3059–3066.

Giles, B., Broomhall, M., Knuiman, M., Collins, C., Douglas, K., Ng, K., Lange, A. and Donovan, R. 2005. Increasing walking: how important is distance to, attractiveness and size of public open space. *American Journal of Preventive Medicine*, 28; 169–176.

Godbey, G., Caldwell, L., Floyd, M. and Payne, L. 2005. Contributions of leisure studies and recreation and park management research to the Active Living agenda. *American Journal of Preventive Medicine*, 28; 150–158.

Goepel, D. 2013. BPMSG AHP Excel template with multiple inputs – version xx, Singapore. Available at: http://bpmsg.com.

Goepel, D. 2013. Implementing the Analytic Hierarchy Process as a Standard Method for Multi- Criteria Decision Making In Corporate Enterprises – A New AHP Excel Template with Multiple Inputs, *Proceedings of the International Symposium on the Analytic Hierarchy Process*.

Kaczynski, T., Luke R. and Brian E. 2008. Association of Park Size, Distance, and Features with Physical Activity in Neighborhood Parks. *American Journal of Public Health*, 98(8); 1451–1456.

Kaplan, R. and Kaplan, S., 1990. *Restorative experience: the healing power of nearby nature*, in Francis, M. and Hester, R. (Eds), MIT Press: Cambridge.

Lo, S., Yiu, C. and Alan, L. 2003. An analysis of attributes affecting urban open space design and their environmental implications. *Management of Environmental Quality: An International Journal*, 14(5); 604–614.

Malczewski, J. and Claus, R. 2015. Desktop GIS-MCDA. *Multicriteria Decision Analysis in Geographic Information Science*. Springer Berlin Heidelberg; 269–292.

Maruani, T. and Amit-Chen, I. 2007. Open space planning models: A review of approaches and methods. *Landscape and Urban planning*, 81; 1–13.

McCormack, G., Rock, M., Toohey, A. and Hignell D. 2010. Characteristics of Urban Parks Associated with Park Use and Physical Activity: A review of qualitative research. *Health Place*, 16(4); 712–726.

Mendoza G. and Prabhu, R. 1999. Multiple criteria decision-making approaches to assessing forest sustainability using criteria and indicators: A case study. *Forest Ecology and Management*, 131; 107–126.

Powell, K., Martin, L. and Chowdhury, P. 2003. Places to walk: convenience and regular physical activity. *American Journal of Public Health*, 93; 1519–1521.

Rabi'ah, A. and Matori, M. 2015. Sustainable Indicator for Feature Attributes Assessment of Urban Green Space. *Proceedings of Postgraduate Conference on Global Green Issues (Go Green)*, UiTM: Malaysia; 417–424.

Rogers L., 2001. Pegger & Roadview – A New GIS Tool To Assist Engineers in Operations Planning. *The International Mountain Logging and 11th Pacific Northwest Skyline Symposium*, Seattle: University of Washington; 177–182.

Rogers, R. 1999. *Towards an urban renaissance: final report of the urban task force*, Department of the Environment, Transport and the Regions, London.

Saaty, T. 1980. *The Analytical Hierarchy Process*. New York: McGraw-Hill.

Saaty, T. 1982. The analytic hierarchy process: A new approach to deal with fuzziness in architecture. *Architectural Science Review*, 25(3); 64–69.

Shen, S., Wang, H., Wen, J., Wang, S. and Fan, C. 2012. Study on Urban Green Space System Planning based on Ecological Suitability Evaluation: A case of Luancheng city in China. *Advanced Materials Research*, 368(373); 1788–1793.

Effects of DEMs from different sources in deriving stream networks threshold values

Nor Azura Ishak, Mohd Sanusi S. Ahamad & Sohaib K.M. Abujayyab
School of Civil Engineering, Engineering Campus, Universiti Sains Malaysia, Nibong Tebal, Penang, Malaysia

Aminuddin Ab. Ghani
REDAC, Universiti Sains Malaysia, Nibong Tebal, Penang, Malaysia

ABSTRACT: Digital elevation model (DEM) is the most common surface topography data used to derive river network. DEM comes with different resolutions and can generate various topographic and hydrological features. This study investigates the effects DEM's threshold values from different sources in deriving stream networks using Geospatial Hydrologic Modelling Extension (HEC-GeoHMS). DEMs of 30m resolution were acquired from SRTM, ASTER, and NEXTMap data. In addition, topographic DEM were derived from contour data of 20 m interval to generate the stream network on the sub-basin of Jawi river, Penang, Malaysia. Subsequently, spatial comparison was made with the existing drainage networks derived from the Malaysian Department of Irrigation and Drainage (DID). The analysis reveals that the drainage network from SRTM with threshold values of 40 was closest to the referenced drainage. The result has signifies the most appropriate values for an effective stream threshold network to be considered suitable for future river sub-basin analysis.

1 INTRODUCTION

The advent availability of the Digital Elevation Models (DEMs) sources and advances in computer processing has resulted in the development of digital mapping or derived channel networks from DEMs (Tarboton, 2003). DEM is the main component of hydrologic modeling and drainage network for river morphological research (Thomas et al. 2014). Topographic maps have been widely used in decades for the estimation of topographic attributes as well as in the delineation of stream networks, which is very time-consuming to produce (Zevenbergen & Thorne, 1987). Generally, the development of Digital Terrain Analysis and DEM have been simplified through the usage of GIS and remote sensing application software such as ArcGIS instead of measuring and digitizing from topographic map (Tarboton et al. 1992). The DEM itself have been widely used in GIS processing as they automatically process the hydrological features such as terrain slope, drainage networks, drainage divides and catchment boundaries thus, reducing the time consumed to analyze the study area (Vaze et al. 2010). In spite of that, DEMs with different resolution and sources generated can lead to different varied topographic and hydrological features.

A study was carried out to investigate the effects of DEMs on deriving topographic and hydrological attributes onto two small forested watersheds located in northern Idaho, USA (Zhang et al. 2008). Six DEMs were used and the result shows that DEMs with different resolution generates varying watershed shapes and structures that gives different extracted hillslope and channel length.

This paper presents a study on the effects of DEMs from different sources in deriving stream networks' threshold values. The stream network analysis applies several DEMs data sources of different resolution in evaluating the hydrological characteristics. The DEMs sources used are namely Shuttle Radar Topography Mission (SRTM) in 30 m resolution, Advanced Spaceborne Thermal Emission and Reflection Radiometer (ASTER) in 30 m resolution, NEXTMap World 30 in 30 m resolution and DEM (TopoRaster) in 20 m resolution.

In signifying the sensitivities of different DEMs sources, tests on the hydrological application were made to delineate drainage network on a specific part of Jawi River, Penang. Subsequently, the delineated drainage river networks were benchmarked with a reasonably accurate data source provided by the Malaysian DID. Lastly, the results were compared with an independent river network study made by (Poggio & Soille, 2011).

2 STUDY AREA

Jawi River is located in the district of South Seberang Perai, about 315 km from the national capital, Kuala Lumpur (Figure 1). It covers a total land area of 243 km^2, stretching between latitude 5° 10'N and 5° 20'N and longitude 100° 25'E and 100° 35'E with the elevation ranges between 1 m and 273 m above

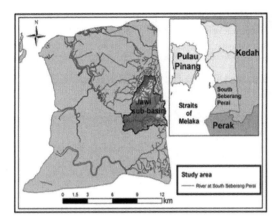

Figure 1. The study area.

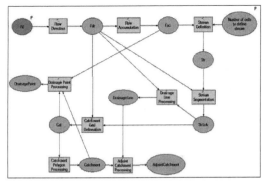

Figure 2. Flow chart showing the entire process of creating watershed delineation using HEC-GeoHMS.

mean sea level. Jawi River originates at Duri river and flowing through of Jejawi river near the coast before discharging into the Malacca Strait. There are five major tributaries of the Jawi river system namely, Bakap river with a drainage area of 65.3 km^2, Duri river with 1.54 km^2, Baung river with 6.8 km^2, Bakar Arang river with 8.5 km^2, and Badak Mati river covering 9.49 km^2. The mean temperature in the region is 27°C with high humidity varying from 50–70%. The study area has two typical monsoons, such as the northeast monsoon and southwest monsoon and sometimes experienced massive change of heavy rainfall.

3 METHODOLOGY

There are several methods of DEM interpolation in Geographic Information System (GIS). However, methods of resampling do not influence the quality of resampled DEMs significantly (Tamiru & Rientjes, 2005). Therefore, 'Topo to Raster' method was used as an interpolation technique where the tool was designed for the creation of correct hydrological DEMs as it is based on the ANUDEM program. The program was developed by Michael Hutchinson and the method was applied in ArcGIS version 10.2 for the hydrological application (Gallant & Hutchinson, 2011). The research applies different sources of DEMs with several levels of spatial resolution. River networks were individually extracted and subsequently compared with the data from Malaysian DID.

The Geospatial Hydrologic Modelling Extension (HEC-GeoHMS) was developed as a hydrology toolkit for engineers and hydrologist to extract physical parameters from DEM. Hydrology module HEC-GeoHMS was used to extract river networks from different DEMs. HEC-GeoHMS is a reliable model developed by the US Army Corps of Engineers and usable for many hydrological simulations (Halwatura & Najim, 2013). The HEC-GeoHMS model are frequently used to analyse urban flooding, flood frequency, flood warning system planning, reservoir spillway capacity and stream restoration (U.S. Army Corps of Engineers, 2008). Flow direction and sinks were identified in each DEM and if sinks existed, the process will filled to revise the flow direction. The automatic process of filling the sinks uses a function that identifies the pits and raised the terrain in order to have a smoother surface that allows the water to flow freely without forming ponds (Jenson, 1991). After correcting the flow direction, the grid cells will identify the streams based on the flow accumulation threshold. The threshold will affect the density of drainage network as well as the size of the watershed area. The higher threshold value will result in a less dense stream networks with fewer internal watersheds as compared to lower threshold (Chang, 2012). By default, the stream threshold value will be 1% of the maximum flow accumulation. Thus, the different threshold numbers were applied to all data sources to facilitate comparison and to generate sufficient stream network.

4 RESULTS AND DISCUSSION

Stream threshold is one of the most important parameters that must be considered in the stream networks. Drainage networks were first extracted from each DEMs using the methodology as described in Figure 2. The number of cells for stream threshold varies based on the spatial resolution of DEM used. The linear parameters such as stream number, stream length were taken into consideration in conducting the comparison study. The test was conducted by changing the threshold number at a constant value, since the 1% of the stream threshold does not replicate all sufficient tributaries on the watershed. The effects of different threshold number are then plotted. Figures 3–6 shows the variation in stream length of each stream order at different thresholds.

The results shows variation in the number of streams in each order as it depends on the threshold value. The number of thresholds varies from 20, 40, 60, 80 and 100 for each data source. However, due to the pixel's resolution, there was split on the stream networks image when the number of thresholds increased to

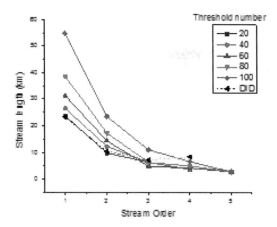

Figure 3. Graph analysis stream order for Jawi sub-basin for SRTM 30 [m].

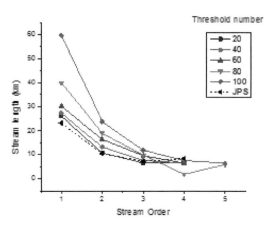

Figure 4. Graph analysis stream order for Jawi sub-basin for ASTER 30 [m].

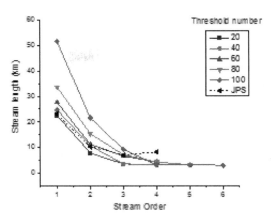

Figure 5. Graph analysis stream order for Jawi sub-basin for World 30 [m].

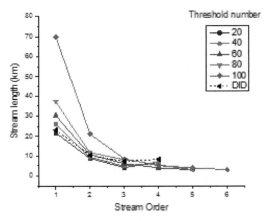

Figure 6. Graph analysis stream order for Jawi sub-basin for TopoRaster at 20 [m].

above 100. Therefore, the optimum threshold number was 100 and the lower is 20 respectively.

The plot on SRTM 30 m (Figure 3) using 20- threshold number matches the line plotted for Malaysian DID reference. The number of stream length was 23.61 km and 9.75 km respectively and achieved at 1st and 2nd stream order. The graph has shown that the stream networks match the Malaysian DID plot at these stream order points. It is found that smaller threshold value will gives denser drainage network, thereby increasing the stream length. Additionally, at stream order 3, the threshold 40 lies on the exact point of Malaysian DID i.e. 6.26 km. The highest stream length recorded was 54.78 km in the 1st order.

As for ASTER 30 m (Figure 4), it shows that the threshold 20 matches the reference point of stream network from Malaysian DID at 2nd stream order. For the 3rd order, the reference stream network from Malaysian DID falls between threshold 20 and 40 (6.44 km and 7.55 km, respectively). However, the Malaysian DID stream network nearly approached the threshold 100 at 7.51 km. The highest stream length from this data source was 59.68 km at threshold 100, while the lowest point was at threshold 80 (1.84 km) at 4th stream order.

Figure 5 shows the graph for World 30 m data source. The graph shows that the threshold 20 lies almost on Malaysian DID point (22.54 km) at the 1st order, but on the 2nd and the 3rd stream order, threshold 40 and 60 matches exactly on the Malaysian DID stream network points (10.75 km and 6.86 km, respectively). The highest stream length on this graph was 51.50 km and the lowest point was 3.24 km at threshold 100.

Lastly, Figure 6 shows the graph for DEM generated from 'TopoRaster' 20 m data. The objective is to identify whether topographic data can yield better result in a stream network as compared to satellite-derived data. The graph lines indicates the points at stream order 1, 2, 3 and 4 matches the line from Malaysian DID stream network. Moreover, threshold 60 lies exactly on the bench mark line at 2nd stream order (10.68 km). The highest and the lowest stream lengths on the graph were (69.87 km and 3.11 km, respectively) with threshold 20.

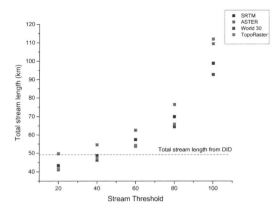

Figure 7. Graph analysis of variation in the total stream length at different threshold with various data sources.

Studying the effect of the threshold number on different sources of DEMs is essential in determining the stream network. A threshold number range 20 to 100 was selected for the study. The effects of threshold number and total stream length are shown in Figure 7.

The graph illustrates a strong positive relationship between ASTER 30m at threshold 20 with SRTM 30 m at threshold 40, having the same stream length as the Malaysian DID (i.e. 49.70 km and 48.65 km, respectively). The total stream length of the sub-basin from Malaysia DID is 49.24 km (represented as a dashed line). From the previous graph analysis, it is discovered that the SRTM 30 m was more significantly near to DID stream data at 1st and 2nd stream order. This result has shown that the threshold 40 in the SRTM 30 m was an effective stream threshold network that can be considered to be suitable for future river sub-basin analysis.

5 CONCLUSION

This study concludes that the accuracy of the automatic stream network derivation depends on the quality and different sources of the DEMs as it has an affect the outcomes of river network extraction. From literature, the SRTM 30 m was expected to be more accurate and reliable and the analysis has shown that SRTM 30 with threshold 40 delivered the best result of river network when compared with the digitized river network from Malaysian DID (i.e. 98% similarity). However, other DEM data sources having same resolution also give similar patterns as indicated in the graphs plot. DEM data from SRTM 30 m has the potential to be used for automated extraction of river networks to less than 1m resolution with superior results.

ACKNOWLEDGEMENT

The authors wish to thank the Universiti Sains Malaysia (USM) for the provision of the Research University Funding (RU Grant No.1001/PAWAM/ 814219 and FRGS Grant No.203/PAWAM/6071258 for the completion of this research.

REFERENCES

Chang, K.T. (2012). Introduction to Geographic. Information Systems. McGraw-Hill Companies.

Gallant, J.C. & Hutchinson, M.F., 2011, 'A differential equation for specific catchment area', Water Resources Research, vol. 47, no. 5, p. W05535.

Halwatura, D. & Najim, M.M.M., 2013, 'Application of the HEC-HMS model for runoff simulation in a tropical catchment'. Environmental Modelling & Software, vol. 46, pp. 155–162.

Jenson, S.K., 1991, Application of hydrologic information automatically extracted from digital elevation mod els Hydrological Processes, 5, pp. 31–44.

Poggio, L. & Soille, P., 2011, 'A probabilistic approach to river network detection in digital elevation models', Catena, vol. 87, no. 3, pp. 341–350.

Tamiru, A. & Rientjes, T.H.M., 2005, 'Effects of LiDAR DEM resolution in flood modelling: A Model sensitivity study for the city of Tegucigal Honduras', pp. 168–173.

Tarboton, D.G., 2003. Terrain analysis using digital elevation models in hydrology. 23rd ESRI International Users Conference, San Diego. July 7–11.

Thomas, J., Joseph, S., Thrivikramji, K.P. & Arunkumar, K.S., 2014, 'Sensitivity of digital elevation models: The scenario from two tropical mountain river basins of the Western Ghats, India'. Geoscience Frontiers, vol. 5, no. 6, pp. 893–909.

Tarboton, D.G., Bras, R.L. & Rodriguez-iturbet, I., 1991, 'On the extraction of channel networks from Digital Elevation Model', vol. 5, no. September 1990, pp. 81–100.

U.S. Army Corps of Engineers, 2008. Hydrologic Mode ling System (HEC-HMS) Applications Guide: Version 3.1.0. Institute for Water Resources, Hydrologic Engineering Center, Davis, CA.

Vaze, J., Teng, J. & Spencer, G., 2010, 'Impact of DEM ac curacy and resolution on topographic indices'. Environmental Modelling & Software, vol. 25, no. 10, pp. 1086–1098.

Zevenbergen, L.W. & Thorne, C.R., 1987, 'Quantitative analysis of land surface topography', Earth Surface Processes and Landforms, vol. 12, no. 1, pp. 47–56.

Zhang, J.X., Chang, K. & Wu, J.Q., 2008, 'Effects of DEM resolution and source on soil erosion modelling: a case study using the WEPP model', International Journal of Geographical Information Science, vol. 22, no. 8, pp. 925–942.

Spatial compliances study of present landfill siting to Malaysian standard

Siti Zubaidah Ahmad, Mohd Sanusi S. Ahamad & Sohaib K.M. Abujayyab
School of Civil Engineering, Engineering Campus, Universiti Sains Malaysia, Nibong Tebal, Pulau Pinang, Malaysia

Mohd Suffian Yusoff
Solid Waste Management Cluster, Science and Engineering Research Centre, Engineering Campus, Universiti Sains Malaysia, Nibong Tebal, Penang, Malaysia

Najat Qader Omar
Civil Engineering Unit, College of Engineering, Kirkuk University, Kirkuk, Iraq

ABSTRACT: The solid wastes in Malaysia are generally disposed through landfills located within the respective states under the control of their local municipalities. The awareness of proper landfill siting has currently become a central issue, where there has been an existence of an inadequate systematic planning decision with respect to locating landfills. This paper highlights a GIS based spatial analytical procedure to investigate the compliances of present landfill location with the criteria set by the by the Malaysian National Strategic Plan for Solid Waste Management (NSPWM). A comprehensive review of the detailed parameters for environmental, physical and socioeconomic (EPSE) criteria set by NSPWM is also discussed. Case studies were conducted on landfill locations in the northern part of Malaysia, viz. Jabi Landfill (Kedah), Pulai Landfill (Kedah), and Alor Pongsu Landfill (Perak). The result has indicated that their locations fulfill the criteria pertaining to sustainable landfill siting set by NSPWM. However, there is non-uniformity with regards to criterion parameters being assessed, where there are evidences of influence from different policies set by the respective state municipalities. The divarication in criterion parameters can create problems in ensuring efficient and effective solid waste landfill management.

1 INTRODUCTION

1.1 Solid waste management

The solid waste generated in Malaysia has increased from 0.5 kg/capital/day in the 1980's to current volume of 1 kg/capital/day, representing an increment of 200% over 20 years (Agamuthu 2001). The abundant of the municipal solid waste is largely due to the urbanization process, increase in population, per capital income and changes in consumption patterns. These factors have not only increased the solid waste volume but also changed the characteristics of the solid waste which have made it more complex for the municipalities to handle.

Over the past years, the country is still battling with the issue of landfill's location even though there have been a total 296 landfills in Malaysia (NSWMD 2012). The disposal of wastes through land filling has become complicated with the landfill sites filling up at a very fast rate within 5–10 years (Ahmad et al. 2014). At the same time, the construction of new landfill sites are heavily influenced by factors such as funding constraint, land scarcity, increase of land prices, swift growth in population, rapid urbanization, lack of public concern and existence of uncontrolled dumping sites (Idris et al 2004, Manaf et al. 2009). If the landfill site can be determined in a proper manner combined with well-managed solid waste collection and transportation system, then the right fundamental level for the solid waste management is achieved (Rylander 1998).

1.2 The national strategic plan for solid waste management

In 2005, the Malaysian Ministry of Housing and Local Government has published a guideline on landfill siting criteria under the National Strategic Plan for Solid Waste Management (NSPSWM, 2005). The NSPSWM acts as strategic plan that provides guidance and an advisory framework to enable the Federal, State and Local Governments to make informed decisions on immediate and long-term management of solid waste as reference to local and private authority. The criteria for setting up of new landfill as stated in NSPSWM guideline is listed in Table 1.

A review study has shown the existence of an inadequate systematic planning decision with respect to locating landfills in Malaysia. The foremost reason is that, the site did not comply with the criteria specified by the guidelines in NSPSWM. Therefore, the scenario in Malaysia requires the usage of special site selection model designed using advance spatial technology such

Table 1. Description of Main and Sub-Criteria of Landfill Siting (DOE 2007, NSPSWM 2005).

Main Criteria	Sub-Criteria	Description
Environmental (E)	Surface Water Body	Criteria that avoid landfill to be situated close to water body including river, lakes and ponds in protecting water body ecosystem and avoiding flood plains.
	Sensitive Areas	Criteria that avoid landfill to be situated close to sensitive areas (i.e. forest reserve).
	Water Permeability (Groundwater)	Criteria that take into account the aquifer potential (i.e. high, medium, and low) around the landfill site to prevent groundwater pollution.
	Climate (Rainfall Density)	Criteria that defines the rainfall density to avoid side effects of drainage and erosion.
	Flood Plain	Criteria that opposed landfill to be situated nearby flood protection embankments.
Physical (P)	Road Access	Criteria that consider an appropriately constructed and maintained access road to and a road system within the landfill site capable of supporting all vehicles hauling waste during the operating life of the landfill to minimize traffic congestion and optimize travelling time and cost.
	Soil Permeability	Criteria that determine the soil permeability classes (i.e. rapid, moderate, and slow) to prevent groundwater pollution from landfill leachates.
	Haul Distance	Criteria that take into account the distances of town centre (collection point) to avoid high transportation costs.
	Wind	Criteria that opposed landfill site to be exposed to wind to control litter and dust.
	Slope	Criteria that select appropriate terrain condition that suitable for construction of landfill site. Presented in slope percentage.
	Geological Fault Properties	Criteria that avoid locating landfill site near to existing fault to prevent the ground motion effect.
	Airport Location	Criteria that determine the landfill site not to be located near to airport area to prevent the birds' disturbance and rising dust from landfill.
	Bedrocks/Lithology	Criteria that determine the landfill site to be situated on the solid unweathered rock that lies beneath the loose surface deposits of soil, alluvium, etc to avoid natural disaster such as earthquake.
Socio-Economic (SE)	Residential Area	Criteria that opposed landfill site to be situated nearby residential area (NIMBY syndrome).
	Urban Area	Criteria that opposed landfill site to be situated nearby urbanized area.
	Land Use	Criteria that determine the land use type that suitable for landfill siting.
	Utilities	Criteria that opposed landfill to be located near or crossing the major lines of electrical transmission, or other utilities (i.e. gas, sewer, water lines).

as GIS to combined spatial criteria that are stated in NSPSWM guideline to new landfill site. The end result will be a composite suitability score map that defines potential candidate sites for landfill. GIS analytical capabilities are useful for landfill site assessment and provide land suitability results in such a good manner with high accuracy (Ersoy & Bulut 2009, Zamorano et al. 2008). GIS can manage large amount of data as well as data changes and very practical in checking spatial parameters (criteria) imposed in landfill selection process (Siddiqui et al. 1996, Vatalis & Manoliadis 2002). GIS spatial analytical tool is useful in checking spatial parameters or the criterion attributes imposed in site selection process (Ahmad et al. 2011).

Locational criteria are important in a new landfill siting. As stated in NSPSWM, landfill cannot be located within certain distance from the aspects such as lakes, ponds, rivers, wetlands, flood plain, highway, critical habitat areas, water supply, well, and airports. Moreover, landfill siting is prohibited in areas where potential contamination of groundwater or surface water body exists. The locational issue will not arise if the site selection process is performed accordingly to the stated criteria and guidelines published by NSPSWM. A proper revision of all restrictions during the preliminary siting process is significant to avoid waste of time and money in evaluating sites that will not conform to the standard requirements (Ahmad et al.

2011). Landfill siting requires a detailed site location investigation which addresses all the issues outlined by NSPSWM that takes into account the environmental, physical and socio-economic (EPSE) parameters including water contamination, air pollution, wildlife conflicts, as well as transportation, economic activity and social processes.

This paper addresses the procedure that investigates the compliances of present landfills locations to the criteria set by the NSPSWM. This procedure can help local authorities in detecting specific work on site selection and in the effort to assure the guidelines for site selection criteria is being followed.

2 RESEARCH METHODOLOGY

2.1 Study area and landfill characteristics

The study investigates three present landfill sites located at different regions in the northern part of Malaysia, viz. Jabi Landfill, Pulai Landfill, and Alor Pongsu Landfill as depicted in Figure 1. Table 2 summarizes the characteristics of the three stated landfills under study.

2.2 Methodology

The research methodology covers field visits to conduct oral interview, discussion and distribution of survey form to landfill operators and maintenance. The task has been difficult when one has to consider the criteria that satisfy stated regulation policy imposed by ruling body. There is no prior record of assessments planning authority per-taining to the suitability of the present landfills under study. No documents or records are accessible. The visit to each landfill site has thoroughly observed and in-vestigates the site condition.

Likewise, the laboratory spatial data extractions are performed using IDRISI GIS analytical tools to determine the parameters of specific criteria such as the bed-rocks/lithology and land use/land cost. The extraction of spatial parameters of present landfill sites are performed using heuristic approach comprised of three main steps:

(i) Preparation of maps pertaining to the present land-fills for different selection criteria such as agricultural activities, road network, conservation land, river network, geology classification and others.
(ii) Extracting the criterion attributes of the present landfill using spatial analysis such as its proximity to existing roads and residential settlement, previous land use type of the landfill, soil and geological properties of the landfill area, and etc.
(iii) Comparing the parameters of the present landfill attributes with the NSPSWM specifications.

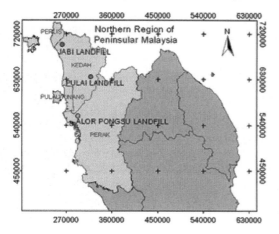

Figure 1. Study Area.

Table 2. Descriptive summary of the landfill sites under study.

Name of landfill	Jabi	Pulai	Alor Pongsu
Local authority	Majlis Bandaraya Alor Setar (MBAS)	Majlis Daerah Baling (MDB)	Majlis Daerah Kerian (MDK)
State	Kedah	Kedah	Perak
Year of start	1985	2001	1986
Year of end	2025	2018	2019
Life span	40 years	18 years	33 years
Area (in hectare)	26.10	6.73	2.4
Coordinate	6°9'9.01"N 100°29'15.8"E	5°38'56.48"N 100°53'11.7"E	5°5'2.89"N 100°35'54.3"E
Land ownership	Government	Kedah Regional Development Authority (KEDA)	Land acquisition made by the government
Site condition	Flat and swampy land	Hilly and swampy land	Flat and swampy land
Amount of waste disposed	300 tone/day	100 tone/day	100 tone/day
Type of vegetation around the site	Rubber tree and paddy field	orchard	Palm oil tree
Vector and animals problem	Under control	Under control	Under control
Complaints	No complaint	Odour problem	No complaint
Land utilization after closure	Recreational park	Recreational park	Recreational park

Table 3. EPSE parameters of present landfills.

Sub-Criteria	Landfill Parameters		
	Jabi	Pulai	Alor Pongsu
Surface Water Body, m	3000	4000	1000
Sensitive Areas, m	No sensitive zone	No sensitive zone	No sensitive zone
Groundwater/Aquifer Potential	No indication of excessive water table rise	No indication of excessive water table rise	No indication of excessive water table rise
Climate/Rainfall Density	Minimal rain	Minimal rain	Minimal rain
Flooding Area,	Not in flood plains	Not in flood plains	Not in flood plains
Road Access (m)	300	850	400
Soil Permeability	Poor draining	Poor draining	Poor draining
Haul Distance (km)	25	25	30
Wind Potential	Calm air	Calm air	Calm air
Slope (%)	<10	<10	<10
Geological Fault Properties	No fault area nearby	No fault area nearby	No fault area nearby
Airport Location (km)	30	No airport area nearby	No airport area nearby
Bedrocks/Lithology	Sedimentary and metamorphic rocks	Intrusive rocks	Unconsolidated deposits
Residential Area (m)	1000	500	500
Urban Area (m)	4000	7000	3000
Land Use/Land Cost	Agricultural land	Agricultural land	Agricultural land
Utility	No major utility line nearby	No major utility line nearby	No major utility line nearby

3 RESULTS AND DISCUSSION

3.1 GIS analysis

From the GIS spatial analysis, the environmental, physical and socio-economic (EPSE) parameters are spatially extracted from the present landfills as shown in Table 3. In brief, the environmental criteria parameters limit the particular geographic areas permissible in protecting the sensitive ecosystem, human health and safety. Likewise, the physical criteria specify the parameter limit to the effect of landfill construction and operation inclusive of technical and operational criteria. Last, the socio-economic parameters take into account the landfill effect towards social aspect such settlement area, urban areas, cultural areas, and visibility and economic aspect such as landfill operating cost, land cost, and construction cost Ahmad et al. 2015). Subsequently, a check list that compares the compliances of present landfills to NSPSWM specifications was made and the summarized result is illustrated in Table 4.

3.2 Discussions

The overall study has shown that the respective landfills in Jabi, Pulai, and Alor Pongsu has generally satisfy the sustainable landfill siting criteria set in the NSPSWM guideline. However, Pulai Landfill and Alor Pongsu Landfill did not satisfy the residential constraint set at 1 km buffer. Overall, the parameters of the respective landfills are common for environmental criteria with the exception of the distance to surface water body i.e. Jabi (3 km), Pulai (4 km), and Alor Pongsu (1 km) respective-ly. Other parameters that vary slightly are road access, haul distance, airport location, and bedrocks/lithology in physical criteria.

The parameters for socio-economic criteria vary with respect to residential area, and urban area proximity. However, this variation may be caused by restriction set by individual local municipalities as they may have their own technical reference and approach in deciding the landfill site suitability that complies with Federal and State regulation. Finding places to put land-fills is difficult. Proper landfill site selection if implemented will guarantee against inconveniences and unfavorable long-term effects. In some cases, the specified criterion imposed in landfill site selection has become subjective through suggestion of some remedial measures. For example, the variation slope stability condition of proposed landfill site can be controlled by improving structure and rock mass strength instead of rejecting areas of steep slopes (Ahmad 2012).

4 CONCLUSIONS

It is evident that the current location of Jabi Landfill Site, Pulai Landfill Site, and Alor Pongsu Landfill Site has satisfied the rules pertaining to sustainable landfill siting criteria. The variation of the criterion parameters from the spatial extraction analysis are being influenced by different policies within respective state municipalities in Malaysia. This cannot be taken lightly since uniformity is been encouraged in NSPSWM guideline for entire Malaysia. Variation of spatial criteria for landfill site se-lection will create problems in managing sustainable landfills. The uniformity will help in providing an efficient and effective solid waste management and public cleansing in the country especially for the site selection process.

Table 4. The compliances of present landfills with NSPSWM specifications.

Sub-Criteria	Landfill Parameters NSPSWM	Jabi	Pulai	Alor Pongsu
Surface Water Body (m)	100	Satisfy	Satisfy	Satisfy
Sensitive Areas (m)	500	Satisfy	Satisfy	Satisfy
Groundwater/Aquifer Potential	No indication of excessive water table rise, springs, or vadose water passages	Satisfy	Satisfy	Satisfy
Climate/Rainfall Density	Minimal rain and rain intensity	Satisfy	Satisfy	Satisfy
Flooding Area (m)	Avoid flood plains	Satisfy	Satisfy	Satisfy
Road Access (m)	500	Accept	Satisfy	Accept
Soil Permeability	Heavy poor draining	Satisfy	Satisfy	Satisfy
Haul Distance (km)	Close to centre of potential service area	Satisfy	Satisfy	Satisfy
Wind Potential	Good air mixing and predominantly downstream of human activities	Satisfy	Satisfy	Satisfy
Slope (%)	<10	Satisfy	Satisfy	Satisfy
Geological Fault Properties, m	100	Satisfy	Satisfy	Satisfy
Airport Location, km	3	Satisfy	Satisfy	Satisfy
Bedrocks/Lithology	Avoid fissured and fractured rocks	Satisfy	Satisfy	Accept
Residential Area, m	1000	Satisfy	No	No
Urban Area, m	1000	Satisfy	Satisfy	Satisfy
Land Use/Land Cost	Avoid areas planned for high value development	Satisfy	Satisfy	Satisfy
Infrastructure, m	Avoid sites of sensitivity	Satisfy	Satisfy	Satisfy

ACKNOWLEDGMENTS

The authors wish to thank the Universiti Sains Malaysia (USM) for the provision of the Research University Funding (RU Grant No. 1001/PAWAM/814219 and FRGS Grant No. 203/PAWAM/6071258) for the completion of this research. This work was also funded by Universiti Sains Malaysia under Iconic grant scheme (Grant No.1001/CKT/870023) for research associated with the Solid Waste Management Cluster, Engineering Campus, Universiti Sains Malaysia.

REFERENCES

Agamuthu, P. 2001. Solid waste: principles and management with Malaysian case studies. Kuala Lumpur: University of Malaya Press.

Ahmad, S.Z. 2012. Spatial Analytical Study on the Effect of Municipal Solid Waste Landfill Siting Using Different Guidelines. *Unpublised Master Thesis*. Universiti Sains Malaysia.

Ahmad, S.Z., Ahamad, M.S.S, and Yusoff, M.S. 2014. Spatial effect of new municipal solid waste landfill siting using different guidelines. *Waste Management and Research*, SAGE Publisher, 32(1), 24–33.

Ahmad, S.Z., Ahamad, M.S.S, and Yusoff, M.S. 2015. A Comprehensive Review of Environmental, Physical and Socio-Economic (EPSE) Criteria for Spatial Site Selection of Landfills in Malaysia. Applied Mechanics and Materials, Trans Tech Publications, Vol. 802 pp 412–418.

Ahmad, S.Z., Ahamad, M.S.S., Wan Hussin, M. A, and Abu Bakar, M. Y. 2011. Application of GIS in Extracting Spatial Criteria from Existing Landfill Site – A Study on Pulau Burung Landfill Site, Presented At 13th International Surveyor Congress (ISC 2011), Putra World Trade Centre, Kuala Lumpur, Malaysia, 22–24 June 2011.

DOE, 2007. Environmental Impact Assessment Guidelines for Municipal Solid Waste and Sewage Treatment and Disposal Projects, Department of Environment, Ministry of Natural Resources and Environment, Malaysia. ISBN 9839119478.

Ersoy, H., and Bulut, F. 2009. Spatial and Multi-criteria Decision Analysis-based methodology for landfill site selection I growing urban regions. Waste Management and Research, 1–12.

Idris, A., Inane, B., Hassan, M.N. 2004. Overview of waste disposal and landfills/dumps in Asian countries. Journal of Material Cycles and Waste Management 6, 104–110.

Manaf L.A., Samah M.A.A. and Zukki N.I.M. 2009. Municipal solid waste management in Malaysia: practices and challenges. Waste Management 29: 2902–2906.

NSPSWM. 2005. *Criteria for Siting Sanitary Landfills, Appendix 6B,* Vol. 3, Local Government Department, Ministry of Housing and Local Government Malaysia.

NSWMD (National Solid Waste Management Department). 2012. The Result of Solid Waste Management' Lab

2012, Ministry of Urban Wellbeing, Housing and Local Government, Putrajaya, Malaysia.

Rylander, H. 1998. Global waste management. In 14th Waste Congress, 12–15 October 1998, Gauteng, South Africa.

Siddiqui, M. Z., Everett, J.W., and Vieux, B. E., 1996. Landfill Siting Using Geographic Information Systems: a Demonstration, Journal of Environmental Engineering, 122(6), 515–522.

Vatalis, K., and Manoliadis, O. 2002. A two-level multicriteria DSS for Landfill Site Selection using GIS: Case Study in Western Macedonia, Greece, Journal of Geographic Information and Decision Analysis 2002, 6(1), 49–56.

Zamorano, M., Molero, E., Hurtado, A. Grindlay,A., and Ramos, A. 2008. Evaluation of a Municipal Landfill Site in Southern Spain with GIS-aided Technology, Journal of Hazardous Material, 160, 473–481.

Monitoring deformation of liquefied natural gas terminal with persistent scatterer interferometry

A.S. Ab Latip & A.N. Matori
Department of Civil and Environmental Engineering, Universiti Teknologi PETRONAS, Seri Iskandar, Perak, Malaysia

D. Perissin
Lyles School of Civil Engineering, Purdue University, West Lafayette, IN, USA

ABSTRACT: The Liquefied Natural Gas (LNG) terminal built on reclaimed land may undergo surface deformation caused by the unconsolidated soil and loading of its heavy structures. Currently, the LNG terminal deformation is monitored using levelling and Global Navigation Satellite System (GNSS) techniques. Physical obstruction and safety issues may prevent the accessibility to the structures of the terminal where the monitoring is required most. This study proposed the state of the art remote satellite-based Interferometric Synthetic Aperture Radar (InSAR) using Persistent Scatterer Interferometry (PSI) technique to perform the deformation monitoring of the above LNG terminal. It does not require intrusion of human and instrument to the monitored structures. The PSI processing results showed the density of persistent scatterer (PS) points extracted from high resolution TerraSAR-X images were high and useful for deformation study of the monitored structures. The deformation trend revealed that the monitored structures experienced relative moderate deformation.

1 INTRODUCTION

Land reclamation from the sea has become the most common way to accommodate the limited land space in coastal areas for development of importance structures such as liquefied natural gas (LNG) terminal. Therefore, the LNG terminal has potential to undergo continuous surface deformation caused by the unconsolidated foundation, its own weight or construction of heavy loaded structures such as plants, storage tanks, large-diameter pipelines and ports on the reclaimed land, which may affect the integrity and safety of the LNG structures. Therefore, continuous monitoring of deformation in the LNG terminal is important to ensure the integrity and safety of the LNG structures as well as to ensure continuous production.

The performance of the Differential Interferometric Synthetic Aperture Radar (DInSAR) technique was limited by the spatial and temporal de-correlation, as well as atmospheric error. To overcome these limitations, the Persistent Scatterer Interferometry (PSI) technique was developed to exploit a long temporal series of SAR images to identify natural or artificial stable scattering targets (persistent scatterers (PS)) over a year or years to reduce the impact of the spatial and temporal de-correlation and effectively remove the atmospheric errors, leaving the surface deformation as the only contribution to the signal phase changes (Ferreti et al., 2001). Comparing with global positioning system (GPS) and levelling techniques, which provide deformation information at the specific observation points, the PSI technique can provide hundreds of points per square kilometre over a widespread area with comparable accuracy (millimetre-level) and extremely lower cost (Sousa et al., 2014; Qin & Perissin, 2015). The use of PSI technique to monitor landslide and volcano could be found in Colesanti et al. (2003) and Hooper et al. (2004). Perrisin used the PSI technique to monitor subsidence of subway tunnel in Perrisin et al. (2012). Works on PSI in oil and gas application were done by Matori et al. (2014) and Ab Latip et al. (2015) for the monitoring of an offshore platform.

In this study, the PSI technique has been applied to monitor the deformation of the small area in the LNG terminal using high resolution TerraSAR-X datasets. The TerraSAR-X datasets were acquired with interval of 44 days from April 21st, 2015 to January 10th, 2016 over the LNG terminal. The average deformation velocity map, deformation time series and three-dimensional (3D) location of PS points were the three outputs from the PSI analysis to reflect the deformation and settlement of the LNG terminal.

2 SAR DATASET AND STUDY AREA

A SAR dataset of seven TerraSAR-X images were acquired with a similar revisiting time of 44 days from April 21st, 2015 to January 10th, 2016 over the LNG terminal (see Figure 1). The images were also acquired at about 6.46 pm each day along the ascending orbit

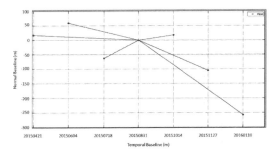

Figure 1. Temporal and normal baseline between master and slave images of TerraSAR-X images used.

(from South to North) with the incidence angles from 49.57 to 51.11 degrees. The acquisitions were in 3 m spatial resolution of StripMap imaging mode and the wavelength of the sensor was 3.1 cm. The coverage of each image was 30 km (width) × 50 km (length) which covered the whole area of the LNG terminal. The image on August 31st, 2015 was selected as a master image among all the available TerraSAR-X images (other images called slave images) to minimize the effect of geometric and temporal de-correlation. The minimum connections of temporal and normal baselines were 44 days and 17 m, while the maximum were 132 days and 257 m, respectively. Normal baseline is the difference in the orbit distance between master and slave images while the temporal baseline is the difference in time.

The study area is the LNG export terminal which it consists of: 1) liquefaction plants for cleaning/treating natural gas and converting natural gas into liquid form, 2) storage tanks for storing liquefied natural gas, 3) pipelines for bringing natural gas from reservoirs to liquefaction plants or transporting liquefied natural gas into storage tanks and then to carrier ships, and 4) port infrastructures. In this paper, small area of LNG terminal was selected for the preliminary PSI data processing, focusing on the storage tanks and large-diameter pipelines as the monitored structures. Small area of PSI processing approach allows focusing on the area of interest, getting familiar with the different strategies in the PSI processing and the results can be analysed deeply. Moreover, this approach did not require the estimation and removal of atmospheric effects as no significant atmospheric effects were expected to influence in small area data processing and also due to the small number of TerraSAR-X images used in this study. The locations of four LNG storage tanks and pipelines of the selected small area on the reflectivity map are shown in Figure 2. The LNG tanks were sequentially marked with A, B, C and D while large-diameter pipelines were marked with E.

3 METHODOLOGY OF PSI PROCESSING

In this study, the small area processing module in Sarproz software was applied for the PSI data processing (Perissin et al., 2011). The persistent scatterers candidate (PSC) points were selected by setting a

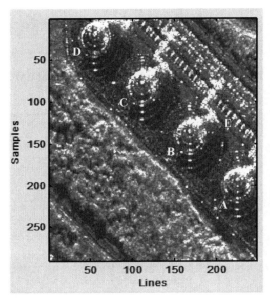

Figure 2. The locations of four LNG storage tanks and pipelines on the reflectivity map.

threshold of 0.80 for amplitude stability index. A reference point was selected based on the PSC who has high temporal coherence value and assumed to be stable. The atmospheric phase screen (APS) error was neglected for small area data processing and the external digital elevation model (DEM) was utilized to remove the topographic error. The linear differential deformation and the differential DEM errors were estimated repeatedly with imposing different parameter ranges. The imposed parameter ranges were judged to be correct from high temporal coherence value and results of the estimated parameters (height and deformation of the targets). The temporal coherence threshold to select the PS points was 0.80 which this threshold level was set to limit the number of potential outliers of PS points.

The geocoding was performed for results visualization and analysis. The PS points were overlaid onto the Google Earth map to clearly identify deformation trends of the monitored structures and 3D function was utilized to combine 3D coordinates and deformation information of the PS points.

4 RESULTS AND DISCUSSION

A total of 653 PSC points which the high density of PSC points observed in the man-made features (i.e., surrounding LNG tanks areas and along the pipeline routes) while those low could be found in the natural features (i.e., along the land road and forested areas) when a threshold of 0.8 was set for amplitude stability index (see Figure 3). The reasonable explanation is the short wavelength of X-band exposed to temporal de-correlation phenomena due to increase sensitivity of

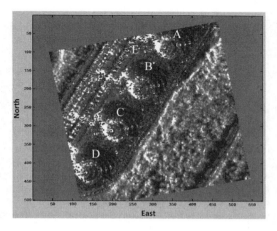

Figure 3. Distribution of 653 PSC points on the reflectivity map.

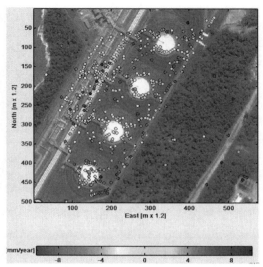

Figure 5. Average deformation velocity map of PS points.

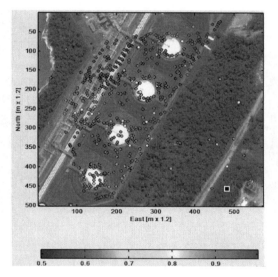

Figure 4. Temporal coherence of PS points.

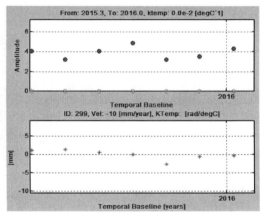

Figure 6. Deformation and amplitude history of a selected PS point (point 299 and its velocity was −10 mm/year).

amplitude values to any changes of natural features during the whole observation period.

Figure 4 shows 577 PS points have been selected when the temporal coherence threshold of 0.8 applied to select only the reliable PS points. Most of the selected PS points were scattered on the LNG tanks and pipeline routes, further utilized for deformation analysis.

The distribution of the PS points and their corresponding average deformation velocity are shown in Figure 5. The average deformation velocity ranging from −10 to 10 mm/year along the sensor-target (line of sight (LOS)) direction with respect to a reference point which the position of reference point indicated by the pink circle. The observed subsidence was indicated by colour coded scale where stable areas were plotted in green colour, maximum subsidence areas in red (−10 mm/year) and maximum uplift areas in blue (10 mm/year). The A, B, C and D of LNG tanks experienced relative moderate deformation rates from −2 to 8 mm/year, −10 to 8 mm/year, −5 to 6 mm/year and −4 to 10 mm/year, respectively. The moderate deformation can also be observed in the E of the pipelines where the rate was from −10 to 7 mm/year.

Figure 6 shows the deformation time series of a selected PS point (point 299) which was on the large-diameter pipeline (see Figure 5). The deformation time series was plotted together with the amplitude of the PS point. The amplitude maintained high values and has small amplitude difference at every acquisition time, showing stable scattering features of the target observed. The deformation time series shows no obvious linear subsidence trend of the selected PS point observed. A slight uplift from first to second data acquisition can be visible. Then, a period of subsidence can be noticed between second and fifth data

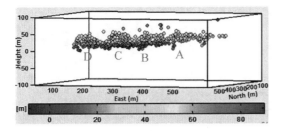

Figure 7. 3D view of PS points (horizontal and height coordinates).

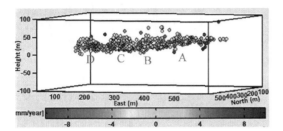

Figure 8. 3D view of PS points combined with deformation trend information.

acquisition, but the movement was moving towards stable again from fifth to seventh data acquisition.

Figure 7 shows the 3D views of PS points for the monitored structures. The colour scale shows the height of the PS points, ranging from −8.5 m (blue) to 90.3 m (red). The top of LNG tanks (roof tanks) was measured to be about 45 m while the bottom was 14 m, thus indicated the height of the LNG tanks was 31 m. The LNG tanks C and D have rather smooth shape compared than A and B due to the sparse PS density can be observed at the top of A and B.

Figure 8 shows the 3D views of PS points where the deformation trend information was available as well. With the combination of the 3D coordinates and deformation information of the PS points, the individual building affected by the deformation can be identified which in this case showed the top parts of the LNG tank C and D have experienced subsidence and uplift, respectively.

5 CONCLUSIONS

The PSI technique has been applied for the first attempt to monitor the deformation of the small area in the LNG terminal, more specifically LNG storage tanks and large-diameter pipelines. The preliminary PSI processing results were processed using Sarproz software showed the density of PS points was high in the LNG tanks and pipelines since the structures were covered by a large number of pixels of high resolution TerraSAR-X data. From the PSI analysis, the monitored structures of LNG tanks A, B, C and D experienced moderate deformation of −2 to 8 mm/year, −10 to 8 mm/year, −5 to 6 mm/year and −4 to 10 mm/year respectively, while E which represented pipelines also experienced moderate deformation of −10 to 7 mm/year. The deformation time series of a selected PS point shows no obvious linear subsidence trend can be observed. Also, the combination of the 3D coordinates and deformation information of the PS points showed the top parts of the LNG tank C experienced with subsidence while D experienced with uplift. To increase reliability and integrity of this finding, ground measurement using GPS and levelling techniques may be suggested to confirm the deformation and settlement observed at the LNG terminal.

REFERENCES

Ab Latip, A. S., Matori, A. N., Aobpaet, A., & Din, A.H.M. 2015. Monitoring of Offshore Platform Deformation with Stanford Method of Persistent Scatterer (StaMPS). *The 2015 International Conference on Space Science and Communication (IconSpace2015), Langkawi, Malaysia, 10–12 August 2015.*

Colesanti, C., Ferretti, A., Prati, C., & Rocca, F. 2003. Monitoring landslides and tectonic motions with the Permanent Scatterers technique. *Engineering Geology* 68: 3–14.

Ferretti, A., Prati, C., & Rocca, F. 2001. Permanent scatterers in SAR interferometry. *IEEE Transactions on Geoscience and Remote Sensing* 39(1):8–20.

Hooper, A., Zebker, H., Segall, P., & Kampes, B. 2004. A new method for measuring deformation on volcanoes and other non-urban areas using InSAR persistent scatterers. *Geophysical Research Letters* 31(23):L23611.1-L23611.5.

Matori, A. N., Ab Latip, A. S., Harahap, I. S. H., & Perissin, D. 2014. Deformation Monitoring of Offshore Platform Using the Persistent Scatterer Interferometry Technique. *Applied Mechanics and Materials* 567:325–330.

Perissin, D., Wang, Z., & Wang, T. 2011. The SARPROZ InSAR tool for urban subsidence/manmade structure stability monitoring in China. *Proc. of ISRSE 2010, Sydney, Australia, 10–15 April 2011.*

Perissin, D., Wang, Z., & Lin, H. 2012. Shanghai Subway Tunnels and Highways Monitoring Through Cosmo-SkyMed Persistent Scatterers. *ISPRS Journal of Photogrammetry and Remote Sensing* 73:58–67.

Qin, Y., & Perissin, D. 2015. Monitoring Ground Subsidence in Hong Kong via Spaceborne Radar: Experiments and Validation. *Remote Sensing* 7:10715–10736.

Sousa, J. J., Hlavacova, I., Bakon, M., Lazecky, M., Patricio, G., Guimaraes, P., Ruiz, A. M., Bastos, L., & Sousa, A. 2014. Potential of Multi-Temporal InSAR Techniques for Bridges and Dams Monitoring. *In: SARWatch Workshop, CENTERIS 2014, Troia, Portugal, 15–17 October 2014.*

Highway and transportation engineering

An evaluation of 85th operating speed and posted speed limit based on horizontal, vertical alignments and traffic conditions: Case study of two lane urban roadway

Azim Muiz Abu Bakar, Mohamad Saharol Nizam Abdul Rani, Rozlinda Mohamed, Nur Suhadah Sani, Nurjannah Jalal, Muhammad Akram Adnan & Tuan Badrol Hisham Bin Tuan Besar
Faculty of Civil Engineering, Universiti Teknologi MARA, Shah Alam, Selangor, Malaysia

ABSTRACT: Operating speed profile analysis is the most used methods in evaluating the consistency of the roadway design where it estimates the operating speed of the vehicle. In the case of roadways for Universiti Teknologi MARA (UiTM) Shah Alam, the horizontal curve was classified as a sub-standard curve and a speed limit of 25 kilometres per hour (km/h) has been enforced along the curve to ensure traffic safety both to vehicle drivers and pedestrians. The 25 km/h posted speed limit was used based on the location of the case study area is within the university. The purpose of this study firstly to evaluate the relationship of 85th percentile operating speed in conjunction with horizontal and vertical alignments at three (3) points which are at the beginning, mid and end of curve. Secondly, to study relationship between turning radius of selected vehicles and the lane width along the curve.

1 INTRODUCTION

Operating speed profile analysis is the most used methods in evaluating the consistency of the roadway design where it estimates the operating speed at which a characteristic vehicle can circulate throughout the segment in study and initiating comparisons with defined criteria [1]. In the case of roadways for Universiti Teknologi MARA (UiTM) Shah Alam, the horizontal curve was classified as a sub-standard curve and a speed limit of 25 kilometres per hour (km/h) has been enforced along the curve to ensure traffic safety both to vehicle drivers and pedestrians. Therefore, evaluation of 85th percentile operating speed for horizontal and vertical alignments was conducted to define the reliability of adopting the 25 km/h of posted speed limit along the curve.

2 PROBLEM STATEMENT

The geometric design of the case study area is a combination of vertical and horizontal curve where the horizontal curve was designed as a sub-standard curve with radius of 50 meter. The 25 km/h posted speed limit was used based on the location of the case study area is within the university. However, mostly the vehicles travel at this case study is 40 km/h. Once, there are case where two buses stuck at the middle curve causing complete standstill to both opposing direction due to the implemented curve was designed under sub-standard curve. This situation becomes dangerous for the safety of vehicles on the both lane. Considering the situations, it is obviously that the road needs to be revised both on operating speed and its geometric elements design such as curve radius, lane width and others.

3 OBJECTIVES

The purpose of this study is to evaluate the relationship of 85th percentile operating speed in conjunction with both horizontal and vertical alignments at three (3) points which are at the beginning, mid and end of curve. In addition, the relationship between turning radius of selected vehicles and the lane width along the curve is also determined. The reliability of adopting the 25 km/h along the curve is also evaluated based on the relationships obtained.

4 SCOPE OF WORKS

In this study, only one site was chosen to conduct the case study which is located in the campus of Universiti Teknologi MARA (UiTM) Shah Alam. For the spot speed study, the data was collected within one day at three point along the curve. During the study, there are two conditions exist on the site which are dry and wet pavement. However, this study only includes the analysis of 85th operating speed for dry pavement. Any equation of relationship defined in this study is only applicable in the selected site of this study that solely describe the physical and geometric condition of the road.

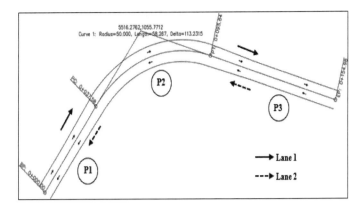

Figure 1. The observation point for the measurement of speed.

5 LITERATURE REVIEW

5.1 Terminology

Speed is an elementary aspect in transportation engineering. It can be defined as rate of motion in distance per unit time. All vehicles travel at a different speed in a moving traffic stream. Therefore, traffic stream has distribution of individual speeds instead of single characteristics speed.

5.2 Previous work on predicting operating speeds at two-lane urban roadway

Research from previous study of different regions and country have developed an operating speed models prediction for urban roadway while a lot of operating speed models have been developed for rural highway. The objective of reviewing these operating speed models is to identify the relationships between the operating speed and principal of geometric design. The guideline for setting up the speed limits while considering the design speed was also provided from previous study. Fitzpatrick et al. [2] [3], found that design speed was the only significant variable in order to predict operating speed on crest vertical curves on sub-urban roadways. The independent variables in the speed models for horizontal curves were approach density and curve radius. After figure out the effect of geometric design, traffic control device factors and roadside on operating speed on four-lane suburban arterials, it is found that posted speed limits was the most significant variables while deflection angle and access density are significant for operating speed on horizontal curves. Poe et al. [8] conducted a study on operating speed at 27 urban collect streets in central Pennsylvania. He found that vehicle speed on low speed urban street was influenced by geometric roadway elements such as lane width, degree of curvature and also grade, access, land-use characteristics and traffic engineering aspect. Free flow speed data was collected by the researchers at specific locations along a corridor. In this study, the researchers also include the effect of vehicles and drivers into speed models.

Table 1. Summary of the Geometrical Characteristics.

Cross Section & Geometry Element	P1	P2	P3
Lane Width (m)	3.25	3.75	3.25
Radius of Curve (m)	–	50.0	–

6 METHODOLOGY

The selected study area for this study is in the Universiti Teknologi Mara (UiTM) Shah Alam, Jalan Ilmu 1/1. The posted speed limit on the case study is 25 km/h. The radius of curve was measured by adopting the method 2 of MIROS [10] with AutoCAD Civil 3D. Study found that the curve radius was designed as a sub-standard where 50-meter radius was used instead of 60-meter based on ATJ 8/86 [7] and REAM guideline. The 16% gradient measured at the site also was exceeding the allowable maximum grade at 12% for U3 design standard. It is also observed that the road shoulder was not constructed along the road. Overall, the road is categorized as U3 where the designed speed is 40 km/hr after consideration of geometry characteristic obtained from the study area.

Three points are selected along the road to collect the speed data as shown in Figure 1. For point 1, the continuous traffic speed data are recorded by using road tube connecting to Trax Appolyon and it is located 20 meter before the beginning of the curve. Besides that, the speed data for 100 vehicles were collected and recorded using laser gun meter detector at point 2 and 3. Table 1 shows the summary of the geometrical characteristic data at the study area.

The turning radius of vehicles are also collected based on the vehicle composition at the study area. The turning radius of vehicles were based on AASHTO 2011 [4] as shown in Table 2.

Table 2. Turning Radius of Selected Vehicles.

Design Vehicle	Type	Length (m)	Maximum Turning Radius (m)
P	Car	5.79	7.75
MH	Medium van	9.14	12.62
City Bus	Bus	12.19	13.71

Table 3. Results of paired t-test of point 3.

Variables	Mean	St. Dev	SE. Mean	P-value
Speed at AC L1	42.805	7.560	0.756	
Speed at AC L2	40.030	4.931	0.493	0.003
Differences	2.775	9.070	0.907	

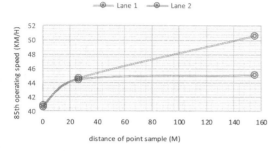

Figure 2. 85th operating speed versus distance of point sample.

7 RESULTS AND ANALYSIS

The 85th operating speed was calculated using Minitab release 17.1.0. The graph of 85th operating speed versus distance of point sample was plotted to show the speed trend of vehicles along the curve as shown in Figure 2.

From Figure 2, it is observed that the speed trend of 85th operating speed at the lane 1 and lane 2 was increase along stretch of the road from point 1 to point 3. However, the Lane 1 has higher 85th operating speed at point 3 compare with the 85th operating speed of lane 2.

From Figure 2, it is observed that the speed trend of 85th operating speed at the lane 1 and lane 2 was increase along stretch of the road from point 1 to point 3. However, the Lane 1 has higher 85th operating speed at point 3 compare with the 85th operating speed of lane 2.

The paired t-test was conducted between speed at point 3 for lane 1 and lane 2 to find if there any significance difference between two samples (speed) mean. [4], The hypothesis for the paired t-test can be stated as:

$H_0 =$ There is no significance difference between two speed means

$H_1 =$ There is strong evidence to indicate a difference between two speed means

From table 3, the p-value is less than 0.05, thus rejecting the H_0 and accepting the H_1. Hence the speed trend shows significant difference between lane 1 and lane 2 at point 3.

To further analyze this situation, this study provides a relationship between 85th Operating speed with vertical gradient and lane width along the curve. To further understand the effect approach using regression had

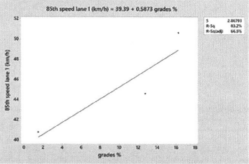

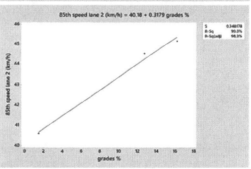

Figure 3. 85th operating speed of lane 1 and lane 2 versus grades.

been performed [4]. One regression model was identified to be correlated with study for predicting 85th Operating speed along 3 point as shown in figure 3 and equation 1 and 2.

For lane 1:
$$V85 = 39.39 + 0.5873 \, GR \quad (1)$$

For lane 2:
$$V85 = 40.18 + 0.3179 \, GR \quad (2)$$

Where,

V85 = 85th operating speed along the 3 points (km/h) GR = Gradient along the 3 points (%)

The equation 1 and 2 shows that the coefficient for the independent variable GR has a positive sign, indicating that increasing in the GR value will lead to an increasing in the 85th operating speed along the 3 points. However, to explain the situation of lower 85th operating speed at lane 2 compare with the lane 1 is due to the uphill condition at the lane 2. Whilst, the gradient at lane 1 is in downhill condition. Thus, the 85th operating speed at lane 2 is lower as the vehicles

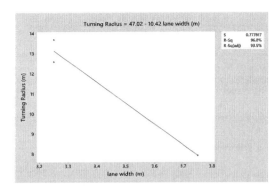

Figure 4. Turning radius of vehicle vs lane width along three points.

required more energy and power to accelerate during travelling at uphill condition.

To define the relativity of lane width with regards the problem of two longer vehicle meet at the middle curve from lane 1 and lane 2. Thus, this study also provides the relationship between lane width and turning radius of the vehicles. The Regression of turning radius versus lane width along 3 point was determined as shown in figure 4 and equation 3.

$$TR = 39.39 + 0.5873LW \quad (3)$$

Where,

TR = Turning Radius of Vehicles Required to travel along the curve (m)

GR = Lane width along 3 points (m)

The equation 1 and 2 shows that the coefficient for the independent variable LW has a negative sign, indicating that increasing in the LW value will lead to decreasing the turning radius required for vehicle along the 3 points. In this case, the maximum lane width adopted at middle curve is 3.75 m where 0.5 m where added in consideration for pavement widening along the curve. Due to the small radius at the curve and long turning radius required for the bus, two busses travelling at the same time from lane 1 and lane 2 will have a difficulty to pass the curve as a bus was not able to yield within the lane width of 3.75 meter at 50-meter radius of curve with normal 85th operating speed.

8 CONCLUSION AND DISCUSSION

The 85th operating speed at before curve, middle curve and after curve was successfully evaluated in this study. From determined 85th operating speed along the three points, the relationship between 85th operating speed and gradient that represent the elements of vertical alignment is successfully developed. Thus, it is observed that the 85th operating speed is higher than the posted speed limit along the curve even though the curve was designed with 50- meter radius. It is observed that the vehicles were speeding due to the uphill and downhill condition along the curve. Thus, the 25 km/h of posted speed limit is not reliable used along the curve as the speed limit on a roadway should be set by an engineer based on the 85th percentile speed even the section of roadway has an inferred design speed lower than 85th percentile speed [9]. However, the standard bus route need to be rearranged to avoid two busses from lane 1 and lane 2 travelling at the same time as the curve is not suitable to be travelled by two longer vehicles at the same time. It is also proposed that the lane width need to be increased or providing the road shoulder to ensure that the longer vehicle with long turning radius will be able to yield within the lane.

ACKNOWLEDGEMENT

The authors would like to thank to the Faculty of Civil Engineering, Universiti Teknologi MARA (UiTM) for the financial support under Ministry of Higher Education- FRGS scheme (FRGS/1/2015/SKK06/UITM/03/2) and also to all individuals and organizations that have made this study possible.

REFERENCES

[1] M. Castro, J.F. Sánchez, N. Ardila and J. Me lo. *Speed Models for Highway Consistency Analysis. A Columbian Case Study*. 4th International Symposium on Highway Geometric Design, Valencia, 2010.

[2] Abbas SKS, Adnan MA, Endut IR Exploration of 85th percentile operating speed model on horizontal curve: a case study for two-lane rural highways. Proc Soc Behave Sci 16:352–363.

[3] K. Fitzpatrick, C.B. Shamburger, R.A. Krammes, D.B. Fambro, Operating speed on suburban arterial curves, Transportation Research Record: Journal of Transportation Re search Board 1579 (1997) 89–96.

[4] Adnan, M.A. "Development of Entrance Ramp merging Density model based on an urban expressway traffic condition." Ph.D thesis School of Civil Engineering, Universiti Sains Malaysia. (2007).

[5] American Association of State Highway and Transportation Officials (AASTHO). *A Policy on Geometric Design of Highways and Streets*. 5th ed. United States of America, 2011.

[6] Road Engineering Association of Malaysia (REAM). *REAM Guideline 2/2002 'A Guide on Geometric Design of Roads'*. Shah Alam, Malaysia, 2002.

[7] Public Works Department Malaysia (PWD). *Arahan Teknik (Jalan) 8/86 'A Guide on Geometric Design of Roads' (ATJ8/86)*. Kuala Lumpur, Malaysia, 1986.

[8] Poe, C.M., Tarris, J.P., and Mason, J.M. (1996). Relationship of Operating Speeds to Roadway Geometric Design Speeds, Pennsylvania Transportation Institute.

[9] Fitzpatrick, K., Blaschke, J. D., Shamburger, C. B., Krammes, R. A., and Fambro, D. B., "Compatibility of Design Speed, Operating Speed, and Posted Speed." Final Report FHWA/TX- 95/1465-2F. Texas Department of Transportation, College Station, TX (October 1995).

[10] Malaysian Institute of Road Safety Research (MIROS). *METRA 'Guidebook for Traffic & Road Safety Audit (part A: Road Alignment and Cross Section)'*. Selangor, Malaysia, 2012.

Mechanical performance of temperature reduced mastic asphalt

M. Dimitrov & B. Hofko
Vienna University of Technology, Institute for Trasnportation, Vienna, Austria

ABSTRACT: Mastic asphalt (MA) mixes are produced at high temperatures (230°C to 250°C) to decrease the viscosity of the bitumen. High production temperatures lead to an increased energy demand and higher emissions of greenhouse gases. This paper presents the results for mechanical performance of temperature reduced MA. A reference MA is compared to temperature reduced MAs: a currently used reduction by modification with wax and an alternative new method by substituting crushed by round aggregates. Both methods realize a maximum temperature drop by 30°C. In addition, a combination of wax modification and round aggregates is investigated, which may achieve a reduction of 50°C. The results show that the resistance to permanent deformation is not decreased by using round aggregates and that it can be doubled by employing wax regardless of the aggregate shape. Resistance to low-temperature cracking is affected by neither of the employed methods for temperature reduction.

1 INTRODUCTION

Mastic asphalt (MA) is an asphalt mix type, which is applied in the field without compaction, it is merely poured. It transfers load mainly by stiff mastic and not by coarse aggregate interaction. MA is applied widely as sealing and surface layer on bridges ...(Widyatmoko et al., 2005, Medani et al., 2007), as road surface layer for city centers where compaction would endanger historic buildings or as surface layer for walk and bike lanes.

The mastic is responsible for the load transfer and therefore usually hard binders are employed for MA. To keep the mix pourable, high temperatures of up to 250°C are necessary for mixing and paving, which makes MA especially energy-intensive in production (Canada, 2005). Also, a number of reports show that workers health is increasingly affected when bitumen is handled at temperatures over 200°C (Hansen, 1991, Ruhl et al., 2007, Kriech and Osborn, 2014). For these reasons, a temperature reduction in MA is seen crucial for enhanced energy efficiency and a healthier work environment. The addition of waxes to bitumen is a state-of-the-art procedure to reduce the mixing and paving temperature of asphalt mixes (Biro et al., 2009, Silva et al., 2010, Rubio et al., 2012, Wu and Zeng, 2012, Rowe et al., 2009, Saboori et al., 2012, Vaitkus, 2009). Due to wax crystallization during cooling of the mix to ambient temperatures, waxes tend to increase the high temperature stability and decrease the low-temperature cracking resistance (Edwards et al., 2006, Cardone et al., 2009, Edwards, 2009, Merusi and Giuliani, 2011, Capitao et al., 2012).

In this paper a wax-modified mastic asphalt is compared to a new innovative mastic asphalt mix. In the new method for temperature reduction, crushed aggregates are substituted by round aggregates within the MA. This substitution brings a significant temperature reduction potential and good mechanical performance.

2 MOTIVATION AND OBJECTIVES

Mastic asphalt mixtures are developed with regard to temperature reduction, mix performance and economic efficiency (Hofko et al., 2015). The following steps taken by the project are presented in this paper:

- Mastic asphalt slabs are produced in the lab using the methods that show best potential for temperature reduction (Hofko et al., 2015). The resistance to permanent deformation by uniaxial cyclic compression tests (UCCT) and to low-temperature cracking by thermal stress restrained specimen tests (TSRST) are assessed and results are compared to results from the reference mix.
- Thus, an optimized mastic asphalt mixture can be recommended with maximum temperature reduction potential while keeping the mix performance at a high level and keeping the material costs low.

3 MATERIALS

A mastic asphalt mixture with a maximum nominal aggregate size of 11 mm (MA 11) was used for the presented research. The filler component was powdered limestone, the coarse fraction was totally crushed (TC) aggregates of porphyritic origin and totally round (TR) limestone, respectively. The binder

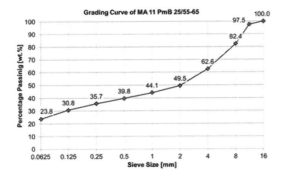

Figure 1. Grading curve MA 11.

was an SBS-modified PmB 25/55–65 (PG 82-16). The binder content was 8,9 wt.% (weight percentage) of the whole asphalt mix mass. The grading curve of the mix is shown in Figure 1.

There was one wax type applied within the test program: amide wax. Amide wax (AW) is a mixture of fatty acid derivatives and is used as a softening-point regulator for binders. Its dropping point is between 139°C and 144°C and is available as a powder or granulates. It can be added previously to the bitumen or directly during the asphalt mix production.

By the substitution of the totally crushed (TC) aggregates by totally round aggregates (TR) within the mastic asphalt, the filler component, the filler content, the binder content and the grading curve remained unchanged for each tested mix.

4 TEST METHODS AND TEST PROGRAM

4.1 Mechanical performance

Two test methods are applied to assess and compare the mechanical performance on asphalt mix level:

- The Thermal Stress Restrained Specimen Test (TSRST) according to EN 12697-46 (CEN, 2012) on prismatic specimens ($50 \times 50 \times 200$ mm) from a starting temperature of +10°C with a cooling rate of 10 K/h. The crack temperature is taken as a benchmark for resistance to low-temperature cracking. The left picture in Figure 2 shows the test setup.
- The Uniaxial Cyclic Compression Test (UCCT) according to EN 12697-25, Part A (CEN, 2005) on cylindrical specimens (148 mm in diameter, 60 mm in height) at a temperature of +50°C with a block-shaped cyclic compressive loading at 0.5 Hz. The permanent, axial strain after 3,600 load cycles is taken as benchmark for resistance to permanent deformation. The right picture in Figure 2 shows the test setup.

There were four mixes tested with UCCT and TSRST:

- MA 11 reference – mastic asphalt mixture with a maximum nominal aggregate size of 11 mm as a reference mix.

Figure 2. TSRST test setup (left), UCCT test setup (right).

- MA 11 + 4 wt.% AW – mastic asphalt mixture with a maximum nominal aggregate size of 11 mm with 4 wt.% amide wax based on the bitumen mass.
- MA 11 100% TR – mastic asphalt mixture with a maximum nominal aggregate size of 11 mm where the crushed aggregates are completely substituted by round aggregates.
- MA 11 100% TR + 4 wt.% AW – mastic asphalt mixture with a maximum nominal aggregate size of 11 mm where the crushed aggregates are completely substituted by round aggregates and 4 wt.% amide wax based on the bitumen mass is added.

5 RESULTS AND DISCUSSION

The resistance to permanent deformation and to low-temperature cracking is investigated for four mastic asphalts. The objective is to keep the mechanical performance at the same or higher level compared to the reference mix.

5.1 Resistance to permanent deformation

Figure 3 shows the results for the resistance to permanent deformation at high temperatures tested in the UCCT at +50°C for 3,600 load cycles. The data is given as mean value from three single tests including error bars that indicate the standard deviation of the results. The reference mix with totally crushed aggregates has an axial strain of −21%. MA 11 + 4 wt.% AW reaches only about of the half permanent deformation of the reference mix (−11%) and MA 11 100% TR for which the crushed aggregates were substituted by round aggregates results in similar deformation as the reference mix (−23%). The combined variant MA 11 100% TR + 4 wt.% AW shows similar results as scenario 1 (−12%).

This indicates that the substitution of aggregates does not change the resistance to permanent deformation significantly and that the addition of AW strongly increases the resistance by about 50% no matter which aggregates are used for the mixture.

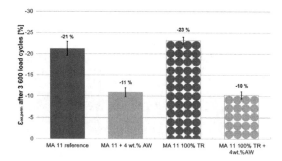

Figure 3. Uniaxial Cyclic Compression Test (UCCT) at +50°C.

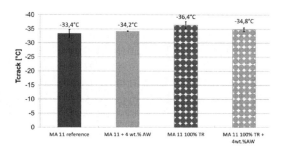

Figure 4. Thermal Stress Retrained Specimen Test (TSRST), Starting temperature: +10°C, Cooling rate: 10 K/h.

5.2 Resistance to low-temperature cracking

Figure 4 presents results of the TSRST for all 4 considered scenarios. The data shown in the diagram are cracking temperatures T_{crack} at which the specimen fails due to the fact that the cryogenic (temperature induced) stress exceeds the tensile strength of the material. Results are given as mean values derived from three single tests together with the scattering in terms of standard deviation.

The cracking temperature T_{crack} of all mixtures ranges between −33°C and −36°C and is especially low compared to other asphalt mix types. A reason for this is the higher bitumen content in the mastic asphalt. Taking the scattering of results into consideration, no significant impact of wax addition or the substation of aggregates can be found.

6 SUMMARY

Optimized mastic asphalt mixes are developed within a recently completed study with regard to maximum temperature reduction and high level of mix performance (Hofko et al., 2015). This paper contains results of the mechanical performance of wax modified mixes. In addition, an alternative approach for temperature reduction by substituting crushed aggregates partly or completely by round aggregates is analysed as well.

The mechanical performance in terms of resistance to permanent deformation at high temperatures was obtained by uniaxial cyclic compression tests (UCCT) according to EN 12697-25, as well as the resistance to low-temperature cracking by thermal stress retained specimen tests (TSRST) according to EN 12697-46. The following conclusions can be drawn from the results:

- The resistance to permanent deformation for the mix with round aggregates is equal to the reference mix. Round aggregates do not have a negative effect on permanent deformation at elevated temperatures (50°C). Addition of 4 M% of AW as a binder modification increases the resistance to permanent deformation by nearly 50% regardless of the aggregate shape.
- The resistance to thermal cracking at low-temperatures is equal for all four tested mixes. All mixes exhibit crack temperatures between −33°C and −36°C. Thus, all mixes can be considered as highly resistant to low-temperature cracking. AW modification does not make mastic asphalt mixtures more prone to cracking.

The results presented in this paper are based on a study which is limited to laboratory investigations so far. The next step of the research includes large-scale production, as well as construction of test fields on the public road network including regular monitoring to validate results found in the lab.

REFERENCES

Biro, S., Gandhi, T. & Amirkhanian, S. 2009. Midrange Temperature Rheological Properties of Warm Asphalt Binders. *Journal of Materials in Civil Engineering*, 21, 316–323.

Canada, N. R. 2005. Road Rehabilitation Energy Reduction Guide for Canadian Road Builders. *In:* CONSERVATION, C. I. P. F. E. (ed.).

Capitao, S. D., Picado-Santos, L. G. & Martinho, F. 2012. Pavement engineering materials: Review on the use of warm-mix asphalt. *Construction and Building Materials*, 36, 1016–1024.

Cardone, F., Pannunzio, V., Virgili, A. & Barbati, S. 2009. An evaluation of use of synthetic waxes in warm mix asphalt. *Advanced Testing and Characterisation of Bituminous Materials, Vols 1 and 2*, 627–638.

CEN 2005. EN 12697-25: Bituminous mixtures – Test methods for hot mix asphalt – Part 25: Cyclic compression test. Brussels.

CEN 2012. EN 12697-46: Bituminous mixtures – Test methods for hot mix asphalt – Part 46: Low temperature cracking and properties by uniaxial tension tests. Brussels.

Edwards, Y. 2009. Influence of Waxes on Bitumen and Asphalt Concrete Mixture Performance. *Road Materials and Pavement Design*, 10, 313–335.

Edwards, Y., Tasdemir, Y. & Isacsson, U. 2006. Effects of commercial waxes on asphalt concrete mixtures performance at low and medium temperatures. *Cold Regions Science and Technology*, 45, 31–41.

Hansen, E. S. 1991. Mortality of Mastic Asphalt Workers. *Scandinavian Journal of Work Environment & Health*, 17, 20–24.

Hofko, B., Dimitrov, M., Schwab, O. & Weiss, F. 2015. High Efficient Low Emission Mastic Asphalt – Final Report. Vienna, Austria.

Kriech, A. J. & Osborn, L. V. 2014. Review and implications of IARC monograph 103 outcomes for the asphalt pavement industry. *Road Materials and Pavement Design,* 15, 406–419.

Medani, T. O., Huurman, M., Liu, X. Y., Scarpas, A. & Molenaar, A. A. A. 2007. Describing the behaviour of two asphaltic surfacing materials for orthotropic steel deck bridges. *Advanced Characterisation of Pavement Soil Engineering Materials, Vols 1 and 2,* 1351–1368.

Merusi, F. & Giuliani, F. 2011. Rheological characterization of wax-modified asphalt binders at high service temperatures. *Materials and Structures,* 44, 1809–1820.

Rowe, G. M., Baumgardner, G. L., Reinke, G., D'Angelo, J. A. & Anderson, D. Warm mix study with the use of wax modified asphalt binders. Binder ETG Meeting, 2009 Irvine, Ca.

Rubio, M. C., Martinez, G., Baena, L. & Moreno, F. 2012. Warm mix asphalt: an overview. *Journal of Cleaner Production,* 24, 76–84.

Ruhl, R., Musanke, U., Kolmsee, K., Priess, R. & Breuer, D. 2007. Bitumen emissions on workplaces in Germany. *Journal of Occupational and Environmental Hygiene,* 4, 77–86.

Saboori, A., Abdelrahman, M. & Ragab, M. 2012. Warm Mix Asphalt Processes Applicable to North Dakota.

Silva, H. M. R. D., Oliveira, J. R. M., Peralta, J. & Zoorob, S. E. 2010. Optimization of warm mix asphalts using different blends of binders and synthetic paraffin wax contents. *Construction and Building Materials,* 24, 1621–1631.

Vaitkus, D. A. The research on the use of warm mix asphalt for asphalt pavement structures. XXVII International Baltic conference, 2009 Riga; Latvia.

Widyatmoko, I., Elliott, R. C. & Read, J. M. 2005. Development of heavy-duty mastic asphalt bridge surfacing, incorporating Trinidad Lake Asphalt and polymer modified binders. *Road Materials and Pavement Design,* 6, 469–483.

Wu, C. F. & Zeng, M. L. 2012. Effects of Additives for Warm Mix Asphalt on Performance Grades of Asphalt Binders. *Journal of Testing and Evaluation,* 40, 265–272.

Influence of factors to shift private transport users to Park-and-Ride service in Putrajaya

Irfan Ahmed Memon
Department of Civil and Environmental Engineering, Universiti Teknologi PETRONAS, Seri Iskandar, Perak, Malaysia
Department of City and Regional Planning, Mehran University of Engineering and Technology, Jamshoro, Sindh, Pakistan

Madzlan Napiah
Department of Civil and Environmental Engineering, Universiti Teknologi PETRONAS, Seri Iskandar, Perak, Malaysia

Mir Aftab Hussain
Department of Civil and Environmental Engineering, Universiti Teknologi PETRONAS, Seri Iskandar, Perak, Malaysia
Department of City and Regional Planning, Mehran University of Engineering and Technology, Jamshoro, Sindh, Pakistan

Muhammad Rehan Hakro
Department of City and Regional Planning, Mehran University of Engineering and Technology, Jamshoro, Sindh, Pakistan

ABSTRACT: Putrajaya in Malaysia is one the finest and intelligent garden cities of the world. It was designed to achieve the target of 70% modal-split towards public-transport and 30% private-transport. The imbalance between public and private-transport usage has been increasing and currently 85% of the travelers' are using private-vehicles. To reduce this imbalance, there is a need to shift car-users towards Park-and-Ride (P&R) service. To shift mode-choice from private-vehicle to P&R service, there is a requirement for travel-behavior survey. Therefore, this research focuses on the influencing factors of employees' mode-choice in Putrajaya. To collect the data, a survey was conducted through a questionnaire. Descriptive and Pearson correlation analysis techniques applied to analyze the data. The research findings were more towards the implementation of parking charges at workplaces to discourage car-users. The research outcomes will support the policy making and provide a base for future study on mode-choice model for P&R service.

1 INTRODUCTION

"A developed country is not a place where the poor have cars; it is where the rich use public transport", these are the famous words of Gustavo Petro, Mayor of Bogota, Colombia.

Vehicle ownership, especially on the private car, is rapidly increasing in Malaysia. According to the Road Transport Department of Malaysia, the registration of cars increased from 90,046 cars in May 2015 to 99,013 cars in June 2015. Averagely, 64,516.93 cars are registered per month from 1988 until 2015. All time high registration in one month was 138,727 cars in March 2015 and the lowest was 9,732 cars in May 1988 (Tradingeconomics.com, 2016).

Registered private cars rapidly increased from 6.5 million in 2005 to 8.5 million in 2009 (Adnan et al., 2012). As of 31st December 2013, around 11.03 million motorbikes, 10.48 million private cars, and 3.2 million commercial vehicles were registered. If motorbikes, cars, and commercial vehicles were included then Malaysia has a total of 23.71 million registered vehicles (Joseph, 2014). This rapid increasing numbers of cars are creating enormous issues in Malaysia. Currently, traffic congestion in Malaysia is not limited to the capital cities only. An even small town like Kluang is often congested due to the exponential growth of private cars. Nationally, from 2007 to 2012 there was a growth of 35 percent in registered vehicles. In the same period of five years, registered vehicles in Wilayah Persekutuan Kuala Lumpur were augmented by 42 percent. Similarly, in that period the presence of the even larger number of registered motorbikes reflects that the car habit in Malaysia is not serving the transportation needs of the poorer segment of the nation (Joseph, 2014).

Similarly, Putrajaya, an administrative city of Malaysia is also facing an increasing number of private car ownerships and their usage. Integrated transportation infrastructure connecting the city with suburban areas and also relatively cheaper housing schemes are the fringes of Putrajaya, which are a mismatch between residential and employment attentiveness (Borhan et al., 2011). Due to the attractiveness of jobs in the city center, commuters' traveling pattern is morning/evening peak hours and it leads to traffic congestion on few major artilleries leading towards and out of the city (Borhan et al., 2011).

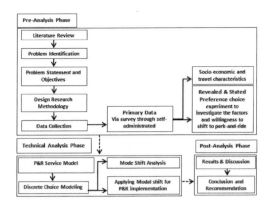

Figure 1. Research process.

2 PROBLEM BACKGROUND

Putrajaya has its own significant status as the world's first intelligent Garden City (Putrajaya, 2013, Memon et al., 2015). It is an inspiring example of sustainable human development design to meet the needs of the growing nation for at least 300 years (Putrajaya, 2013). It is located 25 kilometres away towards the south of Kuala Lumpur. The federal government of Malaysia developed this new administrative city because of the traffic congestion and land limitation inside Kuala Lumpur. The aim of this new city is to have a new well-planned and organized capital and symbol of the Malaysian identity, befitting the nation's aspiration to become a developed nation by 2020. A total number of 25 ministry offices, 51 government offices and also few other companies and financial institutions have moved to Putrajaya and are providing around 250,000 job opportunities (Nor Ghani Md. Nor and Nor, 2006, Borhan et al., 2014).

Putrajaya has a unique nature because of the transport policy perspective, due to its explicit policy goal to achieve a 70 percent share of all travels by public transport to its core precincts (Nor Ghani Md. Nor and Nor, 2006). Customarily, the rapid traffic growth and excessive use of private cars are the most important concern of this city (Borhan et al., 2011). According to the objectives, Park and ride (P&R) service seems to be a good alternative to the car users, however, very few commuters use this P&R service (Borhan et al., 2011). The task is threatening the city authority, which appears impossible because this goal needs a reversal of the current modal split of 15:85 between the public and private transport (Nor Ghani Md. Nor and Nor, 2006). The current modal split is 70 percent for cars, 15 percent motorcycles and 15 percent public transport (Nor Ghani Md. Nor and Nor, 2006).

Currently, the congestion in Putrajaya is the result of people driving their cars to work (Borhan et al., 2011). People cannot be convinced to shift towards public transportation without understanding their travel behaviour (Anable, 2005, Elias et al., 2013, Hamid et al., 2008, He et al., 2009, Ho et al., 2008, Phil Goodwin and Gordon Stokes, 2004, Lindstrom Olsson, 2003, Gurcharan, 1996). The imbalance between the public and private transport has increased nowadays, thus, there is a need to understand how the travel behaviour of private car users can be fit in public transport, such as the P&R service (Qin et al., 2013). P&R service has been extensively used in many countries and proved to be a successful part of the travel demand management in decreasing the congestion and difficulty of finding parking spaces in the urban centres (He et al., 2012).

Therefore, this research intended to understand the car travellers' willingness to use P&R service and factors that influence their mode choice and decision. Thus, to determine safety and security, frequency of bus, comfort preferences and increase parking fees factors that influence traveler's willingness to use P&R service.

3 RESEARCH IMPLEMENTATION AND METHODS

This research was carried out in several phases. The study started with the current scenario of the urban transportation system in the study area. Based on the available literature, this study was divided into three phases as shown in Figure 1. Phase one was pre-analysis which is also known as preliminary, phase two was the technical analysis, and the third phase was the post-analysis phase. This study was based on primary data which was the major and only data source. Primary data was collected through a self-administered questionnaire which is also discussed below.

3.1 Sample and data collection

A self-administered questionnaire was distributed in the ministry offices, governmental departments and private offices in the Putrajaya city centre. The respondents in this study were all working in the above-mentioned offices. A face-to-face approach was adopted. Only the non-users of P&R service, especially those who drive their own mode of transport for office trips were considered for this study. Stratified random sampling technique was adopted in this study and this technique is very common for these types of studies (Borhan et al., 2014), in which car travellers were the focus and they were requested to fill in the self-administrated questionnaires. This approach was adopted to increase the accuracy of the data collected. If any respondent refused to answer, another respondent from the targeted population would be approached for participation. According to the Krejcie and Morgan's well-known sample size calculating method, this study should have 300 samples (Krejcie and Morgan, 1970). According to a recent research in University of Leeds, England, it was suggested that the sample should be large enough so that when it is divided into groups, each will have a minimum sample size of 100 (Pathan and Faisal, 2010). Therefore, 400 questionnaires were distributed and around 326 were returned. Exactly 300 questionnaires were included in the study and the rest 26 were rejected due to invalid responses and incomplete answers.

4 RESULTS AND DISCUSSION

This section includes the analysis of the survey. The survey questionnaire was divided into two sections i.e. personal information and travel characteristics. The first section on personal information was based on revealed preference, whereas travel characteristics were based on both the revealed and stated preferences as shown in Figure 1. Similarly, the sections were also about the descriptive analysis of the survey. The descriptive analysis was based on the personal information (socio-economic and demographic data) and travel characteristics of the respondents from Putrajaya. It was discussed earlier in the research implementation and methodology section that the respondents were selected from the Putrajaya city centre. In Putrajaya, data was collected through a physical questionnaire at the workplaces.

4.1 Respondents' socio-economic and demographic characteristics

This segment elaborates on detailed socio-demographic characteristics and travel patterns of the selected population. In the survey, age was distributed in five different ranges; 21–30, 31–40, 41–50, and 51–60 years old. In overall terms, 51.7% of the respondents were males and the remaining 48.3% were females. 48.7% of the respondents (including females) lay under the age group of 31–40 years old, 42.7% belonged to 21–30 years age group and the remaining 7.3% and 1.7% belonged to the age group of 41-50 and 51-60 respectively. More than half (56.3%) of the employees have completed post-secondary education (Diploma or Bachelor Degree). The majority of employees around 66% were working in the government sector and most of the respondents around 38% were earning RM2000-4000, which counted them into the middle-class income group. Furthermore, almost 79% of the population owned cars and only 21.7% claimed ownership of a bike.

Table 1. Descriptive statistical analysis.

Attributes	N	Mean	Standard Deviation
Safety and Security	300	2.0800	0.98157
Frequency of Bus	300	2.2500	1.09461
Comfort Preference	300	1.5200	0.50043
Increase Parking fees	300	1.4167	0.63593

4.2 Respondents' travel characteristics

Part B of the questionnaire was about travel characteristics of the respondents. The major mode of travelling was by car which was 67%, 31% were using motorbike as a mode of transport, and only 2% walked or cycled to the office. There was also a positive sign that 77% of the people shared their vehicles with colleagues and others.

Consequently, exactly 60% of the population was willing to switch their mode of transport towards P&R service. Around 40% of the respondents did not come directly to the office, and they had to drop their kids or wife at their schools and workplaces. Most of the people (46.3%) had two trips from home to the office and vice-versa in a day. Only 45.3% had one trip per day.

Travel time is one the major factors for the travellers especially private transport users. 34.3% of the people took more than 30 minutes from home to the office and 34% took only 5-10 minutes. Similarly, travel distance also matters and 37% lived more than 20 kilometres away from the workplaces. 34.3% of the employees lived within the range of 10 kilometres. Roughly, 46% spent RM 101-300 per month as travel expenses from the office to home and vice-versa. The total travel time of the respondents per day from home to office and going back was 30 minutes for the majority (44.3%) of the population. The remaining population spent more than 30 minutes for the same reason.

Table 1 shows the factors that can influence the private transport users to remain on their own mode of transport instead of using P&R facility. These factors were asked in questions through different options. Respondents were asked about the four major influencing factors, such as the safety and security at the parking lot and on public transport, frequency of the bus service, comfort levels such as air-conditioned bus service with seat availability or comfort standing, and increased parking cost at the workplace.

Therefore, all the respondents were asked about the attributes that influence them to use P&R service. It was found that almost 50% of the respondents were willing to use P&R service if the frequency of the bus is at every 10 minutes. More than half (52%) of the employees were willing to switch their mode of transport if the safety and security at the parking lot

Table 2. Results of the correlation test.

Attributes	Pearson Correlation	90%
Safety and Security	0.169	0.003
Frequency of Bus	−0.149	0.10
Comfort preferences	0.253	0.000
Increase Parking Prices at Workplace	0.439	0.000

and on public transport are provided at nominal cost. Respondents had no compromise regarding the comfort level. Around 52% of respondents were willing to shift their mode of transport if the assurance of seat availability is given to them. If the government decides to increase the parking cost at workplace then more than 65% of the respondents will switch towards P&R service.

The Pearson correlation method was applied to measure the linear correlation between the willingness to use P&R service (X) and each of the four factors (Y) as shown in Equation 1, and in Table 2. This correlation can measure between −1 (negative correlation) and +1 (positive correlation) range. A value of 0 indicates there is no association between the two variables.

$$r = \frac{\sum_{i=1}^{n}(X_i-\bar{X})(Y_i-\bar{Y})}{\sqrt{\sum_{i=1}^{n}(X_i-\bar{X})^2}\sqrt{\sum_{i=1}^{n}(Y_i-\bar{Y})^2}} \quad (1)$$

The four factors were tested and three of them were found positively correlated with the intention to use P&R service as shown in Table 2. Increased parking fee at workplaces had a medium relationship ($r = 0.439$, $p < 0.000$) as compared to other attributes. Consequently, it was found that safety and security had small relationship ($r = 0.169$, $p < 0.003$). Comfort preference also had small but near to medium correlation ($r = 0.253$, $p < 0.003$), because; 0.10 to −2.9 is small, 0.30 to 0.49 medium and 0.50 to 10 counts in large relationship. The bus frequency had a negative relationship ($r = −0.149$, $p < 0.10$) as low bus frequency was associated with high level of switching towards P&R service.

Therefore, it was found that increasing parking charges at the workplace can motivate people to switch towards P&R service. It was also pointed by Hole (2004) that one of the methods to reduce car number from the CBD is to impose parking charges at the workplace (Risa Hole, 2004).

5 CONCLUSION

Research work to develop a mode choice model for private transport, especially car users, to shift them towards P&R service in Putrajaya was still undergoing. In order to understand the travel behaviour of private transport users, there was a need to explore further on more factors that influence the travel behaviour of the car users. Travel behaviour theory will be adopted to understand and identify the influencing factors of car users. Furthermore, revealed preference and stated preference questions were used together in one questionnaire to reveal the current travel behaviour and future preferences of the respondents. It will be helpful in future to explore the influencing factors of private transport users (Memon et al., 2014)

Hence, this study found that implementing parking charges at the workplace is a major influencing factor and can be very helpful in forcing the middle-income people to use P&R service instead of bringing their cars up to the doors of the offices. Comfort, safety, and security were also influencing factors which can be managed and implemented. This study showed few correlations of some factors which can help policy makers to switch private transport users towards P&R service at Putrajaya.

ACKNOWLEDGMENT

The authors are thankful to the Department of Civil Engineering, Universiti Teknologi PETRONAS, Malaysia for their help and support in providing favourable research environment and facilities.

REFERENCES

Adnan, S., Alyia, S. A., Hamsa, K. & Azeez, A. 2012. A literature review on the parking demand of park and ride facilities. *Conference on urban planning and management in Malaysia*, 8 November 2012, Kuala Lumpur, Malaysia

Anable, J. 2005. 'Complacent car addicts' or 'aspiring environmentalists'? Identifying travel behaviour segments using attitude theory. *Transport Policy*, 12, 65–78.

Borhan, M. N., Rahmat, R. A. A., Ismail, R. & Ismail, A. 2011. Prediction of travel behavior in Putrajaya, Malaysia. *Research Journal Of Applied Sciences, Engineering and Technology*, Vol.3, 434–439.

Borhan, M. N., Syamsunur, D., Mohd Akhir, N., Mat Yazid, M. R., Ismail, A. & Rahmat, R. A. 2014b. Predicting the Use of Public Transportation: a case study from Putrajaya, Malaysia. *The Scientific World Journal*, 2014, 9.

Elias, W., Albert, G. & Shiftan, Y. 2013. Travel behavior in the face of surface transportation terror threats. *Transport Policy*, 28, 114–122.

Gliem, J. A. & Gliem, R. R. Calculating, interpreting, and reporting Cronbach's alpha reliability coefficient for Likert-type scales. 2003. *Midwest Research-to-Practice Conference in Adult, Continuing, and Community Education*.

Gurcharan, S., A/L , Sangkar, Singh. 1996. Factors influencing mode choice for journey to network among government employees case study: Johor Bahru. Master of Science Transportation Planning, Universiti Teknologi Malaysia.

Hamid, N. A., Mohamad, J. & Karim, M. K. 2008. Travel behaviour of the park and ride users and the factors influencing the demand for the use of the park and ride facility. *EASTS International Symposium on Sustainable Transportation incorporating Malaysian Universities Transport Research Forum conference 2008 (MUTRFC08)*. Universiti Teknologi Malaysia.

He, B., He, W. & He, M. 2012. The attitude and preference of traveler to the park and ride facilities: A case study in Nanjing, China. *Procedia – Social and Behavioral Sciences*, 43, 294–301.

He, Z., Ma, S. & Tang, X. 2009. Empirical study on the influence of learning ability to individual travel behavior. *Journal of Transportation Systems Engineering and Information Technology*, 9, 75–80.

Ho, J. S., Sadullah, A. F. & Vien, S. L. L. 2008. Understanding travel behaviour: An important approach to switch private car users to public transport. *EASTS International Symposium on Sustainable Transportation incorporating Malaysian Universities Transport Research Forum Conference 2008 (MUTRFC08)*.

Joseph, J. 2014. *Our love affair with cars can be deadly* [Online]. Free Malaysia Today. Available: http://www.freemalaysiatoday.com/category/opinion/comment/2014/09/27/our-love-affair-with-cars-can-be-deadly/ [Accessed 30 September 2015].

Krejcie, V. R. & Morgan, W. D. 1970. Determining sample size for research activities. *Education and Psychological Measurement*, 30, 607-610.

Lindstrom Olsson, A.-L. 2003. Factors that influence choice of travel mode in major urban areas. Ph.D, Royal Institute of Technology.

Memon, I. A., Madzlan, N., Talpur, M. A. H., Hakro, M. R. & Chandio, I. A. A Review on the factors influencing the park-and-ride traffic management method. *Applied Mechanics and Materials*, 2014. Trans Tech Publ, 663–668.

Memon, I. A., Napiah, M., Talpur, M. A. H. & Hakro, M. R. Mode choice modelling method to shift car travelers towards Park and Ride service. *Malaysian Technical Universities Conference on Engineering and Technology 2015*, 2015.

Nor Ghani Md. Nor & Nor, A. R. M. 2006. Predicting the impact of demand and supply side measures on bus ridership in Putrajaya, Malaysia. *Journal of Public Transportation*, Vol. 9.

Pathan, H. & Faisal, A. 2010. Modelling travellers' choice of information sources and of mode. University of Leeds.

Phil Goodwin, S. C., Joyce Dargay, Mark Hanly, Graham Parkhurst, & Gordon Stokes, A. P. V. 2004. Changing travel behaviour.

Putrajaya, P. 2013. *World's first intelligent garden city* [Online]. Putrajaya: Perbadanan Putrajaya. Available: URL: http://www.putrajaya.gov.my/m_tourist/green_city/ [Accessed 22/10/2013 2013].

Qin, H., Guan, H. & Wu, Y.-J. 2013. Analysis of park-and-ride decision behavior based on Decision Field Theory. *Transportation Research Part F: Traffic Psychology and Behaviour*, 18, 199–212.

Risa Hole, A. 2004. Forecasting the demand for an employee Park and Ride service using commuters' stated choices. *Transport Policy*, 11, 355–362.

Tradingeconomics.Com. 2016. *Malaysia New Vehicles Registered* [Online]. Trading Economics. Available: http://www.tradingeconomics.com/malaysia/car-registrations [Accessed 25 Janurary 2016].

Statistical interpretations about the use of footbridges by diverse groups of pedestrians in Malaysia

R. Hasan & M. Napiah
Department of Civil and Environmental Engineering, Universiti Teknologi PETRONAS, Seri Iskandar, Perak, Malaysia

ABSTRACT: The study presents the results of analyzed data about the characteristics of different groups of pedestrians, and their relation with the use of footbridges. Results showed that the frequency of crossing the street has a positive impact on the use of the footbridge. The level of pedestrian's education, income level and occupation have no influence on the decision of either jaywalking or crossing by the footbridge. The study suggests the well design of the footbridge that complies with the recommended standards to be suitable for the majority of pedestrians groups.

1 INTRODUCTION

Road safety is one of the important issues that concerns transport developers all around the world. This issue is due to the increased vehicular units combined with the increased traveling demand, which resulted in a higher number of traffic accidents in many cities in the world, and have led to a horrible global number of 1.24 million road deaths per year (Organization, 2013). Out of this number, pedestrians formed 22% of annual fatalities, which called for responsible authorities to take the necessary actions in order to secure safety. These actions demonstrated by building friendly and suitable infrastructure for pedestrians, and by applying appropriate procedures to enhance their compliance with safe walking. One of the built structures is the footbridge, which has remarkable characteristics that secure absolute safety for people when crossing the street. Yet, this structure suffers from non-optimal performance resulting from non-compliance with pedestrians to use it. This risky traversing of the street will increase the proportion of accidents and the rate of road deaths, thus it will affect badly the free traffic flow. In this context, Abojaradeh (2013) discussed the importance of safety, where traffic accidents are considered as the second major cause of death in Amman, Jordan. The drawn results from this study stated that footbridges can increase the safety when the awareness of its benefits is available, especially for some pedestrians who do not obey the traffic law, where the severe congested streets lead to a tragically higher rate of danger. To find out the explanation of pedestrians' behavior who dangerously crossed the street, the study of Sabet (2013) tried to test their fear of risk as a motive in preventing them to follow-up to do this behavior in the future. The study explained that the perception of risk was taken into account only when the respondent was not in a rush, where there is a comfort situation to decide which type of crossing he will perform. Therefore, the existence of risk is insufficient reason to urge people to more use of the footbridge when this risk is placed versus the pedestrian's desire of saving time by crossing the street. In some countries, absence of safety culture is the main reason for performing such a risky behavior by pedestrians. This irresponsible behavior did not come from nowhere, but it stemmed from the cultural concepts which rule the house education for raising children before they become independent pedestrians (Arias Gallegos, 2012). This paper endeavors to identify the relation between pedestrians with different socioeconomic characteristics and different purpose of mobility, and their use of the footbridge. This was investigated by carrying out a field data collection from two footbridges in Ipoh city, Malaysia, and by distributing a questionnaire among pedestrians in the same area.

2 METHODOLOGY

Two footbridges in Ipoh city were chosen for this study, where pedestrians can choose between using the footbridge and crossing on the street. A pre-visit to the two sites was conducted in order to obtain the first impression about the pedestrians' walking patterns, and pedestrians' population which is very diverse, and composed of a mix of students, shoppers, doctors, nurses, patients, retail and restaurant employees, and other visitors to the area. The first footbridge No.1 is located in the street of Sultan Idris Shah towards the roundabout of Sultan Ismail, and it services an area which includes two secondary schools, one elementary school, shopping mall, bus stop, car parking, three petrol stations, and some houses and restaurants. The

Table 1. Characteristics of the footbridge.

Bridge No.	Height m	Length m	Width m	Stairways Count
1	8.0	21.8	2.5	2
2	6.7	24.5	1.7	2

Table 2. Characteristics of the street beneath the bridge.

Bridge No.	Width m	Direction Count	Lanes Count	Traffic light Count
1	12	1	4	No
2	17	2	2*2	Yes

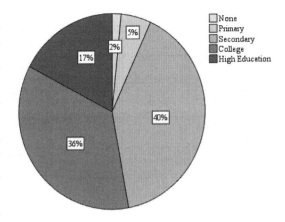

Figure 1. Educational level of participants.

second footbridge No.2 is located in the street of General Hospital towards the roundabout of Sultan Iskandar, and it services an area which includes a general hospital, bus stop, a mosque, two schools and a construction site. These various facilities and amenities were well used by the population in their journeys. The characteristics of footbridges and roads are summarized in Tables 1 and 2. After that, a questionnaire was distributed on the street among the public. The design of the questionnaire contained respondents' socioeconomic characteristics, and questions about their perception on using the footbridge under the impact of various factors. To check out the content validity, that is the extent to which these questions within the questionnaire are relevant and representative of the construct that they will measure; a group of eight expert engineers in the field of traffic and transportation engineering from JKR (Jabatan Kerja Raya), Ipoh city council (Majlis Bandaraya Ipoh), and IEM (Institute of Engineers Malaysia) have reviewed the questionnaire, and they gave comments on the content and the general shape (Buysse et al., 1989, Messick, 1980). The comments were totally taken into account, and minor corrections in the questionnaire were made. In order to check the needed time for the respondents to fill in the questionnaire and to check the clarity of the questions, a pre-test survey was conducted. Quota sampling, which is a non-probability sampling technique, was chosen to be used in order to choose the participants in this study. The population was represented by being divided into groups (strata) according to gender, age, and user/non-user of the footbridge (Moser, 1952). Eventually, the size of the sample was 191 participants of all age groups (age ranged from 11 to over than 59 years, 62.3% male). The Statistical Package for Social Science (SPSS) software V.20 was used in analyzing the collected data and in interpreting the results.

3 RESULTS

3.1 Participants profile

The results from the questionnaire showed that more than half of the participants (52.9%) had a level of education that belonged to the college and higher education, where 62.9% of them were male and 38.1% were female. Figure 1 represents these percentages, and shows a very small percentage of people who are uneducated and are in total three male participants. The Chi square test for association was run to indicate if there is any association between the educational levels of participants and the use of the footbridge. The test indicated that no statistically significant relationship was existed between the educational levels of participants and the use of the footbridge (χ^2 (1, $N = 191$) = 0.002, $p = 0.961$, and Phi & Cramer's $v = 0.004$), and this was confirmed by another study concluded that the level of education is not a factor of influencing the use of the footbridge (Arias Gallegos, 2012). On the other hand, the test showed a significant relationship between the level of education and the gender of the participant (χ^2 (1, $N = 191$) = 4.476, $p = 0.034$, and Phi & Cramer's $v = 0.153$). This means that the higher the level of education, the higher the percentage of men participants to be belonging to this category than women participants. This result was confirmed by the Mann-Whitney U test, where $U = 3463$, $n_1 = 119$, $n_2 = 72$, $p = 0.018$.

In terms of the nature of job and monthly income, the majority of participants were students who belonged to the lowest category of income, while the highest two categories of income shows least number of participants.

Table 3 shows the distribution of participants in terms of gender. Male and female participants, who were observed in the highest income level, belonged to professional and free business categories. Participants who belonged to the lowest level of income were the most users of the footbridge, contrary to what was reported in the study of Manjanja (2013). The Mann-Whitney U test was run, and it showed that the income level for male participants was statistically significantly higher than female participants ($U = 3540.5$, $n_1 = 119$, $n_2 = 72$, $p = 0.032$). In order to find out if there is any association between the income/job observations and the use or non-use of the footbridge, the Chi square test for association was run.

Table 3. Occupation and monthly income distribution according to the gender of participants.

Classification		Monthly income (RM)				
		<1000	1001–2001	2001–3001	3001–4000	>4000
Male	Professional	0	2	11	5	9
	Business	0	2	4	3	4
	Student	42	3	1	0	0
	Unemployed	4	0	0	0	0
	Retired	5	6	1	1	0
	Self-employed	1	6	6	1	2
Female	Professional	0	2	4	3	3
	Business	1	0	0	1	0
	Student	29	0	0	0	0
	Unemployed	5	1	1	0	0
	Retired	5	2	1	0	0
	Self-employed	1	9	2	0	2

The test showed that there was no statistically significant association between either the job occupation or the income level and the use of the footbridge [χ^2_{job} (5, $N = 191$) = 3.505, $p = 0.623$, and Phi & Cramer's $v = 0.135$, χ^2_{income} (4, $N = 191$) = 1.848, $p = 0.764$, and Phi & Cramer's $v = 0.098$]. However, it should be noted that some studies considered that the income level has an impact on the mode of trip and the length of the traveled distance, that is the higher the level of the income, the higher the traveled distance in the trip which will be more likely for entertainment (Mokhtarian et al., 2001).

3.2 Trip purpose and frequency

Participants of this study were asked four questions about their purpose of visiting the area where the footbridge is located, and if they have utilized the bridge while crossing the street in each visit. The most cited reason for visiting the area was for recreation, which was attributed to the available amenities in the area. The frequency of crossing the street in this area was stated as once a week or once a month by the majority of participants who were in turn either students, free-business, professionals, or retired people. The second rank for crossing frequency was cited by students who crossed the street almost every day. Participants' answers about the frequency of using the footbridge when crossing the street were cited as: seldom to use (31.9%), often used (28.8%), always (22.5%), and never used the footbridge (16.8%). A Kruskal-Wallis H test showed that there was a statistically significant difference in the frequency of using the footbridge between the different groups of the frequency of visiting the area (χ^2 (5) = 27.440, $p < 0.005$), with a median rank usage frequency of 3 for crossing the street once a week, 3 for crossing the street almost every day, 2.5 for crossing the street many times a day, 2 for crossing the street once a day, 2 for crossing the street once a month, and 1 for never crossing the street. In other words, the frequency of crossing the same street has an effect on the frequency of using the footbridge in this street.

To indicate which groups of frequencies of visiting the area are statistically different from each other, the data of each two groups were compared by running Kruskal-Wallis H test. Results indicated that the group of people who never or almost never crossed the street was statistically significantly different than the whole other groups in terms of times of utilizing the footbridge. In addition to that, the group of people who crossed the street once a month was statistically different than both of the group of who crossed the street once a week, and the group of who crossed the street almost every day in terms of times of utilizing the footbridge. The remaining groups showed no statistically significant differences between each other. This means that people who used to cross the street once a week and almost every day utilized the footbridge in a higher frequency than other pedestrians, and people who never crossed the street were less in frequency of utilizing the footbridge than other people. This could be attributed to the including of the footbridge in the route of pedestrians who used to cross the street once a week and almost every day as it represents a convenient path for them after they learned how to optimize this route, and contrary to previous studies, visiting the area frequently has increased the usage of the footbridge (Räsänen et al., 2007).

Most of the participants (37.2%) who usually do not use the footbridge blamed the driver for being the cause of their sense of difficulty and loss of time when crossing at the street level when this driver does not give them a way to cross. This situation will mainly occur when there is a high traffic volume associated with a high pressure on the driver to reach at the final destination on time, thus the pedestrian will have no chance to cross safely in front of this hasty driver (Arias Gallegos, 2012). The second reason for facing difficulties when crossing the road was the wideness of the street (27.2%). On the other hand, the rest of participants (35.6%) had no problem while crossing on the road level. Statistically, there was no significant association between either of the four questions of trip purpose and the use or none use of the footbridge.

4 CONCLUSIONS

The study identified some characteristics of different groups of pedestrians and its relation with the decision of using the footbridge. Pedestrians who used to cross the street once a week and almost every day utilized the footbridge in a higher frequency than other pedestrians. The study found that the level of education is not a factor of influencing the use of the footbridge. Statistically, there is no association between either the job occupation or the income level and the use of the footbridge. Reckless drivers and wide streets were the main cause of pedestrians' discomfort when crossing the street on the level. The use of the footbridge will vary according to the diverse characteristics of the pedestrians, therefore the structural design should meet the maximum volume of these people preferences while crossing the street (Räsänen et al., 2007).

REFERENCES

Abojaradeh, M. 2013. Evaluation of Pedestrian Bridges and Pedestrian Safety in Jordan. *Civil and Environmental Research*, 3, 66–78.

Arias Gallegos, W. L. 2012. Motives of disuse of pedestrian bridges in Arequipa. *Revista Cubana de Salud Pública*, 38, 84–97.

Buysse, D. J., Reynolds, C. F., Monk, T. H., Berman, S. R. & Kupfer, D. J. 1989. The Pittsburgh Sleep Quality Index: a new instrument for psychiatric practice and research. *Psychiatry research*, 28, 193–213.

Manjanja, R. 2013. *Non-Usage of Pedestrian Footbridges In Kenya: The Case Of Uthiru Pedestrian Footbridge On Waiyaki Way*. University of Nairobi.

Messick, S. 1980. Test validity and the ethics of assessment. *American psychologist*, 35, 1012.

Mokhtarian, P. L., Salomon, I. & Redmond, L. S. 2001. Understanding the demand for travel: It's not purely 'derived'. *Innovation: The European Journal of Social Science Research*, 14, 355–380.

Moser, C. A. 1952. Quota sampling. *Journal of the Royal Statistical Society. Series A (General)*, 115, 411–423.

Organization, W. H. 2013. *WHO global status report on road safety 2013: supporting a decade of action*, World Health Organization.

Räsänen, M., Lajunen, T., Alticafarbay, F. & Aydin, C. 2007. Pedestrian self-reports of factors influencing the use of pedestrian bridges. *Accident Analysis & Prevention*, 39, 969–973.

Sabet, D. M. 2013. Fear Is Not Enough: Testing the impact of risk on pedestrian behavior in Dhaka, Bangladesh.

Physical and storage stability properties of linear low density polyethylene at optimum content

N. Bala & I. Kamaruddin
Department of Civil and Environmental Engineering, Universiti Teknologi PETRONAS, Seri Iskandar, Perak, Malaysia

ABSTRACT: In this research for selecting best optimum linear low density polyethylene content for bitumen modification. 80/100 Pen bitumen were used as the control sample and linear low density polyethylene (LLDPE) were blended with 80/100 pen bitumen to prepare a modified mix. The concentrations of LLDPE polymer in the blends were 4%, 5% and 6% by weight of bitumen. Conventional test which includes penetration, softening point, ductility, temperature susceptibility tests were conducted using the control 80/100 and LLDPE modified binders. Also storage stability test were conducted to access the stability of the control and LLDPE modified binder. The results indicate that 6% LLDPE modified binder offers the best results compared to control, 4% and 5% LLDPE modified binders when they were investigated in terms of conventional properties. Storage stability test result shows that phase separation might occur in all the polymer modified binders during storage or transportation especially at higher temperatures.

1 INTRODUCTION

Bitumen is generally black or brown viscous solid or liquid comprising of hydrocarbons with their derivatives (Read & Whiteoak 2003, McNally 2011). Presently the estimated consumption of bitumen around the world is nearly 102 million tons yearly in which around 85% were used solely in the construction industries for various types of pavements (Zhu and Brickson 2014).

Pavement performance on different weather conditions, traffic loads and volume increase around the world together with the rise in the cost of bitumen results in urgent need for an improvement in properties of pure bitumen (Sengoz and Isikyakar 2008, Kok et al. 2011).

Researchers places much emphasis in modification of bitumen to achieve higher performance through additive modification, chemical modification and polymer modification and finally results indicates that polymer modification of a bitumen improved much properties compared to other classes of modifiers (Zhu and Brickson 2014).

Brickson et al. (2014) reported that modification of bitumen with polymer started many decades in the past. Polymers used for bitumen modification are classified into classes as follows: thermoplastic plastomers, elastomers, and reactive polymers. Currently 75% of all the polymers used in the modification of bitumen are thermoplastic elastomers, 15% are plastomers and the other 10% are rubber or any other suitable materials (Airey et al. 2003).

Recently, the most typical polymers applied for modification of bitumen includes styrene butadiene styrene (SBS), styrene-butadiene rubber (SBR), polyethylene (PE), ethylene vinyl acetate (EVA) and also ethylene propylene diene monomer rubber (EPDM) [3]. Plastomeric polymers such as polyethylene among thermoplastics are less flexible and have higher crystalline network structure which makes them more resistant to cracks propagation during service life if used as bitumen modifier (Sperling et al. 2015).

Polymer modified bitumen generally have poor storage stability and aging resistance, a phase separation exist in PMBs during storage especially at higher temperatures (Polacco and Berlincioni et al. 2005, Zhang & Yu 2010, Yousefi et al. 2004).

Airey et al (2003) explains that the mechanism of polymer modification with bitumen consist of polymer swelling by absorption of the bitumen light fractions (Maltenes) and an establishment of network (rubber elastic in nature) within the bitumen binder, the modification depends on network formed and also its significance on polymer modification as a function of the nature of the bitumen, contents and properties of the polymer and solubility between the polymer and the bitumen (Perez Lepe and Martinez 2003).

Generally preparation of polymer modified bitumen requires the use of high shear mixer for effective dispersion of polymer which allows more generation of polymer rich microphases (Airey et al. 2002).

Also Airey et al. 2002 found out that, the optimum polymer content in bitumen for best modification ranges between 4%–6% which provides higher enhancement in properties of the modified binder and is considered to be more economical. In majority cases addition of more than 6% by weight of bitumen results only in agglomeration and negative result and as such it becomes only a source of material related problems (Polacco and Stastna 2004, Kalantar and Karim et al. 2012).

Also several standards which includes Indian Road Congress (IRC) and Australian Asphalt Pavement Association (AAPA) recommends the use of between 3%–6% polymer content by weight of the bitumen as the optimum (IRC 1999, AAPA 2004).

The main purpose of this research is to evaluate the stability during storage and physical properties of linear low density polyethylene at optimum polymer content with a view of selecting best optimum polymer content.

2 MATERIALS AND SAMPLE PREPARATIONS

2.1 Materials

The base bitumen grade used for this investigation is 80/100 penetration which were obtained from PETRONAS refinery in Melacca, Malaysia.

Also, the polymer used for the modification is linear low density polyethylene which was supplied by Polyethylene industry located in Kerteh, Malaysia in pellet form.

2.2 Sample preparations

Samples for the investigation were prepared by adding 4%, 5%, and 6% linear low density polyethylene by weight of bitumen to a 80/100 pen grade bitumen using a multimix desk top high shear mixer at shearing rate of 4000 rpm, temperature of $150 \pm 10°C$ for 120 minutes in order to produce homogenous mixture. Also, the base bitumen were subjected to the same shearing rate, shearing temperature and shearing time in order to obtained a uniformity conditions for testing throughout the research.

2.3 Experimental procedures

2.3.1 Penetration test
Penetration test were performed all in accordance to the ASTM D5-97 specifications. Under this method penetration needle were allowed to penetrate for five seconds under load of 100 gm into the bitumen maintained at a temperature rate of 25°C. The distance entered by the penetration needle was recorded as the penetration of the bitumen expressed in decimillimeter (dmm).

2.3.2 Softening Point (Ring and Ball) Test
The softening point or ring and ball test is another consistency test to characterized bitumen, the test were all performed in accordance to the ASTM D 36–97 specifications, softening point temperature is recorded as the temperature to which a sample of a bitumen can no longer withstand the weight of a 3.5 g steel ball when it were heated at a uniform temperature rate of 5°C/min.

2.3.3 Ductility test
Ductility test were performed in accordance to ASTM D 113 specifications. Ductility of bitumen is normally the longest length to which a briquette specimen could be stretched without being breaking at a specified rate and temperature under the influence of increasing tensile forces.

2.3.4 Temperature susceptibility
The behavior of bitumen depends entirely on the temperature and loading rate, the temperature susceptibility of bitumen is referred as the change in rheology of bitumen with changing temperature which is a very influential property for bitumen. Penetrations and softening points are used to obtain penetration index (PI) based on equation 1:

$$PI_{TR\&B} = \frac{1952 - 500\log P_{25} - 20T_{R\&B}}{50\log P_{25} - T_{R\&B} - 120} \quad (1)$$

Where P_{25} = bitumen penetration tested at temperature of 25°C, and $T_{R\&B}$ = softening point temperature.

2.3.5 Storage stability test
Storage stability is used to check the risk of modifier separation from bitumen, the test was performed in accordance to PN-EN 13399 (tube test) by taking the difference between the softening point temperature of the top and bottom samples of bitumen after placing vertically in oven and heating for 48 hours and cooling.

After test samples preparation, the samples for the test were poured directly into an empty aluminum foil tube having diameter of 3 and a height of 16 cm. The foil tubes were immediately closed and stored vertically upward in an oven without interruption for 48 hours at temperature of 160°C. Then the samples were allowed to cool first in to a room temperature which later are cut into three equal sections horizontally. The upper and lower sections after cutting horizontally were used to assess the stability during storage of the base and polymer modified bitumen by determining the corresponding softening points temperatures of the upper and lower sections. The samples are regarded to have excellent stability during storage or transportation if the softening point temperature difference were differed by temperature not more than 2.5°C.

3 RESULTS AND DISCUSSIONS

3.1 Penetration test

Bitumen penetration is referred as a good representation of consistency which fully reflects the rheological properties (flow and deformation) of bitumen, from Figure 1 it can be observed that addition of LLDPE decrease the penetration of the base bitumen and the improvement increased with increase in polymer content. Moreover, 6% LLDPE addition decreases the penetration of the base bitumen by morethan 50% which indicates that 6% has the highest improvement.

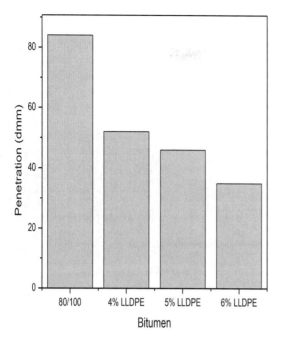

Figure 1. Penetration Test.

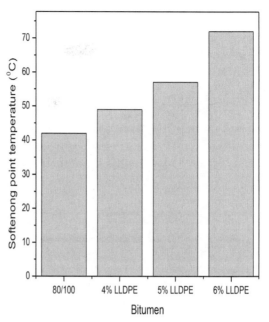

Figure 2. Softening Point Test.

LLDPE addition shows a strong effect on the modified binders through reduction of penetration values and increasing stiffness which improves the binder resistance to high temperatures and less susceptible to deformations.

3.2 Softening point test

Softening point temperature describes the plastic flow of bitumen and also reflects the stability of bitumen under a high temperature condition. In general, higher softening point temperatures of bitumen, indicates a high stability of the bitumen under conditions of high temperatures.

From Figure 2 it can be seen that addition of LLDPE increases the softening point temperature of the base bitumen. Softening point temperature increase in all the LLDPE modified binders implies that the modifies binders ability to soften easily under high temperature condition decreases making it less susceptible to temperature

However, addition of LLDPE further improves the softening point and the improvement increases with an increase in LLDPE polymer content. 6% LLDPE shows the highest enhancement in softening point temperature among the modified binders, this result indicates that 6% LLDPE modification enhances the binder resistance to permanent deformation.

3.3 Ductility test

Ductility of bitumen is a property which presents the maximum elongating ability for the bitumen before it

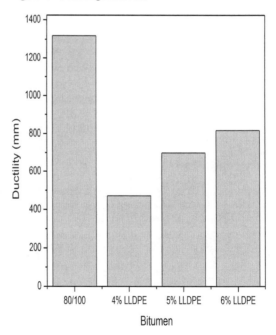

Figure 3. Ductility test.

breaks under an applied tension. All pavement constructed using bitumen with high ductility results in higher durability during service life. Reference to Figure 3, it is observed that LLDPE modified binders can easily results in premature pavement failures such as fatigue cracking due to their poor ductility compared to base bitumen.

Ductility decrease in all the LLDPE modified binders could be attributed due to absorption of

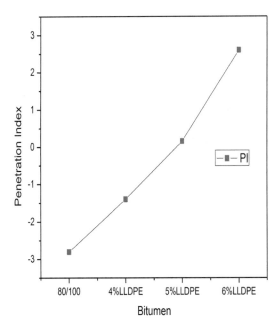

Figure 4. Temperature Succeptibility.

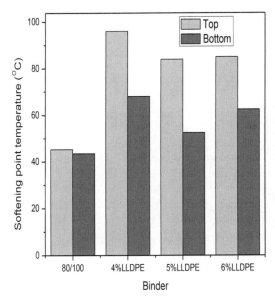

Figure 5. Storage stability test.

bitumen light components by the LLDPE polymer which makes the modified binder more thicker compared to the base bitumen.

3.4 Temperature susceptibility

Bitumen is a material that is sensitive to temperature variations which shows a diverse properties under a various service temperatures. Moreover, in bitumen higher values of penetration index (PI), indicates less sensitivity for the bitumen to variations in temperature changes. The PI value for base penetration bitumens normally in the range of almost −3 (for bitumen with a high temperature susceptibility) and to approximately value of +7 for a highly blown which are low temperature susceptible bitumens (Read & Whiteoak et al. 2003).

From Figure 4 a PI value of −2.8 obtained for base bitumen, indicates that changes in temperature will not have much influence on the properties of base bitumen. After LLDPE polymer modification at different percentages of 4%, 5% and 6%, the PI value is observed to be increasing which fully indicates that temperature susceptibility for all the binders modified were enhanced greatly compared to the unmodified binder. In addition, 6% LLDPE modified bitumen increases further the PI value to 2.27 which indicates higher resistance to temperature susceptibility among all the modified binders.

3.5 Storage stability test

Handling Polymer Modified bitumen both at factories and construction sites required additional attention to ensure that the effectiveness of the polymer modifier were not affected through contamination, high temperature or any other degradation during storage and transportation.

Carcer and Masegosa et al (2014) found out that, polymer modified bitumens are highly susceptible to segregation during transportation or storage if not agitated continuously especially if high amount of polymer were used during modification.

Instability problems of LLDPE modified bitumen during storage or transportation especially at elevated temperatures is an important criterion in the industries producing modified bitumen as the modified binder has to be stored at the factory after production for certain period of time before final application.

Storage instability or phase separation occurs in modified bitumen as a result of large differences in densities, molecular weight, polarity and solubility between polymers and bitumen which are the main keys to improving bitumen modification effects (Liang and Xin et al. 2015).

This phenomenon completely damaged the modified bitumen during storage at factories or construction sites and rendered the binder modified unuseful for paving applications, roofings or any other purpose (Fang and Liu et al. 2014). Therefore, phase separation in this context can lead to great loss to the industries which modified bitumen if not accounted properly to maintained homogeneity.

Poor solubility of polyethylene polymer in bitumen makes the modified bitumen to be too stiff and brittle which in turns increases early chances of premature failures during service life of the pavement (Ahmedzade & Fainleb 2014).

From figure 5 it can clearly observed that phase separation might occur among all LLDPE polymer modified binders during storage or transportation especially at elevated temperatures as a result of much differences in softening point temperatures (above 2.5°C) obtained between the top and bottom parts of the sample, these poor stability during storage were attributed due to differences in solubility (compatibility) and density between the LLDPE polymer and the bitumen which makes the LLDPE polymer that dissolved and dispersed in the bitumen to aggregates together and float in the upper sections of the modified bitumen especially at high temperature conditions and form a stationary state.

Furthermore, the LLDPE polymer addition indicates a distinct effects on the base bitumen as it observed to increase the softening point temperatures of the binders modified, the improved softening point temperature indicates that the binder modified becomes much harder than in the case of the base bitumen.

4 CONCLUSIONS

Based on this research conducted using a linear low density polyethylene and bitumen grade of 80/100 penetration, the following conclusions were observed

(1) Addition of LLDPE polymer with bitumen decreases the penetration of the binder modified which indicates an improvement in the binder rheological properties.
(2) LLDPE polymer addition increases softening point temperature and temperature susceptibility of the binder modified and all the improvements were observed to increases with an increase in LLDPE polymer content.
(3) All LLDPE polymer modified binders shows a poor ductility compared to base bitumen, this situation can easily results in premature pavement failures such as fatigue cracking during pavement service life.
(4) LLDPE polymer modified binders shows poor storage stability due to their poor ability to dissolved and dispersed in the bitumen which leads to phase sepration and to some extent limit their applications.
(5) 6% LLDPE content shows the greatest improvement in properties as such can be selected as the best optimum content.

REFERENCES

Ahmedzade, P., Fainleib, A., Günay, T. & Grygoryeva, O. 2014. Modification of bitumen by electron beam irradiated recycled low density polyethylene, *Construction and Building Materials*, vol. 69, pp. 1–9.

Airey, G. D. 2002. Rheological evaluation of ethylene vinyl acetate polymer modified bitumens, *Construction and Building Materials*, vol. 16, pp. 473–487.

Airey, G. D. 2003. Rheological properties of styrene butadiene styrene polymer modified road bitumens, *Fuel*, vol. 82, pp. 1709–1719.

De Carcer, Í.A., Masegosa, R.M., Viñas, M.T., Sanchez-Cabezudo, M., Salom, C., Prolongo, M.G., Contreras, V., Barceló, F. and Páez, A., 2014. Storage stability of SBS/sulfur modified bitumens at high temperature: Influence of bitumen composition and structure. *Construction and Building Materials*, 52, pp. 245–252.

Fang, C., Liu, P., Yu, R. & Liu, X. 2014. Preparation process to affect stability in waste polyethylene-modified bitumen, *Construction and Building Materials*, vol. 54, pp. 320–325.

Kalantar, Z.N., Karim, M.R. & Mahrez, A., 2012. A review of using waste and virgin polymer in pavement. *Construction and Building Materials*, 33, pp. 55–62.

Kök. B.V. & Çolak. H. 2011. Laboratory comparison of the crumb-rubber and SBS modified bitumen and hot mix asphalt, *Construction and Building Materials*, vol. 25, pp. 3204–3212.

Liang, M., Xin, X., Fan, W., Luo, H, Wang, X. & Xing, B. 2015. Investigation of the rheological properties and storage stability of CR/SBS modified asphalt, *Construction and Building Materials*, vol. 74, pp. 235–240.

McNally, T. 2011. Polymer modified bitumen: properties and characterisation: Elsevier.

Pérez-Lepe, A., Martınez-Boza, F., Gallegos, C, González, O., Muñoz, M. & Santamarıa, A. 2003. Influence of the processing conditions on the rheological behaviour of polymer-modified bitumen, *Fuel*, vol. 82, pp. 1339–1348.

Polacco, G., Berlincioni, S., Biondi, D., Stastna, J. & Zanzotto, L. 2005. Asphalt modification with different polyethylene-based polymers, *European Polymer Journal*, vol. 41, pp. 2831–2844.

Polacco, G., Stastna, J., Biondi, D., Antonelli, F., Vlachovicova, Z. & Zanzotto, L. 2004. Rheology of asphalts modified with glycidylmethacrylate functionalized polymers, *Journal of Colloid and Interface science*, vol. 280, pp. 366–373.

Read, T. & Whiteoak, D. 2003. The shell bitumen handbook: Thomas Telford.

Sengoz, B. & Isikyakar, G. 2008. Evaluation of the properties and microstructure of SBS and EVA polymer modified bitumen, *Construction and Building Materials*, vol. 22, pp. 1897–1905.

Sperling, L. H. 2015. Introduction to physical polymer science: *John Wiley & Sons*.

The Indian Road Congress IRC, 1999. Tentative guidelines on used of polymer and rubber modified bitumen in road construction, *Indian Road Construction special publication 53*, IRC: SP: 53 pp. 4–9.

The Australian Asphalt Pavement Association (AAPA), 2004. Guide to the Manufacture, Storage and Handling of Polymer Modified Binders, *Australian Asphalt Pavement Association publication*.

Yousefi, A. A. 2004. Rubber-polyethylene modified bitumens, *Iranian Polymer Journal*, pp. 101–112.

Yu, R., Fang, C., Liu, P., Liu, X. & Li, Y. 2015. Storage stability and rheological properties of asphalt modified with waste packaging polyethylene and organic montmorillonite, *Applied Clay Science*, vol. 104, pp. 1–7.

Zhang, F. & Yu, J. 2010. The research for high-performance SBR compound modified asphalt, *Construction and Building Materials*, vol. 24, pp. 410–418.

Zhu, J., Birgisson, B. & Kringos N. 2014. Polymer modification of bitumen: Advances and challenges, *European Polymer Journal*, vol. 54, pp. 18–38.

Feasibility of reclaimed asphalt pavement in rigid pavement construction

G.D. Ransinchung.R.N, S. Singh & S.M. Abraham
Civil Engineering Department, IIT Roorkee, India

ABSTRACT: Construction and demolition waste along with waste obtained from reconstruction or resurfacing of pavements produces enormous amount of aggregates. Many studies have been carried out to analyse the use of Recycled Concrete Aggregates (RCA) and Reclaimed Asphalt Pavement (RAP) in flexible and rigid pavement respectively, but very few on each other Due to the benefits offered by rigid pavement, developing countries like India is paying more attention in constructing the same. But the lack of funds and natural aggregates seems to be one of the hurdles, which paves the path for use of RAP in rigid pavement construction. The overall objective of this paper is to bring out the studies conducted on use of RAP in rigid pavements. It is concluded that RAP with or without admixtures can be used in rigid pavements after thoroughly investigating its respective properties.

1 INTRODUCTION

In India, 98% of total pavements are flexible in nature and only 2% are of rigid pavement. Now-a-days more attention is given on construction of rigid pavement because of the advantages being longer life, low maintenance and less life cycle cost compared to flexible. The only hindrance that it faces is the high initial cost contributed by aggregates and cement. In recent past, due to scarcity of natural aggregates, more attention has been given to the utilisation of recycled aggregates in the form of RCA in rigid pavement slabs and RAP in flexible pavement. But still it has not been utilised on a greater scale in India because of following reasons:

- Lack of confidence among designers and contractors
- Limited literature in usage of RAP in Indian conditions
- Common notion that using recycled aggregates will not serve its purpose and rather can be harmful to life (structures collapsing).

Some of the benefits of using RAP in construction are (Carswell et al. 2005)

- Elimination of disposal problem
- Conservation of natural resources
- Saving in land which otherwise would be used for disposal
- Energy saving which would be required for processing of additional virgin aggregates and reduced haulage of materials
- Inconvenience attributed to traffic caused by haulage of materials.

2 PROPERTIES OF RAP

Various researchers conducted studies in order to analyse the characteristics of RAP to bring out the suitability of the same in various pavement layer. Table 1 shows the properties of the RAP material used in one of the study (Thakur et al. 2012). From Table 1 it is observed that the bulk and SSD specific gravity of the RAP aggregates are lesser than the natural aggregate (NA) and percentage of adhered asphalt is around 7%.

This adhered asphalt film is considered to be main reasons for loss in properties of RAP concrete compared to normal. This can be removed by using different methods such centrifuge (ASTM D2172), Ignition (ASTM D6307) and Abson recovery method (ASTM D1856) but are time consuming

Table 1. Properties of RAP.

Aggregate			Test method
Bulk Specific gravity	Fine aggregate	2.48	ASTM C 128
	Coarse aggregate	2.39	ASTM C 127
SSD bulk specific gravity	Fine aggregate	2.56	ASTM C 128
	Coarse aggregate	2.49	ASTM C 127
Uncompacted void content	Fine aggregate	39.15%	ASTM C 1252 (Method B)
Mean particle size (D_{50}) (mm)		2.0	
Binder content	Centrifuge method	6.71%	ASTM D 2172
	Ignition method	6.87%	ASTM D 6307

and highly uneconomical. To make it economical, several researchers are using RAP after sieving and then washing it with water, which not only removes some fraction of film but contaminants also (Bressi et al. 2015; Bressi et al. 2016). This can enhance some of the properties of RAP concrete.

3 USES OF RAP

3.1 *RAP in flexible pavement*

Table 2 shows maximum limits of RAP to be used in flexible pavements specified by different agencies for different conditions. Recently, a tentative guidelines is also published for Indian conditions by Indian Roads Congress (IRC:120-2015) which restrict a limit of 30%.

3.2 *RAP in base and subbase*

There is a general trend of using up to 50% RAP content by weight in virgin aggregate base and subbase layers due to concerns of lower shear strength and excessive permanent deformation resulting in approximately 30% saving in material cost (Hoppe et al. 2015). In order to reduce the aforesaid deformation, geocell-reinforced RAP (Thakur et al. 2012) or blending with other type of aggregates like Construction & Demolition aggregate (D'Andrea et al. 2001). The performance of reinforced RAP bases over weak Subgrade was improved, exhibited a stable response (based on shakedown theory), significant increase in percentage of resilient modulus, reduction in vertical stresses transferred to the Subgrade in comparison to unreinforced base section and concluded that 100% RAP can be utilized as a base course material with geocell confinement (Thakur et al. 2012). Similar performances and elastic moduli values of sub-base layer mixtures with 100% natural aggregates and mixture with 50% natural aggregate and 50% RAP were observed when LWD and FWD tests were conducted, whereas the energy provide to the subbase with RAP is slightly higher than those of natural mix. Homogenous trend of thickness in the test track was observed using GPR test and remained unvaried with time (Montepara et al. 2012). The presence of RAP in mixture enhances resilient modulus and generally improves the drainage characteristics of base and subbase layer mixtures (Kang et al. 2011). The performance of granular base and subbase layers containing RAP materials has been characterized as satisfactory to excellent, evident from the usage of RAP by 13 states in United States (Collins and Ciesielski 1994).

3.3 *Roller compacted concrete pavement*

The feasibility of RAP as replacement of NA in Roller compacted concrete (RCC) by making 4 mixes, RCC 50/50 (i.e. replacement of 50% of coarse NA by coarse RAP aggregates/ replacement of 50% of fine NA by fine RAP aggregates; similarly others) RCC 0/100, RCC100/0 and RCC100/100 was studied (Settari et al.

Table 2. Maximum limits of RAP to be used in flexible pavement by agencies (Thakur et al. 2012).

Highway Agency	Maximum amount of RAP [%]
Alabama	Permeable Asphalt Treated Base – 0 Open Graded Friction Course – 0 Stone Mastic Asphalt – 15 Superpave & Improves Bituminous – 20
Arkansas	30% max
California	15% in new Hot Mix Asphalt (HMA)
Colorado	15% max
Connecticut	40% max
Florida	50% by weight if ESAL < 10000000 30% by weight if ESAL > 10000000
Georgia	40% max for continuous mix type plant 25% max for batch mix type plant
Iowa	≤30% of asphalt binder in a final surface course mixture shall come from RAP

2015). They investigated the mechanical and durability properties of these mixes by performing compressive and split tensile strength, modulus of elasticity, porosity and water absorption by capillary. It was observed that compressive strength at 28 days of RCC decreases by 32.5% and 55% for RCC0/100 and RCC100/0 respectively. Same trend was observed for split tensile strength. Modulus of elasticity decreased by 30% and 44% for RCC50/50 and RCC100/100 respectively. Except RCC50/50, all mixes had an increased water absorption value of about 20–60% compared to reference concrete. Finally it was concluded that all these variation in properties was due to the attached asphalt film on RAP aggregates.

The effects of Rice Husk Ash (RHA) on the mechanical properties of RCC made with NA and RAP aggregates was studied (Mondarres et al. 2014). Cement was replaced by ash by 3% and 5%. 4 mixes were prepared, reference concrete, CA RAP+ fine NA, CA NA+ fine RAP and full RAP concrete. It was observed that adding RHA resulted in high optimum moisture content and lower maximum dry density. The fatigue life of RCC containing RAP was lower than mix with NA. Moreover, mix with RAP CA results in better fatigue life than RAP FA. With increasing the curing time the mechanical properties of RCC with RAP increases and an increase in the energy absorbency value. Adding 3% RHA improved fatigue resistance, materials flexibility and reduced the porosity after 120 days curing.

3.4 *Rigid pavement slab*

Delwar et al.1997, were the first to investigate the feasibility of RAP as aggregate in Portland Cement Concrete (PCC). They found that increase in RAP content, the compressive strength of concrete decreases. High strength concrete made up with RAP requires longer curing periods. High modulus of rupture of RAP inclusive concrete can be used in pavement slabs and was observed that concrete with high RAP contents was more flexible than conventional concrete.

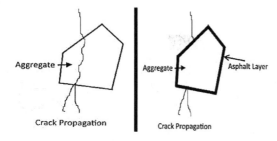

Figure 1. Crack propagation through natural and RAP aggregates in concrete (Huang et al. 2005).

Huang et al. 2005, studied the effect of RAP on toughness and brittleness behaviour of PCC. They performed numerous trials on different arrangements of RAP (coarse and fine) and conventional aggregates. It was observed that RAP aggregate concrete had low slump value but was workable. With increase in RAP content compressive strength and split tensile strength decreased compared to control mix. The toughness of concrete increased tremendously due to the presence of thin layer of adhered asphalt film all around the circumference of aggregates as shown in Figure 1. This layer didn't allow the crack propagation, thereby increasing the toughness of mix.

Huang and Shu 2005, investigated the effect of silica fume and high range water reducing agent (HRWRA) on RAP aggregate concrete. It was observed that there was no effect of silica fume on properties such as compressive strength, split tensile strength and elastic modulus of concrete but HRWRA improved the properties to some extent.

Hossiney 2008, studied the performance of RAP concrete in rigid pavement slab. It was found that the properties like compressive strength, split tensile strength, modulus of elasticity, and flexural strength were inversely proportional to amount of RAP. Coefficient of thermal expansion remained unaffected and only shrinkage decreased with increasing amount of RAP in concrete. One of the major results was that stress ratio decreases with increase in RAP content, which suggest that RAP may perform well as aggregates in rigid pavement slab.

Brand et al. 2012, investigated mechanical and durability properties of RAP concrete by replacing natural aggregate by coarse RAP aggregate. They also observed the same effect on mechanical properties as others researchers i.e. compressive and tensile strength, elastic and dynamic moduli decreases as RAP content increases. There was no significant effect on shrinkage properties, chloride penetration tests and freeze-thaw resistance. Toughness decreases as RAP content increased.

FHWA 2013, investigated the effect on properties of hardened concrete by replacing the natural fine aggregate (FA) up to 50% and coarse aggregates (CA) up to 100% by RAP. They used response surface methodology (RSM) in estimating the influencing parameters on hardened concrete (slump, air content, compressive strength at 7 and 28 days etc). Based on

Table 3. Measured Concrete Properties of Test Slabs (FHWA 2015).

Parameter	High RAP	High strength	Control concrete	MDT Specification
Air Content [%]	4.80	5.00	4.60	5–7
Slump [in]	2.25	1.75	3.50	0.75–2.25
28-days fck [psi]	3001	3960	6000	3000
28-days tensile[psi]	431	670	697	500
365-days fck[psi]	3505	4676	6119	–

RSM, a high RAP (50% fine and 100% coarse RAP) and a high strength (half of RAP used in previous case) were selected and test were done. They concluded that both high RAP and high strength mixes had adequate mechanical and durability properties and can be used for concrete pavements.

4 CASE STUDY

FHWA 2015, in their second phase, evaluated the field performance of high RAP (50% FA &100% CA) and high strength (25% FA & 50% CA) mixes by constructing individual slabs (15x15x10 foot) for these mixes on a roadway near Lewistown, MT. The mixes were processed by using conventional machinery both for laying and finishing. Some obstacles were faced in adjusting water content due to variation in slump values but no major issues were countered. The performance was monitored for 2 years via site visits and internal vibrating wire strain gauges. It was found that no cracking or spalling were present in both slabs. Internal gauges showed that no excessive shrinkage or curling was present but high RAP mix slab had experienced more shrinkage than high strength mix slab which clearly suggest that RAP up to 50% coarse and 25% fine can be used in rigid pavement slab.

It can be seen in Table 2 that, except high RAP's tensile strength, all other properties are within the limits specified by MDT guidelines (MDT 2006). This shows the effectiveness of use of RAP in rigid pavement slab.

5 CONCLUSIONS

From the literature, it is clear that RAP can be utilised in rigid pavement slab up to certain extent. More research should be carried out in order to increase this higher limit such as employing such methods which can remove asphalt completely and be cost effective, mixing some mineral or chemical admixtures to improve its mechanical and durability properties to a greater extent etc. Based on the extensive review on literature for feasibility of RAP in rigid pavement construction following conclusions can be drawn.

1. RAP concrete can be easily batched, placed and finished like normal concrete using conventional equipment's and machinery. Curing period required to achieve same strength is more for RAP concrete than normal concrete.

2. As the percentage of RAP in rigid pavement slab increases, mechanical properties such as compressive strength, modulus of elasticity, split tensile strength decreases. Whereas, no significant effect on shrinkage properties, chloride penetration and freeze-thaw resistance can be seen.
3. Toughness of RAP concrete decreases as RAP content increase. This is due to attached asphalt film all around the periphery of the RAP aggregate which resists the crack propagation. Similarly, modulus of rupture is more for RCA concrete which makes it more flexible than normal concrete.
4. Use of silica fume has insignificant effect on the properties of RCA concrete whereas superplasticizers may results in increasing the properties.
5. RAP can be effectively used in base or subbase of rigid pavement. If it is purposed to be provided over weak subgrade, geogrids can be a successful technique to improve the deformation.
6. By using up to 50% RAP content by weight of virgin aggregate in base and subbase layers, approximately 30% cost savings is present, where as utilizing geocell-reinforced RAP bases, due to advantages of reduced permanent deformation and compatible performance, 100% RAP can be utilized as a base course material. The presence of RAP in mixtures enhances resilient modulus and the addition of RAP generally improves drainage characteristics of base and subbase layer mixtures.
7. From literature it is observed that, RAP can be used in rigid pavement slab up to 50% replacement for CA and 25 % for FA without affecting much properties. Statistical tools such as response surface methodology can help in deciding better replacement proportion of NA by RAP. Similarly, RAP can be used as an aggregate in RCCP but up to 50% replacement of NA (both CA and FA). Use of mineral admixtures such as rice husk ash may results in better properties.

REFERENCES

Active Standard ASTM D1856, Standard Test Method for Recovery of Asphalt from solution by Abson Method, Book of Standards Volume:04.03, United States of America.

Active Standard ASTM D2172, Standard Test Methods for Quantitative Extraction of Bitumen from Bituminous Paving Mixtures, Book of Standards Volume:04.03, United States of America.

Active Standard ASTM D6307, Standard Test Method for Asphalt Content of Hot-Mix Asphalt by Ignition Method, Book of Standards Volume:04.03, United States of America.

Brand A. S., Roesler, J. R., Al-Qadi, I. L., and Shangguan, P., 2012, Fractionated Reclaimed Asphalt Pavement (FRAP) as a Coarse Aggregate Replacement in a Ternary Blended Concrete Pavement, Illinois Center for Transportation Research Report ICT-12-008.

Bressi S, Pittet M, Dumont A.G and Parti M.N, 2016, "A framework for characterizing RAP clustering in asphalt concrete mixtures, Construction and Building Materials, 564–574.

Bressi S, Dumont A.G and Pittet M, 2015, Cluster phenomenon and partial differential aging in RAP mixtures, *Construction and Building Materials*, 288–297.

Carswell, I., Nicholls J.C., Elliot R.C., Harris, J., and Strickland D., 2005, Feasibility of Recycling Thin Surfacing Back Into Thin Surfacing Systems. TRL Report TRL645. Transport Research Laboratory, Wockingham, UK.

Collins R.J and Ciesielski S.K, 1994, Recycling and Use of Waste Materials and By-Products in Highway Construction, NCHRP Synthesis 199, Transportation Research Board of the National Academics, Washington D.C.

D'Andrea. A, Lancieri. F, and Marradi A, 2001 Materiali da Demolizione di Pavimentaziono Stradali Bituminose per la Formazione di Foundazioni e Sottofondi, Proceedings the National XI SIIV Conference, Verona (Italy).

Delwar, M., Fahmy, M., and Taha, R., 1997, Use of Reclaimed Asphalt Pavement as an Aggregate in Portland Cement Concrete. *ACI Materials Journal*, 251–256.

FHWA 2013, Feasibility of reclaimed asphalt pavement as aggregate in Portland cement concrete, FHWA/MT-13-009/8207

FHWA 2015, Feasibility of reclaimed asphalt pavement as aggregate in Portland cement concrete Phase-II: Field demonstration, FHWA/MT-15-003/8207-002.

Hoppe E.J, D.S. Lane, G.M. Fitch, S. Shetty, 2015, "Feasibility of Reclaimed Asphalt Pavement (RAP) Use as Road Base and Subbase Material", Final Report VCTIR 15-R6, Virginia Center for Transportation Innovation and Research, 25–26.

Hossiney, N. J., 2008, Evaluation of Concrete Containing RAP for Use in Concrete Pavement (Master's Thesis). Gainesville: University of Florida.

Huang, B., and Shu, X., 2005, Experimental Study on Properties of Portland Cement Concrete Containing Recycled Asphalt Pavements. Paper presented at the Innovations for Concrete Pavement: Technology Transfer for the Next Generation, Colorado Springs.

Huang, B., Shu, X., and Li, G. 2005, Laboratory Investigation of Portland Cement Concrete Containing Recycled Asphalt Pavements. *Cement and Concrete Research*, 2008–2013.

IRC:120-2015, "Recommended Practice for Recycling of Bituminous Pavements", 2015.

Kang D.H, Gupta S.C, Ranaivoson A.Z, Roberson R and Siekmeier J, 2011, Recycled Materials as substitutes for virgin aggregates in road construction:II, Inorganic Contaminant Leaching, *Social Sciences Society of America Journal*, Vol 75, No.4, 1276–1284.

MDT., 2006, Standard Specifications for Road and Bridge Construction. Helena, Montana.

Modarres A and Hosseini Z, 2014, Mechanical properties of roller compacted concrete containing rice husk ash with original and recycled asphalt materia *Material and Design*, 227–236.

Montepara A, Tebaldi G, Marradi A and Betti G, 2012, Effect on Pavement Performance of a Subbase Layer Composed by Natural Aggregate and RAP", *Procedia-Social and Behavioral Sciences*, Issue 53, 981–990.

Settari C, Debieb F, Kadri E.H and Boukendakdji O, 2015, Assessing the effects of recycled asphalt pavement materials on the performance of roller compacted concrete, *Construction and buiding materials*, 617–621.

Thakur J.K, Han j, Pokharel S.K, Parsons R.L (2012), Performance of geocell-reinforced recycled asphalt (RAP) bases over weak subgrade under cyclic loading, Geotextiles and Geomembranes, Issue 35, 14–24.

Optimal maintenance planning for Trans-West Africa coastal highway infrastructure using HDM-4

O. Oloruntobi, N. Madzlan, K. Ibrahim & O. Johnson
Department of Civil and Environmental Engineering, Universiti Teknologi PETRONAS, Seri Iskandar, Perak, Malaysia

ABSTRACT: The objective of this study was to propose an optimum maintenance standard for Trans-West Africa coastal highway corridor to minimize the life-cycle costs and maximize return on investment of the critical infrastructure. The coastal highway networks have a pivotal impact on the economic prosperities, trades capabilities and social complexities of the entire West African nations via sustainable transportation and logistics services. Dakar-Lagos section of the coastal highway had depreciated in value which depicts significant loss on investment of public infrastructure. The deteriorating condition of the infrastructure and its inability to effectively sustained movement of freight and passengers from Dakar-Senegal to Lagos Nigeria had reflected the poor performance of the transport infrastructure. In spite of this, growing capacity of heavy trucks and transporters across ECOWAS community countries further incapacitate the coastal highway infrastructure condition in the face of poorly defined asset management plans. Unsuitably managed highway asset despite huge public investment continuously increases maintenance frequency and often amount to resource wastage when the right roads were not given the right treatments and at the right time. This present study had proposed possible options to redress Trans-West Africa coastal highway infrastructure using the HDM-4 asset management tool and appraisal software. The deterioration model contained in HDM-4 was calibrated to local condition for development of pragmatic framework to optimally managed Trans-West Africa coastal highway sections. HDM-4 improved analysis models clearly projected appropriate maintenance options for optimal outcome and based on significant information of the transport infrastructure.

1 INTRODUCTION

The vast majority of the West African populace depends on road transportation for domestic and Trans-African trades. Highway infrastructure remains a crucial public asset that boosts commercial activities such as small scale businesses and manufacturing through cost effective short and long distances travel by land. Demographic expansion and exponential growth in transport activities had necessitated the needs for vast development of functional highway networks across West African countries. The Economic Community of West African States (ECOWAS) had benefited immensely from the Trans-West Africa Coastal Highway Infrastructure that had greatly improved the socio-economic of the region and continuously molded unique strategies to enhance highway functionality and long-term performance of the highway corridor. Naturally, the highway pavement surface tends to slowly deteriorates due to temperature differences and accumulative vehicular stresses which increases the pavement's vulnerability to distresses and failure (FHWA, 2003). Trans African trades and movement of heavy truckloads across the coastal highways had contributed to the current worsening condition of the infrastructure, as these loads are repeatedly transported via very heavy trucks and transporters on the pavement surface. Trans-West African coastal highway had failed to function as expected and clearly performed below the standard requirement despite the huge public investment and re-investment in the infrastructure. The infrastructure dynamic loading progression however increases its pavement surface susceptibility to roughness which evidently results in loss of net profit value (NPV) of the critical asset. Managing highway asset using several appraisal software tools have drawn much attention presently than ever, on strategies that would gainfully reinstated highway agencies maintenance skills and commitment to transport infrastructure asset management through systematic and empirical options (FHWA, 2015). Until now, sections of the Trans-West African costal highway corridor are managed by means of non-methodical or native strategies and poor asset management skills leading to the infrastructure deprived condition with defective sections. Environment and physical influences critically induced the highway surface unevenness (roughness), while the ECOWAS region's economies suffers the unfavorable impacts of its coastal highway deterioration of which partly contributed to the hardship and heightened poverty level following ever increasing transportation costs, rising vehicle operating costs and frequent vehicle repairs.

Highway Development and Management (HDM-4) model offers vast options and considerations in making optimal maintenance decisions for highway systems and predicts the consequences of different maintenance options based on highway information (Kerali, 2000). With regards to the study objectives, development of optimal maintenance plan for Trans-West Africa Coastal Highway (functional and structural) condition improvement and prioritization of the highway maintenance options using HDM-4 was proposed for recommendation in this study to optimize the infrastructure condition so as to meet the key performance requirements.

2 COASTAL HIGHWAY ASSET MANAGEMENT

The concept of Trans-West African highway network was a conscious effort to stimulate trades and development in West African countries. The Trans-West African route linked the Senegalese capital Dakar on Africa's western coast with Nigeria's commercial hub Lagos in central-west Africa (Global Development, 2013). Governments in West Africa and international organizations had grieved over the condition of Trans-West African highway network despite intensified efforts and huge investments. The worsening condition of the highway network had poses huge barriers to Trans African trades and commerce, inhibiting growth and tourism.

The concept of highway asset management plan (HAMP) became increasingly relevant to agencies that govern and supervise management of networks of roads. However, highway authorities constantly perform elements of asset management while the service wide application of asset management is a fast developing concept to highway authorities (U.K. Road Liason Group, 2013). Transport asset management (TAM) aims were to optimally maintain existing highway systems, and upgrade management capabilities and to preserve the huge public investments so as to clearly ensure return on investment (ROI) of the asset (Auckland Transport, 2015).

Figure 1. showed to stretch of Trans-West African highway network from Dakar to Lagos; the deteriorated sections of the highway network generally irked the need for optimal maintenance solutions that will enhance the infrastructure performance. An optimal maintenance strategies and measures will improve the infrastructure's worsening condition and downtrend resource wastage on the deprived asset.

3 METHODS AND DATA

Optimal maintenance framework development and strategic asset management planning for the 60.1 km Trans-West African coastal highway section that span from Cotonou-Benin to the central of Lagos-Nigeria. The highway section was investigated in this study and was divided into uniform sections to carefully assess

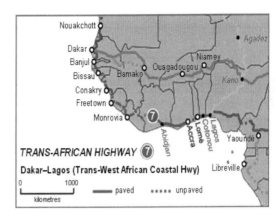

Figure 1. Dakar-Lagos Trans-West Africa Coastal Highway (Sydney 2007).

the condition and data collected from each of the section include the number of lanes, detailed geometry, horizontal curvatures and elevations, and construction materials used along with the material characteristics. The information that comprises climatic variation data (mean monthly precipitation, average temperatures of air and pavement) were appropriately input into the Highway Development and Management (HDM-4) analysis software, the information were significant to calibrate HDM-4 deterioration models to the local condition. HDM-4 systematically address highway maintenance and rehabilitation problems (Hiep and Tsunokawa, 2005), and accurately provides road performance predictions, road treatment programming, funding estimates, project appraisal, policy impact studies, and a wide range of special applications (Jianhua et al, 2004).

Subsequently, exponential increase in road dynamic loading (i.e. from heavy vehicle-to-pavement surface), is a complex interaction that obviously increases highway surface roughness (Misaghi et al, 2012). Thus, 60.1 km Cotonou-Lagos section of Trans-West Africa coastal highway condition is bi-annually assessed and its degrees of roughness from the data obtained were quantified based on the environmental, traffic and structural factors. Response-Type Road Roughness Measuring System (RTRRMS) was used to measure the response of heavy trucks and other vehicles to the highway profile by means of the in-built transducers that accumulate the vertical movement of the axle of the survey vehicle with respect to the vehicle body, the Average Annual Daily Traffic (AADT) or Equivalent Single Axle Loads (ESALs) and structural factors were equally obtained.

3.1 Data collection

The Trans-West African coastal highway deterioration data for this study were the highway pavement condition data collected by the Lagos Metropolitan Area Transport Authority (LAMATA) Data Service Centre (DSC) in collaboration with the Nigeria Federal Roads Maintenance Agency (FERMA) for the

World Bank Group to foster African-wide infrastructure maintenance programs that will enhance community transport, capacity building and accessibility problems minimization. These data was requested and obtained from the LAMATA-DSC, Automatic Road Analyzer Vans (ARAV) was used to photolog the highway sections. The wheel-mounted distance measuring instrument (ARAV) in-built sensors and cameras were used to capture data at intervals of 10 meters and the data collected comprises high resolution pavement images, pothole depth, rut-depth dimensions, roughness values, horizontal and vertical curvature, cross slope and grade.

These data were collected in two broad subdivisions:

1. Inventory data which includes the road details e.g. road nomenclature, road class, carriageway and shoulder width, road thickness and surface types, and pavement layer constituents.
2. Pavement Evaluation which combined both structural and functional evaluation of road pavements.

4 RESULT AND DISCUSSIONS

The highway data were carefully analyzed using the Highway Development and Management (HDM-4) software tool, HDM-4 model conveyed the objectives of the study. Deterioration model contained in HDM-4 was calibrated to the local condition and the highway sections datasets obtained were carefully input into HDM-4 asset management software and processed subsequently. Highway infrastructure roughness was the primary factor in the analysis, the trade-offs that involved highway asset quality versus user cost was characterized by summary of index which are obtained by means of mathematical analysis which reduced the profile data to distinct values and realized at 100 m distance per point. The analysis technique and procedures followed are outlined as follows:

1. Deterioration Models Calibration to Local Condition: The bituminous highway deterioration models incorporate the pothole progression model, plastic deformation coefficients and the standard deviation of rut depth model.
2. Asset Valuation: This was done prior to analysis to properly estimate the highway asset value, the financial and economic value (or level of investment of the infrastructure) was of utmost importance during the analysis.
3. Creation of Project Worksheet: Highway maintenance project worksheet created on the HDM-4 user interface includes titles which are assigned to each project worksheet and specified the highway names to be analyzed.
4. Definition of Project and Tasks: This specified the variables used and clearly defined the set of data that were feed into HDM-4.
5. Analysis Method "Sensitivity Analysis" models contained in HDM-4 version 2.01 software was used is to define sensitivity scenarios and to investigate the influence of disparities in key parameters on the analysis of the data from the selected highway sections.
6. Generation of Reports: Printable format reports and feedbacks were generated.

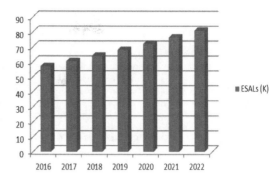

Figure 2. Projected Years Damage from ESAL Progression.

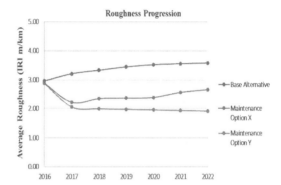

Figure 3. Multi Years Roughness Prediction.

The HDM-4 version 2.01 software run-data produced during analysis were transplanted into a single file in Microsoft Access and Microsoft Excel formats. The HDM-4 appraisal software accurately model future pavement behavior and forecasts the highway pavement performance based on the condition information obtained through various sources and ESAL progression of the coastal highway section under study as shown in Figure 2.

The analysis projected long term deterioration of the 60.1 km Trans-West African coastal highway section from Cotonou-Benin to the central of Lagos-Nigeria, and suggests maintenance options that will inventively preserves and improves the highway condition and enhances its performance through the improved connectivity and data organization models contained in HDM-4.

The multi-years roughness forecast with respect to the varied maintenance options as shown in Fig. 3 justifies the erstwhile maintenance strategies implemented and give in-depth understanding into development of

Table 1. Margin settings for A4 size paper and letter size paper.

Alternatives	Works Standard	Work Description
Base Alternative	Routine Maintenance	Defects repair which include pothole patching and cracks sealing with no major repair work done.
Alternative 'X'	Routine Maintenance Option 'X'	Routine maintenance of the highway infrastructure using 40 mm thick bituminous concrete overlay.
Alternative 'Y'	Routine Maintenance Option 'Y'	Routine maintenance the highway pavement using 40 mm thick natural rubber modified bitumen overlay.

unique framework that will create a balance between comparative cost estimates and economic evaluations of different maintenance options. The multi-year roughness progression with respect to different maintenance alternatives and typical recommendations along with work descriptions are illustrated in Table 1 below, which was proposed in this study to improve the prevailing maintenance strategies for the entire Lagos-Dakar Coastal highways.

The base alternative represents minimum or prevailing bi-annual routine maintenance of the highway section investigated and the depth of highway asset management currently practiced across the West African States, to minimize roughness progression and often yielded little or no results. Alternative options 'X' and 'Y' will inventively preserve the highway infrastructure and recommended based on the asset condition appraisal, and would minimize frequent reconstruction of the critical infrastructure.

5 CONCLUSION

The aims of highway asset management plan are to optimally manage highway infrastructure and provides the basis for an informed understanding of the possible consequences of alternative policies. This strategy in-conjunction with native infrastructure management practices will intensify the on-going efforts to provide a structured approach that will enable West African highway authorities to optimally maintain the existent ECOWAS highway and reinstate the infrastructure condition in a cost effective manner so as to foster African-wide trades among the West African coastal countries.

Optimal maintenance strategies for the Trans-West African coastal highway infrastructure had been lacking, since the road authorities and agencies that are charged with the infrastructure maintenance stick slavishly to the traditional maintenance strategy which neither take the options 'X' nor 'Y' into future consideration. The HDM-4 asset management tool incorporation into the existent highway asset management system has unveil several benefits through cost effective maintenance options for the entire Trans-West Africa Coastal highway network and the highway authorities can therefore manage the critical infrastructure based on the best maintenance decisions against the prevailing and imprecise maintenance option (Base Alternative) that often times result in greater consequences when applied. The analysis result had delivered the tool for ideal implementation of project appraisal and prioritization of multi-year routine highway maintenance program that will deliver the expected level of service of the entire highway network.

It is hereby recommended that West African coastal highway maintenance department should manage (maintain and rehabilitate) the vast highway network based on the infrastructure inspection data and pavement condition index for transport infrastructure optimum performance. Infrastructure condition evaluation and pavement structural condition evaluation is extremely suggested for the infrastructure maintenance and rehabilitation prediction for future management planning.

REFERENCES

FHWA. 2003. A Report on Infrastructure/Pavement Distress Identification Manual for the LTPP (Fourth Revised Edition): Publication Number: FHWA-RD-03-031.

FHWA. 2015. AASHTO Transportation Asset Management Guide: A Focus on Implementation. Publication No: FHWA-HIF-10-023.

Kerali, H.R. 2000. Overview of the HDM-4 System, Volume 1, the Highway Development and Management Series, International Study of Highway Development and Management (ISOHDM), Paris: World Roads Association (PIARC).

UK Road Liason Group. 2013. Highway Infrastructure Asset Management: Highway Infrastructure Asset Management Guidance Document.

Auckland Transport. 2015. Asset Management Plan 2015–2018. Auckland Transport's Asset Management Plan.

Sydney, R. 2007. Map of Dakar–Lagos Highway (Trans-West African Coastal Highway). Self-Made Using Wikipedia Base Map. https://commons.wikimedia.org/wiki/File:Dakar-Lagos_Highway_Map.PNG.

Hiep, D.V., Tsunokawa, K. 2005. Optimal Maintenance Strategies for Bituminous Pavements: A Case Study in Vietnam using HDM-4 with Gradient Methods." ISSN 1881-1124.

Jianhua, L., Muench, S.T., Mahoney, J.P., White, G.C., Pierce, L.M. and Sivaneswaran, N. 2004. Application of HDM-4 in the WSDOT Highway System Prepared for Washington State Transportation Commission in cooperation with U.S Department of Transportation (Federal Highway Administration).

Misaghi, S., Tirado, C., Nazarian, S. and Carrasco, C. 2012. Effect of Road Vehicle Interaction on Pavement Condition Deterioration: Transport Research International Documentation (TRID), 12-2757 – 16p.

Global Development. 2013. Senegal to Nigeria superhighway still stuck in a rut. http://www.theguardian.com/global-development/2013/apr/18/senegal-nigeria-superhighway-stuck-rut

The effect of water on the performance of polymer fibre-reinforced road bituminous mixtures

Ibrahim Kamaruddin, Madzlan Napiah & Mohammed Hadi Nahi
Department of Civil and Environmental Engineering, Universiti Teknologi PETRONAS, Seri Iskandar, Perak, Malaysia

ABSTRACT: The presence of water in the bituminous layers of a pavement structure often produce serious distress, reducing its performance and increasing the maintenance costs. Experience has shown that many factors such as bitumen characteristics, aggregate properties, mixture design, construction procedures, environmental conditions and traffic loading contribute to the severity of the condition. The influence of water on the service performance of bituminous pavements has been the subject of extensive research. This paper is based on a laboratory work and addresses the damaging effect of water in polymer fibre-reinforced bituminous mixtures.

1 INTRODUCTION

The damaging effects of moisture on the physical properties and mechanical behavior of bituminous mixtures have been the focus of study for many years. Many laboratory tests have been developed in order to evaluate and quantify the amount of damage that is caused by water on bituminous mixtures. The most widely used laboratory method in conducting these tests appear to be the immersion-mechanical tests which measures the changes in mechanical properties of the bituminous specimens after exposure to moisture. Typically the results are reported in terms of percentage retained strength of the specimens. This paper is based on some laboratory work and addresses the damaging effect of moisture in polymer fibre-reinforced bituminous mixtures.

2 STRIPPING IN BITUMINOUS MIXTURES

The unfilled void spaces in compacted bituminous mixtures are also able to hold sufficient quantities of water that can cause distress and damage and reduce mix performance. The volume of these voids is a variable dependent on the nature of the voids, characteristics of the mix and the degree to which they are compacted in the pavement (Ahmad, Yusoff et al. 2014, Nahi, Kamaruddin et al. 2014). The phenomenon can be further aggravated by the presence of aggregate surface coatings and by aggregates with a smooth texture surface (Nahi, Ismail et al. 2011, Min, Fang et al. 2015). Stripping is primarily an aggregate problem but that the type of bitumen used is also important (Mehrara and Khodaii 2015). Vacuum saturation of specimens followed by conditioning and testing using the Indirect Tensile test show promise for introducing moisture into the specimens and measuring their strength to predict the moisture susceptibility of the mixtures (Kim, Pinto et al. 2012, Jaskula and Judycki 2014).

3 ADHESION PROPERTIES OF BITUMINOUS MIXTURES

Loss of adhesion is especially common in bituminous mixes which utilizes hydrophilic aggregates i.e. aggregates which have an affinity for water (Nahi, Kamaruddin et al. 2014). An example of this type of aggregate is granite which is widely used in road pavement construction in Malaysia. Ishai and Craus (Movilla-Quesada, Vega-Zamanillo et al. 2013) listed two ways in which modifications are possible should the use of hydrophilic aggregates prove to be unavoidable (Movilla-Quesada, Vega-Zamanillo et al. 2013):

1. modification of the adhesion properties of the bitumen by additives
2. modification of the surface properties of the aggregates by treating them with cement-water solution or hydrated lime-water solution.

4 MATERIALS USED IN INVESTIGATION

4.1 *Mineral aggregates, filler and bitumen*

Limestone aggregates and Ordinary Portland cement (OPC) filler and a binder of nominal penetration Grade 50 were used in this study. Some relevant properties of the material used are shown in Table 1.

4.2 *Synthetic fibres*

Two types of synthetic fibres polypropylene and polyester were used in this study. The fibres were used

Table 1. Properties of the Mineral Aggregates, Filler and Bitumen Used in the Investigation.

Material	Percentage by Weight	Relative Density	Absorption %	BS Specification
Coarse Aggregate	35	2.75	0.47	BS 594: Part 1:
Sand	55	2.65	1.37	1992 Table 3,
Filler (Ordinary Portland Cement)	10	3.15		Type F Wearing Course designation 30/14
	Penetration (0.1 mm)	Softening Point (°C)	Penetration Index, PI	
Bitumen	52	48.5	−0.37	

as a partial replacement of the filler; on an equal volume basis; at two concentrations of 0.5% and 1% by weight of the mix. The chopped fibres were the by-products of the textile industry and thus their potential use was desirable on environmental grounds.

5 WET-DRY INDIRECT TENSILE TEST

The wet-dry indirect tensile test was adopted as a principal measure of the bituminous mix response to moisture damage. Most evaluations of moisture damage have been assessed quantitatively by mechanical tests in which such properties as loss of tensile strength or decrease of resilient and stiffness moduli have been measured. These are then given in the form of a tensile-strength ratio and a moduli ratio, for which the tensile strength and modulus of the dry specimens served as references. The tensile strength ratio (TSR) and modulus of elasticity ratio (MER) are dimensionless numbers used to represent the portion of tensile strength and modulus retained following conditioning. Low values indicate high moisture damage. The use of the static indirect tensile strength test to study the effect of moisture on bituminous mixtures and recommended a minimum tensile strength ratio of 0.7 to differentiate between a stripping and a non-stripping bituminous mix while (Goh, Akin et al. 2011) reported values of between 0.7–0.75. Ishai and Nesichi cited values of 60–75 percent retained stability values for roads and highway pavements and 75 percent for airfield pavements as the quality criteria used in Israel (Ishai and Nesichi 1988).

6 DEGREE OF SATURATION

The degree of saturation gives a measure of the amount of water that is absorbed by the specimen into its permeable voids. The degree of saturation is thus defined as the ratio of the volume of water in the wet specimen to the total volume of voids in the specimen.

A number of researchers have come up with various regimes for vacuuming and saturating the bituminous specimens. Lottman for example used 26-inches of mercury to vacuum the specimens (Lottman 1982). while lower levels of vacuuming (4 and 15-inches of mercury) have been cited in the literature (Xiao and Amirkhanian 2009, Jaskula and Judycki 2014). In addition to vacuuming, Lottman also subjected the specimens to an advanced moisture conditioning in which thermal cycles or a cycle of freezing-soaking was carried out. Ishai and Nesichi subjected the specimens to hot water immersion (at 60°C) for up to 14 days and testing the specimens at different immersion period to determine the retained strength of the specimens while Mehrara, subjected the specimens to the boiling test, a freeze-thaw cycle before conducting the indirect tensile test on dry and conditioned specimens (Ishai and Nesichi 1988, Mehrara and Khodaii 2015).

7 VOID STRUCTURE IN BITUMINOUS MIXTURES

The moisture conditioning process attempts to allow water to penetrate and occupy the air voids in the specimen. An appreciation of the void structure in bituminous mixtures is thus very vital (Kumar and Goetz 1977). The model divides the air voids system into three categories; through passage accessible air voids, dead end accessible air voids and non-accessible air voids (Kamaruddin 1998). The 24 hour soaking allows water to only occupy the through passage accessible voids.

8 EXPERIMENTAL PROCEDURES

As the water susceptibility of the mixes was to be determined, the specimens were tested in both the dry and wet condition. The dry conditioning involves curing the specimens at room temperature for two days prior to testing. The wet conditioning involved subjecting the bituminous samples to a combination of air vacuum, vacuum saturation and static soaking. Air vacuuming is to evacuate all the accessible pores from air and water. The objective of vacuum saturating the specimens was designed to accelerate the moisture damage process. It is extremely important that the process of vacuum saturating the specimens do not result in a degree of saturation greater than 100% which is indicative that more water was introduced into the voids than there are void space, making comparison between samples no longer valid.

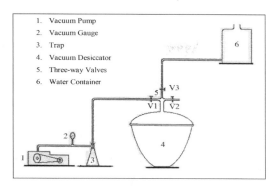

Figure 1. Schematic Diagram of Vacuum Saturation Apparatus.

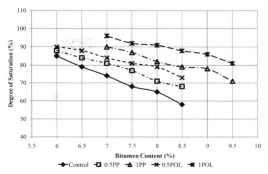

Figure 2. Degree of Saturation vs. Bitumen Content.

Figure 1 is a schematic diagram of the vacuum saturation apparatus. In this study, the specimens were placed in a thick-walled desiccator jar. Valves (W) and (A) are closed and valve (V) which led to the vacuum pump is opened. The air vacuuming process takes about half an hour to drive out all the air trapped in the accessible voids. Distilled water was then used to fill the jar to about 2 cm. above the specimens and about 3 cm. below the top rim of the jar. This was followed by a 1-hour vacuum saturation period at 1 atmospheric pressure which was considered a pre-treatment of moisture conditioning and a means of water-saturating the specimens, during which time the jar surfaces was gently agitated. After the one hour of vacuuming, the vacuum was removed and the inside of the jar was allowed to reach ambient atmospheric pressure and the specimens undergoing static soaking for a period of 24 hours for the purpose of achieving a constant weight of the specimens (fully saturated condition). Earlier studies on moisture damage in bituminous materials reveal that this saturation regime produced a degree of saturation in the region of 75–100% of the bituminous sample voids and was therefore also adopted (Çelik 1996, Sabina, Sangita et al. 2009). After the immersion process, the specimens were weighed in water. They were then weighed in air in the saturated surface dry condition. The volumes of the saturated specimens were determined by subtracting the mass of the saturated specimen from its mass in the saturated surface-dry state. The volume of absorbed water was determined by subtracting the air-dry mass of the specimen from its saturated surface-dry mass. The degree of saturation is given by:

$$DegreeofSaturation(\%) = \frac{VolumeofAbsorbedWater}{VolumeofPores} \times 100\% \quad (3)$$

Alternatively, the degree of saturation can be determined from the following relationship:

$$DegreeofSaturation(\%) = \frac{B-A}{V(B-C)} \times 100\% \quad (4)$$

where: A = dry weight of specimen in air (gm)
B = weight of surface-dry specimen after saturation (gm)
C = weight of saturated specimen in water (gm)
V = porosity of the specimen (%)

The moisture conditioning process using the vacuum saturation technique can also be used to measure the porosity of the specimens which is determined by dividing the volume of absorbed water by the volume of the saturated specimens. Lottman suggested the following equation for the calculation of measured porosity (Lottman 1982).

$$MeasuredPorosity(\%) = \frac{(B-A)}{(B-C)} \times 100\% \quad (5)$$

where: A = weight of dry specimen in air (gm)
B = weight of surface-dry vacuum saturated specimen in air (gm) C = weight of vacuum saturated specimen submerged in water (gm)

9 DISCUSSION OF RESULTS

Figure 2 shows the relationship between the degree of saturation and bitumen content for the control Hot-Rolled Asphalt (HRA) mix in comparison with the fibre modified mixes. The general trend is for the degree of saturation to decrease with increasing bitumen content. The addition of fibres appears to bring about an increase in the degree of saturation, this increase being more pronounced at the higher fibre concentration. This may be the result of the higher porosity that is associated with the higher fibre content mixes. The polypropylene mixes appear to exhibit better result than the polyester fibre mixes. For all the mixes, the saturation lines obtained appear parallel to one another with a somewhat similar slope. This is indicative that the degree of saturation in all the mixes decreases consistently with increasing bitumen content.

Figures 3 and 4 show the variation between the calculated and measured porosity between the control and the polypropylene fibre modified mixes and the polyester fibre modified mixes respectively. The

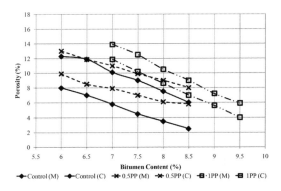

Figure 3. Calculated and Measured Porosity vs. Bitumen Content for Control and PP Fibres.

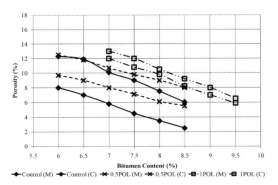

Figure 4. Calculated and Measured Porosity vs. Bitumen Content for Control and POL Fibres.

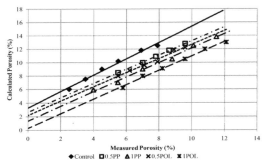

Figure 5. Calculated vs. Measured Porosity for Different Mixes.

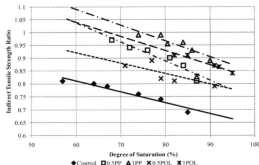

Figure 6. Indirect Tensile Strength vs. Degree of Saturation.

calculated porosity gives a measure of all the voids in the specimens that include both the accessible and non-accessible voids while the measured porosity as was obtained from the moisture conditioning process determined only the accessible voids. The calculated porosity therefore is always greater than the measured porosity as shown in the figure. The general trend is the porosity decreases with increasing bitumen content. The result of adding fibres to the HRA mix resulted in higher porosity in the resulting mix.

The relationship between the measured and calculated porosity is plotted graphically in Figure 5 resulting in a linear relationship between the two. The point of intercept of the lines with the calculated porosity axis gives an indication of the percentage of non-accessible or unconnected pores in the respective mixes. The addition of fibres appears to reduce the percentage of non-accessible pores in the mix, this reduction was seen to be more pronounced in mixes with greater fibre content.

A general trend shown from the figure suggest that increasing the bitumen content resulted in an increase in the unconnected voids as the bitumen fills up the void space or continuous channels in the specimen. Subsequently, this reduces the connectivity of the voids. The presence of the fibres also suggests an increase in the connectivity of the voids as the porous nature of the fibres gave a continuous channel (path) of void space. It must also remembered that the control mix gave lower porosity than the fibre incorporated mixes, thus justifying its behaviour as in Figure 5.

The indirect tensile strength (ITS) ratio is effectively an indication of the amount of strength loss due to the effect of water. The variation of indirect tensile strength ratio and the degree of saturation is shown in Figure 6. The ITS ratio shows it decreasing with increasing degree of saturation for all the mixes. The lines obtained are rather parallel to one another indicative that the decrease is somewhat similar. The fibre-modified mixes exhibited higher ITS ratios of around 11–25% over the control mix.

It is appropriate to be reminded that the fibre incorporated mixes had higher porosity and permeability than the control mixes that will permit easier access to water and increase the potential for stripping. It may thus appear that the more viscous binder of the fibre-incorporated mixes had a better cementing and adhesive property at the binder-aggregate interface that resulted in a reduction in stripping. It is believed that de-bonding may not have been solely responsible for the decrease in wet tensile strength values but other moisture damaging factors such as binder matrix softening may have been responsible as well.

10 CONCLUSIONS

Based on this study, the following conclusions can be drawn:

1. Changes in both the cohesive properties of the bitumen and the adhesion of the bitumen to the aggregate surfaces may occur as a result of exposing the bituminous mixtures to moisture. Polymer fibre incorporation into bituminous mixtures helps reduce the high level of moisture damage that was noted from the control mix. The polyester fibre modified mixes also showed lower moisture susceptibility than those of the polypropylene mixes at the same fibre concentration. However, the 0.5% fibre concentrated mixes showed better resistance to water damage than that at the 1% concentration.
2. It is important to remember that mixes with polymer fibres had greater bitumen content and yet greater void contents than the control mix. Regarding resistance to moisture damage, these two parameters would be expected to oppose each other. Suffice to say that the additional bitumen in the fibre mixes increased the film thickness on the aggregate particles thus affording additional protection from moisture. The 0.5% fibre concentration may have provided enough reinforcement across the plane of failure in the mixtures while the 1% fibre concentration may have far too high void contents that allowed for more water penetration into the mixtures.
3. The incorporation of polymer fibres in bituminous mixtures also acts to decrease the moisture sensitivity of the bitumen to aggregate bonding. This may be due to the strengthening of the wetted binder matrix which promote both adhesion and cohesion retention.

REFERENCES

Ahmad, J., et al. (2014). "Investigation into hot-mix asphalt moisture-induced damage under tropical climatic conditions." Construction and Building Materials 50: 567–576.

Çelik, O. N. (1996). The engineering properties and fatigue behaviour of asphaltic concrete made with waste shredded tyre rubber modified binders, University of Leeds.

Goh, S. W., et al. (2011). "Effect of deicing solutions on the tensile strength of micro-or nano-modified asphalt mixture." Construction and Building Materials 25(1): 195–200.

Ishai, I. and S. Nesichi (1988). Laboratory evaluation of moisture damage to bituminous paving mixtures by long-term hot immersion.

Jaskula, P. and J. Judycki (2014). "Durability of asphalt concrete subjected to deteriorating effects of water and frost." Journal of Performance of Constructed Facilities: C4014004.

Kamaruddin, I. (1998). "The Properties and Performance of Polymer Fibre Reinforced Hot-Rolled Asphalt." Unpublished PhD Thesis, University of Leeds.

Kim, Y.-R., et al. (2012). "Experimental evaluation of anti-stripping additives in bituminous mixtures through multiple scale laboratory test results." Construction and Building Materials 29: 386–393.

Kumar, A. and W. Goetz (1977). Asphalt hardening as affected by film thickness, voids and permeability in asphaltic mixtures. Association of Asphalt Paving Technologists Proc.

Lottman, R. P. (1982). "Laboratory test methods for predicting moisture-induced damage to asphalt concrete." Transportation research record(843).

Mehrara, A. and A. Khodaii (2015). "Evaluation of moisture conditioning effect on damage recovery of asphalt mixtures during rest time application." Construction and Building Materials 98: 294–304.

Min, Y., et al. (2015). "Surface modification of basalt with silane coupling agent on asphalt mixture moisture damage." Applied Surface Science 346: 497–502.

Movilla-Quesada, D., et al. (2013). "Experimental study of bituminous mastic behaviour using different fillers based on the UCL method." Journal of Civil Engineering and Management 19(2): 149–157.

Nahi, M., et al. (2011). "Analysis of asphalt pavement under nonuniform tire-pavement contact stress using finite element method." Journal of Applied Sciences 11(14): 2562–2569.

Nahi, M. H., et al. (2014). "Finite Element Model for Rutting Prediction in Asphalt Mixes in Various Air Void Contents." Journal of Applied Sciences 14(21): 2730.

Nahi, M. H., et al. (2014). "The Utilization of Rice Husks powder as an Antioxidant in Asphalt Binder." Applied Mechanics & Materials(567).

Sabina, T. A., et al. (2009). "Performance evaluation of waste plastic/polymer modified bituminous concrete mixes." Journal of Scientific & Industrial Research 68: 975–979.

Xiao, F. and S. N. Amirkhanian (2009). "Laboratory investigation of moisture damage in rubberised asphalt mixtures containing reclaimed asphalt pavement." International Journal of Pavement Engineering 10(5): 319–328.

Structures and materials

Engineering Challenges for Sustainable Future – Zawawi (Ed.)
© 2016 Taylor & Francis Group, London, ISBN 978-1-138-02978-1

3D nonlinear finite element analysis of HPFRC beams containing PVA fibers

Nasir Shafiq
Department of Civil and Environmental Engineering, Universiti Teknologi PETRONAS, Seri Iskandar, Perak, Malaysia

Tehmina Ayub
Civil Engineering Department, NED University of Engineering & Technology, Pakistan

Sadaqat Ullah Khan
Urban & Infrastructure Engineering Department, NED University of Engineering & Technology, Pakistan

ABSTRACT: In this paper, three dimensional (3D) nonlinear finite element analysis (FEA) results of 04 high performance fiber reinforced concrete (HPFRC) beams containing polyvinyl alcohol (PVA) fibers are presented. The results of FEA are validated by the results of laboratory tests of 04 HPFRC beams having similar rectangular cross-section of 100 × 200 mm and 1500 mm long. Testing of all beams was carried out under three-point bending condition to ensure flexural failure. Among the 04 beams, 01 beam was cast using plain concrete without fibers and used as a control beam. Rests of the 03 beams were reinforced with 1, 2 and 3% of PVA fibers volume. Test results were examined in terms of maximum load and corresponding mid span deflection attained by the beams prior to the failure. The maximum strains were likewise obtained by using strain gages. These experimental results were used to verify the FEA results obtained by simulating all beams in nonlinear finite element program ATENA 3D. Both FEA and experimental results showed good agreement with each other in terms of maximum load, load vs. mid-span deflection patterns and maximum tensile strains.

1 INTRODUCTION

High performance fiber reinforced concrete (HPFRC) is a concrete based composite containing fibers. In HPFRC, steel fibers are the most widely investigated [1]; however, the use of steel fibers is not preferable in environmentally exposed infrastructures (e.g. bridge decks) as they are susceptible to corrosion [2]. Polyvinyl alcohol (PVA) fiber is a well-known fiber for engineered cementitious composites (ECC) [3] and there are no concerns of fire and corrosion with the use of PVA fibers. Furthermore, PVA fibers contribute in the pseudo strain-hardening behavior and resist the crack development under uniaxial tension [3]. Therefore, there is a need of investigating the use of PVA fibers in HPFRC from practical application's point of view. The reason is that the structural applications of ECC, despite exhibiting excellent mechanical properties, are limited due to higher cost resulted by the use of higher cement content and low modulus of elasticity due to the absence of coarse aggregates (e.g. beams, columns, etc.).

In this research, an economical mix design of HPFRC is investigated by using lower cement content, coarse aggregates and PVA fibers. The primary aim was to analyze the structural behavior of HPFRC using finite element analysis (FEA) approach. In this paper, FEA results of 04 HPFRC beams containing PVA fibers are presented using 3D nonlinear finite element (FE) program ATENA 3D. The aim of FEA was to predict the flexural failure mechanism caused by the internal forces when beams were subjected to the concentric load applied at the middle of the beam span. This program had been successfully used for the structural analysis of fiber reinforced concrete members [4, 5].

2 METHODOLOGY

2.1 Finite Element Analysis (FEA)

The 3D nonlinear FEA of HPFRC beams containing PVA fibers was performed using ATENA 3D program. In this program, beams were modeled as 3D solid object characterized by the material model "SBETA" with increased fracture energy resulted by the PVA fibers, which exhibited hardening response in the post peak behavior in compression and tension. The reason for selecting this model was the consideration of the nonlinear hardening and softening behavior of concrete in compression and the fracture of concrete in tension [6]. Thus, the tensile strength of HPFRC was determined through a unique experimental program

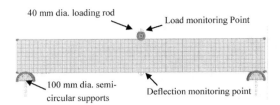

Figure 1. Typical meshing patterns of the beams for finite element analysis.

Figure 2. Typical stress concentration location and crack pat-terns of the beams.

described in [7] and the fracture energy needed was derived from a failure function described in section 2.1.2.3 [6].

2.2 Beam modeling

All beams were modeled in full scale mode by providing actual beam dimensions of $100 \times 200 \times 1500$ mm. It was assumed that HPFRC beams were simply supported at the bottom on semi-circular steel plates. The displacement in the direction of beam length was set as zero (refer to Figure 1). The load at top of the beams was applied through a circular steel rod of 40 mm diameter. The loading was applied at the middle center of the beam in the form of prescribed displacement (i.e. displacement control analysis was performed) and load was monitored as shown in Figure 1 as "Load Monitoring Point".

2.3 Material properties assignment

In ATENA 3D, program calculated material properties (e.g. tensile strength, fracture energy etc.) based on the cube compressive strength were used.

2.4 Mesh generation

3D "brick" element mesh of $25 \times 25 \times 25$ mm size was created (refer to Fig. 1).

2.5 Support restrains

The support restrained were modeled as typical roller and hinged i.e. the movement of the modeled beam was vertically and horizontally restrained.

2.6 Finite Element Analysis (FEA)

Smeared crack approach was adopted for fracture mechanism. Newton-Raphson solution with the concept of internal step by step analysis was used for the FEA of beams [6]. The critical stress in cracked beam is shown in Figure 2.

Table 1. Mix materials and their quantities.

Mix ingredients	Quantity (kg/m^3)
Portland cement	450
Fine aggregate (river sand)	670
Coarse aggregates (<10 mm)	600
Coarse aggregates (10–20 mm)	500
PVA fiber volume, Vf	13, 26 and 39
Water to cement ratio	0.4
Superplasticizer	Variable dosage to maintain a slump of 50 ± 10 mm

Table 2. Physical properties of PVA fibers.

Fiber type	Straw type
Diameter (mm)	0.66
Cut length (mm)	30
Tensile strength (MPa)	900
Elastic modulus (GPa)	23
Specific gravity	1.3
Elongation (%)	7

2.7 Experimental program

2.7.1 Materials and quantities

The detail of HPFRC concrete mix ingredients and quantities and fiber volume percentage is given in Table 1. Kuaray RF 4000 type polyvinyl alcohol (PVA) fibers were used. The physical properties of PVA fibers are presented in Table 2.

2.7.2 Mix design and casting detail

All materials (refer to Table 1) were mixed in a pan mixer according to BS 1881-125: 1986 standard. During mixing of PVA fibers into the concrete mix, extra time was utilized to ensure proper mixing and even distribution of the fibers. Total 04 mixes of HPFRC were prepared using 0, 1, 2 and 3% volume of PVA fibers. After mixing, slump was determined as per BS 1881-102: 1986 standard. Then, $100 \times 200 \times 1500$ mm sized beam (without notch) was cast along with 03 cubes of size $100 \times 100 \times 100$ mm using each of the 04 mixes. The cube compressive strength was required as input for FEA. Formwork of all beams and molds of cube specimens were removed after 24 hours and specimens were fully immersed in a water tank for 28 days.

Figure 3. Typical loading arrangement and testing setup for flexural testing.

2.7.3 Specimens testing detail

After 28 days, all specimens were taken out from water tank and air dried for few hours.

2.7.4 Compression testing of the cubes

Compression test of the three HPFRC cubes of the size $100 \times 100 \times 100$ mm was carried out according to BS 1881-116: 1983 standard using a compression testing machine of 3000 kN capacity. Compressive load was applied using a loading rate of 0.3 kN/sec.

2.7.5 Flexural testing of HPFRC beams

Flexural testing of HPFRC beams was carried using Universal Testing Machine (UTM) of 500 kN capacity (refer to Figure 3). Testing was performed under load control condition at the loading rate of 10 N/sec by applying a concentric loading. Linear variable differential transformer (LVDT) was attached at mid bottom of the beam span to obtain the deflection corresponding to the load. Total 05 strain gages were applied to the beam to obtain the strain measurements before failure. Each of the 02 strain gages were applied on each face and 01 strain gages was applied at the soffit of the beams.

3 EXPERIMENTAL RESULTS AND ANALYSIS

In Table 3, compressive strength, modulus of elasticity and flexural strength are given.

3.1 Failure modes and cracks pattern

Figure 4 shows the failure modes and crack pattern of the tested beams. In all beams, initial crack line appeared in the tension zone, which propagated vertically towards the neutral axis making I-shaped crack pattern finally. This I-shaped crack confirmed the flexure failure mode. The failure of beam B0 was abrupt without any prior failure sign. While, cracking in the beams in which PVA fibers were added was slowly propagated towards the neutral axis. The presence of PVA fibers protected the beams from splitting and the beams exhibited small amount of deflection. Therefore, it may be inferred that the addition of PVA fibers fully transformed the brittle failure mode into the ductile.

Table 3. Compression and flexural test results at 28 days.

Beam ID	CS (MPa)	EM (GPa)	MFL (kN)	DML (mm)	TSL (mm/mm)
B0	88.73	42.03	10.45	0.227	-1.310×10^{-3}
BP1*	94.43	39.62	11.68	0.262	-1.243×10^{-3}
BP2	96.2	38.76	12.10	0.289	-1.286×10^{-3}
BP3	95.08	38.85	10.88	0.282	-1.708×10^{-3}

* "B" represents beam; "P" represents PVA fibers while numeric value "1" represents 1% PVA fiber volume. Similar notations and nomenclature are followed for other beam Ids. And CS: Compressive Strenght, EM: Elastic Modulus, MFL: Maximum Flexural Load, DML: Deflection at maximum load, TSL: Tensile Strain at maximum load.

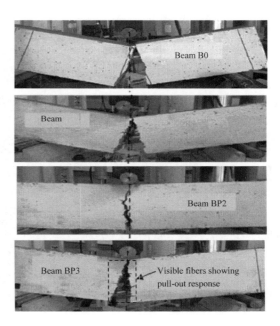

Figure 4. Failure modes and cracks pattern of HPFRC beams where, Beam B0 (without fibers), Beam BP1 (containing 1% PVA fibers), Beam BP2 (containing 2% PVA fibers), Beam BP3 (containing 3% PVA fibers).

3.2 Load vs. deflection response

The load vs. deflection response obtained from the flexural testing of the control and HPFRC beams are shown in Figure 5. The load-deflection response shows that the deflection-hardening response was achieved with 2 and 3% volume of PVA fibers. At 1% PVA fiber volume, toughness (area under the load-deflection curve) was higher than the control beam, but not significant when compared with the load-deflection response of HPFRC containing 2 and 3% PVA fiber volume. Clearly, the load vs. deflection of the control beam "B0" is stiffer than beams containing PVA fibers. Overall, loading capacity and deflection of beams was improved by adding PVA fibers.

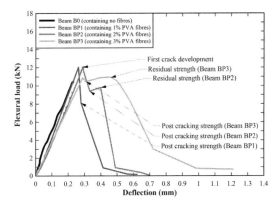

Figure 5. Experimentally obtained load vs. deflection patterns Deflection (mm).

Table 4. FEA results generated from ATENA 3D program.

Beam ID	FEA load (kN)	FEA deflection (mm)	Program calculated modulus of elasticity (GPa)	Tensile strain at maximum load (mm/mm)
B0	10.416	0.2199	43.0	-0.9450×10^{-3}
BP1	11.67	0.2599	38.0	-0.9260×10^{-3}
BP2	11.954	0.2799	36.0	-0.9635×10^{-3}
BP3	10.264	0.2750	32.5	-0.9912×10^{-3}

3.3 Finite Element Analysis vs. experimental results

The finite element analysis (FEA) results of all HPFRC beams with and without PVA fibers are presented in the Table 4, while FEA results are discussed in the next section.

3.4 Crack pattern and failure modes

A typical I-shaped failure pattern of HPFRC beam obtained through FEA is shown in Figure 2, which is similar to the experimental failure mode shown in Figure 4.

3.5 Load vs. deflection response

Figure 6 shows the FEA and experimental load vs. deflection behavior of all HPFRC beams, which is showing good coherence at all fiber volume.

4 CONCLUSIONS

The addition of 1 to 3% PVA fibers slightly increased the loading capacity of HPFRC beams as compare to the control beam from 5.71 to 31.64%. For structural application, use of 2 and 3% volume of PVA fiber seem suitable as deflection-hardening post peak load

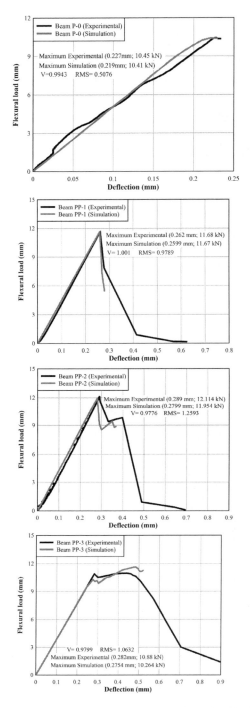

Figure 6. Comparison of experimental and simulated load vs. deflection for HPFRC beams.

vs. deflection behavior was observed in beams PB2 and PB3. The failure mode, crack pattern and load vs. deflection behavior obtained from experiment and simulation are well matched, which depicts that FE model can successfully predict the fracture mechanism in the beams containing PVA fibers.

ACKNOWLEDGMENTS

Authors wish to acknowledge the administration of Universiti Teknologi PETRONAS (UTP), Malaysia for financial support.

REFERENCES

[1] Q. Guan, P. Z., Zhang and X. Xie, Flexural behavior of steel fiber reinforced high-strength concrete beams, Research *J. Appl. Sci., Engg. Tech.* 6 (2013) 1–6.
[2] C. Jiang, C., K. Fan, F. Wu and F. D. Chen, Experimental study on the mechanical properties and microstructure of chopped basalt fibre reinforced concrete, Mater. Design, 58 (2014) 187–193.
[3] J. Zhou, J. Pan and C. Leung, Mechanical behavior of fiber reinforced engineered cementitious composites in uniaxial compression, *J. Mater. Civil Engg.* 27 (2015).
[4] G. Campione, Simplified flexural response of steel fiber-reinforced concrete beams, *J. Mater. Civil Engg* 20 (2008), 283–293.
[5] T. Ayub, N. Shafiq, M. F. Nuruddin and S. U. Khan, Flexural behaviour of high performance Basalt fibre reinforced concrete beams: 3D nonlinear finite element analysis, *IEEE Colloquium on Humanities, Sci. Engg.*, Penang, 2014.
[6] L. J. Vladimír Červenka, L. Jendele and J. Červenka, *ATENA program documentation Part 1 Theory*, March 14, 2012.
[7] S. U. Khan and T. Ayub, Modelling of the pre and post-cracking response of the PVA fibre reinforced concrete subjected to direct tension, *Accepted in Constr. Build. Mater.* (2016)

Displacement-based design for precast post-tensioned segmental columns with different aspect ratios

E. Nikbakht
Department of Civil and Environmental Engineering, Universiti Teknologi PETRONAS, Seri Iskandar, Perak, Malaysia

M.S. Mahzabin
Department of Civil Engineering, Lee Kong Chian Faculty of Engineering and Science, Universiti Tunku Abdul Rahman (UTAR), Malaysia

ABSTRACT: The precast bridge columns have shown increasing demands over the past few years due to their advantages compared to conventional bridge columns. However, there is still lack of confidence on their application in high seismicity zones. The necessity of a comprehensive design method which includes both nonlinear stiffness and self-centring capability of post-tensioned segmental columns is unavoidable. In this study, in order to increase energy dissipation capacity of precast segmental columns continuous mild steel bars are introduced in critical areas. Yielding displacement and the maximum applicable ductility level of post-tensioned segmental columns with various aspect ratio are evaluated. A displacement-based design procedure which allows increasing energy dissipation capacity and keeps the residual drift less than 1.0% is proposed.

1 INSTRUCTION

In recent years, precast segmental bridges have attracted the interest of researchers. The reason for this interest is due to their appropriate performance against severe earthquakes, where they remain functional and repairable with lower amount of cracks and damage. Moreover, these kinds of precast post-tensioned columns are capable of accelerating the construction period, thereby avoiding significant periods of traffic disruption during bridge construction. A number of researchers have investigated unbonded and bonded post-tensioned segmental concrete bridge piers experimentally in recent studies (Ou 2007; Motaref et al. 2014; Sideris et al. 2014; Bu et al. 2013).

Although many research have been conducted on precast segmental columns over the past few years, there is still limitation on application of the segmental columns in seismic regions mainly due to their unknown behavior against earthquake loading. In this study, a design procedure is developed to be implemented for post-tensioned segmental columns in different seismic zones and different displacement ductility demands. In force-based seismic design, the initial stiffness is used to estimate the period of the structure and also the seismic forces, which is applied to distribute the design seismic force between structural elements. However, for the reinforced concrete structures, it is found that the assumption of stiffness being independent of strength is invalid (Priestley 2007). This study, proposes a displacement-based design procedure for segmental columns with various mild steel and aspect ratios. To develop such design procedure, residual displacement, lateral peak displacement and equivalent viscous damping are evaluated.

2 NUMERICAL ANALYSES

For the analysis, 3D nonlinear finite element ANSYS software is applied. The details of the modelling of the samples, including geometry, properties, real constants of materials, modelling of bars and loading program, are reported in earlier works by the author (Nikbakht et al. 2014; Nikbakht et al. 2015a; Nikbakht et al. 2015b). In this study, segmental columns with four aspect ratios of 4.5, 6, 7.5 and 9 are investigated.

3 AXIAL LOADING

It was shown that high initial stress level of PT strands leads to stiffness reduction of self-centring segmental columns at large drift levels. Also, prestressing force level equivalent of 40% of yield stress of PT strands was proposed as a safe force level against severe earthquake loading (Nikbakht et al. 2014). Accordingly, in this study, 40% initial stress level and a normal axial load with coefficient ratio of 25% for the contribution of axial loading and post-tensioning forces (σ) in Eq. 1 is adapted.

$$\alpha(\%) = \frac{P_G + P_P}{f_c' A_g} \quad (1)$$

4 DESIGN PROCEDURE

Based on the numerical results of the parametric study conducted in earlier work by the author (Nikbakht et al. 2015), two empirical expressions for yielding displacement and the limit of displacement ductility level of the segmental columns with various aspect ratios are expressed in Equations 2 and 3, respectively. In Equation 2, yielding displacement of self-centring post-tensioned segmental columns is derived based on the hysteretic performance of the segmental columns with various continuous mild steel and aspect ratio of columns. Also, based on the limitation of 1% residual drift, the maximum allowable displacement ductility of the segmental columns is derived in Equation 3. The proposed displacement-based design procedure is presented as follows.

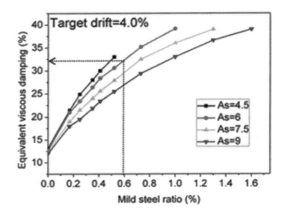

Figure 1. Equivalent viscous damping of various aspect ratio columns at 4.0% target drift.

Steps 1 and 2: Determining yielding displacement according to mild steel ratio ($\rho\%$) and aspect ratio (AR) from the Eq. 2 & finding displacement ductility level.

$$Dy = \frac{1.66 - 2.20(\rho) + 3.24(AR)}{1 - 0.14(\rho) - 0.04(AR)} \quad (2)$$

where; D_y is yielding displacement, ρ is mild steel ratio and AR is aspect ratio. D_y is dependent on variation of mild steel ratio and aspect ratio.

Step 3: Checking the specified displacement ductility level (μ) based on limitation of maximum 1% residual drift.

As mentioned before, the developed Eq. 3 indicates the limitation of displacement ductility of the post-tensioned segmental columns with different aspect ratios.

$$\mu_L = -0.0288 e^{(AR)/2.381} + 5.245 \quad (3)$$

Step 4: Determining equivalent viscous damping.

In this step, the equivalent viscous damping of the column can be specified according to the Fig.1, which is reported in earlier work by the author (Nikbakht et al. 2015a).

Step 5: Selection of a seismic demand and determining the effective period (T_{Aff}) of column based on displacement response spectra (Priestley 2007).

After specifying the equivalent viscous damping, the effective period of the columns can be determined according to Fig. 2.

Step 6: Calculation of base shear force (V_h).

In the final step, the base shear force can be calculated as shown in Eq. 4. (Priestley et al. 1996).

$$V_b = K_e \Delta_d = \left(\frac{4\pi^2 m_e}{T_e^2}\right)\Delta_d \quad (4)$$

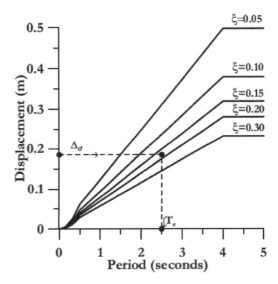

Figure 2. Effective periods for the columns with different equivalent viscous damping (Priestley 2007).

where; K_A is effective stiffness; Δ_d is design target displacement; and T_e is the effective period of structure determined from step 5.

Column design example:

Design a segmental column with aspect ratio (AR) of 6.0, diameter of 610 mm, height of 3660 mm, and mild steel ratio (ρ) of 0.6%. The column is pinned at the top and fixed at the bottom. The target ductility demand is 4.0. There is a gravity load of 500 kN.

From Eq. 2, the yielding displacement $D_y = 35$ mm; and $\Delta_d = 4 \times 36 = 140$ mm.

For checking displacement ductility (μ_r) Eq. 3 is used as follows:

$$\mu_L = -0.0288 e^{6/2.381} + 5.241 = 5.05 > 4 \quad OK.$$

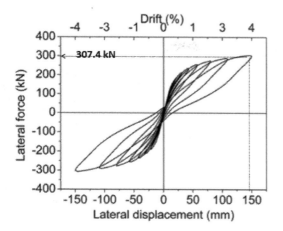

Figure 3. Load-deflection derived by finite element method.

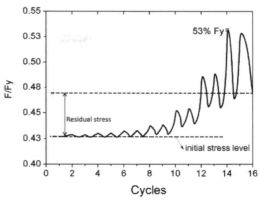

Figure 4. Maximum induced stress in central PT strands.

An equivalent viscous damping of 32% is obtained from Fig. 1. According to Fig. 2, the effective period is determined as: $T_{eff} = 2.95$. And finally, the base shear force (V_b) according to Eq. 4 is:

$$V_b = K_e \Delta_d = \left(\frac{4\pi^2 \times 500}{2.95^2}\right) \times 0.140 = 317.5\ kN$$

From finite element method analysis conducted in this study as shown in Fig. 3, the lateral force at 140 mm displacement (target displacement) is 307.4 kN, which is in good agreement with predicted force by displacement-based design procedure.

5 SAFETY OF DESIGN

As can be seen in Fig. 3, the designed example column based on the proposed displacement-based approach has less than 1% residual drift and possesses a target displacement ductility level of 4.0 which is desirable. Also, Figs. 4–5 show the maximum stress induced in central PT strands. It can be observed in Fig. 4 that the maximum stress level in PT strands is 53% of yielding stress of strands which is in adequate safety. The stress induced in PT strands in every drift levels is also shown in Fig. 5 in which the maximum of 53% stress level in PT strands occurs in 3.0% drift level.

6 CONCLUSIONS

In this study, numerical analyses were performed in order to evaluate the bridge columns seismic demands. Main design parameters such as mild steel and aspect ratio were investigated. Based on the results of parametric study, expressions for estimation of yield displacement and maximum allowed displacement ductility level at different aspect ratio columns were proposed. Finally, in order to achieve less than 1%

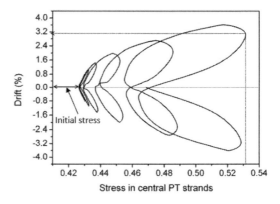

Figure 5. Stress induced in central PT strands at every drift levels.

residual drift and a target displacement ductility level, a displacement-based design procedure was developed for the post-tensioned self-centring segmental columns at different seismicity zones and different displacement ductility demand.

It was shown that the design procedure is capable to predict the ultimate load of the post-tensioned segmental columns. The results of load-deflection response obtained by the proposed design procedure showed good agreement with the numerical results.

REFERENCES

Bu, Z. Y., & Ou, Y. C. 2013. Simplified analytical pushover method for precast segmental concrete bridge columns. *Adv. Struct. Eng.* 16: 805–822.

Motaref, S., Saiidi, M. S. & Sanders, D. 2014. Shake table studies of energy-dissipating segmental bridge columns. *J. Bridge Eng.* 19: 186–199.

Nikbakht, E., Rashid, K., Siti. A.O and Hejazi., Farzad. 2014. A numerical study on seismic response of self-centering precast segmental columns at different post-tensioning forces. *Latin Am. J. solids struct.* 864–883.

Nikbakht, E., Rashid, K., Mohseni, I, & Hejazi, F. Evaluating seismic demands for segmental columns with low energy dissipation capacity. 2015a. *Earthquakes and Structures.* 8:1277–1297.

Nikbakht, E., Rashid, K., Hejazi, F., & Osman, S. A. 2015b. Application of shape memory alloy bars in self-centring precast segmental columns as seismic resistance. *Struct. Infrastruct. Eng.*, 11: 297–309.

Ou, Y. C. 2007. Precast segmental post-tensioned concrete bridge columns for seismic regions. Ph.D. thesis, Dept. of Civil, Structural and Environmental Engineering. State Univ. of New York, Buffalo, NY.

Priestley M. J. N., Calvi, M. C., & Kowalsky, M. J. 2007. Displacement-Based Seismic Design of Structures. IUSS Press, Pavia, 670 pp

Priestley, M.J.N., Seible, F., and Calvi, G.M. Seismic design and retrofit of bridges. 1996. Wiley, New York.

Sideris, P., Aref, A. J. & Filiatrault, A. 2014. Quasi-static cyclic testing of a large-scale hybrid sliding-rocking segmental column with slip-dominant joints. *J. Bridge Eng.* 10, 04014036-1-11.

Carbon emission analysis of double storey reinforced concrete house in Malaysia

M.S.A. Hamid
Universiti Teknologi PETRONAS, Seri Iskandar, Perak, Malaysia
Universiti Teknologi MARA Pulau Pinang, Malaysia

N. Shafiq, N.A. Zawawi & M.F. Nuruddin
Universiti Teknologi PETRONAS, Seri Iskandar, Perak, Malaysia

M.F. Khamidi
University of Reading Malaysia, Johor, Malaysia

M.S.H.M. Shaharmi
Universiti Teknologi MARA Pulau Pinang, Malaysia

ABSTRACT: In developing new sustainable design, low carbon emission design was an important parameter needs to be considering in the design. Furthermore, rapid development in the world was caused increasing of carbon dioxide emission that affected the world environment. Carbon emission analysis has been produced in this study specifically for two types of double story houses. This study was focused on structural elements of the building that were beam, column and slab. An environmental factor was the main parameter that was calculated to determine the amount of carbon emission. This paper showed a comparison of carbon emission by three different types of alternative concrete materials that were normal concrete, fly ash concrete and blast furnace slag concrete. The comparison was made to determine lowest carbon emission for the proposed double story houses. This study will develop new criteria in designing sustainable reinforced concrete structure in term of low carbon emission design.

1 INTRODUCTION

A mission to develop the world was caused a lot of environmental problems such as depletion of ozone layers, global warming, increasing of air temperature and pollution. The sustainable development approached was important to solve the climate change issues that focused on global warming. The global warming was among environmental issues that need to be concern off and it is due to amount of greenhouse gasses that produce from surrounding development activities (Taehoon et al. 2012). IPCC in 2007 was gaves the best estimate of increases in global temperatures over the next century, which ranged from 1.8 to 4.0 degree Celsius. Although this seems like a small change, it was much more rapid than any changes that have occurred in the past 10,000 years (Nordhaus 2008). Thus, it was obviously and acknowledged by the scientific community that temperatures would increase over the coming century up to 3.0 to 4.0 degree Celsius, which is certain to raise people's vulnerability to climatic catastrophes (Nordhaus 2008, Carter et al. 2006, IPCC 2007, Weitzmen 2007). In addition, there will be indirect damages, which are expected to be serious as well (Al-Amin & Filho 2014). Carbon dioxide was a major content of greenhouse gasses that produce from the activities (IPCC 2006, Quasem et al. 2015). Determination of all gasses that produce by concrete mix materials were impossible because of the coefficient factor that has been produced was only equivalent carbon dioxide (Taehoon et al. 2012).

The modeling projections indicate a continued and rapid increase in CO_2 emissions from industry, rising from 188.0 million tons per year in 2010 to 720.5 million tons per year in 2050 and 2024.4 million tons per year in 2105 under the baseline scenario of economic development. However, under the optimal scenario CO_2 emissions from industry will also increase but at a slower pace after 2040 until 2050 before starting to fall sharply from 720.5 million tons in 2040 to 326.1 million tons in 2055 and gradually thereafter (Quasem et al. 2015). The cumulative cost of climatic damage over the period 2010 to 2105 will amount to RM 40,128. 1 billion under the present climatic regime in Malaysia, it will fall to RM 5263.7 billion under the optimal regime. In addition, the government would have collected revenue from carbon taxes amounting to RM 9535.4 billion over the 95 years. The World

Bank Data in 2011 stated the average world carbon emission was 4.84 metric tons and Malaysia has 7.90 metric tons of carbon emission, fall in the rank 33 out of 192 countries which shows quite critical as Malaysia has been developing fast in these nearly centuries. In 2010, 7.99 metric tons has marked the maximum of carbon dioxide emission. Whereas, 1970 marked a minimum of 1.34 metric tons and average of 4.01 metric tons of carbon dioxide is produced throughout 40 years of data (The World Bank, 2011). In cement, carbon dioxide was produced by the carbonate oxidation in the cement clinker production process, which is the main constituent of cement and largest non-combustion sources of carbon dioxide of industrial manufacturing. In iron and steel production, carbon dioxide was generated in the use of coke ovens, blast furnaces and basic oxygen steel furnaces during the making process. The global production of crude steel has increase 3 percent in 2013 compared to 1.4 percent in 2012, on that, Malaysia's crude steel production has increased by 5 percent (WSA, 2014).

By all the reviews, it was found that the important of analyzing carbon emission that produce from construction materials because it gave lots of negative impact to the environment. This study was developed to enhance the reinforced concrete design of building in promoting sustainable design in term of low carbon emission from structural elements such as beam, column, slab and foundation. Furthermore, the development of this formulation in the design was also improved the usage of alternative materials in reinforced concrete structure.

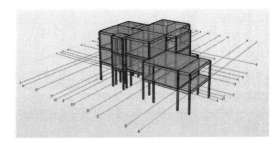

Figure 1. House Type A.

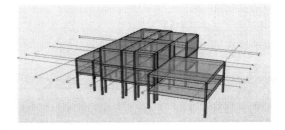

Figure 2. House Type B.

2 DETERMINATION OF CARBON EMISSION

The calculation procedure of the carbon emission amount was using multiplication of the material amount and coefficient of equivalent carbon emission of the material. The amounts of material that were steel reinforcement and concrete were determined by doing reinforced concrete design of the structure and the data was extracted from the results. Furthermore, the coefficient of equivalent carbon emission was referred to Inventory of Carbon and Energy Document Version 2.0. Other than that, a specific coefficient for reinforced concrete structure has been developed in this study to ensure the coefficient represent the reinforced concrete material. Eq. 1 and Eq. 2 shows the equations used to calculate the carbon emission of reinforced concrete structure for original type of reinforced concrete and alternative type of reinforced of concrete (Hamid & Shafiq 2015).

$$CO_{2,ori} = (Q_c + Q_s).EF_{RC} \quad (1)$$

$$CO_{2,alt} = (Q_{ca} + Q_s).EF_{RC} \quad (2)$$

where $CO_{2,ori}$ = carbon emission of normal concrete; $CO_{2,alt}$ = carbon emission of alternative material; Q_c = quantity of concrete; Q_s = quantity of steel; Q_{ca} = quantity of alternative concrete; EF_{RC} = carbon coefficient of reinforced concrete.

Two types of double story houses have been choose in this study as shown in Figure 1 and Figure 2. Furthermore, alternatives materials that have been choose in this study were Fly Ash and Blast Furnace Slag. Comparison was made in order to justify the most effective design that produce less amount of carbon emission.

3 CARBON EMISSION OF DOUBLE STORY REINFORCED CONCRETE HOUSE

Carbon emission was defined as an amount of equivalent greenhouse gasses that produce by the material used toward its implementation. In this study, the carbon emission was focused on reinforced concrete elements that mean the amount of carbon dioxide emission produced from concrete and steel elements. Furthermore, an alternative materials has been used in the design in order to differ the amount of carbon emission so that it may produce alternative approach of materials that can be compose in reinforced concrete structure. The comparison of the carbon emission based on two types of house which were Type A and Type B. A comparison was also made based on three different alternative materials which were normal concrete, fly ash concrete and blast furnace slag concrete. Analysis of most efficient materials took place in order to justify the low carbon emission design.

Table 1 shows the result for total carbon emission of double story house based on structural elements that consists of beam, column and slab for three different types of concrete materials. Table 2 shows the result on

Table 1. Total Carbon Emission of Double Story House Based on Structural Elements.

Elements	Type of Concrete Material	Floor	Material Weight (kg)		Carbon Emission Coefficient (kgCO$_2$/kg)		Total Carbon Emission (kgCO$_2$)	
			Type A	Type B	Type A	Type B	Type A	Type B
Beam	Normal	Foundation	–	–	–	–	–	–
		GF	48504	59582	0.238	0.266	11543	15821
		FF	29185	79307	0.280	0.201	8183	15909
		RF	20447	51977	0.205	0.169	4188	8804
	Fly Ash	Foundation	–	–	–	–	–	–
		GF	48504	59582	0.242	0.270	11737	16059
		FF	29185	79307	0.284	0.205	8300	16226
		RF	20447	51977	0.209	0.173	4270	9012
	Blast Furnace Slag	Foundation	–	–	–	–	–	–
		GF	48504	59582	0.206	0.234	9991	13914
		FF	29185	79307	0.248	0.169	7249	13371
		RF	20447	51977	0.173	0.137	3534	7104
Column	Normal	Foundation	21018	14827	0.215	0.241	4519	3569
		GF	34494	25325	0.200	0.267	6904	6763
		FF	24383	19854	0.179	0.276	4357	5473
		RF	–	–	–	–	–	–
	Fly Ash	Foundation	21018	14827	0.219	0.245	4603	3629
		GF	34494	25325	0.204	0.271	7042	6865
		FF	24383	19854	0.183	0.280	4455	5552
		RF	–	–	–	–	–	–
	Blast Furnace Slag	Foundation	21018	14827	0.183	0.209	3846	3095
		GF	34494	25325	0.168	0.235	5800	5953
		FF	24383	19854	0.147	0.244	3577	4838
		RF	–	–	–	–	–	–
Slab	Normal	Foundation	–	–	–	–	–	–
		GF	35599	34154	0.506	0.525	17998	17939
		FF	20037	21532	0.374	0.409	7502	8796
		RF	–	–	–	–	–	–
	Fly Ash	Foundation	–	–	–	–	–	–
		GF	35599	34154	0.510	0.529	18141	18076
		FF	20037	21532	0.379	0.413	7582	8882
		RF	–	–	–	–	–	–
	Blast Furnace Slag	Foundation	–	–	–	–	–	–
		GF	35599	34154	0.474	0.493	16859	16846
		FF	20037	21532	0.342	0.377	6861	8107
		RF	–	–	–	–	–	–

total carbon emission of double story house based on type of materials used in the concrete mixed. Furthermore, the coefficient of carbon emission was used in the analysis have been determine by Hamid & Shafiq (2015). This coefficient was a new development as a reinforced concrete carbon emission coefficient. Beside that, an analysis of the results was furthered into production of carbon emission per square meter in order to give guidelines in calculation of carbon emission for different type of houses based on area.

4 DISCUSSION

The total amount of carbon emission was referred to the reinforced concrete materials amount. Higher amount of materials will contribute higher amount of carbon emission. Therefore, the reinforced concrete design parameter should be determined accurately in order to maximize the design and produce low amount of carbon emission. Based on data obtained in Table 1, the highest elements that produce carbon emission were beam and slab elements. It is because, beam and

Table 2. Total Carbon Emission of Double Story House Based on Type.

Concrete Type	Total CO$_2$ Emission (kgCO$_2$)	Emission per Square Meter (kgCO$_2$/m^2)	Total CO$_2$ Emission (kgCO$_2$)	Emission per Square Meter (kgCO$_2$/m^2)
	Type A		Type B	
Normal	65194	309	83074	318
Fly Ash	66129	313	84301	323
Blast Furnace Slag	57717	274	73265	281

slab was produced high amount of reinforced concrete materials in the building. It was analyzed that beam element was contributes about 37 percent and slab element contributes about 39 percent of the total amount of carbon emission. While column element contribute about 24 percent only. Therefore, reduction of carbon emission produce by reinforced concrete elements is possible by reduces the elements size in order to reduce the amount of materials weight. It is because the parameter that will affect the total amount of carbon emission is material weight.

Other than that, column elements were produced less amount of carbon emission by comparing with slab elements. Structural elements at ground floor level were contributed highest amount of carbon emission of the building about 48.9 percent of the total amount of the carbon emission. Therefore, minimization of structural elements may reduce amount of carbon emission. Beside that, the amount of CO_2 that was embodied in concrete was primarily affected by the cement content in the mix design. It is important to control the cement content in the mix design in order to control production of carbon emission from the structural elements. In reinforced concrete elements, it is also important to control amount of steel reinforcement due to high amount of steel reinforcement will increased high coefficient that affected the carbon emission of the structural elements. Concrete performs well compared to other building materials but when it comes to sustainable development there is always an opportunity for improvement. As with any building product, concrete and its ingredients do require energy to produce that in turn produces carbon dioxide or CO_2. Furthermore, the amount of carbon emission based on type of concrete mix has been analyze with three different additional materials which were normal concrete, fly ash concrete and blast furnace slag concrete. Based on the analysis, it showed fly ash concrete produce slightly higher of carbon emission compared to normal concrete and blast furnace slag. Data in Table 2 shows the total carbon emission of double story house based on different concrete material as an alternative materials embedded in the concrete mix design. For both cases Type A and Type B, the total carbon emission pattern were same where fly ash concrete will produce the highest amount of carbon emission. Study shows that the coefficient of equivalent carbon emission that produce by fly ash concrete was higher than normal concrete and blast furnace slag concrete (Hammond & Jones 2011). This is because fly ash was a byproduct produced from burning coal in electrical power plants and through the burning produce process that contribute high content of carbon. Blast furnace slag is the waste byproduct of iron manufacture. After quenching and grinding, the blast furnace slag takes on much higher value as a supplementary cementations material for concrete. Blast furnace slag is used as a partial replacement for cement to impart added strength and durability to concrete. Furthermore, by using byproduct in the concrete mix as an alternative material, it will increase total amount of carbon emission and not promoting sustainable structure vision. Therefore, it is important to choose greener alternative material that may produce low amount of carbon emission.

The amount of carbon emission based on square meter of the building has been analyzed in order to ensure the production of carbon emission of the building per square meter. Table 2 shows the total amount of carbon emission that produced by reinforced concrete building elements. Analysis of carbon emission per meter square has been done in order to determine the carbon emission of the building. Based on data obtained in Table 2, the average carbon emission for normal reinforced concrete house was $313.5\,kgCO_2/m^2$. While for fly ash reinforced concrete and blast furnace slag reinforced concrete of carbon emission were $318\,kgCO_2/m^2$ and $277.5\,kgCO_2/m^2$. Therefore, increasing of the building area will increase the carbon emission values because it was reflected by the material amount used in the building. Optimization of the elements size will directly reduced amount of carbon emission because it may reduce the reinforced concrete materials amount. Other than that, better alternative materials is possible to be proposed as a composite material in the concrete that promote low carbon emission in order to reduce carbon emission of the building.

5 CONCLUSION

As a conclusion, the amount of carbon emission is reflected by two main parameters that are coefficient of equivalent carbon emission and amount of materials used in the building. Greener materials that are used as composite or alternative materials in concrete will lead to reduced amount of carbon emission of the building. The amount of greenhouse gasses (GHG) has been established to be expressed as equivalent carbon dioxide (CO_2) values because of 80 to 90 percent GHG is contain by CO_2. Therefore, in this study it was showed the average carbon emission of double story house per meter square was varies about $7\,kgCO_2/m^2$ to $10\,kgCO_2/m^2$. A comparison of three different alternative concrete materials was showed blast furnace reinforced concrete produced slightly lower carbon emission compared to fly ash reinforced concrete and normal reinforced concrete. Production and sources of alternative materials that used as composite materials in concrete will give different coefficient of equivalent carbon emission. Greener materials will produce much lower coefficient and it will produce less amount of carbon emission. Therefore, new alternative green material is recommended to use in reinforced concrete structure in order to promote sustainable design.

REFERENCES

Al-Amin, A.Q. & Leal Filho, W. (2014). A Return to Prioritizing Needs: Adaptation or Mitigation Alternatives? Prog. Dev. Stud. 14 (4): pg 359–371.

Carter, R.M. De Freitas, C. Goklany, I.M. Holland, D. Lindzen, R.S. Byatt, I. McKitrick, R. (2006). The Stern review: a dual critique. *World Econ.* 7 (4): pg165–232.

Hamid, M.S.A & Shafiq, N. (2015). Eco-efficiency Index Model for Reiforced Concrete Structural Design: Malaysia Case Study, *ARPN Journal of Engineering and Applied Sciences*, ISSN 1819–6608.

Hammond, G. & Jones, C. (2011). Inventory of Carbon and Energy Version 2.0, Sustainable Energy Research Team (SERT), Department of Mechanical Engineering, University of Bath, United Kingdom.

IPCC. (2006). IPCC Guidelines 2006a for National Greenhouse Gas Inventories: General Guidance and Reporting, vol. 1. viewed on 12.2.2011.

IPCC. (2006). IPCC Guidelines 2006b for National Greenhouse Gas Inventories. In: Energy, vol. 2. viewed on 6.4.2011.

IPCC. (2007). Climate Change 2007: The Physical Science Basis. Fourth Assessment Report of the IPCC. Cambridge University Press, Cambridge, United Kingdom.

Nordhaus W.D, (2008), A Question of Balance: Weighing The Options on Global Warming Policies, Yale University Press.

Quasem, A.A. Rajah, R. Santha, C. (2015). Prioritizing Climate Change Mitigation: An Assessment using Malaysia to Reduce Carbon Emissions in Future Environmental Science & Policy 50: pg 24–33.

Taehoon, H. Changyoon, J. Hyoseon, P. (2012). Integrated Model for Assessing the Cost and CO_2 emission (IMACC) for Sustainable Structural Design in Ready-mix Concrete, *Journal of Environmental Management* 1013: pg 1–8, Elsevier Ltd.

The World Bank. (2011). World Development Report: Conflict, Security and Development, Washington DC.

Weitzman, M.L. (2007). A Review of the Stern Review on The Economics of Climate Change, *J. Econ*. Lit. 45 (3): pg 703–724.

World Steel Association. (2014). World Steel in Figures 2014, ISBN 978-2-930069-73-9.

Addition of natural lime in incinerated sewage sludge ash concrete

Tan Yeong Yu, Doh Shu Ing & Chin Siew Choo
Faculty of Civil Engineering & Earth Resources, University Malaysia Pahang, Lebuhraya Tun Razak, Gambang, Kuantan, Pahang, Malaysia

ABSTRACT: The reason of carrying out this investigation was to study the ISSA concrete after adding eggshell powder to it. In this study, four different percentages of eggshell powder with respect to cement were added into the concrete mix of Grade 30. The percentage of the ISSA was fixed at 10% as the partial cement replacement with incinerated at 800°C. The materials used were Ordinary Portland Cement, coarse aggregates, fine aggregates, ISSA and eggshell powder. Based on the investigation, all the slump test results of eggshell ISSA concrete were at the desired range of workability from 65–80 mm. Eggshell concrete with 15% achieve the highest compressive strength at 38.96 MPa with was 21% higher than the normal plain concrete. Moreover, addition of ISSA and eggshell powder had shown a significant reduction of the rate of water absorption. Although the trend of the compressive strength decreased when the addition of eggshell powder increased up to 20%, the compressive strength of the specimens still able to be used as structural components.

Keywords: incinerated sewage sludge ash, eggshell powder, compressive strength, water absorption, partial cement replacement.

1 INTRODUCTION

The use of the sewage sludge ash in the concrete production has attracted an international interested because of the significant growth of the sewage sludge. Moreover, Malaysia is one of the largest egg consumers in the world amounting 36.5 million eggs daily (Chong, 2015). Most of the sewage sludge and eggshell are disposed as landfills without go through any pre-treatment (Tsai et al, 2007). The sewage sludge is not recommended to be buried in the soil or used as agricultural fertiliser due to its high heavy metal content (Kartini et al, 2015). Cu, Zn, Pb and Cd are the main elements that often reported to cause contamination of soil and food chain which lead health problem (Al-Musharafi et al, 2013). The number of landfilling increase significantly from 49 in 1988 to 161 in 2002, yet the number is still increasing annually (2010). Moreover, the increasing demand of using cement in construction had made the possible of searching alternatives cement replacement. Sewage sludge ash and eggshell powder are rich in SiO_2 and $CaCO_3$ respectively. High content of SiO_2 and $CaCO_3$ in sewage sludge and eggshell is useful for strength development of concrete. Thus, the combination of sewage sludge and eggshell powder can be possibly potential materials in concrete production.

2 MATERIALS

In this study, sewage sludge ash and eggshell powder were used as cement replacement in concrete production. The quantity of the sewage sludge used as 10%, while the eggshell powder were 5%, 10%, 15% and 20%. Weng et al (2003) recommended the optimum of sludge was 10%. The sewage sludge was collected from STP Megamall located at Kuantan, Pahang, while the eggshell powders were obtained from Eggtech Manufacturing Sdn Bhd located at Puncak Alam. The sewage sludge was incinerated process at 800°C, while the eggshell powder was dried at controlled temperature of 105°C for 24 hours. The incinerated sewage sludge and eggshell powder were ground into incinerated sewage sludge ash (ISSA) and eggshell powder. ISSA and eggshell powder then were sieved using 2.36 mm. and the particles that pass through 2.36 will be used as the partial cement replacement. The size of the concrete cubes used in this study was 100 mm × 100 mm × 100 mm. The other type of materials that used in this study was Ordinary Portland cement, granite, river sand and water. The Ordinary Portland Cement manufactured by YTL Orang Kuat, which is suitable for structural concreting. The granite and river sand are taken form Pancing, Pahang. Figure 1 and 2 shows the sample of the incinerated sewage sludge ash and eggshell powder that were used in this investigation.

Table 1 show the concrete mix design that was used in this investigation. The amount of coarse aggregate and fine aggregate were fixed for all the concrete cube specimens. The changes of the materials were amount of cement, ISSA and eggshell powder. Once the concrete cubes were marked for identification, it de-moulded after 24 hours from casting and placed in water tank for full water curing. The

Figure 1. Incinerated sewage sludge ash (ISSA).

Figure 2. Eggshell powder.

Table 1. Concrete mix design.

Material	Amount (kg/m^3)				
Cement	400	380	360	340	320
Coarse Aggregate	1120				
Fine Aggregate	650				
ISSA (%)	0	10	10	10	10
Eggshell powder (%)	0	5	10	15	20
W/C ratio	0.45				

testing specimens carried out at the age of 1, 7 and 28 days.

3 METHODOLOGY

In the investigation included slump test, X-ray diffraction test, compressive strength test and water absorption.

X-Ray Diffraction was conducted to determine the chemical composition of the sewage sludge and eggshell powder. It is relies on the dual wave or particle nature of X-rays to obtain information about the structure of crystalline materials. Through this test is able to identify the diffraction pattern of the compounds.

Slump test was conducted to determine the workability of concrete during the fresh stage. The result of the slump should be within 75 ± 25 mm in order to maintain the workability of the concrete. The slump test was conducted according to BS 1881: Part 102. In this investigation, 100 mm × 100 mm × 100 mm cube is selected to carry out this test. The test carried out at day 1, 7 and 28. There are three specimens for every type of concrete.

Compressive machine MATEST was used to perform this investigation. The compressive strength test was conducted according to BS 1881: Part 116.

Concrete cubes size 100 mm × 100 mm × 100 mm were selected for this test. The specimens were water cured for 28 days. After the day 28, the specimens need to be oven dry for 24 hours at 105°C. Three cubes for each type of steel fibers to obtain more accurate data. The specimens need to submerge in the water, and the bottom is support by a small aggregate, so that the whole specimens were contact with the water. The weights of the specimens were taken for 3 days.

4 MATERIAL PROPERTIES

4.1 X-Ray Diffraction

The function to carry out X-Ray Diffraction is to identify the chemical composition in the ISSA and eggshell powder. The diffraction pattern of the ISSA and eggshell powder were shown in Figure 3 and 4. From Figure 4, it is observed that the Major component of eggshell powder was calcite ($CaCO_3$) which is the carbonate mineral and the most stable calcium carbonate mineral. The amount of $CaCO_3$ was recommended not to exceed 20% which able to increase the compressive strength of the concrete with curing age (Okeyinka, O.M. and Oladeja, O.A., 2014). From Figure 3, the result from XRD pattern of 800°C of ISSA showed that the main components of ISSA were Silicon dioxide, phosphorous pentoxide and iron oxide. Jamshidi et al (2010) reported that the main phase of the sewage sludge is quartz (Jamshidi et al, 2010). The existing of silicon dioxide and aluminium oxide give pozzolanic properties to the specimens while calcium oxide provides strength to the concrete when under hydration process. Lin et al (2012) revealed that sewage sludge ash had the similar main components with ordinary Portland cement such as C_3S, C_2S, C_3A, C_4AF (Lin et al, 2012). ISSA and eggshell powder can be used as raw material in cement production/ to produce blended cement. Thus, there will be great reduction of using natural materials in producing cement such as clinker.

5 RESULTS AND DISCUSSION

5.1 Slump test

Slump test is to determine the workability of the concrete which is prior to casting during the fresh stage.

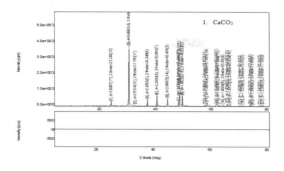

Figure 3. XTD pattern of eggshell powder.

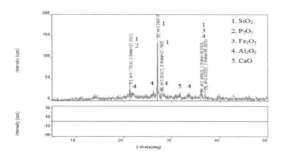

Figure 4. XRD pattern of ISSA at 800°C.

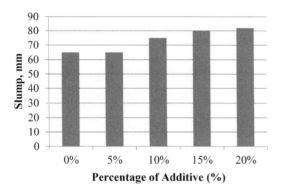

Figure 5. Result of slump.

In this investigation, slump test were carried out with varied percentage of eggshell powder namely 0%, 5%, 10% and 15% and 20%. The results of the slump test for different percentage of the eggshell powder was tabulated in Figure 5. From Figure 3 it can be observed that the slump for all specimens can be group by true slump ranging between 65–82 mm. Concrete CE4 showed the highest slump of 82 mm while both C and CE1 had the lowest slump value of 65 mm. The slump value for all proportions fell into acceptable range of slump which is 75 ± 25 mm. Addition of eggshell powder increase the workability of the concrete which it is still within the range of 75 ± 25 mm. Figure 2 show the summary result of the slump test of the specimens during the fresh stage. Thus eggshell powder is suitable to be used as filler in the concrete mix as it does not absorption water excessively.

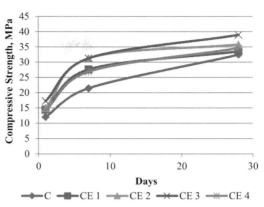

Figure 6. Compressive strength of the specimens.

5.2 Compressive strength

Compressive strength tests are one of the important properties for hardened concrete. In this investigation, concrete of different proportion of eggshell powder were tested at 1, 7 and 28 days. The result of the compressive strength test was tabulated in the Table 3. From Figure 6, CE 3 had the highest initial compressive strength at day 1 which is 17.25 MPa. Wang et al (2009) stated that the early strength of the concrete cube ISSA replacement is reduced because the hydration process is delayed (Wang et al, 2009). In this investigation, addition of the eggshell powder provides extra calcium oxide which is responsible of the strength development of concrete (Yerramal, 2014). As the result, the strength development with addition of eggshell powder is higher than the control specimen. At the day 7, CE 2 and CE 3 had already exceeded the design grade strength G30. At the day 28, CE 3 had the highest compressive strength that was 38.96 MPa with 15% of eggshell powder. The compressive strength of the specimens increases gradually when content of the eggshell powder increase. Although the compressive strength of the specimens decrease when the eggshell powder increased up to 20%, it was still can be used as the structural purpose which it able to gain the strength at 34.6 MPa still higher than the plain concrete. Based on the Figure 7, the optimum of using eggshell powder as partial cement replacement is between the ranges of 10–15%. The reduction of the cement in concrete production can be greatly reduced since the quantity of the ISSA and eggshell powder that used in this investigation was 30% with 20% higher compare to the control specimen in term of compressive strength.

5.3 Water absorption

Figure 8 shows the water absorption rate for every type of the specimens. Based on Figure 8, it showed that the CE 3 possessed the lowest rate of water absorption compare to other specimens with 2.8%. The rate of the water absorption decrease gradually when increase the percentage of the eggshell powder. Moreover, the control specimen with 0% of the ISSA and eggshell

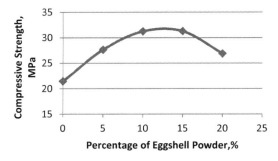

Figure 7. Compressive strength vs eggshell powder content.

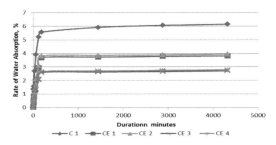

Figure 8. Rate of water absorption.

powder had the highest rate of water absorption due to high amount of permeable voids. Thus, eggshell powder can be used as potential materials to fill up the existing voids in the concrete (Yerramal, 2014). Since the rate of the water absorption of the concrete had been greatly reduced and made the concrete more durable compare to the normal plain concrete. The rate of the water absorption for 5% and 10% were nearly the same at 4% while 15% and 20% showed the lowest rate of water absorption at 2.8 %.

6 CONCLUSION

The ISSA and eggshell powder are the potential materials to use as the partial cement replacement. Since both of them contain silicon dioxide and calcium carbonate respectively, blended cement yet environment friendly cement can be used to partially replace cement in concrete. The emission of the carbon dioxide can be reduced dramatically since there is no incineration process for eggshell powder and only 800°C for sewage sludge to use partial cement replacement up to 30%.

The water cement ratio of 0.45 can produce desirable workability which is suitable for the most common concrete casting. The compressive strength of specimens exceeds the design strength. The highest compressive strength was the specimen was CE 3 with 38.96 MPa at the day 28 days. Although the trend of the compressive strength of the specimens decreases when the eggshell powder reached 20%, it was higher than the control plain concrete. Addition of the eggshell powder of the ISSA concrete can reduce the rate of water absorption of the concrete. This is because addition of eggshell powder able to fill up the existing voids of the plain concrete and improve the performance of the concrete.

ACKNOWLEDGEMENT

The authors would like to express their gratitude to the technicians of Concrete Laboratory, Faculty of Civil Engineering & Earth Resources, University Malaysia Pahang and Concrete, Staff of Indah Water Konsortium, Pahang and Eggtech Manufacturing Sdn. Bhd.

REFERENCES

Al-Musharafi, S.K., Mahmoud, I.Y. and Al-Bahry, S.N. 2013. Heavy metal pollution from treated sewage sludge effluent. *APCBEE Procedia*. 5: 344–348.

BS 1881: Part 102. *Method for determination of slump*. London: British Standard Institution.

BS 1881: Part 116. *Methof for determination of compressive strength of concrete cube*. London: British Standard Institution.

Chong, C. 2015. Electronic sources: The Edge Financial Daily. http://www.theedgemarkets.com/my/article/teo-seng-malaysia%E2%80%99s-edd-cosumption-grow-3-5 (16 November 2015)

Jamshidi, A., Jamshidi, M., Mehrdadi, N. and Shasavandi, A. 2010. Mechanical performance of concrete with partial replacement of sand by sewage sludge ash. *Journal of Construction and Building Materials*. 3–8.

Kartini, K., Dahlia Iema, A.M., Siti Quraisyah, A.A., Anthony, A.D., Nuraini, T. and Siti Rahimah, R. 2015. Incinerated domestic waste sludge powder as sustainable replacement material for concrete. *Science and Technology*. 23(2):193–205.

Lin, Y. Zhou, S., Li, F. and Lin, Y. 2012. Utilization of municipal sewage sludge as additives for production of eco-cement. *Journal of Hazardous Materials*. 214: 457–465.

Okeyinka, O.M. and Oladeja, O.A. 2014. The influence of calcium carbonate as an admixture on the properties of wood ash cement concrete. *International Journal of Emerging Technology and Advanced Engineering*. 4(12):432–437

Tsai, W.T., Hsien, K.T., Hsu, H.C., Lin, C.M., Keng, Y.L. and Chiu, H.C. 2007. Utilization of fround eggshell powder waste as absorbent for removal of dyes from aqueous solution. *Bioresource Technology*. 99(6):1–6.

Wang, C., Xi, J.Y., Hu, H.Y., Arhami, M., Polidori, A., Delfino, R. and Murray, V. 2009. The effects of different types of nanao-silicon dioxide additives on the porperties of sludge ash mortar. *Journal of the Air & Waste Management Association*. 59(4):88–97.

Weng, C.H., Lin, D.F and Chiang, P.C. 2003. Ulitilization of sludge as brick materials. *Advance in Environmetal Research*. 7: 679–685.

Yerramal, A. 2014. Properties of concrete with eggshell powder as cement replacement. *The Indian Concrete Journal*. 94–102.

Zaini, S. 2010. Municipal solid waste management in Malaysia: solution for sustainable waste management. *Journal of Applied Science in Environment Sanitation*. 6(1).

Properties and structural behavior of sawdust interlocking bricks

Bashar S. Mohammed, Muhammad Aswin & Vethamoorty
Universiti Teknologi PETRONAS, Seri Iskandar, Perak, Malaysia

ABSTRACT: Bricks are broadly utilized in construction and building materials. Conventional bricks are created by clay with high temperature kiln firing. Topsoil removal and continuous excavation of clay causes substantial decimation and sustainability of original resources, also burning can evoke environmental matter. To address such problem, this paper presents possibility of interlocking bricks produced using supplementary cementitious materials and sawdust as byproduct, in terms properties and structural behavior. Sawdust combined with cementitious (fly ash and silica fume) can produce the sawdust interlocking bricks utilized as alternative building materials beside conventional bricks. Test results showed that mechanical properties of sawdust interlocking bricks meet ASTM requirements. In addition, structural behavior of prisms tested by using axial compression load has exhibited good performance in range pre-peak and post-peak region. Moreover, stress-strain behavior and failure modes analyzed using VecTor2 FEM is in good agreement with experimental results.

1 INTRODUCTION

Bricks are one of the oldest construction materials, dated back to thousands of years ago, hand-moulded and sun-dried clay bricks were used in Egypt [1]. Nowadays bricks continue to be most popular and leading construction materials, but the use of clay bricks also pose serious problem. Shakir et al. [2] stated that removal of topsoil and continuous excavation of clay causes substantial decimation, scarcity as well as sustainability of original resources. Accumulation of the unmanaged wastes in landfill resulted can evoke the adverse environmental matter. In addition, Goyer [3] reported that many people across several countries uses wood, grass, coal, e.t.c, to ignite the fire used for burning the clay bricks in other to gain strength. Reddy and Jagdish [4] also mentioned that the conventional bricks are produced under high temperature kiln firing and consume significant amount of energy, as well as releases large quantity of greenhouse gases which significantly affect the environment negatively through global warming. Lippiatt [5] justified that firing clay bricks, on average, have an embodied energy of approximately 2.0 kWh and release about 0.41 kg of carbon dioxide per brick.

Related to producing bricks in many countries, a lot of people have no concern on the environmental and sustainability of resources, therefore research on the possible solutions need to be carried out. Zhang [6] explained that waste materials or byproducts can be utilized to produce bricks, for sustainable resources and environmental protection. Many variety of waste materials or byproducts have been studied, involving silica fume, slags, mine tailings, fly ash, construction and demolition waste, wood sawdust, limestone powder, cotton waste, paper production residue, crumb rubber, cigarette butts, craft pulp production residue, waste tea, rice husk ash, cement kiln dust, etc.

Kinuthia and Oti [7] reported the benefits of using the ground granulated blast furnace slag (GGBS) in non-fired clay building material development which reduces the emissions of greenhouse gases and improves durability. Shakir et al. [2] produced bricks with each of the following; fly ash, quarry dust, and billet scale by non conventional way (no require high pressure or high temperature kiln firing), and found out mechanical properties and durability of bricks were promising. Turgut and Algin [8] reported the potential use of limestone powder wastes (LPW) and wood sawdust wastes (WSW) combination to produce a low cost and lightweight composite as a construction material. The test results exhibited good effect of high-composition replacement of WSW by LPW, such as: reduces the unit weight, indicates high energy absorption, introduces smother surface concrete bricks, and shows no sudden brittle fracture even beyond the maximum loads. For its application, this composite can be used also for wooden board substitute, concrete blocks, ceiling panels, sound barrier panels, etc.

Commonly, to construct the brick wall, usual way is to utilize mortar as binder between bricks and at the top surface of bricks, this method needs skilled labor and its time consuming. To overcome these issues, interlocking bricks can be utilized as it requires less time and light labor. Watile et al. [9] stated that utilization of interlocking bricks has achieved rapid popularity in many countries for sustainable housing. Up to present, interlocking bricks is the latest advancement in brick wall construction. These bricks are connected and are joint by their interlocking actions without need of mortar at their top surface, but grouting the two vertical holes with mortar is required. This way is quite easy and simple. In addition, interlocking bricks can form a stable wall and reduce time of construction. Moreover, interlocking bricks can be made up by using non-clay, so it can protect the sustainability of clay resources. In order to reduce the adverse environmental impacts, interlocking bricks also can be produced by using recycled materials or waste materials or byproduct, and without firing.

Mohammed [10] studied mechanical properties of soil cement interlocking brick (SCI). The research revealed a good quality for the SCI with all its mechanical properties accomplished the requirements. Watile et al. [9] investigated the mechanical properties of interlocking bricks, in which the percentages of fly ash, stone dust, and sand with different mix proportion were varied and glass fiber reinforced utilized as strengthening. The test results showed that interlocking bricks were durable in aggressive environments and have adequate strength.

Watile et al. [9] also stated that researchers always face challenge in producing lightweight and low cost interlocking bricks with increased performance. In addition, the interlocking brick if produced has to maintain the sustainability of virgin resources and should not give the adverse effect to the environment.

In order to attain the aim, this paper presents effect of use of supplementary cementing materials and sawdust as byproduct materials on properties and structural behavior of interlocking bricks. Twenty four different mix designs were prepared and each mixture was created by varying the percentage of each material used. In this research, fly ash and silica fume were used as cement replacement materials, whereas sawdust was utilized as filler and as sand replacement material. By conducting the mechanical properties test based on ASTM provision, it will be investigated the possibility of using sawdust interlocking bricks produced using supplementary cementitious materials and sawdust as alternative building materials aside from the conventional bricks. To support this requirement, also need to studied structural behavior of bricks prisms constructed by the sawdust interlocking bricks by applying axial compression load. Moreover, based on the simulation results by VecTor2 FEM, all models have shown almost similar behavior against experimental results.

2 EXPERIMENTAL PROGRAM

2.1 Materials

Materials used for this research consist of ordinary Portland cement (OPC), fine aggregate (sand), fly ash, silica fume, and sawdust. The basic characterization of these materials was carried out in accordance with American Society for Testing and Materials (ASTM) standard. All materials are available locally in Malaysia. OPC manufactured by Tasek Corporation SDN Berhad, conforming to Malaysian Standard MS 522 which is loosely based on European Union standard, EN 196 was used in this research activity. Sand consists of fine grains or particles of mineral and rock fragments. The sand was from a natural or river sand with specific gravity of 2.63. Class F fly ash which conforms to the specifications of ASTM C618 and ASTM C311 and having specific gravity of fly ash is 2.76 was used. Silica fume is a byproduct of the smelting process in the silicon and ferrosilicon industry, conforming to ASTM C1240 and having a specific gravity of 2.24 was used. Sawdust used in the research is obtained from nearby sawmill with a specific gravity of 1.41 and a byproduct of cutting, grinding, drilling, sanding or otherwise pulverizing wood with a saw or other tools.

Table 1. Mix proportion 1–6.

No.	Cementitious			Fine Aggregate	
	Cement	Fly Ash	Silica Fume	Sand	Sawdust
Mix-1	0.8	0.15	0.05	0.95	0.5
Mix-2	0.8	0.15	0.05	0.9	0.1
Mix-3	0.8	0.15	0.05	0.8	0.2
Mix-4	0.8	0.15	0.05	0.7	0.3
Mix-5	0.8	0.15	0.05	0.5	0.5
Mix-6	0.8	0.15	0.05	0	1

Table 2. Mix proportion 7–12.

No.	Cementitious			Fine Aggregate	
	Cement	Fly Ash	Silica Fume	Sand	Sawdust
Mix-7	0.65	0.3	0.05	0.95	0.5
Mix-8	0.65	0.3	0.05	0.9	0.1
Mix-9	0.65	0.3	0.05	0.8	0.2
Mix-10	0.65	0.3	0.05	0.7	0.3
Mix-11	0.65	0.3	0.05	0.5	0.5
Mix-12	0.65	0.3	0.05	0	1

2.2 Mix design

In order to achieve the maximum compression strength of the sawdust interlocking bricks, twenty four different mix designs were derived as shown in Table 1–4. The mixtures were prepared by varying the percentage of each material used such as cement, fly ash, silica fume, sand and sawdust. In this research, the cementitious (fly ash and silica fume) were used as cement replacement materials, and sawdust was utilized as filler or as sand replacement material. Cement and cementitious will be combined with sand and sawdust to produce the sawdust interlocking bricks. There is no clay and without firing in producing the sawdust interlocking bricks. Based on compressive test, the maximum compression strength was achieved by using Mix-1.

Table 3. Mix proportion 13–18.

No.	Cement	Cementitious		Fine Aggregate	
		Fly Ash	Silica Fume	Sand	Sawdust
Mix-13	0.5	0.45	0.05	0.95	0.5
Mix-14	0.5	0.45	0.05	0.9	0.1
Mix-15	0.5	0.45	0.05	0.8	0.2
Mix-16	0.5	0.45	0.05	0.7	0.3
Mix-17	0.5	0.45	0.05	0.5	0.5
Mix-18	0.5	0.45	0.05	0	1

Table 4. Mix proportion 19–24.

No.	Cement	Cementitious		Fine Aggregate	
		Fly Ash	Silica Fume	Sand	Sawdust
Mix-19	0.35	0.6	0.05	0.95	0.5
Mix-20	0.35	0.6	0.05	0.9	0.1
Mix-21	0.35	0.6	0.05	0.8	0.2
Mix-22	0.35	0.6	0.05	0.7	0.3
Mix-23	0.35	0.6	0.05	0.5	0.5
Mix-24	0.35	0.6	0.05	0	1

2.3 Standard test method

The compressive strength test for the brick was carried out according to ASTM: C67-03a and that is representable by applying the bricks in their dry curing condition to the test. The compressive load was applied in the direction of brick width.

For the modulus of rupture, the bricks were tested refer to ASTM: C67-03a as well, where the bricks were in a whole dry, full size-unit and five brick specimens were tested. The brick is supported by using 2 steel rollers, and 1 point load was applied in the direction of brick depth. The average of the modulus of rupture determinations must be nearest to 1 Psi or 0.00689 MPa.

For the initial rate of absorption, the test was conducted also based on ASTM: C67-03a. The bricks were dried in a ventilated room at constant temperature. The bricks were measured to the nearest 0.05 inch length and width of flaws surface of the brick where water was kept in contact to brick surface. The weight of each dry and wet brick was recorded. For the water absorption, the test was also conducted according to ASTM: C67-03a, where it is divided into 5 hour and 24 hour submersion test. The initial weight of each brick was measured and then put them in water tank. The bricks were removed from bath and after 5 minutes each brick is wiped off using a dry cloth and weigh them again.

Meantime, the masonry prism test was conducted based on ASTM: C1314, where the method covers procedures for constructing masonry prisms and testing procedures. In this research, to investigate the structural behavior of sawdust interlocking bricks, it will be conducted the masonry prism test by applying the axial compression load on the top of prisms as shown in Figure 1. There are three different set prisms used, namely Prism 3; Prism 5; Prism 7. Prism 3, 5 and 7 consists of 3, 5 and 7 pieces of sawdust interlocking bricks, respectively, that arranged vertically, so total prism tested is 9 pieces. All prism holes were grouted using mortar without any reinforcement and the top level was screed off. The prisms were cured under room temperature at laboratory. After 2 days of curing white paint was applied on both sides surface area, this is to help monitoring on crack pattern during testing. After painting works, the prisms were marked to attach the strain gages.

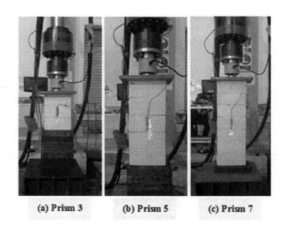

(a) Prism 3   (b) Prism 5   (c) Prism 7

Figure 1. Experimental setup for prisms.

3 RESULTS AND DISCUSSION

3.1 Compressive strength

The compressive strength of sawdust interlocking bricks recorded that the compressive stress in the range of 0.05 N/mm^2 to 10.43 N/mm^2. The maximum compressive strength was achieved by Mix-1 and the minimum one was gained by Mix-18. It was clear that the compressive strength of bricks can be influenced by the cementitious and sawdust content. Increase in sawdust percentage will reduce the compressive strength of interlocking bricks. In addition enhancement of cementitious percentage will increase the compressive strength of interlocking bricks. Based on the test results, the optimum compressive strength (10.43 N/mm^2) attained by Mix-1 was referred for another tests in this research. According to ASTM C67, in general, the maximum compressive strength of conventional bricks is 10.3 N/mm^2. This value is lower than strength achieved by Mix-1, therefore the sawdust interlocking bricks developed in this study can be recommended as alternative brick aside from the conventional one.

3.2 Modulus of rapture

The modulus of rapture was obtained by conducting the flexural test of bricks. The test was performed on five bricks for 7 curing days and Mix-1 was used for this requirement. The test result is ranging from 0.00107 MPa to 0.02128 MPa. ASTM C67 denotes that the modulus of rupture should be determinate to the nearest 1 Psi or 0.00689 MPa, therefore the results obtained show that it is satisfactorily acceptable. Some of specimens reached modulus of rupture over than ASTM provision.

3.3 Initial rate of absorption

The initial rate of absorption for sawdust interlocking brick records almost same for all the five bricks with 7 days of curing tested. Initial rate of absorption is defined as the number of water absorbed in grams and in one minute over the immersed area of brick. According to ASTM C67 acceptable values are range from 10 to 30 grams. Since the test results show 6.55 g/cm^2/min, it fulfills the standard requirement. Adding water to the fresh mix will create pores in the hardened matrix due to bleeding and water evaporation thereby increases the water absorption. The dry brick with an initial rate of absorption more than 30 g/cm^2/min, should be wetted before laying. If the brick is too dry, it can absorb too much water from the applied mortar.

3.4 Water absorption test

Based on the 5-hour of water absorption test, the results are in accordance with the ASTM C67 requirements. The code requires average cold water absorption of less than 8%. Whereas for the 24-hour of absorption test, the results do not meet the code requirements, where its value exceeds and records up to 10% of absorption.

3.5 Masonry prism test and finite element modeling (FEM)

Stress-strain is a unique relationship that represents how the material behaves when subjected to load. Each material will have different stress-strain relationship. According to the curve, it can be determined the characteristic of related materials. The most three common types of materials are brittle, elastic, and ductile material. To investigate the structural behavior of sawdust interlocking bricks, all the masonry prisms were tested by applying the axial compression load on the top of prisms. According to the test results, three graphs of each prism are tabulated as shown in Figure 2–4.

Referring to Figure 2, it can be shown that Prism 3 records higher max stress (namely average stress is 3.45 MPa) compared to the other two prisms, in which for Prism 5; the average max stress is 3.02 MPa, and for Prism 7; the average max stress is 2.21 MPa. As the height of prisms increases from the 3 bricks

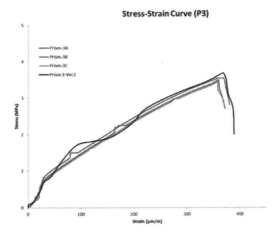

Figure 2. Stress-strain relationship for axial compression load of Prism 3 (P3).

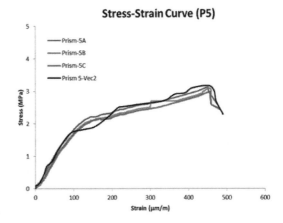

Figure 3. Stress-strain relationship for axial compression load of Prism 5 (P5).

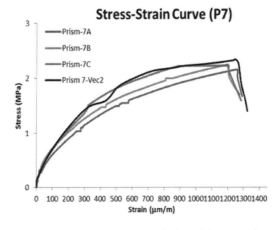

Figure 4. Stress-strain relationship for axial compression load of Prism 7 (P7).

prisms to the 7 bricks prisms, it clearly shows an increment in strain value and decrement in stress value. It can be exhibited that the 7 bricks prisms records higher strain (average max and rupture strain are 1217 and 1283 μm/m, respectively) compared to the other two prisms, in which for the 5 bricks prism (average max and rupture strain are 448 and 479 μm/m, respectively), and for the 3 bricks prism (average max and rupture strain are 354 and 374 μm/m, respectively).

Obviously, the height of prisms provides effect to stress and strain value of the prism. Both values were attributed the compressibility of bricks, in which each brick has same content of sawdust. Sawdust is organic material and has higher compressibility compared to another ingredient materials of brick, such as clay, sand, etc. So if the load is applied on it, then it will compressible and able to experience higher elongation (stretching) compared to other materials of interlocking bricks. Even the sawdust in brick was covered or wrapped by cement and cementitious, the presence of sawdust in brick maintains providing the void in the bricks, so that the compressibility of bricks will give effect against the stress and strain value of bricks. For the 7 prisms, accumulative amount of void is higher compared to other prisms. That is why the 7 prisms have the smallest average stress value and have the greatest average strain value compared to other prisms. The sawdust interlocking bricks exhibit good performance, after beyond the elastic point, the stress still increases until peak condition and experiences the compression softening after peak-post, in addition to, they have quite high strain capacity up to failure. So the sawdust interlocking bricks are in accordance with for buildings located in earthquake region.

According to Figure 2–4, the modulus of elasticity was obtained, which relies in range of 13 to 14 GPa. The obtained modulus of elasticity is higher than normal bricks. Refer to Totoevl and Nichols [11], the elastic modulus of common bricks should be in between 10 to 12 GPa. In this research, it is attained higher elastic modulus owing to use of cementitious materials such as fly ash and silica fume in the mixture of bricks. The particle size of cementitious is smaller than cement. Cementitious can fill in the void in cement paste and increase the bonding strength between matrix and aggregate.

The finite element modeling (FEM) analysis in this study was carried out using VecTor2 [12]. Different amounts of element libraries and models have been adjusted with the characteristic of each test specimens. For the brick prism, the element libraries cover 300 elements which consist of 300 rectangular elements and 341 nodes. Masonry models consists of several types: compression pre-peak (Hognestad: Parabola), compression post-peak (Montoya 2003), compression softening (Vecchio: 1992), tension softening (linear), dilatation (constant Poisson ratio), cracking criterion (uniaxial), crack stress calculation (basic: DSFM/ MCFT), crack width check (Agg/2.5 max crack width), crack slip calculation (Maekawa: monotonic), hysteretic response (Palermo

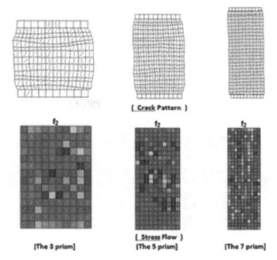

Figure 5. Crack pattern and stress flow of prisms by VecTor2.

with decay: 2002). Lastly, concrete bond (Eligehausen) was adopted as bond model. The materials properties are: compressive strength is 10.4 MPa, tensile concrete strength is 1.1 MPa, Elastic Modulus is 14 GPa, and Possion ratio is 0.25.

Based on the simulation results by VecTor2 FEM, all models have shown the similar behavior with experimental results as shown in Figure 2–4 and in Figure 5. Due to the axial compression load, all prisms experience high stress flow in longitudinal direction. Stress concentration occurs at the top and bottom of prisms. Based on FEM analysis results, it can be exhibited that owing to the sawdust interlocking bricks have adequate compressibility, the prisms experience bulging transversely at the middle height of prisms due to the axial compression load. Finally, if the applying load is increased, all prisms will failure. The stress-strain behavior and failure modes analyzed using VecTor2 are in good agreement with the experimental test results.

4 CONCLUSION

1. Sawdust as sand replacement material combined with cementitious (fly ash and silica fume) as cement replacement materials can produce the sawdust interlocking bricks which can be utilized as alternative building materials aside from the conventional bricks. The mechanical properties of sawdust interlocking bricks meet the ASTM requirements.

2. Use of cementitious in sawdust interlocking bricks can increase the strength capacity, meantime the use of sawdust can improve the compressibility so can perform large elongation or strain up to failure. Structural behavior of prism tested by using axial compression load has exhibited good performance in range pre-peak and post-peak condition.

3. Stress-strain behavior and failure modes analyzed using VecTor2 FEM is in good agreement with the experimental test results.

REFERENCES

Lynch, G.C.J., 1994, Bricks: Properties and Classifications, Structural Survey, Vol. 13, No. 4, 15–20.

Shakir, A.A., Nagathan, S., Mustapha, K.N., 2012, Properties of bricks made using fly ash, quarry dust and billet scale, Construction and building materials.

Goyer, K., 2006, Kiln and brick making, on: www.aprovecho.org/lab/rad/rl/stove-design/doc/14/raw.

Reddy, B.V.V., Jagdish, K.S, 2003, Embodied Energy of Common and alternative building materials and technologies, Journal of Energy and Buildings, Vol.3, 129–137.

Lippiatt, B.C., 2000, Building for Environmental and Economic Sustainability, Technical Manual and User Guide, National Institute of Standards and Technology, USA.

Zhang, L., 2013, Production of bricks from waste materials- A review, Construction and Building Materials, Vol. 47, 643–655.

Kinuthia, J.M, Oti, J.E., 2012, Designed non-fired clay mixes for sustainable and low carbon use, Applied Clay Science, Vol. 59, No. 60, 131–139.

Turgut, P., Algin, H.M., 2007, Limestone dust and wood sawdust as brick material, Building and Environment, Vol. 42, 3399–3403.

Watile, R. K., Deshmukh, S. K., Muley, H.C., 2014, Interlocking brick for sustainable housing development, International journal of science, spirituality, business and Technology, Vol. 2, No. 2.

Mohammed, A.A., 2012, Assessment the Mechanical Properties of Soil Cement Interlocking (SCI) Bricks: A Case Study in Malaysia, International Journal of Advances in Applied Sciences, Vol. 1, No. 2, 77–84.

Totoevl, Y.Z., Nichols, J.M., 1997, A Comparative Experimental Study of the Modulus of Elasticity of Bricks and Masonry, 11th International brick block masonry conference, Tongji University, Shanghai, China.

Vecchio, F.J., Wong, P.S., 2002, VecTor2 & Formworks User's Manual, University of Toronto, Canada.

Development of nano silica modified solid rubbercrete bricks

N. Mahamood, B.S. Mohammed, N. Shafiq & S.M.B. Eisa
Department of Civil and Environmental Engineering, Universiti Teknologi PETRONAS, Seri Iskandar, Perak, Malaysia

ABSTRACT: Poor management of scrap tires in Malaysia have caused the accumulation of non-biodegradable materials in landfills and dumping sites, which contributes to the health issues. This research intends to exploit the scrap tires to produce rubbercrete brick. Utilizing recycled materials in construction industry helps to reduce the environmental impacts. However, inclusion of crumb rubber in Portland cement product lead to reduction in the strength due to the hydrophobic nature of the crumb rubber which repels water and entrap air on its surface. Therefore, in this research, nano silica has been introduced to the rubbercrete to modify its microstructure by densifying the interfacial transition zone and refining the pore system which consequently enhances the compressive strength. In this research, nano silica modified rubbercrete solid brick has been developed with a compressive strength of 15 MPa. The newly developed nano silica modified rubbercrete solid brick can be used for non-load-bearing construction application.

1 INTRODUCTION

In Malaysia, the absence of specific laws or obligations practiced to treat used tires has led to the poor management of these waste materials The result is a large amount of scrap tires, approximately 8.2 million, produced annually (Sandra Kumar a/L Thiruvangodan 2006). The accumulation of these scrap tires causes the occupation of large areas in landfills and dumpsites, giving rise to land pollution. Furthermore, these landfills, if poorly managed, will become a potential breeding ground for mosquitoes, rats and rodents, hence leading to health issues.

According to Ohio Environmental Protection Agency, scrap tires is the unwanted or discarded tires that cease to serve their original purpose (Kasich and Butler 2016). Since scrap tires are classified as non-biodegradable material, the easiest way to dispose them is by burning which inevitably causes air pollution as the activity releases huge amount of smoke to the surrounding. This has raised concerns in environment and health issues in a global scale regarding the disposal of scrap tires. For example, the government of Nepal has banned the uncontrolled burning activities as heavy metals such as iron, zinc, lead, cadmium and chromium that are released by the tires endangers the surrounding living organisms (Shakya 2006). Even in South Australia, the disposal of whole tires in the landfill was banned as its bulky shape uses up huge areas and encourages the breeding of mosquitoes and insects. Furthermore, if the tires are no longer suitable for recycling, it is shredded or chipped into pieces of 250 mm or smaller before being disposed in landfills (Environment Protection Authority 2010). The European Union also took the initiative to overcome the problem related to the disposal of the scrap tires by prohibiting the disposal of whole tires in the landfill. They have also refined their law later by banning the shredded tires from being dumped in the landfill (Mahlangu M. 2009).

In order to cater the environmental and economic issues, many researchers have opted to utilize waste materials such as cigarette butts, fly ash, marble powder, and clay powder in construction materials such as concrete and bricks. Scrap tires are of no exception as the crumb rubber from scrap tires are used as a partial replacement for fine aggregate to produce a new type of concrete known as rubbercrete brick. In comparison with normal concrete, rubbercrete excels in some properties such as higher energy absorbing, lower thermal conductivity, better sound absorption, and higher electrical resistivity (Mohammed et al. 2012; Aliabdo et al. 2015). However, the drawbacks are strengths reduction due to poor bonding between crumb rubber and hardened cement matrix. This is caused by the hydrophobic nature of the crumb rubber as it repels water thus entrapping air on its surface which consequently lead to strength reduction. Figure 1 shows the trapped air attached to the surface of crumb rubber (CR) in the mixture of cement paste, course aggregates (CA) and fine aggregates (FA). Therefore, to enhance the strength of rubbercrete brick, interfacial transition zone (ITZ) has to be densified. One of the method to densify the ITZ is to incorporate nano silica to the rubbercrete brick mixtures. Nano silica will modify the microstructure of rubbercrete through its physico-chemical effects. Whereas, physically, nano silica has a great fill ability which leads to refining the rubbercrete pore system. The porosity of the rubber-cement matrix is reduced with the inclusion of nano silica, hence improving the microstructure of the ITZ

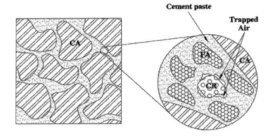

Figure 1. Microstructure of crumb rubber concrete (Mohammed et al. 2012).

Table 1. Mix design proportion for rubbercrete bricks.

Materials	Per mix (kg)
Coarse aggregate	0.688
Cement	0.720
Fly Ash	0.108
Sand	1.032
Crumb Rubber	0.169
Nano Silica	0.036
Water	0.220

(Mukharjee & Barai 2014). While, chemically, nano silica will react with the $Ca(OH)_2$ to produce C-S-H gel which responsible on the strength of the rubbercrete brick (Mohammed et al. 2016).

2 MATERIALS AND PROPORTION

In producing rubbercrete bricks, Ordinary Portland Cement (OPC) Type I conforming to the requirements of ASTM C150 (ASTM Standard 2003) and fly ash class F conforming to the requirements of ASTM C618 (ASTM Standard 2003) have been used. Crumb rubber mesh 30, river sand with maximum size of 5 mm, course aggregates with maximum size of 10 mm and clean tap water were also used in preparing the rubbercrete dry mixtures. The crumb rubber was used as partial replacement by volume to fine aggregates by 25% while 5% of nano silica was incorporated in the mixture. The materials were mixed for about 5 minutes then water (8% of the mixture mass) was added to form rubbercrete. The fresh rubbercrete are then casted under pressure into steel molds with sizes of 210 mm × 110 mm × 65 mm and kept for curing at room temperature for 28 days. The mixture of the rubbercrete brick incorporating nano silica as in Table 1.

3 RESULT AND DISCUSSION

3.1 Dimension

The dimension test was measured based on the overall measurement of 24 bricks which has been placed in a straight line upon a flat surface. The average size of 24 bricks is tested according to BS 3921:1985 (British Standards Institution 1985) and BS EN 772-16 (ASTM Standard 2011). The individual size should not vary greatly from average of 24 bricks and the tolerance of the individual brick should not exceed 6.4 mm on length and 4 mm on width and height. The limitation and measurement of 24 bricks is presented in Table 2, and the measured dimension is within the tolerable limit.

Table 2. Brick dimension.

Size (mm)	Maximum limit (mm)	Minimum limit (mm)	Measured (mm)
210	3054	2945	2954
110	1554	1408	1540
65	936	884	910

Table 3. The compressive strength of individual bricks.

Brick	Top bed area (mm^2)	Bottom bed area (mm^2)	Average bed area (mm^2)	Ultimate load (kN)	Compressive strength (N/mm^2)
1	23104	23103	23103.5	336.8	15.9
2	23103	23100	23101.5	332.6	14.4
3	23100	23101	23100.5	334.3	14.5
4	23101	23102	23101.5	366.4	15.9
5	23102	23102	23102	360.3	15.6
				Mean	15.2

Table 4. Compressive strength of prism.

Brick	Compressive strength (N/mm^2)
Prism 1	8.7
Prism 2	10.4
Prism 3	10.8
Average	10

3.2 Compressive strength

The compressive strength test of rubbercrete brick incorporating nano silica was conducted according to ASTM: C67-03a (ASTM Standard 2000) and BS 3921:1985 (British Standards Institution 1985) at 28 days of curing period. The compressive strength of individual bricks was recorded in the range of 14.4 N/mm^2 to 15.9 N/mm^2, while for the rubbercrete prism ranges between 8.7 MPa to 10.8 MPa as in Table 3 and Table 4. The stress strain curve for the three prisms are illustrated in Figure 2.

3.3 Rupture modulus

The rupture strength test was conducted according to ASTM C67-03a (ASTM Standard 2000) at 28 days of curing period. The average modulus of rupture for five solid rubbercrete bricks specimens is 5.1 N/mm^2 as shown in Table 5. The maximum load is

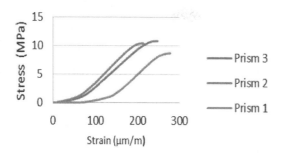

Figure 2. Stress strain curve.

Table 5. Modulus of rupture of rubbercrete bricks.

Brick	Maximum load (N)	Distance between supports (mm)	Distance between fracture & the centre of the brick (mm)	Modulus of rupture (N/mm²)
1	6590	150	22	9.0
2	1830	150	21.5	1.4
3	1790	150	22	2.4
4	830	150	7	3.6
5	830	150	18	9.2
			Mean	5.1

used to calculate the modulus of rupture by using the equation 1 below:

$$S = 3W\left(\frac{l/2 - x}{bd^2}\right) \quad (1)$$

where S = modulus of rupture of the specimen at the plane of failure (Pa); W = maximum load indicated by the testing machine (N); l = distance between the supports (mm); b = net width (mm); d = depth (mm); and x = average distance from the midspan of the specimen to the plane of failure measured in the direction of the span along the centerline of the bed surface subjected to tension (mm).

3.4 Water absorption

The water absorption test was conducted by 24 hours cold immersion method in accordance to BS 3921:1985 (British Standards Institution 1985). The bricks were immersed in the water until saturation to determine the water absorption of nano silica modified rubbercrete solid brick. After removing the bricks from the immersion tank, the specimens were dried in the oven to a constant mass for 48 hours. The specimens were left to cool for 2 to 4 hours before the bricks weight was recorded. The water absorption for

Table 6. Water absorption of rubbercrete bricks.

Brick	Dry mass (g)	Wet mass (g)	Water absorption (%)
1	2730	3022.1	10.69
2	2783	2962.7	6.45
3	2721	2909.4	6.92
4	2743	2956.8	7.79
5	2741	2963.3	8.11
6	2679	2955.1	10.3
7	2824	3022.5	7.03
8	2851	2927.3	2.68
9	2781	2946.3	5.94
10	2736	2879.4	5.24
11	2613	2890.4	10.62
12	2854	3017.3	5.72
13	2772	2955.1	6.61
14	2753	3033.7	10.2
	Mean		7.08

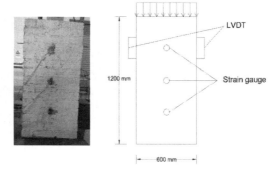

Figure 3. Solid rubbercrete wall set up.

Figure 4. Rubbercrete prism before test.

Figure 5. Failure pattern of rubbercrete prisms.

each specimens is calculated by using equation 2 and tabulated in Table 6.

$$\text{Water absorption (\%)} = 100 \left(\frac{Wetmass - Drymass}{Drymass} \right) \quad (2)$$

3.5 Structural behavior of rubbercrete prisms and wall

A wall and three prism samples were tested to determine the failure pattern and deflection behavior. The set up for the wall is shown in Figure 3. The solid rubbercrete prisms experienced cracking before failure through crushing at 200 kN to 250 kN as shown in Figure 4 and Figure 5. The maximum average load that can be resisted by nano silica modified rubbercrete solid brick prism is 230.6 kN with average strain of 247 μm/m. The inclusion of nano silica in the rubbercrete solid brick reduces the porosity of bricks, hence densifies the interfacial transition zone (ITZ) as agreed by (Mukharjee & Barai 2014). The high pozzolanic activity of nano silica and fly ash and its fine particles size creates denser microstructure.

4 CONCLUSION

The use of waste materials such as scrap tires in civil engineering has gained more interest due to the economic and environmental concern. The nano silica modified rubbercrete solid brick and prism can achieve up to 15 MPa and 10 MPa respectively. From this research, it can be concluded that the newly developed nano silica modified rubbercrete solid bricks can be used for non-load-bearing and load-bearing construction application. The modified nano silica rubbercrete solid bricks are able to undergo large deformation before failure and are also environmental friendly. Recycling the scrap tires as crumb rubber concrete has been successfully proven in various researches and also solve the environmental and health issues due to the accumulation of scrap tires.

REFERENCES

Aliabdo, A. a., Abd Elmoaty, A.E.M. & AbdElbaset, M.M., 2015. Utilization of waste rubber in non-structural applications. *Construction and Building Materials*, 91, pp. 195–207. Available at: http://www.sciencedirect.com/science/article/pii/S0950061815005772.

ASTM Standards (2000) *C67-03a, Standard Test Methods for Sampling and Testing Brick and Structural Clay Tile. Section four: Construction vol. 04.05.* West Conshohocken, PA: ASTM International.

ASTM Standards (2003) *C150-07, Standard Specification for Portland. Section four: Construction vol. 04.01.* West Conshohocken, PA: ASTM International.

ASTM Standards (2003) *C618-03, Standard specification for coal fly ash and raw calcined natural pozzolan for use in concrete. Section four: Construction vol. 04.02.* West Conshohocken, PA: ASTM International.

British Standards Institution. 1985. *BS 3921:1985. Specification for clay bricks.* London: BSI.

British Standards Institution. 2011. *BS EN 772-16:2011. Method of test for masonry units. Determination of dimensions.* London: BSI.

Environment Protection Authority, 2010. Waste Tyres, (15), pp. 1–6.

Mahlangu M., 2009. Waste Tyre Management Problems in South Africa and the, (May).

Mohammed, B.S. et al., 2012. Properties of crumb rubber hollow concrete block. *Journal of Cleaner Production*, 23(1), pp. 57–67. Available at: http://linkinghub.elsevier.com/retrieve/pii/S0959652611004215.

Mohammed, B.S. et al., 2016. Properties of Nano Silica Modified Rubbercrete. *Journal of Cleaner Production*, 119, pp. 66–75. Available at: http://www.sciencedirect.com/science/article/pii/S0959652616001529.

Mukharjee, B.B. & Barai, S. V., 2014. Influence of incorporation of nano-silica and recycled aggregates on compressive strength and microstructure of concrete. *Construction and Building Materials*, 71, pp. 570–578. Available at: http://www.sciencedirect.com/science/article/pii/S0950061814009520.

R. Kasich, J. & W. Butler, C., 2016. *Scrap Tires.* [online] Epa.ohio.gov.Available at: http://epa.ohio.gov/dmwm/Home/ScrapTires.aspx [Accessed 14 Apr. 2016].

Sandra Kumar a/L Thiruvangodan, 2006. *Waste tyre management in Malaysia.*

Shakya, P., 2006. Studies and determination of heavy metals in Waste Tyres and their impacts on the environment. *Pak. J. Anal. & Envir. ...*, 7(2), pp. 70–76. Available at: http://www.ceacsu.edu.pk/PDF file/Journal Vol 7 issue 2/01-PJAEC-01-02-07-09.pdf.

Analysis layers method of strengthening reinforced concrete T-beams wire rope moment negative (attention reinforcement slab)

D.L.C. Galuh & H.P. Riharjo
Sarjanawiyata Tamansiswa University, Yogyakarta, Indonesia

ABSTRACT: Layers is an analysis method used to determine the flexural capacity of a structural element by dividing the cross-section of the structural elements with varying thickness. Three specimens were made that consisted of one controlling beam (BK), one beam reinforced by 4Ø10 mm steel wire in the tensile area (BP1) and one beam reinforced by 4Ø10 mm steel wire in the tensile area and 2P8 mm in the compressive area (BP2) with mortar as the concrete blanket. The diameters of wire rope and reinforcing steel used for the reinforcement were 10 mm and 8 mm, respectively. Results of the analysis showed that the comparative ratios of flexural capacity for BK, BP1, and BP2 crack condition were 0.53, 0.72 and 0.72 respectively. While the maximum condition 1.28, 0.76, and 0.78, respectively.

1 INTRODUCTION

Retrofitting can be done if there are additional load capacity that exceeds the previous plan, or a solution that is used because there are errors in the work field, so it does not correspond to the actual function [1]. The selection of wire rope as a reinforcement material is a wire rope has a light weight with high tensile strength. The objective of this research is to identify the increasing flexural capacity for concrete T-beam. The experiment results were then compared to the results of layers method.

2 LAYERS METHOD

Analysis using layers method can be used to determine the capacity of reinforced concrete beam section is based on the stress-strain diagram of the constituent materials [2]. Formulas in beam material used in analyzing the layers method include [3]:

a. The concrete stress-strain model based on Popovics model (high strength), as shown in the following equations:

$$f_{ci} = -\left(\frac{\varepsilon_{ci}}{\varepsilon_p}\right) f_p \frac{n}{n-1+(\varepsilon_{ci}/\varepsilon_p)^{nk}} \quad \text{untuk } \varepsilon_{ci} < 0 \quad (1)$$

$$n = 0{,}8 + \frac{f_p}{17} \quad (2)$$

$$k = \begin{cases} 1{,}0 & \text{for } \varepsilon_p < \varepsilon_{ci} < 0 \\ 0{,}67 + \dfrac{f_p}{62} & \text{for } \varepsilon_{ci} < \varepsilon_p < 0 \end{cases} \quad (3)$$

$$E_c = \frac{f_p}{|\varepsilon_p|} \cdot \frac{n}{(n-1)} \quad (4)$$

where:
f_p: peak concrete compressive stress (MPa); ε_p: concrete compressive strain corresponding to f_p; f_{ci}: the stress occurs in the concrete (MPa); ε_{ci}: strain occurs at peak; n: curve fitting parameter for stress-stain response of concrete in compression; k: post-peak decay parameter for stress-strain response of concrete in compression; E_c: concrete elasticity modulus (MPa).

b. The reinforced concrete stress-strain model was based on the trilinieri curve models, as shown in the following Equation:

$$f_s = \begin{cases} E_s \varepsilon_s & \text{for } |\varepsilon_s| \leq \varepsilon_y \\ f_y & \text{for } \varepsilon_y < |\varepsilon_s| \leq \varepsilon_{sh} \\ f_y + E_{sh}(\varepsilon_s - \varepsilon_{sh}) & \text{for } \varepsilon_{sh} < |\varepsilon_s| \leq \varepsilon_u \\ 0 & \text{for } \varepsilon_u < |\varepsilon_s| \end{cases} \quad (5)$$

$$\varepsilon_u = \varepsilon_{sh} + \frac{(f_u - f_y)}{E_{sh}} \quad (6)$$

where:
ε_s: reinforcement strain; ε_y: yield strain; ε_{sh}: strain at the onset of strain hardening; ε_u: ultimate strain; E_{sh}: strain hardening modulus (MPa); f_y: yield strength (MPa); f_u: ultimate strength (MPa); E_s: elastic modulus (MPa).

Wire rope stress strain model based on the Ramsberg-Osgood, as shown in the following Equation:

$$f_s = E_s \varepsilon_s \left\{ A + \frac{1-A}{\left[1+(B\varepsilon_s)^C\right]^{\!1/C}} \right\} \leq f_u \quad (7)$$

Table 1. Specifications of specimens off reinforce concrete beam [5].

KODE	L (mm)	b_f (mm)	t_f (mm)	b_w (mm)	Main Reinforcement		Cross Bar Reinforcement	Reinforcement	
					Tensile	Compressive		Tensile	Compressive
BK	2400	400	75	150	3D13	2P8	P8-40	–	–
BP1	2400	400	115	150	3D13	2P8	P8-40	4Ø10	–
BP2	2400	400	115	150	3D13	2P8	P8-40	4Ø10	2P8

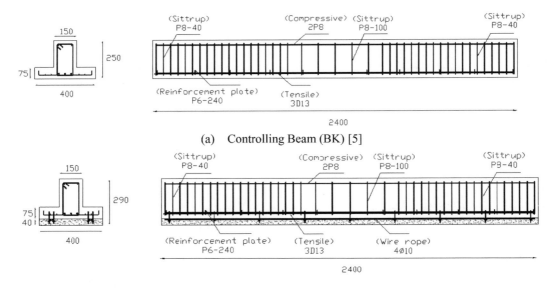

(a) Controlling Beam (BK) [5]

(b) Reinforcing Beam (BP1) [5]

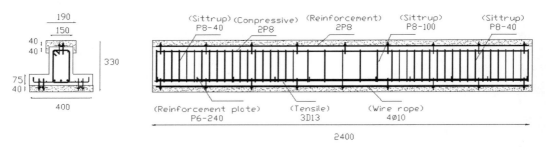

(c) Reinforcing Beam (BP2)

Figure 1. The size and the reinforcement of beam specimen (mm unit).

$$A = \frac{E_{sh}}{E_s} \quad (8)$$

$$B = \frac{E_s(1-A)}{f_s^*} \quad (9)$$

C = transition coefisient (10)

where:
ε_s: reinforcement strain; E_s: elastic modulus (MPa); E_{sh}: strain hardening modulus (MPa); f_u: ultimate strength (MPa); f_s^*: value at which the second linear branch intercepts the stress axis at zero strain; C: 10 [4].

3 BEAM SPECIMENS

There were three Specimens of reinforced concrete beam: the controlling beam (BK), the reinforcing beam type 1 (BP1) and type 2 (BP2). Specifications of reinforced concrete beam specimens are presented on Table 1 and Fig. 1.

Table 2. The capacity of beam load layers method.

No.	Specimen	Load Capacity (kN) experiment		Load Capacity (kN) layers method		Ratio P_{max}	
		crack	max	crack	max.	crack	max.
1.	BK	28.6	88.5	15.17	113.37	0.53	1.28
2.	BP1	31.8	180	22.87	143.07	0.72	0.78
3.	BP2	44.8	259	32.17	208.28	0.72	0.80

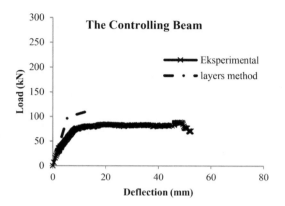

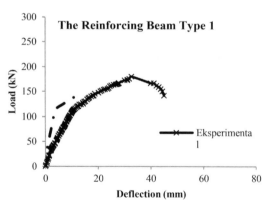

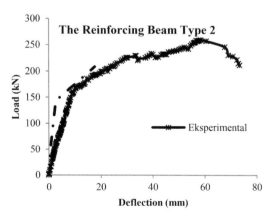

Figure 2. Comparison between analysis and experimental results.

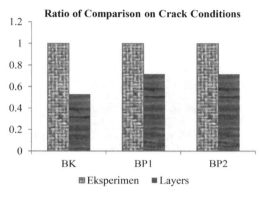

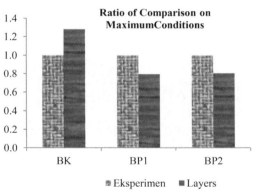

Figure 3. The ratio of load capacity of beams and layers analysis method on condition crack and Maximum.

4 COMPARISON BETWEEN ANALYSIS AND EXPERIMENTAL RESULTS

The load capacity value based on the analysis is presented on Table 2, Fig. 2, and Fig. 3.

Based on Fig. 2 shows that the results of the analysis of the layers method have different initial stiffness to the experimental results. This is caused by differences in the deflection BK, BP1, BP2 happens to load crack first experiments with the analysis were 2 mm, 2.025 mm, 1.955 mm respectively. While BK, BP1, BP2 in the layers method were 0.9 mm, 0.15 mm, 0.12 mm respectively. The maximum load deflection BK, BP1, BP2 results of the experiments also have a longer deflection when compared with the results of the analysis were 48.23 mm, 32.72 mm and 58.52 mm respectively. While BK, BP1, BP2 on layers method were 14.68 mm, 13.32 mm, 18.5 mm respectively.

Based on Table 2 and Fig. 3 on the condition of the crack, it appears that the results obtained in the comparative test piece BK, BP1 and BP2 were 0.53, 0.72, and 0.72 respectively. While in the maximum conditions, the comparative results obtained on the test object were 1.28, 0.78, and 0.80 respectively.

5 CONCLUSIONS

Based on the results of this research and data analysis, the following conclusions can be drawn:

1. The ratio of load capacity of beams and layers analysis method on condition crack BK, BP1 and BP2 were 0.53, 0.72, and 0.72 respectively. While in the maximum conditions, the comparative results obtained on the test object were 1.28, 0.78, and 0.80 respectively.
2. The initial stiffness to the experimental results BK, BP1, BP2 were 2 mm, 2.025 mm, 1.955 mm respectively. While BK, BP1, BP2 in the layers method were 0.9 mm, 0.15 mm, 0.12 mm respectively. The maximum load deflection BK, BP1, BP2 results of the experiments also have a longer deflection when compared with the results of the analysis were 48.23 mm, 32.72 mm and 58.52 mm respectively. While BK, BP1, BP2 on layers method were 14.68 mm, 13.32 mm, 18.5 mm respectively.

REFERENCES

Bentz, E. dan Collins M. P., 2001, *Response-2000 User Manual*, http://www.ecf.utoronto.ca/~bentz/manual2/final.pdf accessed on February 9, 2011.

Park, R. dan Paulay, T., 1975, *Reinforced Concrete Structure*, John Wiley & Sons Inc, Canada.

Triwiyono, A., 2004, *Perbaikan dan Perkuatan Struktur Beton*, Teaching Materials Special Topics, Structural Engineering, Graduate School of UGM, Yogyakarta.

Wong, P.S. dan Vecchio, F.J, 2002, *Vector2 & Form Works User's Manual*, http://www.ecf.utoronto.ca/~bentz/manual2/final.pdf accessed on February 9, 2011.

Y. Haryanto, "Efektifitas *Wire Rope* Sebagai Perkuatan Pada Daerah Momen Negatif Balok Beton Bertulang Tampang T," Dinamika Rekayasa journal, Vol. 8, No. 1, February 2012.

Concrete panel from polystyrene waste

D. Sulistyorini & I. Yasin
Sarjanawiyata Tamansiswa University, Yogyakarta, Indonesia

ABSTRACT: The polystyrene concrete panel consist of cement, polystyrene waste, water and wire mesh placed on the top and bottom panel was compressed 2 MPa. Test specimen for three panels with dimension 80 cm length, 30 cm width, 1 cm thick and three panels with 0.5 cm thick. Each panel use water cement factor 0.4. The cement content 250 kg/m^3, wire mesh diameter of 0.6 mm with grid 6 mm × 6 mm. The test results obtained unit weight for 0.5 cm thickness was 1367 kg/m^3 and flexural strength was 2.3 MPa and than for 1 cm thickness was 1081.9 kg/m^3 and flexural strength was 0.721 MPa. The crack pattern at the midspan, the first occured curve of wire mesh simultanously with polystyrene concrete and than panel failure. Maksimum deflection for 0.5 cm thickness was 72.83 mm and for 1 cm thickness was 15.34 mm.

1 INTRODUCTION

1.1 Background

Styrofoam or expanded polystyrene is environmental waste very bad if one time use for a cup of coffee, marriage decoration and packaging. Because polystyrene is non biodegradable materials so polystyrene waste seen floating in the river and the sea for hundreds years into the future. To prevent environmental pollution, polystyrene is best for long term use, for example in building technology. (b-panel, 2013).

Expanded polystyrene waste in a granular form is used as lightweight aggregate to produce lightweight structural concrete with the unit weight varying from 1600 to 2000 kg/m^3 (Ben Sabaa, R.S. Ravindrarajah, 1997).

The use of polystyrene panels for structural walls has been applied by b-panel. The method used is a sandwich panel which is located amid massive polystyrene wire mesh reinforced with a layer of shotcrete (b-panel, 2013).

1.2 Case study

This study was to utilize the polystyrene waste as a non structural panels that consist of cement, polystyrene waste, water and wire mesh placed on the top and bottom panels is compressed in order to obtain thin panel. Polystyrene panels will be reviewed unit weight, flexural strength and deflection.

2 SPECIMEN INVESTIGATION

2.1 Materials

The cement used was Portland cement type I (40 kg/bags) produced by PT. Semen Tiga Roda,

Figure 1. Polystyrene waste granules.

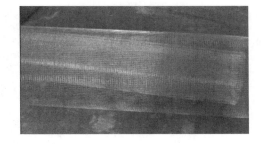

Figure 2. Wire mesh.

shaved polystyrene waste into granules as shown Figure 1, wire mesh 0.6 mm with grid 6 mm × 6 mm cut to panels size as shown in Figure 2.

2.2 Specimen test

The specimen test were six polystyrene panels with two various thickness planned listed in Table 1 and specimen detail shown in Figure 3.

Table 1. Specimen dimensions.

Specimen	Length	Width	Thickness
	cm	cm	cm
T0.5-1	80	30	0.5
T0.5-2	80	30	0.5
T0.5-3	80	30	0.5
T1-1	80	30	1
T1-2	80	30	1
T1-3	80	30	1

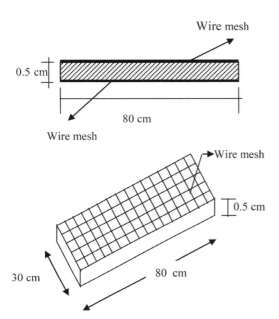

Figure 3. Specimen detail.

Table 2. Composition for polystyrene concrete panel.

Speciments thick	Cement content	Water	Polystyrene	Compressing
cm	kg	liter	kg	MPa
0.5	1.5	0.6	0.1	2
1	3	1.2	0.2	2

2.3 Composition of polystyrene panels

Polystyrene concrete panels was made from a mixture of cement, polystyrene waste, and water.

Wire mesh was placed on the top and bottom polystyrene panel. The panel was compressed at 2 MPa. The cement content 250 kg/m^3, water cement ratio 0.4, unit weight of polystyrene obtained was 15.9 kg/m^3. Table 2 shown the composition for polystyrene concrete after trial mix.

Figure 4. Compressing by hydraulic jack.

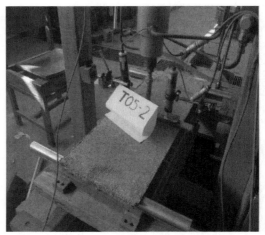

Figure 5. Flexure testing.

Table 3. The test results of the wire mesh.

Diameter	Yield stress, σ_y	Tensile Strength, σ_t	Strain at failure	E_s
mm	MPa	MPa		MPa
0.60	913.52	1085.16	0.095	9649.76

3 RESEARCH METHOD

Materials mix is poured into the mold which covered with wire mesh. After pouring the mixture, put wire mesh on the top and then covered with a cover mold and compressed by hydraulic jack at 2 MPa, shown Figure 4. At the time of compression reaches 2 MPa hydraulic jack is allowed to stand for 15 minutes so that the test object incompressible. The specimens is placed in storage for 28 days without wetted. Flexural testing using simple supported with one point loading, shown Figure 5.

Table 4. Test Results of modulus elasticity of polystyrene concrete.

Cement content	Length	Width	High	Area	Maximum Load	$f'c$	$0{,}4 f'c$	ε	E_p
kg/m³	mm	mm	mm	mm²	N	MPa	MPa		MPa
250	103.55	103.59	205.85	10727.08	63287	5.90	2.36	0.00710	372.99

Table 5. The test results of unit weight of polystyrene concrete panels.

No	Specimen	Early Dimension			Dimension after 28 days			Early weight	Weight after 28 days	Unit weight
		L	W	t	L	W	t			
		cm	cm	cm	cm	cm	cm	Kg	Kg	Kg/m³
1	T 0.5-1	80	30	0.61	80	30	0.61	2.0	2.0	1355
2	T 0.5-2	80	30	0.58	80	30	0.58	1.8	1.8	1293
3	T 0.5-3	80	30	0.57	80	30	0.57	2.0	2.0	1454
	Average						0.59	1.93	1.93	1367
1	T 1-1	80	30	1.25	80	30	1.25	4.0	4.0	1338
2	T 1-2	80	30	1.57	80	30	1.57	4.3	4.3	1139
3	T 1-3	80	30	2.17	80	30	2.17	4.0	4.0	769
	Average						1.66	4.1	4.1	1082

4 RESULTS AND DISCUSSION

4.1 Wire mesh

The test results showed that the wire mesh is classified into high strength steels because it has a yield stress of more than 275–480 MPa (see Table 3). The modulus elasticity of steel in the British Standard was 205000 MPa and the AISC was 200000 MPa, so that for this analysis used modulus elasticity 200000 MPa.

4.2 Modulus elasticity of polystyrene concrete

Modulus elasticity of polystyrene concrete testing is done using specimens 10 cm × 10 cm × 20 cm that has been done previously linked by Ade Okvianti Irlan (see Table 4) (A.O. Irlan, 2014).

4.3 Unit weight of polystyrene concrete panels

The test results showed in Table 5, that the polystyrene concrete panel planned 0.5 cm thick obtained average thickness of 0.59 cm and 1 cm thick plan obtained average thickness of 1.66 cm. The additions were given due to the unstable character (not-masive) of polystyrene panels. The unit weight of 0.5 cm panel was 1367 kg/m³ and 1082 kg/m³ for 1 cm panel The unit weight each specimen as shown Figure 6. The 1 cm panel was lighter than 0.5 cm panel, because weight increased was not comparable with additions thickness.

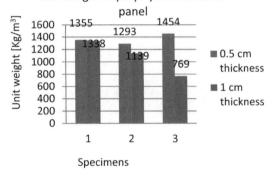

Figure 6. Unit weight of polystyrene concrete panel.

4.4 Flexural strength of polystyrene concrete panel

Flexural strength of polystyrene concrete panel would be calculated as a composites panel, so used comparison between modulus elasticity of wire mesh with modulus elasticity of polystyrene concrete. Maximum flexural strength from experiment result would be calculated as follow (A. Triwiyono, 2000):

$$n = \frac{E_s}{E_p} \qquad (1)$$

$$I_t = I_p + n\, I_s \qquad (2)$$

$$M = \tfrac{1}{8} q L^2 + \tfrac{1}{4} P L \qquad (3)$$

453

$$\sigma = \frac{M \cdot y}{I_t} \quad (4)$$

where n = ratio modulus elasticity, E_s = Modulus elasticity of wiremesh, E_p = Modulus elasticity of polystyrene concrete, I_t = Amount of moment of inertia, I_p = polystyrene concrete of moment of inertia, I_s = wiremesh of momen of inertia, M = maximum momen, q = panel weight per length, L = specimen length, P = maximum load from experiment, σ = maximum flexural strength, y = depth of neutral axis measured from the more highly compressed face.

Table 6. Flexural strength of polystyrene concrete panel.

Specimens	Weight N	Maximum Load N	Moment Nmm	Flexural Strength MPa	Deflection mm
T 0.5-1	20	400	76875	1.97	78.54
T 0.5-2	18	560	106687	2.94	70.26
T 0.5-3	20	370	71250	1.99	69.70
Average				2.30	72.83
T 1-1	40	400	78750	0.86	13.04
T 1-2	43	360	71531	0.59	18.74
T 1-3	40	640	123750	0.71	14.25
Average				0.72	15.34

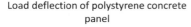

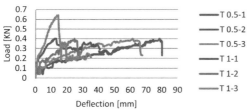

Figure 7. Load and deflection for all panels.

Maximum load and Flexural strength result shown in Table 6. The compression each panel was same at 2 MPa but thickness and maximum load was varying. It is observed that polystyrene waste as agregat have not stability because its not masif material.

4.5 Crack pattern

The crack pattern at the midspan, the first occured curve of wire mesh simultanously with polystyrene concrete and than panel failure. Average maksimum deflection for 0.5 cm thickness panel was 72.83 mm and for 1 cm thickness panel was 15.34 mm (see Fig. 7).

The crack pattern that occurs on wiremesh was crooked and failure, for polystyrene panels crack in the loading position. Figure 8 showed the crack on polystyrene panel after flexural test.

5 CONCLUSIONS

a. The polystyrene concrete panel planned 0.5 cm thick more stable in thickness than planned 1 cm thick but the panel weight increased was not comparable with additions thickness. The polystyrene panels showed unstable character or not-massive materials. Both of them could as lightweight concrete category because the unit weight under 1600 kg/m^3.

b. Flexural strength of polystyrene concrete panel for 0.5 cm thickness greater than polystyrene concrete panel for 1 cm thickness because the maximum load was not comparable with additions thickness. A decline in the flexural strength 68.7%.

c. The crack pattern at the midspan, early occured curve of wire mesh simultanously with polystyrene concrete and than panel failure. Wire mesh and polystyrene concrete could regarded as composit panels.

d. In general polystyrene waste has the potential to be used non structural panels and its expected to reduce polystyrene waste.

Figure 8. Crack pattern.

REFERENCES

A.O. Irlan, Thesis of post graduate student in Gadjah Mada University (2014) 92.

A. Triwiyono, Teaching materials of Mechanics of Materials (2000) 52–53.

Ben Sabaa, R.S. Ravindrarajah : submitted to Symposium MM: Advances in Materials for Cementitious Composites (1997).

B-panel, PT Beton Elemindo Putra (2003).

Fault tree analysis for reinforced concrete highway bridge defect

W.S. Wan Salim
Department of Civil and Environmental Engineering, Universiti Teknologi PETRONAS, Seri Iskandar, Perak, Malaysia
Faculty of Civil Engineering, Universiti Teknologi MARA, Malaysia

M.S. Liew
Faculty Geosciences and Petroleum Engineering, Universiti Teknologi PETRONAS, Perak, Malaysia

A. Shafie
Fundamental and Applied Science Department, Universiti Teknologi PETRONAS, Perak, Malaysia

ABSTRACT: Fault tree method is among the risk assessment tool that is widely used to evaluate the probability of failure of a system. This paper presents the used of Fault Tree Method in qualitative modelling the bridge defect in terms of failure causes and mechanism that can be directly correlated with the proposed bridge management system namely as Bridge Integrity Management System (BIMs) bridge. The minimal cut sets of Fault tree was determined using Boolean algebra formulas which contains 20 double components of cut sets. The results of this qualitative analysis establish the probability expression of bridge defect which may provide guidance for future safety management for bridge failure. Further quantitative analysis will determine the ranges probabilities of bridge defect that will be used for the development of risk matrix.

1 INTRODUCTION

With a fast development of economy since a past few decades, the needs of having a good and efficient transportation system are very demanded, which has resulted a mushrooming of transportation system especially an increase in construction of highway. Bridge is one of the important components of transportation system, which facilitate a vehicle or a pedestrian to get across a river, a road or any other obstacles. Each bridge which is either old or new carries the possibility of failure. Early sign of damage that occurred on the bridges could help engineers to take an action to prevent them from a serious failure in the future because once there is a fault it could lead to structural failure or even collapse and cause a lot of loss in terms of humanity and economic. Hence, it is essential to have a continuous bridge performance assessment with a comprehensive method and management system to detect the early stage of failure or damage to ensure that it is in safe condition for use and to prevent any possible collapse event in the lifespan of the bridge.

Overall, it was reported that there are about more than 10 000 bridges in Malaysia owned by Public Work Department (JKR), Malayan Railways (KTMB), Kuala Lumpur City Council (DBKL) and Malaysia Highway Authority (LLM) which consists of various types of bridge that is concrete bridge, steel bridge, masonry bridge and wooden bridge (Idris & Ismail 2007). These bridges constitute a huge asset to the various agencies as mentioned. These agencies also have an authority to monitor and manage the bridge performance which under their control. With a large number of bridges to be observed and several issues related to the bridge management due to numbers of ageing bridge, the low fund for maintenance and occurrence of catastrophic bridge failure (King 2001), the comprehensive assessment and management system is essential to develop in order to reduce cost related to bridge performance assessment and which bridge should be given the priorities to be maintained first. Furthermore, in the current situation of increasing of traffic demands and ageing bridge, it is significant to have an effective method for deciding which bridges need inspection and maintenance first and which ones need to be prioritized.

At present, Malaysia uses a system known as Bridge Management System (BMS) for measuring and evaluating the bridge performance for further maintenance action. The system is generally based on condition evaluation and cost optimization (Omar et al. 1992, Weng et al. 1991). Moreover, the inspection frequencies of the existing system are generally time-based but some of the bridges may require less frequent of inspection compared to the bridge with high potential of failure. Furthermore, the current bridge management system is more concern on the prediction of the likelihood of failure only rather than the consequences if the failure occurred.

Since repair and maintenance fund is always limited, the new approach of management system is essential to be developed to rank the necessity of inspection for future maintenance by considering both the probability of failure and its consequences during the decision making process. Thus, the cost of inspection of bridges that categorized in a good condition may be reduced by increasing the interval of the next inspection. This approach for the decision making in the asset integrity management system is more recognized as a risk-based approach.

As the risk assessment method has been successfully implemented in structural integrity management of offshore platform structure in order to ensure the fitness of structure for its purpose with long term reduction in operating and assessment cost (O'Connor et al. 2005, Straub et al. 2006), it is possible to develop the bridge management system by using the same approach.

This approach of assessment demonstrates a concept of assessing the probability of system failure by identifying the causes of events that can lead to the failure and its consequences. It can be used both qualitatively and quantitatively to evaluate the failure.

Fault tree method and Even Tree method are among of the risk assessment tool that may be used to evaluate the causes and consequences of failure. These tools depict the failure event through a combination of causes and consequences of event.

However, this paper only focuses on the development and qualitative analysis of Fault Tree method for reinforced concrete (RC) highway bridge defect. This Fault Tree analysis is a part of the evaluation process of Bridge Integrity Management System (BIMs) which was proposed before.

2 BACKGROUND

2.1 Fault tree method (FT)

In every event of failure or accident, investigations of failure causes are very significant for future improvement of the respected system of process. FT can be used to identify the combinations of event which could cause a major failure of any system. Generally, a probability of system failure may determine through a developed model of FT and evaluating each system level of failure probability, by considering an interaction at each component of the system. FT concepts can be explained as a translation of a physical system into a structured logical diagram that shows the relation between system failure and failures of the components of the system which specified causes lead to one specified main event of interest (Aven 2008, Lee et al. 1985). Fault tree analysis (FTA) is considered as a deductive approach of failure analysis starting with a potential undesirable event called TOP event. All the possible way that cause the TOP event occurred is determined using logical gates such as AND/OR gates, which are the most commonly used to combine the causal event.

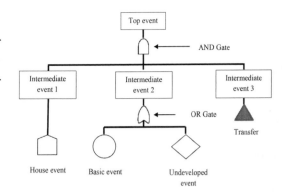

Figure 1. FT concept.

In general, the fault tree model consists of two basic components, 'gates' and 'event' where 'gates' act as connector to show the relationship between of the main 'event' and lower event that contribute to the occurrence of the main event (Top event). The simplified FT model can be explained through FT concept diagram as shown in Figure 1.

FT method has been widely used in risk assessment studies for identifying the causes and probability of failure. For risk assessment studies of bridge structures, McDaniel et al. (2013) has shown that analysis failure through fault tree model can help in identifying countermeasures to minimize risk failures and the study are based on a case study of segmental box girder bridges. Zhu et al. (2008) & Setunge et al. (2015) had uses the fault tree analysis to model the probability of existence of a major distress mechanism to identify the important risks and their relative severity and to rank the performance trends of bridges. While, Johnson (1999) uses the fault tree analysis to show the interactions of the complex processes of erosion at bridge piers and abutments and to determine the overall probability of bridge failure due to scour and channel instability.

2.2 Minimal cut sets

Generally, cut sets which is also known as failure path is defined as a combination of events, including intermediate and basic event which, if all events occur, it will result the failure of the top event (Stamatelatos et al. 2002). While, minimal cut sets are those cut sets with minimum number of basic events which still assures the occurrence of the top event (Stamatelatos et al. 2002).

3 PROPOSED FAULT TREE MODEL OF BRIDGE DEFECT

In this section, an application of the FT method to model failure causes and mechanism of selected bridge components which leading to the defect of bridge as a top event is described. In general, the overall process includes knowledge on the FT concepts, construction

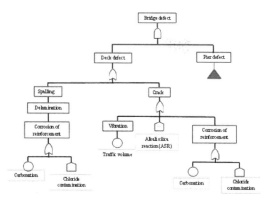

Figure 2. Fault tree model for bridge defect.

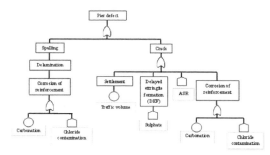

Figure 3. Sub-fault tree model of pier defect.

of the FT model, qualitative evaluation of the minimal cut sets and evaluation of the top event.

In this study, the basic causes of failure considered are limited only to the effect of environmental conditions, chemical reaction in concrete and traffic volume only due to the limited access of existing data. In addition, the fault tree of bridge defect is developed also by assumed the bridge has no major maintenance or repair work performed during their service life. Currently, the existing bridges have gone through a routine inspection yearly.

All the failure mechanism and causes of bridge failure are identified through the compilation of a list of possible causes, failure modes and mechanism that driven to the total bridge defect. This information is gathered from previous related study (King & Ku Mahamud 2009, LeBeau & Wadia-Fascetti 2007, McDaniel et al. 2013, Setunge et al. 2015, Sianipar & adams 1997, Zhu et al. 2008) and also from previous and current damage record of bridge inspection report. However, most of the information for the identification process of the failure causes is collected from an existing of JKR BMS database. For this study, the bridge system also limited to only considering supported by deck and pier for the development of top event of the fault tree. In general, these structural defects are explained through the failure mechanism or mode and further determined by the basic event that causes the failure of each structure as conclude in Figure 2 and 3.

Table 1. Symbols and event description for FT model of bridge defect.

Symbol	Event description
T	Bridge defect
C1	Deck defect
C2	Spalling on deck
C3	Delamination on deck
C4	Corrosion of reinforcement in deck
C5	Crack on deck
C6	Vibration
C7	Pier defect
C8	Spalling on pier
C9	Delamination on pier
C10	Corrosion of reinforcement in pier
C11	Crack on pier
C12	Settlement
C13	Delayed ettringite formation (DEF)
A	Carbonation (Deck)
B	Chloride contamination (Deck)
D	Traffic volume (Deck)
E	Alkali silica reaction (ASR)(Deck)
F	Carbonation (Pier)
G	Chloride contamination (Pier)
H	Traffic volume (Pier)
I	Sulphate
J	ASR (Pier)

Fig. 2 and 3 illustrates the FT model for bridge defect and sub-FT model of pier defect. Based on these figures, Bridge defect considered as the Top Event of the FT. Descending down the tree is a few levels of intermediate event in terms of failure mode or mechanism such as spalling crack etc. Finally, the bottom part of the tree is a set of causes that could give impact on failure of bridges which know as basic events. Connection in between the top events and other event are linked by AND and OR Gate.

4 QUALITATIVE EVALUATION OF FAULT TREE

The bridge defect is the TOP EVENT of the developed FT model. In order to developed algorithm for probability of bridge defect of the RC highway bridge, each event is presented in terms of symbols. The intermediate events are designated as symbol C and the basic events at the lowest level of the FT are designated as listed and described in the Table 1. All the events are assumed to be independent.

Qualitative analysis of the FT involves with obtaining the minimal cut sets. The FT model of Figure 2 and 3 can be written as an expression of Boolean algebra. The analysis of the developed FT begins from the top of the tree with the initial equation as below:

$$T = C1 \cdot C7 \quad (1)$$

Moving down on the FT model, each event is substituted with a detailed expression from a Boolean equation that developed at each level. Based on Figure 2 and 3 and symbols designated in Table 1, the detail Boolean equations of each level are given in following:

$$C1 = C2 + C5 \quad (2)$$

$$C2 = C3 = C4 = A + B \quad (3)$$

$$C5 = C6 + E + C4 \quad (4)$$

$$C6 = D \quad (5)$$

$$C7 = C8 + C11 \quad (6)$$

$$C8 = C9 = C10 = F + G \quad (7)$$

$$C11 = C12 + C13 + J + C10 \quad (8)$$

$$C12 = H \quad (9)$$

$$C13 = I \quad (10)$$

By substituting the expression (3) and (5) in expression (4), the expression of C5 is expanded to

$$C5 = D + E + A + B \quad (11)$$

Then, expression (11) is substituted in the expression (2) which gives

$$C1 = A + B + A + B + D + E \quad (12)$$

By applying the same process to the pier defect, the expression of C7 is given as below:

$$C7 = F + G + H + I + J + F + G \quad (13)$$

In order to simplify both expression C1 and C7, an Idempotent law of Boolean equation is applied to both expressions. The Idempotent law shows that the union or intersection of the same event produces the same original event. Thus,

$$C1 = A + B + D + E \quad (14)$$

$$C7 = F + G + H + I + J \quad (15)$$

Finally, substituting the expression (14) and (15) in expression (1) to obtain an equation for the top event in terms of minimal cut sets.

$$T = (A + B + D + E) \cdot (F + G + H + I + J) \quad (16)$$

Expanding expression (16) produces 20 double components of minimum cut sets for the top event of bridge defect. The respected minimal cut sets are represented by an intersection of basic event as described in the Table 2.

Based on the established expression of the top event of bridge defect, the variations of probability value are determined by substituting a range value of each basic

Table 2. Minimal cut sets.

Minimal cut sets	Description of intersection of basic events
$A \cap F$	Carbonation (Deck) and Carbonation (Pier)
$A \cap G$	Carbonation (Deck) and Chloride contamination (Pier)
$A \cap H$	Carbonation (Deck) and Traffic volume (Pier)
$A \cap I$	Carbonation (Deck) and Sulphate
$A \cap J$	Carbonation (Deck) and ASR (Pier)
$B \cap F$	Chloride contamination (Deck) and Carbonation (Pier)
$B \cap G$	Chloride contamination (Deck) and Chloride contamination (Pier)
$B \cap H$	Chloride Contamination (Deck) and Traffic volume (Pier)
$B \cap I$	Chloride contamination (Deck) and Sulphate)
$B \cap J$	Chloride contamination (Deck) and ASR (Pier)
$D \cap F$	Traffic volume (Deck) and Carbonation (Pier)
$D \cap G$	Traffic volume (Deck) and Chloride contamination (Pier)
$D \cap H$	Traffic volume (Deck) and Traffic volume (Pier)
$D \cap I$	Traffic volume (Deck) and Sulphate
$D \cap J$	Traffic volume (Deck) and ASR (Pier)
$E \cap F$	ASR (Deck) and Carbonation (Pier)
$E \cap G$	ASR (Deck) and Chloride contamination (Pier)
$E \cap H$	ASR (Deck) and Traffic volume (Pier)
$E \cap I$	ASR (Deck) and Sulphate
$E \cap J$	ASR (Deck) and ASR (Pier)

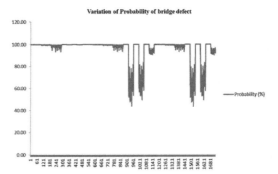

Figure 4. Variation of probability of bridge defect.

event of minimal cut sets. The range value of basic events involves are previously determined by using the Monte Carlo simulation method due to the limited data provided by the bridge stakeholder. This probability variation was developed by using several combination cases of FT basic event, which resulted as shown in the Figure 4.

The result of the graph may consider as a preliminary measure for the determination of the range value of probability of failure in order to develop a risk matrix for bridge failure.

5 CONCLUSION

The study demonstrates that Fault tree model provides an effective option of assessing the possibility of bridge defect through identification of causes of failure. This paper establishes the fault tree model which composed of a top event and a combination of 35 number of basic and intermediate event. However, due to some repeated event, it's reduced to 19 numbers of combination events with 8 numbers of basic events.

Using Boolean algebra formulas, the fault tree model was analyzed qualitatively to determine the minimal cut sets by using Boolean algebra formula. The analysis demonstrates bridge defect as the top event of a fault tree model which contains 20 double components of minimal cut set involving an intersection of basic event. The respected basic events which contribute to the double components of minimal cut sets are carbonation, chloride contamination, traffic volume, ASR and sulphate for both occurring on pier and deck components of bridge.

Based on the previous Monte Carlo simulation data of events, the probabilities of minimal cut sets event are determined. These results were used to calculate the ranges of bridge defect probabilities and for the future development of risk matrix.

ACKNOWLEDGEMENT

The authors are very grateful to the Public Work Department of Malaysia (JKR) for their support in terms of information and bridge data.

REFERENCES

Aven, T. 2008. *Risk Analysis Method. Risk Analysis: Assessing Uncertainties Beyond Expected* Value and Probabilities. England: John Wiley & Sons, Ltd.

Idris, S.M. & Ismail, Z. 2007. Appraisal of concrete bridges: Some local examples. *Jurutera*: 36–41.

Johnson, P.A. 1999. Fault tree analysis of bridge failure due to scour and channel instability. *Journal of Infrastructure Systems*, vol. 5, no. 1: 35–41.

King, N.S. 2001. Key issues in bridge management. *Special Issue of IEM Journal on Bridge Engineering.*

King, N.S. & Ku Mahamud, K.M.S. 2009. Bridge Problems in Malaysia. *Seminar on Bridge Maintenance & Rehabilitation*, 24 February, 2009, Kuala Lumpur.

LeBeau, K.H. & Wadia-Fascetti, S.J. 2007. Fault tree analysis of Schoharie Creek Bridge Collapse. *Journal of Performance of Constructed Facilities*, vol. 21, no. 4: 320–326.

Lee, W.S., Grosh, D.L, Tillman, F.A. & Lie, C.H. 1985. Fault tree analysis, methods and applications – A review. *IEEE Transactions on Reliability*, vol. R-34, no. 3: 194–203.

McDaniel, C.D., Chowdhury, M., Pang, W. & Dey, K. 2013. Fault-tree model for risk assessment of bridge failure: Case study for segmental box girder bridges. *Journal of Infrastructure Systems*, vol. 19, no. 3.

O'Conner, P.E, Bucknell, J.R., DeFranco, S.J, Westlake, H.S. & Puskar, F.J. 2005. Structural integrity management (SIM) of offshore facilities. *2005 Offshore Technology Conference*, 2–5 May 2005, Houston, TX, USA.

Omar, W., King, N.S. & Hashim, S.M. 1992. System approach in bridge management. *7th Road Engineering of Asia & Australasia (RE-AAA) Conference*, 22–26 June 1992, Singapore.

Setunge, S., Zhu, W., Gravina, R. & Gamage, N. 2015. Fault-tree-based integrated approach assessing the risk of failure of deteriorated reinforced-concrete bridges. *Journal of Performance of Constructed Facilities.*

Sianipar, P.R.M. & Adams, T.M. 1997. Fault-tree model of bridge element deterioration due to interaction. *Journal of Infrastructure Systems*, vol. 3, no. 3: 103–110.

Stamatelatos, M., Vesely, W., Dugan, J., Fragola, J., Minarick III, J. & Railsback, J. 2002. *Fault Tree Handbook with Aerospace Applications*, NASA office of Safety and Mission Assurance. Washington, DC: NASA Headquarters.

Straub, D., Sorensen, J.D. & Faber, M.H. Benefit of risk based inspection planning for offshore structures, *OMAE'05 25th International Conference on Offshore Mechanics and Arctic Engineering*, June 4–9, 2006, Hamburg, Germany.

Weng, T.K., King, N.S., Hashim, S.M. & Jasmani, Z. 1991. The JKR bridge management system (JKR BMS). *2nd Regional Conference on Computer Applications in Civil Engineering*, 12–13 November 1991, Johor Bahru, Malaysia.

Zhu, W., Setunge, S., Gravina, R. & Venkatesan, S. 2008. Use of fault tree analysis in risk assessment of reinforced concrete bridges exposed to aggressive environments. *Concrete in Australia*, vol. 34, no. 1: 50–54.

Effect of high volume fly ash in ultra high performance concrete on compressive strength

Norzaireen Mohd Azmee & Muhd Fadhil Nuruddin
Department of Civil and Environmental Engineering, Universiti Teknologi PETRONAS, Seri Iskandar, Perak, Malaysia

ABSTRACT: Technology advances in concrete industry and increasing demand for high strength construction materials have led to the development of Ultra High Performance Concrete (UHPC). Conventional UHPC preparation method requires high cement consumption, high material cost and sophisticated technical procedures that makes it unsustainable and too costly to meet the demand of large-scale constructions. This paper addresses the development of UHPC incorporating high volume of fly ash (FA) of up to 50% as cement replacement material and the use of coarse aggregates in order to reduce the economical and environmental disadvantages of current UHPC. Influence of binder content, water to binder ratio and FA on workability and compressive strength of concrete were investigated. The test results showed that low cement UHPC of 450 kg/m^3 with compressive strength of about 160 MPa at 90 days could be prepared successfully with incorporation of both high volume of FA and coarse aggregates.

1 INTRODUCTION

With a view to reduce the high cost of ultra high performance concrete (UHPC), efforts have been directed to reduce its cost by incorporating high volume of fly ash (FA) as cement replacement and coarse aggregates in the mix. In typical UHPC mix proportioning, the cheapest constituent of conventional concrete is removed or replaced by more expensive components. Removal of coarse aggregates increased the powder usage in the mix. The cement used in UHPC becomes equivalent to the coarse aggregates used in conventional concrete. The mineral component optimization alone resulted to substantial increase in cost making the usage of UHPC very limited. As commonly known, the cement production is said to represent about 7% of the total CO_2 emissions (UNSTATS 2010, Friedlingstein et al. 2010 & Habert et al. 2013). Hence, one of the key sustainability challenges for concrete industry is to produce concrete with lower CO_2 emissions while providing same reliability with an improved durability.

Effective method in having sustainable UHPC mix is by reducing the cement amount without affecting the mechanical properties. Using certain amount of cement replacement material (CRM) in UHPC mixes can lead to a sustainable UHPC mix. Research work reported by Naik et al. (Naik & Ramme 1989) showed that 40% cement replacement by FA improved the concrete strength by 23% and 38% at 28 days and 56 days, respectively. The FA dosage is generally restricted to 15–20% by mass of total cementitious material for economical purposes. However, this small percentage is only beneficial in achieving good workability without any enhancement in durability.

Larger amount of FA that is in the range of 25–60% is needed to improve its durability. Study made by Malhotra (Malhotra 1999) showed that 50% or more cement replacement by FA is possible to produce sustainable, high performance concrete mixtures that show higher workability, higher ultimate strength, high durability with lower thermal and drying shrinkage. John and Ashok (John & Ashok 2014) reported that the use of high volume of FA along with superplasticizer in concrete achieved high strength both at early and later ages with good workability. Siddique (Siddique 2004) carried out experimental investigations on class F FA with percentage replacements of 40%, 45% and 50%. He concluded that partial replacement of FA in concrete mixes resulted in reduction of compressive strength, split tensile strength, modulus of elasticity and abrasion resistance at 28 days of age. However, all these properties of hardened concrete show significant improvement at 90 days and later.

Sujjavanich et al. (Sujjavanich, Sida & Suwanvitaya 2005) investigated the effect of high volume of FA in concrete on steel corrosion and chloride penetration. They concluded that using high amount of FA in concrete has lower down the chloride permeability therefore minimizing the risk of corrosion. Study made by Jerath et al. (Jerath & Hanson 2007) reported that increase of FA from 30% to 45% increased the durability of concrete without any loss of compressive and flexural strength. Halit (Halit 2008) made a study on self-compacting concrete with FA in the range of 30% to 60% together with 10% of silica fume (SF). He reported that addition of 10% SF positively affected both the fresh and hardened properties of high performance high volume FA concrete. From previous results, it can be summarized that fly ash have negative effects on concrete strength which can be contradicting in preparing UHPC. However, when properly designed high volume FA concrete can have adequate early-age strength and high later-age strength.

In UHPC design, the absence of coarse aggregate is considered as the key aspect for the microstructure and performance in order to reduce heterogeneity between cement matrix and the aggregates (Raja & Sujatha 2014). However, due to the use of very fine sand, the cement factor of UHPC can reach as high as 900–1000 kg/m^3 compared to 300–500 kg/m^3 for conventional concrete. Besides the increase in cost, the unusual usage of cement factor in UHPC could also increase its drying shrinkage and creep strain. Collerpardi (Collerpardi 1997) suggested that adding 8 mm natural aggregate into UHPC mix replacing part of the cementitious material does not reduce the compressive strength. The cube compressive strength for tested mix was reported around 160–210 MPa. Research work made by Ma et al. (Ma et al. 2004) using basalt aggregate of 2–5 mm size on UHPC mixes also concluded that the use of coarse aggregate did not affect its compressive strength. The reported cylinder compressive strength for tested mix was around 150–165 MPa.

Recent work reported by Yang et al. (Yang & Abdul Razak 2012) using limestone as coarse aggregate with two particle size ranges from 5–10 mm and 10–20 mm. The aggregate mass proportion used in UHPC mix was 3:7 with crushing index of 4.0%. They reported that the UHPC mix incorporating coarse aggregate could be prepared successfully with the compressive strength of about 150 MPa. Table 2 presents various mix designs for UHPC incorporating coarse aggregate obtained from previous researchers. These results are in contrast to the original concept of Richard and Cheyrezy (Richard & Cheyrezy 1995), which, suggest that coarse aggregate is the weakest link in achieving ultra high strength, and therefore should be excluded from the mix design. Previous findings showed that the use of coarse aggregate in UHPC mixes led not only to decrease in cementitious paste volume fraction, but also to some changes in mixing process and in mechanical properties. UHPC mix containing coarse aggregate was reported to have shorter mixing time, as it was easier to be fluidized and homogenized. The lower paste volume in UHPC mix incorporating coarse aggregate resulted in lower autogenous shrinkage.

The incorporations of high volume FA and coarse aggregate in UHPC are considered as a new type of concrete; therefore, it is important to have a complete knowledge on the behavior of the composite material. In this paper, an effort was made to investigate the possibility of producing low cement UHPC incorporating both high volume FA in the rage of 30%–50% and coarse aggregate without affecting concrete strength at both early and later age.

Table 1. Chemical oxide composition of OPC and FA.

Oxide	Percentage (%)		
	OPC	FA	SF
Na_2O	0.20		
MgO	2.42		
Al_2O_3	4.45	29.1	0.71
SiO_2	21.45	51.7	90.36
P_2O_5	0.11	1.7	
K_2O	0.83	1.6	
CaO	63.81	8.84	0.45
TiO_2	0.22	0.7	
Fe_2O_3	3.07	4.76	1.31
SO_3	2.46	1.5	0.41
MnO	0.20		

Table 2. High FA UHPC mixture proportions.

Mix ID	OPC	CRM		Sand	Coarse Aggregates	W/B
		SF	FA			
0.15F30	630	–	270	620	930	0.15
0.16F30	630	–	270	620	930	0.16
0.17F30	630	–	270	620	930	0.17
0.18F30	630	–	270	620	930	0.18
Reference	900	–	–	620	930	0.16
F30	630	–	270	620	930	0.16
F40	540	–	360	620	930	0.16
F50	450	–	450	620	930	0.16
F30S10	540	90	270	620	930	0.16
F40S10	450	90	360	620	930	0.16
F50S10	360	90	450	620	930	0.16

2 EXPERIMENTAL WORK

2.1 Work program

This work was designed to study the effects of water to binder ratios and the use of high volume FA content on compressive strength of UHPC. The cement was replaced by weight with FA at various fractions of 30%, 40% and 50%. The optimum amount of FA used in UHPC should be established by testing to determine whether the supplementary material indeed improve the properties of UHPC and to determine the right quantity for FA used, as an excessive or inadequate quantity can negatively affect the concrete properties.

In this work, a total of 11 UHPC mixes were prepared and its effect on compressive strength was studied.

2.2 Material properties

FA and SF used as cement replacement materials were locally produced. The chemical oxide composition for ordinary Portland cement (OPC), FA and SF are given in Table 1. River sand with a fineness modulus of 2.8 were used in all mixes. Polycarboxylic Ether based superplasticizer was used to adjust the workability of UHPC.

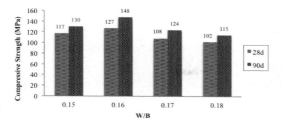

Figure 1. Effect of W/B on compressive strength of UHPC at different ages.

2.3 Concrete mix proportions

Details of the mix proportions are given in Table 2.

2.4 Experimental program

The experimental program consisted of casting and testing 66 cubes (100 mm). Concrete compressive strength of UHPC was determined using cubes tested at 28 and 90 days. Three specimens were tested at each age to compute the average strength.

3 RESULTS AND DISCUSSIONS

3.1 Influence of water binder ratio on compressive strength of UHPC

The influence of water to binder ratio (W/B) on compressive strength of low cement UHPC was investigated only on the UHPC mix with 30% FA content. Four mixes containing 900 kg/m^3 binder material with various W/B of 0.15, 0.16, 0.17 and 0.18 were tested. The results can be seen in Figure 1. It can be concluded that compressive strength of the UHPC decreased with the increasing of W/B. However, the highest strength was achieved with a W/B of 0.16 and not with W/B of 0.15 at later age of 90 days. This is because the unhydrated cement in low W/B had a detrimental effect on the structure of the hydrated material. Previous research made by Wang et al. (Wang et al. 2008) stated that hydration degree of hardened cement paste with W/B of 0.16 is higher than mix with W/B of 0.10 at all ages. Madhavi et. al (Madhavi, Raju & Mathur 2014) also demonstrated similar results and proved that early strength of high volume fly ash concrete could be achieved with W/B of less than 0.3. Beside lower compressive strength, it is observed that mix with a W/B of above 0.18 possessed unstable mix with slight segregation that can be seen during the workability test as shown in Figure 2 indicating poor cohesiveness between mortar paste and aggregate.

3.2 Influence of FA and SF dosage on compressive strength of UHPC

Compressive strength of UHPC influenced by different dosage of FA volume for up to 50% together with or without the combination of SF was investigated.

Figure 2. Workability test (slump flow) for mix 0.18FA30.

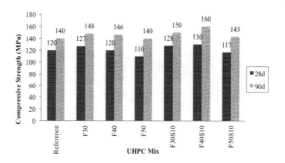

Figure 3. Effect of different mineral admixture on compressive strength of UHPC.

Seven mixes with various FA and SF replacements are listed in Table 2.

The result in Figure 3 indicated that, blending 40% FA together with 10% SF had the best effect on improving the compressive strength of UHPC at both 28 and 90 days. No reduction of strength can be seen for F40S10 mix on all ages. The compressive strength at the age of 28 and 90 days for mix F40S10 showed higher value than the reference mix by 8 and 15% respectively.

Incorporation of 30% FA in UHPC increased the compressive strength at 28 days however, the increased of FA from 30% to 50% resulted to lower 28 days compressive strength by about 15%. Strength development can be seen only after 28 days for all FA mixes. Combining SF in the mix can counteract the reduced strength due to the increase of FA volume. Addition of 10% SF showed positive effect on compressive strength when, combined with FA at both 28 and 90 days except for mix F50S10. The compressive strength was up to 10% higher compared to mixes with FA alone together with 15% improvement on reference. Incorporation of FA reduces the heat of hydration thus slower down the setting time that causes reduction in the early age strength. Higher early strength can be achieved with the combination of SF in the mix as reactive pozzolan and effective void filler to improve cement paste cohesion and aggregate particle adhesion. Xu and Sarkar (Xu & Sarkar 1994) explained that better dispersion of cement grains and development of gel phases are largely responsible for the strengthening effect.

Higher compressive strength of more than 140MPa can be seen at 90 days for all mixes with FA with an improvement of up to 13% compared to reference. Larger FA replacement of up to 40% showed better strength performance at later age for UHPC mix with W/B of 0.16. The strength enhancement due to the increase of FA volume is acceptable when compared to results obtained by Lam et al. (Lam, Wong & Poon 1998). They demonstrated an increase of 28-day compressive strength for high strength concrete mix with large volume of FA replacement of up to 45% if W/B was reduced from 0.5 to 0.3. The high volume of FA with low W/B ratios underwent a lower degree of hydration/reaction. This is because there is less water available for the reaction and less space for the reaction products to form. It is reported that about 40% of the cement and 80% of the FA remained unreacted at the age of 90 days. These unreacted cement and FA particles served as micro-aggregates, which also contributed to the strength of the cementitious material.

4 CONCLUSIONS

Based on the results of this experiment the following conclusions can be made:

1. Low cement UHPC with compressive strength of 130 and 160MPa at 28 and 90 days respectively, could be obtained through 40% FA and 10% SF with W/B of 0.16. UHPC mix with W/B of less than 0.16 possessed unworkable mix (stiff mix) with lower compressive strength at both 28 and 90 days.
2. The negative effect of using high volume of FA for up to 50% on concrete strength of UHPC appeared to be insignificant with strength reduction of less than 3%. The combinations of high volume of FA with low W/B in UHPC mix lower down the hydration process for cement and FA particles which then acted as micro-aggregates to increase the concrete strength.
3. Addition of 10% SF showed positive effect on compressive strength when combined with FA at both 28 and 90 days with improvement of 2% to 15% respectively. SF acted as reactive pozzolan and effective void filler to improve cement paste cohesion and aggregate particle adhesion to form better dispersion of cement grains and development of gel phases.

REFERENCES

Collepardi, S. 1997. Mechanical Properties of Modified Reactive Powder Concrete. *Proceedings of the Fifth CANMET/ACI International Conference on Superplasticizers and Other Chemical Admixtures in Concrete*, Rome, Italy, 173:1–21.

Friendlingstein, P. Houghton, R.A. Marland, G. Hackler, J. Boden, T.A. & Conwat, T.J. et al. 2010. Uptake on CO_2 Emission. *Nat Geosci* 3:811–812.

Habert, G. Denarie, E. Sajna, A. & Rossi, P. 2013. Lowering the Global Warming Impact of Bridge Rehabilitations by Using Ultra High Performance Fiber Reinforced Concretes. *Cement and Concrete Composites* 38:1–11.

Halit, Y. 2008. The Effect of Silica Fume and High Volume of Class C Fly Ash on Mechanical Properties, Chloride Penetration and Freeze Thaw Resistance of SCC. *Construction Building and Materials* 22:456–462.

Jerath, S. & Hanson, N. 2007. Effect of Fly Ash Content and Aggregate Gradation on the Durability of Concrete Pavement. *Journal of Materials in Civil Engineering* 19(5):367–375.

John, J. & Ashok, M. Strength Study on High Volume Fly Ash Concrete. 2014. *International Journal of Advanced Structures and Geotechnical Engineering* 3(2):168–171.

Lam, L. Wong, Y.L. & Poon, C.S. 1998. Effect of Fly Ash and Silica Fume On Compressive and Fracture Behaviors of Concrete. *Cement and Concrete Research* 28:271–283.

Ma, J. Dehn, F. Tue, N.V. Orgass, M. & Schmidt, D. 2004. Comparative Investigations on UHPC With and Without Coarse Aggregates. *Proceeding of The International Symposium on UHPC, Kassel, Germany*, 205–212.

Malhotra, V.M. 1999. Making Concrete Greener with Fly Ash. *Concrete International* 21(5):61–66.

Madhavi, T.C. Raju, L.S. & Mathur, D. 2014. Durability and Strength Properties of High Volume Fly Ash Concrete. *Journal of Civil Engineering Research* 4:7–11.

Naik, T.R. & Ramme B.W. 1989. High Strength Concrete Containing Large Quantity of Fly Ash. *ACI Material Journal* 86(2):111–116.

Raja, L.V.N. & Sujatha, T. 2014. Study on Properties of Modified Reactive Powder Concrete. *International Journal of Engineering Research and Technology* 3(10):937–940.

Richard, P. & Cheyrezy, M. 1995. Composition of Reactive Powder Concrete. *Cement and Concrete Research* 25(7):1501–1511.

Siddique, R. Performance Characteristic of High Volume Class F Fly Ash Concrete. 2004. *Cement and Concrete Research* 34:487–493.

Sujjavanich, S. Sida, V. & Suwanvitaya, P. 2005. Chloride Permeability Corrosion Risk of High Volume Fly Ash Concrete with Mid Range Water Reducer. *ACI Material Journal* 102(3):243–247.

UNSTATS. 2010. Green House Gas Emissions by Sector (Absolute Values). *United Nation Statistical Division* Springer.

Wang, C. Pu, X.C. Chen, K. Liu, F. Wu, J.H. & Peng, X.Q. 2008. Measurement of hydration progress of cement paste material with extreme low W/B. *Material Science Engineering* 6:852–857.

Xu, A. & Sarkar, S.L. 1994. Microstructural developments in high-volume fly ash cement system. *Journal Material Civil Engineering* 6:117–136.

Yang, J. & Abdul Razak, H. 2012. Assessment of effectiveness of CFRP repaired beams under different damage levels based on flexural stiffness. *Construction Building and Materials* 37:125–134.

Behaviour of structural properties of fiber reinforced high strength concrete beams subjected to static loads

Nasir Shafiq, Muhd Fadhil Nuruddin & Ali Elheber Ahmed Elshekh
Department of Civil and Environmental Engineering, Universiti Teknologi PETRONAS, Seri Iskandar, Perak, Malaysia

ABSTRACT: This investigation has been conducted to assess the static structural properties of high strength concrete (HSC) beams utilizing chopped basalt stands fiber (CBSF), hybrid fiber (HYF) of (CBSF + polyvinle fiber (PVAF)) and PVAF as internal strengthening materials. The effect of the three type fibers on the structural properties of HSC beams subjected to static loads has been done using three point's flexural load. The test results showed that the yield and ultimate static flexural loads of fiber reinforced HSC beams were found to increase as compared to that of the control mixes. In addition, the middle-span deflection of HSC beams was found to increase as compared to that of plain HSC.

1 INTRODUCTION

Cracks in HSC concrete can propagate rapidly causing damage at a low ultimate strain without any noticeable deformation. This makes the collapse of HSC without warning deformation, causes a high level of risk to the users of the structures V. Bajaj (2010) and, J. Proulx (2006). That related to that, the changes are mostly related to the progressive development of internal micro-cracks, resulting from a significant increase of strain ACI (1997). The micro-crack would appear when the mechanical properties of the materials of concrete due to the cyclic load are changed such as brittleness and ductility of concrete which effect on the serveavable life of structure. Therefore, researches on concrete structural building are focused towards improving the serviceability and safety of concrete structures V. Bajaj (2010) and S. K. Kaushik (2003).

Fiber reinforced concrete is known as a composite of concrete reinforced by a random distribution of short, discontinuous and discrete fibers of a specific geometry. Fibers have been extensively used to improve the tensile strength, reduce the brittleness and provide ductility to HSC. Fibers have been studied to become an alternative to traditional steel reinforcement of concretes which are utilized as externally and eternally strengthening materials ACI (1997), V. C. Li (1999), H. Hammoud (1998) and J. Walraven (2006).

A number of researches have been conducted to study the effect of discontinuous short fibers on the mechanical and structural characteristics of hardened concrete S. Kannan (2013) and F. Bondioli (2012). However, the application of the used fibers in concrete is still limited due to the environmental unfriendliness of steel fiber, high alkaline condition of the glass and the brittleness and cost of carbon fiber I. De Iorio (2011), N. Kabay (2014), T. M. Borhan (2012), D. Mikulskis (2014) and D. Moon (2005).

Therefore, the natural fiber, which named basalt fiber, is going to be an alternative due to its properties. The effect of CBSF, HYF of (CBSF and PVAF) and PVAF on the structural properties of fiber reinforced beam of HSC are not yet existing in the open published literatures. The main aims of this paper were to determine the effect of the CBF and HYF and PVAF on the structural response under the static loads. The results of the experimental tests of HSC mixes with various contents of the CBSF, HYF and PVAF are presented in this paper.

2 METHODOLOGY

The materials used in this study are cement, fly ash, fine and coarse aggregate, high effective superplasticizer, CBSF, PVAF and water.

The CBSF, HYF and PVAF with different amount were added to the control mixtures 100% OPC and 80% OPC as per 1%, 2%, 3%, and 4% of CBSF, (1.5% CBSF +1.5% PVAF) and (2% CBSF +2% PVAF) of HYF and 3% and % 4% of PVAF by weight of the binder materials as the internal strengthening materials of HSC. Nine beams having a dimension of $100 \times 100 \times 500$ mm were cast and tested to study the static flexural strength of the fiber reinforced HSC beams. Four beams were strengthened with CBSF; two beams were strengthened with PVA fiber; two beams were strengthened with hybrid fiber and control beam. The load was applied from UTM hydraulic machine actuator of the 100KN capacity controlled by Star-II software. Three-point bending tests were applied in the closed loop of the controlled material test for the static

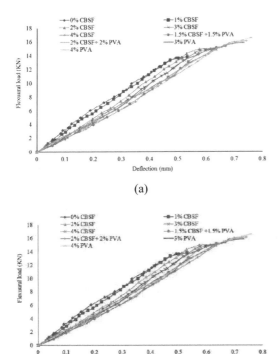

Figure 1. Effect of the fiber on the static flexural load-deflection curve of HSC beams.

and dynamic loading tests. A test of the span beam of 400 mm was used under the capability of the loading frame 1000 KN (20 kips) as per ASTM C1963. The deflection of the middle-span beams was measured using a computer system which is capable to record the stress loads as well.

3 RESULTS AND DISCUSSION

In this study, the effects of the fibers on the structural properties of HSC were investigated. The tests result, analysis and discussion are presented in the following sections.

3.1 Effect of fibers on the static flexural strength of HSC beams

Figure 1 shows the effect of fibers on the load deflection curve HSC beams under static flexural load, which presents the yield and ultimate flexural load as well as yield and ultimate deflection. The following paragraphs discuss the effect of each fiber on the static flexural strength of the HSC.

It was observed that the yield and ultimate load of the fiber reinforced beams utilizing the CBSF was marginally increased as compared to that of the control mixes.

The enhancement was proportionally increased with increasing the CBSF content up to 3% then slowly decreased. The load carrying capacity has been increased even after starting the cracks at middle span. The averages increase of the yield flexural loads of the CBSF reinforced beam was found to be in range of 14 to 24% as compared to the averages of that of the control mix. Similarly, the ultimate flexural of the CBSF reinforced beam shows typical trend, which was found to be in range of 20 to 29% as compared to that of the control mix. CBSF provided more capability to resist the bending flexural strength resulting in an increase in the load carrying capacity after the middle-span cracks start.

It was also observed that the yield and ultimate middle–span deflection of the CBSF reinforced beams were significantly increased with the increase of the CBSF content up to 3% then slowly getting decreased. The improvement of the middle-span deflection at yielding was found to be in the range of 14 to 53% as compared to that of the control mix. Typically, the enhancement of middle span at ultimate load was found to be in the range of 16 to 63%, which could be due to the increased fiber availability of the fiber made it highly efficient in delaying the propagation of micro cracks. In addition, the smaller length of the CBSF can increase the maximum tensile stress capacity resulting to improve in the flexural strength. Additionally, the mechanical properties of the CBSF, such as toughness, tensile strength and flexure strength actually caused the enhancement of the interfacial adhesive bond between the fiber and paste materials.

HYF provided more capability in resisting the bending flexural strength as compared to that of the CBSF. Consequently, there was an increase in the load carrying capacity of the HSC beam after the starting of the bending middle-span cracks. The yield and ultimate static flexural load of the HYF reinforced beams increased proportionally with an increase of the hybrid fiber content. The enhancement of the yield flexural load of HYF reinforced beams were found in the range of 31–33% as compared to that of the control beam. Likewise, the ultimate flexural load of the HYF reinforced beam shows an enhancement as compared to that of the yield flexural load. There were about 29.39% and 33.27% increase in the ultimate flexural load respectively, at the same percentage of the fiber as compared to that of control. Additionally, it was observed that the crack propagation started from the bottom to top which means that the fiber helps increase the load capability of the beams after the cracks start resulting in a delay of the collapse of the beams.

Similarly, yield and ultimate middle-span deflections of the HYF reinforced beams were found to be significantly increased with increasing the HYF content. The enhancement of middle-span deflection at yielding was found to be in the range of 60–70% as compared to that of the control mix. In addition, the enhancement of middle-deflection at ultimate load of HYF reinforced beams was found to be in the rage of 65–80% as compared to that of control. The HYF showed more effect in delaying the propagation of micro cracks than the CBSF, which could be the reason to the increase in the middle deflection of the HYF reinforced beam. Consequently, the PVA showed better bridging of the PVA fiber reinforced concrete cracks, and the interfacial transition bond degradation is a possible source as compared to that of CBSF.

The ultimate flexural load of the PVAF reinforced beam was increased by about 28.18% and 31.48% at 3% and 4% PVAF, respectively, as compared to that of control mix. The PVAF lodged the cracks by widening themselves, thus making an increase in the energy-absorbing mechanism and stress relaxing of the micro-cracked region adjacent to the crack–tip. The crack crossing of the PVAF exceeded the CBSF in the concrete fiber spreading and the tensile strength. Moreover, the flexural load of PVAF was found to be even higher than that of the HYF reinforced HSC beams at the same percentage of fiber. The PVAF have provided a higher toughness and established the post peak-load trend as compared to the control mixes. Similarly, The yield middle-span deflection of the PVA reinforced beams at 3% and 4% of PVAF were found to be increased in the range of, as compared to that of the control mix. The mechanical properties of the PVA, such as toughness, tensile strength and flexure strength, were actually affected to enhance the interfacial adhesive bond between the fiber and paste materials as well.

Furthermore, the yield and ultimate load at middle-span deflection of the three types of fiber reinforced beam were increased using FA together with the fiber as compared to that without FA. That because the FA can increase the distribution of the fiber during the fresh stage of the concrete.

3.2 *Effect of fibers on the of HSC beams*

Figure 2 presents the effect of the CBSF on the ductility of the concrete beams. The addition of 1%, 2%, 3% and 4% of the CBSF was found to increase the ductility index by 12.1%, 24.84%, 33.54% and 26.99% respectively, as compared to that of the 100% OPC beam. This enhancement indicates that the addition of CBSF as strengthening material greatly increased the capability against the deformation before the failure. This enhancement can be explained from the experimental observation as follows: due to the strength of the fiber and bond between the fiber and concrete paste that can help increase the transfer load to concrete after yielding after the initiating of the flexural crack.

It can easily be seen that the ductility of the CBSF reinforced beam with 20% replacement of cement by FA was found higher as compared with that of the CBSF reinforced concrete beams with 100% OPC. That is because of that FA improved the disruption of the fiber inside the beam resulting to increase in capability of concrete against the tensile stress more than that of 100% OPC concrete. As compare to the control beam of the 80%, the increase in the ductility index was found to be by 20.18%, 29.474%, 34.527%

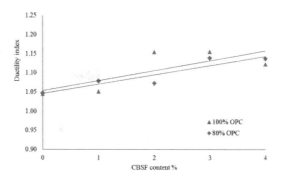

Figure 2. Effect of the BCSF on ductility of HSC beams.

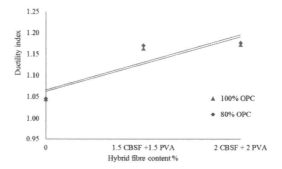

Figure 3. Effect of the Hybrid fibre on ductility of HSC beams.

and 28.53% at the percentage of the fiber of 1%, 2%, 3% and 4%, respectively.

Figure 3 shows the effect of the hybrid fiber which combined CBSF and PVA on the ductility of the concrete beams. The addition of 1.5% CBSF + 1.5% PVA and 2% CBSF% + 2% PVA of the CBSF increased the ductility index by 36.6 % and 41.9% respectively, as compared to that of the control beam of 100% OPC beam. Similarly, the increase in ductility index of the beam was found to be 37 % and 41.6% at the same percentage of hybrid fiber as compared to that of the 80% OPC beam.

This improvement reflects that the addition of hybrid fiber as strengthening materials has a positive effect of increasing the deforming capability at the failure. That is because of the of the PVA fiber contribution in term of strength which is higher than that of the CBSF. That is because of the of the PVA fiber contribution in term of strength which is higher than that of the CBSF.

Moreover, the bond between the fiber and concrete paste has more effect to increase the load transfer to concrete after yielding point after the flexural crack as compared to that of the CBSF. The ductility index of hybrid reinforced of 80% OPC beam was higher than that of the100% OPC beam. That is due to the effect of the FA in enhancing the distribution of the fiber inside the concrete resulting in an increase of capability of concrete against the tensile stress.

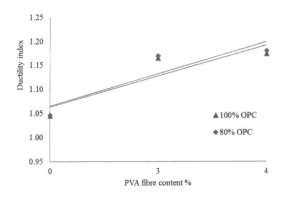

Figure 4. Effect of the PVA fibre on ductility of HSC beams.

Figure 4 presents the influence of the PVA fiber on the ductility of the concrete beams. The addition of 3% and 4% PVA fiber was found to increase the ductility index by 40.99% and 46.21%, respectively as compared to that of the control beam of 100% OPC. Similarly the increase in the ductility index of the beam was found to be 41.7% and 47.2% at the same percentage of hybrid fiber as compared to that of the 80% OPC.

The ductility index of the PVA reinforced beams was found to be higher than that of the hybrid reinforced beams due to the higher strength of the fiber when compared to the CBSF and HYF reinforced HSC beams.

Moreover, PVA fiber has a larger surface area as compared to that of the CBSF and HYF resulting in an increased the bond strength between the PVA fiber and the concrete paste. This makes the beams carrying more loads after yielding point before reach the ultimate load at the failure point. The ductility index of PVA reinforced beams of 80% OPC was higher when compared to the ductility index of PVA reinforced beams of 100% OPC.

The ductility index of the PVA reinforced beams was found to be higher than that of the hybrid reinforced beams due to the higher strength of the fiber when compared to the CBSF and HYF reinforced HSC beams.

Moreover, PVA fiber has a larger surface area as compared to that of the CBSF and HYF resulting in an increased the bond strength between the PVA fiber and the concrete paste. This makes the beams carrying more loads after yielding point before reach the ultimate load at the failure point. The ductility index of PVA reinforced beams of 80% OPC was higher when compared to the ductility index of PVA reinforced beams of 100% OPC.

4 CONCULUSION

The yield and ultimate flexural static loads of CBSF reinforced HSC beams were found to increase compared to that of the control mixes. The average static flexural strengths of CBSF reinforced HSC was in the range of 22.5–38.85% as compared to the control mix. Moreover, CBSF was found to increase the middle – span deflection of the beams. The values of middle-span deflection of the CBSF reinforced beams were found to increase, indicating an improvement in the ductility of the beams by utilizing CBSF as strengthening material.

Bending flexural strengths of HYF and PVAF fiber reinforced HSC beams were higher as compared to that of the CBSF reinforced HSC beams. On average the static flexural strength of the HYF reinforced HSC beams increased in the range of 26.66–36.67% as compared to the control mixes. Similarly, middle-span deflection of PVAF reinforced HSC beams increased in the range of 187–253% as compared to control mixes.

The increase of the values of the middle-span deflection of HYF reinforced beams were in the range of 76–79% for 80% as compared to the control mixes. Addition of PVAF shows higher values of middle-span deflection as compared to that of the CBSF and hybrid fibers which were in the range of 83–88% as compared to that of the control mixes.

ACKNOWLEDGEMENT

The support received from the Universiti Teknolgi PETRONS is sincerely acknowledged. This work at the civil engineering department laboratory could not have been completed without the help of Mr. Johan and Mr. Hafiz, lab technicians, and the authors would like to thank them for their help.

REFERENCES

ACI 215R, Considerations for Design of Concrete Structures Subjected to Fatigue Loading, 1997.

E. Naaman and H. Hammoud, "Fatigue characteristics of high performance fiber reinforced concrete," Cem. Concr. Compos., vol. 20, no. 5, pp. 353–363, Oct. 1998.

C. Gheorghiu, P. Labossière, and J. Proulx, "Fatigue and monotonic strength of RC beams strengthened with CFRPs," Compos. Part A Appl. Sci. Manuf., vol. 37, no. 8, pp. 1111–1118, Aug. 2006.

E. Lappa, C. Braam, and J. Walraven, Flexural fatigue of high and ultra high strength fiber reinforced concrete, no. 1. RILEM Publications SARL, 2006, pp. 509–518.

E. Quagliarini, F. Monni, S. Lenci, and F. Bondioli, "Tensile characterization of basalt fiber rods and ropes: A first contribution," Constr. Build. Mater., vol. 34, pp. 372–380, Sep. 2012.

J. Bošnjak, J. Ožbolt, and R. Hahn, "Permeability measurement on high strength concrete without and with polypropylene fibers at elevated temperatures using a new test setup," Cem. Concr. Res., vol. 53, pp. 104–111, Nov. 2013.

J. Zhang, H. Stang, and V. C. Li, "Fatigue life prediction of fiber reinforced concrete under flexural load," Int. J. Fatigue, vol. 21, no. 10, pp. 1033–1049, Nov. 1999.

J. Sim, C. Park, and D. Moon, "Characteristics of basalt fiber as a strengthening material for concrete structures,"

Compos. Part B Eng., vol. 36, no. 6–7, pp. 504–512, Jan. 2005.

M. Mahalingam, R. Pulipakka, N. Rao, and S. Kannan, "Ductility Behavior Fiber Reinforced Concrete Beams Strengthened With Externally Bonded Glass Fiber Reinforced Polymer Laminates," Am. J. Appl. Sci., vol. 10, no. 1, pp. 107–111, Jan. 2013.

M. M. Kamal, M. a. Safan, Z. a. Etman, and R. a. Salama, "Behavior and strength of beams cast with ultra high strength concrete containing different types of fibers," HBRC J., vol. 10, no. 1, pp. 55–63, Apr. 2014.

M. S. A. Sivakumar, "Mechanical properties of high strength concrete reinforced with metallic and non-metallic fibres," Cem. Concr. Compos., vol. 29, pp. 603–608, 2007.

M. Sinica, G. a Sezeman, D. Mikulskis, M. Kligys, and V. Èesnauskas, "Impact of complex consisting of continuous basalt fibres and SiO2 microdust on strength and heat resistance properties of autoclaved aerated concrete," Constr. Build. Mater., vol. 50, pp. 718–726, Jan. 2014.

N. Kabay, "Abrasion resistance and fracture energy of concretes with basalt fiber," Constr. Build. Mater., vol. 50, pp. 95–101, Jan. 2014.

S. Singh, A. Singh, and V. Bajaj, "Strength and Flexural Toughness of Concrete Reinforced with Steel-Polypropylene Hybrid Fibres," Asian J. Civ. Eng., vol. 11, no. 4, pp. 495–507, 2010.

S. P. Singh and S. K. Kaushik, "Fatigue strength of steel fibre reinforced concrete in flexure," Cem. Concr. Compos., vol. 25, no. 7, pp. 779–786, Oct. 2003.

T. M. Borhan, "Properties of glass concrete reinforced with short basalt fibre," Mater. Des., vol. 42, pp. 265–271, Dec. 2012.

V. Lopresto, C. Leone, and I. De Iorio, "Mechanical characterisation of basalt fibre reinforced plastic," Compos. Part B Eng., vol. 42, no. 4, pp. 717–723, Jun. 2011.

Single strap pull test of carbon fibre reinforced polymer plated steel member under fatigue loading

Norfarahana Osman & Airil Yasreen Mohd Yassin
Department of Civil and Environmental Engineering, Universiti Teknologi PETRONAS, Seri Iskandar, Perak, Malaysia

Ibrisam Akbar
Heriot-Watt University Malaysia, Putrajaya, Wilayah Persekutuan Putrajaya, Malaysia

ABSTRACT: This paper presents the result of the bonded CFRP-to-steel through the single strap pull test survived until 1 million cycles during fatigue test which have not yet failed. The test results demonstrated that the specimen able to withstand approximately 1 million cycles and despite the evidence of some debonding of the specimen was yet to fully debond and fail.

1 INTRODUCTION

Conventionally, steel is used as strengthening material to strengthen structures by welding, bolting and riveting. However this approach would then increase the total weight of the strengthened structure. Alternatively, Carbon Fibre Reinforced Polymer (CFRP) is lightweight thus makes it a good material in strengthening steel structures. CFRP able to resist high pressure and produces small displacement. The good performance of CFRP as strengthening material and method have been reported numerously in the literature [Tavakkolizadeh, M., & Saadatmanesh, H. (2003), Jones, S. C., & Civjan, S. A. (2003), Schnerch, D. (2005), Colombi, P., Fava, G., & Sonzogni, L. (2014), Colombi, P., Fava, G., Poggi, C., & Sonzogni, L. (2014) & Colombi, P., & Fava, G. (2015)].

Fatigue is related to the failure of a structure when subjected to a cyclic load. Crack propagation would occur through the bonded structure when fatigue failure takes place [Kamruzzaman, M., Jumaat, M. Z., Ramli Sulong, N. H., & Islam, A. B. M. (2014)]. Weakest part of the bonded structure of CFRP plate and steel would be the adhesive. There are four types of debonding failure; CFRP and adhesive interfaces failure, steel and adhesive interfaces failure, adhesive layer failure and CFRP delamination [Zhao, X. L., & Zhang, L. (2007)]. In bonded structure, the aim is to ensure that the stress can be successfully transferred from the structure through the adhesive layer to the CFRP plate. To achieve this, knowledge on types of failure modes of CFRP is vital. An effective transfer of stress would be when the CFRP fails by delamination. Studies have found that when the adhesive layer used was relatively thick (>2 mm) the delamination would be the potential failure mode whilst for lesser thickness ($\leq$2 mm) the material failed at the adhesive layer regardless of the types of adhesive used [Xia, S. H., & Teng, J. G. (2005) & Akbar, I., Oehlers, D. J., & Ali, M. M. (2010)].

The present paper is the extension of the work of [Xia, S. H., & Teng, J. G. (2005) & Akbar, I., Oehlers, D. J., & Ali, M. M. (2010)]. The study [Xia, S. H., & Teng, J. G. (2005) & Akbar, I., Oehlers, D. J., & Ali, M. M. (2010)] worked on the same geometrical properties of the specimen and focused on the interfacial shear stress and the bond strength under static loading. This paper presents the result of specimen which currently withstands 1 million cycles of fatigue test and not yet failed.

2 EXPERIMENTAL WORK

2.1 Test method

The single strap pull test was carried out in the present study. This method is an established method to investigate the behaviour of CFRP-to-steel joint [Xia, S. H., & Teng, J. G. (2005), Akbar, I., Oehlers, D. J., & Ali, M. M. (2010) & Yu, T., Fernando, D., Teng, J. G., & Zhao, X. L. (2012)]. This method is suitable in the study of bond-slip relationship since there is only one side of the possible debonding. The specimen was subjected to the static tensile test. The test was conducted using Universal Testing Machine (UTM) with 1000 kN capacity. The CFRP plated steel member was placed

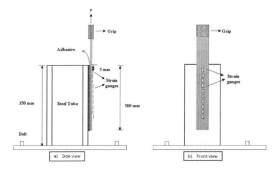

Figure 1. Fatigue test set-up.

Table 1. Material Properties of CFRP.

Thickness (mm)	Tensile strength (N/mm^2)	Tensile E-Modulus (N/mm^2)	Strain at Break (N/mm^2)
1.2	Mean Value: 3100 Min Value: >2800 5% Fractile-Value: 3000 95% Fractile-Value: 3600	165,000	>1.70%

Table 2. Material properties of adhesive.

Adhesive	Tensile strength $\sigma\ max$ (MPa)	Young's Modulus Ea (MPa)	Strain at Break (%)
Sikadur 30	22.53	4013	0.5614
Araldite 420	20.48	10793	0.1898

upward, while the bottom of the steel was connected to the UTM as shown in Figure 1 to minimize plate bending.

2.2 Specimen detail

A single specimen of CFRP plated steel members was tested. The CFRP plate had a bonded length of 350 mm, width bp of 50 mm, thickness tp of 1.2 mm and Young Modulus E of 165 GPa. The details of material properties of CFRP plate are as given in Table 1. The steel block was formed by welding two 10 mm thick steel plates to two (75 mm × 40 mm) rectangular hollow section of 2 mm in thickness. The CFRP plate was glued and placed at the centre of the steel block using adhesive of Araldite 420.

Two surfaces (top and bottom) of the steel blocks were sandblasted and cleaned with acetone to enhance the bonding capability and remove dirt, rust and residues. Yellow tape was applied in lined on the steel surface to ensure the neatness of the steel surface while applying the adhesive. The thickness of adhesive was set approximately to 1 mm. In order to ensure selected thickness been achieved, steel rods with diameter 1 mm were used and glued to the steel surface with a tiny drop of adhesive to keep it even along the bonded length. The mixed of Sikadur 30; Part A and Part B were used as an adhesive for this purpose. A ratio of 3:1 was used accordingly.

Since CFRP plate was too thin to be gripped by the UTM, four aluminium plate (50 mm × 50 mm) with 2 mm thickness were glued at both side of the end of CFRP plate using Araldite 420. This was to provide a rigid grip between the UTM and the CFRP plate. Material properties of adhesive used in the tests are shown in Table 2. Then, CFRP plate was pressed down right after the adhesive was laid out. This procedure aimed to squeeze out excessive adhesive out and provide an even surface and obtain the required thickness of adhesive. Sufficient weight (25 kg–30 kg) was placed on top of the CFRP plate-to-steel for seven days to achieve uniform surface and the required thickness. The process was called curing period.

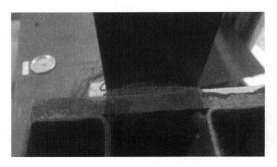

Figure 2. CFRP plate starts to debond.

2.3 Instrumentations

In the fatigue test, ten strain gauges were applied along the bonded length. The first strain gauge was positioned and glued 5 mm away from the load end and followed by the others. Distance between each of the strain gauges was 10 mm. The purpose of the strain gauge is to provide information on the CFRP plates Young's Modulus from the stress-strain curve which can be obtained from the load and strain data. Strain gauges can also detect any bending that may occur on the CFRP plate during experimental test.

2.4 Results

2.4.1 Debonding

Currently, the specimen has not yet failed and the result presented herein was recorded for 1 million cycles. Figure 2 shows the loaded end of the specimen when the debonding initiated after 43,000 cycles with the load range of 10 kN to 16 kN. After 1 million cycles, it was shown that half of the bonded part was debonded as shown in Figure 3. During the fatigue test, slow propagation through the adhesive layer was observed due to low frequency used, 1 Hz.

Figure 3. Half of debond.

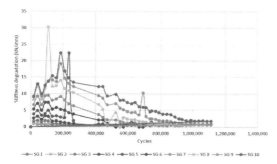

Figure 4. Stiffness degradation of the bonded structure.

2.4.2 Stiffness degradation

Durability of the bonded structure is measured by the stiffness degradation. Figure 4 shows the stiffness degradation throughout the 1 million cycles approximately. Ten strain gauges along the CFRP plate used to measure the elongation of the CFRP plate during the load pull which individual results are plotted in the Figure 4. Stiffness degradation is influenced by the load level, curing process of adhesive and environmental effects [Bai, Y., Nguyen, T. C., Zhao, X. L., & Al-Mahaidi, R. (2013)]. During the test, stiffness degradation can clearly be observed in Figure 4 by the applied cyclic load and the number of cycles.

3 DISCUSSION

Debonding of the CFRP plate from the steel is the main failure mode of the bonded structure. Expected debonding failure for the specimen are combination of cohesive failure and CFRP delamination by referring to the condition of the half debond in the Figure 4. Proper surface treatment need to be done to ensure the expected type of debonding is achieved. Based on the observation of fatigue test and the stiffness degradation in the Figure 5, using the load range of 10 kN to 16 kN, the cycles may continue up to 2 million cycles.

Behavior of the bonded specimen during the test can be observed by refer to the stiffness degradation curve. Later, stiffness degradation will be used to derive the S-N curve.

4 CONCLUSION

The single strap pull test was carried out subjected to the fatigue loading. The specimen survived until 1 million cycles and not yet failed although there was detachment of the CFRP plate from the adhesive layer. Based on the degradation of the stiffness, it is predicted that the full bonding failure will occur after 2 million cycles.

REFERENCES

Akbar, I., Oehlers, D. J., & Ali, M. M. (2010). Derivation of the bond–slip characteristics for FRP plated steel members. *Journal of Constructional Steel Research, 66*(8), 1047–1056.

Bai, Y., Nguyen, T. C., Zhao, X. L., & Al-Mahaidi, R. (2013). Environment-Assisted Degradation of the Bond between Steel and Carbon-Fiber-Reinforced Polymer. *Journal of Materials in Civil Engineering, 26*(9), 04014054.

Colombi, P., Fava, G., & Sonzogni, L. (2014). Fatigue Behavior of Cracked Steel Beams Reinforced by Using CFRP Materials. *Procedia Engineering, 74*, 388–391.

Colombi, P., Fava, G., Poggi, C., & Sonzogni, L. (2014). Fatigue Reinforcement of Steel Elements by CFRP Materials: Experimental Evidence, Analytical Model and Numerical Simulation. *Procedia Engineering, 74*, 384–387.

Colombi, P., & Fava, G. (2015). Experimental study on the fatigue behaviour of cracked steel beams repaired with CFRP plates. *Engineering Fracture Mechanics*.

Jones, S. C., & Civjan, S. A. (2003). Application of fiber reinforced polymer overlays to extend steel fatigue life. *Journal of Composites for Construction*.

Kamruzzaman, M., Jumaat, M. Z., Ramli Sulong, N. H., & Islam, A. B. M. (2014). A Review on Strengthening Steel Beams Using FRP under Fatigue. *The Scientific World Journal, 2014*.

Schnerch, D. (2005). Strengthening of steel structures with high modulus carbon fiber reinforced polymer (CFRP) materials.

Tavakkolizadeh, M., & Saadatmanesh, H. (2003). Fatigue strength of steel girders strengthened with carbon fiber reinforced polymer patch. *Journal of Structural Engineering, 129*(2), 186–196.

Xia, S. H., & Teng, J. G. (2005, December). Behaviour of FRP-to-steel bonded joints. In *Proceedings of the International Symposium on Bond Behaviour of FRP in Structures (BBFS 2005)* (pp. 411–418).

Yu, T., Fernando, D., Teng, J. G., & Zhao, X. L. (2012). Experimental study on CFRP-to-steel bonded interfaces. *Composites Part B: Engineering, 43*(5), 2279–2289.

Zhao, X. L., & Zhang, L. (2007). State-of-the-art review on FRP strengthened steel structures. *Engineering Structures, 29*(8), 1808–1823.

Experimental study for stepped reinforcement concrete beams

Mohamamad Abdalameer Hussain & Nasir Safiq
*Department of Civil and Environmental Engineering, Universiti Teknologi PETRONAS,
Seri Iskandar, Perak, Malaysia*

ABSTRACT: The aim of this study is to presents, testing and analysis result of stepped beams strengthened by carbon fiber-reinforced polymer, subjected to a point load at the center; eight beams have been used to investigate the flexural behavior of reinforced concrete beams which strengthened with carbon fiber reinforced polymer CFRP. The results showed that using of carbon fiber reinforced polymer CFRP it will provides additional strengthening flexural reinforcement, the reliability for this material application depends on how good they are bonded to concrete beam to transfer the stress from concrete to CFRP laminate.

1 INTRODUCTION

Some engineers and system manufactures have recommended that the increase in the load-carrying capacity of a member strengthened with fiber reinforcement polymer FRP system be limited. The philosophy is that a loss of CFRP reinforcement should not cause member failure.

Fire and life safety: CFRP-strengthened structures should comply with all applicable building and fire codes.

The strength of externally bonded CFRP system is assumed to be lost completely in a fire. For this reason, structure member without the CFRP system should possess sufficient strength to resist all applicable loads during a fire. (ACI 440.2R-02, 2002).

To overcome the problems stated above, the future new technique for strengthening the beam with carbon fiber reinforcement polymer CFRP uses different options (bonded surface configurations, end anchor, spacing and fiber orientation) to understand the behavior of strengthened beam with CFRP laminates to improve the understanding of reinforced concrete beams retrofitted with CFRP and this brings new challenges for professionals and who are working in the field of structural and strengthening of reinforced concrete structures and due to the latest technologies in binding the delaminating concept can totally eradicated. In order to achieve the aims of this research study as highlighted in the problem statement. The objectives of this study are to predict the failure mode of stepped reinforcement concrete RC beam at static load point at centre and to evaluate the performance of stepped reinforcement concrete beams with bonded by CFRP fabric in single layer at the (soffit of the beam, side of the beam, and U-shape).

The beam under static loading CFRP rupture failure mode and bonding failure mode. Then to check the adequacy of the chosen the strengthening system (flexible) of CFRP to restore the ultimate capacity of the defected stepped beams.

For the purpose of achieving the desired objectives of the proposed research study, the following scope of work was set:

- To investigate the flexural behavior of stepped reinforcement concrete which strengthened by CFRP, with different angle at the bottom.
- To increase the strength of stepped reinforcement concrete beams that are laminated by CFRP.
- To determine the effects of CFRP strengthening on stepped reinforcement concrete with three different configuration of laminate.
- To determine the mode of failure and type rupture of CFRP.

2 BACKGROUND

Stepped beam is an example of non- prismatic beams that can be used to support a split- level floor. This application is commonly used in theaters and private housing for aesthetic reason. The stopped beam provides additional need for reinforcement detailing to fulfill the stress concentration at the stopped joint.

The target of the researcher was to check the adequacy of the chosen CFRP strengthening configurations to restore the ultimate capacity of the defected stepped beams.

Kachlakeva et al., (2000) Examined four full–scale reinforced concrete beams were replicated from an existing bridge of the four replicate beams, one served as a control beam and remaining three beams were implemented with varying configurations of carbon fiber reinforcement polymer CFRP and glass fiber reinforcement polymer GFRP composites to simulate the retrofit of the existing structure. Test results showed that beams retrofit with both GFRP and CFRP should well exceed the static demand of 658 KNm sustaining up to 868 KNm applied moment. The addition of GFRP to the side face of the beam alone for shear was sufficient to offset the lack of steel stirrups and allow conventional reinforced concrete beam failure by yielding of tension steel.

Grace et al., (2002) Investigated (13) rectangular beams. Two strengthening configuration were used such as strengthening material only on the bottom face beam and strengthening material on the bottom face and extending up to 150mm on both side face of beams. The author stated that the beam strengthened using carbon fiber strengthening system showed lower in yield load than those strengthened with hybrid fabric. The beam strengthened with hybrid fabric system showed no significant loss in beam ductility.

The efficiency and effectiveness of different but very practical FRP scheme for flexure and shear strengthening of RC beam has been studied by Nadeem (2009), six reinforced beams were casted in two groups, each group containing three beams. In each group, one beam was taken as a control specimen and the remaining two beams were strengthened using CFRP strengthening schemes 2.

The results showed that the tension side bonding of CFRP sheets with U-shape end anchorages is very efficient in flexural strengthening, where reinforcement beams are very effective in improving the shear capacity of beams.

Berna et al., (2003), experimentally carried out on twenty rectangular beams. Two beams were used as reference beams and eighteen beams were strengthening using CFRP. The author concluded that by adding transverse straps along the shear span and de-bonding of the longitudinal composites was delayed. The flexural capacity of reinforced concrete beam can be increased by attached CFRP laminated than control beam.

Habibur et al., (2011) tested five rectangular beams in order to evaluate the effect of externally bonded CFRP laminates to the different strengthening scheme for the entire beam length. A total of five reinforced concrete beams have different CFRP configuration. First beam was not bonded with CFRP laminates, three beams were bonded with different layers of CFRP laminates (1, 2 and 3 – layers respectively) and the rest one beam were bonded with one layer CFRP laminates and having one U- shape edge strips.

All these beams specimens were 150×200 mm in cross section and 1900 mm in span length on a simply supported span. They were reinforced with two T10 mm in diameter bars at the bottom and T6 mm in diameter at top. R6 stirrups were placed at a constant spacing of 125 mm throughout the entire of the beams. The stirrups are designed to ensure that none of the beams would fail in shear.

The results shown that un-strengthened control beam failed by yielding of steel tension reinforcement followed by crushing of the concrete directly under the four-point bending test.

The three beams with different levels of CFRP strengthened reinforced concrete beams showed significant increases in flexural stiffness and ultimate capacity as compared to that of control beam.

It is identified that the percentage increase of ultimate load are 54%, 73% and 85% respectively, that mean an increase in stiffness and flexural strength with the increase of CFRP layers as compared to the control beam. The one beam with U-shape warp strips gives an increase of 82% flexural capacity as compared to that of control beam.

Balamuralikrishnan and Antony, (2009) tested ten beams, of which two were control beams, all having size of $150 \times 250 \times 3200$ mm length and designed with reinforced by 2–12 mm at bottom, and 2–10 mm at the top using 6mm diameter stirrups at 150 mm, center to center. To evaluates the performance of RC beams with bonded CFRP fabric in single layer and two layers at the soffit of the beam under static and cyclic loading. The beams were divided in two series of strengthened, first series having four beams with bonded CFRP fabric in single layer which is parallel to beam axis of which two beams were subjected to static loading and remaining two beams were subjected to compression cyclic loading. In second series having four beams with bonded CFRP fabric in two layers which are parallel to beam axis of which two beams were subjected to static loading and remaining two beams were subjected to compression cyclic loading under virgin condition and tested until failure.

The results shows that the strengthened beams exhibit an increase in flexural strength of 18 to 20 percent for single layer and 40 to 45 percent for two layers both static and compression cyclic loading respectively. All the beams strengthened with CFRP fabric in single layer and two layers experience flexural. None of the beams exhibit premature brittle failure, and the CFRP strengthened beam gives appreciable ductility when compared to control beam.

Owners of structures always ask and want in many cases to continue with their activity or service when the strengthening system is applied, and as there is lack of understanding as how cyclic loads during strengthening.

There are three parameters will affect the performance of concrete which laminate with CFRP during apply the materials:

- It is better if the structure can be completely unloaded including the self weight of the structure because this will enhance the utilization of the CFRP contribution before the structure reaches critical stages, yielding of reinforcement, concrete failure and so on.
- Also depends on what kind of structure needs to strengthened and can of course be different for different structures, for structures with high dead loads from furniture and archives it is probable that the structure must not only be taken out of service, the fixtures must also be removed to unload the structure.

- Some structures like the bridges, they have a high amount of self-weight, especially railway bridges, but the frequent live service loads are in comparison quite low, the vehicles that give the highest loads are not so frequent compared to lighter vehicles.

3 METHODOLOGY

This experimental includes fabricating and casted eight beams in two groups to test up to failure.

a. The first group A which called control beams, contained two beams representing un-strengthened beams with different arrangement at the bottom.
b. The second group B contained six beams representing the CFRP strengthened beams. The criteria include the mode of failure, load-deflection relationship, cracking load, ultimate capacity.

All the beams were casted using ready mix fabrication using 30 grade concrete, and slump test 75±, with same curing according to BS 1881: Part 102;1983[8] and techniques, which had 28 days three cube 150 mm × 150 mm × 150 mm, which has been carried out and test by compressive strength at seven days and 28 days, three cubes tested each time and the average strength is calculated and it will be 27.84 N/mm^2 for seven days and 36.73 N/mm^2 for 28 days according to BS 1881: Part 116:1983.[9] The reinforcement concrete stepped beams divided in two groups. First group used as control test beams have two different arrangement by the angle at the bottom, one of them with 45° at the bottom and the other one with 90° at the bottom.

The second group was casting six beams laminated by CFRP, three of them with angle at bottom 45°, and other three with angle 90° at the bottom. All the eight reinforcement concrete stepped beams designed with same internal reinforcement to be strong in shear and weak in flexure beam with cross section 150 mm × 300 mm, and the step at the center. The top reinforcement was deformed 2–10ϕ and the bottom reinforcement was deformed 2–12ϕ, the stirrups 8ϕ mild steel were had two different designs of the shear reinforcement, which can be categorized as stirrups 8ϕ–100 mm with a distance 750 mm from both end, and stirrups 8ϕ–150 mm at the distance 900 mm at the center.

The first activities are to clean the surface by grinded manually to remove all lose particles or grease, blown the dust by pressure air and debris, to make sure the surface was smooth by wiping the surface to develop it for a sound bond adhesive. The two- part of curing epoxy resin (Sikadur "330) part (A) and part (B) with mixing (1:4).

Figure 2 a, b, c shows the reinforcement of the beams.

The epoxy adhesive paste had been used and troweled onto position, and then sheet of CFRP had been applied which prepared early with sufficing length and width on the adhesive. Then Pressed on the CFRP for enough pressure to remove any void or air between CFRP and adhesive and to make sure that CFRP bonded to the stepped reinforcement concrete beam at that area with applied adhesive on it.

A total of eight stepped reinforcement concrete beams were tested. Two beams were designated as reference specimens and six beams were strengthened using carbon fiber reinforcement polymer CFRP wrapping of different configuration and then tested by applying the static load at center until the failure occurred on the beam.

Equation 1 has been applied to determine the ductility of the tested stepped beams:

$$\mu D = (\Delta u / \Delta y) \qquad (1)$$

Figure 1. Stepped concrete reinforcement beam with 90° at the bottom.

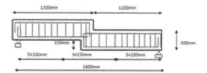

(a) Reinforcement Stepped Beams Stepped Beams with 90° at Bottom

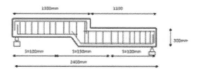

(b) Reinforcement Stepped Beams Stepped with 45° at Bottom

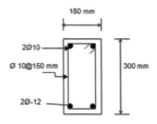

Figure 2. a, b and c. Reinforcement stepped beams and cross section.

Table 1. Test result for the stepped reinforcement concrete beams for Group A 45° at bottom.

Beam name	Failure load (KN)	Mid-span deflection (mm)	Mode of failure
Monitor BEAM 2	13.385	2.518	flexure
B2	19.793	3.509	flexure and CFRP peel at center at the step zone area
B3	20.271	4.982	flexure and CFRP peel from center at the step zone to the end for one wing only
B5	20.672	8.36	Flexure, CFRP peel from center at the step zone to the end, and CFRP rupture

Table 2. Test result for the stepped reinforcement concrete beams for Group (B) 90° at bottom.

Beam name	Failure load (KN)	Mid-span deflection (mm)	Mode of failure
Monitor BEAM 1	12.349	3.133	Concrete crushing
B1	12.473	2.781	Flexure and de-bonding of sheet
B4	12.63	3.172	flexure and CFRP de-bond from concrete to the end for one wing only
B6	12.849	3.628	Flexure and CFRP rupture at the center at the step zone area

where:

μD: Ductility index
Δu: Mid-span deflection at ultimate load (mm) from experimental
Δy: Mid-span deflection at yield load (mm) from experimental.

4 RESULTS AND DISCUSSION

The test were divided in two groups, group A, which were designed as stepped reinforcement concrete beams with bottom by 45° at the step zone, and another, group B were designed as stepped reinforcement concrete beams with bottom by 90° at the step zone.

Table 1 showed the results for the stepped reinforcement concrete beams for group A with bottom by 45° showed, that the service loads B1, B4 and B6 where 12.473 KN, 12.63 KN and 12.849 KN respectively, and the corresponding deflection were 2.781 mm, 3.172 mm and 3.628 mm respectively. The stiffness of beam B1 which laminated by CFRP at the bottom was increased 1.2%, the stiffness of beam B4 which laminated by two side of CFRP was increased by 2.22%, while the beam B6 which Laminated at bottom and two sides (U-shape), was increased by 3.89% compare to control beam BEAM1.

The crack appeared at the mid-span of the beam near to the center and the crack widened and travelled towards the internal support along the bottom face of the beam and upwards towards of the load point.

Loss of bond occurred between the steel reinforcement and concrete which resulted in separation of the concrete cover of the bottom face of the beam.

Table 2, The test results for the stepped reinforcement concrete beams for group B with bottom by 90° showed that the service loads B2, B3 and B5 where 19.793 KN, 20.271 KN and 20.67 KN respectively, and the corresponding deflection were 3.509 mm, 4.982 mm and 8.36 mm respectively. The stiffness of beam B2 which laminated by CFRP at the bottom was increased 32.35%, the stiffness of beam B3 which laminated by two side of CFRP was increased roughly 33.95%, while the beam B5 which laminated at bottom and two side (U-shape), was increased by 35.23%. This mean that the beam with fully sheets wrapped increase the strength more than other beams with side wrapped or at bottom.

Failed as results of CFRP sheet rupturing along with concrete splitting at the bottom face of the beam, and the crack also observed at the top face of the beam which is indicative of spilling failure. It is evident that epoxy resin favored in strengthening and the test results clearly show that CFRP sheets increase the loading carrying capacity.

The relationships between the (load-deflection) for all four flexure-strength of stepped reinforcement concrete beams for group A which are with step 45° at the bottom at the center, plotted in Figure 3, and for stepped reinforcement concrete beams for group B with step 90° at the bottom at the center were plotted in Figure 4.

It was observed that failure for control stepped reinforcement concrete beams was purely due to the flexure as the cracks initiated near the mid-span and propagated almost in vertical direction with increase of applied load until failure.

The reduction in ductility for strengthened beams in reference to the control specimens is not considered to be significant. Therefore, all the strengthened beams were shown to have adequate ductility, which ensured that their failure mode was of ductile nature.

5 CONCLUSION & FUTURE DIRECTIONS

A total of eight stepped reinforcement concrete beams were tested. Two beams were designated as reference

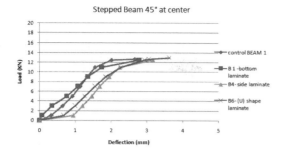

Figure 3. Load deflections curves for stepped beam 45° at bottom.

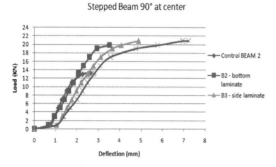

Figure 4. Load deflections curves for steppe beams 90° at bottom.

Table 3. Ductility index of the stepped beam.

Beam ID	Yield deflection (mm) Δy	Ultimate deflection (mm) Δu	Ductility index ($\Delta u/\Delta y$)
Group A with angle 45° at the bottom			
Control BEAM1	2.66	47.47	17.84
B1	2.781	44.09	15.85
B4	3.172	36.63	11.54
B6	3.628	30.517	8.41
Group B with angle 90° at the bottom			
Control BEAM2	2.518	35.76	14.21
B2	3.509	44.09	12.56
B3	4.982	29.49	5.92
B5	8.36	59.388	7.11

specimens and six beams were strengthened using carbon fiber reinforcement polymer CFRP wrapping in different configuration and then tested by applying the static load at center until the failure occurred on the beam.

The summary of the test result shows there are three modes of failures which were observed in this investigation, they called failure mode (I) for flexure cracks due to concrete crushing, failure mode (II) due CFRP peel from center at the step zone to the end, and failure mode (III) due to CFRP rupture.

As expected, the flexural stiffness increased when CFRP was laminated to the stepped reinforcement concrete beams for group B which was with step 90° at the bottom at the center, can be more stiffness with laminated by CFRP than for the stepped reinforcement concrete beams group A which was with step 45° at the bottom at the center.

The strengthening of stepped reinforcement concrete beams with externally bonded by CFRP sheets is effected and lead to increase the flexural strengthening for group A stepped beam 45° at bottom was between percent 1% to 4%, and for group B stepped beam with 90° at bottom was between percent 32% to 36%. This mean that the beam with fully sheets wrapped increase the strength more than other beams with side wrapped or at bottom.

The crack appeared at the mid –span of the beam near to the center and the crack widened and travelled towards the internal support along the bottom face of the beam and upwards towards of the load point and the beams failed as results of (CFRP) sheet rupturing along with concrete splitting at the bottom face of the beam, and the crack also observed at the top face of the beam which is indicative of spilling failure. Loss of bond occurred between the steel reinforcement and concrete which resulted in separation of the concrete cover of the bottom face of the beam. From the above, The CFRP strengthening provides additional flexural reinforcement, the reliability for this material application depends on how well they are bonded and can transfer stress from the concrete component to CFRP laminate and the behavior of CFRP strengthened RC structures is often controlled by the bond strength of the interface between the CFRP and the concrete. In same time this material can be applied while the structure in use and therefore it is expected to replace most of previous existing repairs and strengthening techniques.

REFERENCES

ACI 440.2R-02 Guide for the design construction of externally bonded FRP system for strengthening concrete structures, Report by ACI Committee 440, American Concrete Institute, (2002).

Brena et al, 2003. Cited by Murali G. and Pannirselvam N., 2011, Flexural Strengthening of Reinforced Concrete Beams Using Fiber Reinforced Polymer Laminate: A Review Journal of Engineering and Applied Sciences, 6 (11).

Balamuralikrishnan1 R., and Antony Jeyasehar, 2009. Flexural behavior of RC beams strengthened with carbon fiber reinforced polymer (CFRP) fabrics, The Open Civil Engineering Journal, 3, 102–109. British Standards Institution, BS 1881: Part 102; Testing Concrete, Method for Determination of Slump, 1983, BSI, London.

British Standards Institution, BS 1881: Part 116; Method for Determination of Compressive Strength of Concrete Cubes, 1983, BSI, London.

Kachlakeva, D and Mc Curry, D.D. 2000, cited by Murali G. and Pannirselvam N., 2011, Flexural Strengthening of Reinforced Concrete Beams Using Fiber Reinforced Polymer Laminate: A Review ,Journal of Engineering and Applied Sciences, 6 (11).

Grace et al, 2002. Cited by Murali G. and Pannirselvam N., 2011, Flexural Strengthening of Reinforced Concrete Beams Using Fiber Reinforced Polymer Laminate: A Review Journal of Engineering and Applied Sciences, 6 (11).

Habibur Rahman Sobuz, Ehsan Ahmed, Noor Md. Sadiqul Ha san, and Md. Alhaz Uddin, 2011, Use of carbon fiber laminates for strengthening reinforced concrete beams in bending, International Journal of Civil and Structural Engineering, 2(1).

Nadeem A. and Siddiqui, 2009, Experimental investigation of RC Beams Strengthened with externally bonded FRP composites. Latin American journal of solid and structures. 6:343–362.

Nano silica modified roller compacted rubbercrete – An overview

Musa Adamu, Bashar S. Mohammed & Nasir Shafiq
*Department of Civil and Environmental Engineering, Universiti Teknologi PETRONAS,
Seri Iskandar, Perak, Malaysia*

ABSTRACT: Roller compacted concrete (RCC) when used in pavement is subjected to dynamic fatigue load from moving vehicles which is the major problem. In order to solve fatigue and flexural strength related problems in RCC, incorporating crumb rubber (CR) will improve the ductility, bending deformation and reduces the brittle nature of RCC due to deformation and elastic properties of rubber, but CR will affects the mechanical, durability properties of concrete negatively, this leads to the use of Nano Silica (NS) in RCC to remedy the negative effect of CR by improving the interfacial transition zone (ITZ) between CR and cement paste, densifying the concrete matrix due to NS filler ability, improving strength due to NS high pozzolanic reactivity. Thus, producing a new material for use as RCC pavement of desirable strength, high fatigue performance, higher energy absorption, with lower risk of distress, which is more economical and having a longer service life.

1 INTRODUCTION

According to ACI committee 207, "Roller compacted concrete (RCC) is a concrete of no slump consistency in its unhardened state that is transported, placed and compacted using earth and rock fill construction equipment. Properties of hardened roller compacted concrete are similar to those of conventional concrete" (American Concrete Institute 207). RCC differs from conventional concrete due to its consistency requirements and compaction effort needed, with lower water content, lower paste content, higher fine aggregate content, no entrained air (Mehta and Monteiro 2006). The major advantages of RCC over conventionally placed concrete include high construction speed, reduced construction cost (American Concrete Institute 207, Fuhrman 2000). RCC has similar strength and performance properties as conventional concrete, but with the economy and simplicity of construction as that of asphaltic pavement with lower maintenance needed and longer service life (Shoenberger 1994).

RCC are mainly used in two areas of applications; in mass concrete dams applications, referred to as roller compacted concrete dams and in heavy duty pavement applications which include intermodal yards, port facilities, ware houses and large storage areas, airport service areas, highway shoulders, rural roads, local streets and intersections (Shoenberger 1994, Adaska 2006).

As a result of increased rate of vehicles every day, and with the higher repetitive cyclic axle loads imposed by the heavy vehicles on pavement, the pavement is subjected to cyclic fatigue. To carter for these factors, qualitative, higher deformation resistant pavements with longer design life need to be designed (Sengoz and Topal 2005). Fatigue is one of the commonest defects affecting pavement which affect the cost of maintenance, and shortens pavement design life. It is therefore necessary to device a means of reducing the fatigue effect by delaying the pavement deterioration, and improving its design life. This is possible by incorporating additives such as polymers (crumb rubber) or fibers into the pavement mix where it absorbs the deformations and strain energy cause by heavy traffic (Moghaddam, Karim et al. 2011).

2 CRUMB RUBBER

A Crumb rubber (CR) is another form of scrap tire that is produced by reducing and grinding the scrap tire into smaller uniform granules with the steel, fiber and inert contaminants such as dusk, glass and rocks removed. The size of the crumb rubber ranges from 4.75 mm (No 4 sieve) to less than 75 μm (No 200 sieve) (Siddique 2007).

2.1 Advantages of using CR in concrete

Research shown that CR decreases mechanical properties and durability of concrete, but some of its advantages include improving energy absorption, toughness resistance, brittleness, fatigue performance, thermal insulation and conductivity, electrical resistivity, sound absorption, reduces crack length and width with significant delay in the cracking initiation time of a shrinkage restrained concrete. Similarly, the rate of propagation of cracks significantly reduced for concrete containing waste tire under flexural and tensile loadings. Concrete containing CR shows more ductile failure characteristics, little brittle behavior have higher energy absorption capacity and have higher deformation properties with failure, cracking or breaking not occurring catastrophically

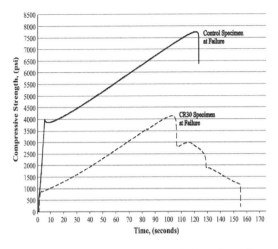

Figure 1. Compressive strength behavior after failure (Kardos and Durham 2015).

but gradually and uniformly under compressive, tensile or flexural loadings Siddique 2007), higher impact resistance and energy absorption (Mohammed, Awang et al. 2016). Atahan, et. al (2012) also found that increasing CR in concrete improves deflection of concrete, increase impact and energy absorption capacity at maximum applied load, which consequently reduces impact effects which makes it suitable for use as safety barriers in highways (Atahan and Yücel 2012). In a similar study, Rubber concrete shows a higher ductile performance with lower brittle index compared to conventional concrete, and even at 15% replacement of coarse aggregate with ground or crush rubber the concrete shows a satisfactory ductile behavior, this gives rubberized concrete a plastic deformation performance at time of fracture due to its high energy absorption capacity (Zheng, Huo et al. 2008), improves deformation, displacement and toughness even at lower modulus of elasticity (Mohammed and Azmi 2011). Incorporating CR into concrete improves and enhances its fatigue performance and fatigue life when subjected to the same level of stress; therefore rubberized concrete can be able to withstand series of cyclic loadings compared to normal concrete. Rubberized concrete also shows higher deformation resistance, seismic performance, and toughness than normal concrete (Liu, Zheng et al. 2013). Fig 1 shows the failure behavior of concrete containing 0% CR and 30% CR

2.2 Roller compacted rubbercrete (RCR)

From the name, an RCR is an RCC in which waste tire particles are used to partially replace either fine aggregate or coarse aggregate. Crumb rubber of size range between 0.075 mm–4.75 mm used as fine aggregate, chip rubber (5 mm–76 mm) as coarse aggregate, and ground rubber (0.15 mm–19 mm), and sometimes waste rubber ash as partial cement replacement (Siddique 2007). Although very few literatures on RCR were found and reviewed below as the topic of RCR is a new research area.

2.2.1 Recent researches on crumb rubber in RCC

The effect of rubber addition on the properties of RCC for pavement application was studied where crushed coarse aggregate was replaced with shredded rubber at 5%, 10%, 15%, 20%, 25% and 30%. The density, unit weight, water absorption all decrease with increase in rubber, but the consistency which is measured using Vebe test was found to decrease with increase in rubber. For 0% rubber consistency was 33 s while for 30% consistency was 23s (Meddah, Beddar et al. 2014). Compressive, tensile and flexural strengths all decreased with rubber addition, but treating the crumb rubber with NaOH solution increased compressive and tensile strengths by 11–28% and 15–20% respectively (Meddah, Beddar et al. 2014, Meddah, Beddar et al. 2014).

2.3 Possible solutions to crumb rubber disadvantage in RCC

The negative effect of crumb rubber when used in RCC can be possibly solved by using Nano Silica due to its pore filling ability, densification of the ITZ between cement paste and crumb rubber, and improving concrete strength due to its high pozzolanic reactivity. Even though nanotechnology in RCC is very new, this research tends to be the-state-of-the-art application of nanotechnology in RCR where Nano silica will be used.

3 NANO SILICA (NS) IN CONCRETE

Nano materials which are ultrafine in size to nanometer (1×10^{-9} m) can totally modify concrete, even though the true knowledge or understanding on the effect of the Nano-sized particles is very little when replaced cement. NS improves concrete mechanical properties, durability, lower the setting time, reduces the overall cost of construction (Givi, Rashid et al. 2010). Also, NS is a highly reactive filler when used in concrete even at lower percentage and it densified the concrete micro structure, increase rate of hydration, reduces bleeding, reduces initial and final setting time in fly ash concrete and improves early strength development in fly ash concrete (Givi, Rashid et al. 2010, Zhang and Islam 2012). The role of NS in concrete is summarized in Figure 2.

3.1 Properties of nano silica

NS is the most commonly used nanomaterial in concrete due to its high pozzolanic effect in addition to its pore filling ability. When used in concrete, it produces a high strength, durable and sustainable concrete, it also speed up the hydration of cement and produces more C-S-H gel as shown in the pozzolanic reactivity (Singh, Karade et al. 2013);

$$Nano-Silica + H_2O \rightarrow H_2SiO_4^{2-} \qquad (1)$$

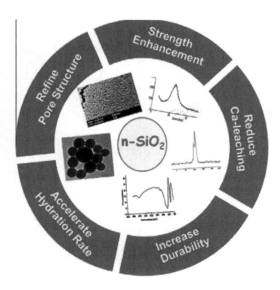

Figure 2. Role of Nano silica in cementitious system (Singh, Karade et al. 2013).

Table 1. Physical and Chemical Properties of NS (Shaikh, Supit et al. 2014).

SiO_2 (%)	Particle Size (nm)	Specific Gravity	Surface Area (m^2/g)	Loss of Ignition (%)
99	25	2.2–2.6	160	–

$$Ca(OH)_2 + H_2O \rightarrow Ca^{2+} + OH^- \quad (2)$$

Then $H_2SiO_4^{2-} + Ca^{2+} \rightarrow C-S-H\_gel \quad (3)$

The physical and chemical properties of NS vary from one another. A typical properties of Nano silica with an average diameter of 25 nm, obtained from Nanostructured and Amorphous Materials Incorporation U.S.A is shown in Table 1 (Shaikh, Supit et al. 2014).

3.2 Mechanical properties of concrete containing NS

Research shows that there was improvement in terms of mechanical properties for cement replacement with NS and Nano Ferrite, at all age of curing with 3% NS and 2% Nano Ferrite been optimum. The highest percentage increase for compressive strength, splitting tensile strength, flexural strength and modulus of elasticity were 21%, 44%, 23% and 25% respectively (Amin and Abu el-Hassan 2015). NS up to 6% replacement by weight of cementitious materials improves strength of concrete which was more observed in the concrete containing 30% fly ash at longer age. The durability of concrete also improved by refining the pore structure, densification of the ITZ, due to its filler effect and consumption of the Portlandites (CH)

(Said, Zeidan et al. 2012). The lower rate of strength development of concrete containing fly ash can be controlled by adding small amount of Nano silica (Singh, Karade et al. 2013); up to 2% by weight of cement in mortar (Shaikh, Supit et al. 2014), up 3% by weight of cement in concrete (Amin and Abu el-Hassan 2015), up to 6% by weight of cement in concrete(Said, Zeidan et al. 2012).

The addition of 4% NS by weight of cementitious materials in high volume fly ash (HVFA) concrete containing 50% fly ash increased the short term strength by 81% when compared with HVFA without fly ash, but the long term strength between 28 days to 2 years of HVFA Nano silica concrete increase was 52.9% lower than HVFA concrete with 99.4% when compared with conventional concrete (Li 2004), 1% NS added to 50% fly ash concrete, accelerates the cement hydration of the paste, both the early strength and long term strengths were improved compared to the HVFA without NS addition(Zhang and Islam 2012). Similarly, 2% and 4% NS was added to HVFA concrete containing 40% and 60% fly ash, there was significant improvement on the early age compressive strengths, where at 3days and 7 days strength increased by 25% and 15% respectively for 40% fly ash 2%NS, but for 60% fly ash and 2%NS significant increase was only observed at 3days but for later ages no significant improvement was found, and for 4%NS similar results were obtained as for 2%NS. This is because NS contributed more to the pozzolanic reactivity at early age than fly ash due to its finer particle size and larger surface area(Supit and Shaikh 2014, Shaikh and Supit 2015), up to 5% NS improves concrete strength, but decreases at 7.5%, which is due to the enhancement of cement aggregate paste and the ITZ microstructure with reduction in the porosity throughout the bulk paste (Shekari and Razzaghi 2011).

Mechanical properties of concrete were improved with NS addition at early and long term ages, although lower particle size NS shows higher increase due to its more pozzolanic reaction. The optimum percentage replacements were 1.0% and 1.5% for 15 nm and 80 nm respectively at 90 days (Givi, Rashid et al. 2010). Adding Nano Silica in concrete reduces its workability due to its larger surface area, but when added in concrete it improves the water permeability by making the paste denser, more uniform and the ITZ between the cement paste and aggregate denser and stronger which in turn improves the strength and durability of the concrete (Ji 2005). Strength loss in CR concrete were mitigated with addition of Nano Silica, this was achieved as a result of enhancement of the microstructure through pore refinement as shown in figure 3 and interfacial transition zone densification as shown in Figure 4 (Mohammed, Awang et al. 2016).

Incorporating NS even at less percentage (1% by weight of cement) improves the fatigue live and the flexural strength of concrete for use in pavement application, with the improvement in the sensitivity of the fatigue life to change in stress and the fatigue following a Weibull distribution function (Li, Zhang et al. 2007).

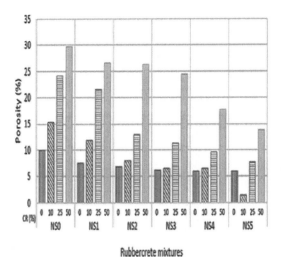

Figure 3. Porosity Versus CR replacement and Nano Silica inclusion (Mohammed, Awang et al. 2016).

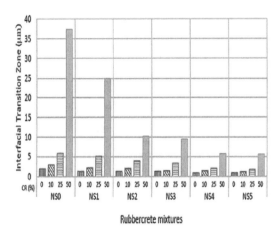

Figure 4. ITZ Versus CR replacement and Nano Silica inclusion (Mohammed, Awang et al. 2016).

3.3 Durability of concrete containing nano silica

The durability effect of 0.3% and 0.9% replacement of cement with NS in concrete were studied, results shows that due to the high pozzolanic activity and Nano filler effect of NS, the paste morphology and ITZ were densified. 0.3% NS shows higher strength and durability. Water depth penetration, chloride penetration and diffusion coefficient were decreased by 45%, 28.7% and 31% respectively for 0.3% Nano silica, and were increased for 0.9% NS due to conglomeration of Nano particles(Du, Du et al. 2014). The durability of HVFA concrete were measured where concrete made with 40% and 60% fly ash and 2% NS. In terms of chloride ion penetration (CIP), results indicated that adding 2% NS to concrete containing 40% fly ash a reduction of 38% CIP was observed, but when 4% NS was added to HVFA with 60% fly ash higher CIP was found which was due to interconnectivity of the pores. Similarly water sorptivity decreases when 2%NS was added to HVFA concrete, for 40% fly ash and 2%NS, a reduction between 27–32% was recorded between 28 to 90 days, while for 60%fly ash and 2%NS, the reduction was the ultimate at 90 days (Supit and Shaikh 2014, Shaikh and Supit 2015).

4 CONCLUSIONS

The following conclusions can be drawn

1) This paper reviews the current state of the field of using crumb rubber (CR) in roller compacted concrete (RCC) applications and the challenges faced i.e. decreasing mechanical and durability properties of RCC which are the most important properties.
2) CR can be used to produce RCC for pavement which is more economical compared to conventional RCC and rigid pavement, with improvement in fatigue performance and bending resistance, thus improving the life time of the pavement
3) The potential of nanotechnology in solving these challenges, which will lead to the development of novel, higher strength, durable, NS modified RCC for pavement application with improved fatigue resistance, energy absorption, crack resistance, longer service life, highly economical when absolutely utilized.

REFERENCES

Adaska, W. S. (2006). "Roller-Compacted Concrete (RCC)." Significance of Tests and Properties of Concrete and Concrete-making Materials: 595.

American Concrete Institute (207). Roller Compacted Mass Concrete. Materials Journal. 85.

Amin, M. and K. Abu el-Hassan (2015). "Effect of using different types of nano materials on mechanical properties of high strength concrete." Construction and Building Materials 80: 116–124.

Atahan, A. O. and A. Ö. Yücel (2012). "Crumb rubber in concrete: static and dynamic evaluation." Construction and Building Materials 36: 617–622.

Du, H., et al. (2014). "Durability performances of concrete with nano-silica." Construction and Building Materials 73: 705–712.

Fuhrman, R. L. (2000). Engineering and design roller compacted concrete, Department of the army US Army Corps of Engineers, EM 1110-2-2006.

Givi, A. N., et al. (2010). "Experimental investigation of the size effects of SiO 2 nano-particles on the mechanical properties of binary blended concrete." Composites Part B: Engineering 41(8): 673–677.

Ji, T. (2005). "Preliminary study on the water permeability and microstructure of concrete incorporating nano-SiO 2." Cement and Concrete Research 35(10): 1943–1947.

Kardos, A. J. and S. A. Durham (2015). "Strength, durability, and environmental properties of concrete utilizing recycled tire particles for pavement applications." Construction and Building Materials 98: 832–845.

Li, G. (2004). "Properties of high-volume fly ash concrete incorporating nano-SiO 2." Cement and Concrete Research 34(6): 1043–1049.

Li, H., et al. (2007). "Flexural fatigue performance of concrete containing nano-particles for pavement." International Journal of Fatigue 29(7): 1292–1301.

Liu, F., et al. (2013). "Mechanical and fatigue performance of rubber concrete." Construction and Building Materials 47: 711–719.

Meddah, A., et al. (2014). "Experimental study of compaction quality for roller compacted concrete pavement containing rubber tire wastes." Sustainability, Eco-efficiency, and Conservation in Transportation Infrastructure Asset Management: 273.

Meddah, A., et al. (2014). "Use of shredded rubber tire aggregates for roller compacted concrete pavement." Journal of Cleaner Production 72: 187–192.

Mehta, P. K. and P. J. Monteiro (2006). Concrete: microstructure, properties, and materials, McGraw-Hill New York.

Moghaddam, T. B., et al. (2011). "A review on fatigue and rutting performance of asphalt mixes." Scientific Research and Essays 6(4): 670–682.

Mohammed, B. and N. J. Azmi (2011). "Failure mode and modulus elasticity of concrete containing recycled tire rubber." The Journal of Solid Waste Technology and Management 37(1): 16–24.

Mohammed, B. S., et al. (2016). "Properties of nano silica modified rubbercrete." Journal of Cleaner Production 119: 66–75.

Said, A. M., et al. (2012). "Properties of concrete incorporating nano-silica." Construction and Building Materials 36: 838–844.

Sengoz, B. and A. Topal (2005). "Use of asphalt roofing shingle waste in HMA." Construction and Building Materials 19(5): 337–346.

Shaikh, F., et al. (2014). "A study on the effect of nano silica on compressive strength of high volume fly ash mortars and concretes." Materials & Design 60: 433–442.

Shaikh, F. U. A. and S. W. Supit (2015). "Chloride induced corrosion durability of high volume fly ash concretes containing nano particles." Construction and Building Materials 99: 208–225.

Shekari, A. and M. Razzaghi (2011). "Influence of nano particles on durability and mechanical properties of high performance concrete." Procedia Engineering 14: 3036–3041.

Shoenberger, J. E. (1994). User's Guide: Roller-Compacted Concrete Pavement, DTIC Document.

Siddique, R. (2007). Waste materials and by-products in concrete, Springer Science & Business Media.

Singh, L., et al. (2013). "Beneficial role of nanosilica in cement based materials–A review." Construction and Building Materials 47: 1069–1077.

Supit, S. W. M. and F. U. A. Shaikh (2014). "Durability properties of high volume fly ash concrete containing nano-silica." Materials and Structures: 1–15.

Zhang, M.-H. and J. Islam (2012). "Use of nano-silica to reduce setting time and increase early strength of concretes with high volumes of fly ash or slag." Construction and Building Materials 29: 573–580.

Zheng, L., et al. (2008). "Strength, modulus of elasticity, and brittleness index of rubberized concrete." Journal of Materials in Civil Engineering 20(11): 692–699.

Contribution of natural pozzolan to sustainability of well cements for oil and gas industry

A. Talah & F. Kharchi
University of Sciences and Technology, Algiers, Algeria

R. Chaid
University of Boumerdes, Algeria

ABSTRACT: In this research study, mechanical and sustainability properties of well cement mixed with natural pozzolan are presented. Well cement's properties in chloride environments are improved by replacing 10% of cement with natural pozzolan of 9600 cm^2/g of fineness. Reduction of water absorption, chlorine ion penetration and oxygen permeability in test specimens put in evidence that the natural pozzolan is convenient for the oil well cement as well as for the mortar cements for oil and gas industry, where, their mechanical properties and sustainability are improved. On the basis of experimental results, it is concluded that the special oil well cement contributes to a sustainable development of drilling oil and gas wells.

1 INTRODUCTION

A typical well can be thousands of feet deep and less than three feet wide, and is constructed by using a metal casing surrounded by a special cement slurry mix that fills the annulus between the outer face of the tubing and the wall formation of the hole. This special cement was replaced by a mortar mixed by sand, ordinary Portland cement and finely ground natural pozzolan.

The worldwide demand for high-performance cement-based materials has increased and predictions are that it will be widely used in construction industry during the early 21st century. Economical and environmental considerations had a crucial role in the supplementary cementing material usage as well as better engineering and performance properties, Badogiannis &. Ramezanianpour (2004, 1995). This study has been exclusively focused on the hardened properties of the high performance cement mortar containing an optimum quantity of natural pozzolan powder under aggressive and normal curing regimes. It is now well established that evaluation of the performance of a mortar mix is not limited to the determination of its mechanical properties since it is of paramount importance to characterize the material in terms of the parameters that rate its durability, Basheer (2001).The higher amount of natural pozzolan additive the longer the setting times and the lower the strength of the specimens for different curing periods, Corinaldesi (2010).

The effects of using natural pozzolan as partial replacement of cement on the mechanical properties of mortar were investigated. Test results indicated that the optimize amount of natural pozzolana as replacement by weight of cement had the best compressive and flexural strengths, Ergun (2011).

Natural pozzolan is pozzolanic material which does react with cement past. Its addition with small mounts to the mortar mix as partial replacement to cement increases the workability in the fresh state. It facilitates the dispersion of the cement past and the compaction which causes the increase in the strength, Talah (2013).

2 MATERIALS

The materials used in this investigation were Portland cement, natural pozzolan, sand, and water. Portland cement (CPA-CEM-I / A 42.5) conforming to the Algerian standard NA 443 and natural pozzolan are utilized as cementitious materials. Sand dune of western erg of Algeria with a nominal size of 400 μm, and a specific gravity of 2.65, is used for the mortar samples. The size grain, the fineness modulus (115 cm^2/g), and the sand equivalent value (SEV=97%) show that sand can be used in developing a high performance mortar (HPMP).

3 FORMULATION, MIXTURES, SPECIMENS AND CURING PROCEDURES

Two formulations of mortar were studied:

- reference mortar (RM);
- high-performance mortar with pozzolan powder (HPMP);

The prepared specimens were stored for one year in an environment containing 5% calcium chloride, (media 1) and drinking water (media 2).

Table 1. Mixture proportions and properties of mortar.

item		MR	HPMP
W/C	(ratio)	0.50	0.50
cement	(kg/m^3)	450	405
NP	(kg/m^3)	0	45
water	(kg/m^3)	225	225
sand	(kg/m^3)	1350	1350
air	(%)	2.0	2.0
slump	(cm)	16	18
density	(kg/m^3)	2280	2300

Figure 1. General view of apparatus.

In order to investigate the natural pozzolan (NP) on the performance properties of mortar, two different mortar mixes were employed, details of which are given in Table 1. The control mix contained only Portland cement as the binder. In the high-performance mortar with NP, Portland cement was partially replaced with, respectively, 10% NP (by weight) obtained by optimization. All cement mortars were mixed in accordance with ASTM C192 standard in a power-driven revolving pan mixer. Mortar cubes of $4 \times 4 \times 16$ mm in size, and cylinders of dimensions $160\emptyset \times 320$ mm were cast in steel moulds for the study of the compressive strengths, rapid chloride permeability test, and oxygen permeability test, respectively. All specimens were cast and compacted by a vibrating table. After casting, the moulded specimens were covered with a plastic sheet and left in the casting room for 24 hours. They were then demoulded and divided into two equal groups and cured under the following conditions: in the first curing condition, the specimens were immersed in water until the age of testing, while in the second curing condition, those were immersed in aggressive water (5% $CaCl_2$) until the age of testing. To ensure a concentration of chlorides constant throughout the tests, the solution in the tanks was regularly checked once a week and changed if the difference between the concentration of the solution and the initial concentration exceeded 5%.

The main tests carried out on the fresh mortar are the workability (slump test), the percentage of air contents determined by the aerometer and the density. The results of these tests are given in Table 1.

4 RESEARCH METHODOLOGY

4.1 Compressive strength

To evaluate strength characteristics of each mixture, the compression test was carried out on mortar 160×320 mm cylinder by a 2000 KN capacity testing machine according to ASTM C39. The strength measurements of mortar were performed at 28, 90, 180 and 365 days of age. The results reported are the average of nine compression tests.

4.2 Chloride permeability

The resistance of the mortar to the penetration of the chloride ions was measured in terms of charge passed through the cement mortar in accordance with ASTM C1202.

4.3 Oxygen permeability

The values of oxygen permeability of mortars were measured in terms of flux passed through the concrete in according with our novel experimental method [7]. The design of the test specimen reflects the actual cases where the permeability plays an important role in the durability, such as shielding petroleum-wells. Figure 2 shows the general view of testing apparatus employed.

5 RESULTS AND ANALYSIS

5.1 Compressive strength

The data regarding the variation of compressive strength with respect to mortar specimens age and curing condition for different types of mortar in the two mediums are shown in the Figure 2. The strength values for the reference mortar and high-performance mortar with NP ranged from 26 to 48 MPa and from 49 to 65 MPa respectively, depending mainly on NP content, curing condition, and mortar age. The result indicates that there was a systematic gain in compressive strength with the NP content. It was observed that the ratio of the compressive strength of the specimens subjected to water curing to those cured under aggressive water for the reference specimen mortar deviated up to 29%. However, this ratio for mortars containing NP lay within a range of 3%, depending mainly on NP content and testing age. This implies that reference mortars are more sensitive to aggressive medium than mortar with NP. Therefore the increase of resistance is remarkable after 28 days, following the pozzolanic property of this addition.

5.2 Resistance to chloride penetration

The effect of curing conditions (up to 28 and 365 days of age) and the partially replacement of cement with NP (from 10%) on chloride permeability of the mortar is shown in Fig. 3. The test results show that the values of the electric charge for HPM with NP are too small. As it is also observed in Fig. 3, the extension of the curing period from 28 to 365 days and the curing conditions applied to the test specimens resulted in a

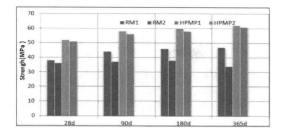

Figure 2. Evolution of compressive strengths at different ages.

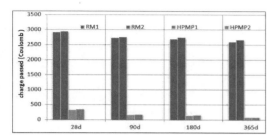

Figure 3. Variation of charge passed at different ages.

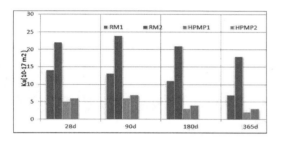

Figure 4. Determination of the apparent oxygen permeability at in inlet 0.3 MPa of mortar samples at different ages.

reduction of the charge passed through the evaluation of the performance of a mortar.

This confirms the contribution of the NP against degradation in a hydrochloric media.

5.3 Oxygen permeability

Results presented in Fig. 4, show that the apparent permeability tends to decrease when the compressive strength increases, which is the most frequent case.

The incorporation of 10% NP is very advantageous; it allows a reduction in oxygen permeability rate from 85% at the age of one year. It is noticed that NP reduced the oxygen permeability during the advancement of age of specimens mortar.

6 CONCLUSIONS

From the results of the tests and analysis carried out in this study, the following conclusion can be drawn.

- The natural pozzolan could be used as partial replacement of Portland cement up to 10% in composite cement.
- Additionally to this, an improvement in durability characteristics is observed; without decreasing the compressive strength of mortar.
- The durability test on the mortar containing NP consisted of immersion in running water, chloride solution, in all cases, structural changes to the samples were noted.
- In all cases the addition had improved the physical characteristics of mortar relatively to the reference mortar sample.
- Finally, it was noted that additive giving a very power of pozzolanicity and effectively contributes to the reduction of chloride ion penetration and oxygen permeability and increase the durability of cement mortar.
- The results show the positive influence of NP on the properties of mortar under hydrochloric mediums.
- This special mortar can be used in petroleum-wells sealing.

REFERENCES

Badogiannis, E.; Papadakis, V.G., Chaniotakis, E. and Tsivilis, S. 2004 'Exploitation of poor Greekkaolins: strength development of metakaolin concrete and evaluation by means of k-value', *Cement and Concrete Research*, vol. 34, pp. 1035–1041.

Basheer, L., Kropp, J. and Cleland, D.J. 2001 'Assessment of the durability of concrete from its permeation properties', *Construction Building Materials* 15 (2–3), pp. 93–103.

Corinaldesi, V., Giacomo, M. and Naik, T.R. 2010 'Characterization of marble powder for its use in mortar and concrete' *Construction and Building Materials* 24, pp. 113–117.

Ergun, A. 2011 'Effects of the usage of diatomite and waste marble powder as partial replacement of cement on the mechanical properties of concrete' *Construction and Building Materials* 25, pp. 806–812.

Ramezanianpour, A.A. and Malhotra, V.M. 1995, 'Effect of curing on the compressive strength, resistance to chloride ion penetration and porosity of concretes incorporating slag, fly ash, or silica fume', *Cement Concrete Composite*, vol. 17, pp. 125–133.

Noha M. Soliman, 2013 'Effect of using Marble Powder in Concrete Mixes on the Behavior and Strength of R.C. Slabs'. *International Journal of Current Engineering and Technology* ISSN 2277–4106 Available online 01 December 2013, Vol. 3, No. 5.

Talah A. and Kharchi F. 2013 'A Modified Test Procedure to Measure Gas Permeability of Hol Cylinder Concrete Specimens' IACSIT *International Journal of Engineering and Technology*, Vol. 5, No. 1.

Costs of urbanization and its impacts on the urban density

B. El Kechebour & S. Haddad
University of Science and Technology Houari Boumediene (USTHB), Faculty of Civil Engineering, Laboratory of Water, Environment, Geomecanique and Works, Bab Ezzouar, Algiers, Algeria

ABSTRACT: The main of this works is the identification of the relation between the costs and the urban density in relation with the urbanization. In this study, only the costs of urbanization and its impacts on the urban density are analyzed. The future owners and the companies specialized in the housing, focalize their choices on the availability of land and its viability with a minimal prices and costs. The study begins by the collection of the square meter costs in relation with the land, the construction in elevation and the viability of the site, then by the identification of correlations between these costs and the urban density. The work finishes by the presentation of the graphs illustrating the relations between the costs and the urban density; and by the process of optimization of the density according to the costs.

1 INTRODUCTION

The Urbanization is considered as a complex system. It consists of numerous interactive sub-systems and is affected by diverse factors including governmental land policies, population growth, and housing market. Land use and the urbanization are considered as the two important subsystems determining urban form and structure in the long term. The Urbanization is a continuous and dynamical phenomenon and its causes and impacts are very varied and very complexes. In this study, only the costs of urbanization and its impacts on the urban density are analyzed. In the practice, the causes of the development of urbanization into a new site are identified. One can cite for example, the accessibility of land, the opportunity of costs of land and viability of the site. Indeed, the future owners and the companies specialized in the housing, focalize their choices on the availability of land and it viability with a minimal prices and costs. The main of this works is the identification of the correlations between the costs and the urban density in relation with the urbanization. The work finishes by the presentation of the graphs illustrating the relations between the costs and the urban density; and by the process of optimization of the density according to the costs.

According to the 2011 Revision of World Urbanization Prospects (UNWUP 2012), the world's urban population increased from 0.75 billion (29.4 percent of the world's population) in 1950 to 3.63 billion (52.1 percent) in 2011 and is expected to reach 6.25 billion (67.2 percent) in 2050. Sprawl is a form of urban growth that happens through rapid suburbanization, especially in North America, with a fragmented form and low density in development (Zhang et al. 2011). As an essential way to learn the urban growth/sprawl phenomena, modeling and simulation is regarded as an efficient way to understand the mechanisms of urban dynamics, to evaluate current urban systems, and to provide planning support in urban growth management (Barthélemy & Flammini 2011). Modeling can either be conceptual, symbolic or mathematical (Courtat et al. 2011). This depends on the purposes of the specific application. Before carrying on the modeling, one has to figure out the driving forces behind urban growth phenomenon (Lechner et al. 2004). It is well known that a city is a complex system (Batty a 2008). It contains various interactive subsystems and is thus affected by a variety of variables or factors. From the perspective of modeling approaches, models evolve from macro simulations to micro-simulations. They are all promising methodologies in a new generation of urban models (Batty 2008a).Cities are considered as being complex systems because there are comprehensive entities coping with infinite variety (Batty 2008b).

2 FORMULATION OF THE PROBLEM

The urban composition operation is realized by using two coefficients: the urban density (d) and the number of stories (n).These coefficients compose a system that defines an urban density. The density shows the number of dwellings setting on a hectare area (10000 m2). In this area there are all spaces with their functions. The density of the dwellings is independent of the organization of spaces, but depends of the number of stories in the vertical plan and the ratios required by the urban and housing rules. The using of these ratios and the number of stories combined with a building typology and a dwelling pattern gives an urban composition. The resolution of problem consists on an analysis of the composition of costs (El kechebour 2009a) relative to each operation during the process of the urbanization. To modelling these components, one must to adopt a mathematical approach (El kechebour 2010b) and (Batty 1992c). This artifice permits to associate every urban function with its corresponding space.

2.1 Formulation of the urban density

The mathematical system composed by the equations 1 to 12 represents an urban composition. The resolution of this system gives an urban density without constraints of costs.

$$d_t = \frac{N*m}{S*m}1000 \quad (1)$$

$$S_f = L*l \quad (2)$$

$$S = Spf*m + Sgs + Sr \quad (3)$$

$$Nr = x*n \quad (4)$$

$$N_{hab} = Spf + S \quad (5)$$

$$Sp = 16*n*m \quad (6)$$

$$S_{gs} = sgs * N_{hab} \quad (7)$$

$$Lr = (17*n^{(-0,77)}) \quad (8)$$

$$W_g = 5 + 3\sqrt{n} \quad (9)$$

$$S_{sr} = Lr*Wr \quad (10)$$

$$S_{pf} = (L + 2*Lo)*(l + 2*lo) \quad (11)$$

$$L_0 = 3,5 + \frac{2,7}{n} \quad (12)$$

Where:
N: number of dwellings in one building.
S: footprint area of one building,
d: urban density (dwelling density),
n: number of stories into one building,
15 m2: parking area for one vehicle,
Sp: total parking area for the plot,
Sgs: total surface of green space for the plot,
x: number of apartments by story,
y: number of habitants par apartment,
Nhab: Number of habitants in the plot,
Lr : total length of road and pavement for the plot,
Wr: width of road and pavement,
Sr: total surface of road and pavement for the plot,
l : building length,
L : building width,
m: number of building in the plot.
Spf: surface of one platform,
Sst: area of one story,
L0: width around every building (for service and maintenance).

2.2 Formulation of the costs and the optimization

The global cost of one building is composed of in three parts: land, building and external works (viability or amenities). In the practice, the cost of the square meter construction in elevation and the cost of the square meter viability are indexed on the cost construction of the first floor. The relations 14 to 19 illustrate these assumptions. The optimization of the elevation of the building (number of stories) is determinate by the comparison of the global cost of every tranche of stories in elevation. For each value of n corresponds one value of density. So, for n optimized corresponds the global optimized cost. The procedure of optimization of cost in the field of build is described by many authors (El kechebour 2010c).

The relations 13 to 19 describe the process of the optimization.

$$Ct(i) = Cl(i) + Cc(i) + Cv(i) \quad (13)$$

$$Cl(i) = Cl(i)\frac{Sf}{n} \quad (14)$$

$$Cc(i) = Cc(1) + i*\Delta*Cc(1) \quad (15)$$

$$Cv(i) = 0008*Cc(1) + i*\Delta v*Cc(1) \quad (17)$$

$$Cav(i) = \sum_{i=1}^{n} \frac{Ct(i)}{n} \quad (18)$$

$$Cav(opt) \leq \min[Cav(1), Cav(2);...;Cav(n)] \quad (19)$$

Where:
i: index representing number i story (i = 1, 2, to n),
Ct(i): total cost for one story in elevation,
Cl(i): cost of land for one story in elevation,
Cv(i): cost of viability for one story in elevation,
Cc(1): cost of square meter construction for the first story (n = 1),
Cl: cost of square meter land,
Δc: additional cost ratio of construction for elevation,
Δv: additional cost ratio of viability for elevation according to the construction cost of the first story,
Cav(i): average cost for one tranche of stories in elevation,
Cav(opt): optimized cost corresponding to n_{opt}.

3 RESULTS AND DISCUSSION

For example, the figures 1 and 2 illustrate a case with following data:
Sgs = 6 square meter/habitant,
X = 2 apartments /story,
Y = 6 persons/ apartment,
Sp = 15 square meter/vehicles
Cc(1) = 300 us dollars/square meter
Cl = 100 us dollars/square meter.

3.1 Results

Many Numerical simulations are realized and the figure 1 shows the variations of costs according to the elevation (n) for urbanization. For example, the figures 2 and 3 illustrate a case with following data:
Sgs = 6 square meter/habitant,
X = 2 apartments /story,

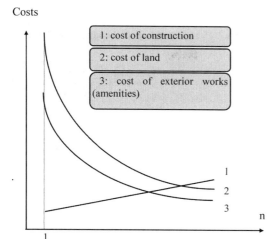

Figure 1. Variations of costs relative to one dwelling according to elevation (n) for urbanization.

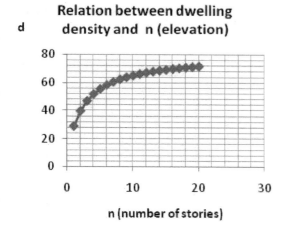

Figure 2. Relation between dwelling density and elevation (n).

$Y = 6$ persons/ apartment,
$Sp = 15$ square meter/vehicles
$Cc(1) = 300$ us dollars/square meter
$Cl = 100$ us dollars/square meter

3.2 Discussion

The density is expressed through the number of dwelling in relation to one hectare area. This assertion is advanced by many authors as Vanegas (Vanegas C. et al. 2009), Zhang (Zhang et al. 2011) and land agencies, but its mathematical formulation is not given. The simulations show that if the cost of land is very low, then the elevation is not significant. It is proposed to find the ratio cost of construction for the first story and the cost of land. In Algeria, this ratio R varies between 0.5 and 3. The cost of land is determinant in the research of the global cost during the urbanization

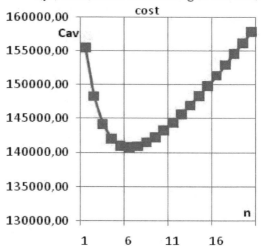

Figure 3. Variation of the average total cost (Cav) of one dwelling in elevation.

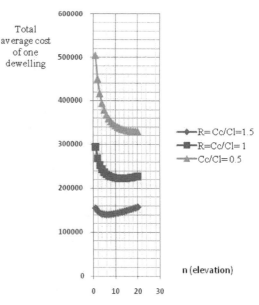

Figure 4. Optimization of n according to the total average cost of one building and the ration R (cost construction/cost of land).

procedure. This observation has been described by many authors as Dempsey (Dempsey 2012) and El kechebour (El kechebour 2014d).

4 CONCLUSION

The modelling approach shows that there is a relation between the cost of urbanization and the urban density. In fact, in the urban zone near to the center, the land is very expansive and the optimized solution gives a high urban density characterized by a high elevation of building and by a very weak green space. The main result concerns the possibility to make simulations on the global cost of urbanization and its impact on the urban density. Every simulation gives a scenario on the urban density according to the costs. This approach permits to gain time and to do the best choice during the design of an urban composition, and confirms that the analysis of the urbanization is a systemic vision.

REFERENCES

Barthélemy, M. & Flammini A. 2009. Co-evolution of density and topologyin a simple model of city formation, *Networks and Spatial Economics*, 9 (3): 401–425.

Batty, M. 2008a. *The Size, Scale, and Shape of Cities*, Science 319: 769–771.

Batty, M. 2007b. Complexity in City Systems: Understanding, Evolution, and Design, Paper 117, March 07, 2007, ISSN 1467-1298.

Batty, M. & Kim K. S. 1992c. Form follows function: reformulating urban population density functions, *Urban Studies* 29: 1043–1070.

Courtat, T. et al. 2011. Mathematics and morphogenesis of the city: a geometrical approach, *Phys. Rev. E*, 83: 036-106.

Dempsey N. 2010. Revisiting the Compact City? *Built Environment*, 36 (1): 05–08.

El kechebour, B. 2009a. Analysis of costs and choice on retaining structures: Land, curtain wall and slope, *International Journal of Applied Engineering Research* ISSN 0973-4562 Volume 4, Number 6, 2009, pp. 845–865.

El kechebour, B. 2010b. *Modélisation de la conception de la trame plane orthogonale (Computer aided design): espace, coûts et site, Doctorat*, Université des sciences et de la technologie Houari Boumediene (USTHB), Bab Ezzouar, Alger, Algeria.

El kechebour, B. 2014c. Optimized Cost of High Performance Concrete in the Build, *Advanced Materials Research* Vol. 911: 479-483, http://www.scientific.net/AMR.911.479.

El kechebour, B. 2014d. Modelling of Assessment of the Green Space in the Urban Composition, Elsevier, *Procedia-Social and Behavioral Sciences* 195: 2326–2335.

Lechner, T. et al. 2004. *Procedural modeling of land use in cities, Technical report*, Northwestern University.

UNWUP 2012 (United Nations World Urbanization Prospects). *The 2011 Revision United Nations (Population Division of the Department of Economic and Social Affairs)*, ESA/P/WP/224, New York.

Vanegas, C. et al. 2009. Interactive design of urban spaces using geometrical and behavioral modeling, *Proc. intern. symp ACM SIGGRAPH Asia, A*rticle No. 111, Session: Urban modeling.

Zhang, et al. 2011. Simulation and analysis of urban growth scenarios for the Greater Shanghai Area, China, *Computers, Environment and Urban Systems*, 35(2), 126–139.

Numerical study on shear and normal stress variation of RC wall with L shaped section

A. Ahmed-Chaouch
*Faculty of Civil Engineering, University of Sciences and Technology
Houari Boumedienne (USTHB-LEEGO), Algiers, Algeria*

ABSTRACT: Numerical investigations have been carried out on reinforced concrete structures in order to evaluate the stress distribution at the base of the L shaped shear walls. Several structural parameters that influence the seismic behavior such the number of stories, length of the shear wall as well as the thickness of the wall have been implemented in our investigation. In total, more than 200 numerical models were crated and analyzed. Based on the analyses results it has been showed that this type of walls have a short shear walls behaviour, so shear cracking turn into flexural cracking once the wall is thick starting from 15 cm, 20 cm and 25 cm, the values of shear span ratio is about 2.00, 2.40 and 2.50 respectively.

1 INTRODUCTION

After the 2003 earthquake that shook the region of Boumerdes located at 60 km east from the capital Algiers with a magnitude of 6.8, the Algerian seismic regulations RPA99 has been revised based on the new seismic zones classification of the different provinces of country. The new version, RPA99/2003, have restricted the use of the RC bare frame systems in moderate and high seismic zones for once the building's height exceed a certain value. In order to overcome this restriction, some construction companies have resorted to insert reinforced concrete shear walls with L shaped section (Oussalam et al. 2003), often at the four edges of the structure, without any numerical investigations on the seismic behaviour of such structural systems. Reinforced concrete structures with L shape walls provide the advantages of open space and flexible architecture modelling. However, until now, seismic behavior of buildings with such configuration is not well known.

Several high rise buildings in the world are made of reinforced concrete shear walls. This type of structures performed very well during the past earthquakes better than reinforced concrete frame buildings. However, some studies were conducted on rectangular reinforced concrete shear walls.

Nonlinear behaviour of reinforced concrete walls such as T, H, and U shape walls have been studied by several researchers. In many cases these walls had different behaviors than rectangular shaped form. Experimental tests were carried mainly on walls with symmetrical flanges. Few studies in the literature treated the behavior of shear walls with asymmetric cross sections, such as L shaped section. Rigidity, strength, and ductility variation of such walls may be completely different in the two directions [8], which may lead to two different failure modes. Only four thorough studies have focused on the seismic behavior of L-shaped walls.

HOSAKA et al. (2008) studied four walls specimens corresponding to the base of a 30-storey building model. At the intersection of the two flanges and at the end of each of these flanges, the concrete was confined by a high density of reinforcement forming columns embedded in the thickness of the walls. Experimental tests were conducted under variable axial and cyclic horizontal loading. Results were compared to a multi-fiber model. The authors observed a localized bending fracture at the base of the wall at the intersection area of the two flanges. Also, the proposed multi-fiber method predicted well the behavior of the tested walls.

INADA et al. (2008) studied the effect of the loading direction and the configuration of a central reinforced concrete core walls on the behavior under cyclic loading. They tested two walls with equilateral flanges with a loading successively perpendicular to one of the flanges and applied at the point of intersection of the two flanges (45° angle), and the third specimen wall having unequal flanges, was loaded toward the intersection point of the two flanges. The results were very interesting especially with respect to the location of the crack inducing failure of the specimen. The first cracks were horizontal due to bending, that were followed gradually by inclined cracks that cause a shear failure of the walls.

WEI et al. (2010) tested six types of L walls under a constant axial load and horizontal cyclic loading. The main test parameter was the wall length. All specimens had the same height of 1400 mm. Based on the test results, the following failure modes were observed:

– For specimen with a shear span ratio relatively low (1.75) shear crack type was observed. First, in the central part of the wall appeared a partial oblique transverse crack, when the load increased, the diagonal crack gradually increased, and the wall surface is divided into a number of lozenge blocks. Several diagonal cracks formed and concrete crashed after buckling of reinforcement bars.

- For specimen with a moderate shear span ratio (2.15) shear-bending cracks were observed. The first cracks were localized at the central part of the perpendicular flange to the loading; the cracks were developed into several smaller inclined cracks. Horizontal bending cracks occurred at the end of the flange. Under the increased displacement loading amplitudes, such horizontal cracks developed faster than the inclined cracks. At the end of the test, vertical cracks appeared with crushing concrete.
- Specimen with a relatively large shear span ratio (2.8) showed a flexural cracking. At the beginning, cracks appeared in the central portion of the wall, and then low inclined cracks appeared in the lower part. Concrete crashing was also observed at the end of the test.

KARAMLOU et al. (2012) were interested in their research work in the seismic behavior of constructive system called R-Iranian ICF panel walls. Four walls with L shaped section have been tested in a combined action, a constant axial load and reversed cyclic lateral loading. Horizontal bending cracks and other angled bending-shear cracks appeared at the end of each flange and at the corner of the two intersection flanges. In addition, at the end of the two flanges, the majority of these cracks propagated horizontally through the wall thickness. These cracks have become more inclined in the inner part of the wall to form diagonal shear cracks which form the majority of the wall cracks.

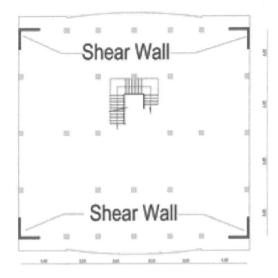

Figure 1. Plan view.

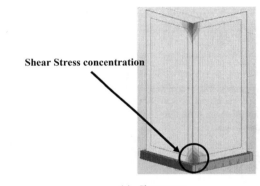

(a) Shear stress

(b) Normal stress

Figure 2. Location of the stress concentration.

2 CONSIDERED BUILDING

To carry out our numerical analysis, we have considered a regular RC building in plan with L shaped section shear walls at the four corners as illustrated in Figure 1.

The building has an equal story height (h) of 3.00 m. The three considered parameter in our analysis are:

- Thickness of the RC walls: t
- Length of the walls: l
- Number of stories: N

In the actual Algerian seismic regulation RPA99/V2003 (2003), reinforced concrete shear wall is defined as an element whose length is at least 4 times the thickness. Three wall thicknesses were considered in our study: 15 cm, 20 cm and 25 cm. The ratio length to thickness (l/t) varied from 4 to 21 for a thickness of 15 cm, from 4 to 16 for a thickness of 20 cm and from 4 to 13 for a thickness of 25 cm.

As for buildings, we considered five buildings with 4, 6, 11, 16 and 21 stories that represent the common buildings that are built in Algeria. Taking into account the variation of the three parameters two hundred (200) models were analyzed.

All buildings were designed according the Algerian seismic regulation RPA99/V2003 (RPA, 2003). We assumed that each building is implanted on S2 soil type (firm soil). Structures were modeled in a 3D using the nonlinear Robot software as illustrated in Figure 1.

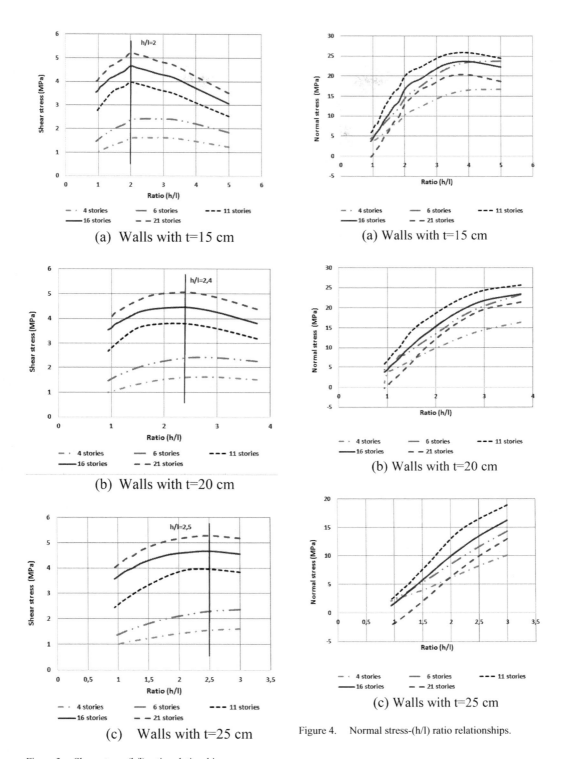

(a) Walls with t=15 cm

(b) Walls with t=20 cm

(c) Walls with t=25 cm

Figure 3. Shear stress-(h/l) ratio relationships.

(a) Walls with t=15 cm

(b) Walls with t=20 cm

(c) Walls with t=25 cm

Figure 4. Normal stress-(h/l) ratio relationships.

3 ANALYSIS RESULTS

Hence, the discussed results and the comparisons involved are those related to the shear stress and the normal stress as illustrated in Figure 2, as an example. The corner location, intersection between the two RC flange is the area where the shear stress is the most important, stress concentration is huge, but for the normal stress, the end of flanges is the most important. This assumption was observed for all models analyzed and also confirmed experimentally by Inada et al. [3–4].

The obtained maximum shear stresses for the different analyzed models are summarized on Figure 3 which shows the shear stress distribution versus the shear span ratio (h/l), for the dead load, life load and earthquake load combination. It is clearly shown that for all buildings the shear stress increased until reaching a peak stress and after that decreased regardless of the walls thickness.

However, for buildings with 4 and 6 stories, It is hardly clear that for walls with a thickness of 15 cm the maximum shear stress is reached once the shear span ratio reached a value of 3. For buildings with 11, 16 and 21 stories this value decreases and becomes equal to 2 (see Figure 3 (a)). For the walls of 20 and 25 cm thickness, regardless of the height of the building, the value of the shear span ratio remains constant and respectively equal to 2.4 and 2.5 (see Figure 3 (b) and (c)).

Figure 4 shows the change in normal stress versus shear span ratio for each of the wall thicknesses. Increased stress is continuous, no peak showing a maximum stress is visible, however, from a ratio (h/l) of 3 a change in the slope of all curves is noticeable.

4 CONCLUSIONS

Numerical investigations have been carried out on L shaped RC shear walls in order to identify the real behaviour of such kind of structural elements.

Main results of those numerical analysis on shear stress variation were discussed and presented. The analysis was mainly focused on 5 RC building with 4, 6, 11, 16 and 21 stories. For each building 3 wall thicknesses were selected respectively 15, 20 and 25 cm. In total more than 200 models were crated and analyzed. It was shown that the shear stress increased until reaching a peak stress and after that decreased independably of the walls thickness.

The first important remark that we carried on is the location of the most important shear stress at intersection between the two RC flanges, and for the normal stress, the end of flanges is the most important.

The study of the variation of the shear stress depending of shear span ratio (h/l) showed that for a certain number of buildings, the increase of the length of the flange increase the value of the shear stress, after that decreased independably of the walls thickness.

The study of the variation of the maximum shear stress and maximum normal stress according to the shear span ratio (h / l) suggests that this type of walls have a short walls system behavior, so shear cracking, turn into flexural cracking when reached respectively for 15 cm, 20 cm and 25 cm thicknesses, the values of shear span ratio are respectively 2.00, 2.40 and 2.50.

REFERENCES

Oussalem, H & Bechtoula, H. 2003. Report on the damage investigation and post-seismic campaign of the 2003 Zemmouri earthquake in Algeria", a full report of 138 pages concerning the earthquake that shook the north part of Algeria with a magnitude of 6.8, published by the ERI, the University of Tokyo and produced by Ohbunsha Press, Japan.

Hosaka, G, Funaki, H, Hosoya, H. & Imai, H. 2008. Experimental study on structural performance of RC shear wall with L-shaped section. The 14th World Conference on Earthquake Engineering. October 12–17, 2008, Beijing, China.

Inada, K., Chosa, K., Sato, H., Kono, S. & Watanabe, F. 2008. "Seismic performance of RC L-shaped core structural walls". The 14th World Conference on Earthquake Engineering. October 12–17, 2008, Beijing, China.

Inada, K. 2008. Horizontal force of resistance mechanism in RC core wall structure ; Study on the model. Master thesis Department of Urban and Environmental Engineering, Graduate School of Engineering, Kyoto University.

Wei, LI, & Qing-ning, LI. 2010. Seismic performance of L-shaped RC shear wall subjected to cyclic loading. Struct. Design Tall Spec. Build. 21, 855–866 Published online 1 December 2010 in Wiley Online Library (wileyonlinelibrary.com). DOI: 10.1002/tal.645.

Karamlou, A. & Zaman kabir, M. 2012. Experimental study of L-shaped slender R- ICF shear walls under cyclic lateral loading. Engineering structures 36 :134–146.

Effects of alkaline solution on the microstructure of HCFA geopolymers

M.F. Nuruddin, A.B. Malkawi, A. Fauzi & B.S. Mohammed
Department of Civil and Environmental Engineering, Universiti Teknologi PETRONAS, Seri Iskandar, Perak, Malaysia

H.M. Al-Mattarneh
Civil Engineering Department, Najran University, Najran, Saudi Arabia

ABSTRACT: The synthesis of geopolymer mortars based on high calcium fly ash (HCFA) has been studied in order to assess the effects of the alkaline solution on the geopolymerization products and microstructure by means of SEM/EDS and compressive strength tests. The results showed that HCFA geopolymers can develop high strength binders with a dense homogeneous microstructure. The higher sodium silicate to sodium hydroxide (NS/NH) ratio in the range of 1–2.5 has increased the density and homogeneity of the microstructure. Similar behavior was observed for the effect of NH concentration with an optimum concentration of 10M. The microstructure of HCFA geopolymers can be described as porous with nanoporosity size and considerable amounts of unreacted fly ash particles depending on the study parameters. The formation of C–S–H crystals was not detectable. The Si/Al ratios of the produced gels were in the range 1.12-5.55 which was different than the initial mixing ratios.

1 INTRODUCTION

Concrete has an essential role in our daily lives. It can be seen everywhere in our surrounding environment. In fact, concrete is the most used material after water, with an average annual consumption around 2.5 tons per capita (U.S. Geological Survey 2016). Cement is the main binder in the concrete industry, and its production bears several sustainability concerns that recently raised the global awareness. Economic, environmental, and social sustainability form the major concerns. This motivates many researchers towards the finding of new cement replacement materials. Geopolymer binders showed increasing potential as cement replacement materials (Davidovits 2011). These binders have the ability to produce concrete mixtures with superior properties in terms of strength and durability (Law et al. 2014, Bernal et al. 2014). Geopolymer binders can utilize many wastes or byproduct materials such as; fly ash (FA), metakaolin (MK), granulated blast furnace slag (GBFS), rice husk ash, red mud, and other several materials. Actually, any material in which the silicon (Si) and Aluminum (Al) form its main component can be used as a geopolymer source material (Davidovits 2011).

Many researchers have investigated the utilizing of the low calcium content class F FA or/and GBFS (Talha Junaid et al. 2015, Nikoliæ et al. 2015, Mallikarjuna Rao & Gunneswara Rao 2015). However, less research has been devoted to studying the feasibility of using high calcium content Fly ash (HCFA) as a source material for geopolymer binder production (Temuujin et al. 2014, Chindaprasirt et al. 2012). This research investigates the effects of the alkaline solution proportions on the microstructure and compressive strength of geopolymer mortars produced using the HCFA as a geopolymer source material.

2 EXPERIMENTAL

2.1 Materials

The geopolymer mortars were prepared using FA as a source material. The FA was provided from the Manjung power plant, Perak, Malaysia. The chemical analysis showed that this FA has a high calcium content of 20%, and the summation of the Si, Al, and Fe oxides was about 70.5%. Hence, this FA may be classified as a Class F FA according to the ASTM standard specification of the coal fly ash (ASTM-C618 2015) but with high calcium content. The chemical composition using the X-ray fluorescence (XRF) analysis is presented in Table 1. The alkaline solution was prepared using a mixture of sodium hydroxide (NH) and sodium silicate (NS) solutions. The NH solution was prepared by dissolving the solids of NaOH in forms of pellets (98% purity) in potable water. The NS solution was composed of $SiO_2 = 29.43\%$, $Na_2O = 14.26\%$, and water $= 56.31\%$ as mass ratios, with a specific gravity of 1.53 g/cm^3.

Table 1. The chemical composition of the used fly ash.

Component	Weight percent
SiO_2	25.9%
Al_2O_3	12.3%
Fe_2O_3	32.3%
CaO	20.9%
Na_2O	0.26%
MgO	2.08%
SO_3	0.70%
K_2O	2.80%
LOI	3.60%

Table 2. Mixing proportions and parameters of the geopolymer mortars.

ID	FA [g]	Sand [g]	NS [g]	NH [g]	NS/NH	NH Molarity [M]
D01	300	600	82.50	82.5	1.0	8
D02	300	600	99.00	66.0	1.5	8
D03	300	600	110.0	55.0	2.0	8
D04	300	600	117.9	47.1	2.5	8
D05	300	600	82.50	82.5	1.0	10
D06	300	600	99.00	66.0	1.5	10
D07	300	600	110.0	55.0	2.0	10
D08	300	600	117.9	47.1	2.5	10
D09	300	600	82.50	82.5	1.0	12
D10	300	600	99.00	66.0	1.5	12
D11	300	600	110.0	55.0	2.0	12
D12	300	600	117.9	47.1	2.5	12

2.2 Specimens preparation

The NH solutions were prepared before using by 24 hours to give enough time to release the generated heat. Different amounts of the NaOH pellets were dissolved in water to prepare solutions with different concentrations. It was required to dissolve 320, 400, and 480 g of the NaOH pellets to prepare one liter of the NH solution with a molarity of 8M, 10M, and 12M, respectively. The two solutions (NH&NS) were mixed together with the required proportions just before mortar mixing. The ratio of the FA: sand was 1:2 and fixed for all mixes. Also, a constant ratio of alkaline solution to FA of 0.55 was used for all mixes. The mixing procedures were similar to that specified by the ASTM standard for ordinary Portland cement (OPC) mortar mixing (ASTM-C305 2014). The mixing proportions and variables are shown in Table 2.

2.3 Methods

The geopolymer mortars were cast in 50 mm cubes molds and left to rest for one hour. After that, the molds were moved to cure in the oven at 60°C for 24 hours. Then, the specimens were de-molded and the curing continued under ambient conditions until the day of testing. The compressive strength test was carried out according to the ASTM standard method for testing of the OPC mortar compressive strength (ASTM-C109 2013). After compressive strength test, some of the fragments were collected and the microstructure was analyzed using the scanning electron microscopy (SEM). The specimens were analyzed using both of the secondary and backscattered electron detectors. The particular elements proportions were determined by the energy dispersive spectroscopy (EDS). All specimens were analyzed at similar ages.

3 RESULTS AND DISCUSSION

3.1 HCFA geopolymer microstructure

Figure 1 shows the BS-SEM micrograph of the HCFA geopolymer. This micrograph shows the formation of geopolymer matrix in a colloidal form, the matrix contains lots of nanopores in uniform spherical shapes, which is similar to the microstructure described for low calcium FA geopolymers (Torres-Carrasco & Puertas 2015). In all mixtures, few crystalline products were noticed, the main reaction products were in gel form. This may refer to the fast geopolymerization reactions; hence, no enough time was available for the growth of well crystalline reaction products due to the fast harden of the matrix caused by the high calcium content and the application of heat curing, similar results were reported for oven cured geopolymers based on class F FA (Al Bakri et al. 2012).

The presence of considerable amounts of unreacted or partially reacted FA particles can be noted in this figure. This indicates moderate geopolymerization reactions, and this was also confirmed by the presence of incomplete geopolymerization. However, it can be seen that the formed geopolymer gel was enough to bind the constituents. Also, it can be seen that some of the FA particles are firmly connected to the geopolymer matrix; on the other hand, many particles are loosely bonded to the structure and the interfacial boundaries can be easily detected.

The HCFA particles composed of precipitators (solids) and cenospheres, the exfoliation of some cenospheres crust can be noticed, which indicates the occurrence of the dissolution reactions and that the generated product due to the destruction of the FA particles. On the other hand, accumulations of the geopolymerization products can be seen on the surface of some solid FA particles.

It is worth mention that even this type of FA contained high calcium content, the formation of the C-S-H gel was not easy to be detected. This was attributed to the formation of nanosize C-S-H crystals within the geopolymer gel, similar results were reported in the case of the GBFS&FA based geopolymers (Oh et al. 2012).

3.2 Effects of the NH solution concentration

Figure 2 shows the effect of the NH solution concentration on the compressive strength of the HCFA

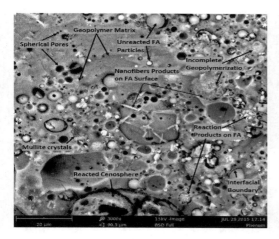

Figure 1. SEM micrograph showing the microstructure details of the HCFA geopolymer mortar.

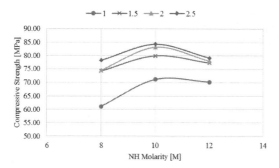

Figure 2. Effect of NH molarity on HCFA geopolymer compressive strength at different NS/NH ratios.

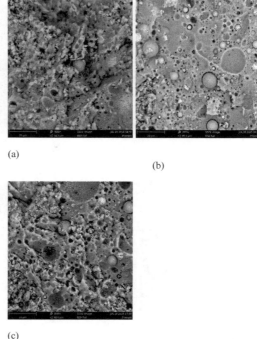

Figure 3. SEM micrographs of the HCFA at NS/NH=2.5 for (a) 8M, (b) 10M, & (c) 12 M.

geopolymer mortars. This figure displays the existence of an optimum NH solution concentration at 10M. Increasing the concentration from 8 to 10M has resulted in higher compressive strength; however, further increasing to 12M has reduced the strength.

Figure 3 shows the SEM micrographs of D4, D8, and D12 mixes representing the mixes of 8, 10, 12M NH solution concentrations, respectively. This figure displays that the mixture of 10M NH solution (Fig. 3b) has the highest density microstructure, where enough geopolymer gel has been created to join the constituents, and this correlates with the highest compressive strength (84MPa) of this mix.

Figure 3a shows a heterogeneous matrix of partially reacted materials, and less geopolymer gel has been created during the reaction period with higher irregular voids. Compared to Figure 3c, the later has a denser matrix with higher regular spherical pores which shows a more homogeneous microstructure. This may be attributed to the changes in reaction rate and the degree of geopolymerization.

The Alkalis content controls the dissolution and leaching of the Si&Al ions from the FA surfaces. The higher alkali content the higher the dissolution rate (Nurudin and Memon 2014). Increasing the NH solution concentration has increased the alkali content; hence, higher geopolymerization rate and more gel has been created. However, the excess content of alkalis (in the case of 12M solution) has increased the rate of the geopolymerization reactions to a level, where there was no enough time for the reaction products to disperse in the matrix and it deposed on the surface of the FA particles which hindered further dissolution reactions and hence less geopolymer gel was produced. This can be verified by the SEM/EDS analysis of the produced geopolymer gel. The SEM/EDS analysis showed that increasing the NH solution molarity resulted in higher aluminum content in the produced gel, the Si/Al ratios for the mixtures shown in Figure 3 below were 5.55, 5.02, and 2.03 for the mixes with NH solution molarity of 8, 10, and 12, respectively. This is a result of the higher Al content in the aqueous solution. This indicates higher FA reactions for the higher NH solution concentration; since the FA is the only source of Al in the mix, where there was no soluble aluminate provided by the alkaline solution.

3.3 *Effects of the NS/NH solution ratios*

Figure 4 shows the effect of the NS/NH solution ratios on the compressive strength of the HCFA geopolymer mortars. This figure indicates that the higher the ratio, the higher the strength. However, increasing the ratio further than 2 has less effect, especially in the case of the highest NH solution molarity of 12M.

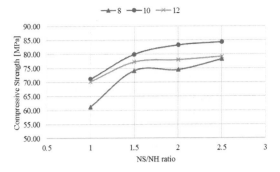

Figure 4. Effect of NS/NH ratios on HCFA geopolymer compressive strength at different NH molarity.

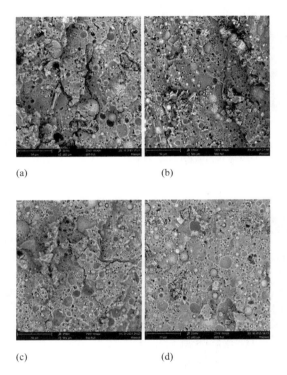

Figure 5. SEM micrographs of the HCFA at NH = 10M for (a) NS/NH = 1, (b) NS/NH = 1.5, (c) NS/NH = 2, & (d) NS/NH = 2.5.

Figure 5 displays the SEM micrographs of the highest strength mixes of the 10M NH solution concentration at different NS/NH ratios from 1 to 2.5. It can be noticed that increasing the NS/NH ratio resulted in a higher density geopolymer matrix. In all mixtures, enough geopolymer gel was formed to fill the space between different particles. However, it is clear that a more homogeneous matrix with a smaller and lower porosity gel was formed for higher NS/NH solution ratio, which resulted in higher compressive strength. This may be related to the amounts of the additional silicates directly available from the sodium silicate solution.

Increasing the NS/NH solution ratio resulted in higher Si/Al ratio, which enhanced the geopolymerization reactions. It is known that the leaching of the Si ions from the FA particles is more difficult than leaching of the Al ions and this refers to the speciation of the atoms. Hence, the presence of the soluble silicate will enhance the geopolymerization processes. In addition, the lower NS/NH ratio indicating higher NH solution content, which results in higher reaction rate and less uniform matrix (Duchesne et al. 2010).

All micrographs showed the presence of considerable amounts of unreacted or partially reacted FA particles; hence, it is difficult to assess the extents of the FA dissolution to form the aluminosilicate geopolymer gel. The SEM/EDS analysis of the produced geopolymer gel showed that the average Si/Al molar ratios are 2.63, 2.87, 4.26, and 5.02 for the 1, 1.5, 2, and 2.5 NS/NH ratio, respectively. This proves that the higher NS/NH solution ratio resulted in higher Si content in the produced gel. This also explains the higher compressive strength, which may be related to the formation of more Si–O–Si bonds, and it is known that this kind of bonds is stronger than the Si–O–Al bonds (Al Bakri et al., 2012).

4 CONCLUSIONS

The results obtained in this study showed that the composition of the produced geopolymer gel based on HCFA was changing based on the alkaline solution proportions. The produced geopolymer gel has different Si/Al molar ratio ranged from 1.12-5.55, and this was different from the initial mixing ratios; where the Si/Al molar ratio varied in the range 2.76-3.05. This can be attributed to the different dissolution level of the aluminosilicate precursors. Moderate geopolymerization level was observed and this was indicated by the presence of considerable amounts of unreacted FA particles in the geopolymer matrix.

An optimum NH solution concentration value of 10M was observed to produce a dense geopolymer matrix which resulted in a high strength mortar. Changing the concentration to lower or higher values resulted in a less homogeneous matrix with a higher porosity which has reduced the compressive strength. Increasing the NS/NH solution ratio resulted in higher Si/Al ratio of the produced HCFA geopolymer gel and increased the density of the produced matrix.

The microstructure of the HCFA based geopolymer did not contain any crystalline phases except some of the mullite crystals which originally existed in the FA particles. Even the high calcium content in this type of FA, the formation of the C–S–H gel crystals was not detected using the SEM-BSE images. This proposing the formation of the C–S–H gel crystals in fine sizes and in a mixed form with the geopolymer gel. However, further investigations are required.

REFERENCES

ASTM C109, Standard Test Method for Compressive Strength of Hydraulic Cement Mortars (Using 2-in. or [50-mm] Cube Specimens), *ASTM International*, West Conshohocken, PA, 2013, www.astm.org.

ASTM C305, Standard Practice for Mechanical Mixing of Hydraulic Cement Pastes and Mortars of Plastic Consistency, *ASTM International*, West Conshohocken, PA, 2014, www.astm.org.

ASTM C618, Standard Specification for Coal Fly Ash and Raw or Calcined Natural Pozzolan for Use in Concrete, *ASTM International*, West Conshohocken, PA, 2015, www.astm.org.

Al Bakri, A. M. M., Kamarudin, H., Binhussain, M., Khairul Nizar, I., Razak, A. R. & Zarina, Y. 2012. Microstructure Study on Optimization of High Strength Fly Ash Based Geopolymer. *Advanced Materials Research* 476–478: 2173–2180.

Bernal, S. A., Provis, J. L. & Green, D. J. 2014. Durability of Alkali-Activated Materials: Progress and Perspectives. *Journal of the American Ceramic Society* 97(4): 997–1008.

Chindaprasirt, P., De Silva, P., Sagoe-Crentsil, K. & Hanjitsuwan, S. 2012. Effect of SiO_2 and Al_2O_3 on the setting and hardening of high calcium fly ash-based geopolymer systems. *Journal of Materials Science* 47(12): 4876–4883.

Davidovits, J. 2011. *Geopolymer Chemistry and Applications*. 3, Saint-Quentin, France: Institut Géopolymère.

Duchesne, J., Duong, L., Bostrom, T. & Frost, R. 2010. Microstructure Study of Early In Situ Reaction of Fly Ash Geopolymer Observed by Environmental Scanning Electron Microscopy (ESEM). *Waste and Biomass Valorization* 1(3): 367–377.

Law, D. W., Adam, A. A., Molyneaux, T. K., Patnaikuni, I. & Wardhono, A. 2014. Long term durability properties of class F fly ash geopolymer concrete. *Materials and Structures* 48(3): 721–731.

Mallikarjuna Rao, G. & Gunneswara Rao, T. D. 2015. Final Setting Time and Compressive Strength of Fly Ash and GGBS-Based Geopolymer Paste and Mortar. *Arabian Journal for Science and Engineering*.

Nikolić, V., Komljenović, M., Baščarević, Z., Marjanović, N., Miladinović, Z. & Petrović, R. 2015. The influence of fly ash characteristics and reaction conditions on strength and structure of geopolymers. *Construction and Building Materials* 94: 361–370.

Nurudin, M. F. & Memon, F. A. 2014. Properties of Self-Compacting Geopolymer Concrete. *Materials Science Forum* 803: 99–109.

Oh, J. E., Moon, J., Oh, S.-G., Clark, S. M. & Monteiro, P. J. M. 2012. Microstructural and compositional change of NaOH-activated high calcium fly ash by incorporating Na-aluminate and co-existence of geopolymeric gel and C–S–H(I). *Cement and Concrete Research* 42(5): 673–685.

Talha Junaid, M., Kayali, O., Khennane, A. & Black, J. 2015. A mix design procedure for low calcium alkali activated fly ash-based concretes. *Construction and Building Materials* 79: 301–310.

Temuujin, J., Minjigmaa, A., Zolzaya, T. & Davaabal, B. 2014. Study of Geopolymer Type Paste and Concrete from High Calcium Mongolian Fly Ashes. *Transactions of the Indian Ceramic Society* 73(2): 157–160.

Torres-Carrasco, M. & Puertas, F. 2015. Waste glass in the geopolymer preparation. Mechanical and microstructural characterisation. *Journal of Cleaner Production* 90: 397–408.

U.S. Geological Survey 2016. Mineral commodity summaries 2016: *U.S. Geological Survey* 1: 196–197. http://dx.doi.org/10.3133/70140094.

Influence of pozzolan on sulfate resistance of concrete

A. Merida & F. Kharchi
University of Sciences and Technology of Algiers, Algeria

ABSTRACT: This paper reports an experimental study the influence of Algerian natural pozzolan of volcanic origin used as partial substitute for Portland cement on the mechanical properties and durability of high performance concretes. The analysis of experimental results on the concrete content of 5% of the natural pozzolan having a fineness modulus of 10000 cm^2/g in a sulfated medium allowed the reduction of pore volume contributing to the increase of mechanical strengths and the durability with respect to migration of chloride ions and oxygen permeability of concrete in sulfated environment. On the basis of the experiments performed, the natural pozzolan is suitable for formulation of high performance concretes and their properties are significantly better compared to the control concrete.

1 INTRODUCTION

The durability of concrete structures is affected by many environmental factors, the sulphate corrosion being one of the most frequent and detrimental processes. Through the capillary pores of concrete due to the concentration gradient, Nadler B and all (2003), Truc O (2000), and react with unhydrated components of the hardened cement paste. In consequence, these chemical reactions may lead to expansive reaction products such as ettringite ($C_3A.3CaSO_4.32H_2O$), Atkinson A and all (1988). In turn, the ettringite may cause the overall expansion of a structural element and its extensive damage progressing from the outer surface towards the specimen inner core, Skalny and all (2002). This process may result in a gradual loss of concrete strength, Tixier R & Mobasher B (2003), accompanied by surface spalling and exfoliation, Bentz DP & Garbicz, EJ (1991).

Pozzolanic materials improve the microstructure of concrete due to their particle size, and may alter chemical composition and hydration reactions. Pozzolan as an amorphous or glassy silicate material that reacts with calcium hydroxide formed during the hydration of Portland cement in concrete. The substance that contributes to the strength of the concrete called calcium silicate hydrates (C-S-H), Bentz DP & Garbicz EJ (1991).Calcium hydroxide will reduce the strength of the concrete.

Pozzolan contains silica that react with calcium hydroxide in concrete to form extra calcium silicate hydrates compound and diminish calcium hydroxide, Shamaran M and all (2007), further strengthening the concrete due to increase of C-S-H compound and making it stronger, denser, and durable during its service life. Many researches on the performance of concretes containing pozzolan in sulfate solution have been performed, Saricimen H and all (2003), Chang ZT and all (2005), Sersale R and all (1998).

The aim of this study is to experimentally investigate the effect of replacing 5% of cement by natural volcanic pozzolan in the mixture of high performance concrete (HPC) on the compressive strength, permeability to the chlorine ions and oxygen permeability of specimens exposed to solutions of 5% sodium sulphate (Na_2SO_4) in comparison with traditional concrete (CC).

The specimens were stored for one year in drinking water (environment 1) and in aggressive solution containing 5% sodium sulphate (environment 2).

2 EXPERIMENTAL PROGRAMME

2.1 Cement

Portland cement (CPA-CEM-I / A 42.5) conforming to the Algerian standard NA 443.

2.2 Natural pozzolan

Natural pozzolan of volcanic origin with a fineness modulus of 10000 cm^2/g.

The chemical compositions of the cement and the pozzolan are shown in table 1.

2.3 Superplasticiser

The study of concrete composition is always to seek simultaneously two essential qualities: strength and workability, but these two qualities are linked to each other but vary in the opposite direction. The idea was to develop a dense concrete from a compact granular skeleton using cement and water and meeting the strength, durability and workability requirements.

The optimized superplasticiser content was 2% at W/C = 0, 3 ratio giving a slump of about 21 cm (table 2).

Table 1. Chemical compositions.

Elements	Pozzolan	Cement
CaO	14.59	63.05
S_iO_2	44.95	21.28
Al_2O_3	16.91	03.85
FeO_3	09.47	04.61
MgO	03.70	01.19
SO_3	0.20	02.54
K_2O_3	01.35	0.80
Na_2O	01.34	0.18
RI	0.56	1.11
PAF	04.30	1.58

Table 2. Optimization of superplasticizer.

	Freshly Concrete			Hardened Concrete
SPa 30%	WC	Slump (cm)	Density (Kg/m^3)	Compressive Strength (Mpa)
1	0.25	5	2524	27
1.5	0.25	9	2536	31
2	0.25	14	2542	33
1	0.3	15	2539	28
1.5	0.3	20	2546	29
2	0.3	21	2549	30

Table 3. Concrete mixtures.

Concrete	CC	HPC
W/C (ratio)	0.5	0.3
Cement (Kg/m^3)	425	430.75
Pozzolan (Kg/m^3)	0	21.25
Water (Kg/m^3)	212.5	107.66
Gravel (3/8) (Kg/m^3)	137	137
Gravel (8/16) (Kg/m^3)	837	837
Spa (%)	0	2
Slumps (cm)	8	21
Density (Kg/m^3)	2430	2596

3 FORMULATION AND CONCRETE MIXTURES

Composition concrete defines optimal dosage of aggregates, cement and water to make a concrete with required qualities: strength and durability.

To investigate the use of natural pozzolan on the performance properties of concrete, two different concrete mixes were employed, details of which are given in table 3. The control mix (CC) contained only Portland cement and mix of HPC (high performance concrete) the Portland cement was partially replaced with 5% natural pozzolan (by weight). All concrete mixtures were prepared according to ASTM C 192 standard. The super plasticizer was added at the time of mixing.

3.1 Curing of specimens

After mixing, concrete specimens were cast into the moulds as specified in EN 12390-2. Concrete cubes of 280 × 70 × 70 mm in size, and cylinders of dimensions 110Ø × 220 mm were used in this investigation. The specimens were kept under laboratory conditions for a day, then removed and transferred to the curing basin. In the first curing condition, the specimens were immersed in water until the age of testing, while in the second curing condition, those were immersed in aggressive solution (5% $NaSO_4$) until the age of testing.

4 TEST METHODS

4.1 Compressive strength

This test was carried out in accordance with ASTM C39. Cylindrical specimens 110Ø × 220 mm were used. The strength measurements of concrete were performed at 28, 90, 180 and 365 days.

4.2 Chloride permeability

The resistance of the concrete to the penetration of the chloride ions was measured in terms of charge passed through the concrete in accordance with ASTM C1202.

4.3 Oxygen permeability

The values of oxygen permeability of concretes were measured in terms of flux passed through the concrete according the method developed by algiers university, Talah A & Kharchi F (2013).

5 RESULTS AND DISCUSSION

5.1 Compressive strength

Figure 1 show compressive strength of the specimens kept in water and aggressive solution. It illustrates the results for compressive strength of concrete versus age. An increasing trend of compressive strength is observed for both specimens. For the specimens kept in water, the increase in compressive strength continuous as the duration of immersion increases. The specimen concretes curing in water, the strength of control concrete increase from 34 to 45 whereas the high performance concrete it increase from 56 to 73 MPa. The results indicates that pozzolan addition helps gain compressive strength. The specimens kept in aggressive solution; the strength of the control concrete is reduced by 17, 77% whereas the high performance concrete the reduction was by 5, 48% only. Compressive strength of samples containing mineral admixture (pozzolan) was greater than control sample. It is evident that the pozzolanic mixture creates a more compact concrete. The density increases when admixture is added because it physically occupies pores in the past by virtue of their particle size. Also, there

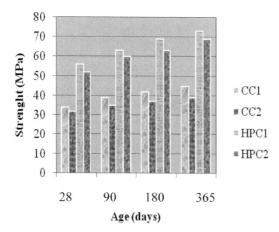

Figure 1. Evolution of compressive strenght at differents ages.

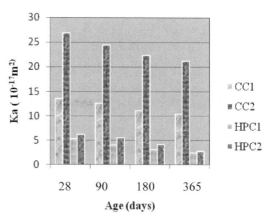

Figure 3. Variation of apparent oxygen permeability at differents ages.

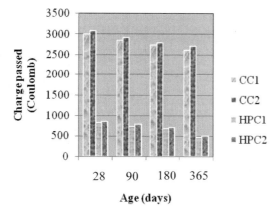

Figure 2. Variation of charge passed at different ages.

is a partial or total pozzolanic reactivity between the particles and the portlandite Ca(OH)2, liberated during the hydration of Portland cement, which results in the formation of an additional C-S-H gel with binding properties similar to those formed in mineral based cements.

5.2 Chloride permeability

Figure 2 gives an indication of the concrete's resistance to the penetration of chloride ion. Concrete specimens were tested at 28; 90; 180 and 365 days. The effect of curing conditions and replacement of pozzolan on chloride permeability of the concrete is presented in this figure.

The specimen concretes curing in water, the charge passed of the control concrete decrease from 3000 to 2614 whereas the high performance concrete it decrease from 825 to 476 coulomb. The specimens kept in aggressive solution, the charge passed of the control concrete is decrease by 4,73% whereas the high performance concrete the decreasing was by 8,3%.

The presence of pozzolan has a very beneficial effect on the chloride permeability of concrete with significant reductions in the charge passed.

This reduction is more at later ages because the pozzolan modifies the microstructure of the concrete in terms of its physical and chemical characteristics. The introduction of this mineral into the cement paste leads to a segmentation of the larger pore and capillaries which reduces the amount of hydrated lime in the cement matrix.

5.3 Oxygen permeability

The effect of incorporation of 5% of cement by natural volcanic pozzolan in the mixture of concrete is very advantageous; it allows a reduction of the oxygen permeation rate in the two preservation solutions at one year age. In water, the HPC permeability coefficient decreases by 52% compared to control concrete, whereas in the aggressive solution, it decreases by 87%.

It is noticed that natural pozzolan reduced the oxygen permeability during the advancement of age of concrete.

6 CONCLUSION

Economic and environmental considerations had a crucial role in the supplementary cementing material usage as well as better engineering and performance properties. In recent years many researches proved that mineral admixtures can be successfully and economically utilized to improve some fresh and hardened concrete properties. The partial replacement of cement in the mixture of high performance concrete by the natural pozzolan gave higher resistance and improved durability against sodium sulfate attack. Compared to conventional concrete (CC), concrete contains pozzolan (HPC) gave higher resistance and durability against sulfate sodium. Indeed, concrete curing during one year in sulphate environment the conclusions are as follows:

– Strength of the control concrete is reduced by 17, 77% whereas the high performance concrete the reduction was by 5, 48%.
– Chloride permeability of the control concrete is decrease by 4, 73% whereas the high performance concrete the decreasing was by 8,3%.
– Compared to control concrete, the HPC permeability coefficient decreases by 87%.

The pozzolan modifies the microstructure of the concrete in terms of its physical and chemical characteristics.

REFERENCES

Atkinson A and all. 1988 "The Chemistry and Expansion of Limestone-Portland Cement Mortars Exposed to Sulphate-Containing Solutions", NIREX Report NSS/R127, United Kingdom.

Bentz DP & Garbicz, EJ. 1991 "Simulation studies of the effects of mineral admixtures on the cement paste aggregate interfacial zone", ACI Mater J, Vol. 88, N°. 5, pp. 518–529.

Chang ZT and all. 2005 "Using limestone aggregate and different cements for enhancing resistance of concrete to sulfuric acid attack", Cement and Concrete Research, Vol. 35, N°. 8, pp. 1486–1494.

Nadler B and all. 2003 "Ionic Diffusion Through Protein Channels: From molecular description to continuum equations", Technical Proceedings of the Nanotechlogy Conference and Trade Show, Vol. 3, pp. 439–442.

Saricimen H and all. 2003 "Durability of proprietary cementitious materials for use in wastewater transport systems, Cement and Concrete Composites", Vol. 25, N° 4–5, pp. 421–427.

Sersale R and all. 1998 "Acid depositions and concrete attack: Main influences", Cement and Concrete Research, Vol. 28, N°. 1, pp. 19–24.

Shamaran M and all. 2007 "Effect of mix composition and water cement ratio on the sulfate resistance of blended cements, Cement and Concrete Composites", Vol. 29, N°. 3, pp. 159–16.

Skalny and all 2002 "Sulphate Attack on Concrete", Spon Press, United Kingdom.

Talah A & Kharchi F.2013 "A Modified Test Procedure to Measure Gas Permeability of Hol Cylinder Concrete Specimens", IACSIT International Journal of Engineering and Technology, Vol. 5, N°. 1.

Tixier R & Mobasher B. 2003 "Modeling of damage in cement-based materials subjected to external sulfate attack – part 1: formulation", ASCE J Mater, Vol. 15, N°. 4, pp. 305–313, England.

Truc O. 2000 "Prediction of Chloride Penetration into Saturated Concrete – Multispecies approach", Chalmers University of Technology, Gutenberg, Sweden.

… Engineering Challenges for Sustainable Future – Zawawi (Ed.)
© 2016 Taylor & Francis Group, London, ISBN 978-1-138-02978-1

Presentation of a helping tool for the sizing and the optimization of 'beam' in reinforced concrete

Ahmed Boukhaled
University of Science and Technology Houari Boumédiène, Algiers, Algeria

Farida Mendaci
University of Jijel, Jijel, Algeria

ABSTRACT: Algeria is witnessing these last years a considerable development in individual constructions of average height (two or four levels). The majority of them are reinforced concrete beam-column framed buildings. This type of structure showed certain fragility during the earthquake of May 23rd, 2003, which shook the region of BOUMERDES. Moreover, in civil engineering the existing design tools do not solve the problem of choosing or optimizing structures. Those tools are often limited to the resisting capacity and the deformation of structure. For all these reasons, the present paper is particularly interested in modeling and constructing the problem related to the optimization of element "*beam*" taking into account the steel situated in the compressed zone of concrete into the global resistance of the section. The resolution of this problem and some illustrative results are presented as well.

1 INTRODUCTION

All The dimensions of the sections of the load bearing structural elements are directly dependent on the applied stresses. These stresses depend in its turn on the actions and dimensions of the structure in plan and in elevation. The approach taken by engineers in the process of structural analysis has as prior the geometrical characteristics of the structural elements and then moves to their verification. The judicious choice of initial parameters (dimensions of the sections) can reduce both the computation time and the volume of used materials. The existing design tools solve neither the problem of choice of materials, nor the system structures, as these tools are limited to traditional reinforcement calculations.

Code specifications such as the French BAEL and the Algerian seismic code RPA regarding the preliminary design of this type of structure gives generally enough wide ranges for the choice of initial dimensions of the sections, which are a necessary precondition for the designer. For these reasons, this work is interesting in the sense that it fits within the general framework of decision support, from the early stages in the operations design and optimization of structural armed concrete. Our work focuses on the study of rectangular beam of constant section.

This work was carried out as part of the preparation of the Magister degree held at the USTHB University, Algiers (Mendaci 2005). It consists in the development of tools for making dimensions and optimization of structural elements: beams, columns.

The objectives of this study are the evaluation of the integration of compressed longitudinal reinforcement section in the overall resistance of the section, and the determination of the optimal cost and the corresponding cross section (Width and Height).

2 INTEGRATION OF THE SECTION OF COMPRESSIVE LONGITUDINAL REINFORCEMENT

2.1 *Justification*

The calculation of reinforced concrete sections often results in a section simply reinforced where only tension reinforcement is required. While in practice, for reasons of assembly or reversibility of seismic effects, another layer of reinforcement named construction reinforcement is introduced. This section of steel works mainly in all cases to compression. To do this, in this work we will be interested to including the reinforcement layer into the overall resistance of a section subjected to bending moment, to assess its impact on the total reinforcement section and therefore the total cost of steel and concrete.

2.2 *Formulation of equilibrium equations*

We will proceed with the formulation of the equations of equilibrium of a rectangular cross-section under flexure at ultimate limit state (ULS) (Guerra & Kiousis 2006, Mougin 1995).

Consider k'_A the ratio of compressive reinforcement expressed by:

$$k'_A = \frac{A_{comp}}{bd} \qquad (1)$$

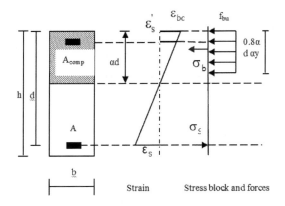

Figure 1. Stresses and strains at ULS.

A_{comp}: section of compressive reinforcement. From the strain diagram we have:

$$\varepsilon'_s = \left(1 - \frac{1}{9\alpha}\right)\varepsilon_{bc} \quad (2)$$

The equilibrium equation is as follow:

$$N_s = N_b + N'_s \quad (3)$$

This gives:

$$\sigma_s A_s = bd(0.8\,\alpha f_{bu} + \sigma'_s k'_A) \quad (4)$$

where N_S, N_b, N_S', σ_s, σ_s', A_s, ε_{bc}: the tensile resulting force in the steel, the compressive resulting force in concrete, the compressive resulting force in the steel, stress in tensile reinforcement, stress in compressive reinforcement, longitudinal steel cross-section and strains of concrete according to the pivots A and B.

The moment of resistance of the section M_{res} is given by:

$$M_{res} = M_b + M'_s \quad (5)$$

where:

$$M'_s = \frac{8}{9} k'_A bd^2 \sigma'_s \quad (6)$$

$$M_b = 0.8\alpha(1 - 0.4\,\alpha)bd^2 f_{bu} \quad (7)$$

The resistant moment done become:

$$M_u = M_{res} = bd^2 f_{bu}\left(0.8\alpha(1 - 0.4\alpha) + \frac{8}{9}\frac{1}{f_{bu}}k'_A\sigma'_s\right) \quad (8)$$

$$\sigma'_s = \frac{E_s}{\gamma_s}\left(\frac{9\alpha - 1}{9\alpha}\right)\varepsilon_{bc} \quad (9)$$

– M_u: acting moment.
– M_{rmb}: moment of the resistance of concrete.
– M'_s: moment of the resistance due to the compression reinforcement.

Knowing the acting moment M_u one can determine the value of α. Equation 8 can be written as follows according to α:

$$0{,}32\alpha^2 - 0{,}8\alpha - \frac{8}{9}\frac{k'_A}{f_{bu}}\sigma'_s + \frac{M_u}{bd^2 f_{bu}} \quad (10)$$

Replacing σ'_{rmS} by its expression, we have two equations of the 3$_{rd}$ degree depending on α, respectively, for the pivots A and B:

$$0{,}32\alpha^3 - 1{,}12\alpha^2 + (0{,}8 + C_1 + C_2)\alpha - C_2 - C_1 = 0 \quad (11a)$$

$$2{,}88\alpha^3 - 7{,}20\alpha^2 - 9C_3\alpha + C_1 \quad (11b)$$

where

$$C_1 = \frac{8}{9}\frac{k'_A E_s}{\gamma_s f_{bu}} \quad C_2 = \frac{10^{-2} M_u}{bd^2 f_{bu}} \quad C_3 = \frac{10^{-3} M_u}{3{,}5 bd^2 f_{bu}}$$

3 CONSTRUCTION OF THE OPTIMIZATION PROBLEM AND RESOLUTION APPROACHES

3.1 Construction of the optimization problem

This section deals with the optimization of a unitary section of beam, the main objective is to determine the optimal dimensions of cross-section, longitudinal reinforcement, and then cost. To achieve this purpose, we propose to consider two levels of reasoning: optimization of longitudinal steel section and the cost of the section while respecting the constraints of size that satisfies BAEL and RPA 2003 codes requirements (Éditions CGS 1999, Éditions CGS 2003).

We build optimization models by the formulation of the factors of the decision: objective function, decision variables and constraints of decision.

3.1.1 Level 1: Optimization of longitudinal steel section

The objective function is defined by A_{tot} (cross-section of longitudinal reinforcement):

$$A_{tot} = A_s + A_{comp} \quad (12a)$$

$$A_{tot} = bd\left(\frac{0.8\alpha(1-0.4\alpha)f_{bu} + k'_A \sigma'_s}{\sigma_s} + k'_A\right) \quad (12b)$$

The decision variables is the compressive reinforcement steel A_{comp}.

Decision constraints apply to both levels of optimization are:

– Geometric constraints formwork:

$$b \geq 20\,\text{cm}, h \geq 30\,\text{cm et } \frac{h}{b} \leq 4 \quad (13)$$

– Constraints related to the percentage of reinforcement:

$$A_s \geq 0{,}23\frac{f_{t28}}{f_e}bh \quad \text{et} \quad 0.5\% \leq \frac{A_{tot}}{bh} \leq 4\% \quad (14)$$

— Constraints related to strength of section:

at pivot A:

$$M_u \leq bd^2 f_{bu}\left(0.8\alpha(1-0.4\alpha) + \frac{16000}{81}\frac{k'_A}{f_{bu}}\frac{9\alpha-1}{1-\alpha}\right) \quad (15a)$$

at pivot B:

$$M_u \leq bd^2 f_{bu}\left(0.8\alpha(1-0.4\alpha) + \frac{56000}{81}\frac{k'_A}{f_{bu}}\frac{9\alpha-1}{1-\alpha}\right) \quad (15b)$$

3.1.2 Level 2: Optimization of the cost

The *objective function* is defined by:

$$Cost = C_b(b\,h + rap\,A_{tot}) \quad (16)$$

Where:

- C_S and C_b: material cost of steel and concrete.
- b and h: width and depth of cross-section.
- A_{tot}: cross-section of longitudinal reinforcement.
- rap: ratio of material cost of steel to material cost of concrete (rap = C_s/C_b).

The decision variables here are three: the depth h, width b of the concrete section and the compression steel area A_{comp}.

The constraints of decision are same at first level.

4 SOLVING APPROACHES

4.1 First level

The problem addressed in this level can be analytically resolved. We can consider K'_A to be fixed; we can than get the value of α from Equation 11a or Equation 11b. So, we obtain equations of 3rd degree with α which we solve by an iterative approach. We use the obtained value of α to compute A_{tot} in Equation 12b. Then, we vary the value of A_{comp} in order to obtain the minimal value of A_{tot}.

4.2 Second level

At first, we present a combinatorial approach based on the method in SPE (Separation and Progressive Evaluation) (Mangin 1989) and an algorithmic approach to achieve an automatic tool.

4.2.1 Branch and bound method (BB)

This method is based on the successive respective application of two principles: a separation principle that can cut a set of solutions D into smaller subsets, and a progressive evaluation that eliminates certain subsets and selects the subset on which we apply the next separation.

We consider initially a fixed A_{comp}, only b and h vary. Dividing the set D into subsets such that the elements of the solution $x_i = (b_i, h_i)$ satisfy the condition:

$$b_i\,h_i = B_i \quad (17)$$

B_i: constant included in bounded set.

Table 1. Database.

Concrete compressive strength	25 MPa
Yield strength of steel:	400 MPa
Limit stress of steel at ULS: σS	348 MPa
limit Stress of steel with SLS	FPN
minimum percentage of steel	0.5%
maximum percentage of steel	4%
Cost ratio: Cs / Cb	52
Unit cost of concrete: Cb	9000 DA

In this way, the cost of concrete is constant, therefore the objective function varies only with A_s. In general, when depth increases, A_s decreases. We can deduce that the cost is minimum for each x_i in which h_i is maximal and consequently b_i is minimal. For each b_{min} and B_i fixed, the optimum depth is equal to B_i / b_{min}. This reasoning leads us to say that the optimal value of b is unique and equal to b_{min}, it remains to find the optimal depth when B_i takes other values. The cost function expressed with depth is:

$$cost\,(b_{min}\,h) = C_b\,(b_{min}\,h) + C_a\,A_s \quad (18)$$

We get an objective function which varies only with depth. Differentiating this function, we obtain the best theoretical value of depth. In pivot A, the cross-section of longitudinal reinforcement is:

$$A_s = \left(\frac{M}{1-0.4\sqrt{1-2\frac{M}{bd^2 f_{bu}}}}\right) \quad (19)$$

$$cost = C_b h b_{min} + C_S \frac{M}{0.9\left(1-0.5\sqrt{1-2\frac{M}{0.81bh^2 f_{bu}}}\right)\sigma_s h} \quad (20)$$

4.2.2 Algorithmic approach

It is preferable to proceed with a comprehensive algorithmic approach which constitutes a structure of a computing tool, which can integrate all the possible constraints (Atabay 2009, Alqedra & Arafa 2011, Kaveh & Sabzi 2012).

5 PRESENTATION AND ANALYSIS OF SOME RESULTS

Some results are presented here to demonstrate the optimization formulation with the use of design examples that study the effects of including the compressive reinforcement on the tensile or total cross-section of steel and finally optimization the cost of beam. The data are:

5.1 Influence of the compression steel area

The result in table 2 presents the variation of the longitudinal steel area according to the compression steel area and the height h of the section.

5.1.1 Reading indications

For each value of h, the first line represents the value of the section of tension reinforcement and the second line the value of the total area of reinforcement. The results of the first column represent the values of the reinforcement sections obtained by a classical calculation, ie., without integrating A_{comp} in the resistance of the section. Fields in bold represent the total optimal sectional area of longitudinal reinforcement.

Table 2. Variation of As in function of Acomp and h (with b=20 cm and M=0.300 MNm).

h(cm)	A_{comp}(cm²)								
	/	3T10	3T12	3T14	3T16	3T20	2T20+2T16	3T25	2T25+2T20
45.	26.56	26.56	26.56	26.56	26.56	20.98	20.57	20.38	20.77
	35.49	35.49	35.49	35.49	35.49	**30.40**	30.87	39.76	36.87
50.	24.75	24.75	24.75	21.93	20.47	18.83	18.74	19.61	19.65
	29.90	29.90	29.90	26.55	**26.50**	28.25	29.04	36.54	35.75
55.	23.31	21.14	19.95	18.91	18.12	17.56	17.67	17.95	17.91
	25.07	23.50	**23.34**	23.53	24.15	26.98	27.97	35.08	34.01
60.	20.97	18.35	17.61	16.99	16.60	16.76	16.70	16.49	16.45
	/	**20.71**	21.00	21.61	22.63	26.18	27.00	34.31	32.55
65.	18.23	16.45	15.99	15.66	15.56	15.47	15.42	15.25	15.21
	/	**18.81**	19.38	20.28	21.59	24.89	25.72	29.95	31.30

5.1.2 Effect of the compression steel area

The results are presented in Figure 2 (variatio of A_s vs A_{comp} and depth). We note that A_s decrease when A_{comp} increase. Also, A_{comp} decreases with increasing depth.

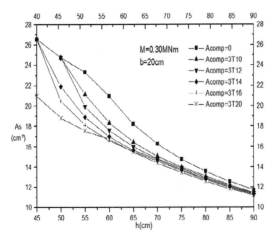

Figure 2. Variation of As in function of Acomp and h (with b=20 cm and M=0.300 MNm).

Table 3. Variation of the total cost depending on b, h and A_{comp}.

b(cm)	A_{comp}								
	0	3T10	3T12	3T14	3T16	3T20	2T20+2T16	3T25	2T25+2T20
20.	2021.41	1938.85	1904.29	1874.93	1837.97	1781.47	1772.43	1742.31	1744.33
	/	2049.30	2062.94	2091.15	2120.17	2222.33	2254.47	2431.67	2497.81
	70	*60*	*60*	*55*	*55*	*50*	*45*	*40*	*40*
	16.27	18.35	17.61	18.91	18.12	18.83	20.56	21.84	21.89
	/	20.71	21.00	23.53	24.15	28.25	30.86	36.54	37.99
25.	2257.60	2174.21	2141.18	2110.84	2070.74	2004.75	2013.13	1958.44	1965.95
	/	2284.66	2299.83	2327.06	2352.94	2445.61	2495.17	2647.80	2719.43
	60	*55*	*55*	*50*	*50*	*45*	*40*	*40*	*40*
	19.39	20.02	19.31	21.07	20.21	21.20	23.78	22.62	22.78
	/	22.38	22.70	25.69	26.24	30.62	34.08	37.32	38.88
30.	2472.94	2390.91	2355.03	2322.41	2284.86	2217.51	2197.68	2172.48	2320.56
	/	2501.36	2513.68	2538.63	2567.06	2658.37	2679.72	2861.84	3074.04
	55	*50*	*50*	*50*	*45*	*40*	*40*	*40*	*40*
	21.11	22.24	21.47	20.78	22.86	24.31	23.88	23.34	26.51
	/	24.70	24.86	25.40	28.89	33.73	34.18	38.04	42.61

5.2 Variation of the total cost depending on b, h and A_{comp}

The variation of the cost depending on the section of compression reinforcements presents a unique optimum global. This optimum is less reduced when height h of the section of the concrete increases. Note that for a constant width b, when increasing the compression steel sections the height corresponding to local optimal cost decreases. The global optimal cost is equal to 2049 AD / m, obtained with the minimum optimum width of 20 cm.

For example, for the minimum optimal width (b = 20 cm) the conventional calculation results in a reduced optimal cost (excluding frames construction) equal to 2021.41DA /m and an optimal height equal to 70 cm

The integration of A_{comp} in calculation enables to have a global optimal cost equal to 1742.31 DA/m and an optimum height of 40 cm. That means a gain in cost greater than 16% and a gain in beam height equal to 42%. We point out that these results are very interesting and open the way for a multi-criteria analysis.

The Indications of reading the contents of Table 3 are as following:

b Cost (bh+A_S) : cost of concrete and longitudinal tension steel
   Cost (bh+A_S+A'_S) : cost of concrete and total longitudinal steel
   h : height of the section
   A_s : longitudinal tension steel area
   A_{tot} : total longitudinal steel area

6 CONCLUSION

This paper presents the necessary stages to optimize a unitary section of beam. We identified and formalized the necessary information (data and knowledge) for this study. We have proposed and built two levels of optimization: the cross-section of longitudinal reinforcement by incorporating the compressed longitudinal reinforcement section in the overall resistance of the section, and the determination of the optimal cost (steel and concrete).

The iterative algorithmic approach adopted for solving the problems was achieved by developing a computer program. The results obtained through the examples are very interesting. They incite in some cases of bending beam (spans and loads large enough) to review the conventional method of calculation of sections. For the optimization approach, the results lead us to conduct a multi-criteria analysis to take into account other criteria such as the beam depth and the weight of the structure.

REFERENCES

Alqedra, M. & Arafa, M. 2011. Concrete beams using genetic algorithms, Journal of artificial intelligence.

Atabay, Ş. 2009. Cost optimization of three-dimensional beamless reinforced concrete shear-wall systems via genetic algorithm. Expert Systems with Applications Journal. Vol. 36 issue 2.

Éditions CGS. 1999. Règles parasismiques Algériennes (RPA).

Éditions CGS. 2003. Addenda (RPA).

Guerra, A. & Kiousis. P.D. 2006. Design optimization of reinforced concrete structures, Computers and Concrete. Vol. 3(5): pp. 313–334.

Kaveh, A. & Sabzi, O. 2012. Optimal design of reinforced concrete frames Using big bang-big crunch algorithm, International Journal of Civil Engineering.

Mangin, J,C. 1989. Méthodes et outils de l'aide à la décision (méthodes d'optimisation monocritère), Université de Savoie. Chambéry.

Mendaci, F. 2005. Élaboration des outils d'aide au dimensionnement et à l'optimisation des Éléments porteurs en béton armé : poutres - poteaux, Magister thesis, USTHB, Algeria.

Mougin, J.P. 1995. Guide de calcul béton armé BAEL91 et DTU associés, Éditions Eyrolles.

Sustainable waste management of bottom ash as cement replacement in green building

Nur Liyana Binti Mohd Kamal, Gasim Hayder, Osamah Abdulhakim Ahmed & Salmia Bte. Beddu
Department of Civil Engineering, College of Engineering, Universiti Tenaga Nasional (UNITEN), Malaysia

Muhd. Fadhil Nuruddin & Nasir Shafiq
Department of Civil & Environmental Engineering, Faculty of Engineering, Universiti Teknologi PETRONAS, Malaysia

ABSTRACT: Since global warming and green buildings have been the concern of the scientific society, efforts were made toward achieving more green and sustainable buildings. One of the efforts was to utilize waste materials such as Coal Bottom Ash (CBA) as cement replacement in the concrete mixture to reduce the harmful effects of cement manufacturing procedure and dumping of coal burning products from utilities industry. This paper studies the effect of replacing cement in concrete mixture by CBA. The porosity of four different concrete mixtures including control mixture was studied using different percentages of CBA, which are 0%, 10%, 20%, and 30% as cement replacement. The results have shown improvement in porosity at 10% and 20% replacement. However, 30% replacement was chosen as the optimum percentage since it showed a 6.825% percentage of porosity which is slightly similar to control mixture, therefore, it utilize larger amount of CBA while providing normal concrete quality, which contribute to the sustainable waste management target of Malaysia.

1 INTRODUCTION

There are a lot of urban areas in Malaysia have been experiencing the problems of waste management because of the growth in population. This increasing in population has enlarged the amount of disposal which will affect the environment and health through the pollution of the environment and disease transmission (Salim et al. 1994). One of the effect of increasing the population is increasing electricity consumption which produce a huge amount of coal burning disposal.

Due to the rising in electricity consumption by the industry, the global warming gained popularity, environmentalists have raised awareness about the issue of coal burning for electricity generation that also affects the environment in many ways, especially the huge amount of Coal Bottom Ash (CBA) waste. It is estimated that 180 tons of Bottom Ash are produced every day in Tanjung Bin power station alone, which end in a disposal ponds beside the power station where the storage has reached alarming level (Abubakar et al. 2012). Since CBA has similar chemical properties of those in cement, the idea of replacing cement by CBA was proposed to utilize it in construction and minimize the disposal problem. Such a solution is expected to utilize huge amounts of CBA that are not going to be burred in the ground as well as reserving the natural resources spent on producing cement and avoiding problems that associated with the disposal in terms of cost and environmental impact (Abubakar et al. 2013).

Concrete is produced using many kinds of cements namely Portland cement, white cement and others types which is usually dry mixed with aggregate and later water is added to trigger the chemical reactions in cement, these chemical reactions binds the components of concrete through a stone-like hard matrix. As concrete has been used for many purposes and under various conditions, chemical additives were introduced to alter its behavior to the required circumstances, which lead to the huge popularity of concrete in the construction industry around the globe. Unfortunately, the production of cement has various harmful effects on the environment, which is represented mostly by emissions from cement plants. The most harmful gaseous emissions are Sulfur Dioxide (SO_2), Carbon Monoxide (CO) and Nitrogen Dioxide (NO_2), these gases have various environmental effects, for example, ground level ozone, acid rain, water quality and global warming by contributing to chemical reactions in the atmosphere which results in high concentrations of Carbon Dioxide (CO_2) and other gases (EPA, 2015, & Mehraj et al. 2014).

Numerous researches have been done on the attempt of utilizing CBA in construction activities. For instance, Denmark is using CBA as sub base in road constructions and as back fills, moreover, there were attempts to replace concrete sand with CBA as well since it is possible to increase the strength of concrete (Leena, 2012). Some studies replaced up to 100% of concrete sand by CBA, the results suggested improvement in concrete durability, drying shrinkage, carbonation depth, and chloride penetration, however, the studied samples showed a decrease in the strength of concrete due to increase in the porosity and water demand that might be solved by using plasticizers. This promises huge benefits of substituting sand by CBA to reserve the natural sand resourced as well as improve the concrete quality, especially when it comes to chemical attacks (Qureshi et al. 2014, Singh et al. 2016, Singh et al. 2014a, Singh et al. 2013, Singh et al. 2014b, & Singh et al. 2014c). Moreover, some researchers replaced cement in the concrete mixture with CBA up to 35% of the cement mass in an attempt to study the environmental and mechanical behavior of the concrete by investigating its leaching under different temperatures (23°C up to 60°C) and various pH values (5 up to 10). The results suggested that the optimum percentages for CBA replacement were 10% and 25% were they performed similar to concrete with only ordinary Portland cement (Menéndez et al. 2014). However, not much research was done regarding the mechanical and physical behavior of the concrete mixture with CBA replacement of cement.

2 EXPERIMENTAL METHOD

2.1 Specimens

The materials used in this paper were Portland cement, aggregate, water, and CBA that was brought from Jimah power plant, Port Dickson. The CBA sample was then grinded to meet the required softness of the cement. In previous studies, the optimum percentage of CBA as cement replacement was 20%. But in this study the percentage was increased to 30%. Four concrete mixtures namely, M1, M2, M3, and M4 with CBA percentage of 0%, 10%, 20%, and 30% respectively, were used to test the effect of different CBA percentages on the porosity of concrete, as shown in Table 1.

2.2 Experimental test

At the age of 28 days, the porosity test was performed according to the following procedure:

The samples were placed in the oven at 105° C for 24 hours to dry and then they were placed in the desiccator to cool down for the next 24 hours. After that, the weight of the samples was recorded and then returned to the desiccator where it was filled with water until the samples were fully submerged and kept under vacuum for 24 hours. Then the desiccator was allowed to equilibrate for the next 24 hours. After that, the weight of the samples was recorded again and then the porosity of each sample was calculated as shown in Equation 1.

$$P = \frac{B-A}{B-C} * 100 \quad (1)$$

Where; P is porosity (%); A is oven-dry weight; B is saturated surface dry weight; C is saturated submerged weight.

Table 1. Concrete mix proportions.

Mixture No	M1	M2	M3	M4
Cement (kg\m^3)	385	346.5	308	269.5
Bottom ash (%)	0	10	20	30
Bottom ash (kg\m^3)	0	38.5	77	115.5
W\R	0.725	0.725	0.725	0.725
Water (kg\m^3)	398.75	398.75	398.75	398.75
Corse aggregate (kg\m^3)	590	590	590	590
Sand (kg\m^3)	728	728	728	728

Table 2. Results of porosity.

Mixture	Sample 1	Sample 2	Sample 3
0%	7.165	6.855	6.664
10%	6.002	5.741	5.702
20%	6.805	6.623	6.432
30%	7.140	6.895	6.755

3 RESULTS AND DISCUSSION

The porosity results of the samples shows that the control mix had 6.89%, which is the highest percentage among the other samples. The porosity in M1 was calculate by taking the average of sample 2 and sample 3 which are 6.855% and 6.664% respectively while the result of sample 1 was ignored since it was far from the other results of the same concrete mixture. This indicates that a human or machine error might have happened during the procedure of the experiment. On the other hand, the lowest value of porosity was 5.721% when the sample was containing 10% of CBA as cement replacement. The result for M2 was also obtained by taking the average of sample 2 and sample 3, which are 5.741%, and 5.702% respectively where the result from sample 1 was ignored for the same reason in M1. The percentage of the porosity increased to 6.62% when the percentage of the CBA was 20% where it was calculated by taking the average of sample 1, sample 2, and sample 3, which are 6.805%, 6.623%, and 6.432% respectively. When the percentage of the CBA was increased to 30%, the percentage of porosity was 6.825%, which was calculated by taking the average of samples 2 and samples 3, which are 6.895% and 6.755% respectively while the result obtained from sample 1 was neglected for similar reason in M1 and M2, as shown in Table 2 and Figure 1.

The results show the porosity is at its maximum in M1 where there is 0% CBA while it is at minimum in M2 where there is 10% CBA, and it kept increasing by rising the percentage of CBA replacement. Therefore, the optimum replacement of CBA is 30% since the porosity is slightly similar to that of the control mix, while the porosity will increase by increasing the CBA more than 30%, which will affect the concrete mix negatively. A thirty percent replacement may not be that significant but it can make a huge difference for the

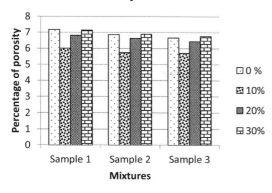

Figure 1. Reults of porosity.

goal of sustainable waste management in Malaysia; in addition, it will reduce the harmful effects of dumping CBA underground.

4 CONCLUSION

In this paper, the porosity of four concrete mixtures that varies in CBA percentage as cement replacement were measured and studied to obtain the optimum maximum replacement of CBA in concrete in order to fulfill the goal of utilizing the maximum amount of CBA in construction. As well as reducing the cement usage in an attempt to decrease the harmful effects of its production. According to the objective of this study and the results of the porosity obtained from the experimental test, the conclusion is stated in following points:

1) The porosity increases when the percentage of CBA is increased in the concrete mixture which affect the quality of concrete negatively and limit the percentage of CBA replacement.
2) The 30% of CBA replacement is the optimum concrete mixture since it has slightly similar porosity to that of control mixture, which means the quality of the concrete produced is similar to that with 0% of CBA replacement.

More research is recommended on the other properties of the concrete produced by replacing cement with CBA to ensure the safety of buildings and the behavior of such mixtures under various climate and working conditions.

REFERENCES

Abubakar, A.U. and Baharudin, K.S., 2013, July. Tanjung Bin Coal Bottom Ash: From Waste to Concrete Material. In Advanced Materials Research (Vol. 705, pp. 163–168).

Abubakar, A.U., & Baharudin, K.S. 2012. Potential use of Malaysian thermal power plants, Vol 3, Issue 2.

EPA, Environmental Protection Agency. 2015. Cement Manufacturing Enforcement Initiative. Information on http://www.epa.gov/enforcement/cement-manufacturing-enforcement-initiative

Leena S. 2012. Characterization and treatment of waste incineration bottom ash and leachate.

Mehraj, S.S. & Bhat, G.A., 2014. Cement Factories, Air Pollution and Consequences. Jammu and Kashmir, India. 30–36.

Menéndez, E., Álvaro, A.M., Hernández, M.T., & Parra, J.L. 2014. New methodology for assessing the environmental burden of cement mortars with partial replacement of coal bottom ash and fly ash, Journal of Environmental Management. 275–283.

Qureshi M.A., & Shakeel, S. 2014. Analysis of Heavy Metals in Sludge and Bottom Ash from a Pharmaceutical Industry.

Salim, M.R., Othman, & F., Marzuky, S. 1994. Problems and challenges of solid waste management: a case study in South Johor, Malaysia.

Singh, M., & Siddique, R. 2013. Effect of coal bottom ash as partial replacement of sand on properties of concrete, Resources, Conservation and Recycling. 20–32.

Singh, M., & Siddique, R. 2014a. Compressive strength, drying shrinkage and chemical resistance of concrete incorporating coal bottom ash as partial or total replacement of sand, Construction and Building Materials. 39–48.

Singh, M., & Siddique, R. 2014b. Strength properties and micro-structural properties of concrete containing coal bottom ash as partial replacement of fine aggregate, Construction and Building Materials. 24–256.

Singh, M., & Siddique, R. 2014c. Microstructure and properties of concrete using bottom ash and waste foundry sand as partial replacement of fine aggregates, Construction and Building Materials. 210–223.

Singh, M., & Siddique, R. 2016. Effect of coal bottom ash as partial replacement of sand on workability and strength properties of concrete, Journal of Cleaner Production. 620–630.

Numerical analysis of the composite "ball-and-socket" carcass design for unbonded flexible pipe

A.F. Billah & Z. Mustaffa
Department of Civil and Environmental Engineering, Universiti Teknologi PETRONAS, Seri Iskandar, Perak, Malaysia

B.T.M.B. Albarody
Mechanical Engineering Department, Universiti Teknologi PETRONAS, Malaysia

ABSTRACT: This paper presents a novel concept of carcass design as an alternative to the existing carcass in unbonded flexible pipe. The conventional carcass made from a flexible metallic layer which has the potential of corrosion and weight issue as the main problems during installation operation. The "ball-and-socket" concept is made from glass epoxy material and with an interlocking design that made from several segment to form a single pipe length. The use of composite has its benefit in term of strength, stiffness and weight as compared with conventional carcass pipe. The results showed that "ball-and-socket" design has solved the second issue of having heavy weight while still maintaining adequate structural strength to hold the hydrostatic pressure.

1 INTRODUCTION

Flexible pipe is currently the new trend for transporting fluid especially in oil and gas industry. Due to the enhance flexibility and stronger configuration compared to traditional carbon steel pipe. Flexible pipe consists of two types; 1) Bonded flexible pipe. 2) Unbonded flexible pipe. This paper will focus on unbonded flexible pipe.

Unbonded flexible pipe consists of few layers that attach together and can slide within layers. The general configuration of unbonded flexible pipe is made up of carcass, internal sheath, armoring layer, and external layer. Currently only few major fabricators for unbonded flexible pipe which are; 1) Technip 2) NKT and 3) Wellstream.

The conventional inner most layer of flexible pipe called carcass is usually made from a flexible metallic structured that prevent collapse due to external hydrostatic pressure and also to protect the pipe from crushing during installation operations. The use of metallic layers in carcass poses a major maintenance issue of corrosion. Furthermore, the weight of stainless steel material is affecting the installation operation which can cause bendability issue. Research has been conducted to make a fully composite flexible pipe to replace steel. Up to date, there is few known to have use fully composite flexible pipe starting from the inner layer up the outermost layer.

The objective of this paper is to present the potential advantages of the "Ball-and-Socket" concept as one of the alternatives that able to replace the existing carcass design in unbonded flexible pipe in term of weight to strength ratio.

Table 1. Layer description of standard flexible pipe API-17B (2002).

Layer	Function	Material used
Carcass	Prevent collapse from hydrostatic pressure	Stainless Steel
Internal polymer sheath	As a leak-proof material	PE, PA, PVDF
Pressure armor	Provide internal pressure hoop resistance, external hydrostatic collapse resistance and radial compression	Steel
Tensile armor	Provide tensile capacity	Steel
Outer layer	Protect from external impact	Carbon Steel

2 LITERATURE REVIEW

2.1 Typical unbonded flexible pipe

The general flexible pipe configuration consists of several layers. The main principle layers of these pipes are shown in Table 1 and Fig. 1.

2.2 Composite material

The most successful offshore application for composites have been in pipework for aqueous liquids (Gibson, 2003). Furthermore, these applications can be accelerated with the use of glass reinforced epoxy. The use of composite as a replacement for the inner

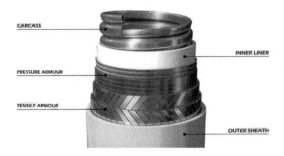

Figure 1. Compression of achievable water depth for different materials (Arbey et al., 2009).

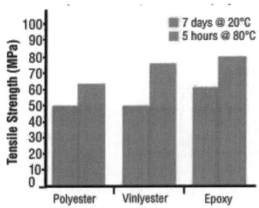

Figure 3. Tensile strength of matrix (Azom, 2001).

Table 2. Effect of ply angle towards stress.

Ply angle	Stress (Mpa)
55	3.3504
45	4.1602
30	5.2047

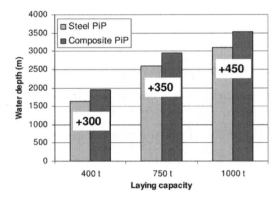

Figure 2. Typical flexible pipe (flexip pipe).

steel has increase the weight reduction of estimated 35%. (Arbey et al., 2009) tested that the composite may enable an increase in achievable water depth. Comparison between steel with composite inner pipe base on the laying capacity shown in Fig. 2.

Despite the benefits of composites, (Gibson, 2003) stated that the initial cost of composites could exceed the use of metallic counterpart. However, due to their ease of handling, the installation cost and the overall cost is often lower than conventional steels. The cost difference is further lowered as composites do not required for expensive corrosion resistance maintenance as compared to steel.

It has been proven that tensile modulus and strength of epoxy resin is much stronger compared to other resin materials (Azom, 2001, Cripps, 2013) shown in Fig. 3.

2.3 Fiber angle orientation

In composite, the fiber can be oriented into specific angle to increase the strength to the needs. Currently the range of a winding angle machine the industry can manufacture is in the range of 30°–60°. According to (Jha et al., 2013), the reinforcement layer usually arranged at 55° as optimum to hold both axial to hoop load in a layer. By applying the fiber angle to be aligned to 55° it has increase the capability of the carcass to hold the tensile force and internal force as to prevent from bursting. A simple simulation using ANSYS Composite PrePost (ACP) was conducted to determine the stress changes with regards on the fiber angle direction shown in Table 2. The result showed angle 55° proven to have lower stress experience compare to other angle configuration.

2.4 Conventional carcass design

The main function of conventional carcass is to hold the external hydrostatic pressure while the tensile and compression force are being hold by the armor and/or tensile layer and also to protect the other layers from contact with the transported fluids. The interlock profile provides the ability for the carcass to flex in addition to prevent the pipe from collapsing. (Rytter et al., 2002) mentioned that to improve the collapse and crushing resistance of the conventional carcass would require an increase of the strip thickness and total height of the inter-locking profile, or replacement of the conventional metallic materials with high strength alternatives.

Fig. 4 shows the conventional interlocked carcass design for flexible pipe that is normally made from folded steel strip.

3 METHODOLOGY

3.1 "Ball-and-socket" geometry

Fig. 5 shows the concept of ball-and-socket design segment. The pipe is made up from a number of composite

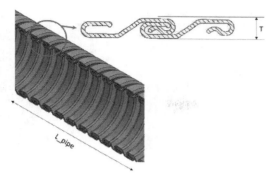

Figure 4. Conventional carcass.

Figure 5. Ball-and-Socket segment.

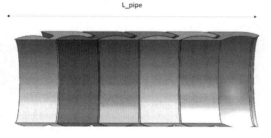

Figure 6. Connected segments.

pipe segments, which are interconnected by rigid ball joints as shown in Fig. 6. The interlocked mechanism allows each segment to be movable to a certain distance thus increasing the bendability of the pipe as a whole. The idea of this concept is based on the normal flexible hose, where it is being modified to be created in segment and connected to each other.

3.2 Design optimization

The optimization of the design consists of two stages.

In the first stage, the internal diameter (ID) of 7 in. (196 mm) is chosen as the base diameter in this research. The next process is to adjust the equivalent thickness (T) of ball and socket to sustain the hydrostatic pressure of 300 m water depth (3 Mpa). The design is first checked visually in the CAD software for any clashing geometry especially in the connected

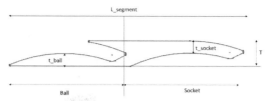

Figure 7. 2D cross section of ball and socket segment.

Table 3. Carcass design criteria.

Pipe Layer	Design Criteria	
Carcass	Stress buckling load	[0.67] for Dmax < 300 m $[(\frac{D_{max}-300}{600}) \times 0.18 + 0.67]$ for 300 m < Dmax < 900 m [0.85] for Dmax > 900 m

*Dmax is the maximum water depth

ball and socket part and the thickness for t_ball and t_socket is adjusted to find the best geometry configuration as shown in Fig. 7.

The next stage is to export the CAD design into numerical simulation to determine the stress capability of the design. Fiber orientation of 55° (optimized angle) was chosen as the base fiber angle. The material used in this research is glass epoxy (40–60) content and the parameter of external pressure is 3 Mpa (for 300 meter water depth). The result is analyzed if the stress value exceeds the ultimate tensile strength of the material.

3.3 Design criteria

Table 3 shows the design standard for flexible pipe base on (API-17J) for internal carcass design that it should be able to sustain at least 0.67 of the maximum pressure applied.

3.4 Material property

The material selected in this research is glass fiber reinforced epoxy with 60% fiber volume ratio. The calculated composite fiber volume ratios based on (Kaw, 2005) shown in Table 4 below.

3.4.1 Finite element model

Table 5 shows the detail of different configuration of carcass geometry that is used in this analysis. The equivalent thickness (T) is set to be 20 mm, and several ball-and-socket geometry configurations were prepared into three different cases. The length of each segment (L_segment) is fixed to be 182.77 mm and the length of the whole pipe (L_pipe) is 768 mm (17 segments).

The model was being tested in ANSYS for the stress experienced under external pressure. For the purpose of this research, ball-and-socket is designed to be used

Table 4. Mechanical Property.

Material		Gl. Epoxy	SS 306L
Young Modulus (GPa)	Longitudinal	52.36	193
Poisson ratio	Major	0.25	0.31
Shear Modulus (GPa)		3.10	73.66
Tensile Stress (MPa)	Longitudinal	954.8	586
Compressive Stress (MPa)	Longitudinal	69.20	207
Density (kg/m^3)		1980	7750

Table 5. Geometry model property (unit in mm).

	Thickness	
Case No.	t_ball	t_socket
1	8	12
2	10	10
3	12	8

Table 6. Conventional carcass geometry.

Equivalent thickness, T	20 mm
Metal sheet thickness	2–4.5 mm
Total length, L_pipe	836.6 mm

Table 7. Stress analysis for "ball-and-socket" carcass.

Case No.	Stress (Mpa)	Mass (kg)	Inertia (kgm^2)
1	1827.08	19.53	1.10
2	833.24	19.63	1.11
3	671.72	19.80	1.13

under shallow water up to 300 m depth with hydrostatic pressure of 3 Mpa. Numerical simulation behaviors of the pipe under external pressure was conducted.

For comparison purpose, a stainless steel conventional carcass (Fig. 4) is simulated to compare the stress experience with the composite ball-and-socket carcass (Fig. 6). Table 6 shows the design configuration of stainless steel conventional carcass.

4 RESULTS AND DISCUSSIONS

4.1 Stress result on ball-and-socket carcass

Based on the three cases used in this paper, Table 7 shows the results of the stress analysis corresponding to the geometry of each case.

It can be seen that Case 3 has the lowest value of stress experienced under hydrostatic pressure compare

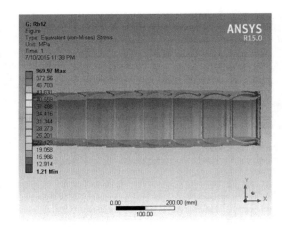

Figure 8. Stress profile of "ball-and-socket".

Table 8. Stress analysis of stainless steel conventional carcass.

Metal sheet thickness.	Stress (Mpa)	Mass (kg)	Inertia (kgm^2)
4.5	530.3	54.61	3.25
3	1233.5	36.29	2.16
2	2669.2	24.14	1.44

to others while the highest stress falls into Case 2. From the above result, it was found out that thickness of ball part is more significant compare to socket. This is probably because the ball part is lying under the socket which serve as the main support to the socket when external pressure is being applied to ball-and-socket segment in relation with the moment of inertia of the geometry of each case.

Fig. 8 shows the stress distribution experienced in the ball and socket where the greatest stress occurred along the intersection between each segment.

4.2 Stress result of conventional stainless steel carcass

A conventional stainless steel carcass was modeled and tested in order to see the differences in structural strength of different material and configuration. Table 8 shows the stress analysis results base on different thickness of metal sheet.

Fig. 9 shows the stress distribution experienced in the conventional carcass where the stress mainly occurred on the interlocking joint.

4.3 Comparison on two models of carcass design

The results on both ball-and-socket and conventional carcass shows significant different in term of structural strength. Table 9 summarizes the comparison of both results.

The results showed that the conventional carcass with stainless steel is better by 27% compare

Figure 9. Stress profile of conventional carcass.

Table 9. Comparison results on carcass models.

Design	Stress (Mpa)	Weight (kg)
Ball-and-Socket	671.22	19.80
Typical Carcass	530.3	48.49
% difference to conventional carcass	26.6%	(59.17)%

to composite ball-and-socket. However, in terms of weight, the ball-and-socket is lighter by 59% due to its lighter composite material compare to heavy stainless steel.

5 CONCLUSIONS

In the present work, the capability of the design and the material selection of glass epoxy with 55° winding angle and equivalent thickness of 20 mm (with Case 3 configuration) is proven to be able to sustain the hydrostatic pressure in shallow water (up to 300 m depth).

The flexibility governs by the design of interlocking mechanism of ball-and-socket which have better interaction compare to a normal standard pipe. And in terms of weight benefit, composite materials can provide a lighter solution for the flexible pipe.

To improve more on the research, model of the fabricated "ball-and-socket" carcass will be experimental tested focusing on Non-Destructive Test (NDT) to further understand the mechanism of the pipe and to validate the result obtained in numerical simulation.

REFERENCES

Api-17b 2002. Recommended practice for flexible pipe. *Iso 13628-11: 2007 identical.*
Api-17j 2008. Specification for unbonded flexible pipe.
Arbey, j., delebecque, l., toussaint, m. & petermann, n. Novel composite pipe-in-pipe solution for ultra deep water field developments. Offshore mediterranean conference and exhibition, 2009. Offshore mediterranean conference.
Azom. 2001. *Resin properties for composite materials* [online]. Sp system. Available: http://www.azom.com/article.aspx?articleid=997 [accessed april 6 2015].
Cripps, d. 2013. *Strength and stiffness of resin systems* [online]. Gurit. Available: http://www.netcomposites.com/guide/strength-stiffness/13 [accessed april 6 2015].
Gibson, a. 2003. *The cost effective use of fibre reinforced composites offshore*, hse books.
Jha, v., latto, j., dodds, n., anderson, t. A., finch, d. & vermilyea, m. Qualification of flexible fiber-reinforced pipe for 10,000-foot water depths. Offshore technology conference, 2013. Offshore technology conference.
Kaw, a. K. 2005. *Mechanics of composite materials*, crc press.
Rytter, j., rytter, n.-j., nielsen, r. & glejbøl, k. A novel compression armour concept for unbonded flexible pipes. Offshore technology conference, 2002. Offshore technology conference.

Milling time influence of ultrafine treated rice husk ash to pozzolanic reactivity in portlandite

Muhd Fadhil Nuruddin, Siti Asmahani Saad & Nasir Shafiq
Department of Civil and Environmental Engineering, Universiti Teknologi PETRONAS, Seri Iskandar, Perak, Malaysia

Maisarah Ali
Material and Manufacturing Engineering Department, Kulliyyah of Engineering, International Islamic University Malaysia, Kuala Lumpur, Malaysia

ABSTRACT: An investigation of pozzolanic reactivity of ultrafine treated rice husk ash (UFTRHA) was evaluated using electrical conductivity measurement. This paper is aimed to identify the effect of grinding duration via high speed milling process to the pozzolanic reactivity level. This pozzolanic reactivity assessment was completed by incorporating UFTRHA powder samples prepared at several milling duration in saturated portlandite solution at 40°C. Structure of these mesoporous substances after grinding process were captured using Field Emission Scanning Electron Microscopy (FESEM) accordingly. Based on the findings, experimental results analysis reveal that UFTRHA ground at 15 min has highest pozzolanic reactivity level among all specimens. Hence, short period of grinding is sufficient to attain huge amount of specific surface area (SSA) by using high speed milling process. Excessive grinding aid i.e. 60 min produced lower specific surface area hence lower pozzolanic reactivity level. Larger amount of specific surface area leads towards vigorous and higher pozzolanic reactivity.

1 INTRODUCTION

Rice is considered one of the most consumed cereal in the world. According to United Nation Food Agriculture Organization (FAO), world paddy milled in 2014–2015 is reaching 500 million tonnes (Food Agriculture Organization (FAO) 2015). Average annual consumption of rice in Asian country alone is around 100 kilogram per capita (Ramziath T. Adjao & Staaz 2011). Hence, the potential of producing ash from husk is about 100 million tons per annum. However, rice husk as the by-product of rice milling process that contributes approximately 20% out of the total rice production worldwide is underutilized. Emerging research and technology unleash the potential of this agricultural waste by specific treatment process. In concrete technology application, silica-rich and amorphous are among the most ubiquitous attribute of pozzolanic activity of rice husk ash (RHA).

It is known that RHA possesses reactive pozzolanic material that makes it possible to be considered as supplementary cementing material or as additive in concrete production. The reactivity is due to high surface area hence provides abundant of nucleation sites for further hydration products precipitation (Singh et al. 2013). This phenomenon leads towards formation of denser paste that eventually increase the strength and durability of the concrete system.

In this paper, pozzolanic reactivity analysis and morphological properties of ultrafine treated rice husk ash (UFTRHA) subjected to grinding duration are highlighted. The pozzolanic reactivity test is completed by measuring electrical conductivity of portlandite-UFTRHA suspensions for several grinding time at 40°C. Pozzolanic reactivity of UFTRHA at various grinding period is compared and deliberated accordingly.

2 METHODOLOGY

2.1 *Preparation of ultra-fine treated rice husk ash (UFTRHA)*

As-received raw rice husk was taken from BERNAS factory in Sungai Ranggam, Perak. Before the rice husk burning process took place, the raw rice husk was undergone pretreatment procedure using diluted hydrochloric acid (HCl). The pretreatment process was done to eliminate metallic impurities i.e. potassium (K), Sodium (Na) and Magnesium (Mg) that exist on the surface of the rice husk. These impurities removal increase purity of silica (SiO_2) content obtained after the incineration process. The specimens were soaked in the acid solution at concentration of 0.1M. After completion of pretreatment process, the samples were washed using distilled water until it neutral pH obtained and dried using laboratory oven at 110°C. It was then burned by using conventional furnace in laboratory. The burning process was done at

700°C for 1 hour. The sample was then ground using planetary ball mill at various grinding durations from 15 to 60 min.

2.2 Chemical composition and surface area analysis of UFTRHA

In order to proceed with the direct pozzolanic reactivity analysis and morphological identification, oxides composition analysis was conducted using X-Ray Fluorescence (XRF) accordingly. The oxides analysis was accomplised using spectrometer of Bruker Axs S4 Pioneer. This test was in accordance to BS EN 12677. On the other hand, specific surface area (SSA) of all UFTRHA powder samples at different grinding time was examined via Brunnet emmet teller (BET) nitrogen adsorption test method. The test was completed using surface area and pore analyzer model micromeritics ASAP 2020.

2.3 Pozzolanic reactivity analysis

As for pozzolanic evalution analysis of UFTRHA samples, rapid evaluation test procedure were conducted. This analysis is known as direct measurement of pozzolanic reactivity. It was completed by recording electrical conductivity decrement value using a conductivity meter at various time interval. In this experimental setup, 200 ml of saturated portlandite solution $(Ca(OH)_2)$ at temperature of 40°C were prepared. Then, the electrical conductivity values before and after addition of 5g of tested UFTRHA samples (i.e. grinding duration of 15 minutes, 30 minutes, 45 minutes and 60 minutes) were recorded until 360 minutes of hydration period. Ratio of liquid to solid (l/s) of the pozzolan in aquoues suspension adopted was 40 as reported elsewhere (McCarter & Tran 1996; Villar-Cocina et al. 2011; Van et al. 2014).

2.4 Morphology characteristics of UFTRHA

For completion of microstructure identification, field emmision scanning electron microscopy (FESEM) model Zeiss Supra 55 VP instrument was used to capture condition of the UFTRHA powder samples with respect to grinding time.

3 RESULTS AND DISCUSSION

3.1 Chemical composition of UFTRHA

Table 1 tabulates the chemical composition determination of UFTRHA. According to the experimental results, it is clear that the preponderant content in the treated rice husk ash is silica (SiO_2). Amorphous silica is vital in promoting higher pozzolanic reaction for additional calcium-silicate-hydrate (CSH) formation in concrete. In addition, alkali metal oxides percentage i.e. potassium oxide (K_2O), sodium oxide (Na_2O) and magnesium oxide (MgO) are recorded less than 0.2%.

Table 1. Chemical composition of ultra-fine treated rice husk ash (UFTRHA) pretreated using 0.1 m hydrochloric acid and incinerated at 700°C for 1 hour.

Chemical composition	Percentage (%)
SiO_2	97
P_2O_5	1.63
Fe_2O_3	0.25
CaO	0.46
Al_2O_3	0.09
K_2O	0.11
Na_2O	–
MgO	–
MnO	–
SO_3	–
Cl	–
ZrO_3	0.25

These percentage reduction amount of metal oxides indicate that metallic impurities on the surface of the raw rice husk grains are eliminated via acid leaching process successfully. Incineration of untreated rice husk ash resulted in early crystallization of silica into cristobalite due to the surface melting of the metallic oxides as reported by in several researches (James & Subbarao 1986; Krisnarao & Godkhindi 1992; Real et al. 1996).

3.2 Rapid evaluation test analysis

Determination of pozzolanic behaviour of a material is essential before incorporating it with cement as supplementary cementing material in concrete production. Fig. 1 illustrates the electrical conductivity readings plotted up to 360 minutes (6 hours) of hydration reaction in portlandite-pozzolan aqueous suspension at four different grinding period i.e. UFTRHA-15, UFTRHA-30, UFTRHA-45 and UFTRHA-60. From the analysis, it can be seen that the electrical conductivity values of portlandite-UFTRHA-60 suspension are higher as compared to the portlandite-UFTRHA-45 at the beginning of 160 minutes. However, the effect is changed at the hydration beyond 200 minutes. Overall, portlandite-UFTRHA-15 suspension has the lowest electrical conductivity values than those of other samples.

On the other hand, Fig. 2 shows the electrical conductivity decrement of portlandite-UFTRHA suspension together with specific surface area values (BET-SSA) at different grinding period. From the analysis, the reduction in electrical conductivity at 0 to 2 min of portlandite-UFTRHA-15 suspension possesses the highest value than other samples i.e. 1.93 mS/cm. It is also noted that the decrease in electrical conductivity of all the portlandite-UFTRHA suspensions reduced as the grinding duration increased.

In addition, the BET-SSA values are inversely proportional to the grinding time increment as well. Reduction in BET-SSA values with the addition of grinding period reveals that the material agglomerate and hence lesser surface area obtained

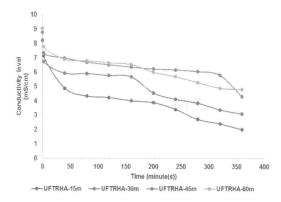

Figure 1. Effect of grinding duration from 15 to 60 minutes to the pozzolanic reactivity level via rapid evaluation method test.

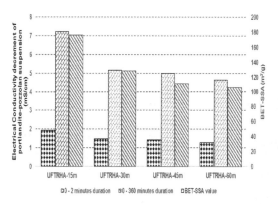

Figure 2. Electrical conductivity (EC) reduction of portlandite-pozzolan in aqueous suspension for 0 to 2 minutes and 0 to 360 minutes together with BET-SSA analysis of the UFTRHA samples.

(Opoczky 1977). This phenomenon justifies higher electrical conductivity value with shorter grinding duration. According to the classification of rapid evaluation method founded by Luxan et al. (1989) (Luxan et al. 1989), a material is deemed as good pozzolanic substance when its electrical conductivity value is greater than 1.2 mS/cm. Therefore, all of the UFTRHA samples ground at 15 to 60 min are considered to have high pozzolanic reactivity.

3.3 Morphological characteriztion of UFTRHA

For further understanding of UFTRHA morphological properties, field emission scanning electron microscopy (FESEM) images for all four samples that have been ground from 15 to 60 min were captured. Fig. 3 (a) to 3 (d) display the FESEM images accordingly.

Based on investigation of the FESEM images, agglomeration of UFTRHA particles are detected. Therefore, it is proven that the agglomeration of particles increase with greater grinding duration. These clump of particles occur due to excessive grinding procedure (Xu et al. 2015). Long duration grinding process leads towards low specific surface area value and hence should be avoided.

Figure 3. FESEM images of UFTRHA after grinding process (a) 15 min (b) 30 min (c) 45 min and (d) 60 min.

4 CONCLUSION

The results obtained indicate that, duration of milling affect the specific surface area value of the substances. High specific surface area is essential in order to boost up pozzolanic reactivity of pozzolan materials. This will ensure high quality supplementary cementing material (SCM) for concrete technology application. Based on the SSA-BET analysis, 15 min of milling period is sufficient to ascertain the highest specific surface area value of 180.72 m^2/g. Meanwhile, from the rapid evaluation test, the decrement of electrical conductivity at 0 to 2 min and 0 to 360 min are 1.93 mS/cm and 7.05 mS/cm respectively. Therefore, it is proven that high specific surface area promotes higher pozzolanic reactivity level.

ACKNOWLEDGEMENT

The author would like express profound gratitude to the Ministry of Higher Education of Malaysia (MoHE) for funding the research work under MyRa Grant Scheme. The author also wishes to thank to the staffs of Department of Civil Engineering and Department of Chemical Engineering, Universiti Teknologi PETRONAS for their kind assistance in this research.

REFERENCES

Food Agriculture Organization (FAO), 2015. *Rice Market Monitor April 2015*, Available at: www.fao.org/economic/RMM.

James, J. & Subbarao, M., 1986. Characterization of silica in rice husk ash. *American Ceramic Society bulletin*, 65(8), pp. 1177–1180.

Krisnarao, R.V. & Godkhindi, M.M., 1992. Studies on the Formation of SiC Whiskers from Pulverized Rice Husk Ashes. *Ceramics International*, 18, pp. 35–42.

Luxan, M.P., Madruga, F. & Saavedra, J., 1989. Rapid Evaluation Of Pozzolanic Activity Of Natural Products. *Cement and Concrete Research*, 19, pp. 63–68.

McCarter, W.J. & Tran, D., 1996. Monitoring pozzolanic activity by direct activation with calcium hydroxide. *Construction and Building Materials*, 10(3), pp. 179–184.

Opoczky, L., 1977. Fine grinding and agglomeration of silicates. *powder technology*, 17, pp. 1–7.

Ramziath T. Adjao & Staaz, J.M., 2011. The Changing Asian Rice Economy and its Implications for the Development of the Rice Subsector in West Africa., pp. 1–10.

Real, C., Alcala, M.D. & Criado, J.M., 1996. Preparation of Silica from Rice Husks.pdf. *Journal of the American Ceramic Society*, 79(8), pp. 2012–2016.

Singh, L.P. et al., 2013. Beneficial role of nanosilica in cement based materials – A review. *Construction and Building Materials*, 47, pp. 1069–1077.

Van, V.T.A. et al., 2014. Pozzolanic reactivity of mesoporous amorphous rice husk ash in portlandite solution. *Construction and Building Materials*, 59, pp. 111–119.

Villar-Cocina, E. et al., 2011. Pozzolanic behavior of bamboo leaf ash: Characterization and determination of the kinetic parameters. *Cement and Concrete Composites*, 33(1), pp. 68–73.

Xu, W. et al., 2015. Effect of rice husk ash fineness on porosity and hydration reaction of blended cement paste. *Construction and Building Materials*, 89, pp. 90–101.

Mix design proportion for strength prediction of rubbercrete using artificial neural network

Aznida Awang, Bashar S. Mohammed & Muhammad Raza Mustafa
Department of Civil and Environmental Engineering, Universiti Teknologi PETRONAS, Seri Iskandar, Perak, Malaysia

ABSTRACT: Data on the mix design of rubbercrete experiments are available throughout the literature and utilized in this paper to provide a platform for prediction of strength to obtained predetermined mix design. Using artificial neural network (ANN), the strengths of rubbercrete are predicted using literature data with water-cement ratio, percentage of CR, cement, fine aggregates, coarse aggregates and water as inputs. The desired output are identified as the compressive strength, flexural strength, splitting tensile strength and modulus elasticity of rubbercrete. From the result, it is concluded that different data set, different neural network parameters are required. The overall regression plot for the prediction achieved a correlation coefficient, R of 0.99157. With this prediction tool, the neural network can be used as mix design for selection of rubbercrete mix proportions to facilitate the application and utilization of rubbercrete, not only the academic field, but also in the industry.

1 INTRODUCTION

1.1 Rubbercrete

Incorporation of tire in concrete has started back in 90's as researchers figure out the alternatives of recycling scrap tire as aggregates within concrete as an act of conserving the environment (Mohammed 2010). The scrap tire went through several process to separate the steel wire from the rubber chip and reduce its size to smaller bits called crumb rubber (CR) and added into concrete mixture as partial or full replacement of coarse or fine aggregate (Mohammed et al., 2012). The resulting concrete is called rubberized concrete or rubbercrete.

CR is known to have low specific gravity and non-polar or hydrophobic surface due to application of zinc stearate formulation during tire manufacturing (Youssf et al. 2014; Sukontasukkul & Chaikaew 2006). Rubbercrete is observed to posed numerous advantages compared to ordinary Portland cement concrete (OPCC) which are lower thermal conductivity an better ductility, electrical resistivity, durability, toughness, impact resistance, damping ration, noise reduction factor, chloride penetration, energy dissipation and plastic capacity (Onuaguluchi & Panesar 2014; Shu & Huang 2013; Bravo & De Brito 2012; Mohammed et al. 2012; Mohammed, Bashar S.; Azmi 2011; Mohammed et al. 2011; Ganjian et al. 2009; Li et al. 2004). Such features enables rubbercrete to be used in impact resistance structure such as road bumper and sound barriers (Topçu & Demir 2007).

Despite such advantages, the reduction in the compressive strength of rubbercrete with the increase in CR addition is observed throughout the literature, hindering the application of rubbercrete in the construction industry especially in load bearing structure. This reduction in strength known to be caused by the weak bond between cement matrix and the CR due to the hydrophobic properties of its surface (Turatsinze & Garros 2008). It repels water and entraps air causing the formation of micro air pockets between the CR and cement, thus thicken the interfacial transition zone (Shu & Huang 2013). This weak chain within concrete leads to micro cracks due to stress concentration and in time, induces structural failure (Li et al., 2004). This properties also arrest water diffusion within cement matrix and deterrent cement hydration (Chou et al. 2010). Chemical and mechanical surface treatment method on the CR, pre-coating, and usage of supplementary cementitious materials has been proposed to overcome this problem and has shown improvement in strength of rubbercrete. Specific amount of CR within cement matrix has also shown to fabricate a rubbercrete of desired advantages and optimum compressive strength. Hence, a rubbercrete mix design is very important to predetermine the required strength of rubbercrete without going through multiple experiments on rubbercrete mixture proportions.

1.2 Neural network

Introduced back in 1980, Artificial Neural Network (ANN) is a simple imitation of the transmission and transformation of information within the neural system of our brain. The similarity between the brain and ANN is due to the capability of ANN to acquire knowledge through learning process and stored it in synaptic weight, an interneuron connection weight (Özcan et al. 2009). The main concept of ANN process is producing outputs resulting from evaluation of a weighted sum of inputs by a non-linear function (Molero et al. 2009). Due to its optimal characteristics and ease of application, ANN has been used in various field and in civil engineering, ANN have been utilized to detect structural damage, identify structural system, predict material behaviour and prediction studies, optimize and control structure, monitor ground water, and evaluate concrete mix (Liu et al. 2002).

Significant amount of research have been carried out to predict the strength and other properties of various type of concrete using ANN. The effect of additives and admixtures e.g. fly ash, silica fumes, metakaolin, blast furnace slag and recycled aggregates on the compressive strength of concrete has also been studied using this network. In term of rubbercrete, two researches has been known to study and predict the effect of CR on rubbercrete. Eldin & Senouci (1994) determine the relationship between size and percentage CR on the engineering properties of rubbercrete and developed two models of back-propagation neural network to predict the compressive and tensile strength of rubbercrete respectively. Topçu & Saridemir (2008) predicted the unit weight and flow table of rubbercrete using feed forward, back-propagation ANN and fuzzy logic using data from experimental works executed. Both papers has highlighted the effectiveness of ANN in predicting the properties of concrete and focusing on the effect of CR on the concrete but less attention on the rubbercrete mix proportion. Hence, it can be concluded that neural network has been around as a data analysis method for prediction of PCC properties albeit its application on rubbercrete is rather limited. The gap within prediction of rubbercrete through neural network is still at large and there is a vast opportunity for research in term of rubbercrete mix design proportion and its materials prediction using ANN.

2 METHODOLOGY

2.1 Neural network

2.1.1 Theoretical structure

Figure 1 shows the basic feed-forward back-propagation neural network. Neural network consist of 5 main parts which are the input, weight, sum function, activation function and output (Ghafari et al. 2015). In a feed forward neural network, the input is pass to the output layer through the hidden layer in forward manner, without any loop. The number of hidden layer and the number of neurons within the layer depends on the architecture of the system. There can be more than one hidden layers and usually one hidden layer is already sufficient to support the network while the number of neurons must be determine whether by trial of error or through experience and trials as there are no fixed method to determine it (Saridemir 2009).

During training phase, the neuron receive input from the previous neuron layer which then is multiplied with weight and added with a bias value. The net

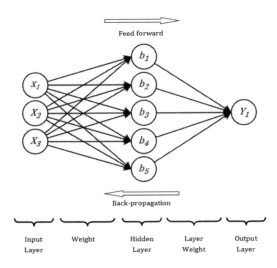

Figure 1. shows basic architecture of feed-forward neural network with back-propagation training algorithm

sum of the weighted input with bias, net_j is:

$$net_j = \sum_{i=1}^{A}(x_i w_{ij} + b_j) \qquad (1)$$

where letter A, B and C represent the number of neurons within the input, hidden and the output layer respectively. While subscript $i, j,$ and k represent the i_{th} ($1 < i <$ A), j_{th} ($1 < j <$ B) and k_{th} ($1 < k <$ C) of neurons within respective layers. x represent input, w represent the weight and b represent bias.

Then, the net_j is processed by a transfer/activation function to yield output in the hidden layer, y_j. The function is expressed as:

$$y_j = f(net_j) = f\left(\sum_{i=1}^{A}(x_i w_{ij} + b_j)\right) \qquad (2)$$

The output from the hidden layer is then feed to the output layer as input. The same process within hidden layer is repeated in the output layer. The net sum of the weighted and bias added input in the hidden layer, net_k is:

$$net_k = \sum_{j=1}^{C}(f(y_j)w_{jk} + b_k)$$
$$= \sum_{j=1}^{C}\left(w_{jk}f\left(\sum_{i=1}^{A}(x_i w_{ij} + b_j)\right) + b_k\right) \qquad (3)$$

The input from the hidden layer is then processed with a linear function in the output layer before passing them into the external world as the anticipated value. The output from the output layer, y_k is then

Table 1. shows the functions/algorithm used in the Neural Network model.

Training function	Learning function	Performance function
trainlm	*learngd*	*mse*
traingd	*traingdm*	*msereg*
traingdm		*sse*
trainscg		

calculated by:

$$y_k = f(net_k)$$
$$= f\left(\sum_{j=1}^{C}(f(y_k)w_{jk} + b_k)\right)$$
$$= f\left(\sum_{j=1}^{C} w_{jk} f\left(\sum_{i=1}^{A}(x_i w_{ij} + b_j)\right) + b_k\right) \quad (4)$$

The error, *e* is then calculated as the square difference between the target, z_k and output values, y_k is expressed as:

$$e_k = (z_k - y_k)^2 \quad (5)$$

2.1.2 Training algorithm

The aim of training the model is to minimise the differences between the output, *y*, and the target value, *z*, hence minimising the error of prediction (Mustafa et al. 2013). The back propagation algorithm passd back the calculated differences between the output and target value from the output layer to the input layer. This differences alter the value of weight and bias according to learning strategies (Topçu & Saridemir 2008a). This global error or known as error function is expressed as:

$$E = \frac{1}{2}\sum_{p=1}^{P}\sum_{k=1}^{C}(z_{pk} - y_{pk})^2 \quad (6)$$

where P is the number of input. Hence, in the next iteration, the new input data will be calculated with different weight and bias value and the error would be reduced and this will continue until it reach a satisfactory convergence, hence the training process has completed (Saridemir 2009). The value of the weight and bias are captured by the trained network and used in testing phase.

2.1.3 Model development

The development of this neural network model for this study is through these processes:

a. Identifying input and output

The parameters which affect the performance of rubbercrete were selected as input consist of water-cement ratio, percentage of CR, cement, fine aggregates, coarse aggregates and water. The desired output identified as the compressive strength, flexural

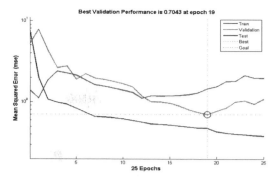

Figure 2. shows the performance for neural network with 5 hidden layer neurons.

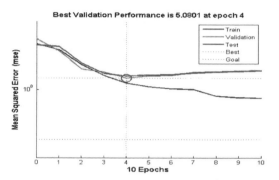

Figure 3. shows the performance for neural network with 10 hidden layer neurons.

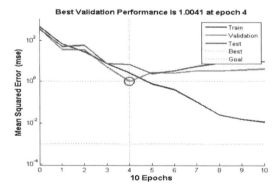

Figure 4. shows the performance for neural network with 15 hidden layer neurons.

strength, splitting tensile strength and modulus elasticity of rubbercrete. Data from experimental work by Mohammed & Azmi (2014) is used.

b. Preparing the training and testing examples

The models were trained with 70% data of experimental results, 15% for validation and the remaining were used only as experimental input values for testing.

c. Neural network architecture

The network's architecture has 6 input nodes to accommodate the 6 input variables, one hidden layer, and

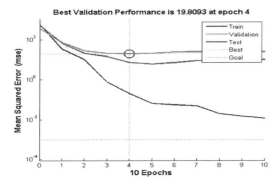

Figure 5. shows the performance for neural network with 20 hidden layer neurons.

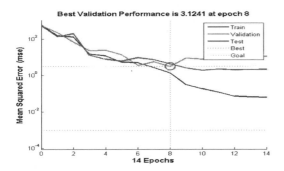

Figure 6. shows the performance for *mse* performance function.

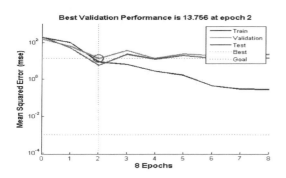

Figure 7. shows the performance for *learngdm* learning function.

3 output node. The number of hidden neuron is determined by trial-and error. The number of hidden neurons will be increased by 5 at a time (5, 10, 15, 20) until the network converge to the desired level of accuracy. The training, learning and performance functions are identified, selected and manipulated to obtain the best validation performance with error goal of 0.001 and epoch number of 1000 as shown in Table 1.

d. Validating (testing) the network

The testing data which has not been introduce (15%) to the neural networks since the data were not used

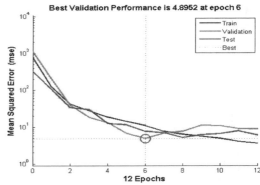

Figure 8. shows the performance for neural network with *trainscg* training function.

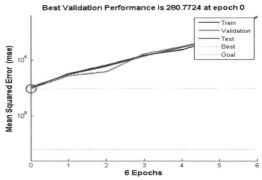

Figure 9. shows the performance for neural network with *traingd* training function.

during the training phase. The error between the target value and the output value will be determined.

3 RESULT AND DISCUSSION

3.1 *Number of neuron in hidden layer*

The performance, learning and training function is set to *msereg, learngd* and *trainlm* to determine the best number of neuron in the hidden layer. Having an optimum number of hidden neurons is crucial as it must be sufficient for correct modelling but sufficiently low to ensure network generalization (Atici 2011; Özcan et al. 2009). An increased number of neurons can give more flexibility to the network when dealing with new data but high quantity of hidden neuron will also cause under-characterized network where the network must optimise more parameters than the data available to constraint them (MathWorks 2016). The performance or network global error of the network for 5, 10, 15 and 20 hidden layer neurons are 0.88054, 5.0801, 1.0041 and 19.8093 respectively as shown in Figure 2–5. Even though the performance of network with 5 and 15 hidden layer neurons are better than 10 neurons, Figure 3 shows similarity between the training, validation and

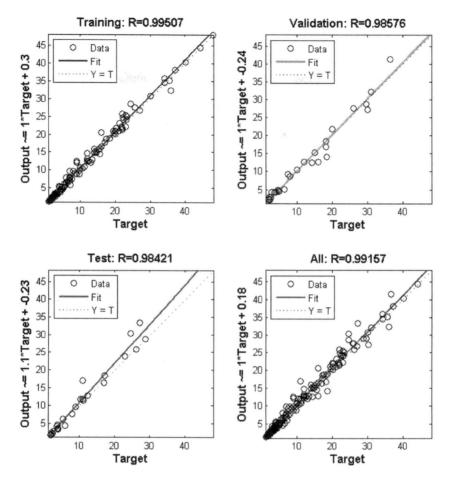

Figure 10. Shows regression plot for the chosen neural network parameter.

testing graph showing that the network have good generalization (MathWorks 2016). Hence, 10 hidden layer neuron is selected.

3.2 Selection of performance function

With 10 hidden layer neurons, the performance function is changed to *mse* while the learning and training function remain as *learngd* and *trainlm*. The performance of the network produced are 3.1241 with 14 epochs shown in Figure 6. Comparing Figure 3 and Figure 6, performance function *mse* give better performance compared to *msereg*. Hence, performance function *mse* is selected.

3.3 Selection of learning function

The learning function is changed from *learngd* to *learngdm* as shown in Figure 7.The resulting network is shown to have poorer performance with 13.756 at shorter epochs of 8. The number of iteration is important because it will affect the fault tolerance of the network. The average iteration taken for curve stabilisation is still lower compared to *learngd* which is at 8 as in Figure 6 and 7. The network convergence speed was also too fast (the number of iteration is only 2) which might cause accuracy of the network to decrease. Hence, learning function *learngd* is selected.

3.4 Selection of training function

The training function selected are *trainlm, trainscg, traingd* and *traingdm*. These are the most commonly used training function (Mustafa et al. 2013). The performance for respective functions are 3.1241, 4.8952, 280.7724 and 451.718 as shown in Figure 3 and Figure 8–9. The error goal for the performance is 0.001 while training function *traingd* and *traingdm* have poor performance compared to *trainlm* and *trainscg*. As *trainlm* function have higher accuracy, the function is selected.

3.5 Neural network modeling and prediction

The selected parameters are used for neural network modelling for prediction of strength for rubbercrete.From Figure 11, it can be seen that the output value is very similar to the target value. The correlation coefficient, R value is more than 0.9 for each phase (training, validation, test and overall) meaning very good generalization between the input and output variable for prediction by the network. Even though the performance of the network is slightly higher than the error goal, the network can still give good prediction proving the capability of the network to deal with noisy data. The scatter plot is very useful to shows that some data have very poor fit/outlier. As there are no outliers in the regression plot, this mean the data presented are precise.

4 CONCLUSION

Throughout this paper, it can be concluded that neural network have a strong ability to be used as a prediction tool for rubbercrete properties. From the result, it is concluded that different data set, will require different neural network architecture. With this prediction tool, the neural network can be used as mix design for selection of rubbercrete mix proportions to facilitate the application and utilization of rubbercrete, not only the academic field, but also in the industry.

REFERENCES

Atici, U., 2011. Prediction of the strength of mineral admixture concrete using multivariable regression analysis and an artificial neural network. *Expert Systems with Applications*, 38(8), pp. 9609–9618.

Bravo, M. & De Brito, J., 2012. Concrete made with used tyre aggregate: Durability-related performance. *Journal of Cleaner Production*, 25, pp. 42–50.

Chou, L.-H. et al., 2010. Improving rubber concrete by waste organic sulfur compounds. *Waste management & research: the journal of the International Solid Wastes and Public Cleansing Association, ISWA*, 28(1), pp. 29–35.

Eldin, N.N. & Senouci, A.B., 1994. Measurement and prediction of the strength of rubberized concrete. *Cement and Concrete Composites*, 16(4), pp. 287–298.

Ganjian, E., Khorami, M. & Maghsoudi, A.A., 2009. Scrap-tyre-rubber replacement for aggregate and filler in concrete. *Construction and Building Materials*, 23(5), pp. 1828–1836.

Ghafari, E. et al., 2015. Prediction of Fresh and Hardened State Properties of UHPC: Comparative Study of Statistical Mixture Design and an Artificial Neural Network Model., 27(11), pp. 1–11.

Li Stubblefield, M. A., Garrick, G., Eggers, J., Abadie, C., & Huang, B., G., 2004. Development of waste tire modified concrete. *Cement and Concrete Research*, 34(12), pp. 2283–2289.

Li, G. et al., 2004. Waste tire fiber modified concrete. *Composites Part B: Engineering*, 35(4), pp. 305–312.

Liu, S.W. et al., 2002. Detection of cracks using neural networks and computational mechanics, 191, pp. 2831–2845.

MathWorks, 2016. Function Approximation and Nonlinear Regression. Available at:http://www.mathworks.com/help/nnet/function-approximation-and-nonlinear-regression.html.

Mohammed, Bashar S.; Azmi, N.J., 2011. Failure Mode and Modulus Elasticity of Concrete Containing Recycled Tire Rubber. *Journal of Solid Waste Technology and Management*, 37(1), pp. 16–24.

Mohammed, B.S. et al., 2012. Properties of crumb rubber hollow concrete block. *Journal of Cleaner Production*, 23(1), pp. 57–67.

Mohammed, B.S., 2010. Structural behavior and m-k value of composite slab utilizing concrete containing crumb rubber. *Construction and Building Materials*, 24(7), pp. 1214–1221.

Mohammed, B.S. & Azmi, N.J., 2014. Strength reduction factors for structural rubbercrete. *Frontiers of Structural and Civil Engineering*, 8(3), pp. 270–281.

Mohammed, B.S., Azmi, N.J. & Abdullahi, M., 2011. Evaluation of rubbercrete based on ultrasonic pulse velocity and rebound hammer tests. *Construction and Building Materials*, 25(3), pp. 1388–1397.

Molero, M. et al., 2009. Sand/cement ratio evaluation on mortar using neural networks and ultrasonic transmission inspection. *Ultrasonics*, 49(2), pp. 231–237.

Mustafa Ph.D., M.R. et al., 2013. Evaluation of MLP-ANN training algorithms for modeling soil pore-water pressure responses to rainfall. *Journal of Hydrologic Engineering*, 18(1), pp. 50–57.

Onuaguluchi, O. & Panesar, D.K., 2014. Hardened properties of concrete mixtures containing pre-coated crumb rubber and silica fume. *Journal of Cleaner Production*, 82, pp. 125–131.

Özcan, F. et al., 2009. Comparison of artificial neural network and fuzzy logic models for prediction of long-term compressive strength of silica fume concrete. *Advances in Engineering Software*, 40(9), pp. 856–863.

Saridemir, M., 2009. Prediction of compressive strength of concretes containing metakaolin and silica fume by artificial neural networks. *Advances in Engineering Software*, 40(5), pp. 350–355.

Shu, X. & Huang, B., 2013. Recycling of waste tire rubber in asphalt and portland cement concrete: An overview. *Construction and Building Materials*, 67, pp. 217–224.

Sukontasukkul, P. & Chaikaew, C., 2006. Properties of concrete pedestrian block mixed with crumb rubber. *Construction and Building Materials*, 20(7), pp. 450–457.

Topçu, İ.B. & Demir, A., 2007. Durability of Rubberized Mortar and Concrete. *Journal of Materials in Civil Engineering*, 19(February), pp. 173–178.

Topçu, I.B. & Saridemir, M., 2008a. Prediction of mechanical properties of recycled aggregate concretes containing silica fume using artificial neural networks and fuzzy logic. *Computational Materials Science*, 42(1), pp. 74–82.

Topçu, I.B. & Saridemir, M., 2008b. Prediction of rubberized mortar properties using artificial neural network and fuzzy logic. *Journal of Materials Processing Technology*, 199(1), pp. 108–118.

Turatsinze, A. & Garros, M., 2008. On the modulus of elasticity and strain capacity of Self-Compacting Concrete incorporating rubber aggregates. *Resources, Conservation and Recycling*, 52(10), pp. 1209–1215.

Youssf, O. et al., 2014. An experimental investigation of crumb rubber concrete confined by fibre reinforced polymer tubes. *Construction and Building Materials*, 53, pp. 522–532.

Roof insulation material from low density polyethylene (LDPE), kapok fibre and silica aerogel

N.H.A. Puad, M.F. Nuruddin, J.J. Lian & I. Othman
Department of Civil and Environmental Engineering, Universiti Teknologi PETRONAS, Seri Iskandar, Perak, Malaysia

ABSTRACT: Malaysia is experiencing hot weather mainly during daylight and mostly affects building occupants. This research aims to develop effective roof insulation materials using three types of composite roof insulation samples from aluminium foil, kapok fibre, and stone wool. Based on the simulation, all of the composites had maintained the indoor temperature at lesser than 34°C which is the acceptable thermal comfort temperature limit. Kapok fibre and silica aerogel incorporated with low-density polyethylene (LDPE) has the lowest thermal conductivity which is 0.485 W/mK in the analysis of hourly temperature and shows better thermal performance in comparison with other composites. Tensile and thermal properties of the composite were dominated by 94% of LDPE which lead to small difference in tensile strength and thermal performance of the composites. Kapok fibre with silica aerogel and LDPE shows the best performance and has potential to be adopted as roof insulation compared to the other composites.

1 INTRODUCTION

Rapid development in Malaysia has increased the energy demand in order to achieve optimum thermal comfort among building occupants. Heating, ventilation and air-conditioning (HVAC) systems has covered almost 50% of building's total energy consumption in most developing countries amid distinct building services. According to Pérez-Lombard (2008), the trend of air-conditioning application will keep inclining soon because of the high demand to achieve indoor thermal (Pérez-Lombard et al. 2008). In Malaysia, 70% of heat gains in most low rise buildings come from roof (Vijaykumar et al. 2007). During the hot day, the heat is absorbed by the roof and transmitted to the attic causing the hot ceiling when the heat remains trapped. The building occupants will experience the heat radiated from the hot ceiling. Thus, occupants depend greatly on fans and air-conditioning in order to acquire thermal comfort and afterward upsurge the electrical energy consumption required for cooling. Therefore the application of insulation can retard the heat transmission through the roof and act as a radiant and conduction heat barrier (Ong, 2011).

Thermal performance of the roof is widely studied in recent years. Many researches have been done to improve the thermal performance of the roof insulation material in order to reduce the air-conditioning load and at the same time provide a satisfying indoor thermal comfort for the occupants. Henceforth, this paper focuses on developing effective composite roof insulation materials from kapok fibre and silica aerogel together with the matrix, low density polyethylene (LDPE). Two most commonly used roof insulation in Malaysia which was aluminium foil and stone wool also was incorporated with LDPE and silica aerogel for comparison. The thermal properties and tensile properties were tested. Also, their respective thermal properties will be used to simulate the thermal performance of the thermal insulated experimental houses in Autodesk Ecotect. The outcome of hourly temperature graph illustrated the outdoor and in-door temperature of three modelled houses installed with three different composite roof insulation materials. The outdoor-indoor temperature difference graphs will be plotted to determine the most effective roof insulation materials with the highest performance to insulate heat from outdoor to indoor of the modelled house in simulation.

2 PROBLEM FORMULATION

2.1 Thermal comfort in tropical country

Malaysia is a country that located near to the equator. The weather in Malaysia is mostly hot and humid the whole year articulated with low daily temperature variation. Ong (2011) claims that the ambient air temperature and relative humidity of Malaysia as a tropical country fall in between 26°C–40°C and 60%–90% respectively. In such condition, it is unfavorable and uncomfortable to the occupants and reduces the productivity of the daily activities of occupants accordingly. The occupants spend a significant amount of time staying in a building and they will opt to use air-conditioners and fans frequently for long duration to make the environment in the room cooler and more hospitable. Thus, both applications of air

Table 1. Comparison of thermal conductivity for conventional and state-of-the-art thermal insulation material.

Thermal insulation	Thermal conductivity (mW/m K)
Conventional	
Mineral wool	30–40
Expanded polystyrene (EPS)	30–40
Extruded polystyrene (XPS)	30–40
Cork	45
Polyurethane (PUR)	35
Cellulose	35
State-of-the-art	
Vacuum insulation panels (VIP)	8
Gas-filled panels (GFP)	4
Aerogels	14

Table 2. Comparison of physical properties for Maerogel and silica aerogel.

Property	Silica Maerogel	Silica Aerogel
Apparent density	0.03–0.35 g/cm^3	0.03 g/cm^3
Internal surface area	600–1000 m^2/g	800–900 m^2/g
Mean pore diameter	∼20 nm	20.8 nm
Thermal tolerance	To 500°C, mp >1200°C	To 500°C, mp >1200°C
Thermal conductivity	0.02 W/m K	0.099 W/m K

conditioning and long hours usage of fans with regards to reduce the indoor temperature consume a lot of energy in their processes and in return contribute to high percentage of greenhouse gases emission. Hence, buildings' capability to insulate heat is the main factor in reducing heat to ensure good indoor thermal comfort. Building insulation is generally referring to wall insulation and roof insulation. However, roof insulation is more critical whereby 60% to 70% of thermal transfer occur at the roof (Vijaykumar et al. 2007; Soubdhan et al. 2005).

2.2 Traditional insulation material

Conventional insulation material has widely used in industry. Common insulation materials for building such as mineral wool, expanded polystyrene (EPS), extruded polystyrene (XPS), cork and polyurethane were among the most used materials applied to building. However the performance for traditional insulation material is not as good as state-of-the-art thermal building insulation material which has better thermal conductivity (Jelle, 2011). Table 1 shows the comparison of thermal conductivity for conventional with state-of-the-art thermal building insulation material. Furthermore, the conventional insulation like polyurethane, polystyrene, and polyethylene (PE) were being used in thicker or multiple layers which result in more complex building details, an adverse net-to-gross floor area and possible heavier load bearing construction (Lyons, 2007).

2.3 Silica aerogel

In 2007, silica aerogel directly prepared from rice husk was produced and known as silica maerogel (Hamdan, 2005). Rice husk is one of the most abundant waste materials in Malaysia. By using waste product as raw material, it becomes more cost effective in production and application. The most remarkable properties of silica maerogel compared to aerogel are non-toxic and environmentally friendly amorphous material which can be the best candidate to replace aerogel in the future. The properties of silica maerogel are mostly identical to silica aerogel and have good thermal conductivity. Being an inert, non-toxic and environmentally friendly amorphous material, silica maerogel possesses established physico-chemical properties which can be modified for specific applications. Table 2 shows the comparison between silica maerogel and silica aerogel.

Numerous kind of study has been done to improve the main concern of silica aerogel which is very brittle. It is believed that the silica aerogel granules can be directly used as super insulating filling materials by exploiting its eccentric properties. Occasionally, the addition of filler in the insulating material can modify or improve its quality. In order to efficiently improve thermal performance of silica aerogel, smaller particle size is recommended to (a) increase surface area and (b) to penetrate through fibrous structures that may also be incorporated into the silica Aerogel.

2.4 Low density polyethylene (LDPE)

Low density polyethylene or LDPE is a thermoplastic that is made from the monomer ethylene. Generally, LDPE is manufactured in the industries by using high pressure process through free radical polymerization. LDPE has low density which is about 0.92 g/cm^3 and thermal conductivity of 0.33 W/m.K. The common applications of LDPE are manufacturing of variety types of containers and molded laboratory equipment due to its good chemical resistance properties. While in construction, LDPE is normally manufactured as sealing membrane for water pond and waterproof membrane for protection barrier on green roofs.

A study was conducted to determine the effect of crystallinity and irradiation on thermal properties and specific heat capacity of LDPE by Zarandi & Bioki (2012). In their findings, they found out that at a prescribed temperature range, the thermal conductivity of LDPE is reducing by increasing the dose of irradiation. Also, as the weight fraction of crystallinity increase, the thermal conductivity of LDPE is increased whereas the specific heat capacity of it is reduced.

2.5 Composite insulating material

Silica aerogel has low strength because of its high porosity and low density, which makes it hard to be used directly as insulating material. The addition of reinforcing fibers into aerogel can enhance the poor strength of silica aerogel. Heat conduction and thermal radiation will be affected through this method. Generally for silica aerogel composite insulating materials, there exist nanoscale pores and matrix together with microscale additives such as fibres and opacifiers (Tao et al. 2013). There is limited systematic study on association of fibre with silica aerogel. The association between fibres and aerogel particles to each other and with the presence of binder and sometimes incorporating them in binder matrix give a mechanically stable material as well as low thermal conductivity (Frank & Zimmermann, 2003). PE has lower cost compared to other types of polymer and it is efficient for condensation polymerization step (Aerogel.org, 2014). However, it has low mechanical and thermal properties, high thermal expansion, subject to stress cracking and low strength/stiffness (Orient HDPE, 2012). The addition of silica aerogel as a filler to the PE and strengthen its properties through reaction with hydroxyl group in crosslinking process can enhance its properties.

3 HYPOTHESIS AND OBJECTIVES

The following hypotheses are created in order to clarify the above issues:

- Varying the mix proportion of silica aerogel, fibre and LDPE in order to get the optimum proportion of thermal conductivity and the best mechanical properties.
- The addition of fibre will enhance the mechanical properties of the composite but it will affect the thermal properties of the composite.

The objectives involved in this study are:

- To determine the tensile property and effectiveness of the three types of composite roof insulation materials which composed of:
  a) Aluminum Foil + LDPE + silica aerogel
  b) Stone Wool + LDPE + silica aerogel
  c) Kapok Fibre + LDPE + silica aerogel
- To evaluate the thermal performance of these composite roof insulation materials by using Autodesk Ecotect.

4 METHODOLOGY

4.1 Preparation of LDPE/kapok/silica aerogel composite

Different percentages of samples were prepared. LDPE was prepared as a control for comparison. Hot compression method was adopted to prepare the mixtures. The mixtures were hot pressed at 150°C under 12.5 ton pressure to obtain the composite.

4.2 Fiber treatment

The kapok fibre was washed (100 mL of water per gram of fiber) in distilled water for 30 minutes. The fibers were dried in an oven with ventilation for overnight at 60°C.

Table 3. Composition of samples for fabrication.

Sample	Materials	Matrix (LDPE) (wt. %)	Aluminum Foil (wt. %)	Stone Wool (wt. %)	Kapok Fibre (wt. %)	Aerogel (wt. %)	No of Samples
A	Aluminum Foil + LDPE + Aerogel	94	4	-	-	2	5
B	Stone Wool + LDPE + Aerogel	94	-	4	-	2	5
C	Kapok Fibre + LDPE + Aerogel	94	-	-	4	2	5

4.3 Characterization

Tensile tests were performed using a Universal Testing Machine, EMIC®model DL 3000, according to ASTM D638. Six samples were tested for each composite. Test was conducted at a crosshead speed of 20 mm/min. Five samples were tested for each composite.

4.4 Fabrication and optimization

Silica Aerogel was being mixed with fiber and binder at different ratio. The optimum ratio masses of silica Aerogel-composite will be determined by using thermal analyzer. Table 3 shows the composition of samples for fabrication.

4.5 Thermal conductivity test

Thermal conductivity test was conducted for all types of composite roof insulation samples by using TCi Thermal Conductivity Analyzer. TCi Thermal Conductivity Analyzer provides simple, highly accurate thermal characterization for lab, quality control and production environments. This instrument has broad testing capabilities of 0.0 to 220 W/mK in a wide range temperature from −50°C to 200°C.

4.6 Simulation

Autodesk Ecotect was used to model three experimental houses in the dimension of 3m width, 5m length and 3m height as shown in Figure 1. Table 4 shows the details and materials assigned to the model. The modelled houses were designated with the same design parameters of standard houses where concrete bricks for walls, plaster board for ceiling and concrete tiles for roof material. The input data for the field study modelling in Ecotect were building construction materials and local climate data of Kuala Lumpur, Malaysia.

5 RESULTS AND DISCUSSION

5.1 Tensile properties

Results from tensile tests show that the fibers and silica aerogel loading are significantly affect the mechanical properties of the composites. Table 5 shows the tensile strength of each composite.

Aluminium foil composite shows the highest tensile strength which is 2.456 Mpa, followed by stone wool

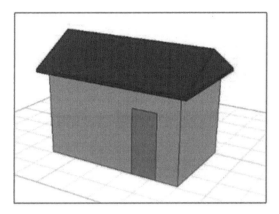

Figure 1. House model for simulation of indoor and outdoor temperature.

Table 4. Materials and details used for house model.

Object	Materials & Details
Wall	• 130 mm Brick plaster wall (100 mm brick with 10 mm plaster either side.
	• 100 mm thick high density polystyrene foam at inner layer of wall (to resist lateral heat transfer of wall)
Door	• 40 mm thick Hollow_core plywood door
Ceiling	• Plasterboard ceiling
	• Composite roof insulation materials
Roof	• Clay tiled roof
	• 30° pitched shaped roof 1 feet overhang

Table 5. Tensile strength of composite samples.

Composite sample	Tensile strength (MPa)
Neat LDPE	12.00
Aluminium+LDPE+Aerogel	2.456
Stone wool+LDPE+Aerogel	2.438
Kapok fibre+LDPE+Aerogel	2.411

Table 6. Thermal properties of composite roof insulation materials.

Sample	Label	Thermal Analyzer Code	Thermal Conductivity, k (W/mK)		Specific Heat Capacity (J/kg/K)	
Aluminium Foil + LDPE + Maerogel	A1	126	0.857	0.824	755.496	749.659
	A2	127	0.729		733.186	
	A3	122	0.864		756.636	
	A4	123	0.758		738.594	
	A5	128	0.914		764.384	
Stone Wool + LDPE + Maerogel	S1	114	0.719	0.668	731.406	720.453
	S2	131	0.631		713.13	
	S3	116	0.622		711.142	
	S4	118	0.767		740.224	
	S5	119	0.6		706.362	
Kapok Fibre + LDPE + Maerogel	K1	102	0.502	0.485	682.55	677.529
	K2	104	0.452		668.812	
	K3	105	0.562		697.374	
	K4	106	0.435		664.032	
	K5	108	0.473		674.878	

Kapok fibre + LDPE + silica aerogel has the lowest thermal conductivity which is 0.485 W/mK while aluminium foil composite has the highest thermal conductivity which is 0.824 W/mK. In general, thermal conductivity of a material is referring to the ability of a material to conduct heat. Heat transfer occurs at higher rate across the material of higher thermal conductivity than the lower thermal conductivity materials. Therefore, kapok fibre composite is the best composite roof insulation materials among the three composites in insulating heat from outdoor to indoor of buildings. Nevertheless, when comparing the thermal conductivity of the composite materials with the raw materials, it is obvious that the composite materials have higher thermal conductivity. There would be high chances of weak interfacial bonding between the materials with the matrix, LDPE. Initially, the thermal conductivity for LDPE is 0.33 W/mK. Although we incorporate LDPE with other materials with lower thermal conductivity (both kapok fibre and silica aerogel have thermal conductivity as low as 0.03 W/mK), instead of reducing the thermal conductivity, it goes the other way.

composite and kapok fibre composite with 2.438 MPa and 2.411 MPa respectively. Overall, there was no distinct significant in tensile strength for all composites. The main reason is due the high percentage of matrix (94% of LDPE) in all samples. Earlier, kapok fibre and silica aerogel are brittle and fragile substance with low tensile strength and this condition promotes the degradation of tensile strength for neat LDPE. The inclusion of of fibres and into the matrix leads to poor interfacial bonding between matrix and fibres due to insufficient matrix to wet the fibres completely. Thus, it can be concluded that LDPE has contributed most in enhancing the tensile strength of the composites.

5.2 *Thermal conductivity test*

Thermal conductivity tests are conducted to determine the thermal properties of the composite roof insulation samples. The results are tabulated in Table 6.

5.3 *Hourly temperature graph (simulation)*

The hourly temperature difference graphs of the composites are calculated by subtracting the indoor temperature from outdoor temperature ($T_{out}-T_{in}$). The purpose of plotting the hourly temperature difference graph is to determine the thermal performance of the composite roof insulation materials. Generally, the higher the positive difference between the outdoor and indoor temperature at hottest hour of the day, the better the thermal performance of the roof insulation materials. Figure 2, Figure 3, and Figure 4 illustrate the hourly temperature difference respectively and Figure 5 shows the combination of hourly temperature difference.

Figure 2 and Figure 3 show that the model house installed with aluminium foil composite and stone wool composite roof insulation has high temperature

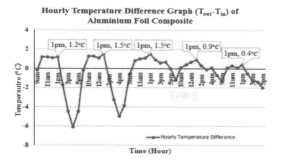

Figure 2. Hourly temperature difference graph ($T_{out}-T_{in}$) of Aluminium Foil Composite.

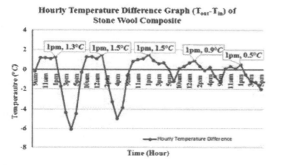

Figure 3. Hourly temperature difference graph ($T_{out}-T_{in}$) of Stone Wool Composite.

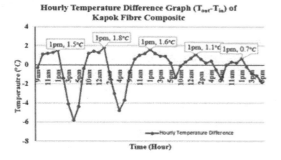

Figure 4. Hourly temperature difference graph ($T_{out}-T_{in}$) of Kapok Fibre Composite.

difference for outdoor and indoor at 1 p.m on the first day. However, kapok fibre composite has obtained the highest temperature difference at 1 p.m on the 1st day which is 1.5°C whereas aluminium foil composite and stone wool composite have temperature difference of 1.0°C and 1.3°C respectively. This means kapok fibre composite performs better in insulating heat from outdoor to indoor. Thermal performance of stone wool composite has about the same as aluminium foil. This scenario remains the same for the next four days. Besides, it is debatable that the thermal performance of these composite are about the same as the difference in temperature difference of the composites are in the range of ±0.3. It is most probably because of the high content of PE in the composites and also it

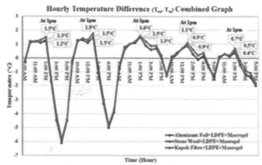

Figure 5. Hourly temperature difference ($T_{out}-T_{in}$) for all composites.

was supported with the tensile strength of all composites has about the same tensile strength. Nevertheless, kapok fibre + LDPE + silica aerogel has lower thermal conductivity, slightly better thermal performance which is comparable to the common thermal insulation material, aluminium foil and stone wool. Therefore, kapok fibre and silica aerogel has a big potential to be developed as a good material to be used as thermal insulation.

6 CONCLUSIONS

Aluminium foil composite has the highest tensile strength which is 2.456 Mpa. However, due to the weak interfacial bonding between matrix and aluminium foil, Kapok fibre composite shows better performance in terms of thermal conductivity with 0.485 W/mK compared to aluminium foil composite and stone wool composite. Kapok fibre composite also shows a great potential to reduce the house model temperature with the highest temperature difference recorded with 1.8°C at 1.00 p.m. This study is still in progress and more tests need to be conducted to support these findings. The combination of silica aerogel with compatible materials can resolve the issue of brittleness of silica Aerogel. These reinforcements may useful for future enhancement to get the best insulation material.

ACKNOWLEDGMENT

A special thanks to MOSTI for the financial support through Geran Penyelidikan Nanoteknologi provided during preparing this research proposal.

REFERENCES

Aerogel.org. 2014. http://www.Aerogel.org/?p=1058. [9] Roof Insulation | Cool Roof Malaysia. [cited 2014 27 October]; Available from: http://thermoshield.com.my/.

Borhani Zarandi, Amrollahi Bioki, Effect of crystallinity and irradiation on thermal properties and specific heat capacity of LDPE & LDPE/EVA. Appl Radiat Isot, 2012. 70(1): p. 1–5.

Frank D. and Zimmermann, A. 2003. Composite Aerogel Material that Contains Fibres, DE Patent.

Hamdan, H. 2005. Nanomaterials as catalysts in the production of fine chemicals, *Akta Kimindo*: 1, pp. 1–10.

Jelle, B. P. 2011. Traditional, state-of-the-art and future thermal building insulation materials and solutions – Properties, requirements and possibilities, *Energy and Buildings*: 43, pp. 2549–2563.

Lyons, A. 2007. 13 Insulation Materials, in Materials for architects and builders, Third ed Butterworth-Heinemann, Oxford.

Ong, K.S. 2011. Temperature reduction in attic and ceiling via insulation of several passive roof designs. *Energy Conversion and Management*: 52(6): p. 2405–2411.

Orient HDPE. 2012. Properties of HDPE, http://www.powerinsulator.com/list1.asp?id=350.

Pérez-Lombard, L., Ortiz, J., and Pout, C. 2008. A review on buildings energy consumption information, *Energy and Buildings*: 40(3): p. 394–398.

Soubdhan, T., Feuillard, T., and Bade, F. 2005. Experimental evaluation of insulation material in roofing system under tropical climate. *Solar Energy*: 79(3): p. 311–320.

Tao, X., Ya-Ling, H., and Zi-Jun, H. 2013. Theoretical study on thermal conductivities of silica Aerogel composite insulating material, *International Journal of Heat and Mass Transfer*: 58, pp. 540–552, 2013.

Vijaykumar, K., P. Srinivasan, and S. Dhandapani. 2007. A performance of hollow clay tile (HCT) laid reinforced cement concrete (RCC) roof for tropical summer climates. *Energy and buildings*: 39(8): p. 886–892.

Application of new codes for fatigue assessment of older bridges

A.Q. Ayilara & T. Wee
*Department of Civil and Environmental Engineering, Universiti Teknologi PETRONAS,
Seri Iskandar, Perak, Malaysia*

M.S. Liew
*Faculty of Geosciences & Petroleum Engineering, Universiti Teknologi PETRONAS,
Seri Iskandar, Perak, Malaysia*

ABSTRACT: Generally, structural design and assessment documents are continually improved upon. The most common challenges that often arise with the introduction of a new document are its application and interpretation. This paper presents the results of evaluation of the use of the Eurocodes for bridge fatigue assessment of existing steel bridges designed to the specifications of older standards. The rate of fatigue damage in truss components of a Pratt through truss steel bridge is evaluated using the nominal stress approach. Comparison is made between the results obtained using the increased loadings of the BS Eurocodes as against using the old BS 5400. The results show that using the new code for assessment is beneficial only when its loadings are not directly employed in the assessment of in-service bridges, as the rate of damage to member details can be 150% higher when the loadings of the new code are directly employed.

1 INTRODUCTION

Generally, structural design and assessment documents (standards, specifications and guides) are continually improved upon, just as steadfast as research breakthroughs. The revised documents are tailored to improve on the overall quality of works and economy. Most developed nations have their own documents and sometimes, a unified standard is adopted by member countries of a union. An example is the European Union, where the Eurocodes are used. In the UK, the introduction of the Structural Eurocodes led to the replacement or withdrawal of previous British Standards. However, revision of established documents or introduction of new ones, come with various challenges.

The most common challenges that often arise with the introduction of a new document are its application and interpretation. For structural bridge standards, it is common for higher loadings to be introduced in newer codes while properties of structural materials like steel are improved. The details of existing steel bridges will not conform to the new specifications. As such, the application of the increased loadings for the assessment of older steel bridges must be executed with caution (Cooper, 2009). Fatigue plays a major role in the deteriorating life cycle of steel bridge details. In order to avoid using excessive loadings in carrying out fatigue assessment, it is pertinent to evaluate the impact of using the recommendations of the new bridge Eurocodes for the assessment of existing steel bridges designed to the requirements of BS 5400.

2 HISTORY OF BRIDGE LOADINGS

2.1 Standards for vehicular loadings on bridges

Standardization of vehicular loadings on bridges was first introduced by the British Standards Institute (BSI) for Great Britain in 1922, with BS 153 part 3 being the first standard containing vehicular load requirements (Childs, 2013, BS 153, 1972). In the development of the bridge loadings (both highway and railway) contained in BS 5400 part 2 (1978), considerations were given to a wider range of factors such as nominal loading effects, combination of loadings, external actions, vehicle configurations (normal and abnormal), traffic volume, width of carriage way, number of lanes, order of passing vehicles, braking/traction force, impact etc. BS 5400 part 2 has been widely employed in the design and assessment of highway and railway bridges within the UK and internationally.

Presently, BS EN 1991-2 (2003) is the recommended specification for traffic load on bridges by the BSI in the UK. The new code co-existed with BS 5400-2 until the later was withdrawn in 2010. The new structural Eurocodes were prepared based on recent material testing, advanced non-linear analysis, non-linear finite element parametric studies, as such; many provisions of the code are more economical than those of the older codes (Hendy et al., 2011). But, most existing highway bridges were designed with the load specifications of BS 5400 as such; BS 5400 still enjoys broad usage amongst design and assessment engineers within and outside the UK.

2.2 Loadings for fatigue verification

Highway bridges are subjected to fluctuating loads from vehicular traffic all through their service life, this makes fatigue an important consideration during design and assessment, especially in the case of steel bridges, since it is important for the design engineer to avoid structural details that are susceptible to fatigue issues during design (Al-Emrani and Kliger, 2009). For short-span bridges, the ratio of the fatigue-design load to the strength-design loads can be large enough that fatigue may control the design of much of the structure while the deck, stringers and floor beams of long-span bridges are sensitive to fatigue, as vehicular live loads are primarily carried by these members (Dexter and Fisher, 2000). National standards often recommend loadings for fatigue verification. These loadings vary among different specifications, typically in terms of truck weight and amount of dynamic impact (Zhou, 2006).

The Eurocode specification recommended for bridge fatigue verification by the BSI is BS EN 1993-1-9 (2005). But older BS 5400-10 (1980) has been the most widely employed reference code for fatigue assessment of highway and railway bridges. It is even thought of as the most comprehensive code in the world for bridge fatigue (Ogle and Chakrabarti, 2010). BS 5400-10 comprehensively contains recommendations for bridges with welded and non-welded connections whereas the new BS Eurocodes do not have a distinctly assembled code of practice for fatigue assessment of bridges. BS EN 1993-1-9: fatigue, is generally employed for fatigue in structural steel works while the recommendations for fatigue design and assessment of bridges are spread through several other parts of the Eurocodes, national annexes (NA) and published documents (PD). As such, bridge engineers will need to study not only multiple parts of the Eurocodes but also NA's and PD's in the process of carrying out assessments of bridges using the new code.

2.3 Fatigue load models

Fatigue load models which are a representation of effects from standard load spectrums, are usually covered in bridge fatigue assessment documents. For highway bridges, the old BS 5400-10 contains standard load spectrum of commercial vehicles and a standard fatigue vehicle (SFV) while five fatigue load models (FML) are covered in BS EN 1991-2. Single vehicle fatigue models are simplified models recommended for direct fatigue verification of highway bridges while applying the traffic data of the bridge's specific location. They represent real fatigue problems especially when evaluating short span bridges (Chatzis et al., 2012, BS 5400, 2000, BS 5400, 1980, BS EN 1991-2, 2003, BS EN 1993-2, 2006). The SFV of BS 5400-10 and FLM 3 of BS EN 1991-2 are single vehicle fatigue load models. These vehicles are conventionally configured with four axles as shown in Figure 1. In terms of weight, SFV of BS 5400-10 has an absolute weight

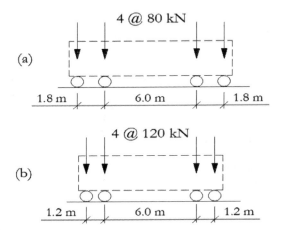

Figure 1. Single vehicle fatigue models: (a) SFV of BS 5400-10; (b) FLM 3 of BS EN 1991-2.

of 320 kN while FLM 3 is configured with an absolute weight of 480 kN representing a 50% increase.

3 BRIDGE ASSESSMENT

3.1 Routine inspection

Bridge inspection is undertaken annually or periodically, depending on type, properties, age and condition rating of critical structural members. The average service life of bridges is between 50 and 100 years (Morales and Bauer, 2006) even though most bridges are designed for 120 years. Also, structural components of a bridge are subjected to different load effects and environmental hazards which influence their rates of deterioration, as such; service life of bridge components are often much less than 50 years. Before the introduction of the Eurocodes, it was recommended that bridges be designed to require marginal level of inspection using the *safe life method* only, for a design life of 120 years while another method, *damage tolerant method*, which requires regular routine inspection is now recommended in the new code. For the assessment of older bridges, the Eurocodes should be employed with caution, as existing bridges are beyond the scope of the new code (Hendy et al., 2011, Cooper, 2009).

3.2 Fatigue assessment

The accuracy of fatigue assessment of steel details is strongly connected with the loadings used in carrying out the assessment. Four methods of assessment can be employed for steel bridge details. These are nominal stress method, structural hot spot stress method, notch stress method and fracture mechanics. National documents are tailored towards use of nominal stresses. For the British Standards, the process of fatigue life assessment of highway bridges essentially involves: identifying the critical details and joints; obtaining

Figure 2. Case study Pratt through truss bridge.

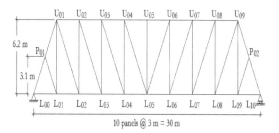

Figure 3. Numbering of truss joints.

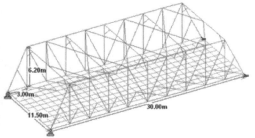

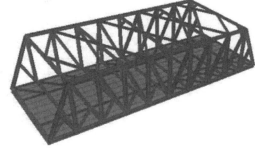

Figure 4. Three-dimensional finite element model of bridge.

necessary loading effects taking into consideration the influence lines of details; streamlining the loadings into ranges of stress, $\Delta\sigma$, with equivalent number of cycles, n, using cycle counting procedures; developing stress spectrums by ranking the stress ranges and cycles in descending order; classifying the detail under investigation according to the code of practice; selecting the appropriate *S-N curve* and calculating the remaining life using Palmgren-Miner's damage summation (also referred to as Miner's rule). While fracture mechanics is recommended for details where cracks greater than or equal to 5 mm have been detected.

4 FATIGUE ASSESSMENT OF A STEEL TRUSS BRIDGE

4.1 Case study bridge

A steel truss bridge located in Ulu Bongawan, a town near Papar in Sabah, East Malaysia is selected for the purpose of carrying out this study (Fig. 2). The highway bridge is a single span Pratt truss steel bridge with a through-truss deck arrangement. The truss components are hot-rolled H-sections with bolted connections and the reinforced concrete deck slab is 0.225 m thick with asphaltic concrete topping of 0.05 m. The bridge is a two lane, all-purpose single carriageway with ten panels of 3 m spans making up a total length of 30 m. It is 11.5 m in width and 6.2 m high. Joints of the vertical truss arrangement are numbered as shown in Figure 3.

4.2 Finite element modelling

Models of the truss bridge were developed in a commercial structural analysis and design software, STAAD.Pro V8i (Fig. 4). Beam and shell elements were used in developing the three-dimensional finite element models (3D-FEM) of the bridge using the member specifications contained in Table 1. The web and chord members of the main vertical arrangements are treated as trusses and structural beams respectively. Shear connectors were not included in the models while all other connection points are treated as fixed, as recommended in (Siriwardane et al., 2007, BS 5400, 2000, Wang et al., 2013).

4.3 Analysis and development of stress spectrum

The axles and vehicle dimensions of the single vehicle fatigue models are configured using vehicle definitions as shown in Figure 5. In order to obtain the most adverse effects, the defined vehicles are positioned 0.15 m away from the edge of the pedestrian walkway and traversed along the entire length of each adjacent lane, moving at 0.1 m intervals for each computational step. The traffic flow of the two adjacent lanes is in opposite direction and each lane of the carriageway is numbered as shown in Figure 6. The passage of one fatigue vehicle on a single lane is taken as one loading event.

Linear elastic analysis was carried out at serviceability Limit State to develop the stresses. Only vertical effects from the fatigue vehicles were considered in determining the stresses since the bridge details are non-welded. Consideration was given to passage of the fatigue vehicles on each lane and both lanes. Other

Table 1. Components of the steel truss bridge.

Component	Cross-section	Steel	Grade
Truss		All are 50B	steel grade
Upper Chord	UC 305×305×118 kg/m		
Lower Chord	UC 305×305×118 kg/m		
Vertical	UC 305×305×118 kg/m		
Diagonal Deck	UC 305×305×118 kg/m		
Floor beam	FB 625×310×(182~165)kg/m		
Stringer	UC 254×254×73 kg/m		
Lower lateral Bracing	UC 254×254×73 kg/m		
Portal			
Portal Bracing	UC 254×254×73 kg/m		
Upper lateral bracing	UC 203×203×46 kg/m		
Edge bracing	UC 254×254×73 kg/m		

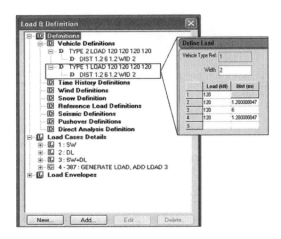

Figure 5. Fatigue vehicle definition for FLM3.

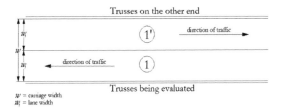

Figure 6. Lane numbering and directions of traffic flow.

fatigue associated factors like corrosion, wind actions and temperature effects were ignored. However, due to the possibility of stiffness in joints, effects of secondary stresses, k, was determined using the UIC scheme (Kühn et al., 2008). Also, mean stress effects were compensated for in determining the stress range values of members subjected to complete stress reversals by adding only 60% of the compressive side of the stress range to the tensile side, as recommended in (BS 7608, 1993, BS 5400, 1980, BS EN 1993-1-9, 2005). While dynamic effects which have already been factored into the fatigue vehicle models (BS 5400, 1980, BS EN 1991-2, 2003), were not further adjusted.

The n values are computed by using the reservoir method with traffic data in the region of the bridge's location (Ministry of Transport, 2013). The steel details in the truss arrangement conform to class D of table 17(a) in BS 5400-10 (BS 5400, 1980) while they agree with detail category 90(10) of table 8.1 in BS EN 1993-1-9 (BS EN 1993-1-9, 2005). The number of cycles to failure, N, for each stress range under SFV model of BS 5400 was estimated using Equation 1, according to clause 11.2 of (BS 5400, 1980).

$$N \times \sigma_r^m = K_2$$
$$\text{Log}_{10} N = \text{Log}_{10} K_2 - m \, \text{Log}_{10} \sigma_r \quad (1)$$

where $K_2 =$ parameter defining the S-N relationship for two standard deviations below the mean line; and $m =$ inverse slope of S-N curve on a logarithmic plot. Values for K_2 and m are given in table 8 of (BS 5400, 1980).

While, the values of N for FLM3 of BS EN 1993-1-9 were estimated using Equation 2, based on the recommendations of clause 7.1 (3) of (BS EN 1993-1-9, 2005).

$$\Delta\sigma_R^m N_R = \Delta\sigma_C^m 2\times 10^6 \text{ with m=3 for } N \leq 5\times 10^6$$
$$\Delta\sigma_R^m N_R = \Delta\sigma_D^m 2\times 10^6 \text{ with m=5 for } 5\times 10^6 \leq N \leq 10^8 \quad (2)$$

Finally, the rate of damage of the truss members under each fatigue vehicle loading is determined using Miner's rule expressed in Equation 3.

$$D = \frac{n_1}{N_1} + \frac{n_2}{N_2} + \cdots\cdots + \frac{n_k}{N_k} = \sum_{i=1}^{k} \frac{n_k}{N_k} \quad (3)$$

where $n_i =$ number of cycles occurring at a stress range, $\Delta\sigma$; $N_i =$ number of cycles corresponding to the fatigue strength of the detail category on the S-N curve; and $D =$ accumulated damage.

4.4 Interpretation of results

The most adverse effects were obtained from lane 1, followed by the combination of lane 1 & 1' while lane 1' gave the least effects. As such, the results presented are of the effects generated from lane 1. The effects generated by each fatigue vehicle for members of the truss after factoring the secondary stress effects are presented in Table 2. It is observed that the vehicle configuration has negligible impact on the shape of the stress spectrum, only the axle loadings mattered. Member L02U01 is most sensitive to fatigue while members L01P01 and L05U05 bear little or no consequences for both fatigue vehicles. The number of cycles for each detail for most elements was found to be 105,000 cycles except for the lower chord members. The rate of damage to each truss member for a year using the recommendations of either BS 5400-10 or the BS Eurocodes with their respective S-N curves and both fatigue vehicles are presented in Tables 3&4 respectively. The results show that the rate of damages is higher with using FLM 3 than SFV for both cases. The percentage difference in damage to member L02U01 is 146.3% and 152.5% for BS 5400-10 and BS Eurocodes respectively.

Also, some members whose fatigue effects would have been negligible under SFV had significant fatigue effects using FLM 3. While, the benefits of the new code with its much improved S-N curves can be observed in the number of members with negligible fatigue effects, these benefits are only observable using lower loadings as is the case of SFV. Also, the amount of fatigue life expended for a service life of 120 years using the two specifications and both fatigue vehicles are plotted in Figures 7-9. The plots show that

Table 2. Stress range and number of cycles.

Member	$\sum \Delta\sigma \cdot k$ (N/mm^2)		$\sum n_i$
	SFV	FLM 3	
L02U01	71.16	83.01	105,020
U04U05	68.25	78.98	105,020
U03U04	64.92	75.25	105,020
L00U01	60.90	70.83	105,020
U02U03	58.37	67.91	105,020
L03U02	53.72	64.45	105,020
L00L01	45.81	54.42	210,040
U01U02	43.68	53.27	105,020
L02U02	40.87	52.27	105,020
L04U03	40.85	51.48	105,020
L04L05	34.74	46.62	210,040
L01L02	34.72	46.49	210,040
L03L04	33.56	44.20	210,040
L05U04	32.57	42.02	105,020
L03U03	29.66	40.51	105,020
L02L03	29.41	38.79	210,040
L04U04	20.00	27.30	105,020
L01U01	16.33	21.04	105,020
L01P01	0.00	0.00	0
L05U05	0.00	0.00	0

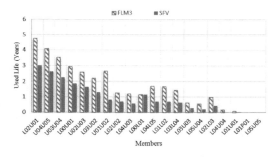

Figure 7. Assessment using recommendations of BS 5400.

Table 3. Results obtained using recommendations of BS 5400-10.

Member	$\sum n_i$	$\sum N_i$		D		Difference (%)
		SFV	FLM 3	SFV	FLM 3	
L02U01	105,020	4,218,676	2,657,031	2.489×10^{-2}	3.953×10^{-2}	146.3
U04U05	105,020	4,780,637	3,085,086	2.197×10^{-2}	3.404×10^{-2}	120.7
U03U04	105,020	5,555,711	3,566,903	1.890×10^{-2}	2.944×10^{-2}	105.4
L00U01	105,020	6,728,331	4,277,568	1.561×10^{-2}	2.455×10^{-2}	89.4
U02U03	105,020	7,644,616	4,853,829	1.374×10^{-2}	2.164×10^{-2}	79.0
L03U02	105,020	9,803,790	5,677,203	1.071×10^{-2}	1.850×10^{-2}	77.9
L00L01	210,040	15,810,017	9,433,172	6.643×10^{-3}	2.227×10^{-2}	38.0
U01U02	105,020	18,242,087	10,057,968	5.757×10^{-3}	1.044×10^{-2}	41.1
L02U02	105,020	22,273,169	10,644,806	4.715×10^{-3}	9.866×10^{-3}	47.1
L04U03	105,020	22,300,124	11,140,351	9.419×10^{-3}	9.427×10^{-3}	128.5
L04L05	210,040	36,249,101	14,997,809	5.794×10^{-3}	1.400×10^{-3}	82.1
L01L02	210,040	36,309,124	15,127,958	5.785×10^{-3}	1.388×10^{-3}	81.0
L03L04	210,040	40,203,393	17,600,886	5.224×10^{-3}	1.193×10^{-3}	67.1
L05U04	105,020	43,980,177	20,492,650	2.388×10^{-3}	5.125×10^{-3}	22.0
L03U03	105,020	58,274,069	22,869,890	1.802×10^{-3}	4.592×10^{-3}	33.2
L02L03	210,040	59,759,327	26,041,171	3.515×10^{-3}	8.066×10^{-3}	45.5
L04U04	105,020	∞	74,672,159	0	1.406×10^{-3}	14.1
L01U01	105,020	∞	∞	0	0	0.0
L01P01	0	∞	∞	0	0	0.0
L05U05	0	∞	∞	0	0	0.0

Table 4. Results obtained using recommendations of BS Eurocodes.

Member	$\sum n_i$	$\sum N_i$ SFV	$\sum N_i$ FLM 3	D SFV	D FLM 3	Difference (%)
L02U01	105,020	4,046,599	2,548,652	2.595×10^{-2}	4.121×10^{-2}	152.5
U04U05	105,020	4,585,637	2,959,247	2.290×10^{-2}	3.549×10^{-2}	125.9
U03U04	105,020	5,430,624	3,421,411	1.934×10^{-2}	3.069×10^{-2}	113.6
L00U01	105,020	7,472,442	4,103,088	1.405×10^{-2}	2.560×10^{-2}	115.4
U02U03	105,020	9,244,353	4,655,844	1.136×10^{-2}	2.256×10^{-2}	112.0
L03U02	105,020	13,993,952	5,629,991	7.505×10^{-3}	1.865×10^{-2}	111.5
L00L01	210,040	31,033,988	13,123,408	3.384×10^{-3}	1.600×10^{-2}	38.1
U01U02	105,020	39,391,989	14,603,854	2.666×10^{-3}	7.191×10^{-3}	38.8
L02U02	105,020	54,944,015	16,051,413	1.911×10^{-3}	6.543×10^{-3}	60.6
L04U03	105,020	55,054,881	17,316,037	3.815×10^{-3}	6.065×10^{-3}	121.9
L04L05	210,040	∞	28,422,575	0	7.390×10^{-3}	73.9
L01L02	210,040	∞	28,834,842	0	7.284×10^{-3}	72.8
L03L04	210,040	∞	37,111,448	0	5.660×10^{-3}	56.6
L05U04	105,020	∞	47,820,478	0	2.196×10^{-3}	18.3
L03U03	105,020	∞	57,419,202	0	1.829×10^{-3}	22.0
L02L03	210,040	∞	71,293,709	0	2.946×10^{-3}	29.5
L04U04	105,020	∞	∞	0	0	0.0
L01U01	105,020	∞	∞	0	0	0.0
L01P01	0	∞	∞	0	0	0.0
L05U05	0	∞	∞	0	0	0.0

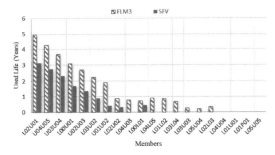

Figure 8. Assessment using recommendations of BS Eurocodes.

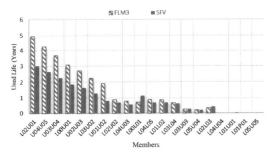

Figure 9. Both codes and their respective fatigue vehicle model.

assessment using the new BS Eurocodes is beneficial to more truss member details but the FLM 3 loading gives excessive used life while the SFV loading is more realistic for both assessment codes.

5 CONCLUSION

Based on the results obtained, it is established that the use of the new BS Eurocodes for the assessment of older steel bridges is beneficial but the benefits are outlived when its fatigue vehicle is directly employed. The axle loading of the single vehicle fatigue model of the BS Eurocodes is excessive for the assessment of older steel bridges. While the loadings contained in BS 5400 may be suitable, it is recommended that efforts should be made to obtain traffic load history of the bridge under investigations. The use of the S-N curves of BS 5400 for assessment purpose should be discontinued as these curves are not beneficial when compared to the fatigue curves of BS Eurocodes. Also, there is need for an assessment version of the BS Eurocode.

REFERENCES

Al-Emrani, M. & Kliger, R. Fatigue Prone Details in Steel Bridges. Nordic Steel Construction Conference, September 2–4 2009 Malmö, Sweden.

BS 153 1972. Parts 3B & 4: Specification for Steel Girder Bridges. British Standards Institution, London, UK.

BS 5400 1980. Part 10: Code of Practice for Fatigue. British Standards Institution, London, UK.

BS 5400 2000. Part 3: Code of Practice for Design of Steel Bridges. British Standards Institution, London, UK.

BS 7608 1993. Code of Practice for Fatigue Design and Assessment of Steel Structures. Incorporating Amendment No. 1 ed. British Standards Institution, London, UK.

BS EN 1991-2 2003. Eurocode 1 – Action on Structures – Part 2: Traffic Loads on Bridges. British Standards Institution, London, UK.

BS EN 1993-1-9 2005. Eurocode 3 – Design of Steel Structures – Part 1-9: Fatigue. British Standards Institution, London, UK.

BS EN 1993-2 2006. Eurocode 3 – Design of Steel Structures – Part 2: Steel Bridges. British Standards Institution, London, UK.

Chatzis, M. G., Tirkkonen, T. & Lilja, H. 2012. Calibrating Eurocode's Fatigue Load Model 3 to Specific National Traffic Composition. *Structural Engineering International*, 22, 232–237.

Childs, D. 2013. *HA and HB Type Loading* [Online]. Available: http://bridgedesign.org.uk/tutorial/tuha.html [Accessed 2 July 2014].

Cooper, D. Highway Traffic Load Model for Bridge Design and Assessment. 7th AUSTROADS Bridge Conference, 2009 Auckland, New Zealand.

Dexter, R. J. & Fisher, J. W. 2000. *Fatigue and Fracture*, CRC Press.

Hendy, C. R., Sandberg, J. & Shetty, N. K. 2011. Recommendations for Assessment Eurocodes for Bridges. *Proceedings of the ICE-Bridge Engineering*, 164, 3–14.

Kühn, B., Lukić, M., Nussbaumer, A., Günther, H.-P., Helmerich, R., Herion, S., Kolstein, M. H., Walbridge, S., Androić, B., Dijkstra, O. & Bucak, Ö. 2008. Assessment of Existing Steel Structures: Recommendations for Estimation of Remaining Fatigue Life. *In:* Sedlacek, G., Bijlaard, F., Géradin, M., Pinto, A. & Dimova, S. (eds.) *Joint Research Centre-European Convention for Constructional Steelwork Report*. Luxembourg.

Ministry of Transport 2013. Statistik Pengangkutan: Transport Statistics Malaysia, Pengangkutan Darat: Land Transport. Malaysia.

Morales, M. & Bauer, D. Fatigue and Remaining Life Assessment of Steel Bridges more than 50 Years Old. 7th International Conference on Short and Medium Span Bridges, 2006 Montreal, Quebec, Canada.

Ogle, M. & Chakrabarti, S. 2010. The UK National Annex to BS EN 1993-1-9:2005 and PD 6695-1-9:2008. London, UK: Institution of Civil Engineers (ICE).

Siriwardane, S., Dissanayake, R., Ohga, M. & Taniwaki, K. 2007. Different Approaches for Remaining Fatigue Life Estimation of Critical Members in Railway Bridges. *International Journal of Steel Structures*, 7, 263–276.

Wang, C.-S., Wang, Q. & Xu, Y. 2013. Fatigue Evaluation of a Strengthened Steel Truss Bridge. *Structural Engineering International*, 23, 443–449.

Zhou, Y. E. 2006. Assessment of Bridge Remaining Fatigue Life Through Field Strain Measurement. *Journal of Bridge Engineering*, 11, 737–744.

Author index

Ab Latip, A.S. 371
Abd Manan, T.S. 299
Abraham, S.M. 401
Abubakar, U.A. 215
Abujayyab, S.K.M. 361, 365
Adamu, M. 483
Adnan, M.A. 377
Ahamad, M.S.S. 361, 365
Ahmad, R. 355
Ahmad, S.Z. 365
Ahmed, M.O. 15
Ahmed, O.A. 183, 517
Ahmed-Chaouch, A. 497
Ajibike, M.A. 215
Akbar, I. 473
Al-Emad, N. 177
Al-Ghuribi, T.M.Q. 81
Al-Mattarneh, H.M. 327, 501
Al-Yacouby, A.M. 21
Alaloul, W.S. 149, 169
Alazaiza, M.Y.D. 289
Albarody, B.T.M.B. 521
Ali, M. 527
Ali, Z.M. 241
Aminu, N. 267
Arockiasamy, M. 45
Aswin, M. 437
Awang, A. 531
Ayilara, A.Q. 543
Ayoub, M.A. 81
Ayub, T. 417
Azizli, K.A.M. 193
Azmee, N.M. 463

Babangida, N.M. 293
Baharom, S. 27
Bakar, A.M.A. 377
Bala, N. 395
Bashir, M.J.K. 221
Beddu, S.B. 517
Besar, T.B.H.B.T. 377
Billah, A.F. 521
Bob, M.M. 289
Boukhaled, A. 511

Chai, K. 335
Chaid, R. 489
Chen, K.W. 263
Choi, H.S. 107, 113, 125
Choo, C.S. 433
Chung, E.S. 311, 317

Dahim, M.A. 327
Dimitrov, M. 381

Eisa, S.M.B. 443
El Kechebour, B. 493
Elshekh, A.E.A. 467
Ezechi, E.H. 207, 211, 279

Fai, L.C. 345
Farhan, S.A. 193
Fauzi, A. 501
Fazli, H. 63

Galuh, D.L.C. 447
Gardezi, S.S.S. 187
Ghani, A.Ab. 273, 361
Gohari, A. 55
Gupta, S. 137

Haddad, S. 493
Hajihassani, M. 323
Hakro, M.R. 385
Hamid, M.S.A. 427
Harahap, I.S.H. 27, 345
Hasan, R. 391
Hashim, A.M. 131, 137
Hassan, S.S. 87
Hayder, G. 183, 235, 517
Heng, G.C. 263
Hofko, B. 381
Hussain, M. 283, 305
Hussain, M.A. 385
Hussain, M.A. 477

Ibrahim, A.E. 33
Ibrahim, F.B. 215
Ibrahim, K. 405
Ibrahim, Z. 331
Idichandy, V.G. 21
Imran, M. 27
Ing, D.S. 433
Isa, M.H. 211, 221, 225, 229, 241, 251, 263, 267, 293
Ishak, D.S.M. 9
Ishak, N.A. 361
Ishwarya, S. 45
Ismail, A. 215
Ismail, M. 245
Ismail, R.M.A. 327

Jaafar, A. 101
Jalal, N. 377

Jamal, M.H. 9
Jameel, M. 33, 119
Jie, L.C. 193
Jiwa, M.Z. 107
Johnson, O. 405
Jumaat, M.Z. 33
Jusoh, H. 279, 335

Kamal, A.R.M.M. 59
Kamal, N.L.B.M. 517
Kamaruddin, I. 395, 409
Kamaruddin, S.A. 289
Kamarulzaman, N. 257
Kassim, K.A. 331
Kazmi, D. 27
Khamidi, F.B. 187
Khamidi, M.F. 199, 245, 427
Khan, S.U. 417
Khan, T. 229
Kharchi, F. 489, 507
Khouna, K.M. 225
Kim, D.K. 93, 107, 113, 125
Kim, J.M. 143
Kim, S.H. 311, 317
Kim, S.U. 311
Kim, Y.T. 113
Klufallah, M.M.A. 155, 199
Kurian, V.J. 3, 21, 39, 51, 59, 69, 87
Kutty, S.R.M. 207, 211, 251, 267
Kwon, T.-H. 351

Laghari, A.J. 241
Laghari, S.M. 241
Lee, J.R. 93
Lee, S.-H. 317
Lian, J.J. 537
Liew, M.S. 21, 51, 69, 81, 149, 169, 457, 543

Madzlan, N. 405
Mahamood, N. 443
Mahzabin, M.S. 423
Makhtar, A.M. 323
Malakahmad, A. 163, 211, 299
Malkawi, A.B. 501
Marto, A. 323
Matori, A.N. 55, 355, 371
Memon, I.A. 385
Mendaci, F. 511
Merida, A. 507
Mohamad Fu'ad, N.F.S.Bt. 235
Mohamad, H. 339

Mohamed, R. 377
Mohammed, B.S. 437, 443, 483, 501, 531
Mohammed, N.I. 245, 257
Muhammad, M.M. 215, 273
Musaad, A. 15
Mustafa, M.R. 229, 251, 273, 283, 293, 305, 531
Mustaffa, Z. 3, 107, 521
Myint, K.C. 55

Nagapan, S. 173, 177
Nagum, A.T.A. 207
Nahi, M.H. 409
Namazi-saleh, F. 3
Napiah, M. 155, 385, 391, 409
Ng, C.Y. 39, 75
Ng, S.J. 75
Ngien, S.K. 289
Nikbakht, E. 423
Noh, D.-H. 351
Noor, Z.M. 245
Nuruddin, M.F. 155, 163, 187, 199, 327, 427, 463, 467, 501, 517, 527, 537

Olisa, E. 211, 225, 279
Oloruntobi, O. 405
Omar, N.Q. 365
Osman, N. 473
Othman, I. 155, 199, 537
Oyejobi, D.O. 119

Paik, J.K. 93
Pakpahan, E.N. 251
Park, H.K. 143
Park, K.S. 93, 113, 125
Perissin, D. 371

Perumulselum, P.A./L. 235
Philip, A.K. 101
Poh, B.Y. 93
Puad, N.H.A. 537

Qasim, S. 27

Rahman, I.A. 173, 177
Rahmat, N.I. 173
Rani, M.S.N.A. 377
Ransinchung.R.N, G.D. 401
Rashidi, A.H.M. 101
Riahi, A. 225
Riharjo, H.P. 447
Rini, N. 51, 69
Ryu, J.W. 143

Saad, S.A. 527
Safiq, N. 477
Salihi, I.U. 163, 267
Sani, N.S. 377
Sapari, N. 225, 279
Selia, S.B.C. 183
Senthil, R. 45
Seo, J.H. 125
Shafiai, S.H. 101
Shafie, A. 457
Shafiq, N. 163, 187, 193, 251, 327, 417, 427, 443, 467, 483, 517, 527
Shaharmi, M.S.H.M. 427
Shahruzzaman, D.B. 131
Shin, S.M. 143
Singh, B.S.M. 225
Singh, S. 401
Sivapalan, S. 299
Sohaei, H. 323
Son, M.W. 311

Song, J.Y. 317
Soon, F.K. 193
Sulaiman, R.B.R. 137
Sulistyorini, D. 451
Sulong, N.H.R. 119
Sutrisno, W. 159
Syed Osman, S.B. 335

Talah, A. 489
Teh, H.M. 87, 101, 137
Teo, W. 63
Tuhaijan, S.N.A. 39

Umar, M. 331
Umar, U.A. 163

Vethamoorty 437

Wahab, A.K.A. 9
Wahab, M.M.A. 59
Wahab, S.N.A. 245
Wan Salim, W.S. 457
Wee, T. 543
Whyte, A. 51, 69
Won, K.J. 311
Wong, L.-P. 221

Yasin, I. 451
Yassin, A.Y.M. 473
Yu, S.Y. 113, 125
Yu, T.Y. 433
Yusof, K.W. 225, 273, 283, 293, 305
Yusoff, M.S. 365

Zahari, N.A.M. 225
Zawawi, N.A. 81, 149, 169, 187, 257, 427